David Dressler
Cambridge, 1975

Principles and practice of experiments with nucleic acids

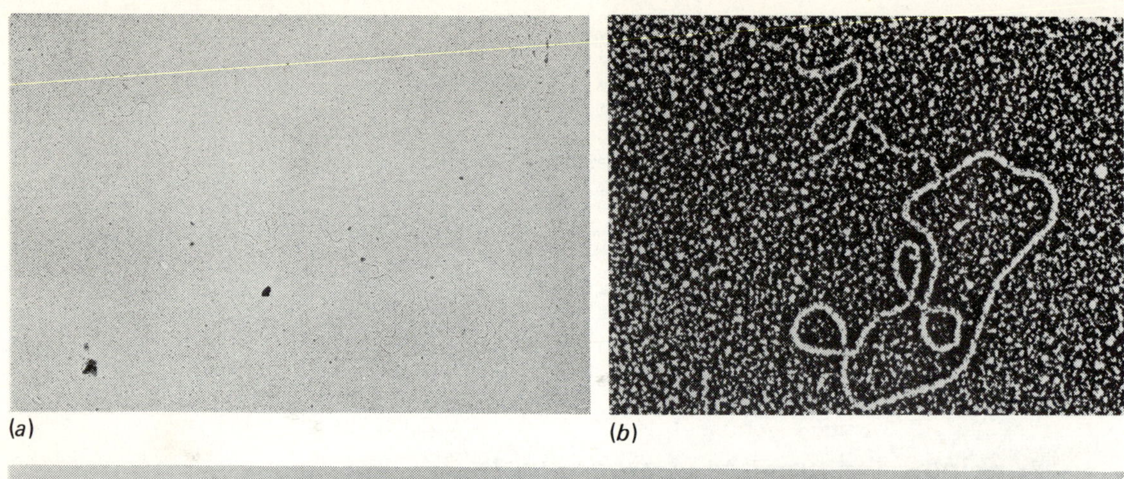

Frontispiece: Examples of the visualization of nucleic acids under the electron microscope. (*a*) A picture of a hybrid λ DNA molecule shadowed with uranium oxide. Most of the length of this molecule is double stranded. However the *b*2 region is missing from one strand so that there is a loop of single-stranded DNA in one chain and there is non-complementarity in the *b*5 region and two single-strand stretches can be seen. See p. 189 for a discussion and the methods. Taken from Westmoreland, Szybalski and Ris (1969). (*b*) A rolling circle of φX 174 DNA shadowed with uranium oxide. See pp. 187 and 189 for discussion. Taken from Dressler (1970). (*c*) Portion of 'nucleolar core' from the oöcyte of *T. viridescens* (see p. 189). The preparation was stained with phosphotungstic acid. Taken from Miller and Beatty (1969).

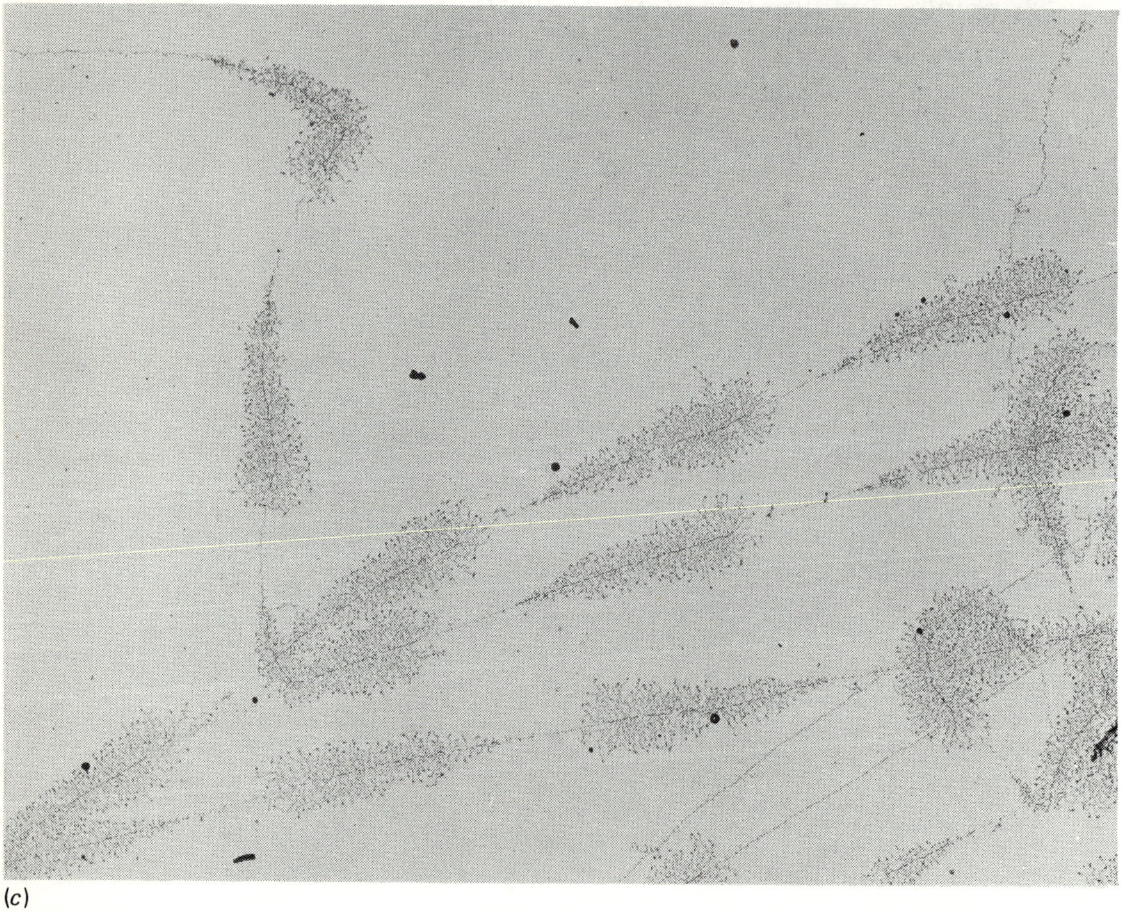

John Howard Parish

Principles and practice of experiments with nucleic acids

Longman

LONGMAN GROUP LIMITED
LONDON
Associated companies, branches and representatives throughout the world

© Longman Group Limited 1972

First published 1972

ISBN 0582 46285 1

Printed in Great Britain by
William Clowes & Sons Ltd
London, Colchester and Beccles

To Margaret

Contents

Preface

This book is largely about chemistry and it is written mainly for biologists. The role of nucleic acids is essential for any overall view of the control of metabolism or genetics. Similarly a precise description of the molecular events involved in such processes as differentiation, carcinogenesis and memory will undoubtedly require an understanding of reactions involving nucleic acids (although we cannot delineate these reactions with certainty at present). Consequently workers in a wide variety of disciplines find their research requires a working knowledge of nucleic acid methodology and the necessary theoretical background. In this book, I have described this background and have exemplified procedures with reference to experiments regarding a variety of biological problems. I hope the choice of examples is balanced rather than random; it is not however, in any sense, comprehensive. The book presents a view of the nucleic acids for those with a real interest in planning experiments or following the literature. It is not a textbook of molecular biology.

Certain topics (such as the isolation and fractionation of nucleic acids) have been described in considerable experimental detail. The reason is that I hope to have presented enough information for the reader to, at least, get started with experiments of his own devising. From such a start, I hope the references are sufficiently comprehensive to cover the more important techniques.

On the other hand, certain topics are almost out-of-bounds to the average biologist. The main example is the technique of X-ray crystallography. For this reason the account of data deduced from diffraction studies is extremely fragmentary. It is important to stress this imbalance as any integrated view of our knowledge of nucleic acid structure would emphasize the enormous (and continuing) contribution from crystallographers.

Any systematic reference to the historical aspects of the subject are also omitted. This omission is regrettable as the elucidation of the basic skeletal structure of RNA and DNA is one of the great achievements of organic chemistry. The chemistry of the nucleic acids that is described here is all relevant to procedures employed in present-day research.

Although nucleic acid research is largely in the province of biochemists, I am aware of the wish of some chemists to change their interest to an aspect of biochemistry—and the study of nucleic acids is a branch of natural product chemistry. Moreover there are several unsolved problems, alluded to in the following chapters, which require the special expertise of physico-organic chemists. It is largely for the reader with an essentially chemical (or physical) academic background, that the first chapter is written. In that chapter, I delineate the outlines of nucleic acid structure and metabolism as a starting point for topics in the rest of the book.

Note

Supplementary notes, signalled in the text by a superscript numeral, are to be found at the end of each chapter.

Acknowledgements

I am extremely grateful to Dr. O. L. Miller, Dr. H. Ris and Dr. D. Dressler for generously supplying the photographs used in the frontispiece and to Dr. J. R. B. Hastings, Dr. I. H. Maxwell, Dr. W. Fuller, Dr. R. M. Malbon, Mr. H. A. Foster and Miss M. Brown for access to unpublished results and for discussions.

Work from the author's laboratory has been supported by the Science Research Council.

Publishers' Acknowledgements

We are grateful to the following for permission to reproduce illustrations:

Academic Press Inc. (London) Ltd: *Journal of Molecular Biology* for Figs. 4.14, 4.19, 4.22, 4.23, 10.4, 10.5, 11.4a, 11.4b, 11.5, 11.6, 12.2; Academic Press Inc. New York: *Advances in Virus Research* for figs. 16.4, 16.7; *Analytical Biochemistry* for fig. 4.7; *Methods in Enzymology* for figs. 4.5c, 4.20; *Molecular Associations in Biology* for fig. 11.10; *The Nucleic Acids* for fig. 2.5; *Progress in Nucleic Acid Research and Molecular Biology* for figs. 1.6, 12.3; American Association for the Advancement of Science: *Science* for Plates 1a, 1c, fig. 12.5; American Chemical Society: *Biochemistry* for fig. 5.12; *Journal of American Chemical Society* for fig. 2.26; American Society for Microbiology: *Journal of Bacteriology* for fig. 5.7; Biochemical Society: *Biochemical Journal* for fig. 4.22; Butterworth & Company Ltd: *Subcellular components* for figs. 4.16, 4.18; Cambridge University Press: *Symposium of the Society for Experimental Biology* for figs. 12.8a, 12.8b, 13.15; J. & A. Churchill: *Cell differentiation–CIBA Foundation Symposium* for fig. 4.17; Cold Spring Harbor Laboratory: *Symposium on Quantitative Biology* for figs. 4.22, 5.5, 12.3, 12.12; Elsevier Publishing Company: *Biochimica et Biophysica Acta* for figs. 2.21, 15.3; *Chemico-Biological Interactions* for fig. 8.9; Genetics Society of America: *Genetics* for fig. 12.9; Macmillan Journals Ltd: *Nature* for figs. 5.8, 11.16, 11.17, 12.10, 16.6, 16.15; National Academy of Sciences: *Proceedings of the National Academy of Sciences* for Plate 1b, figs. 2.27a, 4.22, 4.24, 5.3, 5.9, 5.11, 11.12, 16.2, 16.3.

1   **Introduction**

What are nucleic acids and what are they for? They are polyelectrolytes of high molecular weight, they are the chemicals that genes are made of and they direct the biosynthesis of proteins. I naturally assume that the reader is capable of amplifying these statements in some sort of detail. However, an introductory chapter can serve two useful purposes; it is a polite way of drawing attention to the background material I am assuming and it is the most digestible context for presenting the conventions and abbreviations which must be used throughout the rest of the book. I have omitted any references from the bulk of this chapter as to include them would make it unwieldy. However, an excellent and very readable book which describes the whole background to the subject and its many ramifications is J. D. Watson's *The Molecular Biology of the Gene* (2nd. ed., 1970), Benjamin.

Primary structure

The backbone of a nucleic acid molecule is a chain of sugar moieties linked by phosphodiester groups. As phosphoric acid is tri-basic, phosphodiesters are mono-basic, and each phosphate residue in nucleic acids has one ionizable hydrogen. The pK_a value for this ionization is around 1·0 (or just less). Nucleic acids are not chemically stable at very low pH and consequently they are always handled as their salts. So whenever we say nucleic acid we mean the 'nucleate' anion. The cations present might typically be sodium, potassium, ammonium (or of course a mixture of more than one type). Attached to the sugar residues are heterocyclic bases. Various types of fragment can be obtained from nucleic acids. The base attached to its sugar, but lacking phosphate is called a nucleoside, the base-sugar-phosphate is called a nucleotide so nucleic acids are polynucleotides. The converse of this statement is not true; polynucleotides include in addition to the naturally occurring nucleic acids a number of synthetic polymeric substances. The nucleotides themselves are phosphomonoesters (dibasic acids). Short chains of nucleotides (obtained either synthetically or by partial degradation of polynucleotides) are known as oligonucleotides. The smallest nucleic acids contain about eighty nucleotide residues and the largest of the order of hundreds of millions.

There are two main sugars found in nucleic acids, D-ribose and 2-deoxy-D-ribose.[1] With minor exceptions which are mentioned later, any one nucleic acid molecule contains only one type of sugar. Thus there are two types of nucleic acid (and consequently two families of nucleosides and nucleotides) those in which the sugar is ribose (ribonucleic acid or RNA with ribonucleosides and ribonucleotides) and those in which the sugar is deoxyribose (deoxyribonucleic acid or DNA with deoxyribonucleosides and deoxyribonucleotides).[2] Shortened words for nucleosides and nucleotides are riboside (for ribonucleoside), ribotide (for ribonucleotide) and deoxyriboside and deoxyribotide.

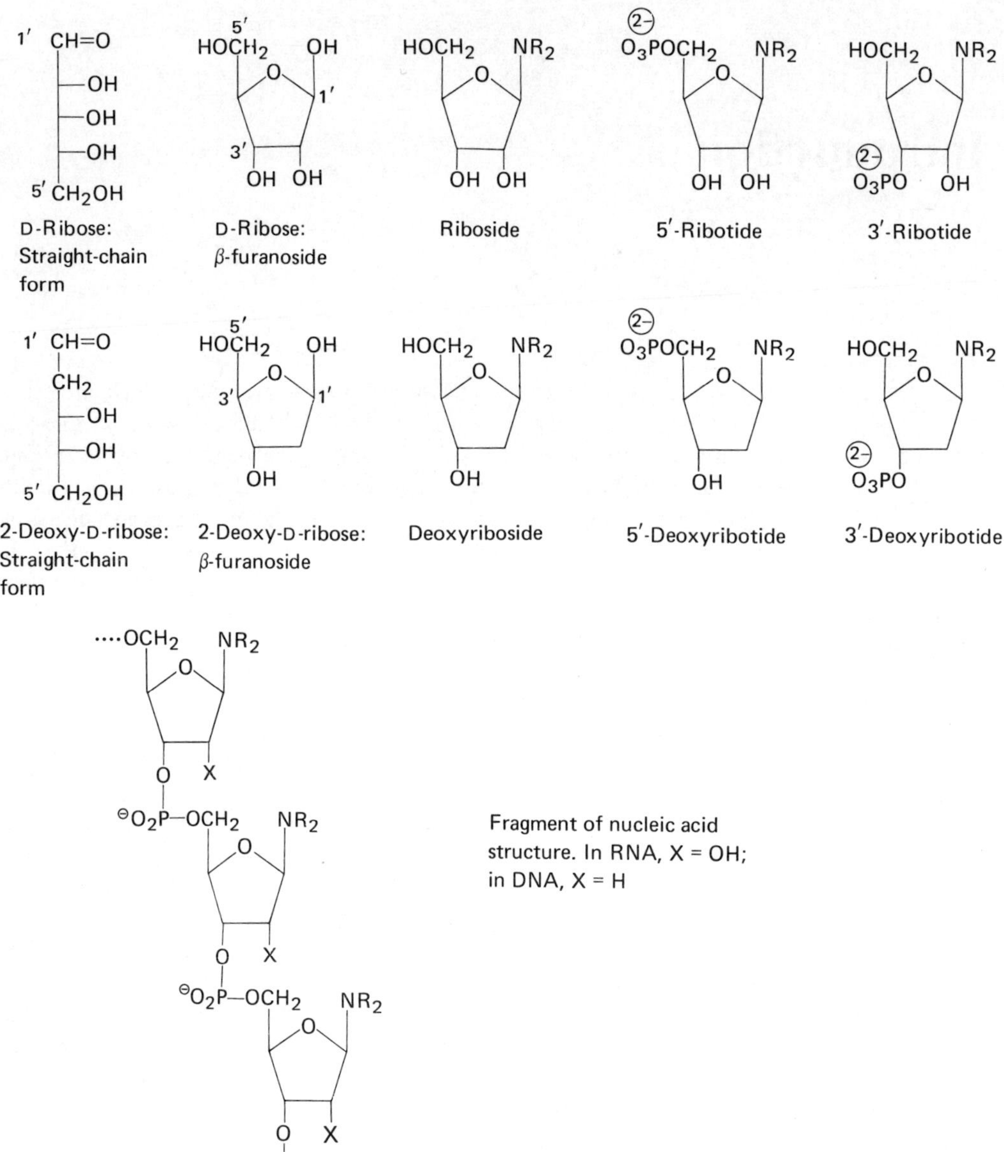

FIG 1.1   Sugars, nucleosides, nucleotides and nucleic acids. In these formulae, the heterocyclic cyclic amines would be represented as R_2NH.

The remaining details of the outline structure of nucleic acids are as follows. The bases are all secondary amines and are attached to the sugars by *N*-glycoside bonds, the sugars being in the β-furanose configuration. The phosphodiester groups are attached to carbon atoms 5 and 3 of the adjacent sugar residues. These carbons are normally designated 5′ and 3′ (pronounced 'five primed' and 'three primed'). The 'priming' is to distinguish reference to positions in the sugar moieties to ring atoms in the heterocyclic bases. Thus, neglecting for the moment the nature of the bases, we can summarize the structures mentioned so far (Fig. 1.1). Note that a nucleic acid can be regarded as a polymer of either 5′- or 3′-nucleotides.

The details of the chemical degradation of nucleosides, nucleotides and nucleic acids are discussed later in the book, but we may note here that in general *N*-glycosides are relatively stable to alkali but are hydrolysed by acid. For a nucleoside, the sugar and the free base would be the products of such a reaction. Ribose itself is fairly stable in acid but in hot strong acid it is dehydrated (as are all pentoses) to form furfural; this forms the base of the orcinol test for RNA. Deoxyribose, on the other hand is less stable in acid and rearranges to form δ-hydroxylaevulinic acid; this is the basis of the diphenylamine test for DNA.

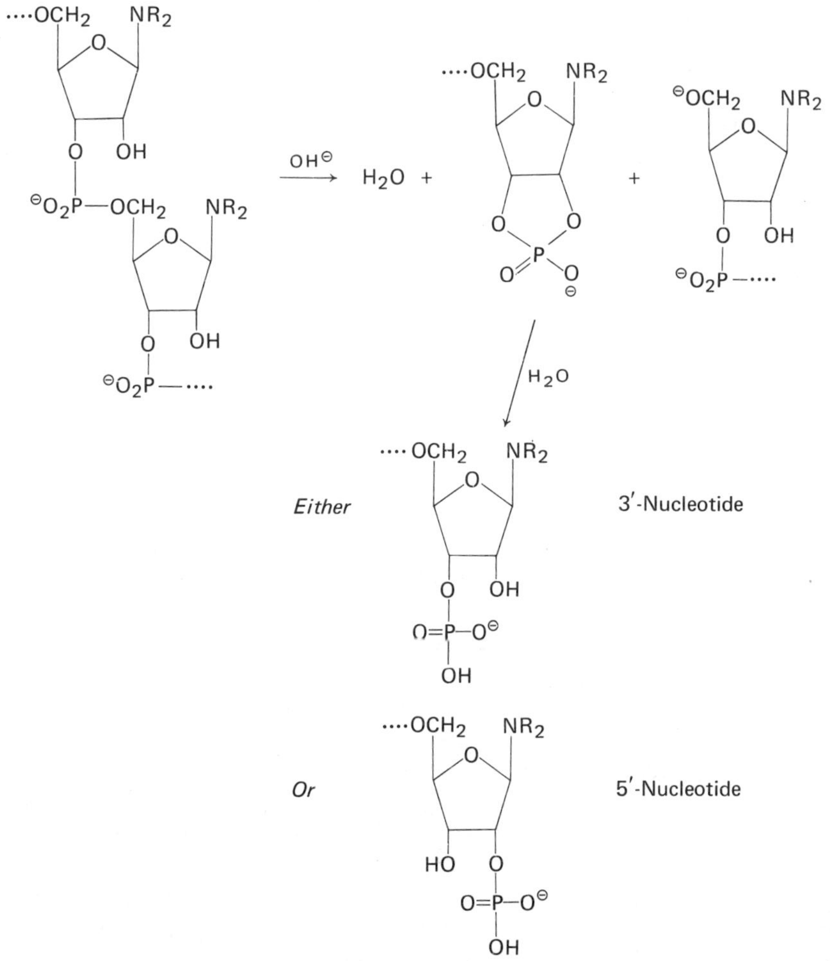

FIG 1.2   Involvement of the 2′-OH groups in the alkaline hydrolysis of a 5′-phosphate ester bond in RNA.

Phosphodiesters are in general fairly stable to hydrolysis. RNA however, is an exception to this general rule. In this case the 5′-phosphate ester bond is relatively easily hydrolysed by alkali. The reason is that the free 2′-hydroxyl group is involved in the reaction to produce a cyclic phosphate intermediate (Fig. 1.2). This type of acceleration of the reaction of one group by the participation of another one in a different part of the same molecule is known as anchimeric assistance. In the present case there are two consequences of this anchimeric effect. One is that when compared to most phosphodiesters (including DNA) RNA is relatively easily hydrolysed by alkali; the other is that the products are mixtures of 3′- and 2′-nucleotides owing to the alternative modes of hydrolysis of the cyclic phosphate intermediate (see Fig. 1.2). This simple observation (that alkaline hydrolysis of RNA yielded both 2′- and 3′-nucleotides) led to a good deal of confusion during the early chemical work on the elucidation of the outline structure for RNA.

The bases present in nucleic acids are substituted pyrimidines or purines. In Fig. 1.3 are the structures of the parent heterocyclic compounds (pyrimidine and purine themselves) together with the conventional numbering of the atoms in the rings. This is the current convention for numbering and is employed throughout this book. However, many earlier publications use alternative conventions and it is extremely important to establish the convention from the context of the writing in reading any paper dating before about 1965. Obviously pyrimidine itself is not a secondary amine and consequently could not form an N-glycoside except as a quaternary ammonium salt. The pyrimidines which occur in nucleic acids are invariably

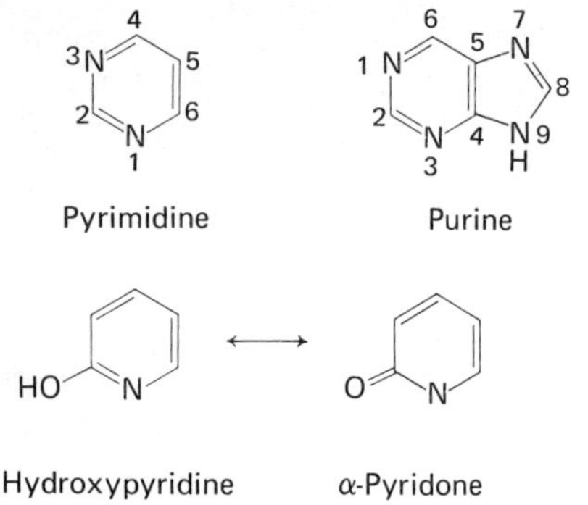

Pyrimidine   Purine

α-Hydroxypyridine   α-Pyridone

FIG 1.3   Conventions for numbering pyrimidine and purine and an example of a tautomeric equilibrium between a hydroxyimine (lactim) and a lactam.

substituted with an oxygen on C-2. A secondary amine group is then generated at N-1 as a consequence of a lactam/lactim equilibrium (sometimes erroneously referred to as a keto/enol equilibrium) of the type illustrated (Fig. 1.3) for the simpler case α-pyridone.

The major bases in RNA are cytosine and uracil (pyrimidines) and adenine and guanine (purines). In DNA, the uracil is replaced by thymine (5-methyluracil); the other three bases are as in RNA. The structures of these bases and the names of the corresponding nucleosides and nucleotides are given in Fig. 1.4. In the pyrimidine nucleosides N-1 is involved in the N-glycoside bond linking the base to the sugar; in the purines it is N-9. Note that the prefix 'deoxy' is dropped from the names of deoxyribonucleoside and deoxyribonucleotides of thymine. This convention originated at a time when it was assumed that thymine only occurred in DNA. It is now known that some RNA molecules contain thymine as a 'minor base'. In this case, of course, it is attached to ribose and the nucleoside has to be called thymine riboside.

Several, but not all, nucleic acid molecules contain minor components. In RNA the minor components are 2′-methylribose, bases methylated, acetylated, reduced or otherwise modified and a nucleoside in which uracil is attached to the ribose via a carbon-carbon bond (C5-Cl′) rather than an N-glycoside. This last nucleoside is known as pseudouridine. In DNA the only minor components identified to date are in the form of substituted bases (certain of which, in some viral DNAs contain oligosaccharide side chains).

The question of shorthand forms of representing these components of nucleic acids is somewhat confused. In laboratories using these compounds (and in many publications) the capital initial letters C, U, T, A and G are used to refer to the bases or nucleosides or nucleotides depending on the context. The conventional abbreviations currently advocated by the advisory committee of IUPAC-IUB[3] are to use three letter abbreviations for bases and ribosides as follows (base first and then riboside): for C Cyt and Cyd, for U Ura and Urd, for T Thy and Thd, for A Ade and Ado, for G Gua and Guo. Pseudouridine becomes Ψrd. Certain minor bases and their ribosides have similar abbreviations (see chapter 2 for the structures). These include xanthine and xanthosine (Xan and Xao), hypoxanthine and inosine (Hyp and Ino) and orotic acid and oritidine (Oro and Ord). In nucleosides in which the sugar is a pentose other than ribose, the three-letter abbreviation for the nucleoside is prefixed by a small letter: d

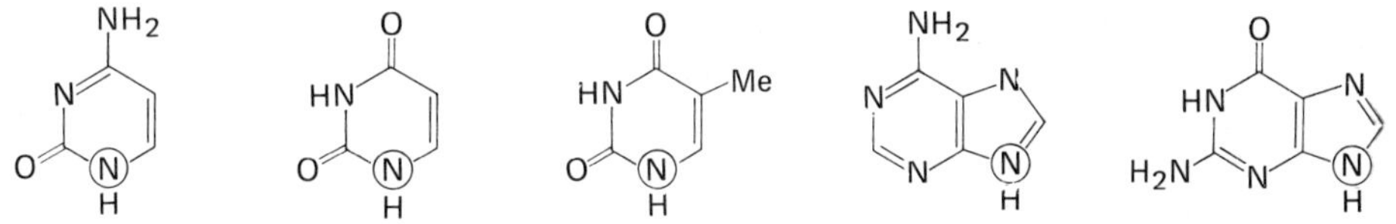

BASE	Cytosine	Uracil	Thymine	Adenine	Guanine
RIBONUC-LEOSIDE	Cytidine	Uridine		Adenosine	Guanosine
RIBONUC-LEOTIDES (2'-, 3'- and 5'-)	Cytidylic acids OR cytidine monophosphates (CMPs)	Uridylic acids OR uridine monophosphates (UMPs)		Adenylic acids OR adenosine monophosphates (AMPs)	Guanylic acids OR guanosine monophosphates (GMPs)
DEOXYRIBO-NUCLEOSIDE	Deoxycytidine		Thymidine	Deoxyadenosine	Deoxyguanosine
DEOXYRIBO-NUCLEOTIDE (3'- and 5'-)	Deoxycytidylic acids OR deoxycytidine monophosphates (dCMPs)		Thymidylic acids OR thymidine monophosphates (dTMPs)	Deoxyadenylic acids OR deoxyadenosine monosphosphates (dAMPs)	Deoxyguanylic acids OR deoxyguanosine monophosphates (dGMPs)

FIG 1.4   Structures of common bases and names of their derivatives. A circle has been drawn round the nitrogen atom attached to the sugar (via an N-glycoside bond) in the nucleosides.

(deoxyribose), a (arabinose), x (xylose) or l (lyxose). Note that according to this convention, thymidine is dThd and thymidine riboside is Thd. Another example would be the drug known as cytosine arabinoside (the same as cytidine but with arabinose replacing ribose) which means aCyd. The conventions shown in Fig. 1.4 for the nucleotides are allowed by the IUPAC-IUB rules. As an example the two deoxyguanylic acids are 3'-dGMP and 5'-dGMP and the thymidylic acids 3'-dTMP and 5'-dTMP.

There are several conventions for representing nucleic acid sequences. They all depend on the assumption that each individual nucleoside is 'viewed' from the same vantage point as is implied by the structures in Figs. 1.1 and 1.2, that is to say that 5' lies to the left of 3'. The shortest convention accords with the following rules. (1) Do not distinguish between RNA and DNA, the nature of the sugar will be apparent from context. (2) Use a single capital letter to identify the major nucleosides. (3) Use a hyphen to indicate a phosphate group. (4) Modifications of bases are shown by small letters preceding the capital initial letter, superscript numbers indicating position and subscripts number or substituents in that position if necessary. (5) Modifications to sugars are represented by small letters following the symbols for nucleosides.

From rules 2 and 3 we could represent $5'$-UMP as -U and $3'$-UMP as U-. A-U would mean a dinucleoside monophosphate which consists of an adenosine with a free $5'$-OH group (no phosphate) a phosphodiester (linking $3'$ of the A to $5'$ of the U) and a uridine with a free $3'$-OH. The dinucleotide -A-U has a phosphomonoester group attached to $5'$ of the A and is otherwise the same. The dinucleotide A-U- has $3'$-phosphomonoester group on the U. A relatively minor additional point is that A-U! is used to indicate the same dinucleotide in which a $2',3'$-cyclic phosphate is present on the uridine. A more extended oligonucleotide would be represented as A-A-G-A-C-C-U-. In such a molecule the 'left-hand' end is referred to as the $5'$-end (in this case an adenosine with a free $5'$-OH) and the 'right-hand' end as the $3'$-end (in this case a uridine bearing a $3'$ phosphate). An alternative to a hyphen for representing phosphate is p. In this book, I only use p when several phosphates are linked by phosphoric anhydride links. Thus pyrophosphate is pp, ATP is pppA and an RNA molecule with a $5'$-triphosphate group at one end would be written pppN-N-N-N. . . .

In addition to the letters C, U, T, A and G the following are recognized I (inosine), Ψ (pseudouridine), X (xanthosine), R (unidentified or non-specified purine), Y (unidentified or non-specified pyrimidine), N (completely unidentified or non-specified nucleoside). In this book, all these abbreviations are employed except R and Y. In my view these are extremely confusing as R means 'alkyl' and Y has a quite specific meaning in sequences of tRNA (see p. 301). Therefore I use Py for an unidentified pyrimidine and Pu for an unidentified purine. Examples of prefix letters are m (methyl, ac (acetyl), h (hydrogenated), s (thio). Examples of the use of these abbreviations are contained in the oligonucleotide C-hU-A-m$_2^6$A-I-ac$_4$C-Gm-G-, which has a free hydroxyl at the $5'$-end and the residues C, 5,6-dihydroU, A, 6-dimethyl A, I, 4-acetyl C, $2'$-methyl G and G (with a $3'$-phosphomonoester).

Synthetic polynucleotides are represented in the following ways. A homopolymer (one in which every base is the same) is represented simply as, for example, poly A (poly adenylic acid) or poly dA (polydeoxyadenylic acid). A polymer with a short, defined, repeating sequence is represented either as poly d (A-T) or d(A-T)$_n$ (for -A-T-A-T-A-T- . . . , all deoxyribosides). Random polymers are molecules of undefined sequence but known overall composition; these are represented as, for example, $(A_3C_2)_n$ for a random co-polymer of adenylic and cytidylic acids with the molar ratio of A and C as indicated (3:2 in this example).

Another set of conventions used a vertical line to represent the side view of the ring of the sugar molecule. These representations are most easily understood by reference to a particular example. Consider the trinucleotide -A-G-U-. This could be represented as:

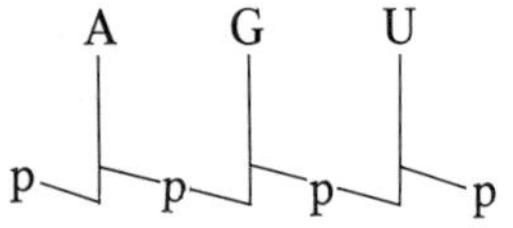

$$\text{A} \qquad \text{G} \qquad \text{U}$$

or alternatively, if we wished to make it quite clear this was an RNA fragment (and not a DNA fragment) we could include the $2'$-OH groups:

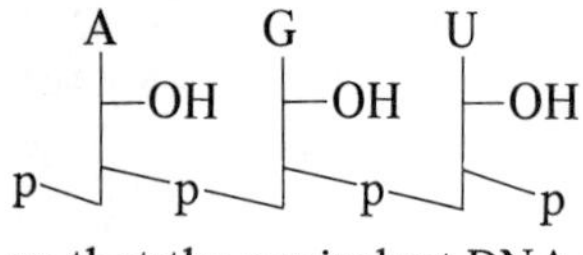

$$\text{A} \qquad \text{G} \qquad \text{U}$$

so that the equivalent DNA fragment would become:

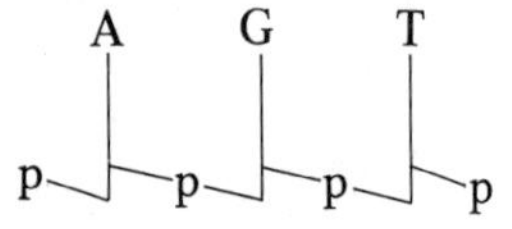

$$\text{A} \qquad \text{G} \qquad \text{T}$$

The main advantage of this convention is that it is immediately clear which are the 5′ and 3′ ends as we could write the deoxytrinucleotide given above equally well upside down thus:

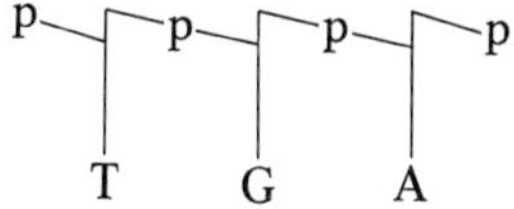

With the simpler form this cannot be done of course as -T-G-A- is a completely different substance to -A-G-T-, so that if we want to draw two chains of nucleotides running 'in opposite directions' as we frequently do, some sort of additional information has to be included either by labelling the ends thus 5′-A-G-T-3′ (which is identical to 3′-T-G-A-5′ but different to 5′-T-G-A-3′) or alternatively by putting in an arrow to indicate the 5′ to 3′ direction.

Before leaving the question of conventions there is just the usual word of warning. The earlier literature is full of different conventions, especially for the minor nucleosides; hU appears variously as DiHU and UH_2, m_2^6A as diMeA, Gm as 2′-GOMe and so on.

Secondary structure

There are two major types of secondary structure (conformation) found in nucleic acids—double helices (or duplices) and base stacking.

Secondary structure of DNA

The double helix was proposed by J. D. Watson and F. H. C. Crick to account for the secondary structure of DNA. Their model accommodated two separate sets of data, X-ray diffraction patterns, obtained from fibres of DNA by M. H. F. Wilkins, and generalizations about the relative amounts of nucleotides in different DNA samples recognized by E. Chargaff. DNA dissolved in water produces an extremely viscous solution. If an organic solvent, typically ethanol, is added to such a solution the DNA is precipitated as material which looks rather like soggy cotton wool. X-ray diffraction patterns can be obtained from fibres picked out of this precipitate and from the pattern it is possible to deduce the parameters of regular repeating structures in the molecules aligned within the fibre. Insufficient data were obtainable from such diffraction patterns to unambiguously point to a specific secondary structure.

The generalizations of Chargaff, which gave Watson and Crick the other clue they needed, are usually referred to as Chargaff's rules. These rules (they are two in number) were based on a massive series of experiments by Chargaff in which DNA from a wide variety of animal and microbial sources was hydrolysed in acid to produce a mixture of the four bases (Gua, Ade, Thy and Cyt) and the bases were quantitatively analysed to produce normalized percentages known as the 'base ratios'. Each base ratio is represented by the capital letters G, A, T and C and are defined as:

$$G = \frac{[Gua]}{[Gua] + [Ade] + [Thy] + [Cyt]} \times 100 \text{ etc.}$$

where square brackets denote the molar concentrations in the hydrolysate. Chargaff's rules state that in any DNA:

$$A + C = G + T$$
and
$$G + A = C + T$$

In words, the first rule can be stated in the form that if we consider the substituent 'at the top of the pyrimidine ring' (position 4 in pyrimidines and position 6 in purines), the total amount of carbonyl equals the total amount of amino and the second rule is simply 'total purines equal total pyrimidines'. In elementary biochemistry text books, Chargaff is frequently credited with a third rule, viz.

$$\frac{G}{C} = \frac{A}{T} = 1$$

However, this is merely an algebraic consequence of the first two rules (together with the trivial equation, $G + A + C + T = $ a constant (100)).

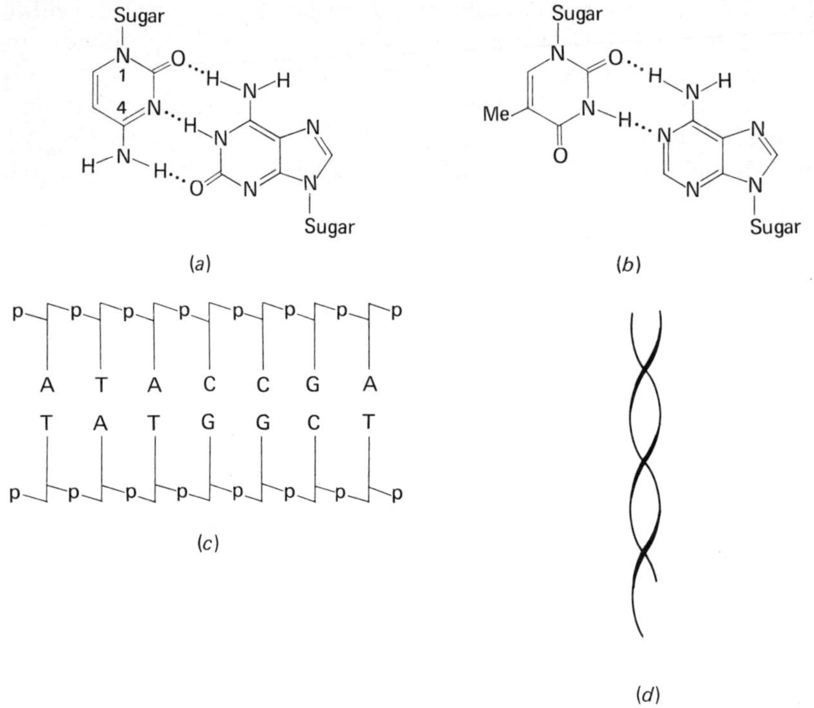

FIG 1.5   Base pairs C and G (a) and T and A (b) and also two anti-parallel complementary DNA strands (c) and a right-handed double helix in which each line represents a sugar-phosphate 'backbone' (d).

A consequence of Chargaff's rules is that if they are strictly obeyed the base ratios for any DNA are defined by the value of any one of the quantities (G, A, C or T) in the equations. In practice base ratios can be measured in two ways, either directly as was done originally by Chargaff, or indirectly for several physical properties of DNA are functions of the base composition. If the first method is employed the base ratios should be calculated as a mean of several determinations and the standard errors of the mean for each base ratio should be quoted. If the base composition is deduced indirectly, the normal practice is to quote the value of $(G + C)$, the 'GC-value'. For example, a GC-value of 40% (approximately true for all mammalian DNA) implies that $G = 20\%$, $C = 20\%$, $A = 30\%$ and $T = 30\%$.

To return to the double helix, Watson and Crick's model of DNA is that it is a double molecule with two strands related to each other by the fact that for every A in one strand there is a T in the other and for every G in one strand there is a C in the other. This model accounts for Chargaff's rules. The bases A and T are said to be 'complementary' as are the bases G and C and also the two DNA strands. The bases are associated with one another by hydrogen bonding to form planar structures. The sugar-phosphate 'backbones' of the two complementary DNA strands are wound round another to form a right-handed double helix. The outlines of the structure that have been described so far are shown in Fig. 1.5. Notice that the two complementary strands are shown with their 5'-3' directions opposed. This conformation is called 'anti-parallel'. The alternative, which does not occur in any polynucleotide structure studied so far, would be called 'syn-parallel'.

There remains the question of the actual geometry of the double helix. Fibrous DNA can exist in one of two conformations known as A-DNA and B-DNA. The simpler structure to visualize is B-DNA, the form for which Watson and Crick originally proposed a model and which is generally assumed to be the predominant structure in solution. In B-DNA, the base pairs are parallel to one another and lie in planes at right angles to the helix axis. The spacing between the layers of this infinite sandwich of base pairs is 3·4 Å, and the number of pairs per complete turn of the helix is ten. Hence the repeat length along the helix is 34 Å. The

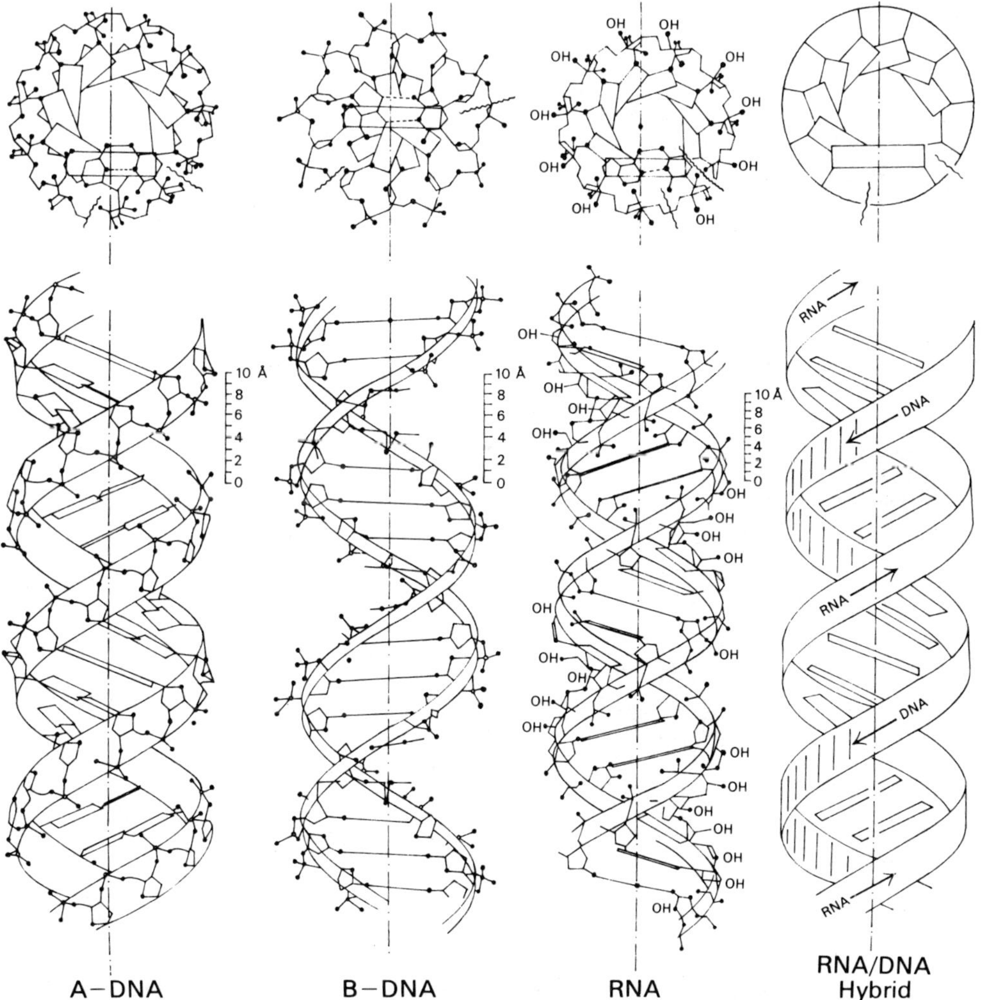

FIG 1.6   Double helical structures of DNA, RNA, and DNA/RNA hybrids. The sugar-phosphate backbone is represented thus ⌂⌂ ..., and base pairs as ▭. Re-drawn from Yang and Samejima (1969).

phosphorus atoms are 9 Å out from the helix axis. In the B-form, the base pairs are tilted at 20° to the planes normal to the axis and the number of base pairs per complete turn is eleven. Models of these two structures are shown in Fig. 1.6.

Note that in the B-structure, the helix produces two grooves of unequal size and that when the projection of the B-structure into a plane perpendicular to the axis is studied, the central area is covered with base pairs. In other words, if you were to look down the helix you would see a tube of sugar-phosphate structure apparently full of base pairs. In the A-structure this is not the case, instead you would get a foreshortened view of base pairs which would only be co-planar as each pair was compared with another 11 pairs further along and down the middle there is a space covered by no bases at all.

The double helix consists of two completely independent, covalently linked molecules to distinguish them from one another (although some wits refer to them as Watson and Crick). The sequence of one strand determines the sequence of its complementary fellow, but there is no reason why the overall base compositions of the two need be the same; to take an extreme case all the G and T could be in one molecule and all the C and A in the other. Thus Chargaff's rules only apply to the overall composition of the pair.

It is possible (by raising the temperature or having extremes of pH) to denature solutions of DNA. This process consists in molecular terms of unzipping the base pairs and unwinding the helix. The result of this process is to produce in solution a mixture of the two complementary molecular species in a denatured random coil conformation (as opposed to the native, double helical conformation). This denaturation is associated with large changes in the viscosity and optical properties of the solution. In the case where the two separate complementary strands are of different overall base composition, the two molecular species can be separated, as the density of DNA is a function of base composition, and macromolecules of different density are separable (see chapter 4). In these cases there *are* names for the two complementary molecules, the denser partner is known as the heavy strand and the less dense one as the light strand.

Secondary structure of RNA

Most RNA species do not exist in the form of two complementary molecules. Certain forms of viral RNA are an important exception to this generalization. However, many single-stranded

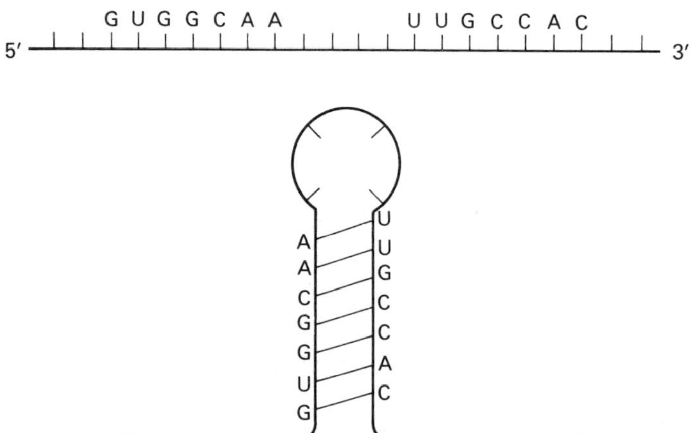

FIG 1.7   Schematic demonstration of the formation of an A-type double helical region in an RNA molecule with intramolecular complementary base sequences. Notice that the strands in the double helix are antiparallel.

RNA molecules do have within them regions of intramolecular complementarity so they can fold back on themselves and produce short sections of Watson-Crickery as illustrated in Fig. 1.7. In all cases of double helical structure in RNA, whether it is in a viral complementary pair of molecules with total helical conformation or a short loop as in Fig. 1.7, the helix is invariably of the A-type with the base pairs inclined to the helical axis. However the repeating unit of eleven base pairs is not universally found.

Base stacking

Base stacking is the description of the other type of secondary structure found in polynucleotides. Just because the nucleotide residues are not paired up with their complementary fellows does not mean that the orientations of the bases with respect to geometric parameters of the chain is not ordered or restricted. In particular, the bases may tend to stack, that is to say lie in a particular orientation roughly parallel to their neighbouring bases. The bases in a loop of the type illustrated in Fig. 1.7 would typically be stacked to some extent. Methods of estimating helical content and extents of base stacking in polynucleotides are described in chapter 11.

Metabolism of nucleic acids

General features

The processes involved in nucleic acid metabolism and the biosynthesis are briefly summarized in Fig. 1.8. DNA, being a double molecule can direct the synthesis of more of itself as each strand contains the necessary 'information' required for the synthesis of its complementary fellow. DNA is thus the ultimate source of information for the whole living system and this

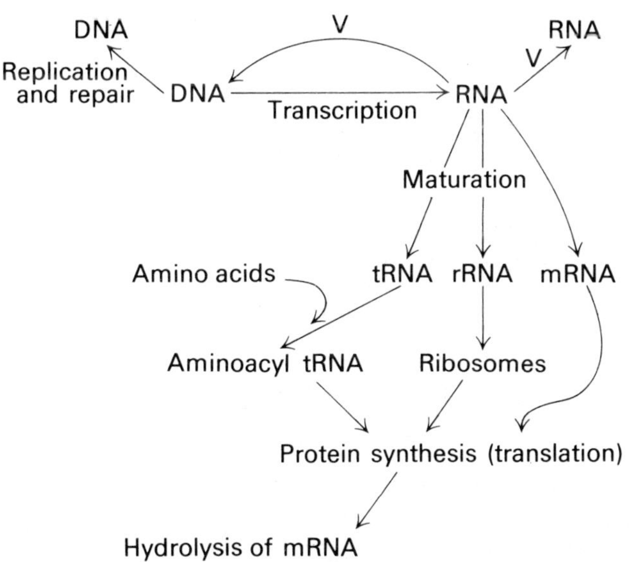

FIG 1.8   Relationship between nucleic acids and protein synthesis. The process marked 'V' (RNA-directed synthesis of RNA and DNA) probably occurs only in cells infected with certain viruses. tRNA = transfer RNA, rRNA = ribosomal RNA and mRNA = messenger RNA. mRNA is a direct product of transcription but tRNA and rRNA are modified by 'maturation' enzymes.

complementary copying is involved in two separate processes. One is repair of regions of DNA in which, accidentally, a base or bases have become altered. The other is the replication of all the DNA in the cell which is a necessary preliminary to cell division. DNA is also involved in the process of RNA synthesis (transcription) in which a molecule of RNA is complementary to a portion of one of the DNA strands ('the transcribing strand'). RNA then is the immediate template molecule required for the direction of 'translation', that is the ordered incorporation of amino acids into protein molecules.

The enzymes required for these two processes in which DNA acts as template are DNA polymerase and RNA polymerase. Both these enzymes have as substrates four triphosphates

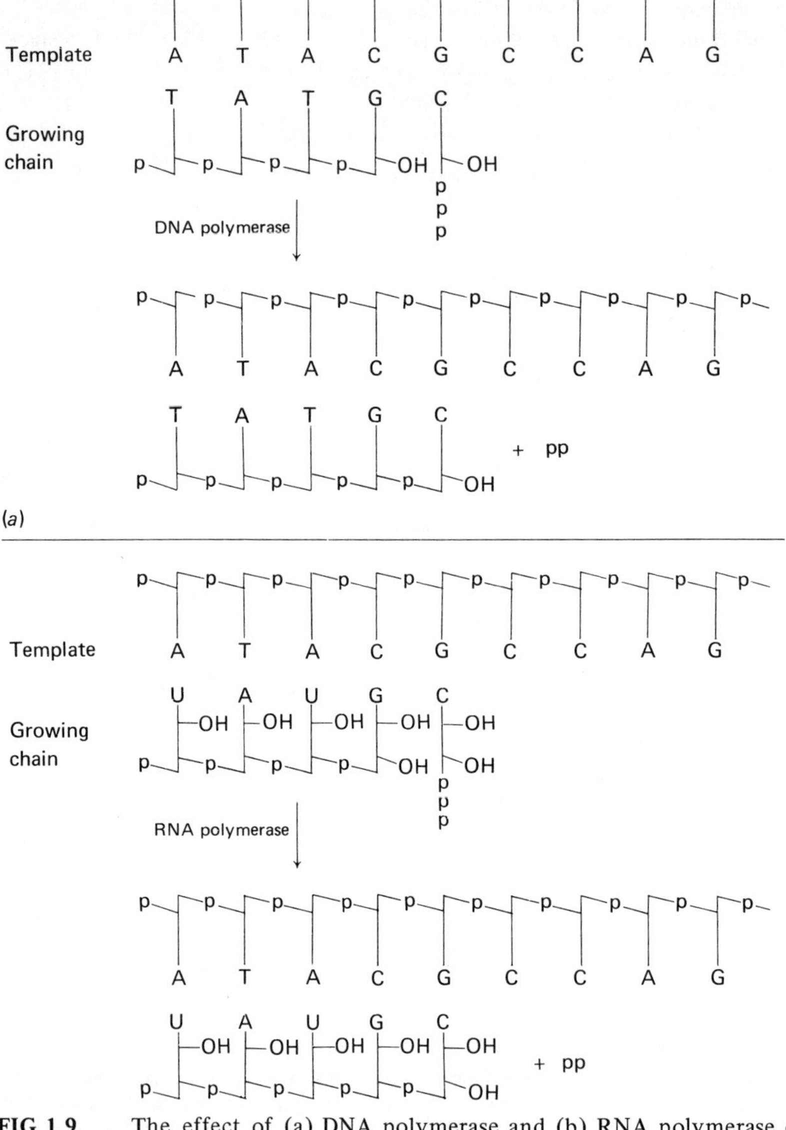

FIG 1.9   The effect of (a) DNA polymerase and (b) RNA polymerase on production of a molecule complementary to the same DNA template.

and the growing polynucleotide chain. The free $3'$-OH of the growing chain alcoholizes the triphosphate to produce the phosphodiester bond with the production of pyrophosphate. DNA polymerase untilizes the deoxyriboside triphosphates dCTP, dTTP, dATP and dGTP; RNA polymerase utilizes the riboside triphosphates CTP, UTP, ATP and GTP. Like all enzymes for which triphosphates are substrates, these polymerases have an absolute requirement for a divalent cation, normally Mg^{2+}. The polymerases are unique however in the role of the DNA template in the process. The incoming triphosphate is selected by the template DNA strand so that a complementary molecule is assembled. Notice that for both enzymes the newly-synthesized complementary molecule is antiparallel to the template and it is extended from the $3'$-end. The reactions are summarized in Fig. 1.9.

The products of transcription are classified according to their metabolic roles. The three classes are (1) messenger RNA (mRNA), (2) transfer RNA, tRNA for short, referred to as soluble RNA (sRNA) in some of the older literature and (3) ribosomal RNA (rRNA). The sequence of nucleotides in mRNA determines the sequence of amino acids to become incorporated in a new protein molecule. Messenger RNA is thus the template for protein biosynthesis. The role of rRNA is structural, it forms a sort of skeleton for the assembly of the ribosomes which are small particles consisting of rRNA and specific ribosomal proteins, and which are the sites at which protein biosynthesis occurs. Transfer RNA plays a dual role in the intermediary metabolism of protein biosynthesis which is described in the next paragraph; tRNA molecules are the smallest naturally occurring nucleic acids (about eighty nucleotides). Messenger RNA is probably a direct product of transcription (at least in bacteria). The other two are not. RNA polymerase synthesizes precursors of rRNA and tRNA. These precursors are 'matured' by specific enzymes which are of two types. There are nucleases which hydrolyse specific phosphodiester bonds to cut down the size of the precursors (which are invariably longer than the final tRNA or rRNA) and there are enzymes which modify certain nucleosides to produce the minor components. The immediate products of any reactions catalysed by the polymerases contain only the four major nucleosides (ribosides of C, U, A and G, or deoxyribosides of C, T, A and G) and all the minor components originate from maturation processes. There is an exception to this statement for certain phage specific DNA polymerases (see p. 269).

The process of protein synthesis consists of repeated sequential synthesis of peptide bonds. The reacting species are aminoacyl and polypeptidyl esters of tRNA. The aminoacyl tRNA synthetases (or 'amino acid activating enzymes') are the enzymes responsible for making the aminoacyl tRNA molecules from the amino acid and a tRNA specific for that amino acid. All tRNA molecules have a common sequence at their $3'$-end, namely . . . -C-C-A. There is an aminoacyl tRNA synthetase specific for each amino acid which recognizes a nucleotide sequence in a part of the tRNA molecule and selects the correct amino acid. The reaction in which aminoacyl tRNA is synthesized involves the overall hydrolysis of one mole of ATP for every mole of aminoacyl tRNA synthesized. The reaction is summarized in Fig. 1.10: the abbreviations for specific tRNA molecules (both 'free' and 'charged') which are introduced in this figure are employed throughout the book.

The reaction between aminoacyl tRNA and the growing end of the peptide chain, which is also attached to a tRNA molecule, is illustrated in Fig. 1.11. Notice that this reaction, which lies at the centre of the processes of protein synthesis, gene expression and the whole subject of molecular biology, is in fact an example of a sixth-form chemistry truism, 'amine plus ester equals amide plus alcohol'. In this case the amine is the amino group of aminoacyl tRNA, the ester is the ester bond in polypeptidyl tRNA, the amide is the new peptide bond and the alcohol is the free tRNA. This reaction occurs without being coupled to the hydrolysis of a triphosphate as the reaction is associated with a significant negative free energy change on its own. The reason is that the amino group in aminoacyl tRNA is 'activated' and reacts readily with the ester bond. An ATP molecule was hydrolysed in the 'activation' (i.e. formation of aminoacyl tRNA). Peptide bond formation occurs on the ribosome and the enzyme required to catalyse it (aminoacyl tRNA peptidyl transferase) is one of the ribosomal proteins.

The other aspect of translation is that the appropriate aminoacyl tRNA molecules are selected to result in the synthesis of a protein of defined amino acid sequence. Twenty amino

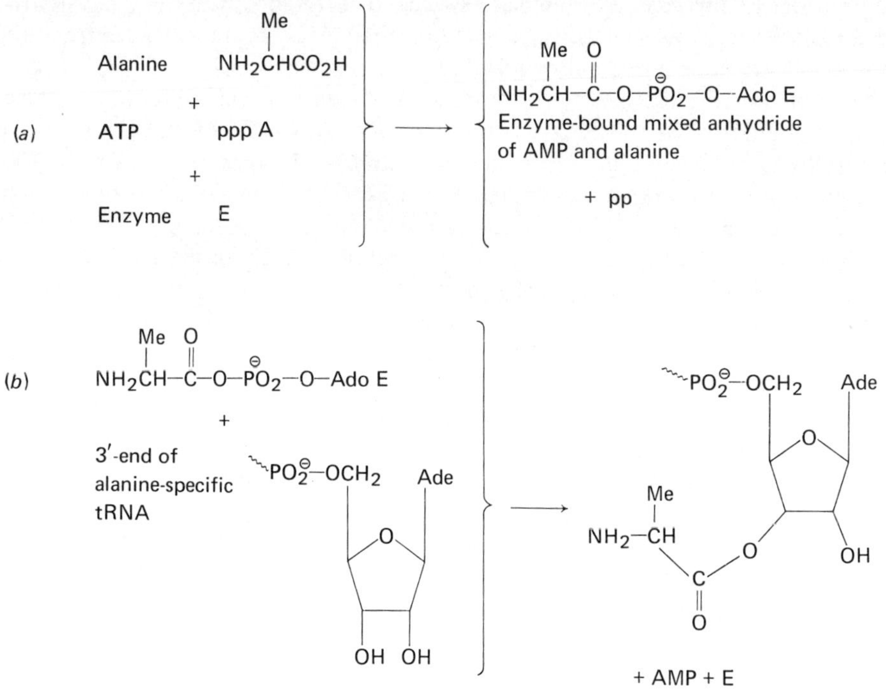

Abbreviated form for these reactions:

(a) ala + ATP + Enz $\rightarrow$ ala-AMP-Enz + pp

(b) ala $-$ AMP $-$ Enz + tRNA^{Ala} $\rightarrow$ ala tRNA^{Ala} + Enz + AMP

FIG 1.10　　Formation of aminoacyl tRNA illustrated with reference to alanine.

acids are incorporated into proteins by this process. With the exception of glycine which is not an optically active molecule, they are all in the L-configuration. These twenty (with their three-letter abbreviations) are as follows: alanine (ala), arginine (arg), asparpartic acid (asp), asparagine (asn), cysteine (cys), glutamic acid (glu), glutamine (gln), glycine (gly), histidine (his), leucine (leu), *iso*leucine (ile), lysine (lys), methionine (met), phenylalanine (phe), proline (pro), serine (ser), threonine (thr), tryptophan (trp), tyrosine (tyr) and valine (val). Any other L-amino acids found in proteins (e.g. cystine, hydroxyproline, glycosylated threonine, serine phosphate) arise from subsequent modification of the peptide chain after peptide bond formation. Polypeptides containing such amino acids as diaminopimelic acid, sarcosine or D-amino acids (examples of such polypeptides are found in the structures of some antibiotics and bacterial cell walls) are invariably synthesized by a process independent of the ribosomes and not directed by mRNA. The mRNA is bound to the ribosomes and the sequence of its nucleotides determines the sequence of amino acids in the protein. The relationship between these two sequences is known as the genetic code. The general features of this code are that groups of three adjacent nucleotides called either triplets or codons correspond to individual amino acids. As there are four nucleotides, there are 4^3 or 64 possible codons. Of these, 61 correspond to amino acids; there are only 20 amino acids and hence most amino acids have more than one codon. This is what is meant when the genetic code is spoken of as being degenerate.

The protein chain is synthesized in the direction amino to carboxyl and the genetic 'message' is 'read' in the $5'$ to $3'$ direction. It has already been stated that the only point in the sequence of reactions in which there is a recognition of a nucleotide sequence by a protein is in the synthesis of aminoacyl tRNA. On the ribosome the selection of the aminoacyl tRNA is by complementary, antiparallel base-pairing between the codon (in mRNA) and a tri-nucleotide

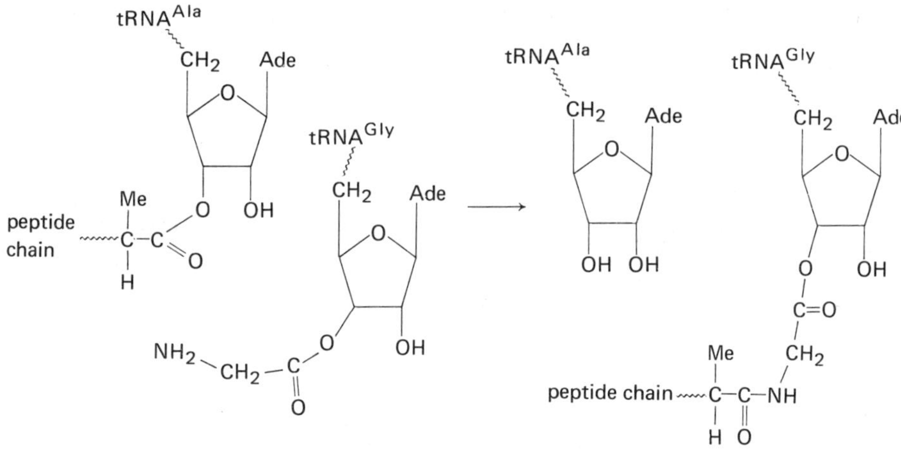

FIG 1.11   Peptide-bond formation. In this example, the growing chain is shown terminating in an alanine residue. At the next amino acid, whose incorporation is shown here, is glycine. After the reaction, the peptide chain is longer by one amino acid residue and ends again attached to a tRNA molecule, albeit a different one. The process is repeated for the incorporation of the next amino acid (when tRNA^{Gly} will be released).

sequence in the tRNA called the anti-codon. In Fig. 1.12, the processes of transcription and translation leading to the incorporation of glycine, following alanine (previously shown in chemical outline in Fig. 1.11) are summarized to emphasize the antiparallel and complementary relationships between the various nucleic acid molecules. The diagram is not meant to imply any particular secondary structure for the nucleic acids, although tRNA has a characteristic secondary structure which is now known. As is correctly implied by this figure, GCA is a codon for alanine and GGU is one for glycine.[4] The two nucleosides marked with asterisks (those in the anti-codons corresponding to the third base of the codon) would typically be minor components (modified bases or sugars) with the same base-pairing properties as the U and A in question. The reason for this is explained on p. 23. After the peptide bond (between ala and gly) had been formed the tRNA^{Ala} would be discharged from the ribosome, and the mRNA would move along to bring the next codon (NNN) into the appropriate binding site on the ribosome to select the next aminoacyl tRNA (one with an anti-codon complementary to NNN).

So much for the simple mechanistics of the protein synthetic process, but what of the energetic aspects? It has been stated above that the actual peptide bond forming reaction is not coupled to the hydrolysis of a triphosphate as it is a thermodynamically favourable reaction as it stands. However for each round of incorporation of a new amino acid into the chain, two molecules of a triphosphate (GTP) are hydrolysed. One is required for the binding of the aminoacyl tRNA molecule to the ribosome. To biologists lacking the advantages of a classical (i.e. chemical) education this may seem puzzling at first as no covalent bond has to be formed or broken during this binding process and the codon/anti-codon interaction appears to serve as a sensible and selective association. The important point is that it is a selection; that is to say out of the population of the many types of aminoacyl tRNA available, we have to choose just one type. Such an event is associated with a negative entropy change (we are creating order out of chaos) and although the codon/anti-codon interaction presents a mechanism whereby it can be achieved, the process only becomes thermodynamically feasible if it is linked to a process associated with a large negative change in free energy (and consequently a large positive entropy

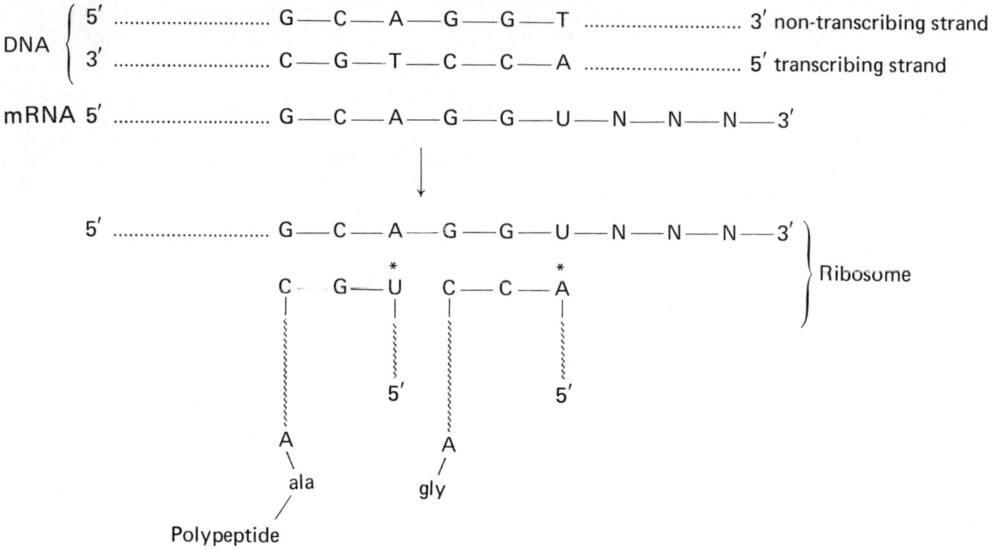

FIG 1.12   Relationship between DNA, mRNA and tRNA for the incorporation of glycine into a growing protein chain. See text for discussion.

change) to swamp the consequences of the unfavourable selection. The other hydrolysis of GTP provides the free energy to drive the mRNA through the ribosome to bring the next codon into a conformation for the next round of aminoacyl tRNA binding and peptide bond formation.

The remaining part of this chapter is devoted to filling in some of the detail in the processes described above. The details have been most precisely elucidated in one particular single-celled, prokaryotic[5] organism, the bacterium, *Escherichia coli*. Therefore I have chosen to go through the various processes as they obtain in *E. coli* and then indicate the differences in other forms of life.

DNA synthesis in *E. coli*

The DNA of *E. coli* consists of a continuous 'circular' molecule referred to as a chromosome. This usage has been criticized, as it is not quite the same as is meant by a eukaryotic chromosome but no alternative nomenclature has yet been universally accepted. In a rapidly dividing culture of *E. coli* the cells contain several copies of the molecule which are all being replicated. DNA synthesis (i.e. chromosome replication) and cell division are synchronized with one another. The mode of chromosome replication is that it is semi-conservative, progressive and starts and stops at a specific point known as the origin of the chromosome. The term 'semi-conservative' refers to the fact that if the two complementary strands are referred to as X and Y, the products of replication are two daughter molecules, one of which consists of the parental X strand and a newly synthesized Y strand and the other of the converse (parental Y and a newly synthesized X). Thus half of the original double-stranded structure is conserved in each of the daughter molecules.

Models for chromosome replication have to account for several features of the process. The two biggest problems are: how does DNA polymerase succeed in copying both strands together and what sort of degree of freedom is introduced into the continuous double stranded helical

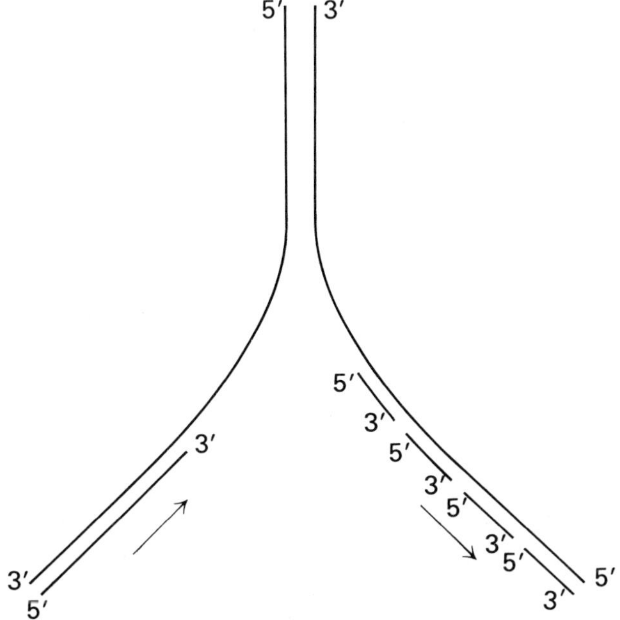

FIG 1.13   Action of DNA polymerase in the replication of double-stranded DNA. The arrows indicate the direction of synthesis.

structure, to account for the unwinding of the helix required for strand separation in the parent molecule? Here I shall only consider the first of these problems. As illustrated in Fig. 1.9(*a*) (p. 12) DNA polymerase is only capable of lengthening a growing DNA chain in one direction ($5'$ to $3'$). In order for such an enzyme to replicate a double-stranded structure semi-conservatively, the synthesis must be discontinuous as illustrated in Fig. 1.13.

Replication of the left-hand molecule could be continuous but the right-hand cannot be. It is known that the overall result is a progressive movement of this type of structure (known as the replication fork) along the molecule and that some of the very recently synthesized DNA does consist of fragments of the type shown on the right. However, the final product contains two continuous strands and DNA polymerase has no activity to join up the breaks, which must be left following the discontinuous synthesis of the right-hand half. The chemical structure of such a break is shown below.

There must be another enzyme in the region of the replication fork that follows the polymerase and patches up these single-strand breaks. An enzyme with the necessary activity is DNA ligase. The enzyme in *E. coli* has as its co-factor nicotinamide adenine dinucleotide (NAD^+) which contains a pyrophosphate bond and (uniquely in this particular reaction) acts as a 'high-energy-phosphate' compound. Normally of course NAD^+ is a proton acceptor in oxidoreductase reactions. The stoicheiometry of the DNA ligase reaction is as follows (only the broken DNA strand is shown).

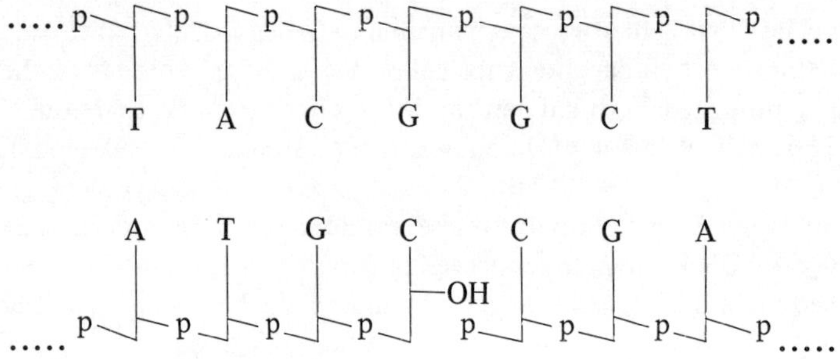

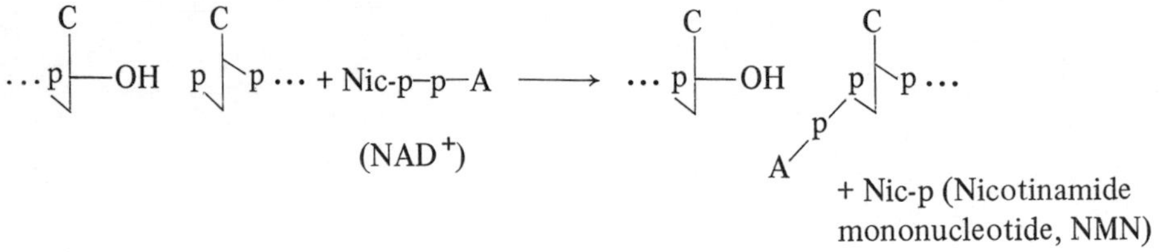

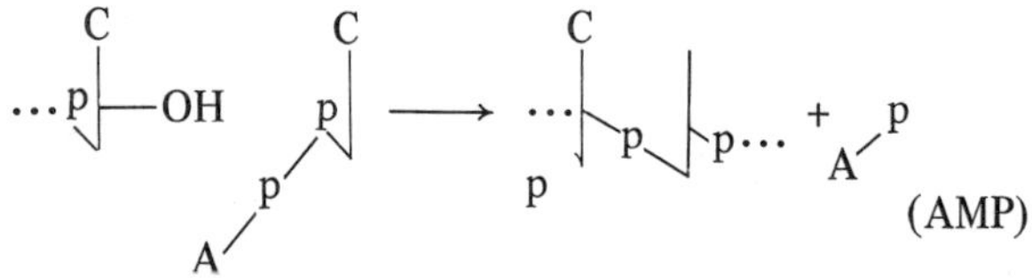

The other process in which DNA polymerase is involved is DNA repair. The main features of the process, which occurs when a base has become accidentally and wrongly modified in DNA, are:

(a) recognition of the 'lesion' by a protein or proteins so far not identified which are apparently scanning the sequence for irregularities,
(b) excision of the damaged base and the surrounding nucleotides,
(c) 'erosion' which consists of a further degradation of the damaged strand and
(d) the synthesis of a new strand to complete the double-stranded structure.

To achieve this last result the two enzymes required for replication, DNA polymerase and DNA ligase, are involved. In this case the polymerase begins at the free 3'-OH at one end of the eroded region and, using the intact strand as template, the polymerase extends the chain in a 3' to 5' direction. Obviously when it gets to the end of the single-stranded region, it leaves a gap exactly of the same type as that produced during the discontinuous synthesis of one of the daughter molecules during chromosome replication. The ligase heals the breach in the same fashion as before.

Details of experimental evaluation of more detailed models for these two processes and the properties of the enzymes involved are discussed later in this book.

General features of transcription and translation in *E. coli*

Chromosome replication is a process which starts at a specific point and is finished when the polymerase returns to the same point, or more properly when the origin has returned to the enzyme if the polymerase is membrane bound and if the nucleic acid is moving. With the processes leading to synthesis of a protein, there is the problem of transcribing and subsequently translating specific regions of the DNA. Moreover the processes have to be exactly controlled as the proteins required by the organism vary with its conditions of growth. The *de novo* synthesis of new proteins in response to changing environmental factors is termed 'induction'. The opposite of induction is termed 'repression'. So the processes of transcription and translation have to cope with two types of refinement. One is that specific regions of DNA have to be transcribed to furnish specific mRNA molecules and the other that there must be a mechanism for induction and repression.

A stretch of DNA whose sequence directs the synthesis of one polypeptide chain is termed a cistron or alternatively a structural gene. The very simplest model for protein biosynthesis would be that transcription is a process which requires specific starting and stopping events (signalled by special nucleotide sequences) and that the product of this process would be an mRNA of exactly the right length to specify the polypeptide in question. As the genetic code is a triplet one this would imply that for a protein of say 200 amino acid residues the mRNA would be of exactly 600 nucleotides. The other point is that of course there are two DNA strands of complementary, not identical sequence. So one of them and not the other must be the transcribing strand. But then the DNA is all part of one giant molecule so perhaps just one strand is the transcribing one for the whole chromosome. To say the very least of it, the truth is somewhat more complex than these simple possibilities. There is no such thing as one transcribing strand for the whole chromosome, for some cistrons one of the strands is transcribing, for others it is the other one. And mRNA is probably never exactly the equivalent size of the protein being made and in many instances it is 'polycistronic', that is to say it has encoded sequences for several different polypeptide chains, so there must be starting and stopping signals in the mRNA as well. The origin of many of the complexities derives from the mechanism of induction and repression in the organism so we shall briefly consider the situation which obtains for the synthesis of a number of polypeptides which are not required all the time but when they are required are required all together. To simplify it as much as possible let us just think of two polypeptides. They could be the two chains of an enzyme with two dissimilar sub-units or they could be two enzymes occupying adjacent positions in a metabolic pathway.

The position is summarized in Fig. 1.14 which also defines some of the neologisms so dear to the hearts of molecular biologists. The only preliminary point to clarify is that the promoter is the binding site for RNA polymerase. Thus in one strand of DNA (the transcribing strand) in the promoter there is a characteristic sequence of bases which bind the enzyme. If the operon is repressed, that is to say transcription is inhibited, the RNA polymerase cannot move and effect the transcription because there is a protein molecule called 'repressor' bound to the operator. The repressor itself is a gene product (i.e. a protein made by transcription and translation) of the gene *i*. Thus in the repressed state there is no transcription of the operon because there is a supply of repressor from the *i* gene which binds to *o* and blocks the polymerase. The *o* and *p* genes are not distinguishable for many operons and it is assumed in these cases that the binding sites for repressor and RNA polymerase are identical.

There are two possible modes of turning on transcription. In one case an inducer molecule, which would typically be a substance of low molecular weight which is required to be metabolized by polypeptides A and B, binds to the repressor and alters the conformation of the repressor protein molecule so that its binding to *o* is prevented; *o* is then exposed and RNA polymerase can move off down the operon transcribing a messenger RNA molecule complementary to the whole stretch of DNA, *o-a-b*, with any 'non-informational gaps' (such as

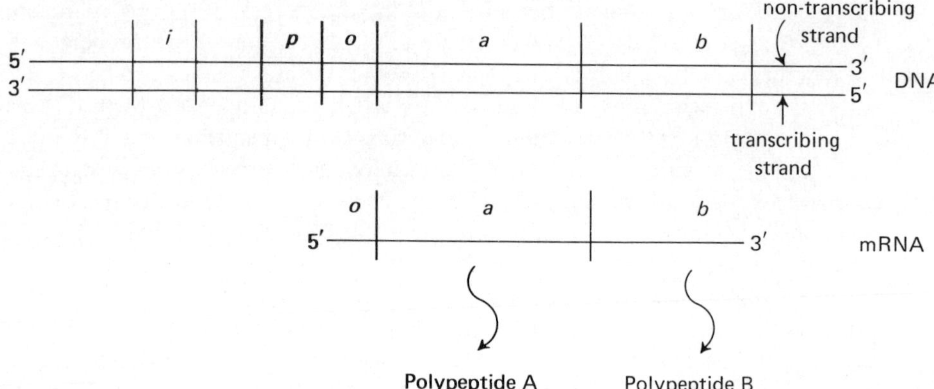

FIG 1.14 Scheme for the synthesis of polypeptides A and B (see text). *i* = regulator gene (*i* stands for 'inducible'), *p* = promoter, *o* = operator, *a*, *b* = structural genes (cistrons). The region of DNA from *p* to *b* inclusive is called an operon.

the one between *a* and *b*) included in the mRNA sequence. When the polymerase reaches specific sequence which means 'stop', the mRNA molecule is discharged. The alternative process operates if the repressor will only bind to *o* when it is associated with a small molecule the 'co-repressor' (typically the end product of the metabolic pathway whose reactions are catalysed by A and B). This substance is bound to the repressor to establish the conformation essential for binding to the *o* gene. It thus has the opposite role to an inducer. When the population of co-repressor molecules is diminished, the repressors change to a conformation in which they cannot bind to *o* any more and transcription can proceed as described above. The first of these processes should strictly be referred to as induction, the second should be referred

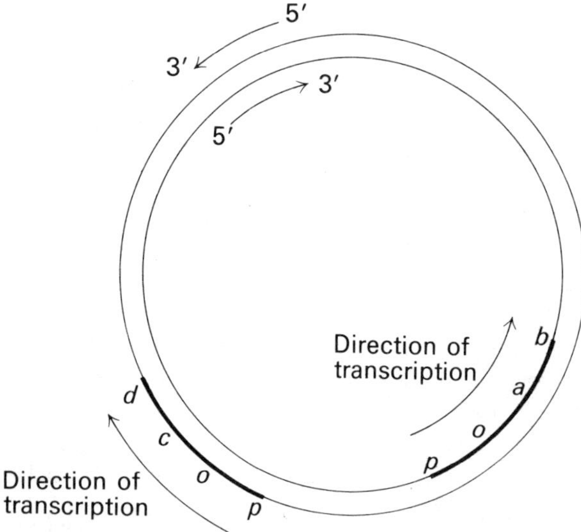

FIG 1.15 Different polarities of two operons in *E. coli*. The whole chromosome is shown in outline as a continuous double molecule. The helical content is not shown. The transcribing strand for each operon is shown as a thickened line (*a*, *b*, *c* and *d* are structural genes).

to as de-repression. A good example of an inducer is lactose which induces the operon containing the structural genes for the enzymes required for the metabolism of lactose (the *lac* operon) in *E. coli*. An example of a co-repressor is histidine which co-represses the operon containing structural genes for the biosynthesis of histidine (the histidine operon) in the very closely related bacterium *Salmonella typhimurium*.

A point that arises from the fact that not all the operons use the same transcribing strand is that they have different directions of transcription with reference to the parameters of the chromosome. This difference is referred to by saying that they are of opposite polarity. This is shown schematically in Fig. 1.15.

As each mRNA contains more nucleotides than those that correspond to the amino acids of single peptide chain, there are sequences in mRNA which signal the starting and stopping of translation, just as there are sequences in DNA which signal the starting and stopping of transcription.

Transcription in *E. coli*

RNA polymerase is a complex enzyme consisting of a 'core' and a protein required to initiate the specific binding of the enzyme to promoters. The core contains three types of sub-unit (α, β and β'); there are four sub-units in the core, which has the composition $\alpha_2\beta\beta'$. The protein required for binding of the core to the promoters is called σ. It is possible that there are different 'σ-factors' for different groups of operons. The nature of the sequences which are recognized by σ is unknown. The first triphosphate to be bound by the enzyme is a purine and then transcription proceeds; σ is discharged during the early stage of transcription. When a termination sequence occurs (say at the end of gene *b* in Fig. 1.14) the polymerase stops. For some operons the actual release of the RNA molecule is effected by another factor ρ. Several other proteins have been implicated in the transcription process. One that is well established for transcription of operons directing synthesis of inducible enzymes is the CAP-factor. The enzymes in question are those whose synthesis is inhibited by catabolites (such as glucose). Catabolite inhibition operates independently of induction and repression. The CAP-factor binds to the promoter. Its binding is required for the binding of σ (and assembly of RNA polymerase). Binding of CAP requires $3',5'$-cyclic AMP. In the presence of excess catabolites, the intracellular level of cyclic AMP is reduced and CAP binding is diminished. The molecular weights of the various proteins referred to above are as follows: α 39 000, β 155 000, β' 165 000, σ 90 000, CAP 45 000 (a dimer), ρ 200 000.

An interesting question is the nature of the secondary structure of the ternary complex consisting of two DNA strands and one of the newly formed RNA. If the choice of complementary nucleotides is via the formation of an orthodox duplex between the transcribing strand of DNA and nascent RNA then clearly the two DNA strands must be denatured in the region of the polymerase (see p. 437).

Translation in *E. coli*

Before outlining the intermediary metabolism of this process there are three groups of data which must be summarized: the structure of ribosomes, the genetic code and the structure of tRNA.

Ribosomes   The ribosomes and some of their components are commonly referred to by their S-values. These are approximate values of the sedimentation coefficients as measured in the ultracentrifuge and expressed in the usual Svedberg units. Ribosomes themselves are 70 S; they are ribonucleoprotein particles of particle weight ('molecular weight' in some publications) about $2·8 \times 10^6$. The diameter is about 180 Å. Under certain ionic conditions each ribosome dissociates to form two dissimilar subunits, 50 S ($1·8 \times 10^6$) and 30 S ($1·0 \times 10^6$). They contain up to four times their own weight of water. Of the dry weight about 60% is RNA and the remainder protein. The ribosomes contain Mg^{2+} ions; the ratio of Mg^{2+} to RNA-phosphate is of the order of 1:5. They also contain spermidine ($NH_2CH_2CH_2CH_2NHCH_2CH_2CH_2\cdot CH_2NH_2$) and other polyamines. Each ribosome contains three RNA molecules, 23 S and 5 S (in the 50 S subunit) and 16 S (in the 30 S subunit). The complete sequence of 5 S rRNA is known. It contains 120 nucleotides, none of which are minor components. The other molecules which are of high molecular weight ($1·1 \times 10^6$ for 23 S and $0·56 \times 10^6$ for 16 S) contain several modified bases.

The remainder of the ribosome consists of ribosomal protein. There are two sorts of ribosomal protein molecules, 'unitary' and 'transitory'. The unitary proteins are present in the ratio of one molecule per ribosome and the transitory in amounts less than this, for not every ribosome contains one at any one time. The 30 S subunit contains twenty proteins of which 12 are unitary and 8 transitory; the 50 S subunit contains thirty (28 unitary, 2 transitory). All fifty ribosomal proteins are different to one another.

The genetic code   The genetic code is usually presented in the form originally recommended by Crick shown in Fig. 1.16. This presentation emphasizes that the first two nucleotides of the codon play a greater role in determining the specificity of a codon than does the third nucleotide. Thus if we just concentrate on the sixteen boxes (each containing amino acids

	2 U	C	A	G	3
1	Phe	Ser	Tyr	Cys	U
U	Phe	Ser	Tyr	Cys	C
	Leu	Ser			A
	Leu	Ser		Trp	G
C	Leu	Pro	His	Arg	U
	Leu	Pro	His	Arg	C
	Leu	Pro	Gln	Arg	A
	Leu	Pro	Gln	Arg	G
A	Ile	Thr	Asn	Ser	U
	Ile	Thr	Asn	Ser	C
	Ile	Thr	Lys	Arg	A
	Met	Thr	Lys	Arg	G
G	Val	Ala	Asp	Gly	U
	Val	Ala	Asp	Gly	C
	Val	Ala	Glu	Gly	A
	Val	Ala	Glu	Gly	G

FIG 1.16   The genetic code. To look up the amino acid for a codon, read off the bases in the order 1, 2, 3. For example, CAU = histidine, AGC = serine, GGG = valine.

corresponding to four codons), in eight of them the amino acid is determined solely by the first two nucleotides. Of the remaining eight it is almost true to say that the only determining fact in the third nucleotide is whether it is a pyrimidine (U or C) or a purine (A or G). Thus if we consider the 'rule' that the genetic code is of the form of three nucleotides for which there are four alternatives for the first two letters and only two (purine of pyrimidine) for the third one (giving us a maximum of 32 different codon types), the only exceptions are AUG (met), GAG (glu) and the three codons for which there is no amino acid equivalent (UAA, UAG and UGA). These three codons signify full stops in the translation process and are known to everyone who pretends to be a molecular biologist by the trivial names ochre (UAA), amber (UAG) and umber or opal (UGA).

Transfer RNA   Several transfer RNA molecules have had their sequence completely worked out; they all contain of the order of 80 nucleotide-residues and intramolecular complementarity results in the assumption by the molecule of a characteristic secondary structure (shaped like a clover-leaf). An example of such a sequence is in Fig. 1.17. The molecule chosen (tRNA^{Ala} of yeast) was the first nucleic acid to be sequenced (p. 283). The anti-codon is the trio of nucleotides at the 'bottom' of the molecule . . I-G-C . . . The codons for ala are of the form GCN (Fig. 1.16). As the codon/anti-codon interaction is antiparallel, the C of the anti-codon recognizes the G of the codon, the G of the anti-codon the C of the codon and the third base of the codon corresponds to the 'odd nucleoside' I. It is frequently found that the nucleoside in the anti-codon corresponding to this third base of the codon is a modified one. It is supposed to have more flexible base-pairing possibilities than the formation of Watson–Crick partners. This flexibility is referred to as 'Crick's Wobble Hypothesis'. The hypothesis postulates that for this particular association (involving position 3 of the codon) that in addition to the Watson–Crick pairs (G:C and A:U) the following pairs are possible, viz. G:U (also Gm:C and Gm:U), I:C, I:U and I:A. Thus we can see that this molecule (tRNA^{Ala}) can recognize three of the alanine codons (GCU, GCC and GCA). Although most amino acids have more than one tRNA, it is clearly not necessary to imagine 61 different tRNA species (one for each codon).

Other features common to all tRNA sequences are discussed on pp. 305 and 419. Here we may note that the sequence . . T-Ψ-C . . (on the 'right') is common to all these sequences and this loop is called the 'T-pseudo U-C loop'. Many tRNA sequences contain the hU-residue (on the 'left') and this loop is referred to as the 'dihydro U loop'. The 3′-terminal sequence (. . -C-C-A) is common to all tRNA molecules; these three nucleotides have a much shorter half-life than the rest of the molecule. During the course of metabolism, the 3′-end becomes (apparently accidentally) hydrolysed and the last three nucleotides are stuck back on again by enzymes not relying on DNA or any template. Notice that the 5′-end has lost the triphosphate group which must have started the transcription event to the molecule's synthesis but that there is a 5′-phosphomonester group attached. (The 5′-G- terminus is another feature common to many tRNA structures.)

Now we can return to polycistronic mRNA and describe its translation. The molecule contains specific initiating sequences. The precise nature of these is not known but the first codon to be translated is always AUG (met) and the first amino acid to be incorporated is *N*-formylmethionine (fmet). Thus the ribosome 'finds' the initiating AUG (i.e. an AUG in the initiating sequence probably defined by a specific secondary structure in the mRNA) and then translates the message until it comes to a terminating (or nonsense) codon UAA, UAG or UGA, when protein synthesis stops and the completed polypeptide chain is discharged. The three stages of the process are known as initiation, elongation and termination. The intermediary metabolism of these processes is well established; the precise role of the ribosomes and its component parts is less well understood. In particular the reference about to be made to two 'sites' on the 50 S subunit is no more than the simplest rationalization of the available data and they may not represent two distinct and geometrically different holes or locks in the structure; it is possible that this part of the jargon will need revision when the conformational changes which occur in ribosomes during translation have been identified.

The mRNA binds to the smaller of the ribosomal subunits (30 S) and the tRNA to the larger (50 S) and the codon/anti-codon interaction is presumed to occur at the interface between

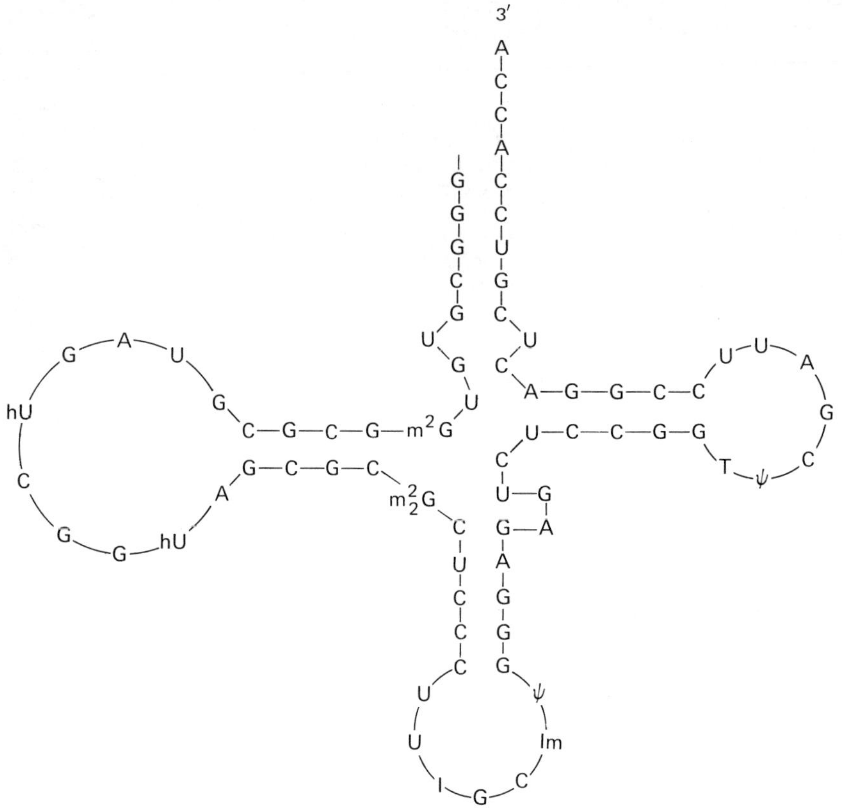

FIG 1.17   Alanine tRNA of yeast. The dotted lines represent H-bonds (three between G and C, two between A and U). The identity of the minor nucleosides is given on pp. 000 and 000.

these subunits. There are apparently two binding sites on the 50 S subunit called the P-site (to which polypeptidyl tRNA is bound during elongation) and the A-site (to which aminoacyl tRNA is bound during elongation). The components of the system are various factors and enzymes to be described shortly, ribosomes, mRNA, the aminoacyl tRNA molecules already described and also the special initiating N-formylmethionyl tRNA. This last species comes into being as follows: There are two methionine-specific tRNA species called $tRNA_F^{Met}$ and $tRNA_M^{Met}$. Both are charged with methionine by the same methionyl tRNA synthetase but, of the two products of this reaction, only met-$tRNA_F^{Met}$ (and not met-$tRNA_M^{Met}$) is a substrate for a transformylase which transfers a formyl group to the aminoacyl tRNA to yield fmet-$tRNA_F^{Met}$. (As with all formylations, the co-factor is tetrahydrofolic acid.) The only aminoacyl tRNA capable of initiating protein synthesis is fmet-$tRNA_F^{Met}$.

The synthesis of a protein is summarized in Fig. 1.18. We start with a 30 S subunit (I) which binds two proteins (initiation factors F_2 and F_3) and mRNA to produce state II, to which fmet-tRNA$_F^{Met}$ and GTP bind to produce the 30 S initiation complex (III). Now another initiation factor (F_1) binds together with a 50 S subunit to give IV. The fmet-tRNA$_F^{Met}$ is now specifically located in the P site on the 50 S subunit. This event is accompanied by hydrolysis of GTP and the release of the factors. The GTPase is F_2 but it is only active in the presence of the other factors and the 50 S subunit. We now have the 70 S ribosome complete with an initiating AUG in the mRNA H-bonded to the anti-codon of tRNA$_F^{Met}$ (V). The next amino acid is bound to the A-site as a complex (VI) consisting of the aminoacyl tRNA, GTP and the 'unstable' transfer factor (T_u). The GTP is hydrolysed and a complex of T_u with GDP is released. This complex is phosphorylated by a GTP phosphotransferase ('stable transfer factor' or T_s) to yield a T_u-GTP complex which can pick up another aminoacyl tRNA and complete the elongation cycle. Concomitantly, peptide bond formation occurs, catalysed by aminoacyl tRNA peptidyl transferase (one of the proteins of the 50 S subunit) to yield VII. This is followed by movement of what is now dipeptidyl tRNA to the P-site (the translocase reaction), discharge of a free tRNA and movement of the mRNA to bring the next codon into the orientation needed for its recognition by an aminoacyl tRNA in the A site. Thus we are back to the equivalent of V and the reactions between V and VIII are repeated as each new amino acid is added to the chain. When the last amino acid (z in Fig. 1.18) has been incorporated, the codon opposite the A site will be one for 'nonsense', i.e. no tRNA will have an appropriate anti-codon. The ribosome in this state (IX) binds release factor (R), the binding of R is apparently dependent on another protein (α). The effect of R is to alter the specificity of the peptidyl transferase enzyme and turn it into an esterase which hydrolyses the ester bond attaching the last tRNA ('tRNA^z' in Fig. 1.18) to the polypeptide chain. The complex now drops to bits; releasing R, α, tRNA^z and free 70 S ribosomes. The ribosomes converted into mixtures of subunits by the dissociation factor (DF) which re-enter the initiation cycle.

There is a degree of heterogeneity among the initiation factors which probably reflects a degree of translational control for various classes of mRNA molecules. Also there are two release factors (R) which differ in the codons for which they are specific (both recognize UAA but R_1 also recognizes UAG but not UGA and R_2 also recognizes UGA and not UAG). The role of the transitory ribosomal proteins in the control of protein synthesis is still largely unknown although there is evidence that one of them associated with the 30 S subunit is only present during the elongation process so it seems likely that it displaces the initiation factors and is itself displaced by the release factors.

One final point about the translation process is that the immediate product of protein synthesis according to this scheme is a polypeptide chain with an N-formyl-methionyl residue at the N-terminus. Most *E. coli* proteins do not have this feature and either the formyl group or the first one, two or more amino acids are removed from this end of the chain by specific proteases to produce the final protein molecule.

Clearly the control of protein synthesis by the feedback system requires that mRNA itself has a short lifetime. Indeed mRNA *is* short-lived compared with rRNA and tRNA. It is degraded by a specific nuclease, ribonuclease V which is only active when bound to 70 S ribosomes. The mRNA is degraded in the 5′ to 3′ direction (the same direction as translation). Another important feature of protein synthesis is that any one mRNA will accommodate a number of ribosomes all progressing in the same direction. Thus if this system is frozen at a moment in time (as it is when a cell-free preparation is made in the laboratory), the structure containing the mRNA will contain a lot of ribosomes, strung along the mRNA. Each ribosome will have attached polypeptidyl tRNA and the polypeptides will be progressively longer as the ribosomes are closer to the end of a cistron. This type of structure is called a polyribosome, or more simply as a polysome.

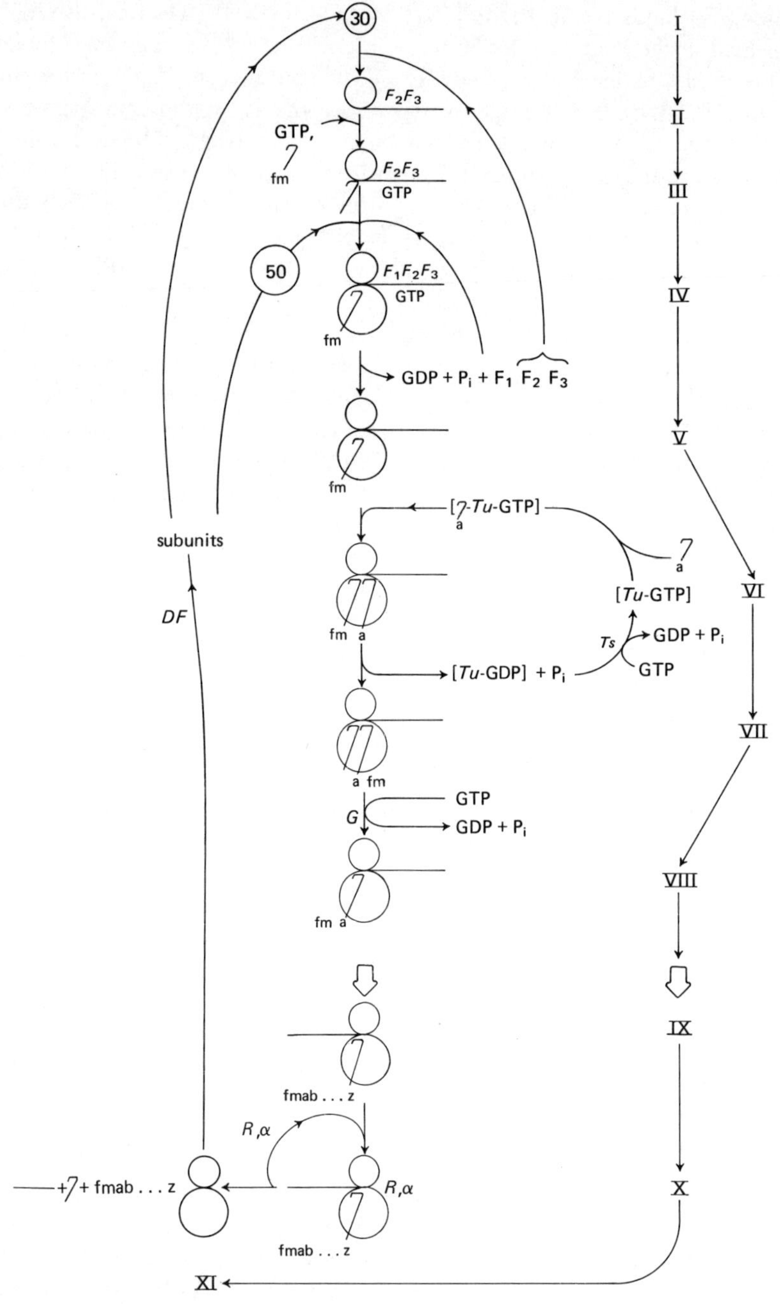

FIG 1.18   Protein synthesis in *E. coli*. The amino acids are fm (fmet) and a, b, c, . . . , z. Transfer RNA is represented by $\nearrow$, with the anti-codon at the top of the bond and the 3′-end at the end of the longer 'tail'. In the ribosomes, the P site is on the left of the 50 S subunit, the A site is on the right. Other abbreviations are defined in the text.

Some polysomes are probably associated with membrane in the intact organism and some of them are certainly associated with DNA. This last association occurs when transcription and translation are coupled. This last phrase refers to the situation which arises when translation of mRNA starts before transcription is complete. This is clearly quite possible as the mRNA is both synthesized and translated from the same ($5'$) end. In fact there is no reason why mRNA need ever exist as a discrete entity. Transcription of a long operon might commence and as the RNA polymerase moves along the DNA, lengthening the RNA chain, ribosomes can start threading on the initiating and elongating chains. At the end of the little train of ribosomes moving along in this way there can be one carrying ribonuclease V and progressively hydrolysing the mRNA from $5'$-end before the polymerase has finished the $3'$-end. This is definitely the case for at least one operon in *E. coli* and the possibility of it happening elsewhere is a cautionary warning for those optimists who are trying to isolate mRNA.

Other nucleic acid systems in *E. coli*

Before briefly surveying this subject there are a few terms that need defining or explaining.

Viruses that infect bacteria are known as bacteriophage, or phage for short. Phage are nucleoprotein particles of varying complexity. Any one type of phage contains either RNA or DNA (never both). The phage attaches to the outside of the cell, the nucleic acid enters the cell (leaving its protein complement outside) and once inside, the viral nucleic acid is replicated; it directs the synthesis of its own proteins and also a protein (such as a lysozyme) required for the release of the phage particles from the cell. Complete progeny phage are assembled inside the cell and the cell then bursts in most phage systems (through the action of the phage-induced lysozyme) and the new phage flood out and are ready to infect new host cells.

Certain strains of *E. coli* contain in addition to the chromosome, other bits of DNA which are normally replicated synchronously with the chromosome but which are capable of passage from one cell to another. These bits of DNA, capable of a sort of semi-independent existence are called plasmids or episomes.

Genetic recombination occurs between DNA molecules of homologous sequence in the same cell. The molecular events behind this process are not precisely defined but schematically it involves the two double helices lying side by side: two equivalent scissions are made in both molecules and the molecules rejoin to produce molecules in which the DNA content has been scrambled as shown in Fig. 1.19. With *E. coli* it is possible to obtain cells containing two chromosomes (or a fragment of one chromosome and another complete chromosome) from two different cells into the same cell as a result of events accompanying either phage-infection or episome-transfer. The results of the recombination which occurs in such a cell can be analysed genetically. If for example the letters *a, b, c, d* and *e* in Fig. 1.19 could represent genes for enzymes (A to E) which are required to enable the organism to grow in the absence of substances α, β, γ, δ, ε, and if the 'primed DNA' came from an organism in which there are mutations in genes *a* to *e* so that this bacterium had an absolute requirement for α, β, etc., to grow we would say this organism was α^-, β^-, γ^-, δ^-, ε^- and the wild-type (upper non-primed DNA) would grow in the absence of substances α to ε and would be said to be α^+, β^+, etc. The progeny from the cross would be α^+, β^-, γ^-, δ^-, ε^+ and α^-, β^+, γ^+, δ^+, ε^- and these measurable consequences of the change in 'genotype' (the measurable description is called the 'phenotype') could be determined by observing the ability or otherwise of the progeny to grow on selective media in which certain of the requirements α to ε were absent. Of course the power of this system is not that recombination can be demonstrated but that it can be quantified. There are several ways of doing this but let us imagine that recombination could be demonstrated in the

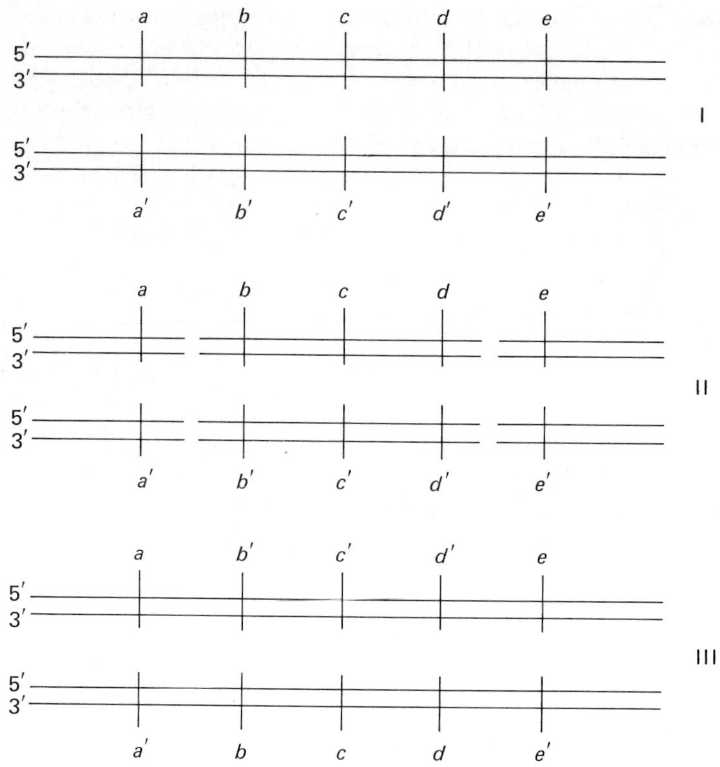

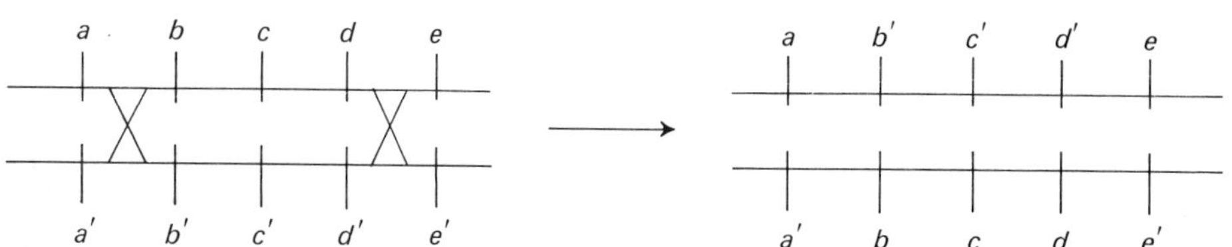

FIG 1.19   Recombination between DNA molecules. (I) *a, b, c, d* and *e* are just characteristic positions in one molecule; *a′, b′, c′, d′* and *e′* are the equivalent positions in another molecule. Scissions in the two molecules (II) and recombination gives the mixed molecules (III). The process is conventionally summarized as in (IV), in which each horizontal line represents a double helix.

system described in Fig. 1.19 and that we just concentrated on those progeny which had one cross-over point between *a* and *b* (i.e. all the α^+, β^- or the α^-, β^+ phenotypes). Among these organisms will be ones which the second cross has occurred not only between *d* and *e* as shown but also between *c* and *d*, to the right of *e* and so on. The order of the genes *a, b, c, d, e* will determine the frequency with which transfer of gene *c, d* and *e* accompanies the transfer of *b*. The power of bacterial and phage systems for genetic analysis lies in the fact that genetic analysis is capable of very high resolution. Thus *a, b, c, d* and *e* do not need to be genes. A general term to describe differences other than gene-differences in genotype is to call the differences 'markers'. So *a, b, c, d* and *e* may all be markers in one structural gene. Clearly certain patterns of recombination between two strains, both with different faults in the same gene could lead to recombinants in which the lesion has been healed. By fine structure analysis, in favourable systems it is possible to obtain the relative ordering of markers (a procedure called 'genetic mapping') when the markers are of the order of a nucleotide distant from one another.

A special type of recombination occurs between DNA molecules which have only a small region of homology between them. This is shown schematically in Fig. 1.20. There is in this

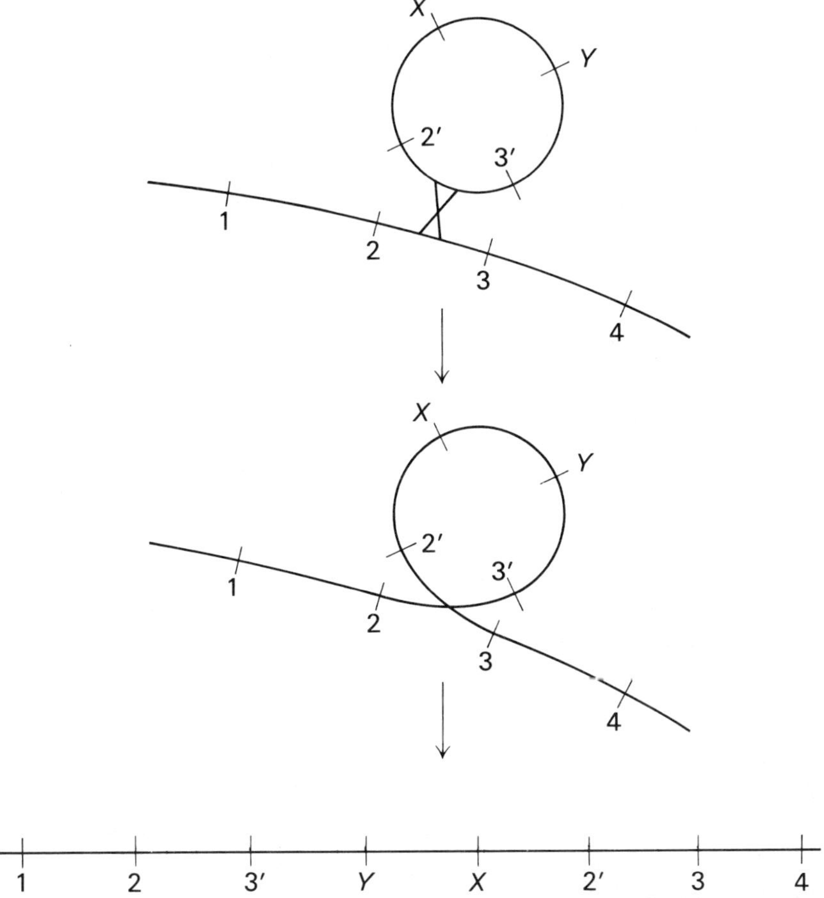

FIG 1.20   A small DNA molecule (with markers 2', X, Y and 3') is incorporated into a large one (markers 1, 2, 3 and 4) by a single cross-over.

case homology between a small DNA molecule and a big one between the markers 2 and 3. A single cross-over will result in incorporation of the little molecule into the big one.

RNA phage in *E. coli*   RNA phage are typically very simple with a coat containing only one type of protein subunit and a single-stranded RNA molecule containing only three or four genes. When the RNA has entered the cell it acts as a messenger molecule and is translated by the cell's ribosomes. The translation products are phage coat protein, RNA replicase (RNA-dependent RNA polymerase) and a maturation protein (or proteins) required to let the complete phage out of the cell. The replicase is responsible for the synthesis of more phage RNA. This is a two-stage process as the immediate product of such a process will not be phage

RNA but rather its complementary copy. The phage RNA is known as + strand and the complementary copy as the − strand. The − strand is the template used for the synthesis of more + strand.

DNA phage in *E. coli*   DNA phage are classified according to the physiology of their infective processes as either 'virulent' or 'temperate'. All DNA phage are more complex than the RNA phage and most are considerably more complex. Most contain double stranded DNA although a few contain single stranded DNA (which is replicated by the two-stage 'plus and minus' procedure outlined for RNA phage).

Typical double-stranded virulent phage have a complex morphology with many proteins in their structure.[6] The DNA enters the cell and is replicated and transcribed. The extent to which the phage relies on the host cell RNA polymerase varies somewhat, and in at least two cases transcription is controlled partly by the fact that genes due to be transcribed late in the phage's timetable of infection are not transcribable by the cell's polymerase but require a phage-specific polymerase which is one of the phage's own gene products. These phage rely on the cell's DNA polymerase but carry the gene for their own ligase, for which the high-energy phosphate co-factor is ATP (not NAD^+ as in the case of the cellular enzyme). Two types of genetic recombination occur in this type of phage infection. If the cells are mixedly infected with two different strains of phage, recombinants are found among the phage progeny. Also, it sometimes happens with certain phage that hydrolyse the cellular DNA as part of their infective cycle that bits of cellular DNA are incorporated into the newly assembled phage particles. If such a particle now infects a new cell of a different strain, genetic recombinants can be obtained between the two strains of *E. coli*. This method of gene transfer between different bacterial strains is known as generalized tranduction.

Temperate phage effect a different type of gene transfer. They are all double-stranded DNA phage; in the phage the DNA is in a linear form. However, on entering the host cell the ends join up by means of a process explained later in this book to form a circle. The circle is then incorporated into the chromosome by the type of single cross-over event illustrated in Fig. 1.20. The cells are not apparently affected and the chromosome (together with its incorporated phage DNA) is replicated and the cells divide. However, the cells which are produced are known as lysogenic, for following exposure to ultra-violet light or certain chemicals the phage DNA is excised from the chromosome by a reversal of the process in Fig. 1.20 and phage replication and cell lysis occur. The mechanism for lysogeny is that while the phage DNA is incorporated in the DNA it produces a gene product which is a repressor for the genes involved in the excision process. Ultra-violet radiation and other agents which effect lysogeny inactivate this repressor protein so it can no longer bind to DNA.

Genetic recombination can be observed in this system because sometimes the reversal of the process in Fig. 1.20 occurs erroneously so that the 'loop' that is formed and nipped off in this process forms in the wrong place leaving a bit of phage DNA left in the chromosome and taking a bit of chromosomal DNA out instead (see marker 2 in Fig. 1.20). When this DNA is introduced into a second host strain which is different in this region, genetic recombination can occur by the introduction of the phage DNA, now carrying a part of the DNA from the 'donor' bacterium into the chromosome. This process is called specialized transduction; 'specialized' because only genes in the vicinity of the phage attachment site are transducible.

Plasmids and episomes in *E. coli*   Plasmids and episomes are extrachromosomal circular DNA molecules in bacteria. They can be transferred from cell to cell, but unlike phage they are not associated with a protein coat and they do not kill the cells receiving them.

Certain plasmids carry genes for drug resistance and others for a class of antibiotics (proteins) called colicins. The former class are of enormous practical importance because they are a method of passing drug resistance through a population of bacteria. At first sight this might appear to be of minor importance in *E. coli* which is a harmless inhabitant of the human colon. However plasmid transfer can occur between bacteria of different species so that patients who have been exposed to antibiotic therapy may have among their gut flora *E. coli* carrying resistance to the antibiotics and this resistance can than be passed on to pathogens introduced in subsequent infection. The plasmids are of genetic interest as they can be incorporated and excised from the chromosome by the processes of Fig. 1.20 and, as with temperate phage, the excised DNA can differ from that that was introduced but with one important difference. As the plasmid does not have a coat protein to accommodate there is not the same restriction on size for viable excised DNA. Thus the plasmids can come out of the chromosome simply bigger, carrying chromosomal genes in addition to the full complement of characteristic plasmid markers.

There is a special class of extra-chromosomal DNA molecules, usually called episomes, which includes the fertility factors. Bacteria carrying the fertility factor (symbol F) are called male. Those lacking this factor are called females. Alternatively male and female are called F^+ and F^- respectively. When F^+ and F^- strains are mixed, providing certain conditions of growth are fulfilled, the cells pair up, a protein bridge is built between them, the F factor is replicated and its copy is transferred to the F^- recipient cell. This process is referred to as bacterial conjugation.[7]

F DNA can be incorporated into the chromosome by the process of Fig. 1.20. However, in this case there is a vitally important difference. With 'normal' F^+ cells a very rare event is that during conjugation the episomal DNA is incorporated into the DNA of the donor cell, so that the whole chromosome behaves as a sex factor; that is it is replicated and the whole lot is transferred (in a linear form) to the recipient F^- cell, the F region going in last. With certain F^+ strains the event occurs frequently during conjugation; these are called 'high frequence recombinant' or Hfr strains. The enormously detailed genetic map of the *E. coli* chromosome has been largely deduced from experiments involving conjugation using Hfr donors.

Bacteria other than *E. coli*

The genetic code and the modes of DNA replication, transcription and translation in other bacteria is the same as in *E. coli* as far as is known. Not all bacteria have all of the same genetic systems as *E. coli* and for most of the species listed in compendia of bacteria, gene transfer systems have not been described. However it is probably because they just have not been looked for. With certain bacterial species it is possible to obtain genetic recombination by a more direct method than those described above. This is the process known as 'transformation' in which chemically purified DNA is isolated from the donor strain and is inoculated into a living culture of recipient cells. DNA enters the cells and recombination occurs within the cell with a proportion of the recipient cells becoming transformed, that is to say assuming the genotype and phenotype of the donor culture.

Eukaryotic cells

Eukaryotic cells are more complex in several respects. For one thing the DNA is organized in a different way. Within the nucleus it is present as a nucleoprotein complex called chromatin, and the distribution of the protein in the chromatin plays some part in exercising a sort of coarse control over transcription. Within the nucleus there is another organelle called the nucleolus in which the genes for rRNA are concentrated. Protein synthesis occurs in the cytoplasm, so clearly the mRNA has some sort of journey to make from the nucleus to the cytoplasm where the ribosomes are to be found. Moreover the ribosomes are larger than bacterial ones (80 S, subunits are of 40 S and 60 S). The large rRNA molecules are larger and the ribosomes also contain proportionally more protein. The polysomes are largely present in a form in which the ribosomes are bound to a membrane, the rough endoplasmic reticulum. Transfer RNA and 5 S rRNA are the same size as in bacteria.

The size of eukaryotic DNA is totally unknown. The general mode of DNA synthesis and repair appears to be the same but no enzymes have been purified. Transcription is effected in a similar fashion too. There are different RNA polymerases in the nucleolus and in the rest of the nucleus. These enzymes have been purified from certain animal tissues; the subunit structure is different and no σ factors or their equivalent have been identified so far.

Translation in eukaryotic cells appears to follow the same lines as in prokaryotes with relatively minor differences. The genetic code and the overall intermediary metabolism of protein synthesis are apparently universal. There is no evidence for translation of polycistronic mRNA in eukaryotes but broadly speaking the same type of initiation and termination seem to occur, with one interesting difference. N-formylmethionine is not the first amino acid to be incorporated, rather it is unmodified methionine. Nevertheless there are two specific $tRNA^{Met}$ molecules. The one essential for initiation is called $tRNA_{F*}^{Met}$. It has this name because although the aminoacyl tRNA derivative Met-$tRNA_{F*}^{Met}$ is not formylated in its own cell, it is formylated by the *E. coli* transformylase enzyme, suggesting an astonishing conservation of tRNA structures during evolution.

A complication that exists in all eukaryotic cells is that there are extranuclear DNA molecules which are responsible for 'cytoplasmic inheritance'. The best known examples are the DNA molecules that occur in chloroplasts and in mitochondria. These molecules contain genes for some (but not all) of the proteins of these organelles. In addition chloroplasts and mitochondria have small ribosomes (somewhat analogous to the 70 S ribosomes of bacteria) and their protein synthesis is inhibited by drugs which inhibit bacterial protein synthesis and do not affect cytoplasmic eukaryotic protein synthesis. The precise relationship of the nucleic acid systems to one another is far from clear.

Some eukaryotic organisms are infected with viruses. Relatively few fungal viruses are known. Most plant viruses studied so far contain single-stranded RNA. Animal viruses are a very diverse group and the examples of all four nucleic acid types (single- and double-stranded DNA and single- and double-stranded RNA). The majority are either double-stranded DNA or single-stranded RNA. However within these two groups there is enormous variety, some are large and contain lipid and membrane fragments from host cells. Others are very small. Very small animal DNA viruses are called picodna viruses; the very small RNA viruses are called picorna viruses.

All eukaryotic viruses differ fundamentally from phage in their mode of infection. The whole virus particle is absorbed by the host cell and the release of nucleic acid occurs within the cell. All plant RNA viruses and most animal RNA viruses replicate their RNA by a process analogous to that described for RNA phage. Some oncogenic (cancer producing) RNA viruses produce an RNA-dependent DNA polymerase which results in a copy of the viral genes being locked into the cellular DNA. There is no direct evidence that nucleic acid from any DNA animal virus is incorporated in the same way as a temperate phage but there is good reason to suppose an analogous process is possible for some viruses. Cellular gene-transfer has not been demonstrated so far by any eukaryotic virus (see, however, p. 190).

The genetic systems in eukaryotic organisms are outside my scope. I assume the reader is aware of the consequences of the facts that the genetic material is organized into chromosomes which can be visualized during mitosis and that many eukaryotic cells are diploid, i.e. carry two (not normally identical) copies of the same genes. The word used to describe the total genetic complement of a cell or a virus is genome.

I have said nothing of the roles of DNA replication; transcription and translation in the control of such processes as morphogenesis, differentiation, embryology, ageing and neoplasia. These are the vital growth points for the present phase of nucleic acid research. It is largely to inform those who wish to exploit these areas that this book is written. In the following pages I have chosen the exemplify techniques by considering as wide a variety of biological problems as possible.

Notes to chapter 1

1   These sugars are universally referred to as ribose and deoxyribose. Actually deoxyribose is a naughty word according to the bodies that pontificate on questions of chemical nomenclature. The reason is that you should never base the name of one compound on another one which is more complex than the compound itself. In the present case ribose has one more centre of asymmetry than has deoxyribose. Another way of approaching the alleged problem is to say that 'deoxyribose' could equally be called 'deoxyarabinose'. Just for the record the correct name for deoxyribose is 2-deoxy-β-D-erythropentose.

2   In the earlier writings of E. Chargaff and in works edited by him (see references) RNA appears as PNA (pentose nucleic acid). Chargaff allowed DNA but for him it stood for deoxypentose nucleic acid.

3   International Union of Pure and Applied Chemistry and the International Union of Biochemistry.

4   It is the usual convention to identify the codons by the three initial letters without any punctuation but the convention 'left to right = $5' \rightarrow 3''$ still applies.

5   Prokaryotic. For the non-biologist: there are two sorts of organism, prokaryotic and eukaryotic. The cells of prokaryotes have a simpler ultrastructure. The defining criterion is the nuclear membrane (absent in prokaryotes, present in eukaryotes). However there are many other differences. Eukaryotic cells have many intracellular organelles, examples of which are mitochondria, lysozomes, chloroplasts, etc. Not every eukaryote has every type of organelle of course. In prokaryotes the specialized functions associated with these organelles are found in invaginations in the plasma membrane ('mesosomes') and other 'more primitive' structures. Bacteria and some algae are prokaryotes, everything else is eukaryotic. Not all organisms consist of cells (at least not at all times during their life cycle). The alternative to cells is mycelium, filamentous segments containing the DNA, enzymes, etc., equivalent to the complement of many cells. The prokaryotic/eukaryotic division still exists between different types of mycelium. Prokaryotic mycelium is found in filamentous bacteria (and certain algae); eukaryotic mycelium is found in the mycelial fungi.

6   The best known virulent DNA phage in *E. coli* are the seven 'T-phage', isolated originally by J. Lederberg. By a remarkable coincidence those with even numbers (T2, T4 and T6) are similar and are referred to as the 'T-even phage'. Similarly the 'T-odd phage' (T1, T3, T5 and T7) have certain features in common with one another.

7   Conjugation is sometimes referred to as 'mating'. This anthropomorphic view of *E. coli* has a certain charm but is confusing. Mating in more familiar circumstances does not result in the females being converted into males.

2 Nucleic acid components

This chapter is to summarize the properties of pyrimidines and purines and their derivatives for two purposes. On the one hand it is impossible to understand the structures and properties of nucleic acids without knowing something of the properties of their component parts. The other point is more pragmatic: isolation, fractionation and structural determination on new pyrimidines and purines is an active and important part of current research. New compounds of these types are discovered both as minor components of nucleic acid structures and also as other classes of natural products, notably the nucleoside antibiotics. These form an important group, not only because of their potential use as drugs but also because many of them are highly selective inhibitors of nucleic acid metabolism and protein biosynthesis and are of great value in probing the mechanisms and control of these processes.

Pyrimidines and purines—a brief survey

Chemical synthesis

The chemical synthesis of pyrimidines and purines is rather outside the scope of this book as it is largely of historic interest for the average reader and I have specifically eschewed such material. However *de novo* chemical synthesis is sometimes required and the general methods are as follows.

Pyrimidines Most chemical synthesis of pyrimidines are based on the reaction of an amidine with a bifunctional carbonyl or nitrile as outlined below.

The group $-R$ can be -alkyl, $-H$, $-O$-alkyl, $-OH$, $-SH$, $-S$-alkyl, $-NH_2$, etc. Groups X and Y can be $-CHO$, $-CO$-alkyl, $-CO_2Et$ or $-CN$. The strong base (B^-) is usually ethoxide or hydroxide. The nature of x and y is determined by X and Y. (As an example, if Y is $-CN$, y will be $-NH_2$.)

Two examples of this process will suffice. First there is the classical synthesis of thymine by Wheeler and Merriam (1903).

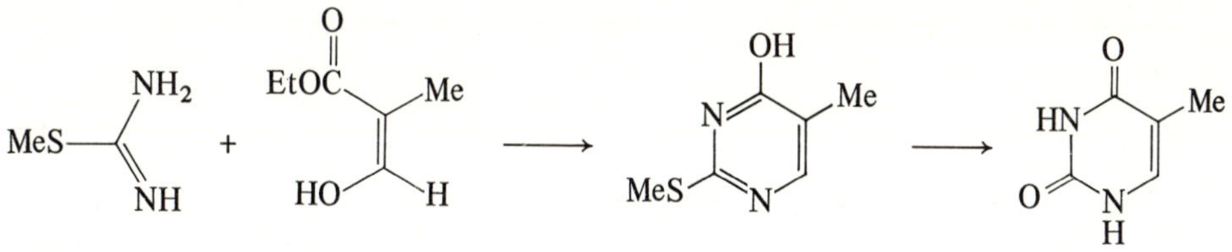

The initial condensation (of *S*-methyl thiourea with ethyl formylpropionate) was catalysed in this case by aqueous potassium hydroxide. The hydrolysis of the intermediate thioether was effected with concentrated aqueous hydrochloric acid.

A more up-to-date example is the synthesis of 5-fluorouracil. This substance is a potent inhibitor of RNA metabolism in mammalian tissues and has been tried as an anti-cancer drug. It has been discontinued on account of its very high toxicity but is of experimental interest as it inhibits transcription, produces errors in translation and is also a teratogen.[1] Unlike the other 5-halouracils it cannot be synthesized from uracil by halogenation and must be synthesized as follows. The procedure is that of Duschinsky, Plevin and Heidelberg (1957). Ethyl fluoroacetate and formic acid react in the presence of potassium metal to produce ethyl formylfluoropropionate. The condensation of this substance with ethyl thiourea (below) is entirely analogous to the reaction of Wheeler and Merriam.

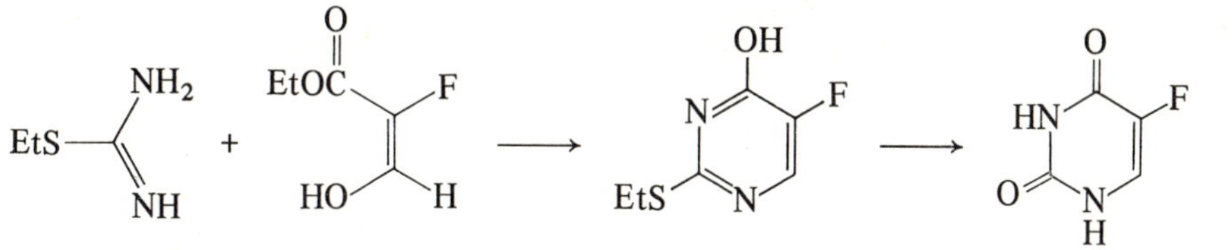

In this particular case potassium metal was used as base for the condensation.

Purines   Chemical syntheses of purines are achieved by ring closure of two amino groups with formic acid so that formylation of one amino group is associated with formation of an internal

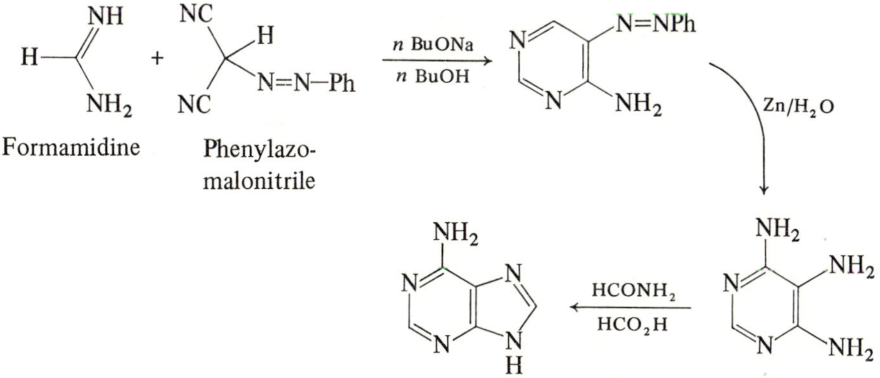

Formamidine

Phenylazo-
malonitrile

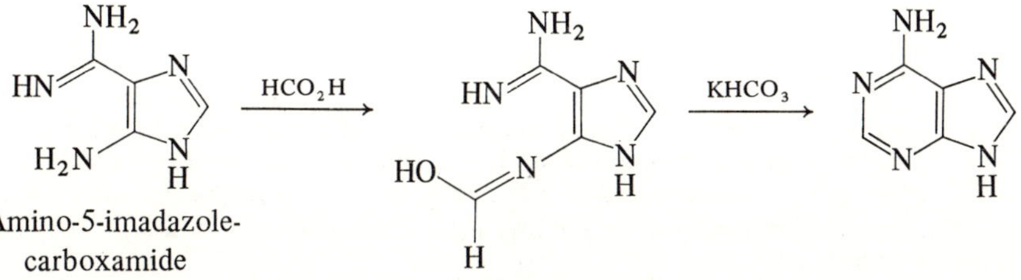

4-Amino-5-imadazole-
carboxamide

Schiff's base between the formyl group and the other amino group. The diamine chosen can be either a diamino pyrimidine or an amidino-imidazole. The two approaches can be exemplified by two different routes to adenine. The first is due to Cavalieri, Tinker and Bendich (1949) and the latter to Shaw (1950).

An alternative synthetic route to purines starts with hydrogen cyanide. Although this procedure is of no preparative value it could be of evolutionary significance. Hydrogen cyanide itself can be made from purely inorganic precursors. Adenine (molecular formula $C_5H_5N_5$) can

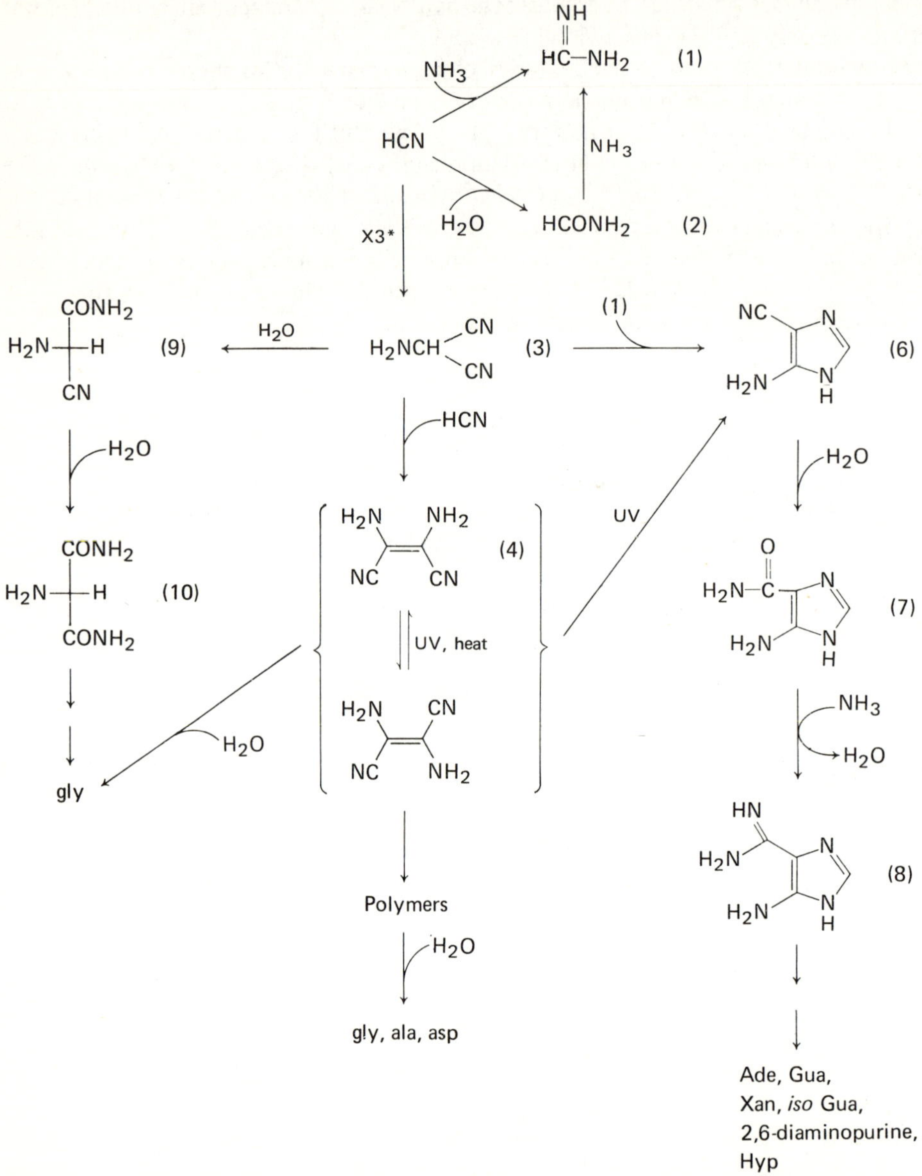

FIG 2.1   Photochemistry of aqueous hydrogen cyanide. (1) Formamidine, (2) formamide, (3) aminomalononitrile, (4) diaminomaleonitrile, (5) diamino fumaronitrile, (6) 4-amino-imidazole-5-carbonitrile, (7) 4-aminoimidazole-5-carboxamide, (8) 4-aminoimidazole-5-carboxamidine, (9) aminomalononitrile monoamide, (10) aminomalonodiamide. * The dimer $HN = CH - CN$ is an intermediate.

be regarded as a pentamer of hydrogen cyanide. It is reasonable to regard adenine as one of the 'primitive' molecules formed at an early stage of pre-biotic evolution (i.e. the production of organic substances prior to their association to form macromolecules and their aggregation to produce the earliest living cells). Adenine, other purines and certain amino acids can be synthesized from hydrogen cyanide. The intermediates and the course of the reaction have been elucidated by L. E. Orgel. The reactions are summarized in Fig. 2.1 (from Sanchez, Ferris and Orgel, 1967).

Naturally occurring pyrimidine and purine compounds

The intention of this brief survey is to indicate the variety of substances of this type. References have been omitted as this compilation does not pretend to be complete and all the compounds can be quickly looked up in *Chemical Abstracts*.

Intermediates in the biosynthesis and breakdown of nucleosides and nucleotides   Many of these compounds are discussed later in the book (see pp. 176 and 177). The purines, xanthine and uric acid are metabolites in the major pathway for nitrogen excretion in 'uricotelic animals' (birds and some reptiles). The intriguing N-3 riboside of uric acid (Fig. 2.2) is found in the blood of some mammals.

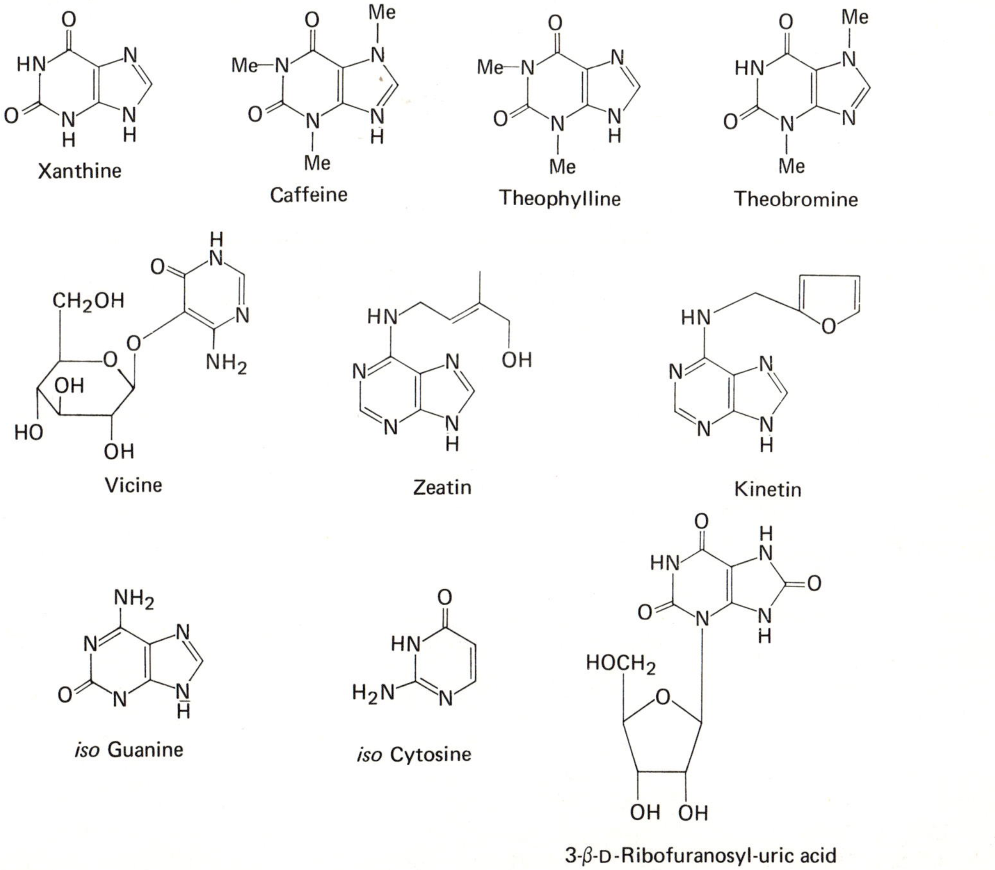

Xanthine

Caffeine

Theophylline

Theobromine

Vicine

Zeatin

Kinetin

iso Guanine

iso Cytosine

3-β-D-Ribofuranosyl-uric acid

FIG 2.2   Miscellaneous purines and pyrimidines mentioned in the text. Zeatin and kinetin are examples of cytokinins.

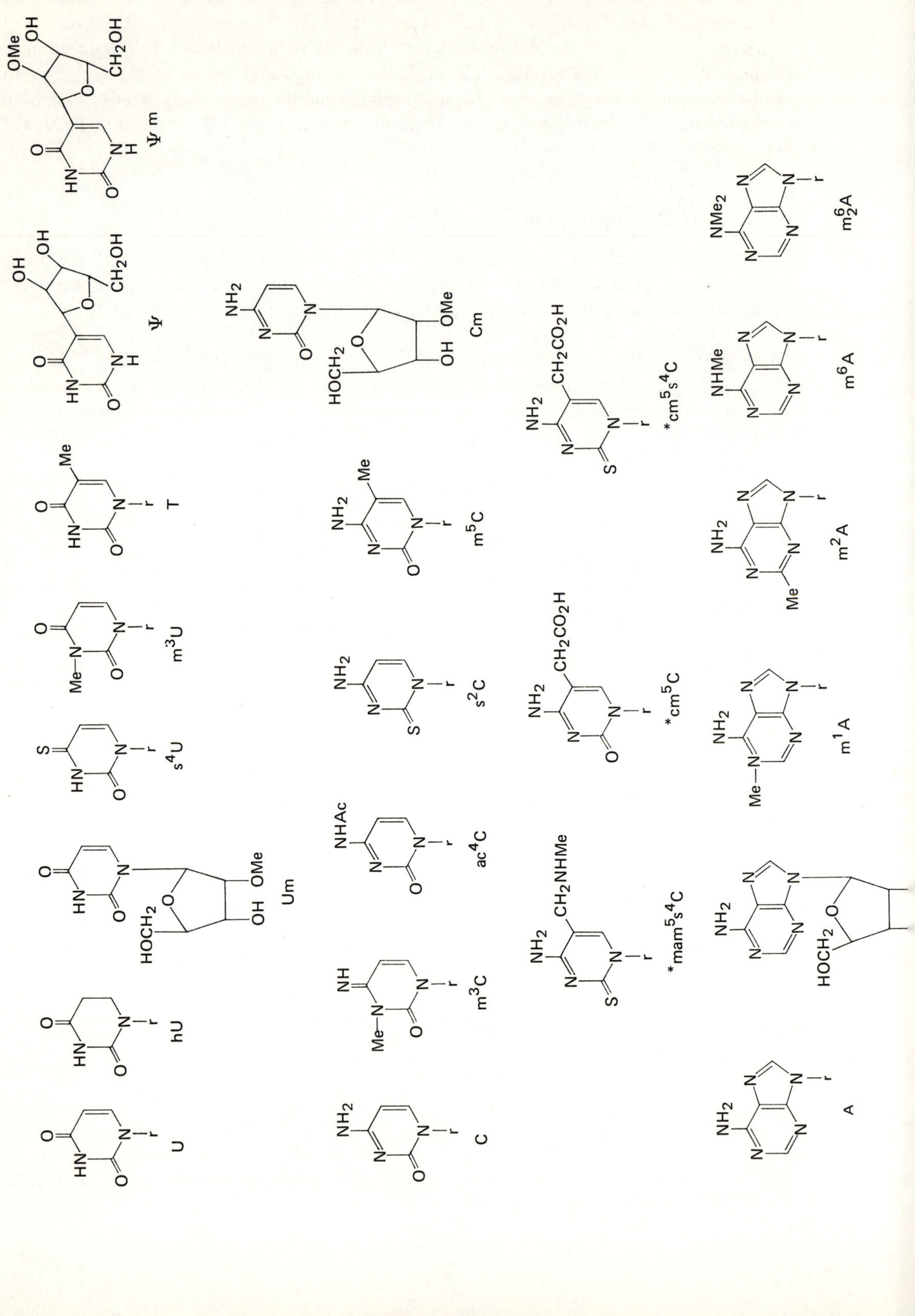

Ψm

Ψ

Cm

m$_2^6$A

m^{6}A

T

m^{5}C

*cm^{5}s^{4}C

m^{2}A

m^{3}U

s^{2}C

*cm^{5}C

m^{1}A

s^{4}U

Um

ac^{4}C

*mam^{5}s^{4}C

hU

m^{3}C

U

C

A

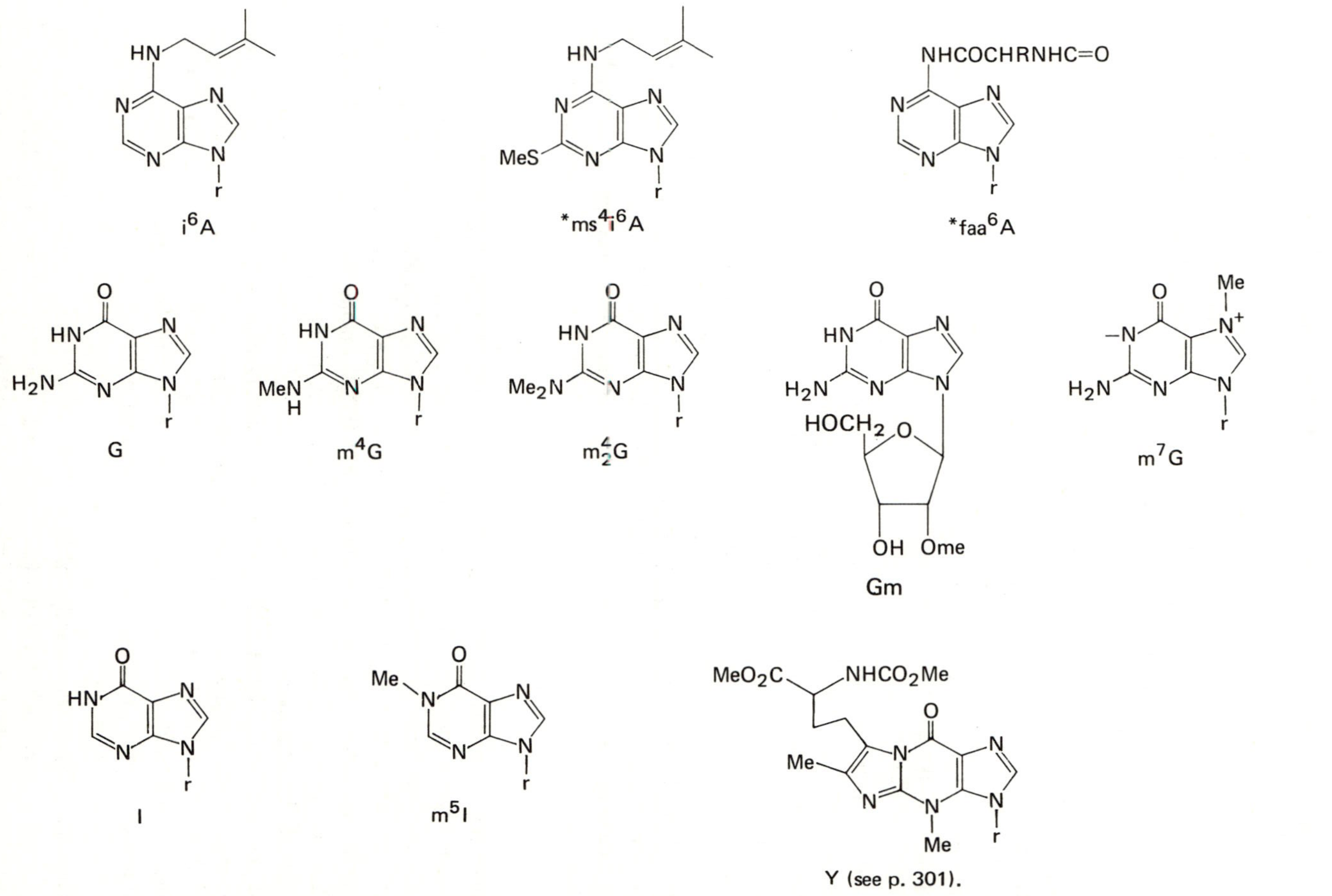

FIG 2.3   Minor nucleosides occurring in RNA. The abbreviations follow IUPAC-IUB recommendations (p. 5) except those marked * which are my suggestions.

Phosphate esters and pyrophosphate derivatives of nucleotides   This group comprises some of the most familiar co-factors in biochemistry. In addition to the di- and tri-phosphates (and the mixed anydrides of adenylic and sulphuric acids which are the 'high energy compounds' of certain bacteria), there are the various nucleoside diphosphate sugars of which uridine diphosphate glucose (UDPG) is the most familiar to mammalian biochemists. Additionally there are the oxidized and reduced forms of nicotinamide adenine dinucleotide (NAD^+ and NADH) and the equivalent forms of nicotinamide dinucleotide phosphate ($NADP^+$ and NADPH), co-enzyme A and S-adenosyl methionine. The nucleotide $3',5'$-cyclic AMP has been mentioned in chapter 1 (pp. 21 and 98). In addition to roles in nucleic acid metabolism it is involved in the initiation of many hormonally controlled processes in mammalian biochemistry. It is apparently a universal allosteric modifier of protein conformation. The nucleotide ppGpp is a naturally occurring substance and its role is discussed on p. 226. A curiosity in this group is the nucleotide pAp which is a co-factor for bioluminescence in the sea pansy (*Renilla reniformis*).

Other co-factors   Pyrimidine rings occur in the structures of vitamin B_1 (thiamin), vitamin B_2 (riboflavin) and the co-factors derived from them (thiamin pyrophosphate TPP, flavin mononucleotide FMN and flavin adenine dinucleotide FAD). An unusual nucleoside, $5'$-deoxyadenosine, is bound to the cobalt atom in B_{12} co-enzyme by a $5'$-C–Co bond. This ligand is replaced by cyanide in vitamin B_{12} itself.

Plant natural products   There are two groups of compounds in this class, the purine and pyrimidine alkaloids and the group of plant hormones known as cytokinins. The purine alkaloids are methylated xanthines. Caffeine occurs in tea and coffee, theophylline in tea and theobromine in cocoa. An example of a pyrimidine alkaloid is the O-glucoside vicine found in vetch seeds. The cytokinins are all N^6-substituted adenines. Structures of representatives of these groups are shown in Fig. 2.3. Comparison with Fig. 2.2 will show that two of the minor nucleosides in RNA (i^6A and ms^4i^6A) are in fact cytokinins. Cytokinins only occur in tRNA. The significance of this fact is not certain but as hormones and modified bases in tRNA might both be expected to control translation, there may be a metabolic relationship as well as a chemical analogy. Also present in Fig. 2.3 are the substances known as isoguanine (whose ribonucleoside is called 'crotonoside') and isocytosine.

Unusual nucleosides in nucleic acids   The only unusual residues identified in DNA molecules are in the bases. In certain phage DNA molecules a base (either C or T) is completely replaced by one of the unusual bases, 5-methylcytosine, 5-hydroxymethylcytosine (frequently glycosylated), uracil or 5-hydroxymethyluracil (or 'hydroxythymine'). In addition alkylated and otherwise modified bases can be introduced into DNA chemically and such bases as 5-bromouracil can be incorporated into DNA by living cells if the cells are starved of thymine and supplied with this analogue.

Some of the nucleosides identified so far in RNA are illustrated in Fig. 2.3 together with their abbreviated form for use in describing sequences (see p. 7). The only categories of bacterial RNA in which any of these nucleosides are found are tRNA and 16 S and 23 S rRNA. In eukaryotic organisms the equivalent statement is almost true but there are many types of RNA in the nucleus some of which contain minor nucleosides.

Spongonucleosides   This term strictly applies to the 'odd' purine nucleosides found in the Caribbean sponge *Cryptotethya crypta* (and presumably in related species). Spongosine is O^4-methylguanine. The other compounds are the 1-β-D-aribinofuranosyl derivatives of certain bases, that is they have the same structure as the ribonucleosides but with arabinose replacing ribose; spongothymine is aThd, spongouridine is aUrd. Another compound of the same type is the synthetic cytosine arabinoside, aCyd.

Nucleoside antibiotics   These compounds are secondary metabolites of certain species of the actinomycetes[2] and a few fungi. These substances are of great interest as they are potentially useful as drugs, as inhibitors of protein synthesis and as a variety of compounds with unusual or specifically orientated configurations and conformations. So far only a small minority have been fully exploited in these respects; however the total group is one of tremendous diversity, and as the number employed in biological research is likely to increase, it is worth trying to classify the various groups. Structures referred to are drawn in Fig. 2.4.

The simplest groups are those in which 'odd bases' are attached to ribose. Of these, nebularine (from the actinomycete *Streptomyces yokosukaensis* and also the mushroom *Agaricus nebularis*) is simply purine ribonucleoside. Of the others toyocamicin (from *S. toyocaensis*) and the closely related sangivamycin (not shown in the figure but with an amide group in place of the nitrile in toyocamycin), tubercidin (from *S. tubercidicus*), formycin and formycin B (both from *Nocardia interforma*) are all analogues of purine ribosides whereas showdomycin (from *S. showdowensis*) has a certain analogy with pseudouridine.

The next group are those in which adenine (or a very closely related purine) is attached to an 'odd' sugar. The simplest compounds of this type (not shown in Fig. 2.4) are $3'$-deoxyadenosine (cordycepin from *Cordyceps militaris*) and $2',3'$-dideoxyadenosine (a synthetic compound, an inhibitor of DNA synthesis). Another simple group are the angustomycins (produced by *S. hygroscopicus* var. *angustomyceticus*). In angustomycin A or psicofuranine, adenine is attached to D-psicose and in angustomycin C, or decoyinine, it is attached to D-angustose, both in the β-furanoside configuration. Angustomycin B turned out to be adenine. Ariskreomycin (from *S. citricolor*) does not contain a sugar at all but rather a deoxycyclitol with the same geometry as ribofuranose. The most important group of substances in this group for biochemists are those in which adenine is attached to 3-deoxy-3-amino ribose, that is ribose in which the substituent on C-$3'$ is $-NH_2$ not $-OH$. Deoxy-amino ribosyl adenine itself and its amides of lysine and homocitrulline ($NH_2CONH(CH_2)_4CH(NH_2)CO_2H$) are found in *Cordyceps militaris*. However, the most familiar of these is puromycin (from *S. alboniger*) in which the amino acid is O-methyltyrosine and also the base is N^6-dimethyladenine. The structure of puromycin is given in Fig. 2.4. The most complex of the adenine antibiotics is septacidin (from *S. fimbriatus*) in which the N-glycoside involves N-6 and the sugar is an aminoheptose. Septacidin is actually a mixture of compounds in which the fatty acid side chains vary from C_{12} acid (shown) to C_{18}.

The cytosine antibiotics are exemplified by the amicetins (produced by *S. plicatus* and other related species), gougerotin (from *S. gougeroti*) and blasticidin S (from *S. griseochromagenes*). Amicetin itself (Fig. 2.4) is the most complex; plicacetin lacks the amino acid (D-α-methyl-serine) and bamicetin has only one N-methyl substituent in the 'left-hand' amino sugar (amosamine).

The polyoxins (from *S. cacaoi* var. *asoensis*) are derivatives of 5-hydroxymethyl uracil. Polyoxin A (Fig. 2.4) is the most complex; in polyoxin B the substituted azacyclobutaine is missing to leave a free acid group and is polyoxin C. Not only is this acid free but so is the amino group on C-$5'$ of the nucleoside sugar. In other words polyoxin C is the same as 5-hydroxymethyl uridine but with C-$5'$ being substituted thus $-CH(NH_2)CO_2H$ rather than $-CH_2OH$.

Streptothricin is one of a very large group, the 3-deazatetrahydropurines produced by many species of *Streptomyces*.

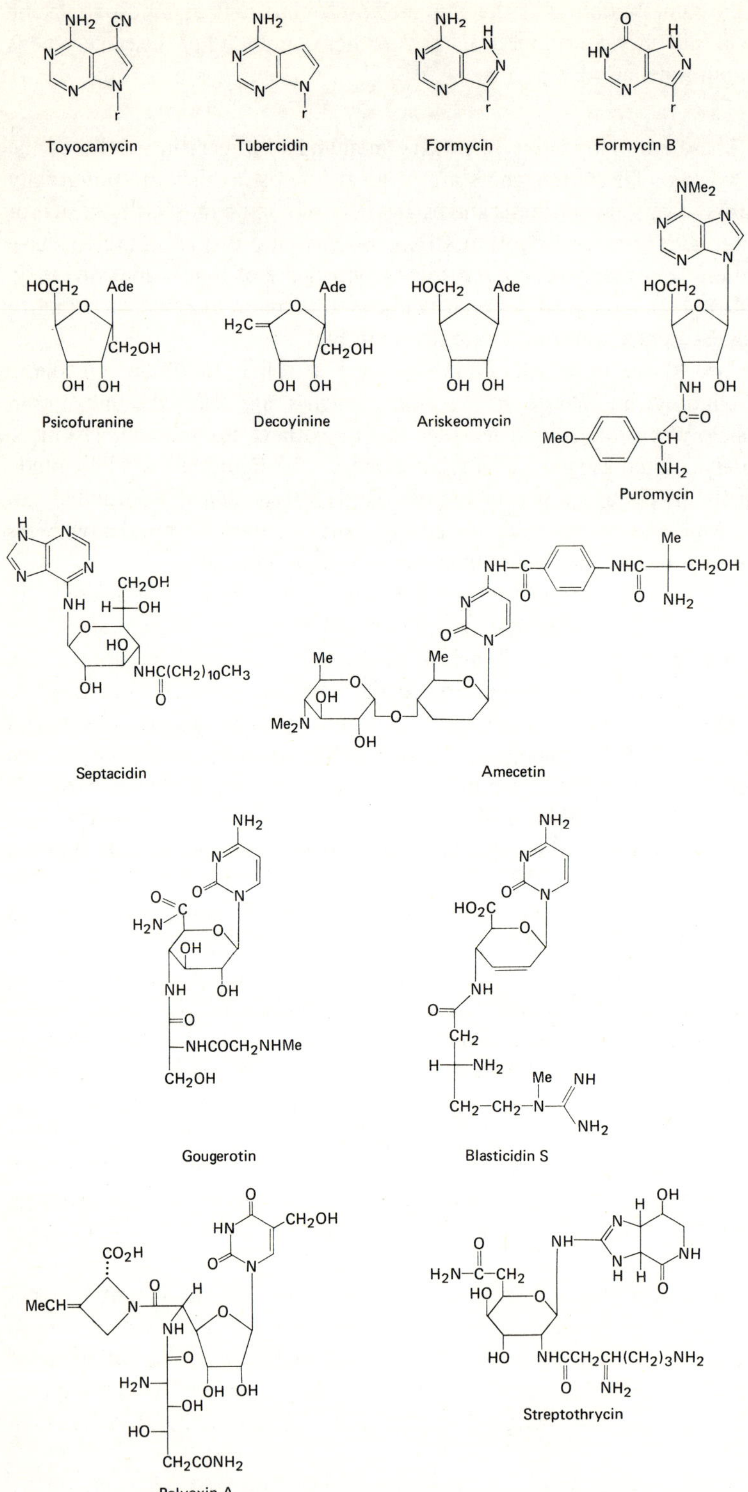

FIG 2.4   Nucleoside antibiotics (*r* = ribose).

Ionic properties and separation methods

Not every separation method is based on differences in ionic properties. However, it is impossible to discuss separations of nucleic acids or their components without a knowledge of the ionic properties of the low-molecular weight components and therefore I have compiled a brief (unreferenced) survey of these ionic properties.

Ionic properties

Dissociation constants (always expressed as pK-values) for the dissociating groups are calculated either from titration curves or from spectroscopic data. In the latter case the value for the dissociation constant is obtained from the following equation (as an example an acid dissociation constant is considered).

$$pK_a = pH - \log \frac{\epsilon_{HA} - \epsilon}{\epsilon - \epsilon_A}$$

The molar absorptivities (see p. 84) are measured at a wavelength at which the values for the undissociated form HA (ϵ_{HA}) and the anion A^- (ϵ_A) are different. The measurement is made at a pH at which both forms are present. It is important to remember that all the pK values quoted here and elsewhere are only 'apparent'. To calculate corrected pK-values from titration curves we need to know the values of all the liquid junction potentials; to calculate them spectrophotometrically we need the activity coefficients for all the ionic species involved. These data are not available for nucleotides and the other compounds under discussion here. However, there is in general good agreement between the pK-values obtained by both methods and the values quoted below are probably reasonably 'true'.

In Table 2.1 there is a summary of the pK-values for ring nitrogens and substituent hydroxyl and amino groups in several pyrimidines, purines and their nucleosides. In all the nucleosides, there is a sugar hydroxyl dissociation with a pK of about 12·5. This is not listed in the table. Note that only the ionization of one pyrimidine ring nitrogen and one imidazole ring nitrogen is measurable and that in a pyrimidine nucleoside no ring-nitrogen ionization can be measured.

The following generalizations should be born in mind. The ionizations of the ring nitrogens are very markedly influenced by substituents in the rings. The dissociations of the amino group in pyrimidines (cytosine and its derivatives) is fairly invariant but in the aminopurines there is great variation in the pK for such groups. The hydroxyl dissociation constants are affected by the position of the lactam/lactim equilibrium (see p. 4). Thus in the case of 1,3-dimethyluracil such an equilibrium is totally precluded and no such ionization can take place at all. However steric factors also play a role in determining the ease with which the lactim can be formed. Such an effect is just apparent when 1-methyluracil and 1-ribosyluracil (uridine) are compared; a much more striking effect is seen by comparing barbituric acid (which with a pK_a of 3·9 is a stronger acid than simple carboxylic acids) with barbital. Another interesting point is illustrated by comparing xanthine and xanthosine. Xanthosine is a significantly 'stronger' acid than its parent base. The reason is of course that the conventional drawing of a purine with a tertiary amine at N-7 and a secondary one at N-9 is only one of a tautomeric pair of structures (the alternative one having the NH at N-7 and the =N− at N-9). In the riboside there is never a hydrogen atom attached to N-7. The presence of a 'pure' −N= group at N-7 favours the stability of a lactim form for the oxygen at C-6 thus leading to easier dissociation and a lower pK.

Finally there are the phosphate dissociations and the influence of phosphate substituents on

TABLE 2.1   pK-values for pyrimidines, purines and nucleosides. pyr-N = pyrimidine ring nitrogen, imi-N = imidazole ring nitrogen

	pK_b for pyr-N	pK_b for imi-N	pK_b for $-NH_2$	pK_b for $-OH$
Pyrimidine	1·30	—	—	—
Uracil	0·5	—	—	9·45, >13
Thymine	0	—	—	9·82, 13
1-Methyluracil	—	—	—	9·95
3-Methyluracil	—	—	—	9·75
1,3-Dimethyluracil	—	—	—	—
Uridine	—	—	—	9·17
Deoxyuridine	—	—	—	9·3
Pseudouridine	?	—	—	9·0
Thymidine	—	—	—	9·8
Barbituric acid	—	—	—	3·9, 12·5
Barbital	—	—	—	7·85, 12·7
Orotic acid	?	—	—	9·45, 13, 2·4*
Cytosine	?	—	4·60	—
5-Methylcytosine	?	—	4·60	—
Cytidine	—	—	4·22	—
Deoxycytidine	—	—	4·3	—
5-Methylcytidine	—	—	4·28	—
Deoxy-5-methylcytidine	—	—	4·4	—
Purine	2·39	8·93	—	—
Adenine	<0	9·8	4·22	—
2-Aminopurine	−0·28	9·93	3·80	—
2,6-Diaminopurine	<1	10·77	5·09	—
Adenosine	?	—	3·45	—
Deoxyadenosine	?	—	3·45	—
Hypoxanthine	1·78	12·1	—	8·94
Inosine	1·2	—	—	8·75
Guanine	—	12·2	3·3	9·20
*iso*Guanine	—	?	4·5	9·0
Guanosine	—	—	1·6	9·16
Deoxyguanosine	—	—	1·6	9·2
Xanthine	0·8	?	—	7·44, 11·12
Xanthosine	?	—	—	5·75 ~ 13
Uric acid	—	—	—	5·4, 10·3

* Carboxyl dissociation. The structures of most of the bases and nucleosides have been introduced already either in chapter 1 or in Fig. 2.2. Hypoxanthine is the name of the purine component of inosine. In addition:

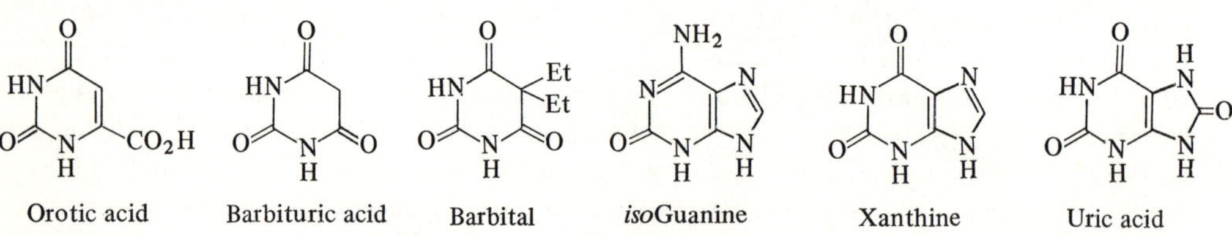

ring substituents. The phosphate dissociation constants for some derivatives of phosphoric acid are given in Table 2.2. Notice that the dissociation constants for esters of phosphoric acid depend quite a lot on the nature of the alcohol used in forming the ester and that sugar phosphates are on the whole stronger acids than phosphoric acid itself.

Phosphate substitution has no significant effect upon the dissociation of hydroxyl groups in nucleotides. This is not surprising as the pH range in which this dissociation occurs (in all but the exceptional cases of barbituric acid and a few other compounds) is so high that phosphate groups will be completely ionized. However the effect of the phosphate group and amino groups on one another is quite profound as can be seen from Table 2.3 in which sets of

TABLE 2.2   Dissociation constants for some esters of phosphoric acid

	pK_1	pK_2
Phosphoric acid	1·97	6·82
Methyl phosphate	1·54	6·31
Propyl phosphate	1·88	6·67
α-Glycerophosphate	1·40	6·44
β-Glycerophosphate	1·37	6·34
Glucose-6-phosphate	0·94	6·11
Glucose-3-phosphate	0·84	5·67
AMP	~0·9	6·1
CMP	~0·8	6·3
GMP	~0·7	6·1
UMP	~1·0	6·4
Dimethyl phosphate	1·29	—
Dipropyl phosphate	1·59	—
RNA	~1	—

TABLE 2.3   Dissociation constants for some bases, nucleosides and nucleotides

	pK_{NH_2}	pK_2 (phosphate)
Adenine	4·22	—
Adenosine	3·45	—
2'-AMP	3·80	6·15
3'-AMP	3·65	5·88
5'-AMP	3·74	6·05
Cytosine	4·60	—
Cytidine	4·22	—
2'-CMP	4·36	6·17
3'-CMP	4·28	6·0
5'-CMP	4·5	6·3
Guanine	3·3	—
Guanosine	1·6	—
2'- and 3'-GMPs	2·30	5·92
5'-GMP	2·4	6·1
Uracil, uridine	—	—
2'- and 3'-UMPs	—	5·88
5'-UMP	—	6·4

compounds are compared. One consequence of these figures is that 2′- and 3′-nucleotides can be separated by ion-exchange chromatography. The first nucleotide pairs of this type to be separated were 2′-AMP and 3′-AMP. The structures of the components were established by synthesis by Cohn (1949). Shortly afterwards the separations were applied to other nucleotides by Loring, Luthry, Bortner and Levy (1950).

Detection of purine and pyrimidine compounds

All these compounds, with the exception of dihydrouracil and similar substances, have strong ultra-violet absorption spectra. They are normally identified in solution (such as the eluate from chromatography columns) by monitoring the absorbance at the wavelength of the absorbance maximum (see p. 85). Spots on paper (following chromatography or electrophoresis) can be visualized by observing the paper under ultra-violet light; the spots appear dark against a light background except for Gua and its derivatives which fluoresce (violet) if the paper has been in acid. If the derivatives have been chromatographed in a basic or neutral solvent, the spots all look dark. In this case the Gua fluorescence can be developed by making the paper slightly acid. This is most easily done by waving the stopper from a concentrated HCl bottle around the paper. If the substances are radioactive they can be identified either by autoradiography or by cutting the paper up and counting it (see p. 178).

Electrophoresis

Electrophoresis is the most versatile method for the fractionation of nucleotides and polynucleotides (including nucleic acids). It is not in general a satisfactory method for bases and nucleosides. The separation of mononucleotides is necessary for the identification of

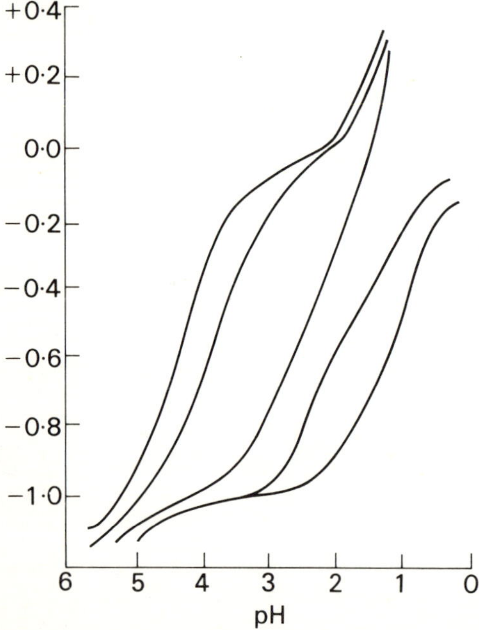

FIG 2.5   Net charge-molecule as a function of pH for various nucleotides. The curves were calculated by Cohn (1950) from the data of Levene and Bass (1931).

components and measuring their relative amounts in alkaline digests of RNA (see p. 204). The great importance of the procedure derives from the fact that certain RNA molecules can only be obtained in very small radiochemical quantities and the easiest general method of labelling RNA is with ^{32}P in the phosphate moieties. In this case the molecules have to be degraded to their nucleotide in order that the radioactive phosphate (the only group whereby the substances can be identified) is not lost. It is also used for the nearest neighbour frequency analysis of DNA (p. 271). The optimum pH for electrophoretic separation is clearly the one at which the nucleotides carry charges as different to one another as possible. From the variation of net charges per molecule with pH (see Fig. 2.5) it is clear that they should separate at pH 3·5 but that there is not much latitude on either side. The separations are normally performed on paper in either a flat-bed electrophoresis apparatus or, preferably, in a tank under white spirit or carbon tetrachloride. Further details of design of apparatus for paper chromatography are to be found in the excellent compendium edited by Smith (1967).

The mobilities of several nucleotides at low pH are summarized in Table 2.4. In the legend to this table are details of the normal conditions employed for such separations.

There is a very good separation of the major nucleotides and several minor nucleotides at pH 3·5 and (as predicted from Fig. 2.5) there is a poor separation of the major nucleotides at the lower pH (2·1). However, this sytem is valuable for separating certain substituted purine nucleotides.

TABLE 2.4   Electrophoretic mobilities of nucleotides. Migration is in the direction of the cathode in all cases. For the ribonucleotides they are all quoted relative to UMP (=1·00) and for deoxyribonucleotides they are quoted relative to dTMP (=1·00). The potential gradient was of the order of 20 V/cm in each case; the fastest moving component moves at about 8 cm/h. The pH 3·5 system was 0·05 M ammonium formate and the data are those of Markham and Smith (1952),[1] Sanger, Brownlee and Barrell (1965),[2] Littlefield and Dunn (1958).[3] The pH 2·1 system[4] is 0·05 M phosphate buffer (Smith, 1967).

Compound	pH 3·5[1]	pH 3·5[2]	pH 3·5[3]	pH 2·1[4]
UMP (Up)	1·00	1·00	1·00	1·00
CMP (Cp)	0·41			0
AMP (Ap)	0·50		0·50	0
GMP (Gp)	0·88	0·73		0·38
hUp		1·00		
Tp		0·98		
Ψp		0·98		
IMP (Ip)		0·90		
m^{7}Gp		0·98		
m^{1}Gp				0·3
m^{2}Gp				0·3
m$_2^2$Gp		0·76		0·25
m^{6}Ap			0·46	
m$_2^6$Ap			0·42	
m^{2}Ap		0·25	0·19	
dTMP				
dCMP				
dAMP				
dGMp				

Electrophoresis at higher pH is useful for certain specific jobs. Obviously electrophoretic separation of sets of compounds such as AMP, ADP and ATP is easy provided a pH is chosen in which the number of phosphate dissociations is different (typically pH 5 to 7 is the chosen range). Substances such as ATP are very strong chelating agents and results are sometimes irreproducible on account of varying amounts of bound divalent cations. For this reason it is preferable to use as buffer a strong chelating agent (such as citrate, Markham, Matthews and Smith, 1954). Ribonucleotides (5′ only) can be separated from either deoxyribonucleotides or 2′-methyl-ribonucleotides in borate buffer at pH 9·2. The separation is based on the fact that *cis* glycols form complexes with borate ions. In this case only the 5′-ribonucleotides have a *cis* glycol in them and therefore only they will form such a complex. As the pK_{a1} for boric acid is 9·14, the borate complex carries about half an additional negative charge and will migrate faster than their methylated or deoxyribotide analogues (Jaenicke and Volbrechtshausen, 1952).

Electrophoresis is not in general a satisfactory method of analysing mixtures of nucleosides. However the technique of making borate complexes of ribose derivatives mentioned above works well for separating ribosides from others. Certain derivatives of nucleosides can be separated by electrophoresis if there are strongly ionizing substituents. The isonicotinic acid hydrazide derivatives of oxidized ribosides (see p. 74) can be separated at pH 3·0. Of more importance are the aminoacyl derivatives of adenosine. These compounds are the 3-terminal nucleosides of aminoacyl tRNA. They migrate towards the cathode at pH 3 owing to the positive charge on the amino group of the amino acid (Zachau, Acs and Lipmann, 1958).

Paper chromatography and thin-layer chromatography

Paper chromatography remains the simplest method for quantitative analysis of mixtures of the normal bases for the purpose of measuring base ratios. It is also the simplest method of analysing mixtures of nucleosides. An excellent review of solvent systems for the paper chromatography of nucleic acid components is by Wyatt (1955). However for the preparative fractionation of new or minor nucleosides paper chromatography has been really superseded, although for confirmation of the identity of nucleosides it remains a useful analytical method. A comprehensive list of R_f-values for many of the minor nucleosides of tRNA in five solvent systems is given by Hall (1967).

For a separation of the four bases derived from DNA an acid system is the best. In all these systems the order of migration (in the direction base line to solvent front) is Gua, Ade, Cyt and Thy. A positive suggestion of mine is to use the method of Kirby (1957) in which the solvent is a mixture of methanol, concentrated hydrochloric acid and water (7:2:1 by volume). Use ascending chromatography with Whatman no. 1 paper and run the solvent across the machined direction of the paper. Allow it to run overnight (about 20 cm) and use about 0·2 mg hydrolysate per spot. For acid hydrolysates of RNA it is probably easiest to use conditions which yield a mixture of the pyrimidine nucleotides and purine bases (p. 203) and to analyse the mixture in the same way but using as solvent system a mixture of methanol, ethanol, concentrated hydrochloric acid and water (50:25:6:19 by volume). In this case the order of elution is Gua, Ade, CMP and UMP (Kirby, 1956). Another useful application of paper chromatography is for separating a set consisting of a base, nucleosides and phosphates such as Ade, Ado, AMP, ADP and ADP. The best system using ordinary chromatography paper is probably that of Turba, Pelzer and Schuster (1954), 0·1 Mtrisodium citrate, pH 8·45 saturated with *iso*amyl alcohol.[3] Alternatively mixtures of this type can be separated on ion-exchange papers such as diethylaminoethyl cellulose paper (DEAE-paper). Morrison (1968) recommends ammonium formate buffer pH 3 of molarities depending on the nature of the nucleotide mixture to be analysed. Finally a recent chromatographic system with great resolving powers for a variety of types of mixture is that of Reeves, Seid and Greenberg (1969). The solvent is a mixture of the following components the relative proportions by volume given in parentheses: *iso*butyric acid (160), water (22), 0·1 M EDTA monosodium salt (3), concentrated aqueous ammonia (2) and toluene (20). The R_f-values of the common bases, nucleosides and nucleotides in this system are summarized in Table 2.5.

TABLE 2.5   Relative mobilities (R_f-values) of nucleic acid components in the paper chromatographic system of Reeves, Seid and Greenberg (1969)—see text

	C	U	T	A	G
Base	0·46	0·32	0·49	0·81	0·35
Riboside	0·20	0·08	0·16	0·50	0·09
Deoxyriboside	0·31	0·19	0·28	0·61	0·19
5′-Ribotide	0·05	0·03	—	0·08	0·01
5′-Deoxyribotide	0·07	0·03	0·08	0·16	0·02

There is not, so far as I am aware, any simple paper chromatographic system for the analysis of mixtures of nucleotides such as AMP, UMP, CMP and GMP. However it is possible to separate them by chromatography on DEAE-paper (such as Whatman DE 11). The method worked out in the author's laboratory by R. M. Malbon is to use a long development (ascending chromatography, 15-20 h, 50 cm travel by the solvent front) using as solvent an aqueous solution of triethylamine (0·6% v/v) and glacial acetic acid (5·0% v/v). The pH is 3·4. R_f-values are GMP 0·17, UMP 0·27, AMP 0·35 and CMP 0·63. With small spots and runs as described above, the UMP and AMP are completely resolved.

Thin-layer chromatography of nucleic acid constituents requires an aqueous developing system and absorbant microgranular cellulose. An extremely well-informed review with much useful technical advice is by Randerath and Randerath (1967). According to Randerath and Randerath the optimum thickness for the layer is 0·5 mm of the wet slurry. They emphasize the importance of thoroughly degreasing the glass or plastic plates prior to layering of the absorbant. The absorbant is either unmodified cellulose or a weak anion exchange cellulose. For the former method, microgranular cellulose powder (20 g/130 ml water) is made into an even slurry by homogenization in an electric mixer followed by degassing; in the latter the microgranular cellulose is dispersed in 1% polyethylene imine hydrochloride pH 6 (22 g in 145 ml) and homogenized and degassed as above. It is recommended that both types of plates be allowed to dry at room temperature. Randerath and Randerath (1967) have tabulated the R_f-values of a number of bases and nucleosides (including many halogenated derivatives) using both acidic (methanol, concentrated hydrochloric acid and water, 70:20:10) or basic (*n*-butanol, methanol, water and concentrated ammonia, 60:20:20:1) systems to develop chromatograms on unmodified cellulose. The acidic system gives the better resolution for the separation of the common bases (in order from the base line, Gua, Ade, Gyt, Ura and Thy). Randerath and Randerath (1967) have also tabulated R_f-values for a variety of nucleotides on polyethylene imine-cellulose plates. For the normal nucleoside monophosphates the best solution for development appears to be 0·25 M lithium chloride (the order is GMP, AMP, CMP and UMP).

An alternative procedure for the thin-layer chromatography of 2′- and 3′-nucleotides on polyethylene imine-cellulose plates is that of Greenman, Huang, Smith and Farrow (1969). They prepare the plates using a 6:1 mixture of commercially available polyethylene imine cellulose powder with untreated cellulose powder; the chromatogram is developed in several stages with formic acid. The first phase consists of development with 0·1 M formic acid, this is repeated several times under the leading spot as visualized under UV (CMP) is well up towards the solvent front. The appearance of the plate (under the UV) is that near the base line is an unresolved mixture of UMP and GMP followed by three spots, the ones nearest the base line are 2′- and 3′-AMP, the one furthest is the mixture of 2′- and 3′-CMP. The plate is then developed with 5 M formic acid and stopped as the solvent front approaches the lower AMP spot. The UMP (nearer base line) and GMP are then resolved. This type of procedure, which would be immensely time-consuming for paper chromatography, is quite feasible for thin-layer chromatography where development times of less than an hour are typical.

Ion-exchange chromatography

Ion-exchange resins consist of ionizing groups covalently bound to an insoluble matrix suitable for packing into a chromatography column. If the ionizing groups are acidic (i.e. the matrix is an insoluble polyanion) the resin is cation exchanger; if the ionizing groups are amino (i.e. the matrix is an insoluble polycation) the resin is an anion exchanger. For the purpose of the present discussion there are three types of matrix, cross-linked polystyrene beads, cellulose and polysaccharide gels. The advantages of the first of these types is that strongly ionizing groups can be incorporated into the resin. Thus the benzene rings can be sulphonated to yield a cation-exchange resin with acid groups with very low pK-values or alternatively quaternary ammonium groups can be introduced to yield an anion-exchange resin with very strongly basic groups. The other types contain weakly ionizing groups, usually either carboxymethyl (CM) for a cation exchanger or diethylaminoethyl (DEAE) for an anion exchanger. In the case of the modified celluloses the groups are attached to either fibrous or microgranular cellulose particles. In the case of the polysaccharide gels the groups are attached to cross-linked dextran (Sephadexes).[4]

Cation-exchange resins Mixtures of bases or nucleosides or nucleotides are easily resolved by chromatography on cation-exchange resins. The eluent can be a constant concentration of an increasing concentration of either acid or salt. As a rough guide the diameter to column length should be in the ratio of between 1:10 and 1:20; the bed volume should be of the order of 10 ml per mg of the most concentrated component of the mixture to be separated and flow rate should be less than 1 ml per min per 10 ml of bed volume. Examples of the method are summarized below, the substances being given in the order of elution.

Cohn (1949): Dowex 50 (sulphonated polystyrene type) in the hydrogen form, eluted with 2 M hydrochloric acid—Ura, Cyt, Gua and Ade. A modification of this procedure is useful for separating alkylated guanines (see p. 60).

Crampton, Frankel, Benson and Wade (1960): Dowex 50 in the sodium form, eluted with a gradient of 0·2 M sodium citrate pH 4 to 1 N unbuffered sodium citrate (i.e. a gradient of increasing ionic strength and pH)—Ura, Thy, Gua, Acyt, Ade, 5-methylCyt.

Cohn and Uziel (1967): Amberlite CG-120 type III 400-600 mesh (sulphonated polystyrene resin) in the ammonium form, eluted with 0·4 M ammonium formate, 0·4 M formic acid—Urd, Guo, Ado, Cyd.

Crampton and Rodeheaver (1967): Dowex 50-X2 in the sodium form, eluted with a solution of 0·1 M citric acid, 0·2 M sodium hydroxide, 0·13 M hydrochloric acid pH 3·25—TMP, dGMP, dCMP, dAMP.

Anion-exchange resins   Anion-exchange chromatography is a versatile method for fractionating nucleic acid components, oligonucleotides and nucleic acids (see p. 160). Examples of the method are as follows:

Cohn (1949): Dowex-1 (quaternary base) 300 mesh chloride form, eluted with 0·2 M ammonium hydroxide, 0·025 M ammonium chloride—Cyt, Ura, Thy, Gua; the ammonium chloride is then stepped to 0·1 M to elute Ade.

Cohn and Volkin (1951): Dowex-1 X8 200-400 mesh formate form eluted with (1) 0·02 M formic acid—5′-CMP, 2′-CMP, 3′-CMP; (2) 0·15 M formic acid—5′-AMP, 2′-AMP, 3′-AMP; (3) 0·05 M sodium formate, 0·1 M formic acid—5′-UMP, 2′-UMP, 3′-UMP; (4) 0·1 M sodium formate, 0·1 M formic acid—5′-GMP, 2′-GMP, 3′-GMP.

The second of these examples is from the work of W. E. Cohn on the 2′- and 3′-nucleotides (see p. 46). Other specialized applications of Dowex-1 chromatography by Cohn are the use of borate buffers (see p. 48) to separate various derivatives or uridine and pseudouridine (Cohn, 1960) and elution with hydrochloric acid to separate the degradation products of dihydrouridylic acid (see p. 69).

DEAE-cellulose is not routinely used for fractionation of nucleotides and other low molecular-weight derivatives; the relatively slow flow rates of this type of column detract from its use when other fast flowing columns (such as the synthetic resins) are effective. Chromatography on DEAE-cellulose is an important technique for fractionation of oligonucleotides and nucleic acids (see pp. 161 and 292) and has been used for mononucleotides in certain special circumstances. An example is the separation of 4-thiouridylic acid from the nucleotides obtained by alkaline hydrolysis of tRNA from *E. coli*. At high pH the thiol group is ionized, so the thiouridylic acid molecule carries three negative charges (two on phosphate, one on sulphur) whereas the other nucleotides just carry the two on phosphate. In order to eliminate association of these charged nucleotides by hydrogen bonding the eluting solution contains 7 M urea (Lipsett, 1965). The salt in this solution is a gradient of 0·05—0·25 M ammonium bicarbonate pH 8·6.

A recent suggestion (Miller and Kirkpatrick, 1969) is to employ polyethylene imine-cellulose for the column chromatography of nucleotides. According to them the column packing should contain a 2:1 mixture of polyethylene imine-cellulose for thin-layer chromatography ('Cellex-PEI') and technical grade 'Avicel'. The affinity of nucleotides for the resin are very weak, for example dTMP, dCMP, dAMP and dGMP (in that order) or UMP, CMP, AMP and GMP are eluted with a linear gradient of 0—0·01 M citrate buffer pH 7·0.

As an alternative to ion-exchange celluloses there are the ion-exchange crossed-linked dextrans ('DEAE-Sephadex'). On the whole, there is everything to be said in favour of the Sephadexes and everything to be said against the celluloses. The Sephadexes are beads, they do not break up or produce fines, they are supplied in a very clean form by the manufacturers and do not need elaborate recycling procedures to rid them of impurities which can be eluted from other types of resins to contaminate the fractionated products and confuse spectrophotometric monitoring of column effluent. The only disadvantages of the Sephadexes are that the columns are more slow-running and the resin is expensive.

Three examples of the use of DEAE-Sephadex A-25 resin (medium grade) are as follows. Maddison and Holley (1965) used a linear gradient (0—0·16 M ammonium carbonate) to elute nucleotides obtained by the alkaline hydrolysis of tRNA and isolated a fraction containing phosphorus with no absorbance in the UV. This fraction was shown to be the minor nucleotide dihydrouridylic acid, not suspected of being a component of RNA before that time. Using a similar system but with a carefully defined concave gradient of ammonium bicarbonate, Rossett and others (1970) have described a separation of 3′,5′-cyclic AMP, AMP, ADP and ATP of value·

in following the degradation *in vivo* of adenosine nucleotides. Finally there is the remarkable separation of a very large number of nucleotides including the co-factors NAD, NADP and FAD by Caldwell (1969). In this case three successive exponential gradients of triethylammonium acetate pH 4·70 followed by strong triethylammonium formate to elute the GMP were used to develop the column.

Partition chromatography in columns

Partition chromatography is the process whereby the components of a mixture are separated on the basis of their different partition coefficients in two immiscible fluid phases. We shall discuss the use of gas-liquid chromatography shortly. In liquid-liquid chromatography there are several applications of the system. The most straightforward theoretically (but not experimentally) is counter-current distribution. Although used for nucleic acid separations (see p. 168) it is of no practical importance for mixtures of nucleotides, nucleosides and bases. The most familiar type of liquid-liquid chromatography is paper chromatography, where the stationary phase is water (present in the paper) and the moving phase is the solvent system used to develop the chromatogram. The technique can be scaled up by using two immiscible phases, one of which is absorbed into an inert support material which is then packed into a column, the other is used to develop the chromatogram. The method has been extensively exploited by R. H. Hall for the separation of minor nucleosides. A review of the method which includes useful technical details has been written (Hall, 1967). One example of the method which is the first fractionation in Hall's scheme is as follows. Nucleosides obtained from the hydrolysate of 5 g tRNA were fractionated. The inert support is diatomaceous earth. It is washed in 3 M hydrochloric acid, the acid is washed out with water and ethanol and it is thoroughly dried. A 9:1 mixture of two grades ('Celite-545' and 'Microcel-E' respectively) is recommended. A two-phase solvent system (ethyl acetate, 2-ethoxyethanol and water in the ratio 4:1:2 by volume) is thoroughly equilibrated and allowed to separate. The lower phase (308 ml) is mixed with 690 g of support material. The liquid is absorbed (in a mechanical shaker) to produce an even 'dry' powder. This is packed into a big chromatography column and rammed down one layer at a time to produce an evenly compacted column. The nucleoside mixture in 35 ml of the same phase is mixed with 80 g of support and packed on top. The column is then developed with 3·5 l of upper phase to elute (in this order) N^6-methyladenosine, adenosine and uridine. The remaining nucleoside (a mixture of methylated guanosines, guanosine and cytidine) are eluted with a further 3·5 l of the upper phase of a different system (ethyl acetate, butan-1-ol and petrol 66-75°C boiling range, 4:1:2). Further fractionation of the nucleosides on smaller columns using different systems is employed for further purification. In particular the two phases of the mixture of butan-1-ol, concentrated aqueous ammonia and 3·8% aqueous sodium borate ($Na_2B_4O_7 \cdot 12H_2O$) in the proportions 3:0·05:1 is used for the separation of $2'$-methyl ribosides. Borate ions which complex the *cis*-glycol of the ribosides are confined to the aqueous (stationary) phase so they are retarded on the column with respect to the $2'$-methyl nucleosides which are eluted first.

An alternative method is to trap the aqueous phase in a porous gel such as polyacrylamide. In the method of Kull and Sodak (1968) isobutyric acid, concentrated aqueous ammonia and water (66:1:33) are used to elute Ado, Cyd, Gua and Urd (in that order) from Biogel P-6. The corresponding nucleotides were separated in the same way using Biogel P-2.

Gel-absorption chromatography

Porous gels are normally employed, not as supports for partition chromatography as above but for the process known as exclusion chromatography. This is a standard method for fractionating macromolecules on the basis of their molecular weight (see p. 167). Large molecules are excluded from the interstices of the gel and are hence eluted from the column before smaller ones, which can enter the interstices. However the gels of this type which are designed for excluding all but the smallest macromolecules (ones with small pores or low exclusion limits) will also fractionate certain classes of small molecules on account of a different type of interaction. The most frequently used resins of this type are the cross-linked dextrans (Sephadex). The glucose moieties of dextran molecules adopt conformations such that the hydroxyl substituents are exposed on one face of a block of glucose units to leave hydrophobic regions (groups of C–H bonds). Low-molecular-weight molecules have differing affinities for these regions. Such a column can be regarded as a partition chromatography column with water or buffer as the moving phase and hydrophobic regions in the gel as the stationary phase. A comprehensive discussion of the facts influencing the chromatographic behaviour of bases, nucleosides and nucleotides on Sephadex gels is by Gorbach and Henke (1968). The gel with the smallest pore size (G-10) is used for bases, pyrimidines are eluted before purines; thymine is retarded (methyl groups add to hydrophobic character) relative to uracil. Nucleosides and nucleotides are fractionated on Sephadex G-15 or G-25. Altogether the technique has much to offer; I have already extolled the general virtues of Sephadex products (p. 51). An additional advantage of the Sephadex gels which contain no ion-exchange groups, is that there is no tailing of the peaks, recovery is quantitative and, in the event of your analysing a mixture which may contain an unknown or unsuspected component, there is no danger of its being left bound to the column. Everything is always eluted from Sephadex eventually as interactions between solute and gel are all very weak. A specific application of this type of chromatography is to the resolution of mixtures of methylated purines on Sephadex G-10 (Sweetman and Nyhan, 1968). These compounds are obtained by hydrolysis of methylated nucleic acids (see p. 239).

An alternative type of gel-adsorption chromatography employs polyacrylamide gels. Gels such as Biogel P-2 will resolve complex mixtures of nucleosides, isomeric nucleotides and di- and tri-phosphates. The separations are most effective at elevated temperatures (John, Skrabei and Dellweg, 1969).

The most sophisticated way of obtaining inclusion of purines in hydrophobic regions of sugar structures is to use cyclodextrins. The techniques require considerable chemical know-how. For the fractionation of several adenine derivatives, Hoffman (1970) employed cyclopeptaamylose cross-linked to form a gel matrix.

Gas-liquid chromatography

Gas-liquid chromatography is a very specialized technique and there is no space here to discuss the apparatus in any detail. Basically it is partition chromatography in which the moving phase is a gas and the stationary phase in a non-volatile liquid. As an analytical technique is has two advantages over other forms of chromatography; extremely long columns can be constructed and the resolution of components which differ only very slightly in partition coefficients is possible. The other advantage is that the monitoring of column effluent is nowadays usually effected by ionization detectors of one sort or another—these are orders of magnitude more sensitive than the spectrophotometric techniques employed in liquid chromatography systems. Gas–liquid chromatography can be scaled up to provide a preparative fractionation system. This is a costly and complex operation and is not suited to nucleic acid derivatives for a reason just to become apparent. The obvious limitation to gas–liquid chromatography is that the components of the mixture to be analysed must be sufficiently volatile to partition significantly into the vapour phase. With bases, nucleosides and nucleotides this implies a temperature at which they will be pyrolysed. Therefore volatile derivatives have to be prepared prior to the chromatography. The delay in the widespread introduction of this, the most sensitive analytical chromatographic system of all, into nucleic acid research has been due to the perfection of techniques for quantitative conversion of the components of such a mixture into volatile derivatives.

An ingenious technique for the bases Cyt, Gua, Ade and Ura has been developed by MacGee (1966). The tetramethylammonium salts of the bases (in solution in alcoholic 1 M tetramethylammonium hydroxide) were injected into the inlet heater of the chromatography apparatus at 360°C. At this temperature the salts decompose to produce N-methyl derivatives of the bases which pass into the column (1·8 m long, 0·4 cm internal diameter) packed with acid-washed glass beads (80–120 mesh) coated with 0·5% polyethylene glycol ('Carbowax 20M'). The carrier gas was argon and the apparatus was programmed so that the overn temperature rose from 143 to 221°C. The flow rate of gas dropped from 139 to 107 ml/min during the run. The method gave good analytical figures for ratios of the bases but suffered from the disadvantage that all the bases except Ura and Thy gave multiple products and hence several peaks had to be integrated for each component.

Compounds containing sugar residues can be analysed by gas–liquid chromatography if they are first of all converted into trimethylsilyl ($-SiMe_3$) derivatives. A simple and effective method for nucleosides is to treat a dry mixture of nucleosides with a 100-fold molar excess of N,O-bis (trimethylsilyl) acetamide (structure below)

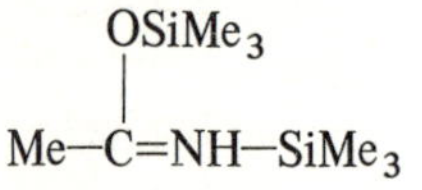

$$OSiMe_3$$
$$|$$
$$Me-C=NH-SiMe_3$$

These derivatives form single peaks for each nucleoside when injected (inlet temperature 260°C) into a column (60 cm long, 0·4 cm internal diameter) packed with 4% phenyl-methyl-silicone gum ('OU-17') on a standard diatomaceous earth support. The carrier gas is argon, the flow rate 75 ml/min and the temperature programmed to rise at 2 deg/min between 160 and 240°C (Jacobson, O'Brien and Hedgcoth, 1968). The order of elution is Ψrd, Urd, Ado, Guo and Cyd.

An alternative method of making trimethylsilyl derivatives is to use hexamethyldisilazane ($Me_3SiNHSiMe_3$) and/or trimethylchlorosilane (Me_3SiCl). A mixture of these two (2:1 by volume in favour of the former) will form trimethylsilyl esters of phosphates and trimethylsilyl nucleotides formed in this way can be separated by gas–liquid chromatography (Hashizume and Sasaki, 1966).

The stationary phase was 5% of a methyl silicone fluid ('DC 430') on a diatomaceous earth. Steel tubes (75 cm) were used for the column; the inlet temperature was 320°C, the column temperature 266°C, the carrier gas helium and the flow rate 75 ml/min. The common nucleotides including 5′- and 2′,3′-isomers were resolved. Two peaks were obtained from 2′- and 3′-UMP, otherwise the 2′/3′ pairs were unresolved.

Chemical properties

In this section there is a brief survey of some of the important chemical properties of nucleotides, nucleosides and bases. The intention is twofold, to summarize those properties which are of importance in characterizing the degradation products of nucleic acids and to explain the basis of those chemical properties which are also of importance in nucleic acids in the simpler context of compounds of low molecular weight. Reference to experimental procedures has largely been omitted except for the few reactions which are commonly employed for preparing reference compounds for work with nucleic acids.

Bases, nucleosides and nucleotides—degradation and synthesis

The simplest method of removing the phosphate group from nucleotides is with a phosphomonoesterase. These enzymes are on the whole of very limited specificity and quantitatively convert nucleotides to nucleosides. The usual enzyme to choose is the alkaline phosphatase from *E. coli*. The enzyme can be cheaply purchased or very easily made. It is induced in *E. coli* grown in a medium in which all the phosphate is present as a phosphomonoester, such as β-glycerophosphate. It is located between the cell wall and the plasma membrane and if the cells are converted into protoplasts (see p. 125) or freeze-dried and extracted with buffer the enzyme appears in the supernatant. The pH-optimum for enzyme activity is about 8·6. Although it is considerably activated by strong tris buffer, for the purpose of converting nucleotides into nucleosides, it is not necessary to optimize everything provided the enzyme is in excess. A typical recipe would be to use about 5 mg of enzyme per mg of nucleotide and to perform the hydrolysis in any suitable buffer of pH 8·6 in the presence of 5 mM magnesium salt and to incubate it overnight at 37°C. It is essential that no inorganic phosphate is present in the buffer as phsophate is an inhibitor of the enzyme.

Hydrolysis of nucleosides The *N*-glycoside bonds in nucleosides are on the whole resistant to alkali. The only exceptions are found in those nucleosides in which the sugar ring is strained or locked in a particular conformation (e.g. 3′,5′-cyclic AMP).

On the other hand *N*-glcosides are, in general, hydrolysed by acid. Among the nucleosides there is a very great variety in the case of acid hydrolysis. Deoxyribosides are significantly more easily hydrolysed than the ribosides and for any one class the purine nucleosides are significantly more easy to hydrolyse than the pyrimidine nucleosides. Thus the pyrimidine ribosides (Cyd and Urd) arc almost impossible to hydrolyse except by such reagents as strong perchloric acid which oxidize the sugar. The mechanism of hydrolysis or nucleosides proposed by Kenner (1957) is shown below.

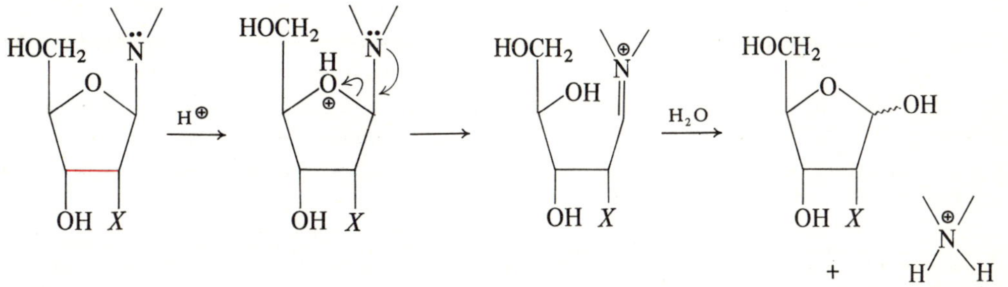

The factor which determines the ease of acid hydrolysis is the readiness with which the ring oxygen can become protonated. In the deoxyribosides ($X = H$) it is apparently easier for this protonation to occur than in the ribosides, which contain two hydroxyl substituents in the furanose ring. It is possible to rationalize the ease of hydrolysis of nucleosides of different bases. The one which hydrolyse easily are those in which there is a nitrogen atom, protonated in acid which, on one conformation of the nucleoside, is close enough to the ring oxygen of the furanose for proton-transfer to occur. The protonated form of three easily hydrolysed nucleosides (purine riboside, 2-aminoimidazole riboside and isocytidine) are shown below.

Pseudouridine is not an N-glycoside and might be expected to be inert to acid. Moderately strong acid (1 M hydrochloric acid at 100°C) does cause isomerization so that an equilibrium mixture of α and β anomers is slowly produced (Cohn, 1960).

Conditions for hydrolysis of nucleic acids to produce the bases are given on p. 203. The same conditions will hydrolyse the corresponding nucleosides (or nucleotides).

Enzymic hydrolysis of nucleosides can occur by two methods. The nucleosidases are enzymes which hydrolyse nucleosides to yield free base and free sugar. They are a relatively poorly characterized group. Considerable base specificity is demonstrated by the purine ribonucleosidases from *Aspergillus foetidus* and related fungi (Reese and Maguire, 1968). A better characterized group are the nucleoside phosphorylases. These are enzymes for which the substrates are nucleoside and inorganic phosphate and the products are free base and sugar-1-phosphate. The enzyme from *E. coli* is of some interest as it has very little base-specificity (the base can be one of a range of pyrimidines, imidazoles and purines) and the sugar can be either ribose or deoxyribose. However the nucleosides must be β-anomers. It has therefore been proposed by Lodemann and Wacker (1967) that the enzyme could be used to determine the anomeric configuration of nucleosides; the preparation and spectrophotometric assay of the enzyme are very simple and α-anomers do not react at all.

Synthesis of nucleosides   The enzymic routes for the synthesis of nucleosides are summarized on p. 175. Methods for chemical synthesis are mostly based on the use of a derivative of the sugar known as a 'halogenose'. Halogenoses are prepared from derivatives of monosaccharides by treatment with hydrogen chloride under rigorously dry conditions. In the halogenose the anomeric hydroxyl on C-1 is replaced by a halogen (usually Cl). Halogenoses are always anomeric mixtures. The simplest procedure for the synthesis of a nucleoside is to treat an acylated halogenose with mercuri-derivative of the base (see p. 403). The reaction takes place in hot dry solvent (usually an aromatic hydrocarbon). The acyl groups are removed from the sugar by hydrolysis. The anomeric configuration of the product is determined by the substituent on C-2′. The base ends up *trans* to it. Thus β-ribosides are synthesized. Mixtures (α and β) of deoxyribosides are produced. The technique as applied to the synthesis of the deoxyriboside of 5-fluorouracil (see p. 35) is summarized in Fig. 2.6*a* (described by Wempen and Fox, 1967; the original is in the U.S. Patent literature by M. Hoffer). The β-deoxyriboside is obtained from the final mixture by fractional crystallization. Obviously β-arabinosyl nucleosides cannot be made in this way. The alternative is to use benzyl derivatives of halogenoses; the stereochemical restriction appears not to operate in this case. The reaction is performed in absolutely dry methylene chloride. The mercuri-derivatives are not required but the base (which must be a pyrimidine) is completely alkylated or (in a more recent modification) silylated (see p. 54).

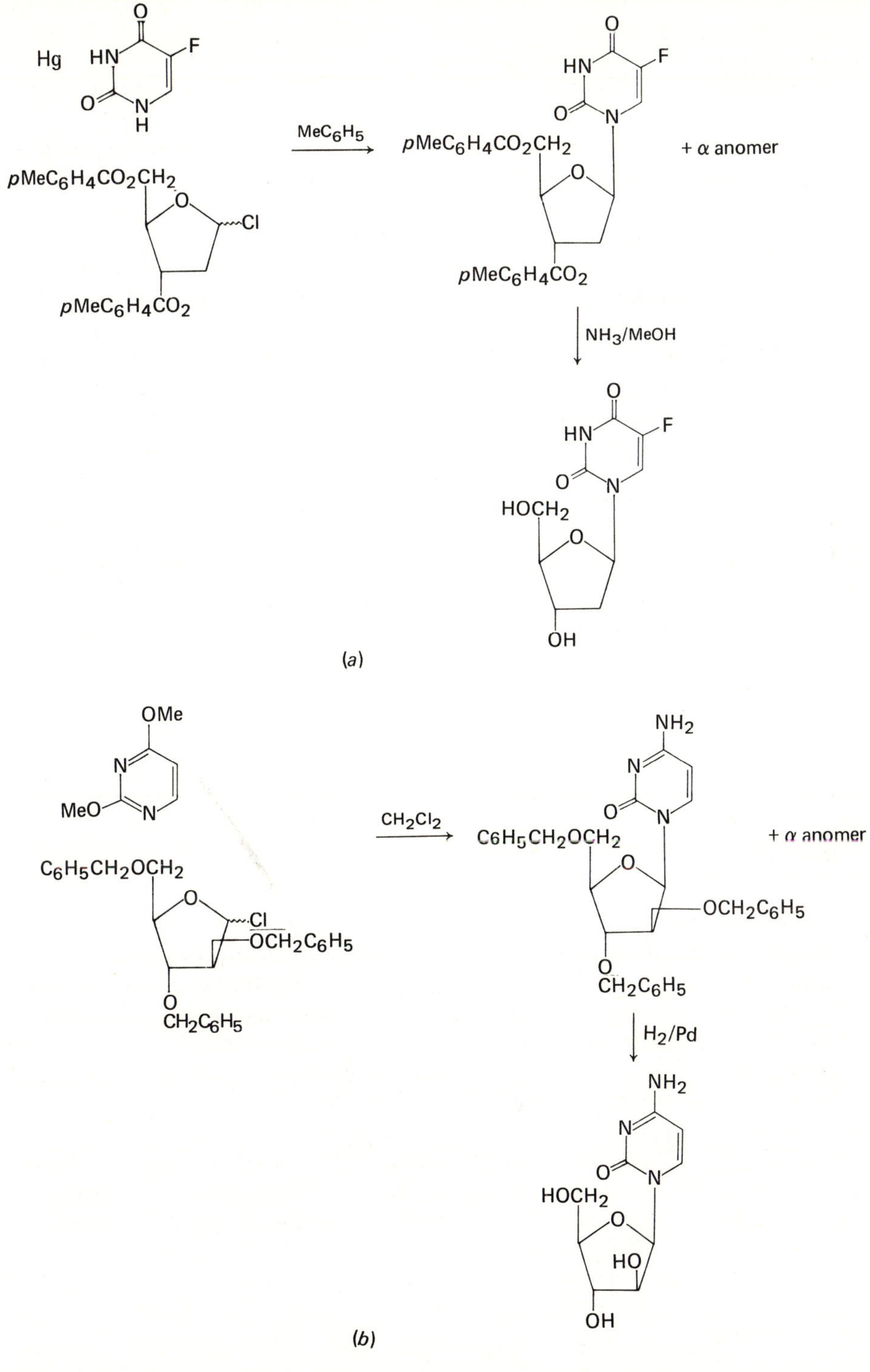

FIG 2.6 Synthesis of nucleosides exemplified by (*a*) 5′-fluorodeoxyuridine ('FUDR') and (*b*) 'cytosine arabinoside'.

The application of the method to synthesis of 'cytosine arabinoside' (see p. 41) is given in Fig. 2.6(*b*) (Shen, Lewis and Ruyle, 1965).

Synthesis of nucleotides   Nucleosides can be phosphorylated either enzymically or by chemical methods.

If it required to obtain [^{32}P] nucleotides, there is no doubt that the simplest thing to do is to let *E. coli* make them for you. If the bacteria are grown in a medium lacking inorganic phosphate (see p. 55) and exposed to carrier-free inorganic [^{32}P] phosphate for several generations (say a few hours) the RNA and DNA will become extremely radioactive. If nucleic acids are extracted (see pp. 107 and 111) and separated, the nucleotides can be obtained by degradation. Methods for obtaining the 5'- or 3'- (and 2'-) nucleotides are described on pp. 203 and 208.

Chemical synthesis of the 5'-nucleotides (from nucleosides) is fairly straightforward. The most popular phosphorylating agent is probably 2-cyanoethyl phosphate. This substance is made by condensing 2-cyanoethanol with phosphoric acid in the presence of trichloroaceto-nitrile. This procedure is easily adapted to synthesis of 2-cyanoethyl[^{32}P] phosphate (Pfitzner and Moffatt, 1964). The conditions for the phosphorylation of a nucleoside are condensation with dicyclohexylcarbodiimide (DCC, see p. 214). The reaction is described in detail by Moffatt (1967). The other hydroxyl (or hydroxyls) must be protected—see p. 213.

The other type of synthesis is the conversion of a 5'-monophosphate into the triphosphate. The classical method of doing this is the procedure of Moffatt and Khorana (1961). It is based on the reaction of the 5'-phosphomorpholidate with tributylammonium pyrophosphate using DCC as condensing agent. The phosphomorpholidate (structure below for the AMP compound) is itself made by condensation (with DCC) of morpholine with the nucleotide.

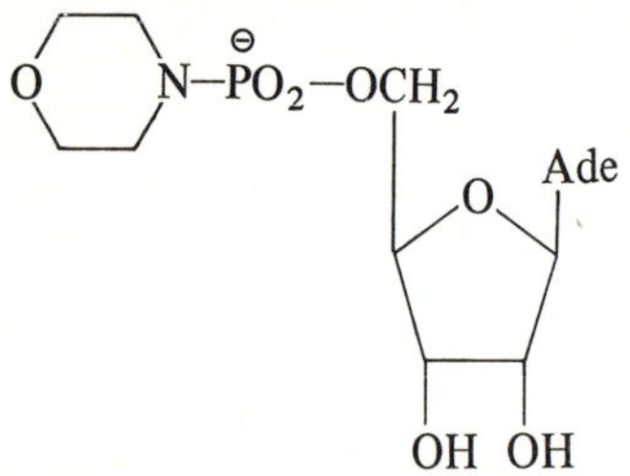

A simpler recipe for the deoxyribotides is that of Ott, Keir, Hansbury and Hayes (1967). The tributylammonium salt of dAMP and tributylammonium pyrophosphate are heated in solution in a mixture of phosphoric triamide and 1,1'-carbonyldiimidazole. The products (dAMP, dADP, dATP and deoxyadenosine tetraphosphate) are separated by elution from a column of DEAE-cellulose (see p. 51). The same reaction can be used to put a triphosphate group on to the 5'-end of an oligomer such as —T—T—T—T.

The 2',3'-cyclic phosphates (see pp. 3 and 221) are synthesized by the dehydration of the 2'/3'-ribonucleotides. The dehydration can be achieved either with DCC or with ethylchlor-formate in the presence of an amine such as tri-*n*-butylamine. An example of a recipe for the former procedure is Smith, Moffatt and Khorana (1958); the latter method is due to Michelson (1959).

Substitution reactions

Substitution at carbon   The carbon atom 5 in pyrimidine has 'aromatic properties'; that is to say the hydrogen attached to it can be replaced by —NO_2, —halogen, etc., by the usual electrophilic reagents employed in benzenoid chemistry. In most naturally occurring pyrimidines this type of reaction is limited by the reactivity of such reagents with other functional groups in the molecule. However most pyrimidines and their nucleosides (other than 5-substituted compounds such as thymine) can be halogenated. Thus uracil, uridine or deoxyuridine will react with chlorine (in acetic acid as solvent) or with bromine or iodine (in dioxan as solvent and with a trace of nitric acid as catalyst) to yield the 5-halo-derivatives (Michelson, Dondon and Grunberg-Manago, 1962). This is the way that 5-bromouracil and 5-bromodeoxyuridine (see p. 154) are made. Fluorination cannot be performed in this way so 5-fluorouracil and 5-fluorodeoxyuridine are synthesized the hard way (see pp. 35 and 57). The 5-halouracils are of interest in that they can become incorporated into DNA and RNA.

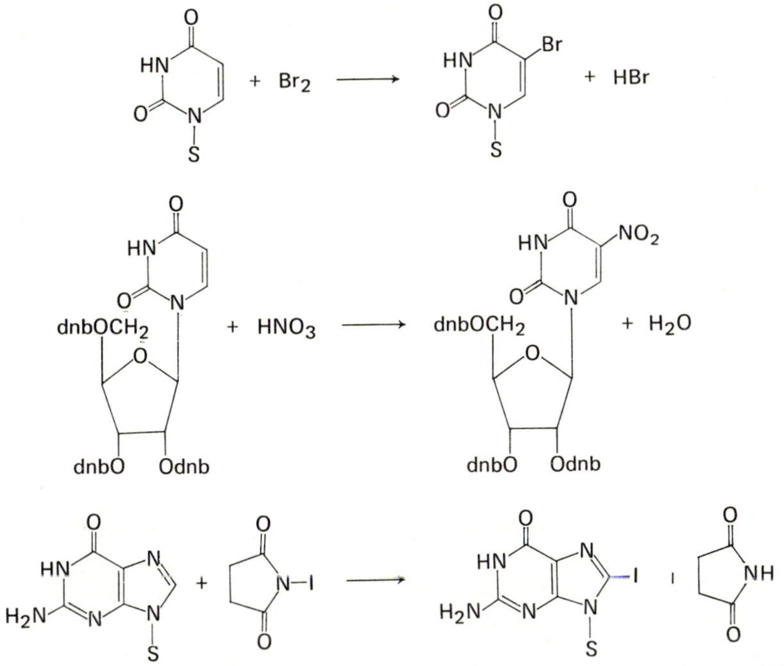

FIG 2.7   Examples of halogenation and nitration reactions of pyrimidines and purines. s = sugar (ribose or deoxyribose); dnb =

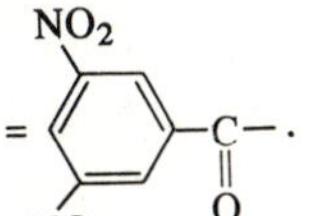

The paper by Michelson and others (above) also describes a way of nitrating a protected derivative of uridine (5′,3′,2′-tri-(3,5-dinitrobenzoyl) uridine)) to yield 5-nitrouridine.

In purines the equivalent atom (C-5) is substituted. Certain purines can be halogenated however at C-8 in the imidazole ring. Thus 8-iodoguanosine can be synthesized by treatment of guanosine in solution in dimethyl sulphoxide with *N*-iodosuccinimide. A dialkyl disulphide is required as catalyst (Lipkin, Howard, Nowotny and Sano, 1963).

These reactions are summarized in Fig. 2.7.

Substitution at nitrogen   The reaction of nucleophiles with pyrimidines and purines to form *N*-alkyl derivatives has been extensively studied because these reactions occur with nucleic acids both *in vitro* and *in vivo*. The reactions with DNA *in vivo* are exploited experimentally in producing mutations in micro-organisms and introducing faults into DNA as a means of studying repair of DNA (see p. 262). It is also likely that DNA-alkylation can lead to carcinogenesis in animals.

First of all let us consider the nature of these nucleophiles (or 'alkylating agents'). They fall roughly into three groups, (i) those that produce as reaction intermediates carbonium ions, (ii) diazoalkanes and their precursors and (iii) substances which produce strained-ring cyclic ammonium of sulphonium ions as intermediates (these reagents are known as 'mustards' for a reason that will soon be obvious—they are in no way related to components of the condiment of the same name). In the following paragraphs there is a summary of the patterns of reaction with various groups of nucleosides and a few illustrative examples.

(a) The classical electrophilic alkylating agents are the alkyl esters of strong acids, preferably ones whose acid anions form good 'leaving groups'. Examples in the case of methylating agents are methyl tosylate ('tosylic acid' is p-toluene sulphonic acid), dimethyl sulphate and methyl methane sulphonate. Dimethyl sulphate is experimentally the most popular of these for the alkylation of nucleosides as the reaction will occur at around neutral pH. The most reactive nucleoside for this reaction is Guo, followed by Ado and Cyd. The same is true of the deoxyribonucleosides. There is no reaction with Urd, dUrd and dTyd (Lawley, 1961). The products of such methylations are 7-methylguanosine, 1-methyladenosine and 3-methylcytidine (see Fig. 2.8(a)). Other alkylating agents which produce carbonium ions form similar products; the reaction of olefin oxides with guanosine (Brookes and Lawley, 1961) is shown in Fig. 2.8(b). The corresponding reactions of the bases are described by Brookes and Lawley (1960).

Other alkylating agents of type (a) are alkyl halides. Methyl iodide (in alkali) converts GMP to 1-methyl-GMP (Pochon and Michelson, 1967). The alkylated products obtained by the pyrolysis of the tetramethylammonium salts of the bases (p. 54) have not been characterized except for the cases of Ura and Thy (which yield their 1,3-dimethyl derivatives) and Xan (which yields caffeine).

(b) Diazomethane (CH_2N_2) is a reagent with different specificity. The order of reactivity of the ribosides is Urd, Guo, Cyd and Ado. In this sequence there is a big jump between Guo and Ado, that is Urd and Guo are very reactive, Cyd and Ado much less so (Haines, Reese and Todd, 1964). The products of methylation of Guo, Cyd and Ado with diazomethane are the same as those shown in Fig. 2.8(a) for the reaction with dimethyl sulphate. The product of the reaction of diniomethane with Urd (also true of dThd) is the 3-methyl derivative (as for Cyd). These data all refer to experiments with diazomethane itself free in solution. Such solutions are obtained by treating a diazomethane precursor such as N-methylnitrosourea or N-methylnitrosourethane with alkali and extracting the liberated diazomethane in ether. The diazomethane precursors themselves will alkylate nucleosides and nucleotides. In these reactions the specificities are different. For example the substance N-methyl-N-nitroso-N'-nitroguanidine ('NMG'), which is the most commonly used chemical mutagen for bacteria and is also an extremely potent carcinogen, has little reaction with GMP but with AMP it produces the 1- and 3-methyladenylic acids (Rau and Lingens, 1967; see Fig. 2.8(c). This reactivity is in contradistinction to the reaction of NMG with DNA *in vivo* (see p. 259). For another type of alkylation effected by the dimeomethane precursors see p. 68.

(c) The other important alkylating agents are bis-β-chloroethyl sulphide (sulphur mustard or mustard gas), bis-β-chloroethyl methylamine (nitrogen mustard or HN2) and certain derivatives. The reactive intermediate in these substances is a cyclic cation. The reaction of guanosine with mustard gas, when the cyclic intermediate is a sulphonium ion is summarized in Fig. 2.8(d) (Brookes and Lawley, 1961). Notice that the mustards are 'two-armed' alkylating agents. Thus once a 7-alkyl guanosine has been produced it still possesses a reactive group ($Cl-CH_2CH_2-S-$) for which either water or another guanosine are competing nucleophiles.

Internal alkylation A special type of alkylation of nucleosides occurs when the 'leaving group' of an ester-type alkylating agent is attached to one of the oxygen atoms of the sugar moiety of the nucleoside. In this case an internal substitution occurs to produce a new ring system. An example of such a reaction is the internal substitution of a protected (isopropylidene) derivative of adenosine 5′-tosylate (Clark, Todd and Zussman, 1951) illustrated in Fig. 2.8(e).

Arylation reactions Arylation is a process of possibly great significance in certain types of carcinogenesis (p. 264). An example which has been well studied is the reaction of the carcinogen N-acetoxy-N-(2-fluorenyl) acetamide[5] with GMP. The product is an 8-arylamidate of GMP (Fig. 2.8(f), Kriek, Miller, Johl and Miller, 1967; see also p. 249).

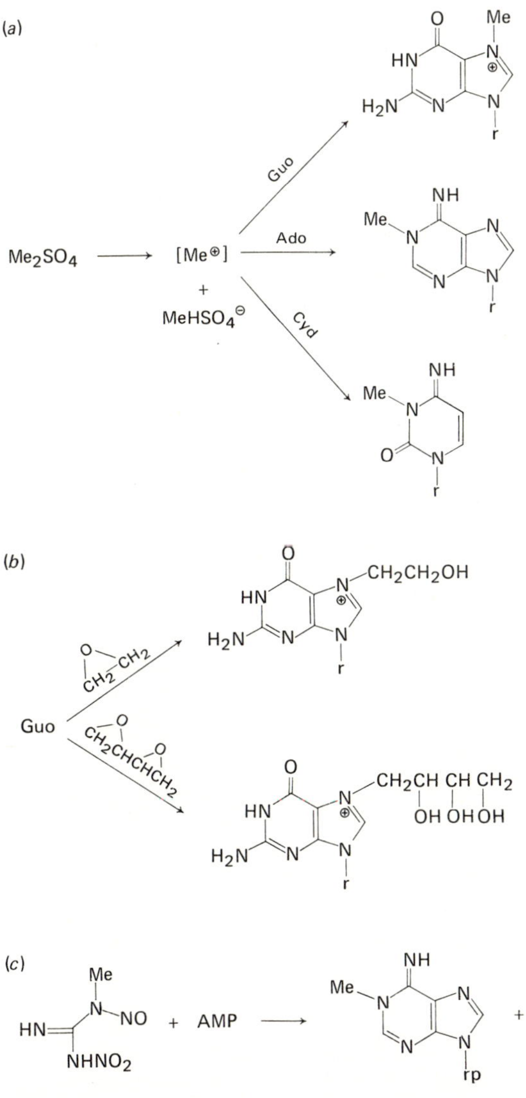

FIG 2.8 Substitution reactions of nucleosides and nucleotides. (r = ribose, Ts = tosyl). (a) Reaction of Me_2SO_4 with Guo, Ade and Cyd. (b) Reaction of ethylene oxide and butadiene diepoxide with Guo. (c) Reaction of 'NMG' with AMP.

(*d*)

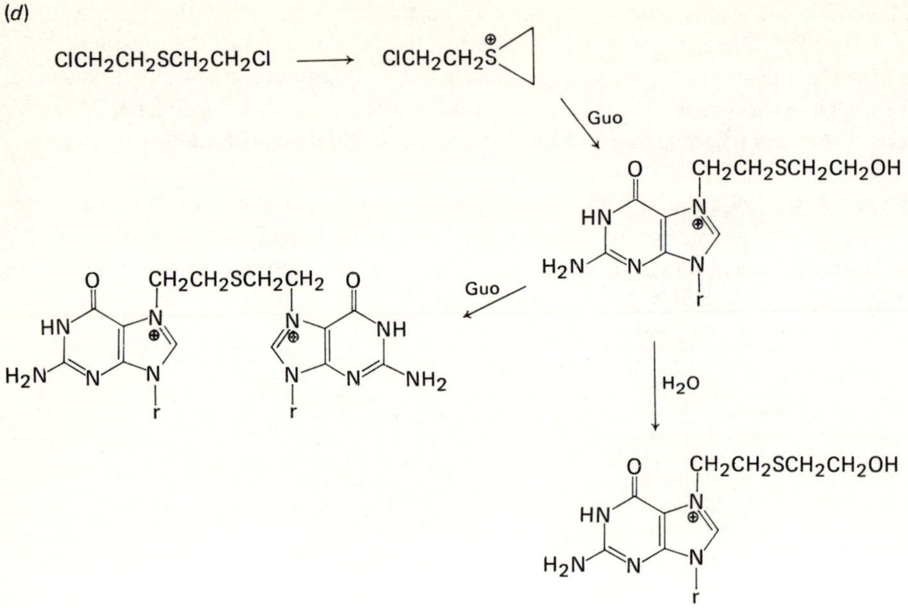

(*e*)

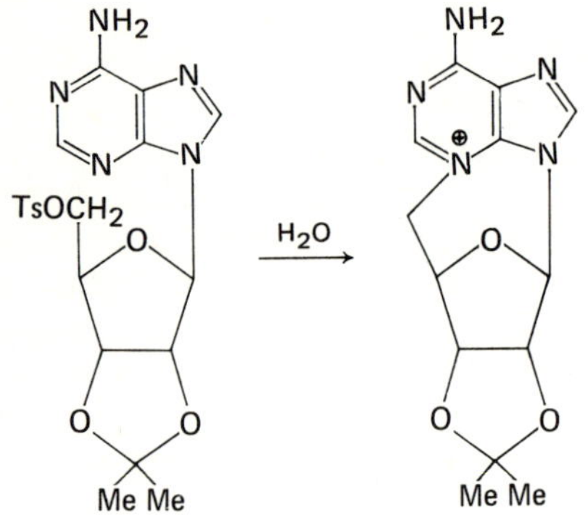

(*f*)

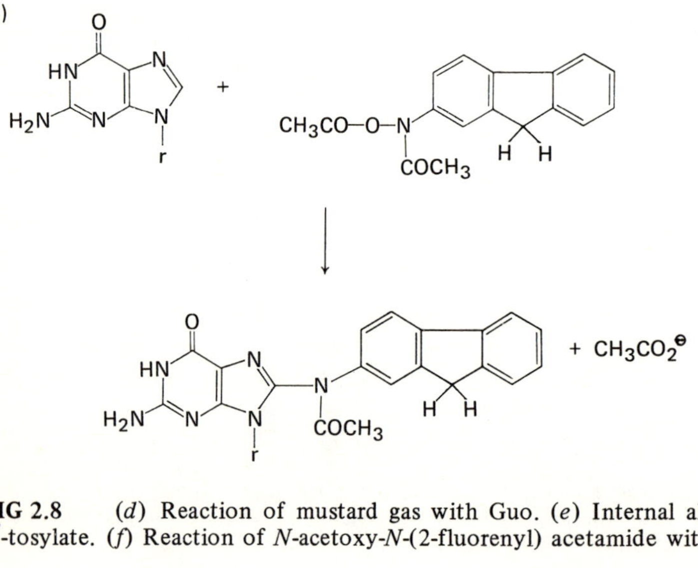

FIG 2.8   (*d*) Reaction of mustard gas with Guo. (*e*) Internal alkylation of a nucleoside-5'-tosylate. (*f*) Reaction of *N*-acetoxy-*N*-(2-fluorenyl) acetamide with GMP.

Addition reactions

The products of these reactions are shown in Fig. 2.9.

Addition of hydrogen   The 5,6-double bond in pyrimidine nucleosides can be reduced to form the saturated 5,6-dihydro-compounds (Fig. 2.9(a). The process is usually effected by hydrogenation in aqueous solution with a rhodium catalyst (Cohn and Doherty, 1956). No comparable reaction occurs with purine nucleosides. An alternative method for the synthesis of dihydrothymidine is via the intermediate diimide (see p. 70).

A particular class of purines, the 7-alkylguanosines, can be reduced. The reducing agent is sodium borohydride and the reduction occurs in the imidazole ring (Pochon, Pascal, Pitha and Michelson, 1970, Fig. 2.9(b)).

Addition of oxygen   Oxidation of N-3-substituted uracils with osmium tetroxide (OsO_4) yields the 5,6-saturated diol. The reaction is most easily demonstrated in N-3-methyluracil (an analogue of uridine; Burton, 1967).

This type of reaction cannot be demonstrated in purines although purines are susceptible to N-oxidation. Adenosine (but no other common nucleoside) is slowly, but quantitatively converted into adenosine-N-1-oxide by reaction with hydrogen peroxide in acetic acid solution (Stevens, Magrath, Smith and Brown, 1958, Fig. 2.9(c)).

The synthesis of other purine-N-oxides is of considerable importance as the use of these substances for cancer therapy has been extensively investigated. Brown (1968) has reviewed the subject.

Addition of halogen compounds   Bromine water reacts quantitatively with uridine in a reaction that proceeds in three stages (Wang, 1957). The reactant is hypobromous acid (HOBr) present in the bromine water. The immediate product is 5,6-dihydro-5-hydroxy-6-bromo uridine, which loses water to produce 6-bromouridine; a second molecule of HOBr reacts to yield the final dibromo-product (Fig. 2.9(d)).

Addition reaction of amides and carbodiimides   Amides form addition reactions with a group of compounds known as the water-soluble carbodiimides. Of these substances, the most commonly used is 'CMCT' (for the record, N-cyclohexyl-N'-(β-morpholinyl-4-ethyl) car-bodiimide methyl-p-toluene sulphonate). This reaction is shown in Fig. 2.9(e). Several nucleosides (Urd, Guo, Ψrd, Ino) contain amide groups (C=O next to NH) in the heterocyclic base and this reaction will occur (Ho, Uchida, Egami and Gilham, 1969). Notice that the reaction is reversed by treatment with mild aqueous alkali. The corresponding reaction with these residues in RNA molecules is of importance in sequencing methods (see p. 304).

Photoaddition reactions   Light catalysed addition reactions of the 5,6-double bond in pyrimidines and their nucleosides are of two types. The simpler reaction is the addition of a small molecule such as the reversible hydration of cytidine (Deboer, Klinghoffer and Johns, 1970, Fig. 2.9(f)).

The more complex one is the self-addition of two pyrimidines to produce a tricyclic adduct containing a cyclobutane structure. The most reactive base for this reaction is thymine. There are four possible dimers which could be produced depending on the orientation of the thymine molecules at the time of reaction (see Fig. 2.9(g). If the reaction is performed by irradiating frozen aqueous solutions of thymine with ultra-violet light the major product is the *cis-syn*

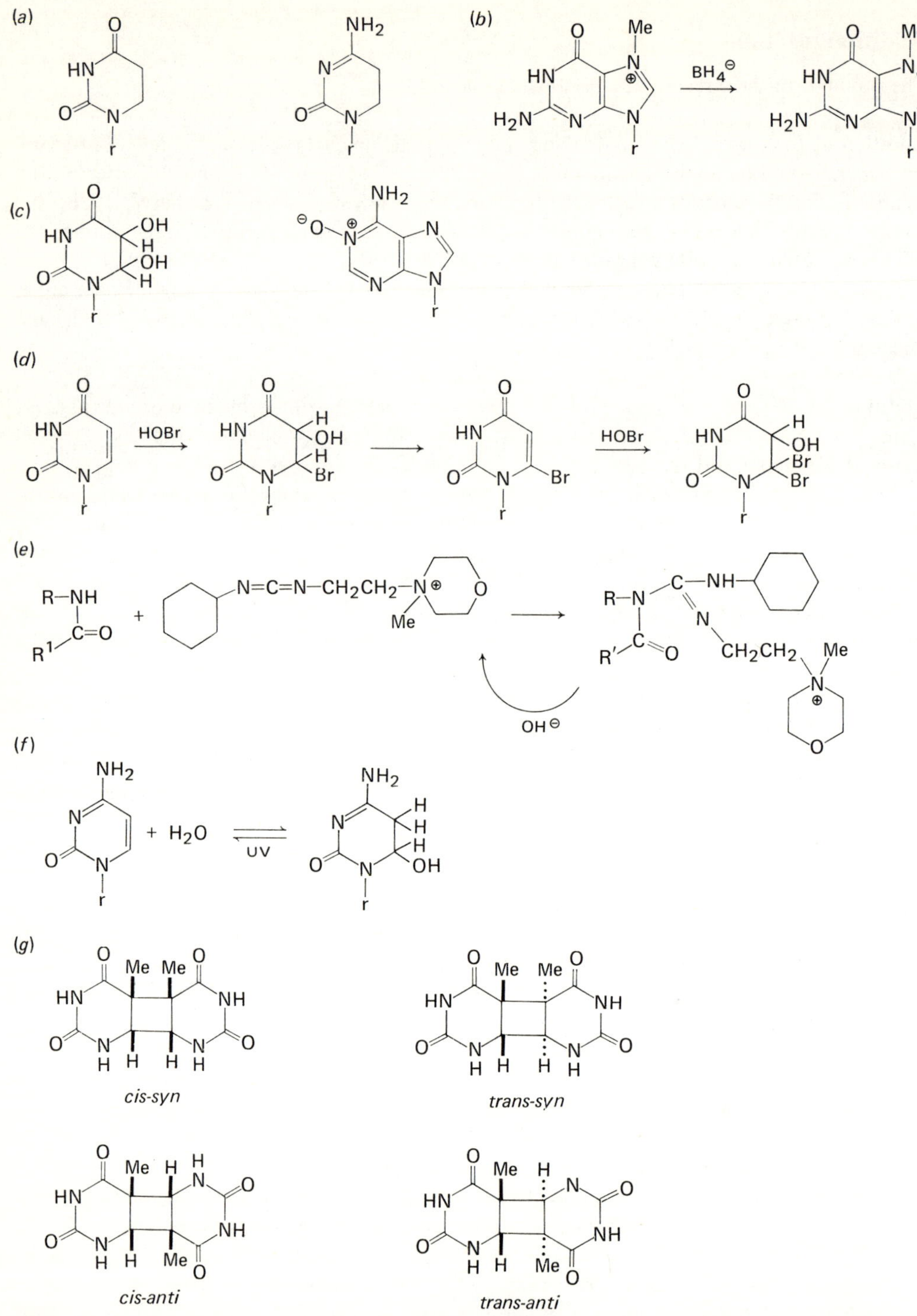

FIG 2.9   Products of addition reactions of bases and nucleosides (r = ribose). (a) 5,6-Dihydrouridine and 5,6-dihydrocytidine. (b) Reduction of 7-methylguanosine. (c) Oxidation products of Urd and Ado. (d) Reaction of bromine water with Urd. (e) Reaction of CMCT with an amide. (f) Photohydration of Cyd. (g) Thymine dimers.

isomer (Blackburn and Davies, 1965), however there are more complex products of this reaction (other than thymine dimers; Varghese, 1970).

Reactions of functional groups

Displacement reactions   This class of reactions includes the route to the synthesis of many 6-substituted purine nucleosides. The starting point is a protected inosine derivative (usually $2',3',5'$-tri-O-benzoyl inosine) which can be chlorinated with phosphorus oxychloride to yield the corresponding 6-chloro compound (Chamberlin, Zambito, Sohar and Jenkins, 1963). This compound can be converted into the corresponding thio-derivative (with thiourea) or can be reduced (hydrogenation with palladium catalyst; Brown and Weliky, 1953) to yield the protected nebularine. Alternatively the nebularine compound can be obtained by desulphurization (Raney nickel; Fox, Wempen, Hampton and Doerr, 1958). The protecting groups can be removed from any of these compounds with alkali. The reactions are summarized in Fig. 2.10(a).

With pyrimidines the oxygen substituents can be replaced with sulphur by direct thiation using phosphorus pentasulphide. Any thio-group can be replaced (in either pyrimidines or purines) with an amino group by nucleophilic displacement with ammonia. The 'rare' DNA nucleoside 5-methyl-deoxycytidine may be synthesized from $3',5'$-di-O-benzoyl thymidine by selective thiation at position 4 and treatment of the 4-thiothymidine derivative with ammonia (Wempen, Ueda and Fox, 1963; Fig. 2.10(b)).

Reactions of amino groups   The amino groups of the bases react reversibly with formaldehyde in the manner familiar to anyone who has done a formol titration of the free amino groups in a protein. The principle of the method is that the formyl derivatives have lower pK-values than have the free amino groups. Hoard (1960) attempted formol titrations of cytidine nucleotides in an attempt to adjust the pK of the amino-dissociation so that the second phosphate dissociation in CMP (or the terminal phosphate in oligonucleotides) could be unambiguously titrated. The quantitativity of his titration was disappointing, probably because the formylation of purines and pyrimidines is a complex reaction involving the heterocyclic imino nitrogens (Eyring and Opengard, 1967). More satisfactory aldehyde reagents are the geminal carbonyl aldehydes glyoxal (CHOCHO) and kethoxal (3-ethoxy-2-oxobutyraldehyde CH_3- CHOEt$-$CO$-$CHO) which form stable cyclic adducts with guanosine (Shapiro and Hachmann, 1966; see Fig. 2.11(a)). These reactions are useful methods of modifying RNA for sequence studies (see p. 304).

The amino group of cytidine has characteristic reactions with hydroxylamine (and the closely related methoxyamine), hydrazine and semicarbazide. The reaction with hydroxylamine occurs at pH 6; the sequence is somewhat complex and has been the subject of careful analysis by Brown and Phillips (1965) and Lawley (1967). The reactions (with 1-methyl cytosine an analogue of cytidine chosen because it and its reaction products have simpler spectral properties than the nucleoside) is summarized in Fig. 2.11(b). It is not absolutely certain whether the initial addition product shown in this figure is an intermediate in the formation of the N-hydroxy product. The reaction is important in the light of the mutagenic activity of hydroxylamine (p. 353). Hydrazine reacts with cytosine at pH 6 to form the two hydrazine derivatives shown in Fig. 2.11(c) (Lingens and Schneider-Bernlöhr, 1965). A more selective reaction (see p. 69 for other reactions of hydrazine with bases and nucleosides) is that with the substituted hydrazine semicarbazide ($H_2NNHCONH_2$). In this case the reaction product (from 1-methylcytosine) is as shown in Fig. 2.11(c) (Takeisha and Ukita, 1966).

Acylation of amino substituents can be achieved with normal reagents. In the special case of the synthesis of the N^6-aminoacyl adenines (which are naturally occurring compounds of uncertain role) synthesis involves condensation of a carbobenzoxyamino acid with adenine in the presence of dicyclohexyl carbodiimide (DCC) in solution in dimethyl sulphoxide. The protecting group is removed with HBr (Brink and Schein, 1963; Fig. 2.11(*d*)).

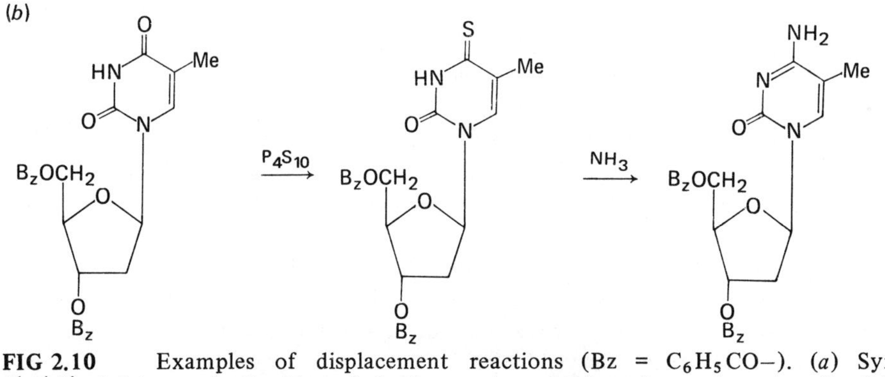

FIG 2.10   Examples of displacement reactions (Bz = C_6H_5CO-). (*a*) Synthesis of 2′,3′,5′-tri-*O*-benzoyl ribosides of 6-chloropurine, 6-thiopurine and purine (in this case the product is 2′,3′,5′-tri-*O*-benzoyl nebularine). (*b*) Synthesis of 3′,5′-tri-*O*-benzoyl 5-methyl-deoxycytidine.

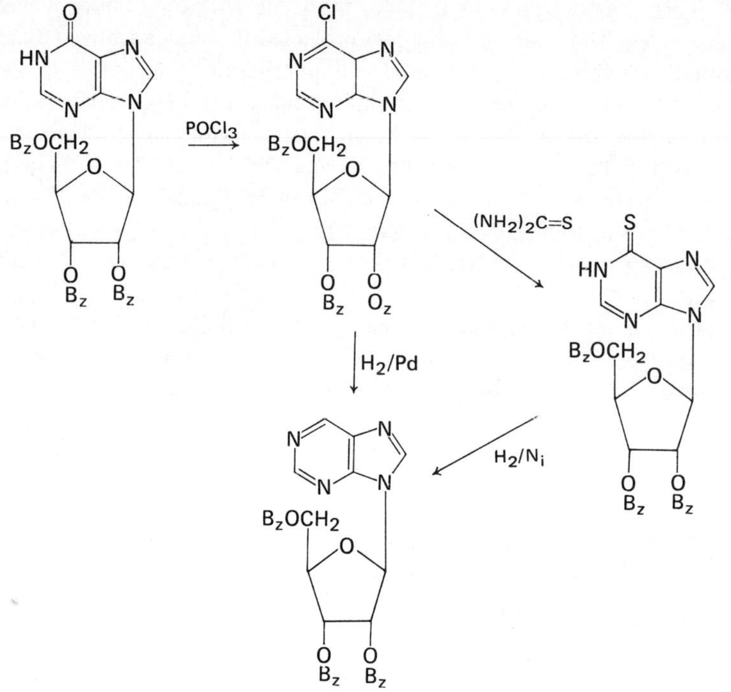

The amino substituents in nucleosides can be replaced by hydroxyl groups following reaction with nitrous acid. Thus Cyd can be converted to Urd and Ado to Ino. The reaction between nitrous acid and guanosine is more complicated. The products of the reaction are xanthosine (straight replacement of $-NH_2$ with $-OH$) and a small amount of 2-nitroinosine (Shapiro, 1964). If this observation is correlated with an earlier observation (Montgomery and Hewson, 1960) that 2-fluoroinosine is one of the products of the reaction of guanosine with nitrous acid in the presence of fluoroboric acid, it looks as though the reaction between HNO_2 and the $-NH_2$ of Gua is to form a diazonium ion which will then react further with water, nitrous acid or fluoroboric acid to yield the substituted products (Fig. 2.11(e)).

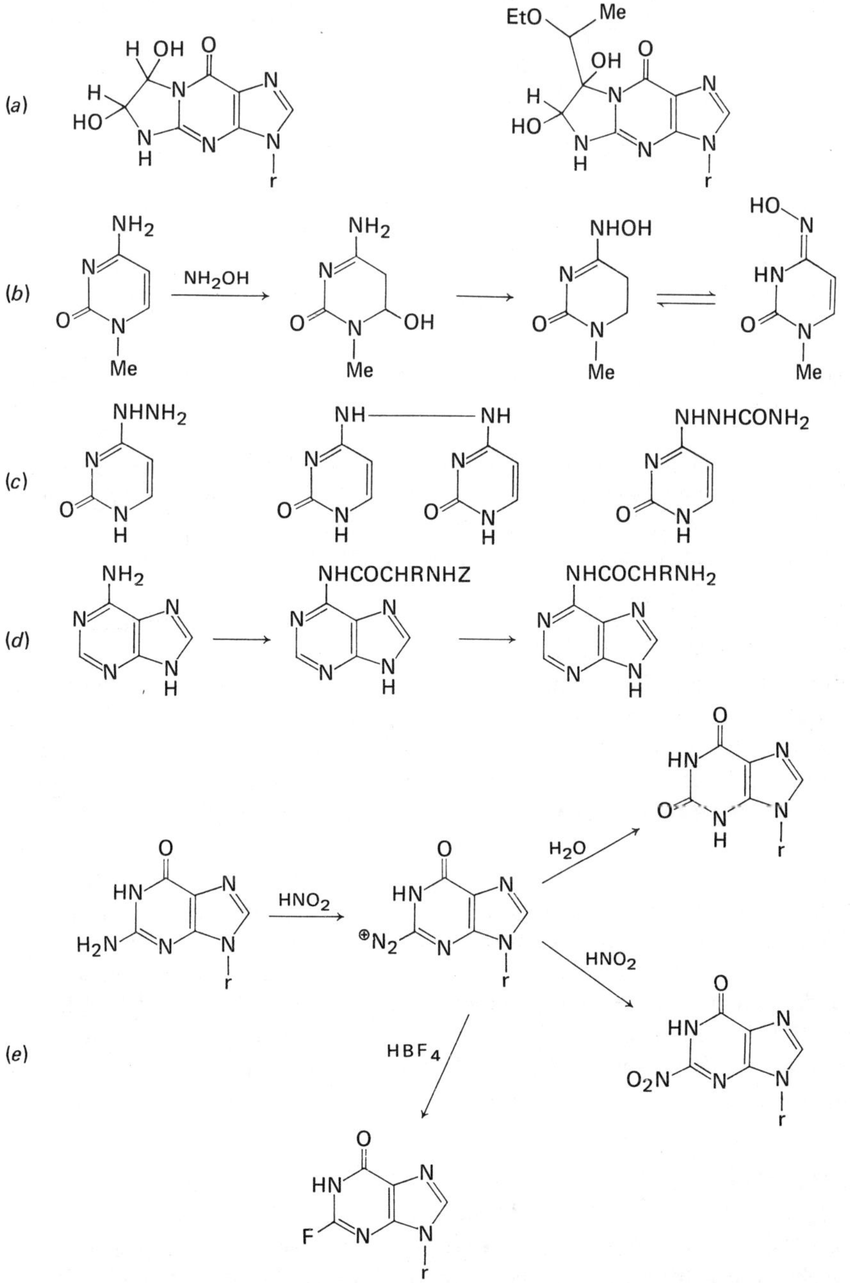

FIG 2.11   Miscellaneous products of side-chain reaction (r = ribose). (a) Products of the reaction between Guo and glyoxal (left) and kethoxal (right). (b) Reaction of 1-methylcytosine with hydroxylamine. (c) Products of the reactions between Cyt and hydrazine (left-hand and middle structures) and semicarbazide (right-hand structure). (d) Synthesis of an N^6-aminoacyl-adenine (Z = carbobenzoxy). (e) Proposed formation of a diazonium derivative of Guo and its subsequent reaction with H_2O, HNO_2 or HBF_4 (see text).

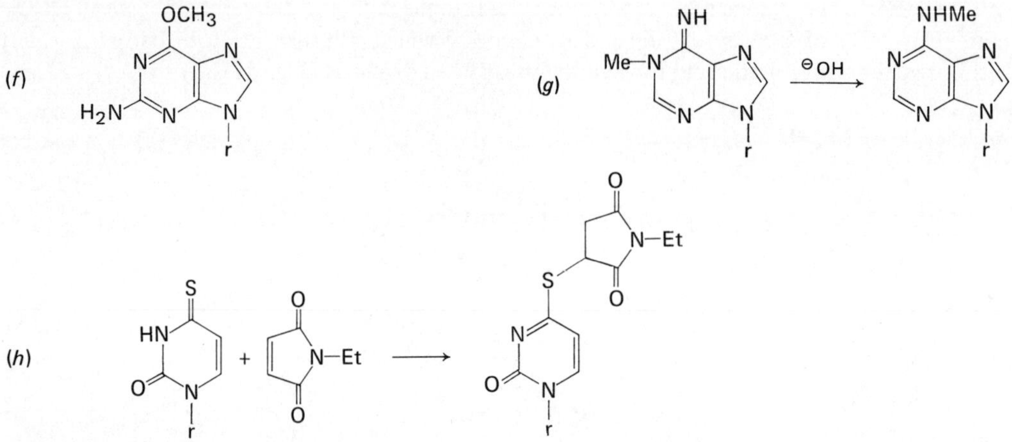

FIG 2.11   (*f*) *O*[6]-methylguanosine. (*g*) Re-arrangement of 1-methyladenosine. (*h*) Reaction of thiouridine with NEM.

With certain reduced aminonucleosides the deamination is achieved simply by hydrolysis. Thus although dihydrocytidine (see p. 63) can be obtained by catalytic hydrogenation of cytidine, the product always contains dihydrouridine due to nucleophilic replacement of $-NH_2$ by $-OH$ (from water) occurring during the hydrogenation reaction (Green and Cohn, 1957).

Alkylation of substituents   It has been recognized recently that the alkylation of guanosine by the alkylnitrosamides and other diazomethane precursors (see p. 60) does not yield exclusively 7-methylguanosine but in addition O^6-methylguanosine (Fig. 2.11(*f*); Loveless, 1969). This discovery has probably profound consequences in the biological effects of alkylating agents (see p. 261. Another interesting methylation reaction is the internal methylation (rearrangement) which occurs when 1-methyl adenosine (see p. 61) is heated in alkali; the methyl group migrates to yield N^6-methyladenosine (Jones and Robins, 1963; Fig. 2.11(*g*)).

Reactions of thiol groups   We have already considered one reaction of thionucleosides, namely their reduction with Raney nickel (p. 66). The other characteristic reactions of thiols are also performed by these nucleosides, e.g. reaction with another molecule or with another mercaptan to form disulphides in the presence of a weak oxidizing agent (normally iodine), formation of mercuri-compounds with such reagents as *p*-chloromercuribenzoate (PCMB) and addition to *N*-ethylmaleimide (NEM). These last two reactions can be used to quantitatively titrate the thiolated nucleosides in RNA molecules (p. 242). The most useful of the reactions for application to RNA is the addition of NEM (Carbon and David, 1968; Fig. 2.11(*h*)).

Ring-opening reactions

Pyrimidine compounds   Certain bases and nucleosides contain amide groups (see p. 5). At first sight one might expect these groups to be readily hydrolysed. In fact the presence of the double bonds in the rings results in electron delocalization with a consequent loss of many characteristic amide reactivities. However in the case of dihydrouridine and its derivatives this is no longer the case and the ring is easily opened. The instability of uridine compounds and uridine residues in RNA following catalytical hydrogenation was one of the key observations of Levene in his classical work on nucleoside structure. The alkaline hydrolysis of hUMP to form a derivative of β-ureidopropionic acid has been studied systematically (Fig. 2.12(a); Green and Cohn, 1957). An analogous reaction with $NaBH_4$ yields the corresponding alcohol (Cerutti and Miller, 1967).

Pyrimidines are ring-opened by hydrazine and hydroxylamine under vigorous conditions. The reaction of hydrazine is an important one in probing nucleic acid structure (see p. 276).

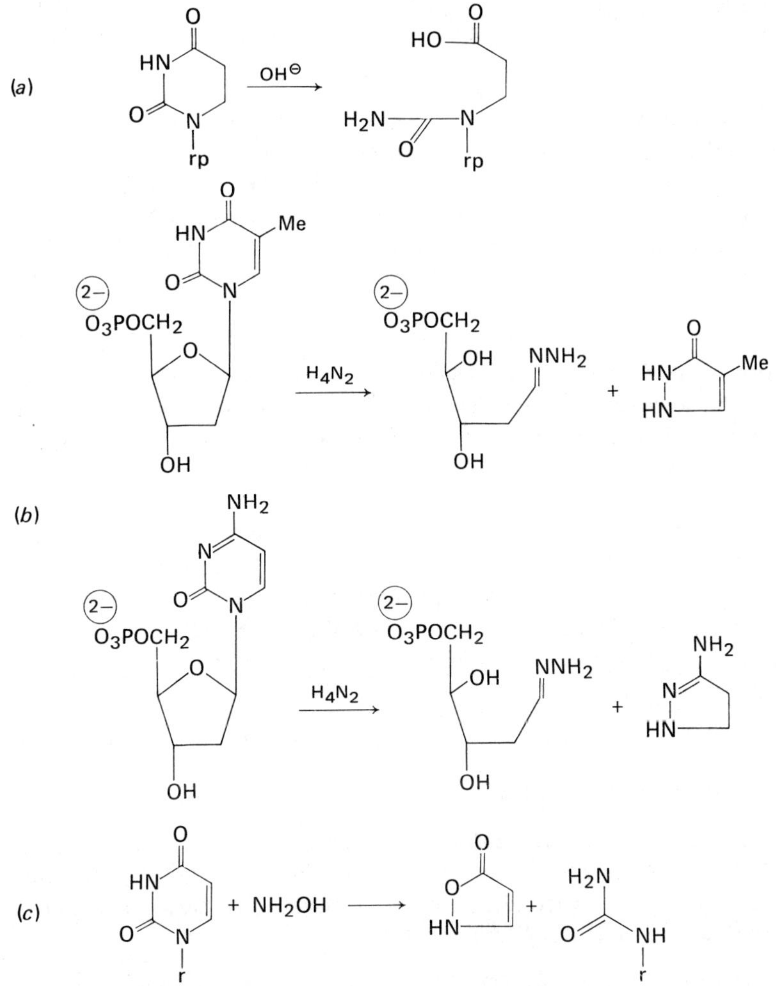

FIG 2.12   Pyrimidine ring-opening reactions (r = ribose, rp = ribose phosphate). (a) Alkaline hydrolysis of 5,6-dihydro UMP. (b) Reaction of hydrazine with 5$'$-dTMP and 5$'$-dCMP. (c) Reaction of hydroxylamine (at pH 10) with cytidine.

Although there is a very extensive literature on the subject, the main work has been done by Chargaff, who first exploited the reaction. A comprehensive survey is in the paper by Chargaff, Rüst, Temperli, Morisawa and Danon (1963). The reaction of 5$'$-dTMP and 5$'$-dCMP are shown in Fig. 2.12(b). The conditions for this reaction are neat hydrazine hydrate or even anhydrous hydrazine. The sugar, free of hydrazide substituents (deoxyribose-5-phosphate in these two instances) can be released by treating the sugar hydrazide with benzaldehyde (Baron and Brown, 1955). At pH 6 there is a different reaction with cytosine and its derivatives (see p. 65). Yet a third type of reaction of hydrazine is with thymine in the presence of oxygen and a copper catalyst. Under these circumstances diimide is formed as an intermediate and the thymine is reduced to 5,6-dihydrothymine (Hünig, Müller and Thier, 1965; see p. 63).

A reaction analogous to the ring-opening by hydrazine is the ring-opening of uridine (but not cytidine) by hydroxylamine at pH 10. This is contrast to the quite different reaction of hydroxylamine at pH 6 where there is a reaction with cytidine but not with uridine. The reaction at pH 10 is shown in Fig. 2.12(*c*) (Verwoerd, Zillig and Kohlhage, 1963). Methoxyamine which will effect the analogous reaction at pH 6 (p. 65) has no reaction with uridine at pH 10.

Purine compounds   In general purines do not readily ring-open unless they are *N*-substituted, in which case the ring containing the substituents is made relatively labile. Here we shall consider just two types of reaction, both of which are important in studies on nucleic acids.

The first type is the opening of the pyrimidine ring in adenines substituted at N-1 or N-3. In straightforward alkyl derivatives the reaction consists of a loss of formic acid (Shaw, 1958). The examples given (Fig. 2.13(*a*)) are the alkaline hydrolysis of 1-benzyladenosine and of the cyclic alkylated structure whose synthesis has been referred to on p. 62. In the case of the reaction of diethyl pyrocarbonate (EtO—CO—O—CO—OEt; see also p. 112 for its reaction with RNA) the intermediate (whatever it might be) is ring-opened and deformylated straight away. The product (Fig. 2.13(*b*) can be converted into adenine again by treatment with ammonia under pressure at 120°C (Leonard, McDonald and Reichmann, 1970). A similar reaction occurs when adenine-1-*N*-oxide (p. 63) is hydrolysed (0·05 M HCl at 100°C). The final product is a carboxamide (Stevens, Smith and Brown, 1958).

The second type of reaction is exemplified by the imidazole ring-opening in the 7-alkyl-guanosines. The reaction has been known for some time but has recently been thoroughly re-investigated by Pochon, Pascal, Pitha and Michelson (1970). Above pH 9·3 the zwitter ion is destroyed and if the solution is irradiated with UV this product is de-formylated (Fig. 2.13(*c*). This reaction is important when the hydrolysis of nucleic acids containing 7-alkyl guanosine residues is undertaken (p. 248).

Reactions of sugars and phosphate residues

The reactions of the sugar moieties in nucleosides and nucleotides are important in three respects, the protection of sugar residues for other chemical reactions of nucleosides, alteration of the sugar for the synthesis of 'odd nucleosides' and the synthesis of aminoacyl adenosines (the type of linkage present in aminoacyl tRNA, see p. 14).

Protection of hydroxyl groups   This is a large area of carbohydrate chemistry and any sort of systematic account of it is completely outside the scope of this book. The references already given to reactions of protected nucleosides (see pp. 58-213) give examples of the experimental procedures actually employed to prepare these derivatives of nucleosides. The normal repertoire of protecting groups for hydroxyl groups in carbohydrates are acetyl, CH_3CO- (reagent is acetyl chloride or acetic anhydride); tosyl, $p\text{-}CH_3C_6H_4SO_2-$ (reagent is tosyl chloride); trityl, $(C_6H_5)_3C-$ (reagent is trityl chloride) or the formation of ketals. This last reaction is performed by *cis* glycols (such as ribose) with ketones. An example of such an acetonide (reagent is alkaline solution of acetone) is given on p. 61; another example is discussed in the next section (p. 72).

An important procedure in nucleotide synthesis is the protection of specifically one or two of the hydroxyls. Clearly acetonide formation is a reaction which selectively protects the 3'-

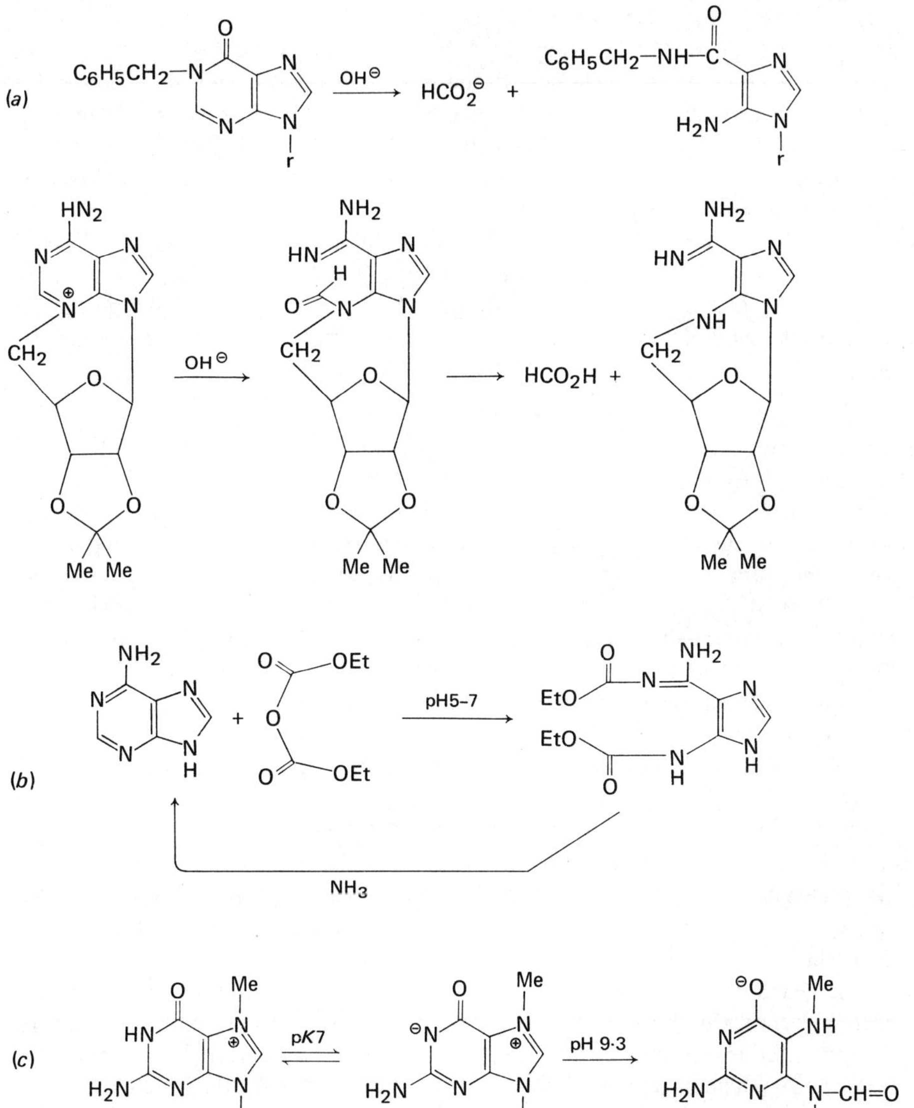

FIG 2.13   Purine ring-opening reactions (r = ribose. (*a*) Alkaline ring-opening of adenosine substituted at N-3; an analogous reaction occurs if the substituent is at N-1. (*b*) Reaction of diethylpyrocarbonate with adenine. (*c*) Alkaline ring-opening of 7-methylguanosine.

and 2'-hydroxyls in ribosides and leaves the 5'-hydroxyl free. Considerable selectivity is shown by acetylation of nucleosides. The reactions have been carefully examined by Lord Todd (Hayes, Michelson and Todd, 1955; Michelson, Szabo and Todd, 1956). To briefly summarize their findings, partial acetylation of deoxyribosides yields the 5'-acetate; less selectivity is shown by ribosides but an acetate residue is less stable when adjacent to *cis* hydroxyl (the same is true of phosphate p. 55). A more usual reagent for selective protection of 5'-hydroxyl is trityl; an example of this reaction is given below.

The use of silylation as a method for rendering nucleosides and nucleotides volatile and suitable for gas–liquid chromatography has been mentioned (p. 54). The silylation method of Hashizume and Sasaki (1966) quantitatively forms silyl esters of phosphate groups in nucleotides (and incidentally of phosphoric acid itself) in addition to form silyl ethers with the sugar hydroxyls.

Alteration of the sugar   An example of the use of protected sugars is in the synthesis of β-furanosyl adenosine (McCarthy, Robins, Townsend and Robins, 1966). A 5'-trityl deoxyadenosine was tosylated (at 3'); the 5'-protecting group was removed to yield 3'-tosyl deoxyadenosine; elimination of this tosylated derivative (sodium ethoxide in ethanolic solution) yields the furanosyl compound. Catalytic hydrogenation yields 2',3'-dideoxyadenosine (Fig. 2.14(*a*)). An alternative route to dideoxyadenosine does not employ the furanosyl compound as intermediate. Rather the 3'-tosylate is treated with a thiol in alkali to yield a thiated nucleoside (Fig. 2.14(*a*); Robins, McCarthy and Robins) which can be desulphurated with Raney nickel to produce the desired dideoxy nucleoside. This particular nucleoside has not so far been identified as a natural product; however it results in the termination of DNA synthesis in *E. coli* and should prove to be a valuable antimetabolite (Toji and Cohen, 1970).

Another very ingenious use of sugar derivatives is in the synthesis of spongouridine (see p. 41). This is a sequence of reactions in which the ribose moiety in a nucleoside is converted to arabinose. The principle is to put a leaving group on to the 2'-hydroxyl and then to dissolve the product in alkali; O-2 now attacks C-2' to displace the leaving group to yield the cyclic anhydrouridine which can be subsequently hydrolysed to form the aribonsul derivative (Fig. 2.14(*b*). The key reaction is the formation of the cyclic anhydronucleoside. One method (Yung, Burchenal, Fecher, Duschinsky and Fox, 1961) is to treat 5'-trityluridine with limiting amounts of tosyl chloride. In this way 2'-hydroxyl is selectively tosylated, reflecting the relatively greater nucleophilic character of 2'–OH. This tosylate ($X = p\text{-}CH_3C_6H_4SO_2$ in Fig. 2.14(*b*) then undergoes the necessary reaction. An alternative method is to treat the 5'-trityluridine with thiocarbonyldiimidazole (Fig. 2.14(*c*)). In this case the initial ketal derivative is so unstable that it is only a chemical intermediate and the anhydronucleoside is formed in one step (Fox and Wempen, 1965).

The synthesis of the aminoacyl adenosines consists of the reaction of a suitable *N*-protected amino acid with a 5'-protected adenosine in the presence of dicyclohexylcarbodiimide. A recent paper on the subject which describes versatile conditions suitable for the synthesis of a wide variety of compounds of this type, and also a comprehensive bibliography is by Chládek, Pulkrábek, Sonnenbilcher, Žemlička and Rychlík (1970). The products of such a reaction (Fig. 2.14(*d*)) are the 2'- and 3'-derivatives and the *bis*aminoacyl compound. The 2'/3' mixture, the unreacted nucleoside and the *bis* compound can be separated by thin-layer chromatography. The 2'/3' mixture cannot be resolved although there is a spectroscopic method for assessing the ratio of isomers.

Finally there is the oxidation of sugars in nucleosides. These reactions have been employed for 'end group methods' in nucleic acids (p. 287). All ribosides are oxidized with periodic acid

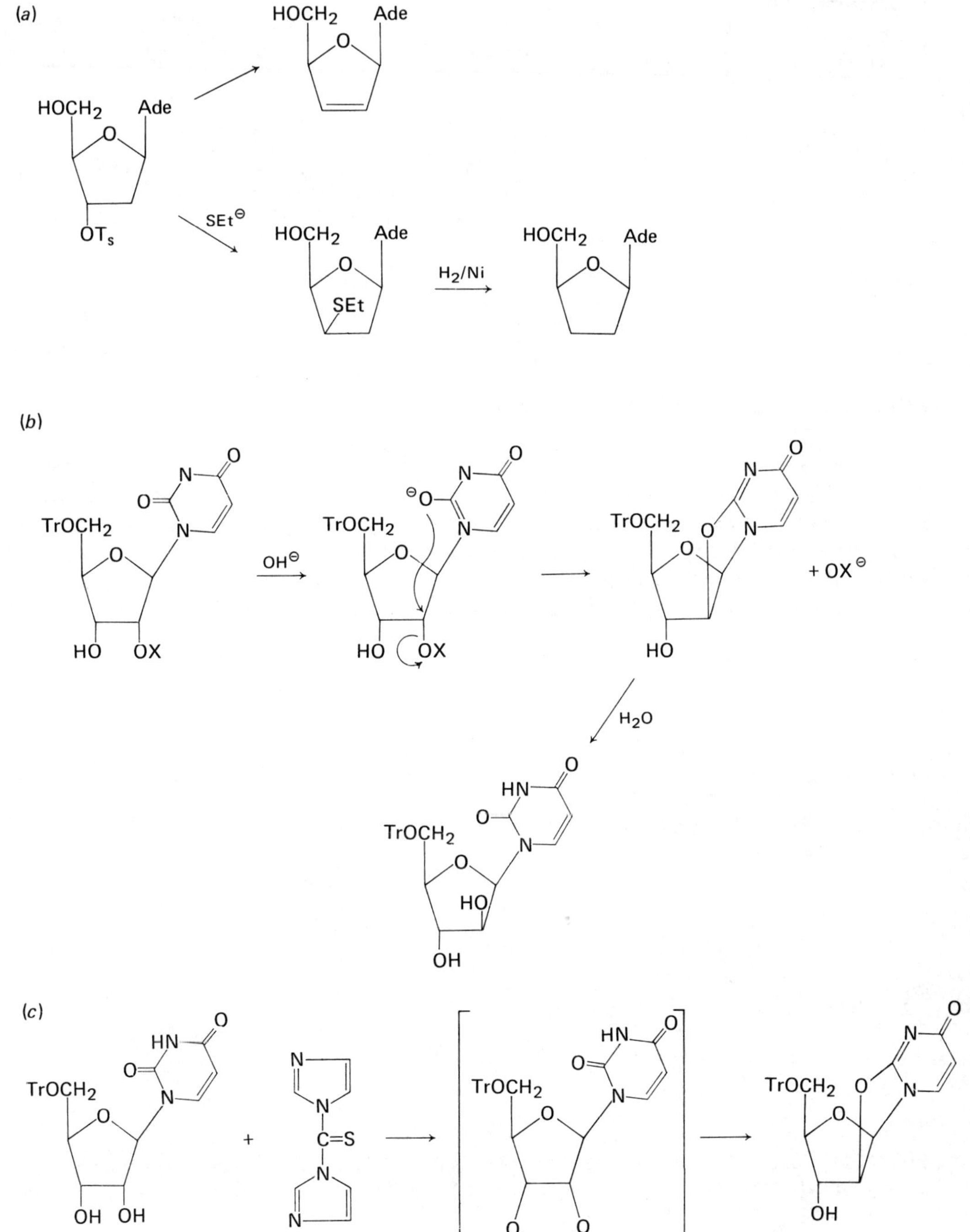

FIG 2.14   Reactions of sugars in nucleosides. (Ts = tosyl, Tr = $(C_6H_5)_3C-$, Z = $C_6H_5CH_2OCO$, DCC = dicyclohexylcarbodiimide, R = a side chain in an amino acid and X = a leaving group). (*a*) Synthesis of $2',3'$-dideoxyadenosine. (*b*) Synthesis of spongouridine. (*c*) Alternative synthesis of anhydrouridine.

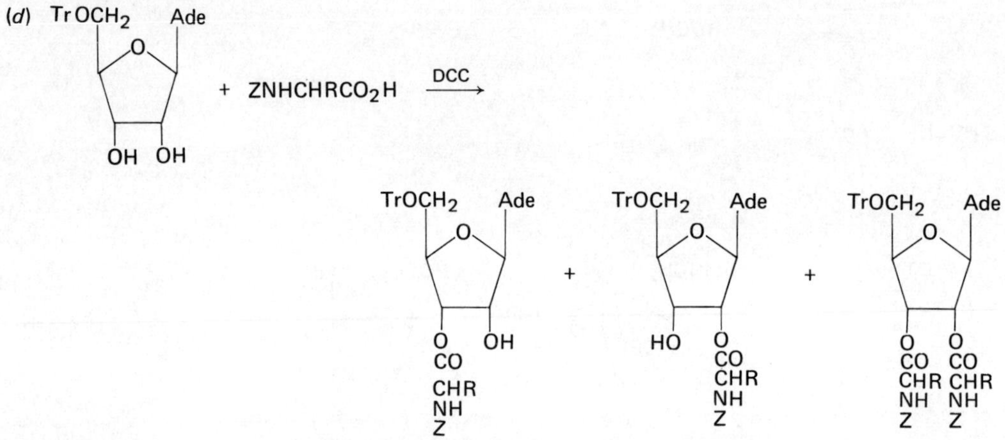

(d) Synthesis of aminoacyl adenosines.

FIG 2.14(d)   Synthesis of aminoacyl adenosines.

(in practice sodium metaperiodate in a solution adjusted to pH 4) to form the dialdehyde. These dialdehydes can be characterized as their derivatives with a hydrazine compound. Isonicotinic acid hydrazide is normally employed; only one molecule reacts to form the derivative shown in Fig. 2.15(a). The stereochemistry of these compounds has not been established. Deoxyribosides can be oxidized to the deoxyriburonosides with hydrogen peroxide in the presence of suitable catalysts (Vizsolyi and Tener, 1962; Fig. 2.15(b)).

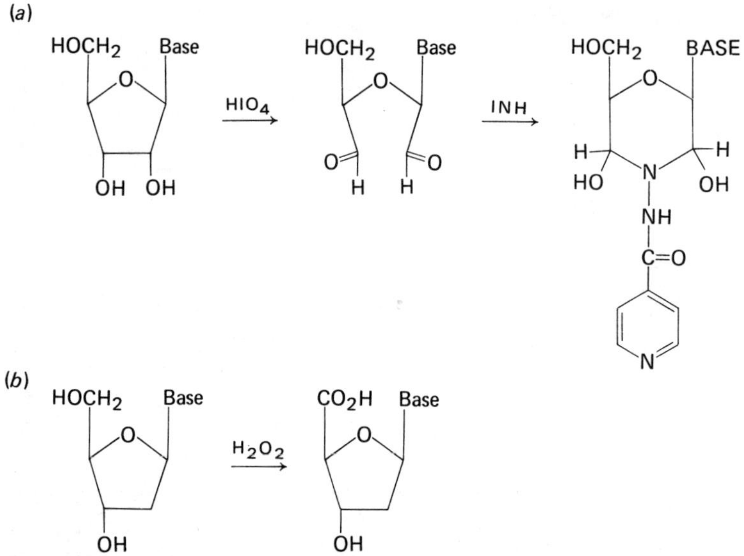

FIG 2.15   Oxidations of the sugars in nucleosides (INH = isonicotinic acid hydrazide). (a) Oxidation of a riboside with periodic acid and formation of a derivative. (b) Oxidation of a deoxyriboside with hydrogen peroxide.

Structural and spectroscopic aspects

I quite realize that the title of this section is hardly likely to thrill the average biologist. But a chemist would say that a thorough understanding of this subject is the only natural starting point for any study of nucleic acids. There is certainly no space to be thorough so let us just be pragmatic. The questions to consider are these. What spectroscopic procedures are suitable for the assay of pyrimidines and purines? What methods are suitable for identifying the structures of new compounds of these types? What are the real electronic structures of these substances? What are the molecular conformations of nucleosides and their derivatives? Is there any real physico-chemical basis for the apparent affinity between G and C and A and T? And what methods suitable for the study of nucleic acid structure require a prior knowledge of data relating to the low-molecular weight structures? What follows is a very fragmentary account of certain methods (starting with theoretical ones) with a statement of their value in present-day research.

Electronic structure

Wave mechanics is a complex and seemingly abstruse subject. The complexity derives not so much from the basic postulates (which can be understood as applied to simple model systems by anyone whose mathematics stretches to partial differentiation), but rather from the fact that the equations which derive from it have no absolute solution for the arrangement of electrons and atoms in molecules. Thus the postulate of Schrödinger, which in 1926 heralded the dawn of modern chemistry, states that the fundamental wave equation:

$$\nabla^2 \psi = -\frac{4\pi^2}{\lambda^2} \psi$$

can be applied to all particles (electrons or other sub-atomic particles or photons). In this equation ψ is the amplitude function for three orthogonal co-ordinates x, y and z, and ∇^2 (pronounced 'del squared') is a shorthand representation of the Laplacian operator:

$$\frac{\partial^2}{\partial x^2} + \frac{\partial^2}{\partial y^2} + \frac{\partial^2}{\partial z^2}$$

Schrödinger himself recognized that the form of the equation that applies to a single particle when wavelength (λ) is replaced by h/p (where h = Planck's constant and p = momentum; this relationship was established by de Broglie) could solve several problems in atomic physics. However to solve the problem of electron distribution and hence the electronic structure of complex molecules with many atomic centres, and large numbers of electrons, the equations have to be applied to approximate models.

The particular problems that apply to wave-mechanical calculations on purines and pyrimidines have been most extensively studied by Bernard and Alberte Pullman and their review of the subject is to be strongly recommended (Pullman and Pullman, 1969). Before briefly stating some of the conclusions of their remarkable work let us just be clear that in this (and other work of a similar nature) the calculations that follow from different models can lead to considerably different conclusions. The Pullman's work is so distinguished because they have always critically evaluated calculations with experimentally determined parameters and have consistently attacked problems which give an insight into biologically important problems.

All the calculations with pyrimidines and purines employ the molecular orbital (MO) treatment in one form or another. It is often imagined that molecular orbitals are some sort of clouds, balloons or sausages floating around the molecule with the electrons in some way confined within them. Strictly they are 'eigenfunctions' (i.e. proper or allowed forms for the wave functions, ψ) which arise from calculations in which the behaviour of electrons are considered from the point of view of individual electrons in the field of all the nuclei and all the other electrons in the molecule. Of the methods actually employed for pyrimidines and purines, probably the most important are one application of the Self Consistent Field idea (SCF).

Without discussing this topic at all, we merely point out that in general it is an incredibly involved method for complex molecules and is simplified by either considering the σ and π electrons separately (this is called $\sigma + \mathrm{SCF}\ \pi$ for short) or one of the CNDO methods for all the valency electrons (CNDO stands for 'complete neglect of differential overlap').

Total net charges The purpose of this calculation is to rationalize the chemical reactivities of the bases, especially with regard to substitution reactions and to deduce the polar state of the molecules, which is important when considering interactions between bases (such as those interactions which occur in nucleic acids). Several calculations of these values have been made, using different molecular orbital treatments. In Fig. 2.16 there is a summary of the calculations of Berthod, Giessner-Prettre and Pullman (1967) which use the $\sigma + \mathrm{SCF}\ \pi$ method. The numbers in this figure represent the fractional charges present on each atom. In the lower part of the figure there is a definition of the angle used to define the sense of the dipole moment, together with the values of the angle and the dipole moments calculated for the net charges calculated for the structures above. These values of the dipole moments can be checked against experimentally determined values to evaluate the accuracy of the method. Satisfactory experimental values for the Gua and Cyt are not available but the experimental values for Ade (3·0) and Ura (3·9) are in good agreement with the values in Fig. 2.16.

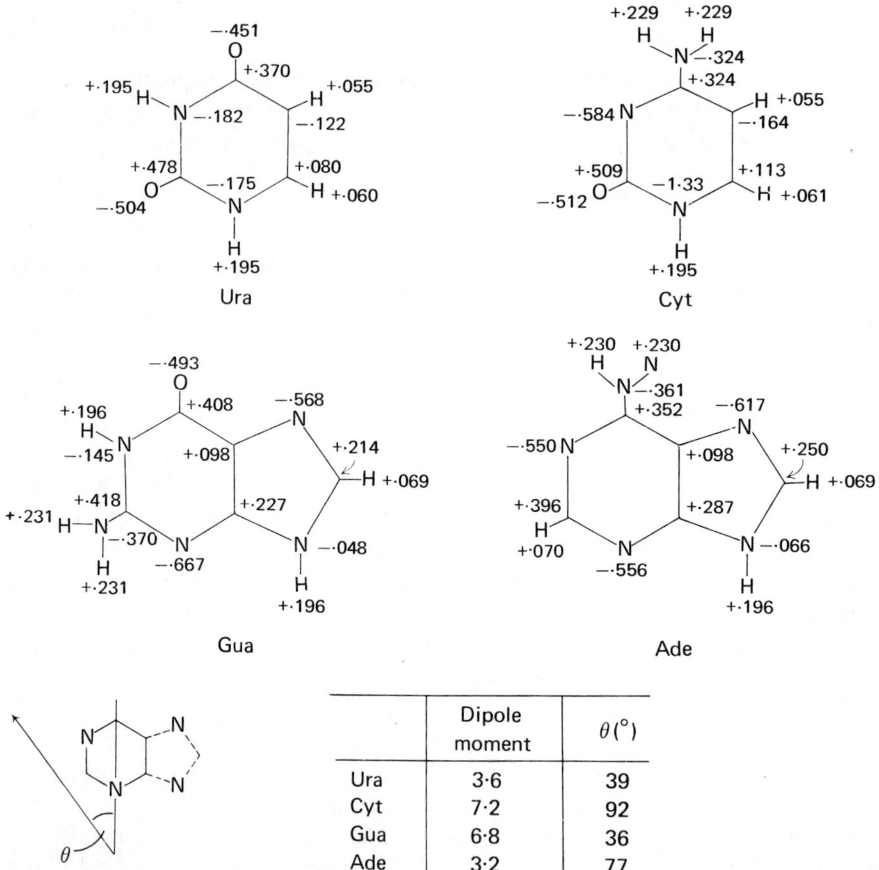

	Dipole moment	$\theta(°)$
Ura	3·6	39
Cyt	7·2	92
Gua	6·8	36
Ade	3·2	77

FIG 2.16   Total net charge distribution on Ura, Cyt, Gua and Ade and dipole moments (in Debye units) calculated by the $\sigma + \mathrm{SCF}\ \pi$ method (see text).

Interactions between bases   This subject is central to understanding the secondary structure of nucleic acids and the bases for the 'recognition processes' that occur in the reactions catalysed by the nucleic acid-dependent nucleic acid polymerases and the codon/anti-codon interaction. It is the aspect of the quantum mechanical work on pyrimidines and purines which has proved most illuminating. First of all we must consider the problem which is to be resolved. If you read an elementary biochemistry text-book, you may get the impression that the base pairs are 'held together by hydrogen bonds' and that stacks of bases are stabilized by quite different factors variously referred to as hydrophobic interactions and van der Waals forces. Such a statement is frankly meaningless. The forces which contribute to in-plane interactions (e.g. those between the two components of Watson−Crick base pairs) can be resolved into polar and dispersive components; so can the forces between adjacent bases in a stack. It is true that in the in-plane interactions the polar interactions are the greater and in the base-stacked interactions the dispersive component is the most important.

To calculate the interaction energy for two such molecules there are two mathematical approaches. The more obvious one is the 'dipole approximation'. In this one the total interaction E_D (D standing for 'dipole approximation') is regarded as the sum of three terms thus:

$$E_D = E_{\mu\mu} + E_{\mu\alpha} + E_L$$

In this expression $E_{\mu\mu}$ is the dipole-dipole interaction, $E_{\mu\alpha}$ (the dipole-induced dipole interaction) is the interaction between one dipole with the dipole that it induces in the other molecule, and E_L is the London dispersion interaction energy. Mathematical definition of these terms are as follows:

$$E_{\mu\mu} = \frac{\mu_A\mu_B - 3(\mu_A\,\mathbf{R}_{AB})(\mu_B\,\mathbf{R}_{AB})/R_{AB}^2}{R_{AB}^3};$$

$$E_{\mu\alpha} = \frac{\alpha_A\left\{3\left(\dfrac{\mu_B\mathbf{R}_{AB}}{R_{\mathbf{AB}}}\right) + 1\right\} + \alpha_B\left\{3\left(\dfrac{\mu_A\mathbf{R}_{AB}}{R_{AB}}\right) + 1\right\}}{2R_{AB}^6};$$

$$E_L = \frac{3I_A I_B\,\alpha_A\alpha_B}{2(I_A + I_B)R_{AB}^6}.$$

In these expressions, A and B are dipoles in the two interacting molecules, R_{AB} is the distance between them, $\mathbf{R}_{AB}$ is the vector from one dipole to the other, μ's are the dipole moments, I's the ionization potentials and α's the polarizabilities.

An alternative approach is the monopole-induced dipole approximation. In this case the total interaction energy (E_M) is considered as being the sum of three terms. One of them (E_L) is the same as above but the others are different:

$$E_M = E_{\rho\rho} + E_{\rho\alpha} + E_L$$

Here the all positive and negative charges are regarded as being separate (monopoles) and $E_{\rho\rho}$ is the monopole-monopole interaction; $E_{\rho\alpha}$ (the monopole-induced dipole term) is the interaction energy between the monopoles and the dipoles which they induce in the other molecule. Clearly the mathematical definition of $E_{\rho\rho}$ and $E_{\rho\alpha}$ are more complex in this case as we have to consider the sum of all the individual monopoles in the molecules. A monopole is simply a charge (ρ) not a vector (and hence not written in heavy type). Now we must name our atoms. If there are N atoms in A let them be $a_1, a_2, \ldots, a_n$ and use the expression $a \in A$ to mean 'for all the a's in A'. Similarly the atoms in B are b_1, b_2, etc. As before R's are distances. The energy of the monopole-monopole interactions is given by:

$$E_{\rho\rho} = \sum_{a \in A} \sum_{b \in B} \frac{\rho_a \rho_b}{R_{ab}}$$

To define the monopole-induced dipole interaction energy, we must use $\mu_{A \to B}$ for the dipole induced in molecule B by all the monopoles in A. The other symbols are $\mathbf{E}_{A \to B}$ for the field induced in B by the monopoles in A at the place in molecule B in which the induced dipole is located and $\mathbf{R}_{aB}$ is the vector between atom a (in molecule A) and this position of the induced dipole. We can now define $E_{\rho\alpha}$ as:

$$E_{\rho\alpha} = -\tfrac{1}{2}[\mu_{A \to B}\,\mathbf{E}_{A \to B} + \mu_{B \to A}\,\mathbf{E}_{B \to A}].$$

$$\mathbf{E}_{A \to B} = \sum_{a \in A} \frac{\rho_a}{(R_{aB})^3}\,\mathbf{R}_{aB}$$

and

$$\mathbf{E}_{B \to A} = \sum_{b \in B} \frac{\rho_b}{(R_{bA})^3}\,\mathbf{R}_{bA}$$

The first of these equations can be re-written with only one class of vectors. This is achieved by replacing $\mu_{A \to B}$ by $\alpha_B \mathbf{E}_{A \to B}$ and $\mu_{B \to A}$ by $\alpha_A \mathbf{E}_{B \to A}$ so we obtain:

$$E_{\rho\alpha} = -\tfrac{1}{2}[\alpha_B(\mathbf{E}_{A \to B})^2 + \alpha_A(\mathbf{E}_{B \to A})^2]$$

An evaluation of the relative merits of these types of calculation was made by Nash and Bradley (1966). They considered the theoretical experiment of fixing an adenine molecule in space and rotating a uracil around it in plane (like a planet going round the sun). They then calculated the varying interaction energy between the molecules as the uracil assumes its different orientations. Using the monopole-induced dipole approximation, it was found that minima were generated corresponding to the hydrogen-bonding possibilities. With the dipole-induced dipole approximation these minima do not appear. It is therefore assumed that monopole approximation is the more reliable.

Considering just these in-plane interactions the calculation has been applied to two separate problems. Choosing the energy minima in these calculations (and neglecting any particular hydrogen-bonded structure as being the 'optimal'), Pullman, Claverie and Caillet (1966) calculated the interaction energies of all the various pairs of bases (there are ten possible combinations). The results (Table 2.6) are very striking to a biologist seeing them for the first time for although G:C is the most stable base pair possible there is nothing remarkable about A:T at all. On the other hand, of the heterologous pairs (G:C, A:T, G:T, A:C, C:T and G:A) only the 'correct' pairs G:C and A:T are more stable than either of the homologous pairs formed by the components of the heterologous pair.

TABLE 2.6 Total interaction energies (kcal/mole of base pair) (E_M) for pairs of bases in the most stable in-plane orientations. (From Pullman, Claverie and Caillet, 1966)

	A	T	C	G
G	$-7 \cdot 5$	$-7 \cdot 4$	$-19 \cdot 2$	$-14 \cdot 5$
C	$-7 \cdot 8$	$-6 \cdot 5$	$-13 \cdot 0$	
T	$-7 \cdot 0$	$-5 \cdot 2$		
A	$-5 \cdot 8$			

This table gives the values for the most stable orientations for in-plane associations; as explained above, these correspond to hydrogen-bonded structures. However the Watson–Crick configurations are not the only possible hydrogen-bonded structures. Structures are possible in which the pyrimidine is swung round (known as the 'reversed' orientation). Also there are structures in which N-7 of the purines is hydrogen-bonded. These are known as the Hoogsteen (and reversed Hoogsteen) orientations. The reason for associating these structures with Hoogsteen's name is given on p. 103.

In Table 2.7 there are diagrams of the four possible hydrogen-bonded structures for the G:C and A:U pairs together with the interaction energies as calculated by Pullman and Pullman (1968). Notice that large negative values of E_M for the Watson–Crick and reversed Watson–Crick structures for the G:C base pair is due in the main to the large contribution of the $E_{\rho\rho}$ terms and of these two it is the greater negative value for the same component that results in the Watson–Crick structure being favoured. With the A:U pair notice that the 'most stable form' used in calculating the data of Table 2.6 relating to A:T is not in fact the Watson–Crick structure but the Hoogsteen.

TABLE 2.7   Interaction energies (kcal/mole of base pair) for hydrogen-bonded structures for the base pairs A:U and G:C. (Data of Pullman and Pullman, 1968)

Structure	Base pair	Name	$E_{\rho\rho}$	$E_{\rho\alpha}$	E_L	E_M
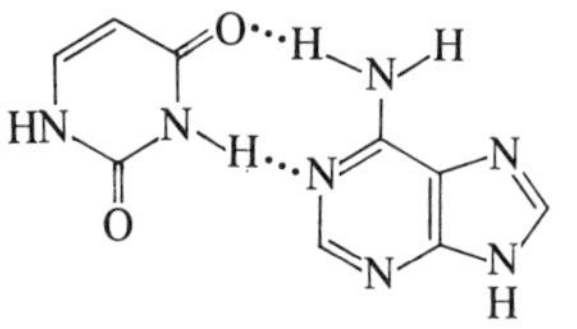	A:U	Watson–Crick	−4·64	−0·25	−0·69	−5·58
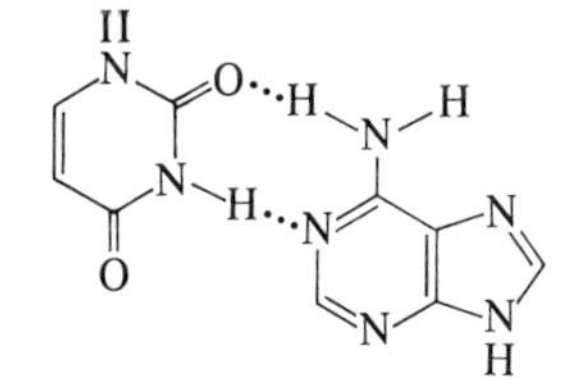	A:U	Reversed Watson–Crick	−3·98	−0·25	−0·71	−4·95
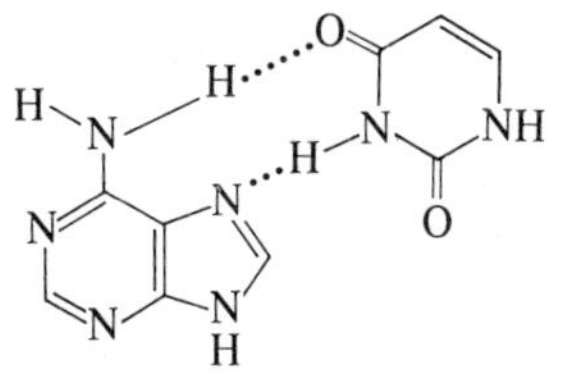	A:U	Hoogsteen	−5·86	−0·22	−0·88	−6·96
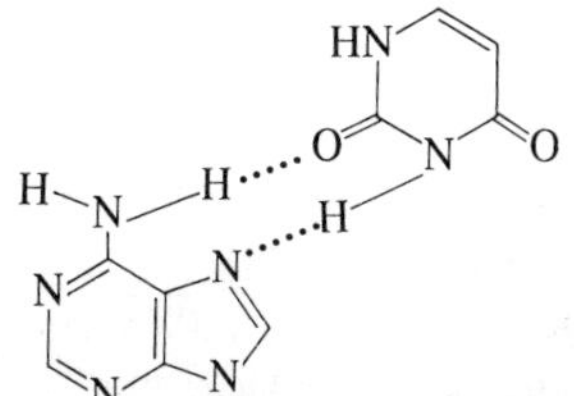	A:U	Reversed Hoogsteen	−5·63	−0·17	−0·94	−6·74

TABLE 2.7–continued

Structure	Base pair	Name	$E_{\rho\rho}$	$E_{\rho\alpha}$	E_L	E_M
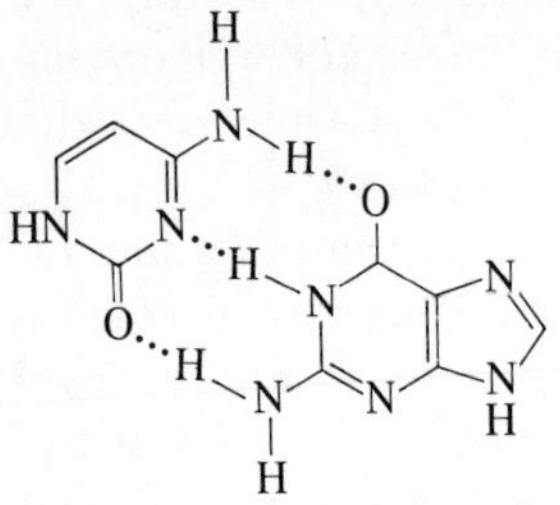	G:C	Watson–Crick	−15·91	−2·02	−1·25	−19·8
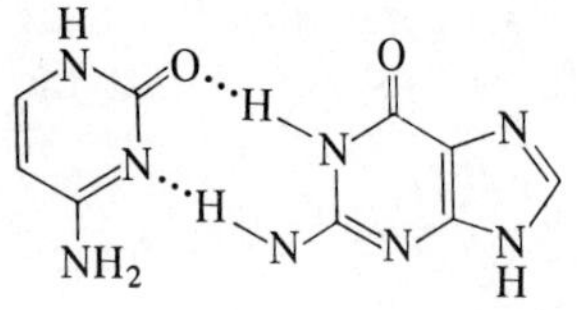	G:C	Reversed Watson–Crick	−10·24	−1·99	−0·62	−12·85
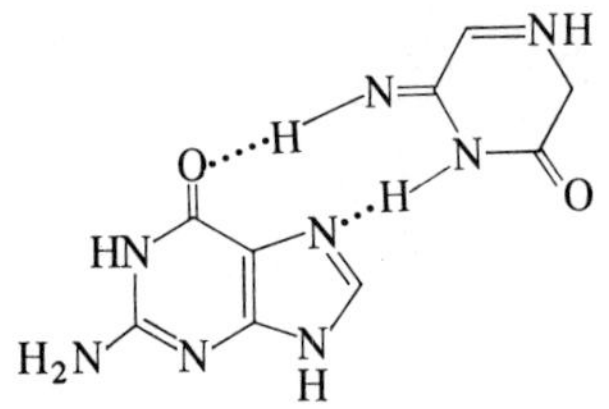	G:C	Hoogsteen	−3·98	−1·33	−0·44	−5·75
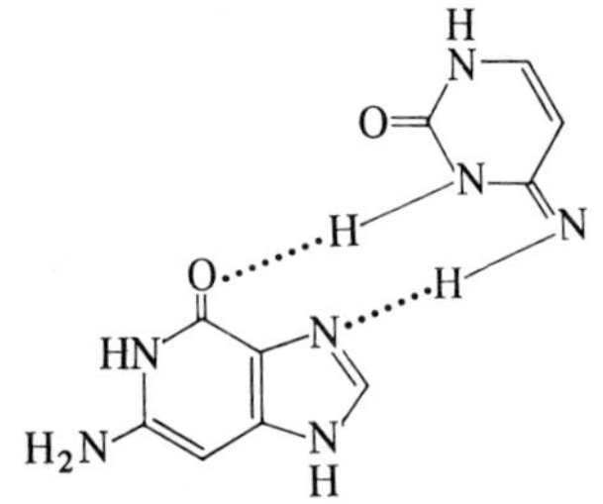	G:C	Reversed Hoogsteen	−2·42	−0·91	−0·34	−3·67

So much for the in-plane interactions; but what are the effects of base stacking on the stability of the bases in nucleic acid conformations? These calculations are extremely complex. The purine dimer (stacked) was considered by Pullman and Pullman (1968) who determined the parameters for the most stable orientation, one of which is 3·4 Å for the inter-base distance. In this stable orientation the total interaction energy (E_M) was calculated to be −4·9 kcal/mole. This value can be experimentally tested and is in agreement with the heat content change for the self-association of purine determined by van Holde and Rossetti (1967).

Pullman and Pullman (1969) have extended these calculations to the far more involved case of the bases in pairs attached to the backbone of B-DNA. The data were summed to give values corresponding to the total interaction energy *in vacuo*. Their results are presented in Table 2.8. The 'average content of in-plane interactions' is simply the average value for E_M for the two Watson–Crick structures using the values of −19·2 kcal/mole for G:C and −5·5 kcal/mol for A:T. The point that emerges from these values is that the total interaction energy in DNA is determined not only by the base ratios but also by the sequence.

TABLE 2.8   Nearest neighbour base:base interactions in the DNA double helix *in vacuo*. Units are kcal/two mol of base. In the sequences in the first column the left-hand doublet is 5′ to 3′ reading upwards and the right-hand one is 5′ to 3′ coming down. Thus in the fourth example the two DNA sequences are . . G—A . . and . . T—C. (Data of Pullman and Pullman, 1969)

	$E_{\rho\rho}$	Vertical interactions			Average content of in-plane interactions	Total
		$E_{\rho\alpha}$	E_L	E_M		
C G / G C	+0·9	−2·0	−10·2	−11·3	−19·2	−30·5
G C / C G	−1·6	−2·5	−4·0	−8·5	−19·2	−27·7
G C / G C	+2·6	−2·0	−8·3	−7·7	−19·2	−26·9
A T / G C	+1·2	−0·8	−10·3	−9·9	−12·2	−22·1
A T / C G	−0·6	−1·7	−4·9	−7·2	−12·2	−19·4
T A / C G	−0·1	−1·7	−5·2	−7·0	−12·2	−19·2
T A / G C	+1·8	−1·0	−7·8	−7·0	−12·2	−19·2
A T / A T	+0·5	−0·5	−2·4	−7·4	−5·5	−12·9
A T / T A	+0·4	−0·3	−6·2	−6·1	−5·5	−11·6
T A / A T	+1·5	−0·7	−5·8	−5·0	−5·5	−10·5

So far as the structure of nucleic acids is concerned there is a third factor contributing to the stability of the secondary structure. This is the effect of solvation. This point is discussed on p. 345.

Tautomerism   So far we have neglected the question regarding the relative stabilities of the lactam and lactim or the amine and exoimine structures (see p. 4) although it has been generally assumed that lactams and amines are the more stable forms (although an imine structure is implicated in the Hoogsteen configurations for G:C, see Table 2.7, p. 80). The quantitative evaluation of the relative stabilities of these forms requires the study of the effects of changing from one tautomer to the other on the 'resonance energy'. Consider for example the structures for adenine shown in Fig. 2.17. We have drawn the amino form with two different distributions of the double bonds in the pyrimidine ring. Structures which differ in this way are referred to as 'canonical formulae'. There are not two similar canonical forms for the imino tautomer. Thus we might expect the amino form to be more stable as it is capable of 'resonating' between these two canonical structures. The situation is more complex than this as there are actually more canonical structures for the tautomers of adenine than those of Fig. 2.17.

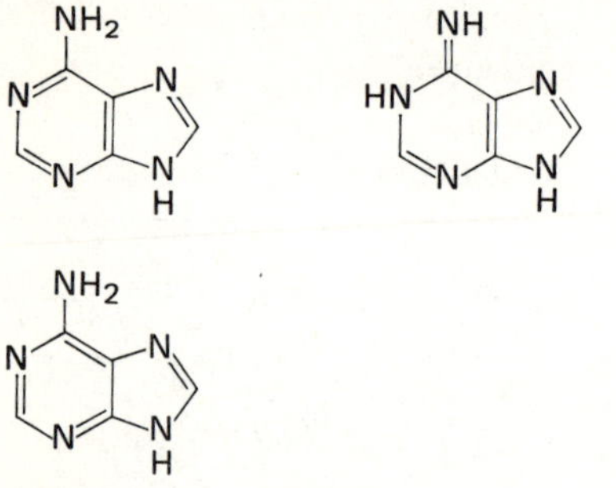

FIG 2.17   Difference between tautomerism and resonance illustrated by reference to adenine.

The resonance energy for delocalized molecules such as benzene, pyrimidines or purines can be computed in two ways. Either the canonical structures can be regarded as having different molecular orbitals, in which case the resonance energy is expressed in terms of the molecular orbital exchange integrals (symbol β) or alternatively the double bonds can be regarded separately and then the resonance energy is expressed in terms of the localized-pair exchange integrals (symbol α). Pullman and Pullman (1969) have calculated the change in resonance energy during the tautomeric transformations for the DNA bases employing the molecular orbital method. Their results are reproduced in Table 2.9. The data cover both the ground state of the molecules and the first excited state. Note that although the lactams are on the whole even more favoured than the lactims in the first excited state, the imine forms approach the amine forms in stability in this state. 'Chemicals in bottles' are essentially in their ground states. However ionizing radiation will produce excited states and could affect the tautomeric equilibria in Cyt and Ade. The possible consequences of this in DNA are discussed further on p. 352.

TABLE 2.9   The variation of resonance energy (ΔR) associated with tautomeric transformations in bases. (Data of Pullman and Pullman, 1969)

Tautomeric transformation	Compound	ΔR in the ground state	ΔR in the first excited state
Lactam–lactim	Ura	$0.22\,\beta$	$0.43\,\beta$
Lactam–lactim	Thy	$0.22\,\beta$	$0.40\,\beta$
Lactam–lactim	Gua	$0.32\,\beta$	$0.36\,\beta$
Amino–imino	Cyt	$-0.13\,\beta$	$-0.01\,\beta$
Amino–imino	Gua	$-0.27\,\beta$	$-0.14\,\beta$

Types of spectroscopy

The remainder of this chapter is about the spectroscopic properties of purines and pyrimidines (together with a brief reference to X-ray diffraction).

The simplest sort of spectroscopy consists of measuring the absorption of electromagnetic radiation by a substance (normally in solution). The type of radiation absorbed can be characterized in a number of ways. For much spectroscopy it is identified by its wavelength (λ) which can be measured in metres, symbol m, or centimetres, symbol cm, or fractions of these. These smaller units are: micron, symbol μ, 10^{-6} m (or 10^{-4} cm); the nanometre, symbol nm, 10^{-9} m (or 10^{-3} μ—hence millimicron, mμ, is not used in this book); and the angstrom unit symbol Å, 10^{-8} cm (or 0·1 nm). Another method of characterizing radiation is not by its wavelength but rather its wave number, i.e. not by the number of units of length per wave but rather the number of waves per unit of length. Thus wave number, symbol ν, is the reciprocal of λ and is always expressed in the same units, cm^{-1}. Thus for radiation of wavelength 400 nm (= 400 x 10^{-7} cm) the wave number is 25 000 cm^{-1}. It is important not to confuse wave number with frequency which is often given the same symbol (ν). The units of frequency are reciprocal seconds (s^{-1}). Frequency is calculated by dividing the velocity of light (3 x 10^{10} cm/s) by the wavelength (in centimetres).

The energy associated with radiation is proportional to the wave number. The precise relationship is Einstein's equation:

$$E = h\nu/c$$

where E is energy, h is Planck's constant and c is the velocity of light. The nature of the molecular perturbation which absorbs the radiation is dependent on this energy term. Radiation of extremely high energy (high wave number or very small wavelength) produces the stripping of electrons from atoms: this is atomic absorption spectroscopy and is produced with low wavelength X-rays. The lowest wavelength (highest energy) radiation useful for molecular spectroscopy is that in the ultra-violet and visible range. This results in the electrons in the molecular orbitals being pushed from their 'ground states' into higher energy levels or 'excited states'. The excited states are indicated by asterisks. There are three types of ground states for such electrons, σ, π and n, which correspond to the electrons in single bonds, multiple bonds and non-bonding electrons (e.g. the 'spare pair' on nitrogen in the ammonia molecule) respectively. The transitions normally encountered in the electronic absorption spectra of organic molecules are n to π^*, π to π^*, σ to π^*, and σ to σ^*. The excited states are examples of 'anti-bonding orbitals'. The next band of the electromagnetic spectrum is the infra-red in which the radiation is absorbed as vibrational and bond-stretching frequencies in the molecule. The two ranges of radiation of next longer wavelength are the far infra-red and micro-wave regions in which energy is absorbed as rotational frequencies (this type of spectroscopy is not of importance for us). At very long wavelength, the energy is so low that no immediately obvious absorption can be measured. However certain atomic nuclei behave as magnetic moments when they are placed in high magnetic fields. The quantized orientations of these little magnetic moments can be shifted into higher energy stakes and absorption due to this phenomenon can be obtained in the radio-wave region of the spectrum; the spectroscopic technique is nuclear magnetic resonance. In addition to these absorption effects various types of emission spectroscopy (fluorescence) will be considered.

Reference will also be made to the diffraction of X-rays by solid lattices of atoms in crystals and also to mass spectrometry—the precise analysis of the products following fragmentation of molecules in the vapour phase.

Electronic absorption spectra

Measurement of absorption in the ultra-violet is the most commonly used spectroscopic technique employed in biochemical research on pyrimidines and purines and their derivatives including the nucleic acids. It is used for three types of experiment, quantitative assay of the compounds in solution, identification of structures and determination of details of molecular structure in solution. It is in fact one of the least versatile spectroscopic techniques for this last purpose, but the most useful for the first two. Definitions of spectroscopic terms used in this section (and throughout the book) are given in Fig. 2.18.

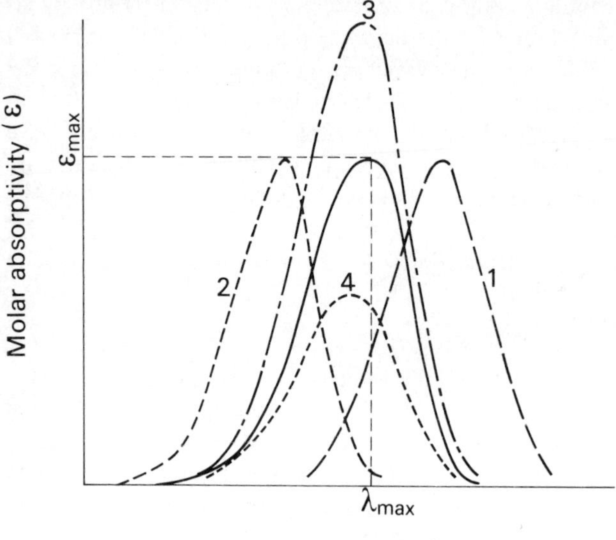

FIG 2.18   Definitions of spectroscopic jargon. The continuous curve is an absorbance band for a substance obeying Lambert's and Beer's laws. This means that if the intensity of incident light is I_0 and the intensity of transmitted light is I, then for a solution of a substance of concentration c mole/l in a light path of l cm:

$$\log_{10} (I_0/I) = \varepsilon c l,$$

where ε is a constant (for the substance in solution) called the molar absorptivity. $\log_{10} (I_0/I)$ is called absorbance (symbol A). If the effect of a chemical substituent or environmental change is to alter the absorbance band to curve 1, it is said to produce a bathochromic shift or red shift. The other changes are referred to as follows: to 2 a hypsochromic (or blue) shift; to 3 a hyperchromic effect and to 4 a hypochromic effect. In some publications A is referred to as extinction (symbol E) or optical density (O.D.) and ε is referred to as the molar extinction coefficient.

The positions and intensities of the UV-maxima of pyrimidines and purines agree quite well with transitions predicted from the SCF calculations and these calculations form the basis for a systematic classification of the spectra (Clark and Tinoco, 1965). However for routine analysis of spectra a qualitative description is more helpful. In all these discussions, water is the solvent.

Pyrimidine itself has three bands, $n \to \pi^*$ (λ_{max} 267 nm) and two $\pi \to \pi^*$ (^1L_b, λ_{max} 240 nm and ^1L_a, λ_{max} 188 nm). In pyrimidines in which there is lactam structure, that is one with an oxygen substituent on any carbon atom other than C-5 (and consequently all those of biological interest), the $n \to \pi^*$ is not observed and instead there are just the two $\pi \to \pi^*$ bands. Typically the first one has a λ_{max} around 260–270 nm and the second one around 210–220 nm and thus forms the low-wavelength absorbance in the region of the cut-off on most spectrophotometers. What then are the factors that determine the position and absorptivity of the first $\pi \to \pi^*$ transition? For convenience, the derivatives are referred to as substituted pyrimidines (uracil as 2,4-dihydroxy uracil) even though these oxygen-substituted compounds and their thiolated analogues exist as lactam or thiolactam tautomers. The 'basic compound' is

2-hydroxypyrimidine. For it λ_{max} is 298 nm and ε_{max} is 4750. Additional substitution with $-OH$ or $-NH_2$ in the ring results in a hypsochromic shift and a hyperchromic effect. In these respects $-OH$ is more effective than $-NH_2$. Substitution with $-SH$ produce a strong bathochromic shift. The shift is produced in both $\pi \rightarrow \pi^*$ bands so that the spectrum of 4-thiouracil does contain a band in the typical pyrimidine region (λ_{max} 249 nm) but in this particular case it is the second $\pi \rightarrow \pi^*$ transition; the first one is at 330 nm. The spectra of all these bases are unchanged by acid as there is no acid dissociation (p. 44), but at high pH the dissociation of $-OH$ (p. 43) constrains the molecule in a lactim configuration with a concomitant bathochromic shift. Substituents at carbon atoms other than those considered so far have very little effect on the spectra (thus the spectra of Thy and Ura are almost superimposable), but substitution at N (which includes the nucleosides) produces a bathochromic shift. The result of this shift is that the other bathochromic shift produced by alkali is less marked. In fact in certain pyrimidine nucleosides there is no detectable change in λ_{max} following titration of $-OH$; this is not to say that pK-values cannot be determined this way however, as the ionization is associated with a big hyperchromic effect. All these generalizations are exemplified in the summary of ultra-violet spectroscopic data for pyrimidines given in Table 2.10.

TABLE 2.10   Summary of ultra-violet absorption spectra of pyrimidines and purines.

In addition to standard abbreviations, $2O$-Pyr = 2-hydroxypyrimidine (lactam), 5mCyt = 5-methylcytosine (similarly the deoxyriboside d5mCyd), 4sUra = 4-thiouracil (and the riboside 4sUrd), Uri = uric acid, Urr = uric acid riboside, 7mGua = 7-methylguanine. In the body of the table, the notation 227/1·8 means at a wavelength of 227 nm, the molar absorptivity is $1·8 \times 10^3$. If the column 'other data', 280/260 means the ratio of molar absorptivities at 280 nm and 260 nm (likewise 250/260); i means inflexion.

Compound	pH	Minimum	Maximum	Minimum	Maximum	Other data
Pyr	7		240/3·2	?	267/0·4	
2O-Pyr	7		212/10·8	?	298/4·75	
Ura	2	227/1·8	259/8·2			280/260 = 0·10
Ura	12	235/2·2	282/6·2			280/260 = 1·40
Urd	2	230/2·0	262/10·2			280/260 = 0·35
Urd	12	241/2·2	262/8·5			280/260 = 0·29
Oro	2		205/10·9	240/0·9	280/7·5	250/260 = 0·57
Oro	12		?	243/3·4	288/12·6	250/260 = 0·80
Ord	2			234/2·6	267/9·5	250/260 = 0·65
Ord	12			250/5·6	267/7·5	250/260 = 0·82
Thy	2	230/2·2	264/7·9			280/260 = 0·53
Thy	12	244/2·2	292/5·4			280/260 = 1·31
dThd	2	235/2·2	267/9·6			280/260 = 0·72
dThd	13	240/4·6	267/7·3			280/260 = 0·65
Cyt	1		210/9·7	237/1·3	277/10·0	
Cyt	7			238/4·4	267/6·2	i223/7·7
Cyt	14			251/1·4	282/7·9	i223/8·3
Cyd	1		212/10·2	241/1·7	280/13·4	
Cyd	7	226/8·2	230/8·3	250/6·6	271/9·1	
Cyd	14	226/8·2	230/8·3	251/6·0	272/9·2	
5mCyt	1		211/14·2	241/0·9	273/9·8	
5mCyt	7		211/12·0	253/3·7	273/6·3	

TABLE 2.10—continued

Compound	pH	Minimum	Maximum	Minimum	Maximum	Other data
5mCyt	14		?	253/1·3	283/9·8	
d5mCyd	1		?	242/1·0	287/11·0	
d5mCyd	6		?	255/4·7	278/7·8	unchanged up to pH 12
4sUra	7	249/2·4		?	330/15·5	
4sUrd	2	260/5·5		272/3·0	335/17·0	
4sUrd	12			260/5·0	320/15·0	
Pur	1		?	228/1·2	260/6·1	shoulder at 240 nm—see text for data in unbuffered water
Pur	12		219/8·3	235/1·0	271/7·6	
Ade	2	229/2·6	263/13·1			280/260 = 0·37
Ade	12	237/3·3	269/12·2			280/260 = 0·60
Ado	2	230/3·5	257/14·6			280/260 = 0·22
Ado	11	227/2·3	260/14·9			280/260 = 0·14
dAdo	2	228/3·0	258/14·1			280/260 = 0·24
AMP	3	229/3·2	257/14·4			280/260 = 0·23
AMP	12	227/2·5	259/15·4			280/260 = 0·15
Hyp	1	215/1·8	248/10·8			250/260 = 1·45
Hyp	11	232/3·7	258/14·4			250/260 = 0·84
Ino	3	223/3·4	248/12·2			250/260 = 1·68
Xan	2	218/5·2	230/6·4	239/2·8	267/9·2	280/260 = 0·61
Xan	10		240/8·9	257/5·0	277/9·3	280/260 = 1·71
Xao	2	217/4·0	235/8·4	248/6·5	263/9·0	250/260 = 0·75
Xao	11	223/2·4	248/10·2	264/7·2	277/8·8	250/260 = 1·30
Uri	2		231/8·5	255/3·6	283/11·5	250/260 = 1·05
Uri	11			260/2·7	294/11·6	250/260 = 1·50
Urr	2		240/8·0	259/3·6	286/5·1	250/260 = 1·12
Urr	12	236/8·0	252/10·6	258/3·6	298/8·7	250/260 = 1·57
Gua	2	224/3·7	248/11·5	267/7·2	277/7·4	280/260 = 0·84
Gua	11	234/6·0	245/6·3	255/6·1	273/8·0	280/260 = 1·13
Guo	1	228/2·5	257/12·2	shoulder	278/8·3	280/260 = 0·70
Guo	11	230/4·2	257/11·2	plateau	267/11·2	280/260 = 0·61
7mGua	1	228/?	250/10·6	shoulder	270/6·9	280/260 = 0·79
7mGua	12	257/?	281/7·3			280/260 = 1·9

The spectra of purines are more complex than those of pyrimidines, less easy to simply rationalize but more sensitive to substituents and hence are of great value in structure determination. Purine itself has a low wavelength $\pi \rightarrow \pi^*$ transition (λ_{max} 188 nm) and two bands with maxima at 265 nm and 240 nm (present as a shoulder). In certain purines these bonds are totally unresolved. This region of the spectrum is supposed by Mason (1969) to be due to a drift in electron density to produce a dipole (low electron density in the imidazole half of the molecule and high electron density in the pyrimidine half). A theoretical analysis of the problem has been made by Kwistiakowski (1968). The views of Mason do account for the fact that electron donating ($-OH$ or $-NH_2$) substituents at the ends of the dipole (positions 2 and

8) have large effects on the spectrum (bathochromic shifts) but rather small effects at position 6 (compare the spectra of purine and adenine in Table 2.10). Similarly substitution at N-9 (nucleosides) has little effect on the spectra although something of an exception must be made of Gua and Guo. The titration of groups in purines has a very large effect on the position of this region of the spectrum, as would be anticipated from absorbance due to creation of a dipole. On the whole protonation of $-NH_2$ produces a hypsochromic shift and dissociation of $-OH$ produces a bathochromic shift.

As sugars and phosphates do not have electronic absorption spectra in the region under discussion, sugar derivatives of nucleosides, including the nucleotides, have very similar spectra to the nucleosides themselves. In Table 2.10, this point is illustrated with reference to Ado, dAdo and AMP.

Another application of ultra-violet spectrophotometry is to determine which atoms carry the charge in ionic species. For example, which one (or both) of the oxygen atoms carry the negative charge in alkaline solutions of thymine? This type of problem is difficult to resolve with UV data alone, but in conjunction with infra-red spectra it can be solved. The subject is deferred (p. 94) until the infra-red spectra of pyrimidines and purines have been considered.

Luminescence

The two forms of luminescence (fluorescence and phosphorescence) have not yielded a great deal of useful information regarding the structures of pyrimidines and purines, but are important in probing the interactions between the bases in polynucleotides, and therefore it is important to understand the basic features of luminescence of the low-molecular weight compounds.

Luminescence is the name given to the radiation which is emitted when a molecule reverts from an excited state to its ground state. The simplest example of the phenomenon is excitation and fluorescence. In the region of an electronic absorption maximum, radiation causes excitation of the molecule to form an excitation spectrum; as the molecules return to their ground state, radiation is emitted and the fluorescence spectrum is produced. The fluorescence spectrum is different to the excitation spectrum because the excited molecules lose some of their energy to the surrounding molecules as thermal energy, so the emitted radiation is of lower energy (longer wavelength) than the excitation spectrum. The excitation spectrum is not the same as the absorption spectrum as it is a measure of a different phenomenon. The absorption spectrum is a measure of absorption of radiation of varying wavelength irrespective of what is going to happen to the molecules. The excitation spectrum is a plot of absorption of radiation from a grating characterized by its ability to fluoresce at a measured, longer wavelength. Instead of fluorescence the emission can be slowed up due to the excited molecules making a radiationless transition to a 'triplet state' of intermediate energy. Direct emission due to the triplet-state molecules returning to ground state is called phosphorescence; if the triplet-state molecules retain the excited state through thermal excitation, and then return directly to the ground state, the phenomenon is called delayed fluorescence.

For any type of luminescence, the 'mean lifetime of the excited state' (symbol τ) is defined by the exponential equation which describes the decay (which is a monomolecular process):

$$I_t = I_0\, e^{-t/\tau}$$

t is the time elapsed from cessation of excitation, where I_t is the intensity of luminescence at time t, and I_0 is the intensity at time 0. The quantum efficiency or quantum yield (symbol φ_L)

is the ratio of the rate of emission of quanta of radiation during luminescence to the rate of absorption of quanta. The methods for calculating these quantities and correcting luminescence spectra are outside my scope but I assume that no-one would attempt to use a fluorospectro-photometer without having access to suitable books and expertise regarding their use.

Luminescence of pyrimidines At very high pH, thymine has some fluorescence at ordinary temperature, otherwise the pyrimidines only exhibit luminescence at low temperatures. These spectra are largely of academic interest, although if the *O*-methylated (lactim) and *N*-methylated (lactam) derivatives of Ura are compared it is found that the latter compound has the same characteristics as Ura itself (at pH 11, 77°K: fluorescence maximum at 350 nm, φ_L of 0·0013; phosphorescence maximum at 410, φ_L of 0·0006) confirming the lactam structure for Ura (Longworth, Rahn and Shulman, 1966).

Luminescence of purines Unlike the pyrimidines, the purines are strongly fluorescent compounds at ordinary temperatures. An extensive compilation of the data has been made by Udenfriend and Zaltman (1962). Adenine and its derivatives only fluoresce in acid whereas guanine and its derivatives have strong fluorescence emission spectra at all pH-values. This observation facilitates an easy assay of guanine or its nucleoside(s) or nucleotides(s) in a mixture. The pH is adjusted to 11 and the fluorescence spectrum (maximum at 350 nm) is recorded following excitation at 285 nm.

Börrensen has published interesting analyses of the spectra of purines. He concludes (Börrensen, 1963) that the excited singlet state n, π^* is non-fluorescent and that fluorescence is delayed and follows thermal activation to a π, π^* singlet state. A more interesting conclusion (Börrensen, 1963) relates to the methylation products of Ado and Guo with dimethyl sulphate. These reactions result in a large increase in fluorescence intensity for Guo and (at neutral pH) the appearance of a fluorescence spectrum in methylated Ado where none occurs in the free nucleoside. These spectra resemble the fluorescence of the acidified solutions of the nucleosides. It is therefore concluded that the methylated positions (N-1 in Ado and N-7 in Guo, see p. 60) are the same atoms that are protonated in acidified solutions of the nucleosides.

Optical rotatory dispersion and circular dichroism

These types of spectroscopy are conveniently discussed here as they are based on phenomena related to electronic transitions. As with the preceding discussion of luminescence, I am only giving sufficient detail to enable the reader to follow simple papers on the optical properties of nucleosides, etc., and not any discussion on instrumentation or detailed theory of the subject. Of the two phenomena, optical rotatory dispersion (ORD) is the simpler to record but circular dichroism (CD) yields data which are, on the whole, easier to interpret.

The specific rotation of an optically active substance (symbol $[\alpha]$) is a function of wavelength. A plot of $[\alpha]$ versus λ forms a sort of spectrum known as the optical rotatory dispersion (ORD). The simplest generalization about these curves is that $[\alpha]$ is of opposite sign on either side of an absorption maximum corresponding to an electronic transition). The first theory of rotatory dispersion produced Drude's equation, which in its simplest form is

$$[\alpha]_\lambda = \frac{K}{(\lambda^2 - \lambda_{max}^2)}$$

where k is constant. Obviously this equation implies that $[\alpha]_{\lambda max}$ is infinity. A plot of Drude's equation and what is really observed are shown in Fig. 2.19(a). The cross-over at λ_{max} (i.e. $[\alpha]_{\lambda max} = 0$) is known as the Cotton effect. The equation of Drude can be fiddled by incorporating a 'damping factor' to produce a real curve. The rationale behind this adjustment is to 'recognize' that Drude's equation is derived from premises that imply that an absorption peak would cover an infinitely narrow wavelength band, whereas in fact they are dispersed to produce an envelope covering a real band width.

To understand CD and to put ORD on a more satisfactory basis it is necessary to consider circularly polarized light. Plane polarized light can be resolved into two circularly polarized components of equal amplitude. If x, y and z are three orthogonal directions and circularly polarized light (of frequency ν) is propagated in direction z through medium of refractive n, the amplitude a of the radiation field is described in the equation:

$$a = a_0 [\mathbf{x} \cos 2\pi(t - nz/\nu) \mp \mathbf{y} \sin 2\pi(t - nz/\nu)],$$

where t is time, a_0 a constant and $\mathbf{x}$, $\mathbf{y}$ are unit vectors at right angles to one another and to z. This equation generates a right-handed helical pattern (z is the helix axis) for right-polarized light and a left-handed helix for left-polarized light. If plane polarized light (consisting of a mixture of these two components) passes through a transparent medium, in which the refractive indices for the left and right components are different, emerging light will consist once more of two circularly polarized components of the same amplitude but different phase-difference to yield a plane polarized beam but with the plane rotated. If on the other hand the medium is not transparent (i.e. ν is in the region of an electronic transition) the absorptivities as well as the phases will be different for the two components and the resultant will be elliptically polarized light. If this phenomenon is measured with an instrument only capable of detecting plane polarized light then the effect will be to produce an apparently anomalous rotation (the Cotton effect). If on the other hand the light is monitored with a suitable optical system it is possible to define the parameters of the ellipse. This effect is demonstrated and the parameters defined in Fig. 2.19(b). The reason why circular dichroism is the simpler phenomenon to interpret is that it can be derived quantitatively from a real molecular model. The reason for the dichroic effect in the region of an electronic transition is that in this region the electrical and magnetic dipoles of an asymmetric molecule are of different sense. However we would get out of depth if this point were pursued. The only remaining necessary points are to know the ways in which ORD and CD curves are related to one another and are drawn. First of all the rotation α and specific rotation $[\alpha]$ ($[\alpha] = 100\,\alpha/lc$, where l is length in decimetres and c concentration in g/100 ml) are not normally plotted, but rather specific molar rotation $[M]$.

$[M]$ is obtained from the relationship:

$$[M] = [\alpha](MW)/100 \qquad (MW = \text{molecular weight})$$

Likewise from Ψ we can calculate $[\Psi]$, the specific angle of ellipticity, but more usually we use the specific molar ellipticity $[\theta]$:

$$[\theta] = [\Psi](MW)/100$$

An alternative to using $[\theta]$ is to use the quantity $(\epsilon_e - \epsilon_r)$ which is the difference in the decadic molar extinction coefficients for the left- and right-circularly polarized components. The relationship between the two is:

$$(\epsilon_l - \epsilon_r) = [\theta]/3300.$$

The relationship between ORD and CD spectra for the region of a single electronic transition is shown in Fig. 2.19(c). The ORD trace is the same shape as the first differential of the CD trace. Actual spectra are normally more complex owing to the overlap of the regions of influence of two or more electronic transitions close together. The value $[A]$ in Fig. 2.19(c) is known as the 'amplitude' of an ORD trace. Cotton effects are said to be either positive or negative. A positive effect is one in which $[M]$ is positive on the long wavelength side of the cross-over point and negative on the low wavelength side. Thus if the effects shown

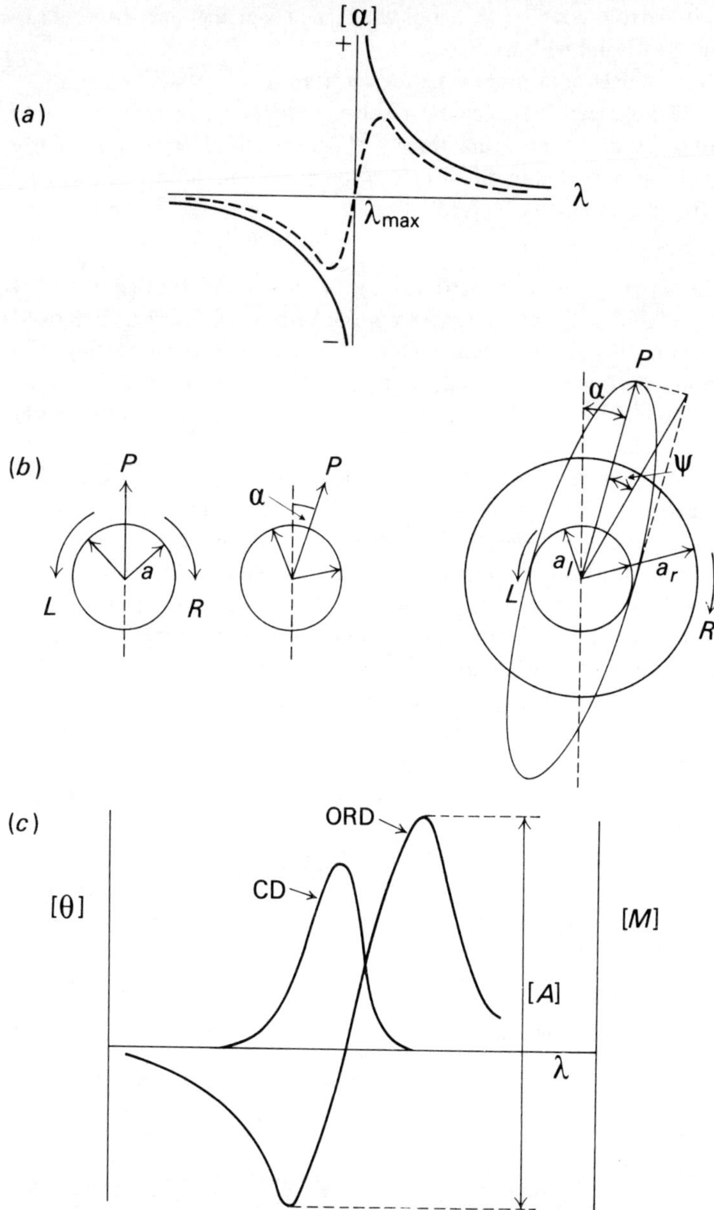

FIG 2.19   ORD and CD. (a) Hypothetical ORD spectrum obeying Drude's equation (———) and the appearance of a real ORD spectrum (- - - - - -). (b) Two circularly polarized components of plane polarized light (the plane polarized light is their resultant (P) emerging from (i) inactive medium, (ii) transparent but optically active medium (a is amplitude, α is angle of rotation) and (iii) from optically absorbing and optically active medium. In (iii), tan $\Psi = (a_r - a_e)/(a_r + a_e)$, where a_r and a_e are the right and left amplitudes. (c) Relationship between ORD and CD spectra of the same substance.

hypothetically in Fig. 2.19 (parts *a* and *c*) are plotted with low wavelength to the left, they are both positive. The opposite type of effect is described as negative. The CD peak corresponding to a negative Cotton effect is negative too (i.e. produces a trough below the base-line rather than a peak above it.

Purines and pyrimidines   As with the technique of luminescence spectroscopy (last section) the real importance of rotatory dispersion and dichroic spectra is for the understanding of the structures of oligo- and polynucleotides. Nevertheless ORD data have been accumulated for several low-molecular weight derivatives, notably by Ulbricht and his collaborators, and have been applied to two problems. One is the determination of the anomeric configuration of uncharacterized nucleosides. The second is the question of the conformation of the molecules about the glycosidic (N—C) bond.

The ORD curves for the common nucleosides and nucleotides were measured by Yang, Samejima and Sarkar (1966). Some of their data are summarized in Table 2.11. Two important generalizations are immediately apparent from these data: the purine nucleosides have negative Cotton effects and the pyrimidines positive Cotton effects; the ORD curves are a far more sensitive monitor of differences in sugar structure than is ultra-violet spectrophotometry (compare the different ORD data for Ado, dAdo, AMP and dAMP with the ultra-violet data for the same set in Table 2.10). A comprehensive survey of many α-anomers of nucleosides has been made by Walton, Holley, Boxer and Nutt (1966). What emerges from these comparisons is that the -anomers have in general Cotton effects of opposite sign to those of the corresponding ('natural') β-anomers. Another way of starting this is in terms of 'Hudson's rule'. Hudson's rule is a generalization about the optical properties of glycosides which considerably pre-dates ORD measurements (Hudson, 1909). The rule is that at the wavelength of Na—D light (i.e. right on

TABLE 2.11   ORD data for common nucleosides and nucleotides. Data of Yang, Samejima and Sarkar (1966). The quantity [*A*] is defined in Fig. 2.19.

	Peak (nm)	$[M] \times 10^{-3}$	Cross-over (nm)	Trough (nm)	$[M] \times 10^{-3}$	[*A*]	UV spectrum λ_{max} (nm)
dAdo	260	−1·55	258	245	+2·25	−3·80	259·5
Ado	275	−3·30	256	245	+1·60	−4·90	260
dAMP	268	−2·49	258	253	+1·30	−3·79	260
AMP	272	−3·00	257	246	+1·80	−4·80	260
dGua	258	−2·40	250	231	+3·65	−6·05	252·5
Gua	258	−2·90	249	230	+1·95	−4·85	253
dGMP	270	−3·35	248	233	+3·10	−6·45	252·5
GMP	281	−1·61	247	237	+0·60	−2·21	253
dCyd	288	+5·85	273	250	−7·20	+13·1	272
Cyd	288	+6·80	273	242	−13·6	+20·4	271
dCMP	288	+6·30	273	245	−9·10	+15·4	272
CMP	288	+7·70	272	240	−13·0	+20·7	270
dThd	290	+2·20	277	260	−4·30	+6·50	267
Urd	283	+4·00	271	253	−9·80	+13·8	261
dTMP	290	+1·50	278	257	−6·20	+7·70	268
UMP	283	+3·50	272	255	−9·70	+13·2	262

the long wavelength side of the Cotton effect in nucleosides) the α-anomer of a glycoside is more dextrorotatory than the β-anomer. The purine nucleosides obey this rule; the pyrimidine nucleosides do not.

The question of conformation of the molecules in solution about this glycoside bond has been examined by T. L. V. Ulbricht using model systems. The two extreme conformations are referred to as *syn* and *anti* and are defined in Fig. 2.20.

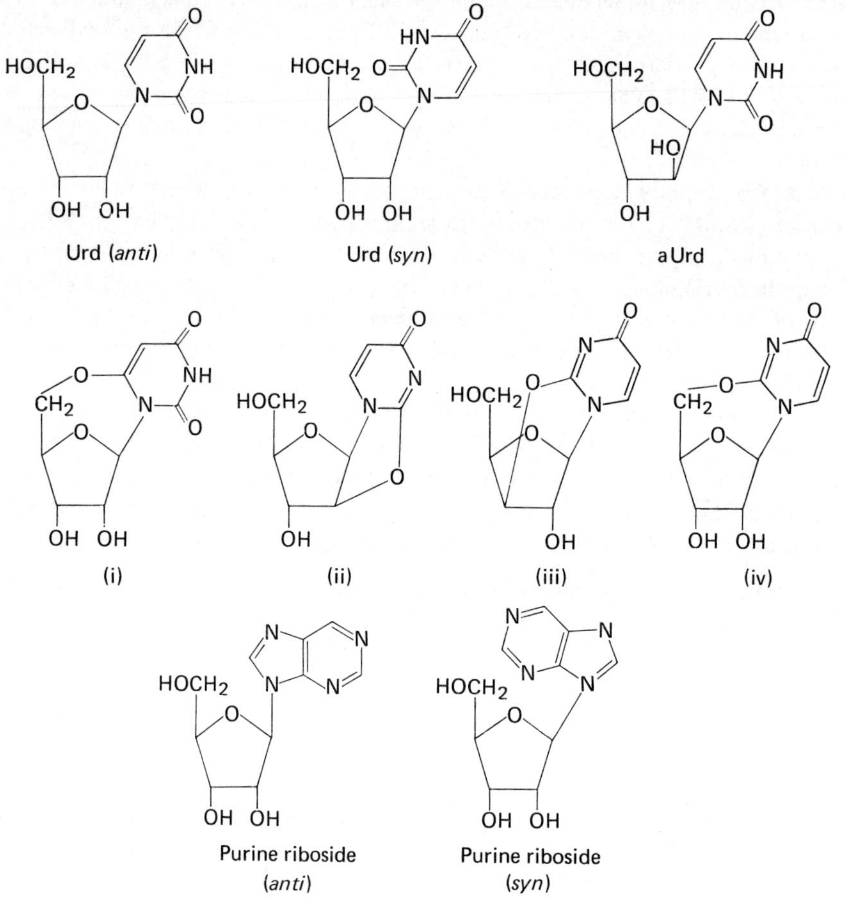

FIG 2.20   Conformations of nucleosides about the *N*-glycoside bond. Structures (i) to (iv) are referred to in the text.

A preliminary deduction (Ulbricht, Emerson and Swan, 1965) was that the positive Cotton effect in pyrimidine nucleosides was due to restricted rotation about the glycosidic bond, as the amplitude of this effect is greater in the arabinsode than the riboside (see Fig. 2.20). A more systematic survey was made using the cyclic nucleosides (i) to (iv) (see Fig. 2.20; Ulbricht, Emerson and Swan, 1966). The amplitudes ($[A]$) of these compounds are (i) $+49 \times 10^3$, (ii) $+2\cdot7 \times 10^3$, (iii) $+4\cdot3 \times 10^3$, and (iv) -37×10^3. Each of these cyclic nucleosides is locked in one particular conformation. The plane of the pyrimidine ring in (i) is taken as a reference plane. The angle of the pyrimidine ring to this plane is $0°$ in (i), $110°$ in (ii), $135°$ in (iii) and $180°$ in (iv). The generalization was that the dipole through the atoms O^4 and O^2 (or O^4 and N^2 in Cyt) must be as far away from C–5$'$ as possible for maximum positivity in the Cotton effect. The value for Urd ($+13\cdot8 \times 10^3$) implies that the anti-conformation is strongly favoured.

With purine nucleosides the situation is less clear but comparison with locked cyclic compounds (such as the compound in Fig. 2.8(*e*), p. 62) implies that the *anti* conformation is weakly favoured. Unfortunately only an incomplete publication of these comparative data has been written (Emerson, Swan and Ulbricht, 1966).

Infra-red spectroscopy

Absorption in the infra-red region, wavelength between about 3 and 15 μ (or wave numbers between 5000 and 600 cm^{-1}), is due to the induction of vibrational frequencies in molecules. Such spectra are on the whole easy to interpret and form excellent 'fingerprints' of molecules as large numbers of well resolved absorption bands can be recorded (see Fig. 2.21). Unfortunately aqueous solutions are useless for this type of spectroscopy, as water has massive absorption throughout the whole of the region of interest. Satisfactory spectra can be recorded in deuterium oxide. This is an expensive experiment, not because of the D_2O which is a

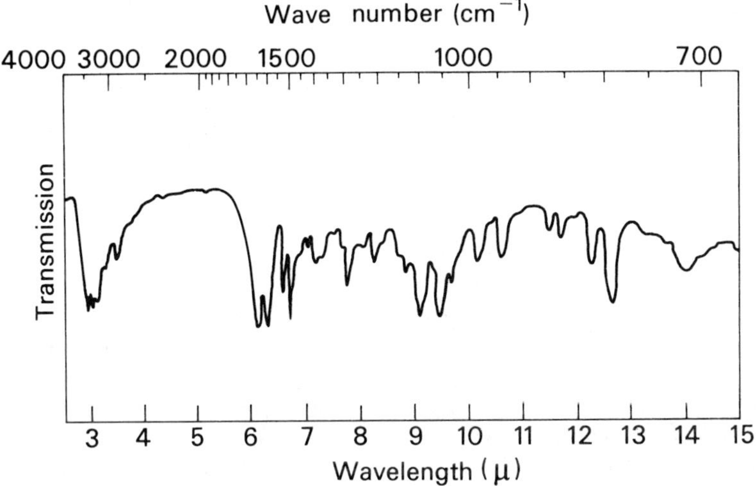

FIG 2.21   Infra-red spectrum of solid Cyd precipitated from neutral solution. Absorbance is 'downwards' (from Tsuboi, Kyogoku and Shimanouchi, 1962).

reasonably cheap material, but because infra-red spectrophotometer cells must be made of a material transparent to infra-red radiation. Glass and quartz are of no use whatsoever and the usual choice is sodium chloride. Clearly such cells are of no use for solutions in D_2O and instead special cells with calcium fluoride windows are used. These are expensive. However satisfactory spectra for certain purposes can be obtained from solid materials which can either be compounded into discs with a suitable solid salt (potassium bromide) or made into an emulsion with a suitable oil and trapped between sodium chloride discs.

The pioneering work on the infra-red spectra of nucleosides is that of H. T. Miles. An example of his work is the experiment (Miles, 1956) which established the tautomeric structure of Urd. His data on part of the spectrum of a variety of model compounds are summarized in Fig. 2.22. The glucoside was chosen for part of the comparison for reasons of availability of derivatives at that time. By analogy it is clear that the two bands in Urd must both be due to C=O stretching vibrations, and the band at 6·14 μ in the third structure must be due to C=N.

A more thorough investigation of these C=O and C=N vibrations in the four nucleosides Urd, Cyd, Guo and Ado was made by Tsuboi, Kyogoku and Shimanouchi (1962). The important conclusions relate to the structure of the ionized structures. It is well established that when an amino substituent (such as $-NH_2$ in aniline) is protonated the ring stretching frequencies are unchanged; on the other hand when a heterocyclic ring nitrogen (such as that in pyrimidine) is protonated these bands shift to higher frequencies. In the spectrum of cytidine at neutral pD (or as a solid precipitated from aqueous solution at neutral pH) there are bands at 1652, 1615, 1582, 1527 and 1504 (all wave numbers in cm^{-1}). The first of these two bands are due to C=O and C=N stretching respectively. In the deuteronated (or protonated) form at low pD (or pH) the bands have all shifted, viz. 1709, 1658, 1591, one missing (corresponding to 1527 above) and 1529. The shift in the C=O and C=N bands indicates a change to a less delocalized, more 'classical', electronic distributions in the molecules in acid. The other shifts indicate that it is a ring nitrogen (N-3) that is protonated. Similarly it is a ring nitrogen in

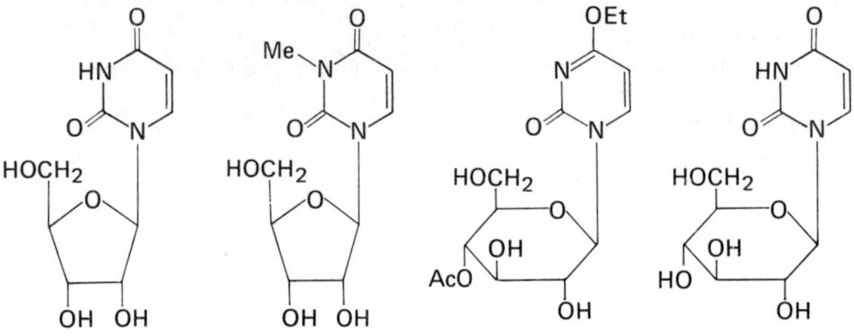

FIG 2.22   Bands in the IR for various uracil derivatives. Figures are wavelengths ( ), $s =$ strong absorption, $vs =$ very strong. The band at 5·73 in (iii) is due to the C=O of acetyl. Data of Miles (1956).

adenosine and guanosine that is protonated in acid. The spectrum of guanosine in alkali has no stretching band of wave number greater than 1952 cm^{-1} indicating (from analogy with model systems) great π-electron delocalization consistent with the predominance of the lactam form (see Fig. 2.23(a)). A similar situation obtains in thymine. The experiments in this case are those of Wierzchowski, Litońska and Shugar (1965). The infra-red data rely on a band at 1654 cm^{-1} common to thymine and 1-methylthymine but absent in 3-methylthymine (see Fig. 2.23b). This implies the presence of a negative charge on N-3 in thymine as this is the only atom capable of carrying the negative charge in the 1-methylthymine molecule. However ultra-violet data in the same paper imply that a proportion of the negative charge is on oxygen so the tautomeric components of the structures are believed to be those indicated (Fig. 2.23(b)).

A different use of infra-red spectroscopy has been to demonstrate 'specific' hydrogen bonding between derivatives of Cyd and Gua (Hamlin, Lord and Rich, 1965). The data are very difficult to evaluate as the measurements have to be performed on derivatives with blocked hydroxyl groups, and in chloroform solution, to 'eliminate' effects of base stacking. The bands studied are those in the regions of 3520 and 3410 cm^{-1} in the derivatives of Cyd and Gua. These two bands are characteristic for the group $-NH_2$. They are the two N–H stretching frequencies (not one for each hydrogen atom but corresponding to antisymmetric and symmetric frequencies for both bonds). In the mixture these bands are weaker and association bands characteristic of hydrogen bonding are produced. The effect is not found in mixtures of Gua and Ado.

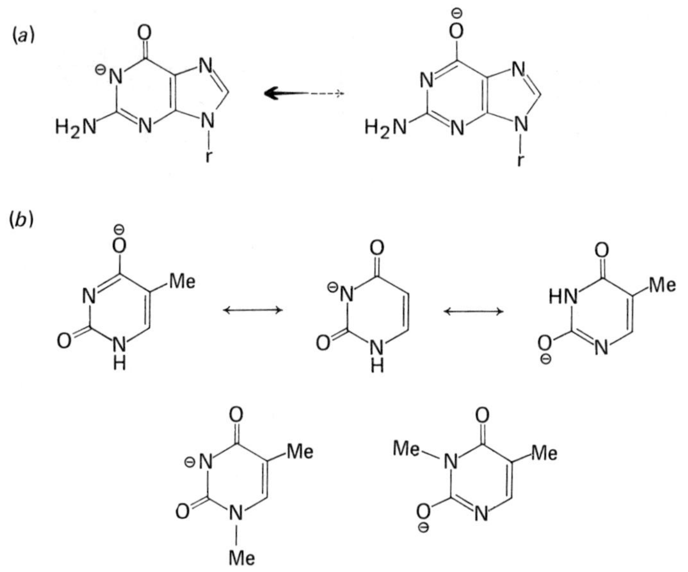

FIG 2.23   Tautomeric structures of anions (*a*) guanosine (*r* = ribose) and (*b*) thymine, 1-methylthymine and 3-methylthymine.

Raman spectra   Molecules in excited vibrational states emit infra-red radiation as they fall back into their vibration ground states. The experiment is performed by irradiating a solution of the sample (which can be in water) with visible light and recording infra-red radiation emitted by the solution. No measurements can be made in the line of fire of the light, so the IR emissions are recorded at right angles to the input beam. Herein lies the technical difficulty of Raman spectroscopy as the input light must be highly directional (well colimated) and the yield of emitted radiation scattered at right angles is very small. Nevertheless the possibility of measuring vibrational transitions in water is obviously of enormous importance in studies on nucleotides and with the availability of laser Raman spectrometers the subject must be one of the growth points in nucleic acid research over the next few years (see p. 342). Raman data on free nucleotides have been published by Malt (1966) and Lord and Thomas (1967).

Nuclear magnetic resonance

Atomic nuclei have the properties of tiny magnetic moments. Consequently they have a quantized angular momentum (**I**) with the value:

$$\mathbf{I} = [I(I + 1)]^{1/2}\, h/2\pi$$

In this equation, I is a quantum number characteristic for each nucleus. The permitted values for I are zero, a positive integer or a half-integer. It is possible to work out the nuclear magnetic moment, which turns out to be a simple function of $[I(I + 1)]^{1/2}$, from which it is apparent that for any nucleus for which I is zero there is no nuclear magnetic moment. However if I is other than zero there is a real nuclear magnetic moment $\boldsymbol{\mu}$. Now if such a nucleus is in a magnetic field it has a potential energy E_H:

$$E_H = -\boldsymbol{\mu}.H_0 \cos\theta$$

where H_0 is the applied magnetic field and θ is the angle between the direction of the field and the quantized angular momentum. This expression can be re-written in a form in which μ is factorized, one of the factors being M_I, a number which can take any of the values $I, I - 1, \ldots, -I$ ($2I + 1$ values altogether). We shall only be concerned with the cases in which $I = \frac{1}{2}$ so that M_I has two possible values ($\frac{1}{2}$ and $-\frac{1}{2}$). Consequently in the field H_0, E_H can have one of two values (there are two energy levels) corresponding to these two values of M_I. In a stable magnetic field transitions between these two levels do not occur. However in an oscillating electromagnetic field such transitions will occur at frequency, defined by:

$$h\nu = E_H' - E_H''$$

where E_H' and E_H'' are the potential energies (in field H) for states defined by $M = \pm\frac{1}{2}$.

So, how is nuclear magnetic resonance (NMR) measured? Compared with other forms of spectroscopy it has the novelty that either the frequency or the energy of the atomic transition (a function of the external field H_0) can be varied by the operator. The design of the equipment involved is obviously outside the scope of this book, but essentially imagine the sample at the centre of three orthogonal co-ordinates x, y and z. The sample is in a little tube rotating on its axis (to even out minor heterogeneities in the magnetic field). In one axis (say x) there is strong magnetic field, in direction y there is a radiofrequency oscillator ('the light source') and the axis of the sample tube is in z. The tube is surrounded by a detector coil to record the emission ('resonance'). The sample can be scanned, either by keeping the frequency constant and sweeping the magnetic field or keeping the field constant and sweeping the frequency. The latter process is more immediately analogous to conventional spectroscopy although the former is the method used in present day NMR machines. When resonance is produced a peak is formed in the detector. In compounds of C, H, O and N present as their normal isotopes, H is the only atom whose nucleus produces NMR. However the spectrum does not consist of just one line. The peaks are resolved as the positions of resonance are affected by the electronic environment of the nuclei. Thus the NMR of hydrogen atoms (proton magnetic resonance or PMR) is a means of studying the different environments of hydrogen atoms in molecules. The two types of effect found in such spectra are best illustrated with reference to a specific example, so let us consider acetaldehyde CH_3CHO. There are two sorts of hydrogen present in a ratio of 3:1. Their positions of resonance are different because of the different efficiencies of their electronic environments in shielding the protons from the applied magnetic field. These shielding effects shift the resonance positions of the protons from that of 'bare protons'. In practice these shifts are related to the resonance position for a hydrogen in a molecule, containing only one sort of hydrogen (e.g. benzene) included in the sample solution. Thus we might expect the PMR spectrum of CH_3CHO to resemble that shown in Fig. 2.24(a) with two peaks the one (corresponding to the protons in CH_3-) three times the size of the other (corresponding to the proton in $-CHO$). In practice the spectrum looks more like that in Fig. 2.24(b) with each peak highly resolved. The reason is that in addition to the effect of electronic shielding, the resonance position for a proton is affected by the energy level of other protons in the same molecule if they are close enough to be 'coupled'. Thus in the case of the methyl protons in CH_3CHO the peak is split into two depending on whether or not the aldehydo proton is in the higher (+) or lower (−) energy level. The splitting of the aldehydo proton resonance by the methyl protons is more complex as there are three protons responsible for the splitting and the quartet pattern is obtained as shown in the figure.

The symbol for the chemical shift is σ and the coupling constant (J) is defined from the statement that the position of the resonance for the components of such a set are given as $H_0(1 - \sigma) \pm J$.

A more complicated situation arises when the value of J is large and of the same order of magnitude as σ. Consider a molecule with the formula $R_2CH–CHR'_2$. This is known as an *AB* system (two coupled protons on different types of group). In this case as *J* increases (in a series of such molecules) the pattern changes from two equal doublets as shown in Fig. 2.24(*c*). Examples of other types of coupled systems are given in the discussion of nucleosides.

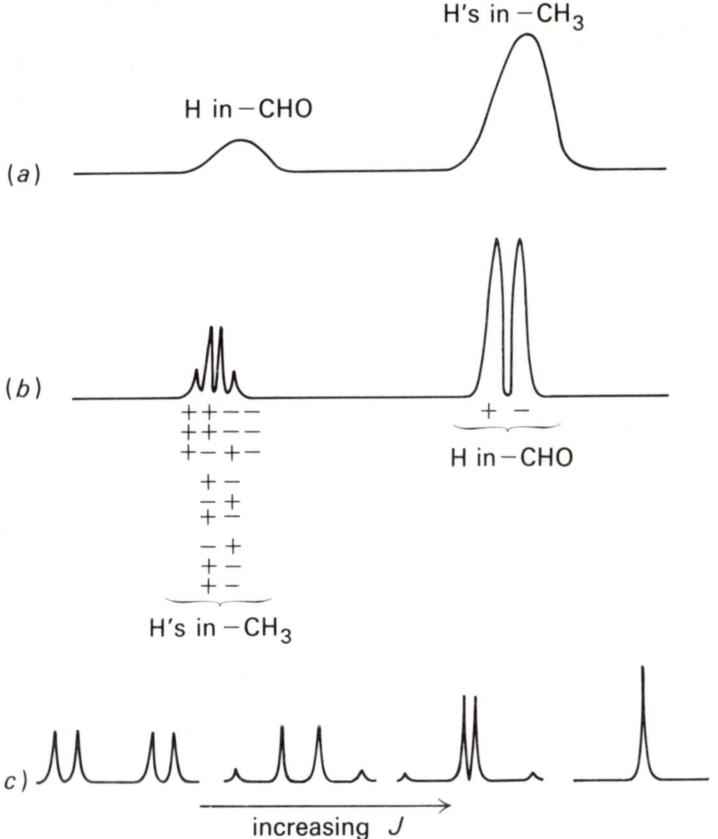

FIG 2.24   Hypothetical PMR spectra of acetaldehyde (*a* and *b*) and for molecules of the type $R_2CH–CHR'_2$ (*c*) See text for discussion.

PMR spectra of nucleosides   Studies on the chemical shifts in bases and nucleosides have been documented by Jardetzkey and Jardetzkey (1960) to correlate the effect of shielding of delocalized π electrons; the discussion has been considerably extended in the case of purine by Schweitzer, Chan, Helmkamp and Ts'o (1964). The paper by Walton, Holley, Boxer and Nutt (1966) referred to in the previous section (p. 91) indicates a method of anomer assignment using PMR.

Several attempts have been made to demonstrate association between bases and nucleosides. NMR is a very sensitive test for hydrogen bonding as (in general) donor protons demonstrate a downfield shift when engaged in H-bonding. The evidence is actually very unsatisfactory, as we should expect from a theoretical consideration of the interactions between these substances (see p. 78). Katz and Penman (1966) were able to demonstrate such associations between Gua and Cyt (and between Guo and Cyd) in an organic solvent but the complexes formed were not of a 1:1 type; no such association could be demonstrated between Ade (or Ado) and Gua (or Guo). A more valuable experiment (Chan, Schweitzer, Ts'o and Helmkamp, 1964) was on the self association of purine in solution (a model of base-stacking) in which it was shown that the data were consistent with models for the calculation of the interaction energy (see p. 78) and also with a value of between 3 and 4 Å for the distance between bases.

An extremely important use of PMR is the work of Christine D. Jardetzky on the conformation of the atoms of the sugars in nucleosides. The central point is that the coupling constant for the interaction between two vicinal H-atoms in a molecule is a function of their dihedral angle. Thus if you imagine two such atoms (H–C–C–H) and look down the axis of the C–C bond, the dihedral angle (ϕ) is the apparent angle between the two C–H bonds. If ϕ lies between 0 and 90° the relationship is:

$$J = 8{\cdot}5 \cos^2 \phi - 0{\cdot}28$$

whereas if ϕ lies between 90 and 180° the equation is:

$$J = 9{\cdot}5 \cos^2 \phi - 0{\cdot}28$$

These relationships were derived by Karplus (1959).

A pyranose sugar forms a saturated six-membered ring of which the simplest model compound is cyclohexane. The stable conformation for a cyclohexane ring is the 'chair' in which ring atoms 1, 3 and 5 are in one plane and ring atoms 2, 4 and 6 are in another parallel plane. Substituents (or H-atoms) are in two classes equatorial (e) and axial (a). The five ring atoms of the furanose ring in nucleosides form a cyclopentane ring. There are several conformations for this ring system. The ones which contain maximum puckering are the 'envelope' and 'half-chair'. There are five types of substituent position in these conformations

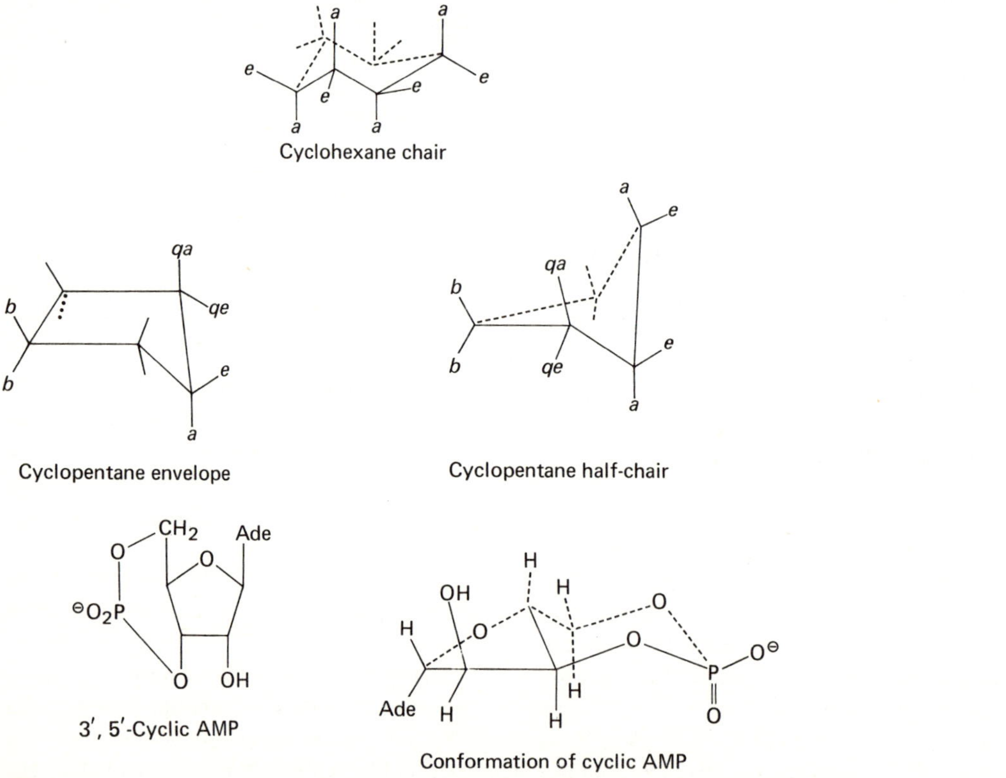

FIG 2.25   Conformations of various cyclic compounds. See text for discussion.

equatorial (*e*), quasi-equatorial (*qe*), axial (*a*), quasi-axial (*qa*) and bisectional (*b*). These conformations are illustrated in Fig. 2.25. Note that the envelope contains four ring atoms in one plane and one out-of-plane. There are four such conformations for nucleosides. They are named after the atom which is out-of-plane which is referred to the side of the plane on which C-5′ is located. Thus C'_2-*endo* is the envelope in which C-2′ is out-of-plane on the same side of the plane (formed by 0, C1-1′, C-3′ and C-4′) as is C-5′. Likewise C'_2-*exo* is the envelope with C-2′ out-of-plane and on the opposite side to C-5′. The other two envelopes are C'_3-*endo* and C'_3-*exo*. There are also two half-chairs, C'_2-*exo*/C'_3-*endo* and C'_2-*endo*/C'_3-*exo*.

Jardetzky (1962) calculated the dihedral angles between H'_1 and $H_{2'}$, between H'_2 and $H_{3'}$ and between $H_{3'}$ and H'_4. She then predicted the coupling constants (from Karplus' equations) for the various conformations and tested the prediction with cyclic AMP (see Fig. 2.25) in which (as with all *trans*-fused medium ring systems) the conformation is locked (chair and half-chair in this case). The agreement between theoretically predicted and experimentally measured spectra was excellent. The coupling constants $J_{1'-2'}$ for a variety of ribosides and ribotides were measured and the data suggested in most cases a predominance of the C'_2-*exo* conformation although with considerable contributions from other structures.

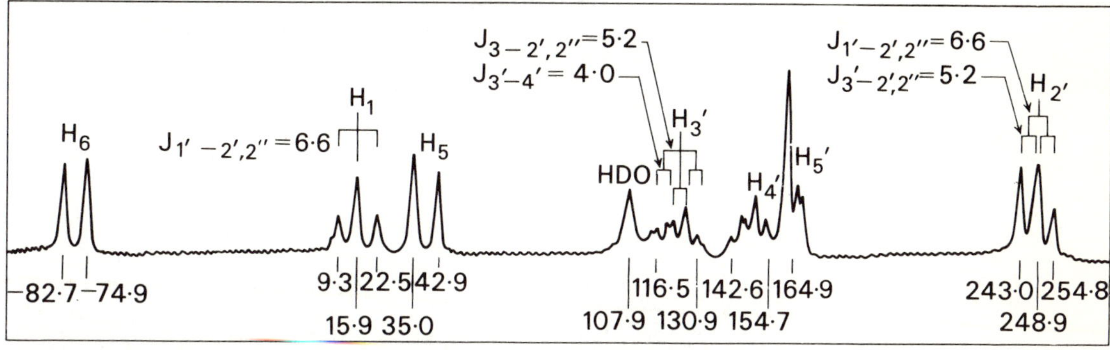

FIG 2.26   PMR spectrum of dUrd (from Jardetzky, 1961).

A more rigorous treatment of the deoxyribosides was possible (Jardetzky, 1961) as it was possible to identify the bands with certainty (Fig. 2.26). The spectrum was analysed on the assumption that the four protons attached to carbon atoms C-1′, C-2′ (two protons) and C-3′ formed an *XABY* system. This is a well-analysed arrangement where *A* and *B* are protons on the same carbon atom. It fitted the assumption that $J_{2'-3'}$ is much greater than the differences in chemical shifts and also greater than $J_{2'-3'}$ or $J_{2'-1'}$. If these assumptions were made and the dihedral angles from various models calculated and inserted in Karplus' equations, the conclusion is that, unlike the ribosides, the deoxyribosides exist predominantly in an envelope in which the O is out of the plane formed by C-1′, C-2′, C-3′ and C-4′ (O-*endo*).

Electron spin resonance

Electron spin resonance (ESR) is a phenomenon analogous to NMR which occurs in radicals (molecules containing one or more unpaired electrons). The spin quantum number (*I*) for an electron is one-half (as with protons). However the resonance is only obtained from ordinary molecules if they are irradiated to yield free radicals. The value of the technique is that it can be employed to determine which atom carries the unpaired electron following irradiation. It is thus a vital probe into the mechanisms of free-radical reactions, such as those that follow the irradiation of nucleosides (see p. 63) which are of such importance in mutagenesis (p. 351).

Spin-spin coupling between the electron and nearby protons in the molecule produces line-splitting. It is the normal practice to record the first (or sometimes second) derivative of the peaks which gives ESR spectra their characteristic appearance (see Fig. 2.27(a)). It is also the practice to scan with the magnetic field and record line-widths in oersteds (Oe). The other figure often quoted in this connexion is the G-value which relates to the efficiency of the irradiation process. G is the number of radicals produced by 100 eV total energy.

Let us consider one particular example, the proof that the free radical produced from thymine has the structure shown (Fig. 2.27(b)). The prediction is that the two protons on C-6 will produce a triplet in the ratio 1:2:1. If we call these two protons a and b and the states of their dipole energy levels + and −, the three bands correspond to the states (a+, b+), (a−, b+ and a+, b−) and (a−, b−). Moreover the three protons on the methyl group should produce a quartet (1:3:3:1). From a knowledge of the field strengths and splittings to be expected the resultant should be an octet (1:3:5:7:7:5:3:1). It is difficult to accurately measure such patterns from ESR spectra (see Fig. 2.27(a)), but by measuring the deuterated compound (Fig. 2.27(c)) in which the three methyl-H atoms are replaced by D, it is possible to predict the simple 1:2:1 triplet. This spectrum was recorded and hence confirmed the structure of the radical (Shields and Gordy, 1959). (There is some peak-broadening but no splitting produced by D).

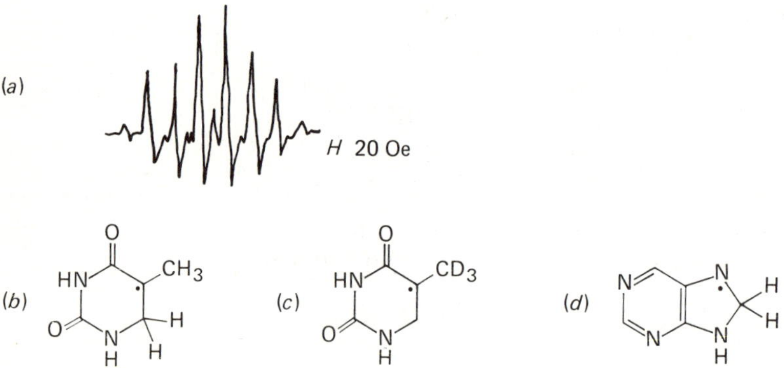

FIG 2.27   Free radicals. (a) ESR spectrum of a single crystal of dThd (from Pruden, Snipes and Gordy, 1965). (b) dThd radical. (c) Deuterated dThd radical. (d) Purine radical.

With irradiated purine simple triplet spectra (1:2:1) are produced confirming the structure of the radical shown in Fig. 2.27(d) (Herak and Gordy, 1965).

The chemical effects of irradiating solutions of molecules may follow the production of solvent radicals (e.g. OH or NH_2) which then attack the purine or pyrimidine. By means of ESR spectroscopy it is possible to establish the structure of the radical so formed. In the case of the two simple radicals mentioned above, the site of attack in purines is C-8 and in uracil it is C-5 (Dertinger and Nicolau, 1970).

Mass spectra

The mass spectrometer is a device for fragmenting molecules to produce charged ions, and to then fractionate and quantitate the fragments on the basis of their mass/charge ratio. As most fragments are single-charged positive ions, it is effectively a method of sorting out the fragments on the basis of mass. The power of the method depends on the fact that with a properly calibrated instrument it is routine to be able to measure the mass numbers of the fragments to within a thousandth of a dalton. It thus provides an extremely accurate technique for elemental analysis and, from the fragmentation pattern, it is possible to make considerable deductions regarding the structure of the original molecule. It has produced the necessary evidence for elucidating the structure of many of the minor nucleosides in tRNA. Examples of this type of structure elucidation are for the prenylamino-purines or cytokinins (see Hecht, 1970 and Armstrong and many others, 1970) and for one of the thiolated pyrimidines (Baczynskyj, Biemann and Hall, 1968).

One of the most interesting structures to be elucidated recently is that of the Y-base. This substituted guanine is a tri-cyclic compound and occurs in $tRNA^{Phe}$ from yeast, wheat germ and rat liver. Its structure has been assigned by a combination of mass spectral and PMR data (Nakanishi, Furutachi, Funamiza, Grunberg and Weinstein, 1970). The riboside is illustrated in Fig. 2.3, p. 39. The free base apparently has a secondary NH at N-7 and imino group at N-9. This is the same tautomer that occurs in purine itself (see below). For a further discussion of the chemistry of Y, see Thiebe, Zachau, Baczynskyj, Biemann and Sonnenbilcher (1971).

X-ray Diffraction

I have already stated in the preface that I must regard X-ray diffraction techniques as largely outside the scope of this book. The principle of X-ray diffraction is simple: any electromagnetic radiation is scattered (diffracted) when it passes through an array of regularly spaced objects (such as lines in a grid, strands in a fabric, etc.) in which the spacings are of the same order of magnitude as the wavelength of the radiation. The spacings between atoms in crystals are of the same order of magnitude as X-rays and hence when a beam of monochromatic X-rays passes through a single crystal it is scattered to produce diffracted beams whose intensities and positions can be measured. The practice is complex: the complexity derives from the fact that there is no direct method of establishing the phasing of the diffracted beams. If such phasings are estimated by a procedure of successive approximations of trial estimates, the relative orientations of the atoms and their spacings can be calculated. Thus provided a substance is available in the form of reasonable-sized single crystals, the molecular parameters (bond lengths and bond angles) can be measured experimentally for the molecules in the crystal. Of course if a molecule has degrees of freedom, the X-ray data only inform us of the conformation that exists in the crystalline state. Spectroscopic methods have to be employed to establish the conformation(s) of molecules in solution. Below are highlighted some of the major results relating to nucleoside structure. For an extensive review with many tables of bond lengths and bond angles see Voet and Rich (1970).

Structures of bases   All the bases are approximately flat molecules except for the saturated molecule dihydrouracil, which has the expected puckered conformation (Rohrer and Sundaralingam, 1968). The others are not absolutely flat (unlike say benzene), apparently because the ring nitrogens are not pure sp^2 hybrids but have some sp^3 character (Macintyre, 1964).

The hydrogens on oxygen and nitrogen atoms can be located and established which tautomer (lactam or lactim; amine or imine) is present in the crystal. In all cases studied (except the unnatural base isocytosine, McConnell, Sharma and Marsh, 1964) stable lactam and amine forms have been found in the crystals. That is not to say that in the free purines, the imidazole N–H is present necessarily on N-9 as is required for the formation of nucleosides. In purine itself, the crystal structure is that of H-bonded chains of molecules with the hydrogen covalently linked to N-9 (see Fig. 2.28, Watson, Sweet and Marsh, 1965).

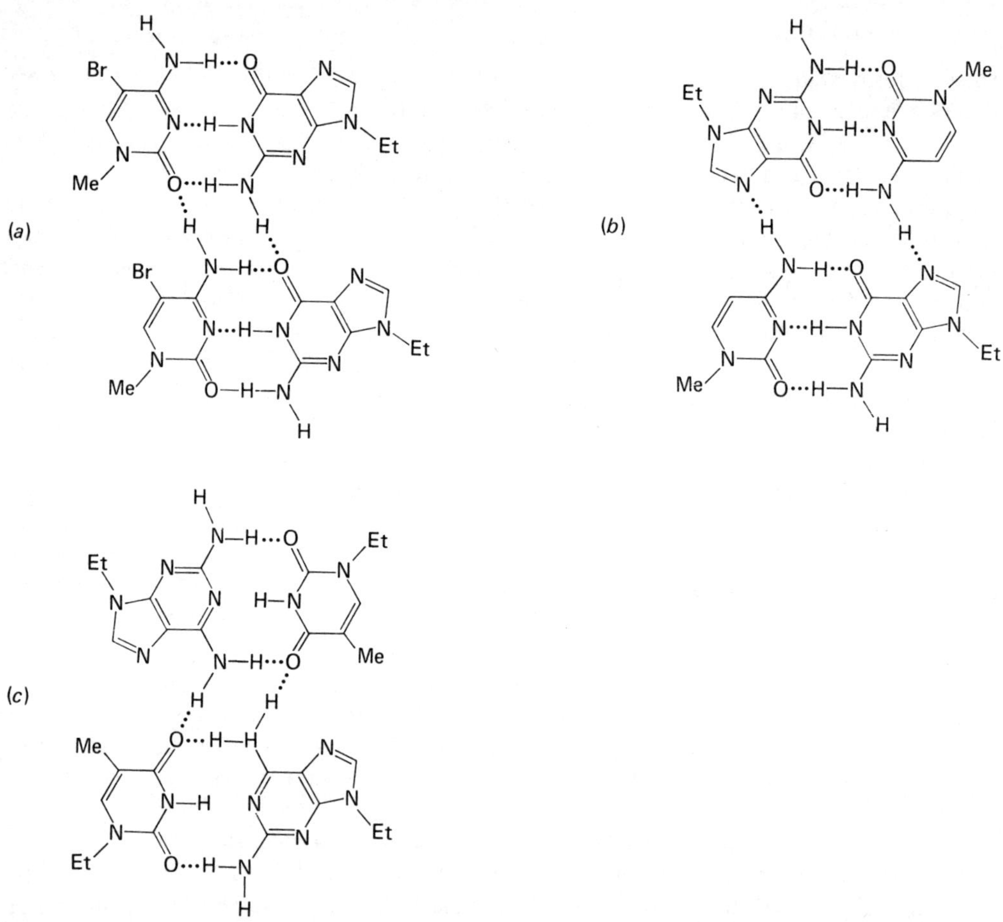

FIG 2.28   Tautomers of purine and hydrogen-bonding in crystals of the free base.

FIG 2.29   Hydrogen bonding in certain mixed crystals. (*a*) 9-Ethylguanine/1-methyl-5-bromocytosine (Sobell, Tomita and Rich, 1964). (*b*) 9-Ethylguanine/1-methylcytosine (O'Brien, 1967). (*c*) 9-Ethyl-8-bromo-2,6-diaminopurine/1-ethylthymine (Simundza, Sakore and Sobell, 1970).

Structures of nucleosides   Data on the conformations of nucleosides in crystals are summarized by Sundaralingam and Jensen (1965). The *anti*-conformation for the arrangement about the glycosyl C–N bond is the norm. Variation is found in the conformation of the ribose ring. For example, in CMP it is C'_2-*endo*, in AMP it is C'_3-*endo* and in dAMP it is C'_2-*endo* (see p. 98).

Association between bases   Co-crystallization of derivatives of Ade and Ura (or Thy) and of Gua and Cyt gives single crystals containing both types of molecule. Such crystals cannot be obtained from 'non Watson–Crick mixtures' (such as Ade and Cyt). At first sight this suggests a simple natural precedent for the relationships between A and T and between G and C. The facts are somewhat more complex. Mixtures of compounds of Ade and Thy in which N-9 in Ade and N-1 in Thy are substituted (i.e. including the nucleosides) co-crystallize with the bases hydrogen-bonded in the most stable, Hoogsteen, configuration (p. 79, Hoogsteen, 1959, 1963). The reversed Hoogsteen configuration is found in the case of complexes involving 5-bromouracil (Matthews and Rich, 1964). With derivatives of Gua and Cyt, only the Watson–Crick configuration is normally produced (Haschemeyer and Sobell, 1965). Again this is the most stable form according to the theoretical calculations (p. 80).

Mention was made above of hydrogen bonding between purine molecules. Other cases of hydrogen-bonded structures involving like bases in non-mixed crystals are reviewed by Pullman, Claverie and Caillet (1966).

Finally there are structures in which hydrogen bonding links bases in configurations other than pairs. Of particular interest (see p. 434) are those in which the bases are arranged in quartets. Three such structures are illustrated in Fig. 2.29. In the crystals, these bonding schemes achieve an infinite array of bases; the grouping of four bases has been illustrated as two of these configurations are discussed in another context later.

Notes to chapter 2

[1]   Teratogen. A substance which, like thalidomide, produces foetal abnormalities when administered to pregnant animals.

[2]   Actinomycetes: members of the order Actinomycetales, mycelial bacteria.

[3]   Ordinary 'amyl alcohol' (at least as sold in the United Kingdom) is in fact largely *iso*amyl alcohol (3-methylbutan-1-ol) and is suitable for this purpose. *n*-Amyl alcohol (not suitable for this purpose) is sold as pentan-1-ol.

[4]   Sephadex is a trade name of Pharmacia (Uppsala, Sweden).

[5]   *N*-(2-fluorenyl) acetamide is often referred to as acetylamino-fluorene ('AAF'). This material is itself a potent carcinogen; the *N*-acetoxy-derivative (in the text) is called 'acetoxy-AAF'.

3   **Isolation**

Introduction

The central technical problem encountered in devising methods of isolating nucleic acids is one of removing protein in such a way that the nucleic acids are not degraded and that one ends up with a pure nucleic acid preparation which can be fractionated. Nucleic acids occur in the cell as nucleoprotein (deoxyribonucleoprotein or DNA variously referred to as chromatin or nuclear material; ribosomes and polysomes sometimes collectively referred to as ribonucleoprotein or RNP). There is one obvious and important exception to this, the tRNA in the 'pool' (not immediately involved in protein synthesis). Viruses are also nucleoprotein particles. The nature of the interactions between protein and nucleic acid in nucleoprotein complexes is still largely unknown. Various models for these complexes are discussed in chapter 16. However all we need to bear in mind in this context is that these interactions are very strong. By utilizing the facts that nucleic acids are soluble in water and can be precipitated by organic solvents, their separation from other types of ellular components is relatively trivial (although polysaccharides can be a bit more of a problem as we shall see later). However, the separation of nucleic acids from protein requires special reagents which will disrupt the nucleoprotein complex and will remove the protein from the system, for if a nucleic acid preparation is contaminated with some residual protein, no amount of precipitation and re-solution, nor in general chromatography, will ever completely eliminate the protein.

We shall only consider in detail chemical methods of deproteinization. Although the use of proteolytic enzymes is a method of achieving partial deproteinization, it is very difficult to remove the residual protein fragments by this method and in any case, proteases are themselves proteins and have to be removed somehow, and we shall see that chemical procedures will satisfactorily deproteinize even whole tissue and the insertion of any stage involving proteolysis is unnecessary.

Before considering the ways in which deproteinization is achieved, let us consider what we are actually trying to make. In the cases of ribosomal and transfer RNA, the aim is to prepare totally undegraded molecules.[1] With ribosomal RNA of high molecular weight, it is important that the molecules are not denatured during the preparation as its renaturation can lead to aggregation. We also aim to obtain a preparation not contaminated with other types of nucleic acid. Similarly with viral RNA, the preparation should yield completely undegraded molecules.[1] As it is now routine practice to prepare undegraded tRNA, rRNA and viral RNA and DNA from a variety of sources, there seems little point in ever preparing degraded material. With mRNA (other than from viruses) the problem is a bit more difficult as in many cases we do not really know what we are looking for. The fractionation and assay of mRNA are discussed in later chapters of this book, but here we may point out that cellular mRNA is a minor component in preparations either of rRNA of high molecular weight or of DNA and clearly no studies on it are worthwhile unless the major fractions of RNA are undegraded. The deproteinization of RNP is not in general very difficult; the art of preparing RNA lies in avoiding the action of ribonucleases. Cellular RNP is almost invariably associated with ribonuclease and it is important that the initial deproteinization is rapid and efficient. It is also important that the final deproteinization is complete, for many ribonucleases are surprisingly resilient enzymes and although they are inactive under the conditions of deproteinization (in which proteins are typically denatured), if there is protein left in the final preparation, the ribonuclease may recover some of its activity with the result that the RNA preparation can be slowly degraded during storage. Apart from the danger of ribonuclease on human fingers it is essential to handle RNA under conditions of the utmost cleanliness. Some biochemists suffer from a ribonuclease neurosis and see the enzyme lurking in every bottle of laboratory reagent. Many of their fears are probably without foundation (it is not after all a particularly difficult enzyme to assay) but their attitude is the right one.

With DNA the problems are rather different. Totally undegraded DNA is obtainable from viruses, from certain bacteria and from organelles (mitochondria and chloroplasts). No one has ever claimed to have a preparation of DNA from the nuclei of eukaryotic cells in which the molecules have not been chopped up to some extent, and consequently we do not know how long (for example) the DNA molecules are in a mammalian chromosome. In general we can say that viral DNA should be intact and cellular DNA as undegraded as possible and should retain its native, double helical structure. From the point of view of preparative methods, deoxyribonuclease is not regarded as such a universal horror as ribonuclease but DNA is exposed to a different hazard, namely being 'sheared'. When solutions of DNA are rendered very turbulent the DNA molecules are fragmented progressively but without losing their secondary structure. I am not aware of any satisfactory theoretical treatment of this effect, but it is presumably due to a cantilever effect during the flexing of a very long and 'stiff' molecule which results in bonds in the phosphodiester bonds of the sugar–phosphate backbone being distorted in such a way that they approach the geometry of the transition state for hydrolysis. The practical consequence of this effect is that methods should be devised for 'soaking' the DNA out of cells without homogenization and solutions of DNA should not be squirted through narrow orifices such as fine pipettes. This latter point is something of a counsel for perfection but should be born in mind if experiments on very high molecular weight DNA are being planned. DNA is also fractured by ultrasonic vibration. We shall see in chapter 9 that this is sometimes a useful technique for isolating specific fragments of DNA, but for preparing DNA it is clear that cells must not be disrupted by ultrasonic methods.

Bearing these general points in mind there are two basic approaches to obtaining nucleic acids. Either we can disrupt the cells (or mash up the tissue) in a system which will effect deproteinization and end up with a 'total nucleic acids' preparation from which we have to sort out the fractions we require, or alternatively we can obtain the sub-cellular fractions and purify them before the deproteinization. For convenience we shall consider methods of deproteinization whole cells first as the problems of deproteinization as such are more difficult in this case and we shall have introduced all the important techniques, before considering the isolation and deproteinization of organelles.

However, we might first consider the question, how do we decide on the appropriate approach for a particular problem? If it is required to obtain one of the major fractions of nucleic acids from a large number of cells or a lot of tissue, there is much to be said in favour of the total deproteinization approach. For example we shall see that it is within the capacity of the average laboratory for anyone to isolate rRNA from several hundred grams of liver or *E. coli* paste, whereas preparing and purifying ribosomes on this scale would be a fairly formidable task. On the small scale there is often a lot to be said for isolating and purifying the organelles first. In particular, the preparation of such minor components as nucleolar RNA or RNA and DNA from mitochondria, usually requires the isolation and purification of the organelles first. The job of fractionating nucleic acids and purifying minor components is a difficult one anyway and identifying the organelle from which such a component was derived is no more than guesswork if the organelles in question have not been thoroughly purified and characterized beforehand.

One final general point. How should purified nucleic acids be stored? DNA is best kept in solution; precipitated DNA tends to denature with time. Some sort of buffer (say 0·1 M sodium acetate) which will also maintain the secondary structure and something to inhibit the growth of micro-organisms (say 0.001 M sodium azide) should be present; 4°C is a suitable storage temperature. RNA of low molecular weight (e.g. tRNA) can be stored as a dried powder (obtained by precipitating RNA from solution with two volumes of alcohol and successive washing with alcohol/water, alcohol and finally ether desiccating). High molecular weight RNA is best kept as a slurry under 75% aqueous alcohol containing 2% sodium acetate at 4°C. The sodium acetate prevents the slight loss of secondary structure which occurs if such precipitates are kept under alcohol or are dried out.

Deproteinization of whole cells

The stages in this process are as follows. (i) Disruption of the cells, (ii) treatment with a solution which will disrupt nucleoprotein and denature proteins, (iii) extraction of the denatured proteins, normally with a suitable solvent and by precipitation, (iv) re-extraction of the nucleic acid fraction to complete the deproteinization, (v) removal of other contaminating materials and (vi) fractionation of the nucleic acids. We whall see in the examples that follow that for soft tissues (such as liver) and certain Gram negative bacteria[2] processes (i) and (ii) can be combined. Also (ii) and (iii) can be linked together by extracting the tissue, etc., with a two-phase system such that the proteins are denatured and extracted in one go. As far as (v) is concerned, the 'contaminating materials' are low molecular weight compounds including reagents used earlier and polysaccharides. Lipids are normally extracted in stage (iii). Low molecular weight compounds can be removed by precipitating the nucleic acids with alcohol and normally are lost during the later processing of the nucleic acids. Various solvent systems which separate nucleic acids from polysaccharide are described in the examples given below and in chapter 4. The question of fractionation (vi) is left until chapter 4 but certain crude fractionations can be achieved during the preparative stages and these are exemplified below.

Before considering a few preparative methods in more detail, we shall briefly consider the ways in which processes (ii) and (iii) may be carried out. The usual reagents for releasing and denaturing the protein during a preparation of nucleic acid are detergents. They are typically used in conjunction with a solvent for denatured protein. Examples of these solvents will be considered later, but if phenol is chosen, it has the effect of achieving deproteinization itself (phenol and water are incompletely soluble in one another so the aqueous layer will contain some phenol). Neither detergents, nor phenol, nor a mixture of both will satisfactorily deproteinize nuclei at room temperature. Therefore for obtaining the nucleic acids from the nuclei of eukaryotic cells, it is necessary to make the conditions for deproteinization more drastic. There are basically two approaches to this; the extraction can be performed at an elevated temperature or alternatively a strong chelating agent can be included in the mixture (both RNP and DNP contain divalent metal ions which contribute to the stability of the nucleoprotein complex). I have delayed a discussion of the 'hot phenol' methods until chapter

4, but we may observe here that there is one possible drawback to this method, namely that there is a danger of the denaturation and consequential renaturation which could produce artefacts during the extraction. Also we know that, at least in the case of DNA, the presence of phenol accelerates the rate of hydrolysis in boiling water (Kidson and Kirby, 1963) and some breakdown might conceivably occur under the conditions of the extraction.

Bacterial DNA

Isolation of DNA by the method of Marmur I have chosen this procedure (Marmur, 1961) as the first example for detailed consideration as it was very carefully thought out and illustrates the way to approach many of the problems of nucleic acid isolation. Moreover it remains one of the most versatile methods of isolating DNA of reasonably high molecular weight. I have included in this, and in subsequent examples, an indication of quantities in a typical preparation, partly to give a more accurate impression of the experiments but also to include enough detail for the reader to have a go for himself without having to resort to the literature.

In this procedure the cells are lysed enzymically or with sodium dodecyl sulphate (SDS, sometimes called sodium lauryl sulphate). Chloroform is used as a surface denaturant and to remove the denatured protein. The second deproteinization is achieved with the chelating agent, sodium citrate. The following recipe (Marmur, 1967) assumes that 2 or 3 grams of packed bacterial cells (obtained by harvesting the culture in a centrifuge) are to be processed.

The cells are washed once in 0·15 M NaCl and 0·1 M EDTA at pH 8 and are then resuspended in about 30 ml of this medium. The EDTA (ethylenediaminetetracetic acid) inhibits deoxyribonuclease. Lysozyme (about 10 mg of the commercial enzyme from hen's egg white) is added and the mixture is incubated at 37°C until the cells lyse (increase in viscosity). Now SDS is added to a final concentration of about 2% and the mixture is incubated at 60°C for 10 min. Note the importance of the relatively high ionic strength of the buffer in maintaining the secondary structure of the DNA during this phase. In the case of Gram negative bacteria (such as *E. coli*, *Salmonella typhimurium* and certain micro-organisms other than bacteria, e.g. *Euglena*, *Chlamydomonas* and *Dictyostelium*) and for some Gram positive bacteria (e.g. *Diplococcus pneumoniae* and certain *Bacilli* and *Clostridia*) SDS alone will lyse the cells and the lysozyme stage is omitted. However with certain other Gram positive bacteria (including *Bacillus subtilis* whose DNA is extensively used in studies on bacterial transformation) the lysozyme stage is essential. After the mixture is cooled (following the treatment with SDS) it is shaken for 20 min with an equal volume of chloroform/*iso*amyl alcohol (24:1); the alcohol is present to prevent frothing and to enable the resulting emulsion to separate more readily. This emulsion is broken up (it produces three phases) by centrifugation in a refrigerated centrifuge (about 5 min at 10 000 r.p.m. at 5°C would by typical conditions in a medium speed centrifuge of the sort found in the average laboratory). The nucleic acids are precipitated from the top layer with two columes of ethanol. As this precipitate contains a lot of DNA it looks like whisps of soggy cotton wool which can be fished out with a glass rod. The precipitate is dissolved in 0·015 M NaCl, 0·0015 M trisodium citrate and extracted repeatedly (by shaking for 10 min and centrifuging) with the choloroform/*iso*amyl alcohol until no more protein can be seen floating at the interface. The nucleic acids, contaminated with capsular polysaccharide, can now be precipitated from the upper layer with two volumes of ethanol. We shall delay discussion of the ways of separating RNA from this type of mixture until the next chapter; however to finish the preparation of DNA, the precipitate is redissolved in the salt citrate buffer, incubated with ribonuclease (say about 50 μg/ml using the bovine pancreatic enzyme p.

206, and about 37°C and half an hour) and again deproteinized with the chloroform/*iso*amyl alcohol mixture. Once again the nucleic acid (now DNA) is recovered by precipitation with alcohol, redissolved in salt/citrate buffer and $\frac{1}{9}$th of a volume of 3 M sodium acetate pH 7·0 containing 0·001 M EDTA is added. *Iso*propanol is added until the DNA just precipitates (about 0·54 volume). Under these conditions the capsular polysaccharide material and any RNA fragments still remaining are not precipitated. The yield of DNA is typically around 50% of the theoretical value (1 g of bacteria wet weight contains of the order of 5 mg DNA). In later chapters we shall discuss methods of characterizing DNA, but we may note here that this preparation yields DNA of molecular weight in excess of 10^7 and containing less than 0·5% protein.

Bacterial spores Bacterial spores[3] are resistant to the action of lysozyme. They can however be rendered sensitive to the enzyme by pre-treatment with very strong reducing agents in the presence of urea (e.g. 50% thioglycollic acid and 8 M urea). Using such reagents it is possible to obtain spore DNA by a modification of the above procedure (Sakakibara, Tanooka and Terano, 1970).

Intact Bacterial chromosomes In order to isolate bacterial DNA in completely undegraded form, it is essential to reduce the handling procedures to a minimum and to lyse the cells by a very gentle method. In experiments by J. Cairns, which are discussed in chapter 13, radioactive DNA was isolated from *E. coli* in a completely undegraded state and autoradiographed. For this purpose, it was essential that the molecules should remain intact and should be deposited on a membrane for the autoradiography. It is less important that the DNA should be totally free of protein, for the DNA had been labelled with a radioactive precursor specific for DNA. The method used by Cairns (1963) was as follows. Labelled bacteria were suspended in 'lysis medium' to a concentration of 10^4 cells per ml. Lysis medium contained 1·5 M sucrose, 0·05 M NaCl, 0·01 M EDTA, 0·01 M KCN and non-radioactive DNA isolated separately (4·7 g/ml). The added DNA was not necessarily of very high molecular weight; its purpose will become apparent shortly. The cyanide is present to stop further replication of DNA in the cells. The suspension was placed in a little cylindrical vessel (20 mm diameter and 2·5 mm deep) with one face consisting of a dialysis membrane in the form of a Millipore VM filter. It was dialysed at 42°C for 45 min versus 1% SDS dissolved in more lysis medium. The combined effect of the SDS and EDTA gently lysed the cells. The osmotic pressure inside the cells was counteracted by the sucrose outside so that at the point of lysis the cells did not 'explode' and create turbulence in which DNA might shear. The SDS was removed by dialysis versus lysis medium and the dialysis vessel was then dialysed against medium in which the NaCl concentration was 4 M and then subsequently against medium in which it was 0·04 M. The strong salt dissociates the DNP; when the ionic strength is lowered again, DNA and protein derived from DNP will tend to reassociate. However the lysis medium contains non-radioactive exogenous DNA in vast excess so that the 'average' DNA molecule (including of course the bulk of the radioactive chromosomes) is more-or-less free of protein. The exposure to 4 M salt lasted 1 hour; the subsequent exposure to 0·04 M salt for 16 h, after which the little cylindrical cell was stood on its side and punctured so that the contents slowly drained out on to a piece of filter paper, leaving a thin film of the contents (including the radioactive chromosomes) to dry on the dialysis membrane.

Another approach to the isolation of undegraded DNA is to use as a strong salt to dissociate nucleoprotein the components of a gradient for buoyant density centrifugation so that the DNA will be released and will 'band'. Alternatively the lysis can be achieved in a layer on top of a pre-formed gradient used for rate-zonal centrifugation. These techniques are discussed on pp. 147 and 150.

Phenol methods

There are numerous recipes in the literature in which phenol is used for isolating nucleic acids. Like chloroform, it acts as a deproteinizing reagent but the reaction occurs in aqueous solution (phenol partitions into the aqueous phase to some extent); furthermore the phenol phase (containing some water) is a solvent for denatured protein. I shall describe here two phenol recipes for obtaining the nucleic acids from rat liver as they illustrate the two types of deproteinization one can do with phenol. If the phenol is used in conjunction with an aqueous solution of a simple salt (present to inhibit denaturation of nucleic acids) the nuclei are not deproteinized but are merely precipitated. The cytoplasm however is deproteinized and cytoplasmic RNA can be made in this way. On the other hand if the phenol is used in conjunction with a strong deproteinizing agent (here we consider a strong chelating agent but hot detergent is another alternative) the whole tissue is deproteinized to yield total nucleic acids.

These two methods are due to K. S. Kirby. Obviously they are not the only approaches to the problem but they do exemplify the precautions which are to be considered during this type of preparation and can be applied to a wide variety of different tissues. Throughout these methods, emulsions of phenol and water have to be separated in the centrifuge. The conditions described on p. 107 for the separation of the chloroform/water emulsion would normally work. The emulsions are less easy to separate in the early stages of the deproteinization as the presence of protein renders the mixture more intractable. In the following procedures, 'phenol mixture' refers to the liquid which is produced if 70 ml m-cresol, 55 ml water and 0·5 g 8-hydroxyquinoline are added to 500 g phenol.[4] The cresol is best distilled to remove the o-cresol which tends to contaminate commercial samples of m-cresol. The ortho isomer has a lower boiling point so that if the first 10% of distillate is discarded and the bulk will be more or less pure m-cresol. (o-Cresol is oxidized with time to produce dark coloured tars.) It is more convenient to start off with the phenol liquefied and the cresol is present as an anti-freeze (the early stages are performed in the cold). The hydroxyquinoline is present to inhibit oxidation of the phenols and to remove di-valent cations as phenol-soluble complexes.

Partial deproteinization   In the first method (Parish and Kirby, 1967) the tissue is removed from the animal and immediately dropped into about 10 volumes of ice-cold 0·5% sodium naphthalene-1,5-disulphonate and about 8 volumes of phenol mixture and homogenized (Waring blendor or equivalent). The particular salt is an effective inhibitor of ribonuclease. The mixture is shaken or stirred at room temperature for about half an hour and the phases are then separated by centrifugation at 0–5°C. Note that the deproteinization occurs most efficiently at room temperature as the phenol is then more soluble in water but the phases separate more readily in the centrifuge if the mutual solubilities of phenol and water are reduced. After the centrifugation there is an upper aqueous layer (slightly cloudy on account of the glycogen) which contains cytoplasmic RNA, a lower phenolic phase which will be almost black (due to iron from haem) and some stuff that looks rather like a thin layer of porridge partly floating between the two liquid phases and partly on the bottom of the centrifuge tube. This material is variously referred to in the literature as 'interface', 'interphase' and 'phenolic nuclei'. We shall leave the interface for the moment. The upper phase is removed and NaCl (3%) and sodium tri-*iso*propylnaphthalene sulphonate (5%) are dissolved in it. The salt is to reduce the solubility of phenol in water and to precipitate some proteins solubilized during the first stage. The tri-isopropylnaphthalene sulphonate is a detergent and a powerful inhibitor of ribonuclease. This mixture is now re-extracted with phenol mixture (half a volume) and the phases are separated as before. The upper phase is again removed (very carefully this time to avoid contamination with any phenol or interface). The tRNA and rRNA are in this phase together with some glycogen. We shall consider the removal glycogen on p. 129.

Total deproteinization   The total deproteinization method is exemplified by the following procedure (Kirby and Cook, 1967). The tissue is homogenized in a solution containing a strong chelating agent (sodium 4-aminosalycilate, 6%), detergent (sodium tri-*iso*propylnaphthalene sulphonate, 1%) and NaCl (1%). The detergent is salted out of solution by the other constituents and therefore this solution should also contain a solvent (either 6% *sec*-butanol or 3% phenol mixture are suitable). Salycilate is a powerful chelating agent but sodium salycilate emulsifies phenol and water to produce a colloidal liquid which, in the presence of protein is almost impossible to separate. The 4-amino derivative produces a mixture which can be separated without too much difficulty. The tissue is homogenized in the solution above (about 10 volumes for liver, more for tissue rich in DNA, e.g. spleen, tumours, etc.) at 0°C. The gentlest method of homogenization that is possible should be employed (such as a Potter or Dounce homogenizer). The homogenate becomes extremely viscous owing to the release of DNA. It is shaken with an equal volume of phenol mixture at room temperature (about half an hour) and the phases are separated by centrifugation. The top layer is removed (it is easier to pour it off than to try and pipette it). An extra 2% NaCl is now added and the mixture is re-extracted with half a volume of phenol mixture and again the phases are separated and the top phase is removed very carefully. This upper phase contains all the cellular nucleic acids and some glycogen.

If the interface obtained during the first of these procedures is put through the second recipe (it does not need homogenizing—it is merely gently shaken with the deproteinizing solution), nuclear nucleic acids can be obtained from it (Parish and Kirby, 1967).

Removal of polysaccharides   We shall consider the fractionation of these preparations in chapter 4 and we shall see then that the final phase of the preparation (removal of glycogen) is normally incorporated into these fractionations. However we may note here that there basically are three approaches to this problem. Certain organic solvents (e.g. *iso*propanol—see p. 108) may be used to precipitate nucleic acids of high molecular weight preferentially. Kirby's suggestion was to use *m*-cresol (10% in 20% sodium benzoate) which precipitates DNA and high molecular weight RNA and leaves glycogen in solution. A second approach is to use strong sodium acetate (3 M, pH about 7) in which glycogen is soluble (so are DNA and low molecular weight RNA—see p. 000). The third suggestion of Kirby is of universal application (Kirby, 1956), provided the nucleic acids (DNA and low-molecular-weight RNA) are not insoluble in strong salt, but does require a little care. The polysaccharide/nucleic acid mixture is in a suitable buffer and to it is added an equal volume of 2·5 M potassium phosphate pH 7. Another equal volume of 2-methoxyethanol ('methyl cellosolve') is added (i.e. we have x ml of nucleic acid, x ml of phosphate and x ml of methyl cellosolve) and the mixture is shaken up and immediately separated in the centrifuge. There are two liquid phases, the upper one much larger than the lower. The nucleic acids are in the upper layer which is poured off. The polysaccharide forms a precipitate floating at the interface. Nucleic acids are precipitated (ethanol equal in volume to the top phase) together with a lot of potassium phosphate which can be removed by dissolving the precipitate in the minimum volume of water and subsequent dialysis. The thing that has to be watched in this procedure is that if it is employed for DNA, the temperature at which denaturation will occur is lowered by the methoxyethanol and moreover all the cellosolves get warm when mixed with water. Therefore the two aqueous solutions should be ice-cold at the start and the methoxyethanol should be pre-cooled to about −20°C. In the case of rat liver preparations there is a fourth method of removing the bulk of the glycogen, namely starving the rats for the night before their execution.

General application of phenol methods   The phenol method can be successfully applied to preparations of nucleic acids from bacteria. With *E. coli* (and presumably most other Gram negative organisms) the second (aminosalicylate) method works well if the bacteria are simply shaken up with about 15 volumes of the deproteinizing solution and the same amount of phenol mixture (Kirby, Fox-Carter and Guest, 1967). With some organisms slight modifications may be necessary. We successfully obtained the nucleic acids from *Serratia marcescens* by stirring the cells up with a paste of tri-isopropylnaphthalene sulphonate, aminosalicylate and a little water and phenol and then adding the rest of the water, salt and phenol mixture to bring it up to about 15 volumes (unpublished work). With *Bacillus subtilis, Clostridium putrificum,* an actinomycete (*Streptomyces chrysomallus*) and presumably many other Gram positive organisms the phenol method works well provided it is preceded by treatment with lysozyme (Kirby, 1968). Attempts at isolating RNA from bacteria and 'leaving the DNA behind' (as in the naphthalene disulphonate method) are not usually very successful. However with *E. coli*, a good starting point for the isolation of RNA is to soak the cells in the aminosalicylate/detergent/salt mixture with only a third of its volume of phenol for $\frac{3}{4}$ of an hour without shaking. If the phases are separated and the upper phase used for the second deproteinization not much DNA will be present in the final preparation (Parish, 1968). This makes the 'salting out' procedure (p. 127) easier to handle.

Provided the tissue can be efficiently disrupted under conditions of deproteinization, the recipes discussed so far are applicable to most of the types of tissue likely to be encountered. I should imagine that anyone interested in nucleic acids from a particular source will be acquainted with the relevant literature. I hope that the above accounts of versions of certain 'classical' procedures may give you a few ideas and introduce suggestions about improving techniques. However, if you are interested in taking the matter further the paper by Loening (1968a) to which we shall be referring later (p. 395) contains descriptions and references to applications of phenol methods to a wide variety of animals, protozoa, algae, fungi, ferns, and higher plants.

It is not necessary to use a two-phase system for phenol-deproteinization. An extremely ingenious density gradient method has been devised by J. R. B. Hastings. A cell lysate is centrifuged through layers of aminosalicylate containing dissolved phenol and the nucleic acids sediment into a gradient suitable for separating DNA and different RNA species. Details are given in the next chapter (p. 146).

Residual protein   These phenol techniques yield DNA preparations which contain 0·01−0·05% protein. Protein in RNA prepared by these methods is undetectable. In chapter 5 we shall consider methods of identifying protein in nucleic acid preparations but we may note here that a very sensitive method is to hydrolyse the preparation with 0·5 M HCl under conditions typical for protein hydrolysis and partially fractionate the amino acids by paper chromatography and develop the spots with ninhydrin. The glycine spot must be cut out and discarded (some glycine is produced by the hydrolysis of adenine) and the remainder are assayed by spectrophotometry of the cadmium complexes (Parish and Kirby, 1967).

Diethylpyrocarbonate

The isolation of undegraded RNA from leaves and setiolated stems of higher plants is allegedly especially difficult. According to Solymosy, Fedorcsák, Gulyas, Farkas and Ehrenberg (1968), extractions with SDS and phenol is unsatisfactory. They do not report results with aminosalicylate but recommend the use of the extremely powerful protein-denaturant and precipitant, diethylpyrocarbonate. In their procedure, 2 g of leaves are ground up with 6 ml of 50 mM tris HCl pH 7·6, 5 mM $MgCl_2$ 1% SDS and 0·2 ml of diethylpyrocarbonate at 4°C. After incubation at 37°C for 5 min, the whole lot is ground up again and centrifuged (8000 g—see p. 131—for 15 min at room temperature). NaCl (0·6 g) is ground into the solution and after incubation (37°C, 5 min) is centrifuged (10 000 g for 20 min at 4°C). Nucleic acids are precipitated (ethanol), redissolved and dialysed against 0·4 M NaCl in dilute phosphate buffer pH 7·4. A precipitate is formed, which is removed by centrifugation and the supernatant (which contains RNA and DNA) is recovered. There are many queries raised by this method. For example, what are the Mg^{2+} ions for? What is the significance in the different temperatures for centrifugation? and what is the final precipitate that is discarded?

The attraction of diethylpyrocarbonate is said to be that it is a powerful ribonuclease-inhibitor. Indeed it is—it is an everything—inhibitor, it even reacts with RNA! It is possible however that the A-residues (Ado is the nucleoside with which it reacts—p. 70) in double-strand sections of RNA are relatively inert to the reagent. Oberg (1970) has shown that single-stranded poliovirus RNA loses its infectivity after treatment with diethylpyrocarbonate but the double-stranded replicative intermediate survives.

Nevertheless this is a fashionable reagent and it has one very useful application in sterilizing solutions (p. 137).

Isolation of organelles and nucleoprotein

The problems involved in the isolation of organelles and tissue-fractionation are central to modern experimental biochemistry. The aim is to disrupt the cells in such a way that the organelle one is interested in is not lysed nor ruptured nor mechanically damaged in any way and then to separate the organelles normally by differential centrifugation. As we shall see later, certain nucleoprotein fractions must be separated by means of zonal ultracentrifugation and we shall discuss this technique in the next chapter. It is outside the scope of this book to consider tissue fractionation in any detail. For more detailed discussions of the aspects particularly relevant to nucleoprotein and nucleic acids refer to the volumes of reviews edited by Campbell and Sargent (1967, 1969); but here we shall consider some procedures which exemplify most of the important techniques.

Nuclei and nucleoli

At first sight, the isolation of nuclei would appear to be a rather simple procedure. In eukaryotic cells, the nucleus is by far the largest organelle and its separation from other organelles by differential centrifugation is trivially easy. In practice good preparations of nuclei require very great care. The major problem is that the nuclear membrane is fragile, and during isolation procedures there is a tendency for material (including enzymes) to be lost from the nuclei. There are basically three approaches to the solution of the problem. One is to homogenize the tissue in a medium containing divalent cations which appear to toughen up the nuclear membrane and also sucrose to prevent damage to the nuclei through osmotic shock. A second approach is to use diluted citric acid as a medium for the extraction. Citric acid is very effective at removing cytoplasm from nuclei and very gentle homogenization can be employed. The third method is to use organic solvents, which are totally immiscible with water, for the homogenization. Here the theory is that no exchange across the nuclear membrane is possible as the water-soluble components are trapped in the nucleus as soon as the disruption has occurred. We shall briefly consider procedures for obtaining mammalian nuclei representative of each of these approaches. The important thing to recognize is that no preparation of nuclei is perfect, in the sense that it does not yield a product which looks like normal nuclei under the microscope *and* has all the metabolic activities expected of nuclei and none of those expected of cytoplasm. The sucrose-divalent cation procedure produces very nice looking nuclei but they have lost a certain amount of DNA-polymerase and have picked up some cytoplasmic enzymes (notably alkaline phosphatase). Citric acid nuclei look a bit the worse for wear. The chromatin appears to be aggregated and under the phase contrast microscope the nucleoli appear less refractile. However according to Higashi, Shankar Narayan, Adams and Busch (1966), who certainly ought to know, there is less autolysis of nuclear RNA of very high molecular weight during the citric acid procedure. The non-aqueous nuclei probably represent the cleanest preparations from the point of view of their integrity and lack of cytoplasmic contamination but morphologically they are very distorted. Incidentally it must be emphasized that no one should try and make nuclei seriously unless he is used to either staining and observing cells and fragments of cells on slides or is used to phase contrast microscopy of tissues. The essential criterion of a nuclear preparation is that it should at least look right (intact nuclei, no cells and no cytoplasm).

An example of the sucrose procedure is the system of Chauveau, Moulé and Rouillet (1956) with technical recommendations of Busch (1967). The tissue is minced and suspended in ten volumes of ice-cold 3·3 mM calcium salt (acetate or chloride) and 60% w/v sucrose. The sucrose concentration needs to be right and is checked with a direct-reading refractometer (see p. 136). The homogenization is performed using a homogenizer of the type in which a pestle is either pushed up and down or is rotated in a tube. Nowadays Teflon-in-glass homogenizers are very popular. The conditions for the homogenization of any given tissue must be perfected by doing a number of trial runs and examining the homogenate under the microscope. To give some impression of the conditions, for rat liver a homogenizer with a clearance of 0·006 in driven by a motor so that the pestle is rotating 12 000 rpm is suitable. For large scale homogenizations a continuous system has been described (Busch and Desjardins, 1965). The homogenate is centrifuged at 40 000 *g* for 60 min at 5°C. Only the nuclei will penetrate the heavy sucrose. The nuclei can be further purified by centrifuging. For example the nuclear pellets can be dispersed in more dilute sucrose which can be layered over 60% sucrose and the centrifugation can be repeated. For large-scale preparations, the centrifugations are performed in continuous-flow centrifuges (Busch and Desjardins, 1965).

The citric acid method is due to Dounce (1955). Following Busch's advice (1967) the tissue is homogenized as above but using as medium 1·5% citric acid. The homogenate is centrifuged at 600 *g* for 10 min. The pellet is suspended (it requires further gentle homogenization) in 0·25 M sucrose and layered over 0·88 M sucrose (both containing 1·5% citric acid) and is centrifuged for 10 min at 900 *g*.

The non-aqueous method is due to Allfrey, Stern, Mirsky and Saetren (1951). The tissue is frozen (liquid air) and broken up roughly (I mean roughly too—like smashing it up with a hammer) and lyophilized in a freeze drier. The powder is dispersed in light petroleum (boiling range around 60°C, 4·5 volumes related to the dry weight) and is then homogenized for about two days in a jar-rolling mill. The homogenate is now filtered through a 0·5 mm mesh and centrifuged for 20 min at about 1000 g. The pellet is dispersed in a mixture of cyclohexane and carbon tetrachloride with a specific gravity of around 1·30 (detailed suggestions for different tissues are given in the paper) and is centrifuged at 2000 g for 40 min. The nuclei sediment and form a pellet and the cytoplasmic material floats. This and all subsequent centrifugations are performed at 2–4°C. There follows a complicated protocol of washing with cyclohexane and chloroform mixtures of varying specific gravity to hit on conditions in which the nuclei are just sedimenting and the cytoplasmic material remains in suspension. The exact conditions vary from tissue to tissue; to give some impression of the technique, for the nuclei of chick erythrocytes the centrifugations are at 2000 g and for 60 min with mixtures of density 1·310, 1·320 and 1·390. In the last of these experiments the nuclei remain in the supernatant and cells are removed as the pellet. The specific gravity is again reduced so that the nuclei are pelleted by a subsequent centrifugation and then the whole procedure is repeated with solvent mixtures closer together in density (1·327, 1·333 and 1·338).

With plant nuclei much the same principles apply. The calcium/sucrose method appears to be the most popular. For example in the procedure for nuclei of pea seedlings due to Rho and Chiphase (1962) the tissue is mashed in 0·003 M tris HCl pH 7·2, 0·003 M CaCl$_2$ and passed through a remarkable machine called a pea-popper, in which a set of knife blades cut it into 1 mm sections which are then ruptured by being squeezed past spring-loaded rollers. The mush is washed off the rollers with a solution of 6 mM tris HCl pH 7·2, 3 mM CaCl$_2$ and 0·2 M sucrose. This suspension is squeezed through fine fabric to filter off the bulk of the unbroken tissue and whole cells and then centrifuged (350 g for 10 min at 2°C). The contents of the centrifuge tube consist of a clear supernatant which is discarded and two layers of sediment. The bottom layer is starch. The upper part of this sediment contains the nuclei and is carefully removed and layered over 1·2 M sucrose and centrifuged at 450 g for 10 min. The nuclei form a pellet. This recipe only yields about 5% of the theoretical (based on the recovery of DNA) whereas up to 50% yields are possible with animal cells. The loss presumably occurs during the separation of nuclei from starch granules.

Nucleoli   Nucleoli are sub-nuclear organelles. Unlike nuclei themselves, nucleoli appear not to have a membrane but are held together by a protein matrix. The principle behind the isolation of nucleoli is to disrupt nuclei and fractionate the nucleoli by differential centrifugation. Superb mammalian nucleolar preparations have been achieved by Busch's group. In the method of Ro and Busch (1964), nuclei (isolated by the calcium/sucrose method) are suspended in 0·34 M sucrose and disrupted at 0°C by ultrasonics. With the Raytheon sonic oscillator, 1 amp for 15–30 s is apparently about right, but it is essential to keep removing a drop, smearing it on a microscope slide and staining with a stain which will dye nucleoli and other nucleoprotein particles, to ensure that the sonication is stopped when the nucleoli are separated from other nuclear material (0·1% Azure C in 0·25 M sucrose is the stain of choice). As an alternative to sonication, the nuclei can be disrupted in the French pressure cell (Somes, Cole and Hsu, 1963). The disrupted nuclei (in 0·34 M sucrose) are layered over an equal volume of 0·88 M sucrose and centrifuged at 2000 g for 20 min. The nucleoli sediment and form a pellet, which is resuspended in 0·34 M sucrose and the centrifugation (over heavier sucrose) is repeated.

Pea nucleoli have been obtained by a similar procedure (Johnson, Setterfield and Stern, 1959). In this case it is apparently essential to remove the calcium before disruption of the nuclei. The presence or absence of calcium during mammalian nucleolar preparations is of no account although it is essential that it should be present for the original preparation of nuclei subsequently used for this purpose.

Mitochondria and chloroplasts

Great interest attaches to the nucleic acids of these organelles. The mere fact of their presence is intriguing in itself, as it makes mitochondria and chloroplasts look like little cells which may at some point have had a life of their own. Moreover there are differences between the nucleic acids of mitochondria and chloroplasts and those of the nucleus and cytoplasm of the cell. We shall be discussing these differences in subsequent chapters. However I regard it as outside my scope to consider the isolation of these organelles in any detail. Without detailed references I shall just sketch in the principles of their isolation for completeness. However an excellent summary of the techniques involved in isolating mitochondria is by Chappell and Hansford (1969).

Bacteria and other prokaryotic organisms (blue-green algae and related forms) do not have either mitochondria or chloroplasts.[5] The principle behind the isolation of these particles from eukaryotic organisms is that the tissues or cells are homogenized in a medium in which the particles are not unduly damaged and the organelles are isolated by differential centrifugation and purified by isopycnic centrifugation. This phrase refers to the process of sedimenting the particles through a gradient of sucrose (see p. 134 for a description of the way in which such gradients may be established) which is of such density that at some point the particles will reach a medium of their own density. At this point they will of course stop. For the purpose of isolating nucleoprotein or nucleic acids from mitochondria or chloroplasts it is essential that the particles are really free from contaminating cytoplasm or nuclear 'debris' and it is especially important that the procedures are performed under sterile conditions to avoid contamination with bacteria which are about the same size, and (unlike mitochondria and chloroplasts) are bursting with nucleic acids.

Mitochondria   To isolate mitochondria, the tissue is homogenized in sucrose (typically 0·44 M) containing EDTA (1 mM, pH 7·6). For mammalian tissue the homogenization is usually performed on a minced preparation using a plunger-type homogenizer with a loose-fitting barrel. With yeast, the most popular method is to use a Nossal shaker in which the cells are shaken very rapidly with small glass beads. Filamentous tissue (e.g. fungal mycelium) and plant tissue are normally ground up with sand. Preliminary centrifugation (say 500 g for 10 min) removes nuclei and debris and unbroken cells and the supernatant is then centrifuged at 15 000 g for 20 min to sediment the mitochondria. The isopycnic centrifugation is performed in swinging bucket rotors of an ultracentrifuge (see p. 135). Typical conditions would be to suspend the mitochondrial pellet in 0·5 M sucrose (EDTA as before) and sediment through a gradient of 0·98–1·66 M sucrose at 64 000 g for 5 h. The mitochondria form a band at their own density (about 1·18) and the gradient can be analysed (see p. 135) and the mitochondrial fraction is recovered. It is normal to identify the mitochondrial band by assaying fractions from the gradient for some typically mitochondrial enzyme (such as succinate–cytochrome c reductase).

Chloroplasts   With chloroplasts much the same principles apply. However in this case the most successful preparations appear to result from the use of magnesium buffers in the presence of a thiol (such as β-mercaptoethanol). Chloroplasts can be successfully purified by centrifugal flotation in heavy sucrose as they are the least dense of the cell organelles. For their work on nucleic acids of chloroplasts, Eisenstadt and Brawerman (1964) used the Dinoflagellate, *Euglena gracilis*. In this case the cells were washed with water and suspended in four volumes of 10% sucrose, 10 M tris HCl pH 7·6, 4 mM $MgCl_2$, 1 mM β-mercaptoethanol and disrupted in the French press. The homogenate was centrifuged at 500 g for 10 min. The pellet contained chloroplasts and much debris. It was dispersed in the original buffer (5 x original volume). After a time a membranous precipitate appeared which was removed by filtration through fabric. The chloroplasts were concentrated (centrifugation and re-suspension) and mixed with two volumes of 75% sucrose. Centrifugation at 23 000 g for 30 min yielded the chloroplasts as a top layer. The flotation was repeated to complete the purification.

Deoxyribonucleoprotein

In this section we are considering attempts to isolate in a chemical pure form the whole of the apparatus concerned with the direction and control of the transcription process. We have already seen that in bacteria we know that the control of transcription is effected by repressors and the process of derepression (p. 19) and that the DNA molecule directs the control of its own transcription. This system has not been directly demonstrated in eukaryotic cells although one would guess that it operates to some extent. What we do know about these cells of higher organisms is that the DNA is only one component of a nucleoprotein complex. The current view is that the protein components of this material is responsible for control of transcription. Perhaps it operates as a sort of coarse control of transcription and that some kind of system of the Jacob and Monod system operates as a fine control of protein synthesis. The coarse control is envisaged (see p. 31) as a 'masking' of sections of the DNA. In fact it is possible that bacteria have such a system as well, as in these organisms too there is a nucleoprotein complex. However here we consider the deoxyribonucleoprotein (DNP) of higher organisms as in some ways we know what we are looking for rather more precisely.

Metaphase chromosomes   Higher organisms contain a number of chromosomes. The chromosomes are made of DNP and people who are fairly confident that their DNP is the stuff that chromosomes are made of refer to it as chromatin. In a diploid cell the chromosomes segregate during mitosis and can be stained and visualized under the microscope when the cells are in the phase of mitosis called metaphase. As we can see the chromosomes during metaphase, the most direct approach to isolating pure chromatin is to try and make metaphase chromosomes in a pure form. The most elegant method is simply to dissect them out of the nucleus. I say 'simply', in fact it is a task that the average biochemist, in his world of tissue-mincers and centrifuges, would not even contemplate. However brilliant and painstaking work by Keyl (1964) has resulted in the isolation of pure and separate metaphase chromosomes from the salivary glands of insects. This particular tissue contains giant (polytene) chromosomes which contain bundles of DNA molecules. Although the yields which can be achieved are minute in chemical terms, very sophisticated ultra-microscale techniques can be applied to study the biochemical properties of these chromosomes (p. 182). For large scale preparations of chromosomes the first requirement is to obtain a large population of cells with a relatively high proportion in metaphase. This can be achieved with cells in tissue culture, for if they are incubated with colchicine (or a derivative) mitosis is arrested in metaphase. Apparently colchicine inactivates or destroys the spindle fibres responsible for the movement of the chromosomes necessary to achieve the next stage (anaphase). In a population of mammalian cells in tissue culture over 5% can be arrested in metaphase by this process. A higher yield can be obtained if the cells are first of all in synchrony (i.e. if the cells are all at the same point of division at the same time). It is outside my scope to discuss methods of achieving synchrony but the principle is to arrest growth at one point of the division of the cycle (for example if they are starved of thymine they will all be waiting at the point before DNA synthesis should begin) and then starting them all off again. Once the cells are in a suitable state (i.e. the maximum number have been arrested in metaphase) they are disrupted in a buffer in which the chromosomes retain their integrity and the chromosomes are obtained by differential centrifugation. The first of these procedures is the critical one. Chromosomes appear to be less liable to fragmentation at low pH. In the procedure of Huberman and Attardi (1966) HeLa cells[6] arrested in metaphase are suspended in 15 volumes of 0·1 M sucrose, 0·7 mM $CaCl_2$, 0·3 mM $MgCl_2$. This and all operations are performed in the cold (0–4°C). The cells swell in this hypotonic medium (it takes them about 5 min). Medium containing dilute hydrochloric acid is added until the pH would theoretically be reduced to 2·6. In fact this amount of HCl reduces the pH to 3·0 as the cells have a buffering capacity. If the conditions are right the chromosomes should appear as refractile bodies in the *cytoplasm* of the cells under the phase contrast microscope. The cells are disrupted in a Teflon-in-glass homogenizer. The homogenate is centrifuged at 900 *g* for 30 min. The pellet contains the chromosomes which are purified by zonal centrifugation at 400 *g* for 20 min through a gradient of sucrose (0·1–0·8 M sucrose) in the same acid medium. The chromosomes remain in the gradient in positions depending upon their size and the pellet (nuclear debris) and the top region (cytoplasmic debris) are discarded. One is a bit doubtful about the fate of DNA that has been exposed to pH 3 (see p. 205) and in an alternative procedure, Lin and Chargaff (1964) use a hypotonic solution (10 mM maleate) at pH 6 for cell lysis. The disruption is completed by a cycle of freezing and thawing and swirling in the same buffer containing 2·5 mM calcium and magnesium chlorides and a non-ionic detergent (0·17% Triton X-100). This method was also applied to HeLa in metaphase arrest. Lin and Chargaff utilized repeated differential centrifugation to isolate and purify the chromosomes.

Chromatin   For most tissues it is not possible to obtain cells in metaphase and an attempt has to be made to isolate interphase chromosomes. As interphase chromosomes cannot be visualized one is largely working in the dark. Whether one refers to the product as DNP or chromatin is something of a matter of choice. Current work on the transcription *in vitro* on preparations believed to be chromatin is outside the scope of this book. However as we shall be considering the structure of DNP in chapter 16, we must discuss the methods for its isolation. As this is a somewhat controversial area, I must exemplify the two basic approaches and also indicate my own preference (or prejudice).

In the method of J. Bonner, whole tissue is homogenized in a Waring blendor and the DNP is isolated by a sequence of procedures including differential centrifugation and further homogenization of pellets of nuclear fragments. For preparations from liver, Maurshige and Bonner (1966) homogenized frozen liver (stored at −80°C) with a solution of EDTA and EDTA at pH 8 in a Waring blendor. A 'nuclear pellet' isolated by differential centrifugation is homogenized and the DNP is isolated by differential centrifugation, first in buffer and then in sucrose solution. For preparations originally designed for pea buds, but subsequently applied to other plant and animal tissues, Huang and Bonner (1965) used unfrozen tissue and a similar protocol to the above except that the medium contained magnesium, and sucrose, but no chelating agent. Bonner and his co-workers refer to their preparations as chromatin.

Busch and his group refer to their preparations as DNP. They start with highly purified nuclei and use two methods for the extraction of DNP. In the first method of Steele and Busch (1963) the nuclei are carefully homogenized in a continuous-flow homogenizer in a medium containing a chelating agent (EDTA or citrate) and tris HCl at pH 7·8 and ionic strength of 0·15 (overall). Apparently these are the conditions most suitable for the inhibition of deoxyribonuclease. The suspension is stirred at 0–4°C for 30 min and centrifuged (6000 *g* for 20 min). The pellet is re-suspended and the centrifugation is repeated. The sediment is very gently homogenized with 0·7 mM phosphate buffer (pH 7) using about 2 ml per gram of original tissue. The stirring and centrifugation are repeated as before except 15 000 *g* is required for this stage. The DNP remains in the supernatant as a very viscous gel. The second method is to use 2 M NaCl to solubilize the DNP. The paper by Steele and Busch (1963) also contains a recipe for this procedure.

It must be emphasized that DNP (or chromatin) is extremely difficult material to handle. The solutions and gels which it forms are extremely viscous. However, in characterizing it there would seem to me to be every advantage in starting with a really clean and well characterized nuclear preparation. Also Busch's emphasis on using very gentle methods for disrupting nuclei would appear to be very important as the danger of shearing DNA (p. 105) is equally important with DNP.

Several attempts have been made to make bacterial DNP. It is a commonplace observation that if bacteria are disrupted the bulk of the DNA sediments rapidly in the 'debris' which sediments at 15 000 *g*. Various methods are available for trying to separate the DNP from this mixture by salt-extraction and differential centrifugation.

Ribonucleoprotein

Ribosomes are very easy to prepare. They are stable particles (provided magnesium is present in the buffer) and their sedimentation properties are very different from other organelles and sub-cellular particles. Polysomes require more care as the messenger RNA strand which holds them together is susceptible to shear and to ribonuclease. Moreover many of the polysomes in eukaryotic cells are attached to a complex membranous system called the rough endoplasmic reticulum. When such cells are disrupted, the endoplasmic reticulum fragments and the fragments invaginate to produce little vesicles called microsomes. Thus in a rat-liver homogenate, for example, the bulk of the ribonucleoprotein (RNP) is present as 'bound' polysomes in the microsome fraction. A few examples of RNP preparations follow to exemplify the conditions that are required. Magnesium is present in all these buffers as ribosomes dissociate into their sub-units in the absence of magnesium. The dissociation is more facile in ribosomes from bacteria than those from other sources. In all these procedures the temperature is kept between 0 and 4°C throughout and the high gravitation fields used for sedimenting ribosomes require the ultra-centrifuge. Ribosomes and polysomes isolated by the methods indicated here still have attached to them various factors (see p. 25) which can be removed by exposing the ribosomes to strong salt.

Bacterial ribosomes   Bacterial ribosomes (as opposed to polysomes) are obtained if the bacteria are disrupted mechanically. The polysomes are nearly always degraded to ribosomes under these conditions, presumably because ribonuclease is released from membranes. A typical preparation from *E. coli* would utilize as buffer 10 mM magnesium (acetate or sulphate), 50 mM ammonium chloride and 10 mM tris HCl pH 7·6. The ribosomes are stabilized by thiols and 6 mM β-mercaptoethanol is included. Cells are harvested, washed with buffer and frozen. This frozen material is ground up with bacteriological grade alumina (2·5 x weight of cells) in a pestle and mortar. It is essential that this stage is performed carefully and thoroughly. The cells are put into the ice-cold mortar. The alumina is weighed out and kept by the side. A little alumina is added to the mortar and is ground up with the cells until it forms an even paste. Then a bit more alumina is added, ground in and so on. As more and more alumina is added the material in the mortar gets more and more sticky until at the end it is very stiff indeed. You now keep on grinding hard for 10 min. If you are not dropping with fatigue at the end, you have not been sufficiently thorough. Now buffer (1·5 x the weight of cells) is added. Again this has to be added slowly and ground in with the pestle to produce an even slurry. For a good yield of ribosomes it should be left for 15 min in the cold room (while you sit down for a rest) and is then centrifuged at 15 000 g for 15 min. The supernatant is noticeably cloudy due to the RNP in suspension. It is carefully removed and centrifuged at 100 000 g for three hours (or an equivalent number of 'g-minutes', 200 000 g for $1\frac{1}{2}$ h, etc.). The ribosomes form a pellet (colourless gel) and are washed by resuspending in buffer and repeating the two centrifugations. If final preparations of bacterial ribosomes are required free of endogenous fragments of mRNA the supernatant from the first centrifugation at 15 000 g can be incubated under conditions of protein synthesis (see p. 194). If it is found that the preparation is contaminated with DNA, the same supernatant is a convenient stage at which to insert an incubation with DNAse.

Bacterial polysomes   To isolate bacterial polysomes the principles are much the same except the cells must be broken open very gently. This is normally done in two stages. First they are incubated with lysozyme to destroy partially the cell wall[7] and subsequently with detergent to solubilize the plasma membrane. It is largely a question of juggling the conditions with any particular organism to overcome the central dilemma, namely that lysozyme is inhibited by magnesium whereas magnesium is essential for maintaining the integrity of polysomes. Thus the exposure to the buffer in which lysozyme is active should be for the shortest possible time. A procedure which is applicable to a wide variety of bacteria (Gram positive and Gram negative) is that of Flessel, Ralph and Rich (1967). The bacteria are inoculated with chloramphenicol to stop protein synthesis (and the consequent loss of polysomes through ribosomes 'running off' during the early stages of the preparation) and are harvested at 0°C. They are suspended in 0·5 M sucrose, 0·1 M NaCl and 0·01 M tris HCl pH 8 containing chloramphenicol (of the order of 1 mg per ml); lysozyme (also about 1 mg per ml) is added and then EDTA to a final concentration of about 10 mM. The cells are swirled gently for 5 min at 0–4°C and then magnesium sulphate is added so that the EDTA is 'titrated' and the magnesium is 10 mM in excess. The 'protoplasts' are harvested by centrifugation. Up to 7 mg bacteria (wet weight) per ml can be used in this preparation. ('One wet *E. coli*' weighs about 10^{-12} g.) The lysis of the 'protoplasts' is effected with a weak detergent. Sodium deoxycholate or a non-ionic detergent such as Brij are suitable. If they are suspended in the 'ribosome buffer' (p. 118), 0·15% deoxycholate is sufficient for *E. coli* (Parish, 1969). The debris is sedimented at 15 000 g for 15 min. Either the supernatant can be treated with DNAse and immediately put on to a sucrose gradient for analysis and fractionation or they can be sedimented through layets of 15% and 30% sucrose at 60 000 g for 2 h to form pellets.

There are alternatives to lysozyme-treatment. Osmotic shock has been used to isolate polysomes from 'fragile' cells of *E. coli* (ones that have divided in abnormal salt concentrations and in which the walls are apparently more easily broken) by Mangiarotti and Schlessinger (1966) and from *Azotobacter vinelandii* by Oppenheim, Scheinbuchs, Biava and Marcus (1968). One final word of warning about bacterial polysomes and ribosomes. It is probable that a proportion of them are membrane-bound. Recent work by van Dijk-Salkinoja, Stoof and Planta (1970) suggests that 96% of the RNP in *Bacillus licheniformis* is membrane-bound. Bacterial polysome methodology will probably need re-thinking during the next few years.

Ribosomes and polysomes from plants and fungi   Plant and fungal ribosomes can be prepared by grinding the tissue or mycelium with sand and extracting the homogenate with a magnesium-containing buffer such as that described already. In general it is very difficult to obtain good polysomes from these tissues.

The levels of ribonuclease are so high that frequently even the tRNA is degraded. The answer is to use a ribonuclease inhibitor (see p. 121). An example of such a procedure is the method of Marcus, Bretthauer, Bock and Halvorsen (1963) for isolating polysomes from yeast (*Saccharomyces Dobzhanskii* or a related species or hybrid). Spermidine ($NH_2(CH_2)_3NH(CH_2)_4NH_2$) and Macaloid gel are used as ribonuclease inhibitors.[8]

The yeast is disrupted by a grinding procedure and the buffer is 5 mM magnesium acetate, 10 mM β-mercaptoethanol, 10 mM KCl, 0·5 mM spermidine and 5 mM tris HCl pH 7·6 containing 0·2% Macaloid gel. The cells are washed in ice-cold buffer, ground in a pestle and mortar with glass beads (120 microns diameter) and extracted with a small volume of buffer. The debris (and beads) are removed by centrifugation (10 000 g for 10 min) and the polysomes are recovered from the supernatant either as a pellet or by centrifugation through a sucrose gradient.

Mammalian RNP   A discussion of mammalian RNP most naturally starts with microsomes. P. C. Zamecnik was the first to establish that the bulk of protein synthesis in rat liver occurs in the microsomes (Zamecnik and Keller, 1964). A typical preparation of rat liver microsomes would be to homogenize liver from starved rats (see p. 111—this is to deplete the glycogen store) in 2·5 volumes of buffer (0·15 M sucrose, 25 mM KCl, 10 mM magnesium acetate and 50 mM tris HCl pH 7·6). The easiest way to homogenize tissue of this sort is first to mince it and then put it in a Potter type of homogenizer. All operations are performed at 0°C. If the homogenate is centrifuged at 12 000 g for 15 min the unbroken cells, nuclei, mitochondria and lysozomes are sedimented. Microsomes can be sedimented from this post-mitochondrial supernatant at 100 000 g in 1 h. The ribosomes can be released from the membrane with sodium deoxycholate or some other detergent. However if this is done with pelleted microsomes all that is obtained are single ribosomes and a few dimers; this is because of nuclease activity associated with the microsomes. However if the detergent is added to the total post-mitochondrial supernatant, polysomes can be isolated by differential centrifugation. We now realize that the reason for this is that there is a ribonuclease inhibitor in rat liver and that this substance is present in the supernatant fraction of homogenates. This substance (a glycoprotein) was discovered by Roth (1956) but it was Lawford, Langford and Schachter (1966) who first recognized its relevance to polysome preparations. The method most commonly used for the release of polysomes from microsomes and their isolation is that of Wettstein, Staehelin and Noll (1963). This group used in their earlier papers (including this one) the term ergosome for polysome. Also this method yields not only polysomes but also single ribosomes. In their paper the authors referred to these 'heavy ribosomes' which they believed (correctly) to be polysomes, or rather ergosomes, as C-ribosomes. This last phrase has stuck and refers to polysomes isolated by this particular method. Post-mitochondrial supernatant is prepared as above but with lower magnesium (5 mM). Sodium deoxycholate is added as a strong solution to a final concentration of 1·3%. The resulting suspension is now carefully layered over sucrose in a centrifuge tube. The sucrose solutions are made up in the same buffer as that used for the homogenization. Typical conditions use two layers of sucrose so that the centrifuge tube contains 3 ml of 2 M sucrose, 4 ml of 0·5 M sucrose and 5 ml of supernatant. Centrifugation at 165 000 g for 3 h results in the C-ribosomes forming a pellet. The membranes are left at the interface between the two sucrose layers. Another advantage of the C-ribosomes procedure is that it removes ferritin from the preparation. Ferritin is an iron-transport protein which has similar sedimentation properties to ribosomes and can easily confuse the appearance of profiles of ribosomes when they are analysed by zonal centrifugation. If the polysomes are not purified by sedimentation through 2 M sucrose (for example if it is required to leave the 80 S ribosomes in the preparation) it is essential to remove the ferritin from a mammalian preparation either by precipitating it with ferritin anti-serum or by precipitating the ribosomes with high concentrations of magnesium salts (50 mM).

Different conditions of disruption are required for other mammalian cells. I shall just mention two important examples. Polysomes are isolated from HeLa cells by allowing the cells to swell in hypotonic buffer (10 mM KCl, 1·5 mM $MgCl_2$, 10 mM tris HCl pH 7·4) and then rupturing them in a tight-fitting homogenizer. The nuclei are removed by d centrifugation (800 g for 10 min). Deoxycholate is added to a concentration of 0·5% and the mixture is centrifuged through sucrose gradients. Presumably mitochondria are removed as the pellet in this gradient (Penman, Scherrer, Becker and Darnell, 1963). With reticulocytes,[9] if the cells are merely soaked in the above hypotonic buffer (10 volumes) they lyse. The polysomes are in the supernatant following centrifugation at 10 000 g for 15 min (Warner, Rich and Hall, 1962).

Ribonucleoprotein from nucleoli   Apart from their DNA complement, nucleoli contain RNP particles. These particles are of great interest as they are involved in the processes involved in the maturation of rRNA (pp. 183 and 396). A paper that describes the isolation of these particles from rat-liver nucleoli, and describes biochemical and electron microscopic criteria for assessing the success of such preparations, is by Shankar Narayan and Birnstiel (1969). Nucleoli were prepared essentially by the methods of Busch (p. 114); it was found to be essential to use 'magnesium nuclei' (not calcium nuclei) for the nucleolar preparation. The nucleoli were suspended in 10 mM tris HCl pH 7·4, 10 mM NaCl, 10 mMEDTA, 10 mM dithioerythritol, in the cold, harvested by centrifugation and resuspended in the same buffer. Brij and deoxycholate were added (each to 0·5%); following centrifugation (25 000 g, 10 min), the RNP was obtained in the supernatant.

Ribonuclease inhibitors   If there is evidence of degradation of RNA during preparations of RNP, the answer may be to include a ribonuclease inhibitor in the preparation. The ribonuclease inhibitors used for this purpose fall roughly into three groups: endogenous inhibitors, certain cations and certain polyanions (or particles with negative ionizing groups). The rat-liver supernatant (p. 120) is an example of the first of these. The cations are supposed to be enzyme inhibitors and the polyanions or negatively charged particles are simply designed to adsorb basic proteins (including ribonuclease). Obviously they have to be used with care as ribosomes themselves contain basic proteins. Although many recipes in the literature incorporate these substances there have been remarkably few systematic surveys of their relative effectiveness. It is important to recognize that the properties of ribonucleases from different sources differ greatly and each new source of ribosomes needs to be examined itself. The best survey of these inhibitors is by Payne and Loening (1970) who were concerned with the breakdown of RNA in microsomes from pea roots. As with much plant tissue, these roots contain high levels of ribonuclease (p. 112).

The assay method exployed by Payne and Loening was to prepare pea-root microsomes in a post-mitochondrial supernatant and either to immediately deproteinize by the method described on p. 110 or to pellet the microsomes and deproteinize the pellets after incubation for varying times. They were able to identify the breakdown products of high-molecular-weight rRNA (18 S and 25 S) and monitor their production during incubation, by polyacrylamide gel electrophoresis (p. 394). The 18 S molecule (MW 0·7 x 10^6) yielded fragments of MW 0·61 x 10^6 and 0·1 x 10^6 and small amounts of 0·45 x 10^6 and 0·37 x 10^6. The 25 S molecule (MW 1·29 x 10^6) had just one breakage point and yielded fragments of 1·04 x 10^6 and 0·97 x 10^6. Using standard incubation conditions they measured the percentage inhibition of the production of these fragments by various inhibitors (or alleged inhibitors).

No evidence for an endogenous inhibitor was obtained, indeed cell sap merely accelerated the breakdown of RNA. Of the cationic inhibitors tried the most effective was 1 mM spermidine (46% inhibition), the diamines putrescine ($NH_2(CH_2)_4NH_2$) and cadaverine ($NH_2(CH_2)_5NH_2$) were less good (both about 30% inhibition). Zinc salts were effective at 2·5 mM (42%) but not at higher (5·0 mM) concentrations (14%).

The polyanions, etc., that were tried included ion-exchange resins as given below: CM is carboxymethyl and SE $-CH_2CH_2-SO_3^-$. Sodium bentonite is made by treating the diatomaceous aluminosilicate, 'bentonite' with EDTA and grading the particle size by differential centrifugation. The treatment with EDTA is required to remove ferric ions from the natural material; a thorough recipe for making sodium bentonite is in the paper by Fraenkel-Conrat, Singer and Tsugitsa (1961). All the substances tried had some inhibition: CM-cellulose, 1 mg/ml, 58%; polyvinyl sulphate, 10 mg/ml, 39%; CM-Sephadex, 0·1 mg/ml, 16% SE-Sephadex, 0·1 mg/ml, 24%; but the really effective one was sodium bentonite: at a concentration of 0·1 mg/ml, 61% inhibition was obtained and at 1 mg/ml, 98–100% inhibition. However (and this is always the trouble with bentonite) about half the yield of ribosomes is lost by adsorption to bentonite at the higher concentration. Payne and Loening also confirmed that the so-called ribonuclease inhibitor 6 mM β-mercaptoethanol had no effect whatsoever.

There is another way of inhibiting the effect of ribonuclease on ribosomes and that is to put some low-molecular-weight commercial RNA into the preparation. As an example, in Payne and Loening's system 0·25 mg/ml of such RNA effected 43% inhibition.

An alternative polyanion to try is the sulphated polysaccharide, heparin. In their work on chick oviduct polysomes, Pallmiter, Christensen and Schimke (1970) made an analysis of the effectiveness of heparin and polyvinyl sulphate and concluded that heparin (which is much cheaper and more readily available) inhibited the endogenous ribonuclease by a factor of 20–40 fold at a concentration of 50 mg/ml.

For general use, sodium bentonite appears to be the best ribonuclease inhibitor. It has two main disadvantages. One is that the preparation is tedious and the exact processing and grading of particles is critical for maximum activity of the preparation (Watts and Matthias, 1967). The other is that low yields of RNP are obtained and a lot of protein is adsorbed to the bentonite. Both these disadvantages are overcome by using 'Macaloid gel' (see p. 120). There is unfortunately no quantitative comparison of the inhibitory effects of the two preparations on ribonucleases. To make Macaloid gels, Macaloid itself is suspended and homogenized in hot 50 mM tris HCl pH 7·6 10 mM magnesium acetate and then cooled and centrifuged. The supernatant is discarded. The pellet consists of a layer of 'ungelled' Macaloid with the gel on top of it. This gel is removed and included in homogenization buffers. It seems to be especially beneficial for certain fungal and plant preparations (p. 120).

By suitable and obvious modifications of these procedures, ribosomes can be isolated from chloroplasts (Eisenstadt and Brawerman, 1964), mitochondria (Küntzel and Noll, 1967) and nuclei (Steele and Busch, 1963). However there are special problems associated with these isolations.

Viruses

I have not described in this book the method of growing or harvesting tissue and cells. No one will be working on any biological material without knowledge and experience of the basic techniques involved. Similarly I give no details nor references for virological techniques. The principle of isolating viruses is that cell lysate (or homogenized plant tissue, etc.) is purified for virus by differential centrifugation. Ultracentrifuges are required to sediment small viruses;

others, including the T-phage of *E. coli*, can be harvested in a medium-speed centrifuge (15 000 *g*).

In certain cases it is possible to precipitate the viruses with ammonium sulphate and harvest the precipitate at lower speeds. Viruses are purified by zonal or buoyant-density centrifugation (see p. 132).

Deproteinization of nucleoprotein and organelles

The techniques already described for the deproteinization of whole tissues are equally applicable to organelles and nucleoproteins. In particular the method described on p. 110 works very well with ribosomes or phage. The nucleoprotein pellet is gently homogenized in the two phases. However in many cases the procedures can be considerably simplified. This is often an advantage when very small amounts of material are being processed. We shall just exemplify this statement with particular techniques.

Detergents   Sodium dodecyl sulphate will totally deproteinize ribosomes and polysomes under suitable conditions. The technique is a most useful one if combined with zonal centrifugation through sucrose gradients for the RNP is treated with SDS and then the various RNA species sediment to form zones and the nucleic acids have been released and fractionated under conditions involving no precipitation or exposure to strong salt nor organic solvents and thus aggregation, which is one of the big problems in handling RNA preparations, is minimized. This technique has been perfected by Noll and Stutz (1968) and its application is discussed in chapter 4. Ribosomes are dispersed in buffer (10 mM NaCl, 5 mM tris HCl pH 7·5). SDS is added as a 10% solution. For ribosomes from the cytoplasm of eukaryotic cells, the SDS is added to a final concentration of 0·5% and the suspension is incubated at 37°C for 10 s and is then layered over the sucrose gradient. For ribosomes from bacteria, chloroplasts and mitochondria the final concentration of SDS is 2% and the incubation time is 1 min.

Strong salts   The use of strong salt to remove selectively and progressively protein from nucleoprotein is a tool for studying the structure and composition of these complexes. However for the preparation of nucleic acids, strong salts alone are not of value. Strong salt solutions do not completely remove protein from ribosomes and in any case rRNA of high molecular weight is insoluble in such solutions. We have already described the use of strong salt solutions to obtain undegraded bacterial chromosomes of value in autoradiographic studies. However there is no merit in this method for other DNA samples. With 4 M CsCl chromatin is dissociated and DNA can be sedimented in the ultracentrifuge leaving the bulk of the protein floating. However such DNA contains up to 4% protein (Huang and Bonner, 1962) and is useless for any chemical or physical studies.

An ingenious way of making the strong salt method satisfactory is due to Cox (1966). It makes use of the fact that RNA is soluble in strong solutions of granidinium chloride; a salt which is also a very effective protein-denaturant. The method is applied to mammalian ribosomes. Guanadinium chloride is added (as a 6 M solution) to a suspension of ribosomes such that the final concentration is 4 M. At −18°C, 50% of ethanol is added and the RNA precipitates. The purification consists of repeated dissolution of the RNA in guanidinium chloride and precipitation as above.

Another effective salt for deproteinization is sodium perchlorate. Freifelder (1966a) has used 7 M perchlorate for isolating viral DNA. The method is probably in general less satisfactory that a phenol method, as residual protein is left with the DNA and DNA is readily denatured in strong perchlorate and so the more effective the solution is for deproteinization (higher temperature and stronger salt) the more risk there is of denaturation of the DNA. However the method is a valuable one in that it can be modified to prepare bacteriophage DNA directly from a cell lysate and eliminate the time-consuming task of first isolating and purifying the phage. The technique has been successfully applied to the T-odd phages (see p. 33) and consists of putting 1 ml of 5 M perchlorate in the bottom of a tube for an ultracentrifuge, overlayering it with a 'spacer' of 0·5 ml 30% sucrose and then filling the tube up with phage lysate. The centrifuge is run under the conditions under which the phage particles would sediment. Bacterial nucleoprotein is unable to penetrate the heavy perchlorate layer. The phage particles are denser and do penetrate and are deproteinized. The perchlorate layer also collects a few large pieces of bacterial debris which are removed by diluting the perchlorate layer and centrifuging at low speed.

Transfer RNA

There are two aspects of transfer RNA which make it worthy of separate consideration. For one thing it is the only nucleic acid fraction which occurs free (not part of a nucleoprotein complex) in the cell. It should therefore be possible to release it under conditions which leave the other RNA and DNA components behind. For another it is often required to obtain a large amount of tRNA, either for use in experiments on protein synthesis *in vitro* (see p. 194), or for fractionation of tRNA and study of the chemical composition and sequence. As far as the sequencing of RNA is concerned micro-scale techniques introduced by Sanger (see p. 291) have largely obviated the requirement for large amounts of RNA but these larger quantities of RNA are required for chemical and physical studies on the minor bases of tRNA.

The simplest way to exploit the unbound character of tRNA in its preparation is to obtain a cell homogenate and to remove all the organelles and nucleoproteins by differential centrifugation. The tRNA can then be isolated from the supernatant by extraction with phenol. An alternative is to separate the tRNA and proteins by ion-exchange chromatography. This has been applied to rat liver by Altener and Maďarič (1966). In their procedure, post-mitochondrial supernatant (p. 120) is centrifuged at 100 000 *g* for 2 h to sediment the microsomes and this post-microsomal supernatant is chromatographed on DEAE-cellulose (p. 161). A column of fairly modest dimensions (3 x 30 cm) will cope with 200 ml of supernatant. A gradient of NaCl in 0·05 M tris HCl pH 7·6 is used for elution. The protein is eluted at around 0·5 M salt; the RNA at about 0·8 M.

With yeast, use can be made of the fact that when yeast cells are treated with phenol (just by soaking them in phenol and water) the cell wall is damaged in such a way that it acts as a sort of molecular sieve, so that tRNA, but no other nucleic acid, will pass through. In this way Holley was able to isolate tRNA from vast amounts of brewer's yeast (45 kg). This can obviously be scaled down and speeded up (as on this scale all the separations must be done by settling rather than centrifugation). For 45 kg, the procedure (Holley, 1963) consisted of soaking the yeast in 43 l of 88% aqueous phenol and stirring in a further 100 l of water. After settling (about a week) the upper phase was siphoned off into a mere 3·5 l more phenol and again stirred up and allowed to settle. The RNA was precipitated from the upper layer by adding 600 ml of 20% of potassium acetate (pH 5·2) and 125 l of alcohol. The precipitate was collected and purified by ion-exchange chromatography (p. 161).

A procedure that has been successfully used in the author's laboratory for the isolation of tRNA from up to 500 *g* of *E. coli* paste is to homogenize (in a Waring blendor) frozen bacterial paste with solid phenol and 0·5 M NaCl. The quantities are unimportant as long as you do not clog the blendor up. The resulting mush is added to more salt solution and phenol so that the final total quantities are about 3 l of salt and 2 kg phenol. This is stirred for about 2 h at room temperature and the phases are separated in a big centrifuge (quite a slow speed is all right—there is surprisingly little protein and DNA in the emulsion). RNA (and a little DNA) is precipitated from the supernatant by addition of two volumes of ethanol (industrial methylated spirits). The precipitate settles down well overnight at 0–4°C. We then treat this precipitate as though it were nucleoprotein and send it through the aminosalyclate phenol treatment (p. 110). The further fractionation of this material (to isolate pure tRNA) is described on p. 129.

However the tRNA that is obtained consists largely of aminoacyl tRNA (p. 14). The amino acids are removed by incubating the tRNA at 37°C for 30 min in buffer (either tris or carbonate) of pH 10. After the incubation, the solution is neutralized and either the amino acids are removed by dialysis or the tRNA is precipitated with two volumes of ethanol.

Finally tRNA from yeast or from certain 'common' strains of *E. coli* (such as K12) are now commercially available. Any commercial sample should be carefully characterized and assayed for amino-acid-acceptor capacity (p. 198) before use.

Notes to chapter 3

1   Chloroplasts from plants in certain phases of growth may contain partially degraded rRNA *in vivo* (see p. 396). Moreover certain viruses appear to contain more than one RNA molecule (see p. 397). With these minor exceptions the above generalizations hold true.

2   For the benefit of physical scientists, Gram negative bacteria are not composed of anti-matter. Gram is the name of the inventor of an important stain used in the microscopic examination of bacteria. After staining with Gram's stain, the bacteria are washed with an organic solvent. Those that are destained are Gram negative; those that are not destained are Gram positive. Although the mechanism of the Gram reaction is complex and incompletely understood, the following generalizations are roughly true. The differences between Gram positive and Gram negative organisms are in the structure of the cell wall. Compared with Gram negative organisms, Gram positive ones have thicker walls containing more hexosamines, less amino acids and less lipid. The Gram positives are more resistant to heat and their walls are more difficult to separate from the plasma membrane, they are more sensitive to lysozyme and cell-wall inhibiting antibiotics such as penicillin. An excellent short review of the Gram reaction is by Stanier, Dourderoff and Adelberg (1968). Gram positivity is unique to certain bacteria (all other cells are Gram negative).

3   Again for non-biologists: Certain Gram positive bacteria, for example members of the genera *Bacillus* and *Colstridium*, have a non-dividing, heat-stable form called spores. The induction of sporulation and the reversion of spores to vegetative cells are excellent systems for studying control of morphogenesis.

4   If high quality 'detached crystal' grade is used, it is not necessary to re-distil phenol.

5   Strictly that statement requires qualifying or at least the terms need defining. Mitochondria are organelles in which the enzymes of intermediary metabolism involved in oxidation and ATP synthesis are located. Bacteria have such particles (called mesosomes) in the region between the wall and the plasma membrane but mesosomes do not appear to have the semi-autonomous characteristics of mitochondria. Similarly photosynthetic bacteria contain their photosynthetic apparatus in 'chromatophores' which are invaginations in the plasma membrane but likewise differ from the chloroplasts of higher plants and photosynthetic protozoa.

6   HeLa is the designation for a strain of cancer cells (originally isolated from a human cervical carcinoma) which have been maintained in tissue culture for many years and are widely used for studies on mammalian tissue culture.

7   Bacterial cells totally lacking the cell wall are called protoplasts (or spheroplasts). I avoid these terms in this context as the lysozyme-treatment is normally so mild for the reason given above that it is extremely doubtful whether the walls are completely removed. There are just sufficient holes in them to allow the detergent to destroy the membrane.

8   Spermidine is a normal constituent of ribosomes although its role in maintaining ribosome structure is not known (see p. 405). Macloid (American Baroid Co.) is an earth containing sodium, potassium, lithium and fluorosilicate. It is discussed further on p. 122.

9   Reticulocytes are immature red blood cells. In mammals the mature cell (erythrocyte) contains no nucleic acids at all. The differentiation leading to these cells is complex but there are to main events. The nucleus is lost leaving a cell with stable mRNA making proteins (largely haemoglobin). This intermediate state is the reticulocyte. Reticulocytes appear in the blood during the pathological condition known as reticulocytosis, which can be induced in animals if they are treated with phenylhydrazine.

4   **Fractionation**

Introduction

In this chapter I only discuss methods for the fractionation of undegraded preparations of naturally occurring nucleic acids. The special techniques which are used for the fractionation of fragments of RNA obtained during sequence studies are mentioned later (chapter 10). In a sense it would have been more logical to discuss the fractionation of nucleotides (chapter 2) then oligonucleotides and finally nucleic acids, but I have opted for the present sequence so that in the early part of the book we deal with Preparation, Fractionation and then Properties of Nucleic Acids in the order used in experimental work.

In the body of the chapter I have grouped the techniques according to the type of apparatus employed (ultracentrifugation, column chromatography, electrophoresis, etc.). However it is worth considering briefly the differences in properties which can be exploited in a scheme for fractionation.

Molecular weight   Molecular weights of nucleic acids vary from around 20 000 (tRNA) to an unknown upper limit for DNA. However we can place an effective upper limit in the range $10^7 - 10^8$ for nucleic acid preparations that are commonly handled. Basically there are two methods of fractionating macromolecules on the basis of differences in molecular weight: hydrodynamic behaviour (sedimentation in the ultracentrifuge) and molecular exclusion chromatography. In both cases, molecular weight is only one factor which determines the behaviour of a particular molecular species. The conformation of the molecule is very important. To exemplify this point, native DNA behaves as a long rod-shaped molecule whereas most RNA molecules appear to be approximately spherical. Thus for DNA and RNA of equal molecular weight (say 28 S rRNA and very degraded DNA, both of around $1 \cdot 5 \times 10^6$) the RNA is a more compact structure and will sediment faster than the DNA, whereas the DNA molecule will be more readily excluded from a chromatographic gel. We shall consider the use of the ultracentrifuge for fractionation purposes in detail later (p. 131). Molecular exclusion chromatography has also proved useful for the separation of relatively small RNA molecules (4 S and 5 S) but it is not generally used for nucleic acids of high molecular weight. The reason is that the gels employed for molecular exclusion in the molecular weight range of the order of several million are difficult to handle and the process is time-consuming when compared with available alternatives. However there is another process which can be employed. When nucleic acids migrate in an electric field through a wide-pore gel the rate of migration is proportional to their sedimentation constants (this statement is an oversimplification which we shall qualify later). Thus fast-sedimenting molecules are retarded relative to slow-sedimenting ones. From the point of view of fractionation, gel-electrophoresis achieves enormously improved resolution over sedimentation methods in the ultracentrifuge and also at a fraction of the cost. The only advantages of the ultracentrifuge (and they are important ones) are that it is possible to work on a much larger scale and it is easier to recover the sample at the end.

Ionic properties   Nucleic acids are polyanions. This is the basis of the fact that electrophoresis will work at all. However there are also ionizing groups in the bases and relative abundance of these bases (nucleotide composition) and their clustering (sequence) should enable nucleic acids of comparable molecular weight to be fractionated by ion-exchange chromatography. Molecular weight itself also affects the ease of elution from ion-exchange columns for a reason we shall discuss later. Altogether ion-exchange chromatography has proved a most powerful tool for the fractionation of nucleic acids especially for transfer RNA. For high-molecular-weight RNA and DNA it is essential to use a very weak anion exchange resin or the binding of nucleic acid to resin is so strong that it is difficult to choose suitable buffers which will effect an elution.

Nucleotide composition   Two important properties of nucleic acids, the density and the ease of denaturation, are functions of nucleotide composition.

DNA samples can be fractionated on the basis of their densities in the ultracentrifuge using the technique known as buoyant density centrifugation. This method is widely used both as a preparative method (this chapter) and also analytically (see chapter 13).

The idea that nucleic acids (either RNA or DNA) can be selectively denatured and the native and denatured molecules fractionated, was the original idea behind counter-current distribution. The actual basis of this type of fractionation is still somewhat obscure. However the approach is to choose a two-phase solvent system between which nucleic acids will partition. If one phase (presumably the one containing more organic solvent) is a solvent for denatured nucleic acid it will preferentially extract the nucleic acid(s) more readily denatured. The counter-current distribution apparatus (see p. 168) is a device for repeated extractions of this type.

Solubility methods   Nucleic acids in solution are rendered insoluble by a variety of reagents (di- and multi-valent cations, certain organic cations, strong salts, organic solvents, proteins). Certain of these reagents are selective for nucleic acids of various types (e.g. RNA as opposed to DNA, denatured as opposed to native structures, or variations in molecular weight). In current methodology these procedures are mainly employed for rather crude fractionations although attempts have been made to achieve more subtle fractional precipitations or extractions. As these methods are frequently employed in the early stages of a scheme for isolation and fractionation, we shall consider them first.

Solubility methods

Strong salt

RNA of high molecular weight is insoluble in strong solutions of salts such as sodium chloride. Thus if salt is added to a solution of total nucleic acids, DNA and low-molecular-weight RNA will remain in solution and RNA of high molecular weight will be precipitated. Conversely if a precipitate of nucleic acids is extracted with such solutions only the high-molecular-weight RNA will remain undissolved. The actual procedure to be adopted depends on the nature of the problem. Thus if DNA is present, it is unwise to attempt the extraction method as the fibrous DNA slowly swells in the salt solution to form a gel, which gradually dissolves to form an extremely viscous solution and it is difficult to tell when the DNA has eventually dissolved. We shall consider the application of these procedures to an actual fractionation scheme on p. 129. However one or two general points of a technical nature should be born in mind. We shall see that for a total preparation of nucleic acid, 4 M NaCl is a suitable precipitant for the high molecular weight RNA. The RNA precipitate is very fine and considerable centrifuge speeds are necessary to sediment the precipitate (about 12 000 *g* for 5 min. If the DNA is of very high molecular weight it may tend to sediment as well. This problem can be overcome by introducing a lower layer of 6 M sodium bromide into the centrifuge tube. The DNA is held up at the 4 M NaCl/6 M NaBr interface, whereas the precipitate forms a pellet at the bottom of the tube (Kirby, Fox-Carter and Guest, 1967). The RNA precipitate can be washed with 3 M sodium acetate (pH 6). This extracts the residual DNA solution from the pellet; it replaces NaCl, which is relatively insoluble in alcohol, with sodium acetate (this is of use if the RNA is to be stored as suggested on p. 106); the 3 M acetate solubilizes glycogen and other polysaccharides which may still be left in the preparation at this stage. The DNA and low-molecular-weight RNA are left in 4 M NaCl (or NaBr). I suggest that this is not a suitable starting material for obtaining tRNA unless the DNA has been only partially extracted (p. 111). The easiest way to remove the salt is to precipitate the nucleic acids. The simplest method is to dilute the solution to about 2 M salt and then add an equal volume of pre-cooled 2-ethoxyethanol ('ethyl cellosolve'—see note on p. 111 regarding the cooling of cellosolves). In this way the precipitation of NaCl is avoided. The precipitate can be dissolved in dilute buffer and treated with ribonuclease (if the DNA is required) and then polysaccharide can be removed by the methoxyethanol/phosphate procedure (p. 110).

Less strong sodium chloride (1 M) is suitable for extracting low-molecular-weight RNA from a total RNA preparation. A 2:1 mixture of 1 M NaCl and 2-ethoxyethanol gives a cleaner separation and is suitable for extracting tRNA from a mixture of bacterial RNA containing a proportion of the DNA (see p. 111, Parish, 1968).

According the McCoy and Carter (1968) there is an advantage in precipitating the rRNA on a column of Sephadex G-75. The sample is applied in 1·5 M sodium chloride (so the rRNA will be out of solution) and the tRNA is eluted at this high ionic strength. A gradient of increasingly dilute salt is then employed. The rRNA is eluted at around 0·1 M; this method is probably useful for salt-fractionation of small amounts of RNA.

Organic solvents

Careful addition of organic solvents to aqueous solutions of nucleic acids will preferentially precipitate nucleic acids of high molecular weight. The procedure is somewhat analogous to the outmoded technique of fractionating proteins with acetone or alcohol. Ethanol is not in general a suitable solvent for the fractional precipitation of nucleic acids. Although high-molecular-weight nucleic acids are more readily precipitated with ethanol than ones of low molecular weight, if the nucleic acids are present in a mixture, an aggregate precipitate of all the nucleic acids is formed as soon as the high-molecular-weight material comes out of solution. However it is possible to use ethanol for the fractionation of RNA if the RNA is slowly eluted from an insoluble state with decreasing concentrations of ethanol. The procedure (Franklin, 1966) consists of trapping a very fine precipitate of RNA in a cellulose column, and eluting the column stepwise with different concentrations of ethanol. The method was successfully used by R. M. Franklin to purify the double-stranded replicative form of R17 RNA (see p. 227). The buffer employed is 0·05 M tris HCl pH 7, 0·1 M NaCl, 0·001 M EDTA. Cellulose powder (Whatman CF 11) is washed in buffer, fines are removed and the optically absorbing contaminants in the powder are removed by washing it with buffer in which the EDTA concentration is increased to 0·01 M and containing 1% β-mercaptoethanol. The slurry is re-equilibrated with ordinary buffer and packed into a column (20 x 1 cm). About 0·7 mg RNA dissolved in buffer is mixed with ethanol (final concentration of ethanol 35%) and poured on to

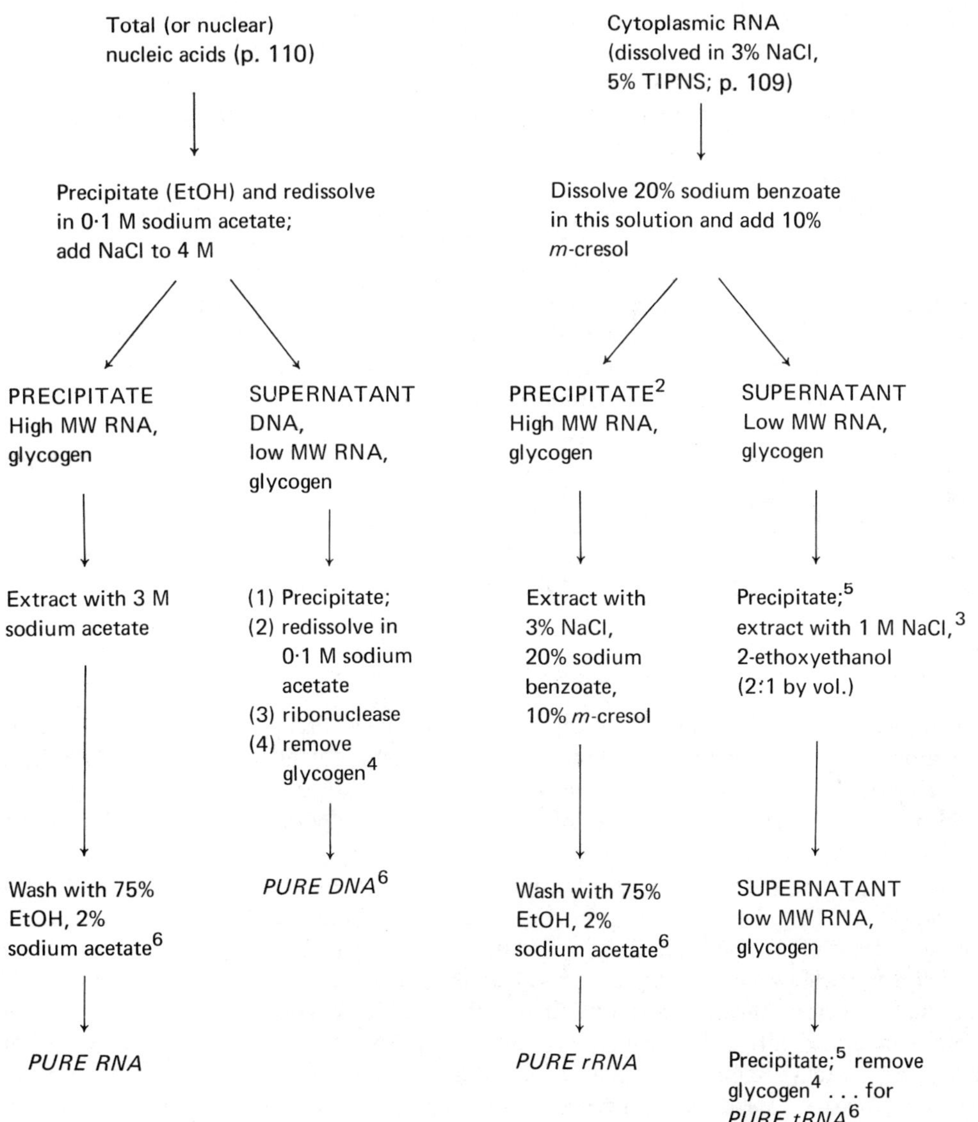

FIG 4.1   Application of precipitation methods to preparation of nucleic-acid fractions from a tissue such as rat liver. The left-hand scheme is suitable too for bacterial preparations within the limitations discussed on p. 000.

NOTES   1   Overnight at about −10°C (ice-making compartment of a refrigerator);
2   This precipitate is a colourless gel; it is easily spun out in the bench centrifuge at 0–4°C
3   Or DEAE chromatography (p. 161)
4   The methoxyethanol/phosphate procedure (p. 110)
5   An equal volume of ethanol
6   Notes on methods of storing nucleic acids are on p. 106.

the column. The column is eluted with 15% ethanol to yield rRNA (from the host bacterium) and single-stranded phage RNA. Replicative form is eluted with between 10 and 0% ethanol. This simple and elegant procedure would seem to have application in any similar fractionation.

Alternatively satisfactory fractionations can be achieved with *iso*propanol (see p. 108), with *m*-cresol or with 2-butoxyethanol ('butyl cellosolve'). These last two solvents are not miscible with water but can be used in conjunction with a salt such as sodium benzoate or sodium butyrate which will solubilize them. (Butoxyethanol is miscible with pure water but not with dilute salt solutions and buffers.) For preparations of bacterial nucleic acids, in which the DNA should be of very high molecular weight (whether the method of Marmur or a phenol procedure is used) the DNA can be precipitated either by the very careful addition of *iso*propanol (Marmur, 1963) or *m*-cresol (Kirby, Fox-Carter and Guest, 1967), in the latter case in the presence of 20% sodium benzoate. In either event it is essential to use the literature recipe only as a guide and to be very careful to observe when the DNA comes out of solution (as a clear gel—not fibres) as the exact concentration of solvent at which precipitation occurs is very dependent on ionic strength and temperature. A more clear-cut application of organic solvent method is to mammalian RNA preparations. In the presence of 20% sodium benzoate and 3% sodium chloride, 10% *m*-cresol reproducibly precipitates only high-molecular-weight mammalian RNA. The other advantage of this particular mixture is that it does not precipitate glycogen.

In Fig. 4.1, I have summarized the application of these procedures to the sub-fractionation of preparations of nucleic acids of rat liver. These schemes are based on those of K. S. Kirby (Kirby, 1965; Parish and Kirby, 1966) and illustrate the use of precipitation procedures.

Protein

Proteins, when associated with nucleic acids, render the nucleic acids insoluble or at least alter their properties very markedly. Protamines (the very basic protein component of DNP from spermatozoa) will precipitate nucleic acids completely. The use of a preparation of protamine (such as 'salmine'—the protamine from salmon sperm) for this purpose is commonplace in the isolation of proteins from tissue when it is required to remove all the nucleic acids. As far as I am aware, no successful attempts have been made to fractionally precipitate nucleic acids with protein. On the other hand attempts have been made at fractional deproteinization. The idea is that the release of nucleic acids from their nucleoprotein complex (deproteinization) can be controlled in such a way that only certain categories of nucleic acid are released. Of course the procedures we have already considered in which only cytoplasmic RNA is extracted from mammalian tissue, only tRNA from yeast and mainly RNA from *E. coli* are all examples of partial and selective deproteinization. A much more ambitious scheme is to attempt to selectively release categories of nuclear RNA from 'phenolic nuclei'. This procedure is particularly associated with G. P. Georgiev. He has reviewed the various techniques employed (Georgiev, 1967) and concludes that the most successful one is to extract the phenolic nuclei ('interface'—see p. 106), with mixtures of phenol and 0·14 M NaCl at successively higher temperatures, viz. 45, 55 and 63°C. Using Georgiev's own terminology, the three RNA fractions are respectively R-RNA, R-RNA/D-RNA mixture, and D-RNA. R-RNA is ribosomal RNA (rRNA elsewhere in this book) and D-RNA is RNA with a nucleotide composition similar to that of DNA. We shall discuss methods of assigning metabolic roles to nuclear RNA elsewhere (chapters 5 and 12) but as far as this procedure is concerned, we have already noted one possible objection to hot phenol methods (p. 107). A puzzling feature of Georgiev's account, is that it is not clear at what point the nuclear tRNA (or its precursors, see p. 184) are extracted.

The ultracentrifuge

Introduction

In a later chapter (13) there is a systematic account of the behaviour of nucleic acids in the ultracentrifuge. Here we are only concerned with the preparative fractionation of nucleic acids. I do include a simple introduction to the ultracentrifuge but most strongly advise anyone who plans to use the instrument to have a working knowledge of it and the physical chemistry that underlies its performance. A useful review is by Berman (1966).

The ultracentrifuge is a device for creating high gravitation fields (up to 500 000 g) in which large molecules will sediment. In this chapter we are only considering applications of the Preparative Ultracentrifuge. By 'preparative' I do not mean that we are confining the discussion to experiments in which it is intended to recover the fractions and use them for subsequent experiments. Some of the applications are powerful analytical techniques. Preparative untracentrifugation rather refers to any process in which the centrifuge is stopped at the end of the run and the contents of the tube are then analysed to determine the nature of the fractionation that has occurred. This is in contradistinction to analytical untracentrifugation in which the sedimentation (or other redistribution) of macromolecules is visualized during the progress of the centrifugation, which is performed in a special instrument (the Analytical Ultracentrifuge) in which the sample is in an optical path such that light passes through the cell and the optical properties of its contents can be recorded.

The Preparative Ultracentrifuge then is a very high speed version of a normal laboratory centrifuge. The rotors normally used in such centrifuges are of either the fixed angle or swinging bucket types. There are however one or two specialized rotors for ultracentrifuges; one of these (the zonal rotor) is discussed on p. 142. The features of an ultracentrifuge are as follows. It is refrigerated and there is some kind of temperature measuring and controlling device. The centrifuge bowl is evacuated and the rotor runs in a high vacuum (of the order of 0.01 mm Hg). The reason for this is that friction has to be minimized to simplify the problem of temperature control at high speeds. The other controls on the centrifuge are a speed selector, a fail-safe overspeed device and a brake. Mechanical braking is clearly impracticable at the speeds achieved by untracentrifuges (up to 70 000 rpm) and the 'brake' consists of a circuit which reverses the magnetic field in the motor so that the rotor then drives the motor as a dynamo and the energy is expended by heating up electrical resistors.

For a molecule in a centrifuge tube, the force upon it is $\omega^2 r$ where ω is the angular velocity (in radians per second) and r is the distance from the molecule to the centre of rotation. It is conventional to express speeds of rotors, not in radians per second but in revolutions per minute. Also it is the convention to express the forces as multiples of the acceleration due to gravity (g) or relative centrifugal force (RCF). To take account of these changes of units the relationship becomes:

$$\text{Centrifugal force (units as } \times g\text{)} = \frac{4\pi^2 \, (\text{rpm})^2}{3\,600} \times r$$

A simple graphical method of solving this equation is to use a nomogram (Dole and Cootzias, 1951).

The important thing to recognize is that the centrifugal force experienced by a molecule increases as it proceeds down the tube. This is perhaps rather an elementary point but not one always grasped by the advertising agents for centrifuge manufacturers who quote (with pride) the 'g_{max}' achieved by their latest rotor. We are hardly likely to be interested in the force experienced by a molecule once it has hit the bottom of the tube. A more useful figure is 'g_{av}' which is the centrifugal force at the midpoint of the tube in a swinging bucket rotor (or the corresponding point on the projection of the axis of a tube in a fixed angle rotor on to the radial axis of the rotor). Elsewhere in this book, I always refer to g_{av} when describing centrifugation at so many 'g'.

We shall be referring to the three following uses of the ultracentrifuge in this book.

Sedimentation velocity centrifugation   This is a long phrase to describe the sort of experiment which is done in the ordinary bench centrifuge. We start with a suspension of particles (or in the ultracentrifgue with a solution of macromolecules) and put them in the centrifuge. The particles sediment to the bottom of the tube and form a pellet. A feature of this type of centrifugation, which is very important when analytical centrifugation is discussed, is that during the process of sedimentation a boundary develops between pure solvent and sedimenting solute molecules (or particles). This boundary is initially coincident with the meniscus and progresses down the tube until all the molecules are at the top. For preparative fractionations, sedimentation velocity centrifugation is only of value when the sedimentation coefficients (see pp. 118 and 119) of the components are so vastly different that we can choose a speed at which

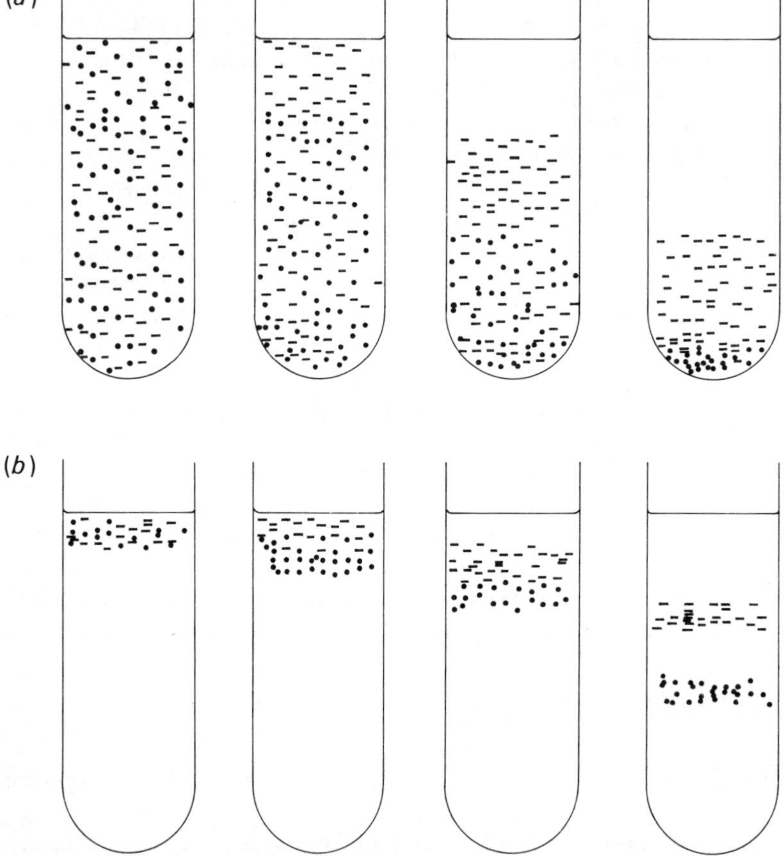

FIG 4.2   Separation of two molecular species by (*a*) sedimentation velocity centrifugation and (*b*) rate zonal centrifugation. The distribution in each case is shown at increasing times from left to right. The dots have a higher sedimentation coefficient than the dashes.

the slower moving component(s) do not sediment at all. This is the process of differential centrifugation which we have frequently referred to in the previous chapter. When the molecules have sedimentation coefficients that do not differ so markedly the technique is of no preparative value. This is illustrated in Fig. 4.2(*a*).

Rate zonal centrifugation   A clear separation of macromolecules such as the hypothetical mixture of Fig. 4.2(*a*) can be obtained if the molecules are confined to a zone at the start of the centrifugation. This is achieved by layering the sample on top of an aqueous solution denser than the sample solution. Typically this dense solution is one of sucrose, and the sucrose solution is present as a gradient of increasing concentration towards the bottom of the tube. The advantage of the gradient is that it stabilizes the contents of the tube against convection currents which would be formed during the inevitable small changes in temperature which occur during the loading and unloading of the rotor and the cycling of temperature which occurs within the centrifuge bowl. A zonal separation is illustrated diagrammatically in Fig. 4.2(*b*).

Buoyant density centrifugation   So far we have only considered uses of the ultracentrifuge in which the centrifugation speed has been fast enough to sediment the molecules completely. There are two forces operating on the molecule in the centrifuge, the centrifugal force which is tending to sediment the molecules and the opposing effect of diffusion. At speeds somewhat lower than those which will achieve a complete sedimentation the molecules will form a concentration gradient which at infinite time will constitute a dynamic equilibrium. 'Infinite time' is effectively only a few days in this context. Observing the equilibrium concentration gradient of proteins gives data from which the molecular weights of certain macromolecules (proteins) can be measured with the analytical centrifuge. As we shall see in chapter 13, this method is difficult for nucleic acids. However, equilibrium–concentration gradients can also be set up in the ultracentrifuge using relatively low-molecular-weight substances such as the salts of metals of high atomic weight. Caesium salts are used for this purpose. The value of the technique depends on the fact that the densities of nucleic acids lie in the range of densities of strong solutions of caesium salts. There is a problem with RNA, as RNA of high molecular weight is insoluble in strong salts but on the analytical scale at least this can be overcome by fractionating the precipitate. However for DNA, which is soluble even in 7 M CsCl, this is the basis for a method of separating molecules on the basis of their densities (and hence their nucleotide compositions—see p. 287). The DNA is dissolved in strong caesium chloride and then centrifuged at a speed such that the caesium chloride forms a gradient. The DNA molecules will move in the high centrifugal field and will all end up in the caesium chloride of their own density. The zone of DNA molecules which is produced is normally known as a band and the technique is often referred to as 'caesium chloride banding'.

In the remaining part of this section, I have illustrated the applications of the ultracentrifuge to a number of fractionations of nucleic acids.

Zonal centrifugation

General technique   Sucrose gradients are formed in the tubes for swinging-bucket rotors for ultracentrifuges. Fixed-angle rotors are not suitable for zonal centrifugation; resolution is impaired owing the the tendency of molecules to 'slip' down the wall of the tube exposed to the greatest force. Tubes for swinging-bucket rotors have capacities of 5–35 ml. Up to 5 mg of

RNA can be fractionated in one tube of the largest rotors. For a theoretical account of the loading capacity of zones see Brakke (1964). The first job is to create the gradient in the tube. In this chapter I shall exemplify several actual separations and leave the discussion of the ideal 'shape' of a sucrose gradient until chapter 13 when we shall consider the use of such gradients

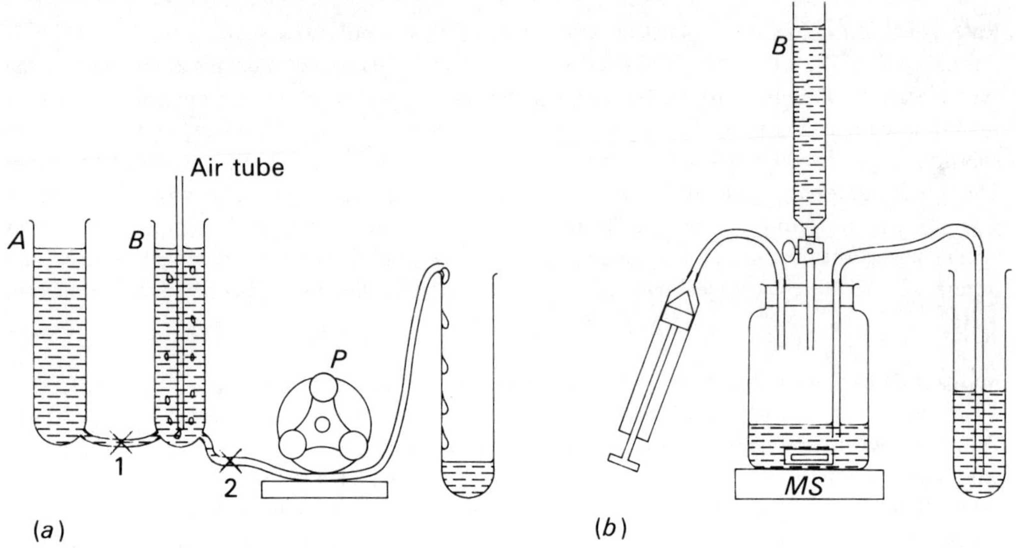

FIG 4.3   Devices for creating sucrose gradients.

(a) This arrangement produces a linear gradient. The reservoir and mixing chamber (A and B) are test tubes with little side arms blown on. The taps (1 and 2) are artery clips. The remaining tube is polyvinyl or silicone rubber, except for the air tube which is a broken syringe needle. P is a peristaltic pump. The heavy sucrose solution is placed in B and the light one in A. Air is gently bubbled through the air tube (from a fish-tank pump) to stir the contents of B. 1 and 2 are opened and the gradient is slowly pumped into the centrifuge tube. If the tube is made of polypropylene it is wise to insert a narrow glass spill in the tube and run the gradient down this, as polypropylene (unlike nitrocellulose and polycarbonate) is not wetted very well by water and the gradient goes in as rather large drops. In either event, the pumping speed should be slow enough to prevent any obvious mixing (shlieren patterns) between the incoming sucrose and the top of the gradient.

(b) This arrangement produces an exponential gradient. The mixing chamber is a little bottle containing the light sucrose. It is stirred with a magnetic stirrer (MS). The heavy solution is in the burette (B). The burette tip and two syringe needles are inserted into the bung of the little bottle. It is essential that this is absolutely air-tight. The other tubes are all narrow pieces of glass tube or syringe needles connected with flexible tubing as above (Fig. 4.3(a)). After the arrangement has been set up, the syringe (on the left) is squeezed so that the light sucrose from the mixing chamber is forced just into the tubing leading to the centrifuge tube. The syringe is clamped in this position and the stirrer is turned on and the burette tap opened. Note that the gradient goes in 'backwards', i.e. increasingly dense sucrose displaces the part of the gradient already formed. At the end the tube is carefully withdrawn from the centrifuge tube.

Clearly the concentration of sucrose at the top of the final gradient (C_t) is equal to that initially in the mixing chamber. Also the final volume of the gradient is that of the heavy sucrose run in from the burette. If we call the concentration of heavy sucrose in the burette, C_h, and the mixing volume V_m, then the concentration C_V when volume V has accumulated in the centrifuge tube is an exponential function of V, viz.

$$C_V = C_h - (C_h - C_t)\, e^{-V/V_m}$$

The application of this equation to idealizing the shape of sucrose gradients is discussed on p. 385. Notice that the shape is exponential convex. If V_m is increased this curvature is flattened; if V_m is decreased it is steepened.

for measuring sedimentation coefficients. However we shall refer to fractionations in both linear and exponential gradients. There are many descriptions in the literature of gradient-making machines. I have chosen to illustrate my own preferences which are for devices so simple that they can be put together from apparatus found in any laboratory (Fig. 4.3). The very ingenious method of creating exponential gradients (Fig. 4.3(*b*)) is due to Noll (1967).

After the gradient is established in the centrifuge tube it is stable for several hours. Gradients are usually stored before use in a refrigerator as they are normally centrifuged at around 4°C. The sample is carefully layered on top of the gradient with a pipette. The volume of sample solution should not be in excess of about 5% of the total volume of the gradient. The tubes are carefully placed in the buckets of the rotor. It is not necessary to be neurotic about handling them; they are reasonably stable but obvious precautions should be taken (such as layering the sample over the sucrose very close to the centrifuge so that the amount of carrying is minimized). The centrifuge is then run for the necessary time. Some typical conditions will be given later in the examples. After the centrifuge has stopped, the tubes are removed and the contents analysed.

The simplest way to sample a sucrose gradient is probably to introduce a syringe needle, attached to a piece of fine flexible tubing into the tube. This is done very carefully to avoid disturbing the sucrose. The tip of the needle should be very close to the bottom of the tube. The gradient is then pumped out (heavy end first) with a peristaltic pump. If the nucleic acids are present in quantities sufficient to detect by their UV absorption at 200 nm (see p. 173) the effluent is best pumped through a flow-through cell in a continuously recording spectrophotometer. The solution emerging from the cell is then collected as fractions using a timed fraction collector (or even by a human being with a stop-clock). Other methods of sampling gradients are to use a similar arrangement to the above to have the sampling tube inserted into a rubber bung which fits tightly into the top of the centrifuge tube. The bung also has a shorter needle inserted in it which ends just above the meniscus. Sucrose solution, denser than the bottom of the gradient, is pumped down the long tube so that the gradient emerges (light end first) through the shorter tube. Yet a third method is to insert the tube into a device called a tube-piercer in which a needle is made to puncture the bottom of the tube and the gradient is pumped out (heavy end first) by pumping water on to the top of the gradient. This technique is the basis for a very elegant design for an automatic recording and fractionating machine for sucrose gradients designed by Noll (1969).[1] For a person who lacks the luxury of a recording spectrophotometer, the optical densities of the samples can be measured after fractionation. If the nucleic acids are radioactive the radioactivity of the individual samples is assayed. The techniques available for this are discussed in the next chapter.

One important point to discuss before considering applications of sucrose gradients (and other methods of fractionation) is the recovery of nucleic acids from the gradient. The simplest method is to add alcohol (two volumes). It is difficult to obtain a visible precipitate as the total amount of nucleic acid in the fraction(s) in question is less than about 0·1 mg. Even with quantities of this order, it is preferable to use pre-cooled alcohol (−10°C). There is a tendency for sucrose to crystallize out. This can be overcome by using the acetone procedure described for the recovery of nucleic acids from very large gradients (p. 146). Rather bulkier precipitates can be obtained with divalent cations (cadmium is a good choice) but then the precipitate has to be dissolved in EDTA and the EDTA chelate must be removed either by dialysis or with a very small column of a molecular exclusion gel such as Sephadex G25. Dialysis of RNA requires great care because of 'finger nuclease' (p. 105). Another precipitant is the cationic detergent cetyl trimethylammonium bromide (known as CTAB, pronounced 'seetab'). CTAB-precipitates of nucleic acids are very bulky. To remove the CTAB and get the precipitated nucleic acid back *into* the sodium form, it can be repeatedly extracted with 75% aqueous ethanol containing 2% sodium acetate. CTAB is soluble in this solvent and an exchange between CTAB and sodium ions occurs in the precipitate. For description of the use of CTAB see Ralph and Bellamy (1964). It must be remembered however that changes in secondary structure will accompany these CTAB manipulations. If the nucleic acids are present in such small amount that no precipitation method will work, then fractions must either be dialysed (or put through Sephadex) or alternatively precipitated with alcohol in the presence of a carrier. For example if it is intended to use the samples for assaying mRNA activity in a cell-free system for protein synthesis (p. 194) it is often convenient to add the tRNA at this stage and precipitate the fraction of 'messenger' together with the correct amount of tRNA.

Sucrose concentrations   There are three methods of expressing sucrose concentration, the percentage composition of the solution expressed as weight/volume (w/v), the molarity and the percentage composition weight/weight (w/w). The relationship between the first two is trivial; the molecular weight of sucrose is 342·30 and consequently a solution $x\%$ w/v contains $10x$ g/l and is $(x \div 34·23)$ M. The relationship between % w/v and % w/w is not a straightforward one as we have to take into account the volume changes which accompany the processes of dissolution and dilution. The advantage in working in w/w values is explained in the context of CsCl solutions (whose concentrations all conventionally expressed as w/w) on p. 149. If the conversion has to be made, the graph in Fig. 4.4 relates the two functions. Also plotted in Fig. 4.4 are the refractive index of sucrose concentrations (as a function of w/w %) and the specific gravity of these solutions. The determination of refractive index is the simplest method of checking (or measuring) sucrose concentration. The specific gravity is not of great importance in the fractionation of nucleic acids as they are all denser than even the strongest sucrose

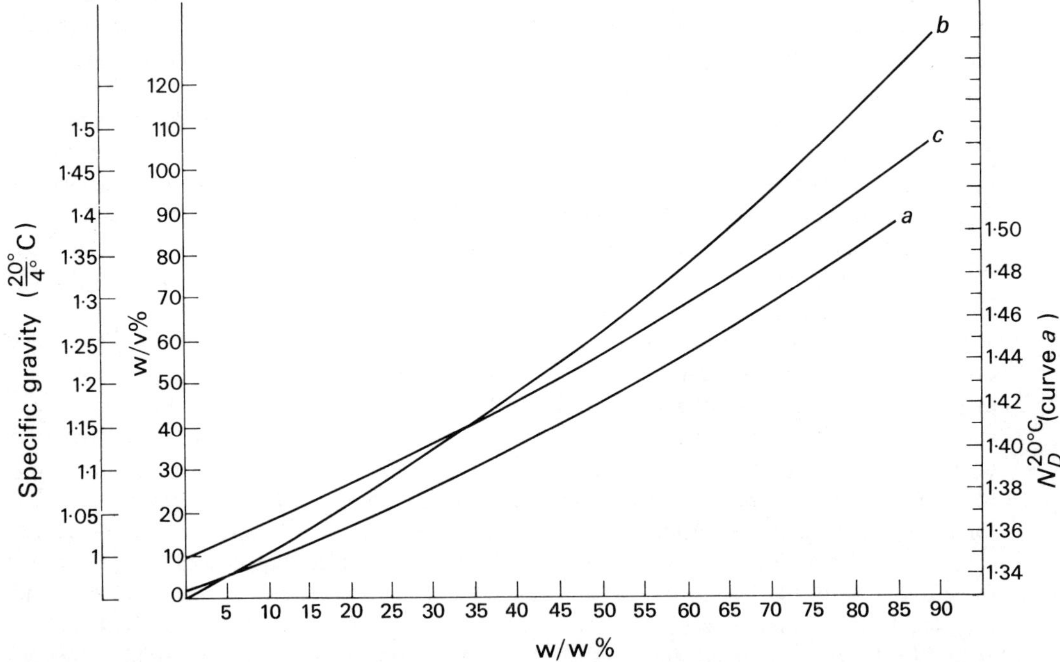

FIG 4.4   Sucrose solutions. Graphs of (*a*) refractive index ($\pi_D^{20^\circ C}$), (*b*) w/v-composition and (*c*) specific gravity as functions of w/w percentage composition for sucrose solutions at 20°C. Low concentrations of slats or buffers (of the order of 0·1 M) can be neglected in reading these graphs. These graphs are plotted from data in the *Handbook of Chemistry and Physics* (1964).

solution so that bouyant-density (or isopycnic) centrifugation is not employed in sucrose. However, these specific gravities are useful to know when centrifuge tubes are balanced and isopynic-sucrose methods are employed in the fractionation of organelles (see p. 115).

Ribosomal RNA   The sucrose gradient is a useful device for separating two RNA species of high molecular weight (23 S and 16 S in prokaryotic organisms, 25–29 S and 18 S in eukaryotic ones). They can also satisfactorily separate these species from RNA of low molecular weight. Examples of the performance of sucrose gradients for these purposes are shown in Fig. 4.5. Technical details which are omitted from the text of the chapter are given in the captions to figures. Note the poorer resolution in the bacterial RNA and the very sharp zones produced in the exponential gradient. The way in which sedimentation coefficients may be calculated from the positions of zones in gradients such as these is described on p. 385.

Degradation and aggregation   Sucrose gradients have been widely used in the past for checking the integrity of preparations of rRNA. It is in fact quicker to use either electrophoresis (p. 156) or the analytical centrifuge (p. 384) for this purpose. However sucrose gradients are still used a great deal for fractionating rRNA and it is worth considering one or two anomalous results.

Something that happens all too frequently is that the pattern of the resolved peaks (Fig. 4.5) turns out to betray some evidence of degradation. Extreme cases of degradation give rise to entirely low-molecular-weight material towards the top of the gradient. Slight degradation yields distribution patterns in which the zones are more diffuse and the ratio of peak heights is odd looking. It is rather commonly imagined that the larger peak should be twice as high as the other. This is not quite true. The area within the envelope of the peaks is proportional to the amount of RNA in the zone (this is only true because the base compositions of the two peaks of high-molecular-weight rRNA are very similar in all organisms studied so far). The two areas should therefore be in the same ratio as the molecular weights. The molecular weights of several rRNA molecules are discussed on p. 395. For bacteria the ratio should be 23 S : 16 S :: 1·96:1 and for mammals 28 S : 18 S :: 2·5:1. If there is evidence of degradation we must look for ribonuclease at some point in the preparation or fractionation. There are three possibilities.

(i) The gradient may be contaminated with exogenous ribonuclease. This is just a case of taking every possible precaution against 'finger nuclease' (p. 105). The centrifuge tubes and other apparatus used in preparing sucrose gradients should be scrupulously clean.

(ii) One of the chemicals used for the gradient may contain low levels of ribonuclease. This is the conclusion most people jump to and sucrose is regarded as the most likely culprit. Ribonuclease-free sucrose specially prepared for gradients is commercially available. I have never found evidence for ribonuclease in Analytical Grade sucrose. It is alleged that beet sugar (as opposed to cane sugar) is less likely to contain ribonuclease. I do not know; we have certainly used commercial beet sugar (of the grade intended for culinary use) for large-scale gradients in the zonal rotor (p. 142) without any evidence of degradation. For those who suspect their sucrose, one procedure is to make up the gradients and then put a layer of an aqueous suspension of Na-bentonite (p. 122) on top of it. The bentonite is then spun through the gradient, allegedly clearing the ribonuclease out by adsorption on the way down. The benetonite pellet is left in the tube and the RNA sample is layered on the gradient and the gradient is spun normally. An alternative is to use diethylpyrocarbonate (p. 112). All the solutions employed in making up the gradient are made 0·6% with respect to diethylpyro-carbonate and are heated to 100°C for 5 min. The ribonuclease in inactivated and the diethylpyrocarbonate decomposes to gaseous products during this procedure (Fedorcsák, Natarajan and Ehrenberg, 1969).

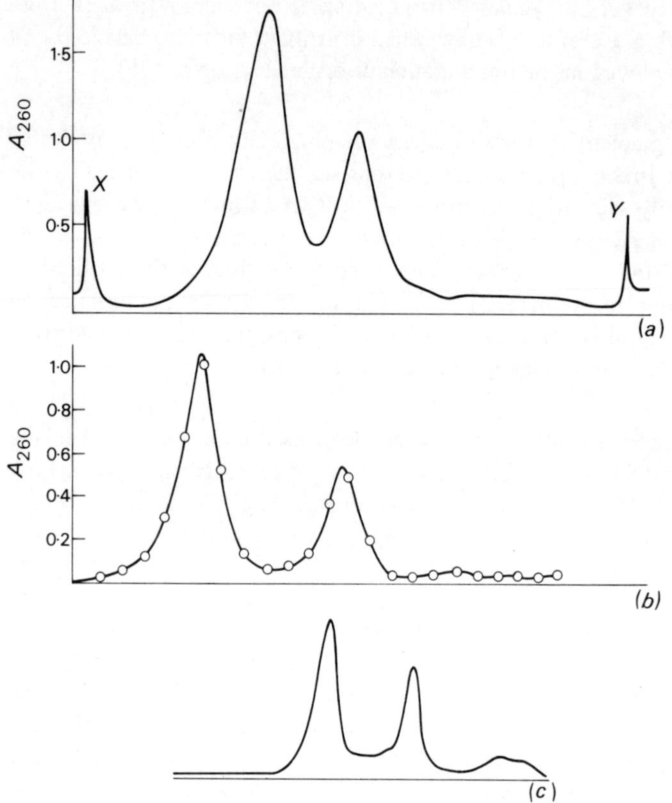

FIG 4.5   Examples of the distribution of high-molecular-weight rRNA through sucrose gradients. In each case the direction of sedimentation is from right to left and the gradient was pumped out so the 'x-axis' is linear with respect to length (and volume) of the centrifuge. In all cases also the RNA was measured by recording the optical density at 260 nm.

(a) RNA from a bacterium (*Myxococcus xanthus*) isolated by the method described from *E. coli* on p. 111 and subsequently salted out with NaCl (p. 128). The gradient was linear (5–20% w/v sucrose in 2% sodium acetate, unbuffered) and was established in a 35 ml tube for the SW 27 rotor of the Beckman model L2-65B ultracentrifuge. Such a rotor will accommodate six tubes. Approximately 1·7 mg RNA in 1 ml of acetate were layered on the gradient and it was spun for $16\frac{1}{2}$ h at 21 000 rpm at 5°C. The gradient was pumped out through the Gilford-Unicam 2000 automatic spectrophotometer and the optical density continuously monitored in a 0·5 cm cell. X and Y are 'spikes' arising from the schlieren effects produced in the cell as the sucrose mixes with water (X) and bubbles appear at the end (Y). They are useful for aligning the profile with fractions. Unpublished data obtained in the author's laboratory by Mr. H. A. Foster.

(b) RNA from rat liver isolated as described on p. 110 and salted out (see above). In this case a linear 7·5–30% (w/v) gradient in 0·1 M sodium acetate was established in a 30 ml tube of the SW 25 rotor of the Beckman model L1 ultracentrifuge. The RNA was introduced as above. The centrifugation conditions were 23 500 rpm for $15\frac{1}{2}$ h at 10°C. In this case, fractions were taken, each diluted to 3 ml and the optical densities recorded manually in 0·5 cm cells.

(c) RNA from rat liver. C-ribosomes were treated with SDS (p. 123) and layered over a 0·44–0·89 M sucrose gradient in 5 mM tris HCl, 0·01 M NaCl in a tube of the 283 rotor of the International B-60 ultracentrifuge. This rotor holds six such tubes (12 ml each). The gradient was exponential, made up in the device shown in Fig. 4.3(b) using 11·8 ml 0·44 M sucrose in the mixing chamber and running in 12·0 ml of 1·15 M sucrose from the burette. This gradient is 'isokinetic' (see p. 385). The centrifuge was run at 40 000 rpm and at 2°C for 11 h. Data of Noll and Stutz (1968).

(iii) The third possibility of course is that the RNA is genuinely degraded. If this is the case it means that the deproteinization process was incomplete or that the preparation allowed for autolysis to occur at some point. The effects of autolysis can be very misleading. The classical example must be regarded as the 'discovery' that the ribosomes of the organism *Rhodopseudomonas spheroides* (a photosynthetic bacterium) had no 23 S RNA, only 16 S. For an account of this story, read the paper by Szilagyi (1968) who showed that by altering the conditions of deproteinization it was possible to minimize the action of a nuclease which was producing the artefactual result. The nuclease in question is apparently specific for one break in the 23 S molecule (compare the case of pea-root nuclease on p. 121 and see also p. 467).

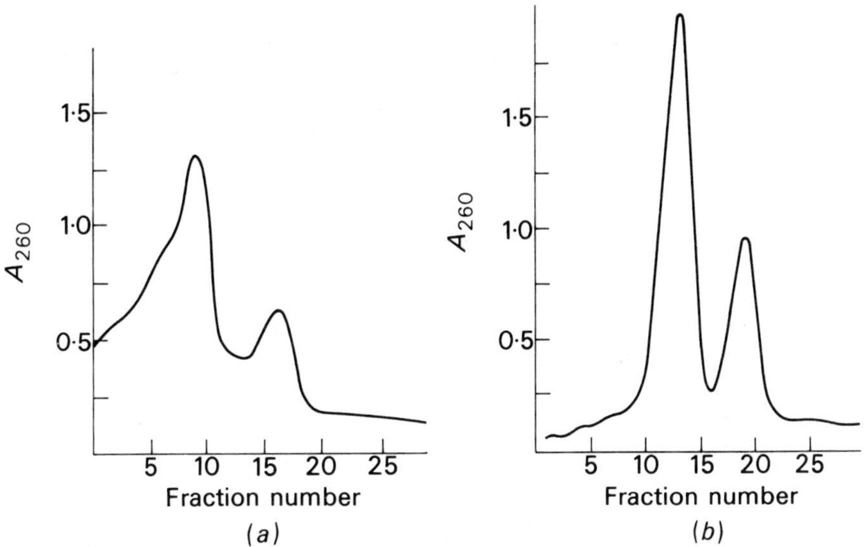

FIG 4.6   Aggregation in mammalian RNA. RNA was isolated from a subcutaneously transplanted rat-kidney tumour by the method of p. 000 followed by salting out (see legend to Fig. 4.8). Both gradients are 5–20% sucrose (linear) in 0·1 M sodium acetate pH 6·5 and were established in tubes for the Beckman SW 25 rotor (see p. 000). In each case the RNA was dissolved in acetate at room temperature. Sedimentation from right to left. In the left-hand diagram the centrifuge had been run at 22 000 rpm at 4°C for 14 h; in the right-hand case it had been run at 19 000 rpm at 18°C for 14 h. Data of J. R. B. Hastings reported in Kirby, Hastings and Parish (1969).

The opposite of degradation is the formation of aggregates in sucrose gradients (Marcot-Queiroz and Monier, 1965; Kirby, Hastings and Parish, 1969). The generalizations that can be made about these aggregates is that their formation is favoured by high ionic strength and low temperature. The most common type of aggregate is a small peak running ahead of the larger rRNA molecule and consisting apparently of one large and one small rRNA. Typically the aggregate peak is no more than a slight indentation in the base line. However in certain cases it is very striking. In Fig. 4.6 there are data, taken from the work of J. R. B. Hastings in which extensive aggregation was found in rRNA obtained from a kidney tumour (originally a spontaneous tumour and subsequently maintained by serial transplantation) in a Wistar rat. Interesting questions are raised by these aggregates. For example, is there some special interaction between the two larger RNA molecules which contribute to the association between the two sub-units of the ribosome? It is to be hoped that further work on the very interesting RNA from the rat tumour will be undertaken to try and identify the nature of the difference between it and other mammalian rRNA and to see whether there are differences in the properties of the ribosomes.

Sucrose gradients have been employed in the past as a method for fractionating the very-high-molecular-weight nuclear and nucleolar RNA of mammalian cells (see for example Busch, 1967). They have been largely superseded by gel electrophoresis for this purpose (p. 158).

The identification of mRNA as minor components in sucrose-gradient fractionation of rRNA is considered on p. 195.

Ribosomes and polysomes   Detailed discussion of the fractionation of ribosomes is outside the scope of this book. However as RNP preparations used as starting material for RNA isolation are also employed for identification of mRNA (p. 194), it is necessary to consider briefly their fractionation.

It is often necessary to obtain purified ribosomal sub-units. A preparation of ribosomes may be dissociated into subunits by lowering the magnesium concentration. For bacterial ribosomes, the dissociation is complete at 10^{-4} M magnesium ions. For mammalian ribosomes, the

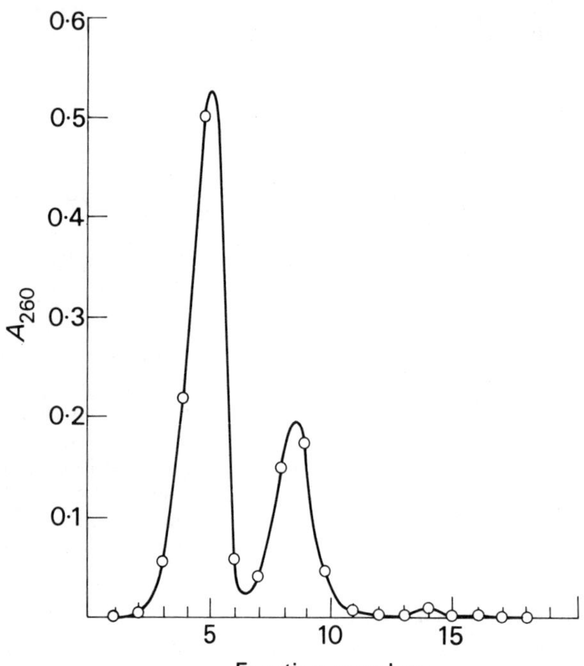

FIG 4.7   Separation of ribosomal subunits in a sucrose gradient. Ribosomes were isolated from *E. coli* MRE600 by the method of p. 118. They were purified by two sedimentations at 160 000 *g* and the pellets were suspended in 10^{-4} M $MgSO_4$, 0·1 M NH_4Cl, 0·01 M tris HCl pH 7·6. The A_{260} was 38. 1 ml of this suspension (containing about 2 mg RNP) was layered over 7·5–30% linear sucrose gradients made up in the same buffer in tubes for the Beckman SW 25·1 rotor (see Fig. 45). The gradients were centrifuged at 24 000 rpm for 15 h at 5°C. The gradients were then pumped out and fractions were collected as described before (Fig. 4.5, p. 138) and 0·1 ml. Aliquots were diluted to 2·0 ml and the absorbances were measured at 260 nm. Fractions 3, 4 and 5 (50 S) and 8·9 and 10 (30 S) were pooled and the subunits were obtained by diluting with an equal volume of buffer and centrifuging. Up to three times as much RNP could have been used without significantly impairing the resolution of the zones. These data were obtained in the author's laboratory by Miss M. Brown.

position is more complicated; however for preparations of C-ribosomes, dissociation is only complete if EDTA is titrated into the suspension such that it is slightly in excess of the magnesium (Stahl, Lawford, Williams and Campbell, 1968). The subunits can then be separated on gradients in which the buffer contains low magnesium concentrations. An example of such a fractionation is shown in Fig. 4.7.

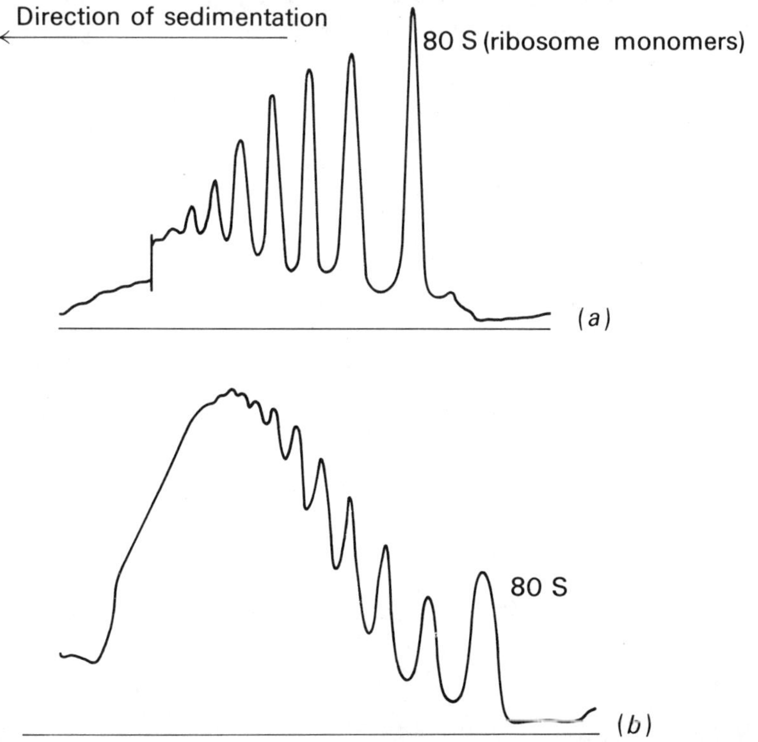

FIG 4.8   Sedimentation profiles of mouse-liver polysomes through exponential sucrose gradients. Exceptional care in monitoring the gradient is required for these remarkable resolutions. In both cases exponential gradients were established using the device shown in Fig. 4.3(*b*) and the gradients were analysed using the apparatus referred to on p. 000. The buffer is not designated in the reference but would typically contain 50 mM KCl, 10 mM MgSO$_4$ and 10 mM tris HCl pH 7·4. In (*a*) the gradient was from 0·3 to 1·1 M in the 14 ml tubes (283) of the International ultracentrifuge (see legend to Fig. 4.6(*c*)) and centrifugation was for 2 h at 37 000 rpm and at 4°C. In (*b*) the gradient (0·39–1·3 M sucrose) was in the 4·8 ml tubes of the SW 65 rotor of the Beckman model L2-65B ultracentrifuge (the rotor holds three such tubes). Centrifugation was for 15 min at 60 000 rpm and at 8°C (Noll, 1969).

As an example of the fractionation of polysomes (Fig. 4.8) in a sucrose gradient I illustrate the sedimentation pattern of C-ribosomes from mouse liver in exponential gradients taken from the data of Noll (1969).

The data of Noll (Fig. 4.8) illustrate the results that can be obtained in the zonal centrifugation of ribosomes if the conditions are optimized. Probably the most important single feature of his method is the use of very high speeds and correspondingly short centrifugation times. Under these circumstances, diffusion is minimized. In recent work, Herzog, Ghysen and Bollen (1971) have shown that centrifugation of *E. coli* ribosomes at 300 000 *g* achieves a resolution of the monomer peak into two components (70 S and 63 S). The 63-S component could not be resolved in more slow running centrifugation. The 63-S ribosome appears to be a conformational varient of the 70-S particle and to be the substrate for the ribosome dissociation factor (p. 25).

The zonal rotor   Before discussing the applications of sucrose gradients to the fractionation of viral RNA and DNA, where the scale is usually very small, we shall briefly consider the way in which fractionations of the sort we have just discussed can be scaled up. To cope with this problem, N. G. Anderson designed the 'Zonal Rotor'.

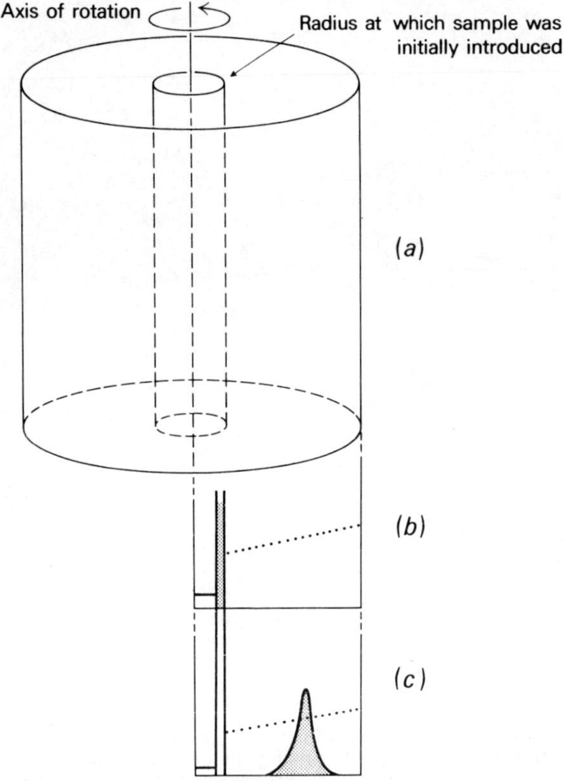

FIG 4.9   Principle of the Zonal Rotor. (*a*) Cylindrical rotor. (*b*) Concentration of sucrose (. . . .) and a hypothetical macromolecule (hatched area) at the beginning of the centrifugation along any radius. (*c*) The same functions after centrifugation.

The zonal rotor is a device specially designed for use in the preparative ultracentrifuge.[2]

The rotor is a cylinder rotating on its axis. The whole volume of the cylinder is full of sucrose solution in the form of a gradient (heavier towards the outside of the rotor). The sample is introduced as an annulus of material at a radius towards the centre of the rotor. The densities of the solution containing the sample and the material between it and the centre of the rotor must be chosen with care to avoid diffusion. Before we consider the design problems posed by such a rotor and the extremely ingenious ways in which they were overcome by Anderson, we shall just consider a diagrammatic illustration of the way in which it works (Fig. 4.9). Note that for the purpose of this illustration we show a gradient which is linear with radius. This is not of course the same thing as a gradient linear with volume in a rotor of this type.

There are two advantages in rotors of this type; one is that larger amounts of RNA can be fractionated as very much larger gradients can be established. The other advantage is that it eliminates a type of diffusion which occurs in a parallel-sided centrifuge tube. This diffusion is illustrated in Fig. 4.10. The zonal rotor (like the sector-shaped cell of the Analytical Ultracentrifuge (see p. 381)) eliminates this effect.

There are two technical problems associated with the design of zonal rotors. One is the tendency for the solution in the rotor to form a vortex as the rotor accelerates. This is eliminated by including four sectors (or 'septae') inside the rotor so that the cylindrical volume of the rotor is divided into four quadrants. The other problem is getting the gradient into and out of the rotor. There are two approaches to this problem. One is use a re-orienting gradient. That is to say that a gradient is made up in the stationary rotor (heaviest material at the bottom). If the rotor is now accelerated the gradient will re-orient so that the heavy end will end up at the wall of the rotor. The geometry for a re-orienting gradient rotor is very critical. In fact all the zonal rotors commercially available for biochemical work use the other principle, namely the gradient and the sample are introduced into the rotor while it is spinning.

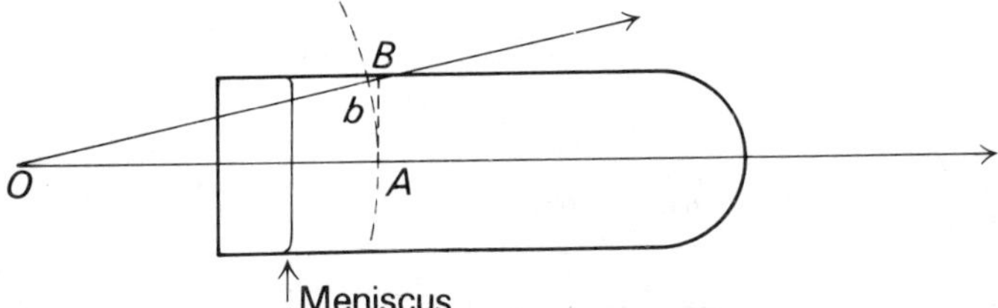

FIG 4.10   Diffusion in a parallel-sided centrifuge tube. A and B are molecules of the same sedimentation coefficient at the same distance from the meniscus (i.e. in a thin zone). O is the centre of rotation. The distance OA is clearly less than OB ($Ob = OA$ and $OB = Ob + bB$). Thus the gravitational force on B is greater than that on A. Moreover B is not travelling in the same direction as A. It is about to 'bounce off' the wall of the tube (edge diffusion).

Anderson and his colleagues at Oak Ridge National Laboratory devised two solutions to the problem. In one version of the zonal rotor (B-IV, Fig. 4.11(a)) the rotor is driven from the bottom and on the top there is a seal with two inlet tubes into the rotor; one leads to the geometric centre of the rotor. Access to the volume of the rotor from this tube is via four holes drilled in the shaft of the assembly which holds the four septae. The other inlet tube has access (via the rotating seal) to the volume over the top of the septae. The centrifuge is run at about 5000 rpm and the gradient is pumped in (light end first through inlet B—see Fig. 4.11(a). The centrifugal force pushes the sucrose to the wall of the rotor and there it is displaced inwards by increasingly heavy sucrose. It is normal to finish the gradient with an underlay of dense sucrose for the following reason. When the rotor is full, the light end of the gradient appears in tube A which is the exit line. The sample is introduced by pumping it in via tube A (part of the underlay is now displaced and emerges through tube B). The last part of the loading procedure is to pump in an overlay (also via A, more underlay emerging from B) to push the sample out from the rotor core so that it starts an even annular zone. The rotor is now accelerated (maximum speed is 40 000 rpm) and run for the time necessary for the separation. After the run, the centrifuge is decelerated to about 5000 rpm again and the gradient is sampled by pumping heavy sucrose (heavier than the underlay) in via B. The gradient (light end first) emerges through A. The design problems encountered in rotors of this type are twofold. The bearing (Fig. 4.11(a)) has to be vacuum-tight and has to put up with speeds up to 40 000 rpm. Also there is a limit to the shape of the rotor. It has to be 'tall and thin' to minimize the wobble which occurs when a centrifuge is accelerated and which would break the seal at the top. The alternative design is that of the B-XIV and B-XV rotors (which differ from one another in capacity). In this case the seal is of a similar type to that in the B-IV type but is removable. The arrangement is more complicated in design but somewhat easier to use; also the restrictions on the shape of the B-IV rotor do not apply and these rotors are of a relatively wider and flatter shape giving a much greater length for the fractionation (i.e. radius) for a rotor of given volume. A simplified diagram of this type of rotor is shown in Fig. 4.11(b). In this case the route to the

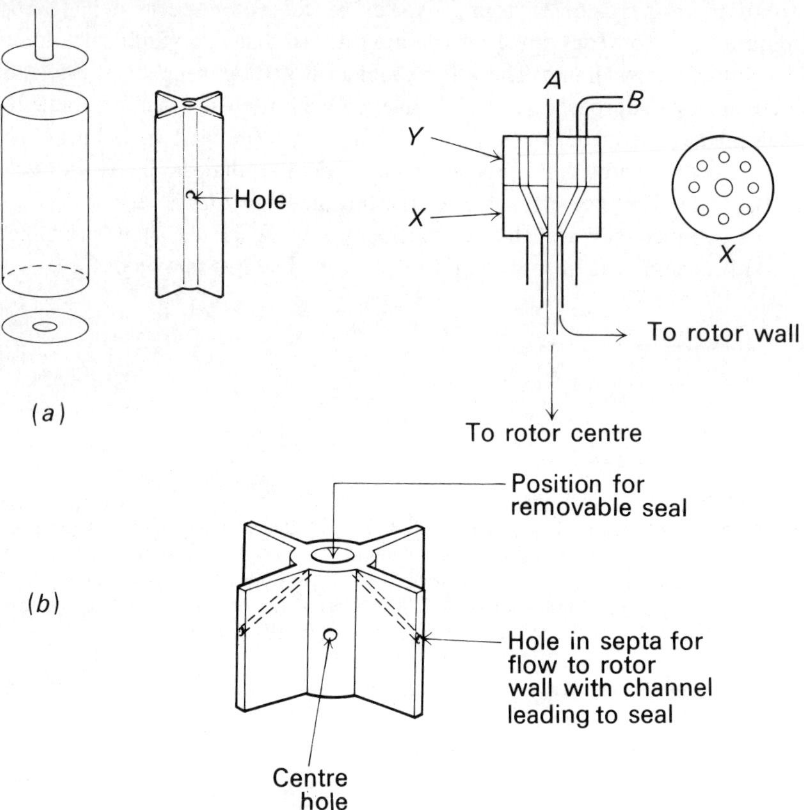

FIG 4.11   Components for zonal rotors for the ultracentrifuge. (*a*) Components of the B-IV rotor. From left to right: The rotor with detachable base plate (including connection to centrifuge drive) and lid. Centre piece with septae; the centre is hollow and there are four small holes (one in each quadrant) giving access from the hollow core to the gradient. Seal unit (on larger scale); the lower seal (X) is attached to the rotor—it is made of Teflon impregnated with jewellers' rouge. The channels in Y (the stationary part of the seal unit) and X are arranged as shown in the top view. Pipes A and B are referred to in the text. In fact there are more connections to the seal unit. There is an inlet and outlet pipe for a cooling water supply to the seal and also pipes for a continuous lubrication system for a second vacuum seal (not shown in this diagram) which allows the top of the rotor centre pipe with the seal attached to project from the vacuum chamber of the centrifuge. (*b*) Centrepiece for a detachable-seal zonal rotor.

wall of the rotor is via holes drilled through the septae in the rotor. The rotor is loaded as for the B-IV type (except that with channels through the septae to the wall, it can be loaded at lower speeds). The seal is then removed (with the rotor still spinning) and it is replaced with a vacuum-tight cap. The lid of the centrifuge is closed, the chamber evacuated and the centrifuge accelerated to speed. At the end of the run, the machine is decelerated to the unloading speed, the vacuum is released and the cap is replaced with the seal and the rotor is unloaded as described before. Technical and design details are available from centrifuge manufacturers; however for an appreciation of the technical problems I cannot do better than refer to the papers by Anderson and his group in the monograph of the National Cancer Institute devoted to the Zonal Centrifuge (Monograph No. 21, 1966). The papers relate to the theory of zonal centrifugation (Anderson, 1966) the design of zonal rotors of different geometries (Barringer, 1966), the B-IV rotor (Anderson and others, 1966), and removable-seal rotors[3] (Barringer, Anderson, Nunley, Ziehlke and Dritt, 1966).

We have not discussed the way in which large gradients can be formed for these rotors. Clearly a scaled up version of the sort of apparatus shown in Fig. 4.3(*a*) is fairly easy to construct from cylinders with a powerful stirrer in the mixing chamber. A big version of the apparatus of Noll (Fig. 4.3(*b*)) has been described by Birnie and Harvey (1968).

Virtually any experiment involving either rate-zonal or isopycnic centrifugation in a sucrose gradient can be performed in the zonal rotor. I illustrate the use of the centrifuge with a fractionation of RNA of high molecular weight from mammalian tissue (Fig. 4.12). In this example the distribution of 'rapidly labelled' RNA in a sucrose gradient is also demonstrated.

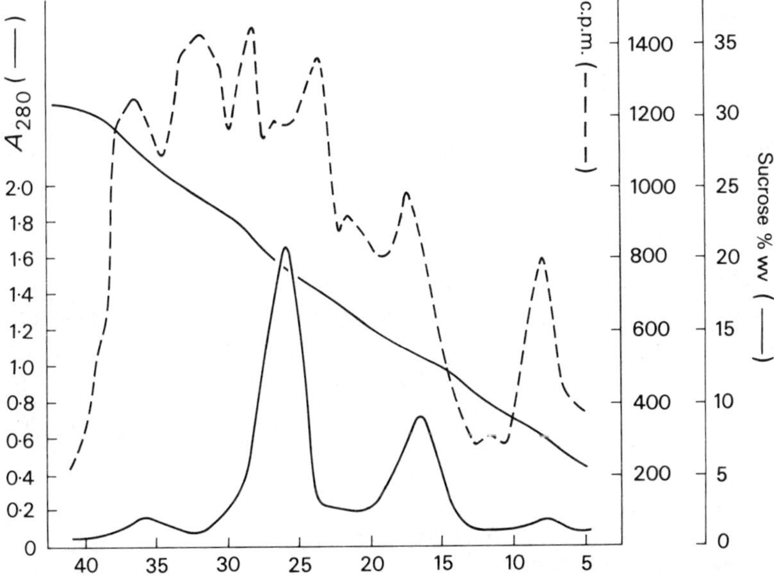

FIG 4.12   Fractionation of RNA using a zonal (B-IV) rotor. RNA was isolated from the livers of rats which had been injected with insulin (80 units per kg body weight) and radioactive orotic acid for 20 min before sacrifice. Forget about the insulin—its effect is not significant. A sucrose gradient was established in the B-IV rotor of the Beckman L4 ultracentrifuge. The gradient was 0·1 M sodium acetate pH 6·5 throughout. There was a sucrose underlay. The sample (40 mg RNA in 40 ml acetate buffer containing 5% sucrose) was introduced and overlayed with 160 ml of 2·5 sucrose (also in buffer). Centrifugation was at 5°C and 40 000 rpm for 6 h. The gradient was pumped out through a Beckman automatic recording spectrophotometer and the optical density was monitored at 280 nm. 40 ml fractions were collected. The concentration of sucrose was worked out by measuring the refractive index of each sample. 2 ml aliquots were removed and radioactivity measured by the method of p. 180. As this method only measures half the radioactivity in the aliquot, the total radioactivity in the samples is 40 x the figures recorded above.

These unpublished data were obtained by the author in collaboration with Mr. E. S. Klucis.

This gradient contains many familiar landmarks. Note the 45 S (small peak) and 35 S (shoulder) both associated with much of the radioactivity. The RNA had been isolated by the 'total' procedure (see p. 110) and most of the labelled material is nuclear. Notice also that the very small amount of tRNA left in the preparation is of high specific activity due to CCA-turnover (p. 228).

The identification of the components of this material will be discussed in the next chapter (p. 183). A suitable method of isolating the RNA from the large volumes of sucrose solution obtained from these gradients is to add strong sodium acetate (pH 6·5) until the final concentration of sodium acetate is about 10%. Ice-cold acetone is added until two phases have just formed (the solution is obviously cloudy—the amount depends on the strength of sucrose but is of the order of 1·8 volumes). The phases settle in an hour or so in the cold. The RNA forms a precipitate in the very small lower layer. The acetone (upper) layer is decanted and the RNA is isolated by adding a drop of alcohol (to produce one phase again) and spinning out the precipitate. It is possible to recover 1 mg of RNA from 100 ml of sucrose solution by this method (Hastings, Parish, Kirby and Klucis, 1965).

One limitation to the application of the zonal ultracentrifuge is that the seal tends to shear very long molecules (DNA). Otherwise the centrifuge is extremely valuable for a variety of large-scale separations. A theoretical discussion of the loading capacities of zonal rotors is given by Spragg and Rankin (1967). As an example of its application to a problem of value in work on nucleoprotein, the large (B-XV) rotors of the removable seal type will fractionate 30 S and 50 S ribosomal subunits (see Fig. 4.7, p. 140) with up to 2 g RNP per run.

Density gradients using materials other than sucrose   Several alternatives to sucrose have been advocated including trimethyl phosphate, cyclic sulphones, urea (Parish and Hastings, 1966) and chloral hydrate (which does not react with nucleic acids, Hastings, Parish, Kirby and Cook, 1969). The main disadvantage of this type of gradient is that there is a danger of fast-sedimenting, non-ribosomal RNA pelleting and (as this last paper points out) this can give rise to artefactual sedimentation profiles, especially if tube-piercing is employed for monitoring the gradient. The advantages of these methods are that certain types of aggregation are eliminated and radioactive material can be easily assayed as these components (with the exception of urea) are not precipitated by liquid scintillant (see p. 179). Although I was partly instrumental in trying out these alternatives to sucrose, I was of the opinion (until recently) that on the whole these gradients are not worth the trouble and we may as well stick to sucrose. However this has completely changed with the introduction by Hastings (1971) of an extremely versatile method of preparing and fractionating all nucleic acids in one go. The technique requires the zonal centrifuge if it is to be performed on a large scale. On a small scale, the cells are radioactively labelled (say with ^{32}P so that RNA and DNA will both be labelled) and are suspended in water (10^7–10^8/ml for bacteria) growth medium or buffer and added to three volumes of a lysing medium consisting of 2% 4-amino salicylate, 1·3% sodium dedecyl sulphate and 4% of phenol/cresol mixture (p. 109). Clearly certain bacteria (see p. 107) must be pre-treated with lysozyme (p. 111). They are left in lysing medium overnight at 4°C and are then layered over the discontinuous gradient illustrated in Fig. 4.13. The material sediments through a band of phenol and is then resolved into nucleic acid peaks as shown. This system is ideally suited to measuring sedimentation coefficients by the calibrated zonal centrifugation technique described on p. 385.

Examples of the use of such gradients are illustrated on p. 184 and p. 429.

All the density gradients described so far (including ones employing sucrose) contain media in which the secondary structure of RNA is essentially native. Attempts have been made to devise gradients in which the RNA sediments in a random-coil (denatured) conformation (p. 339). In the method of Sedat, Lyon and Sinsheimer (1969), the centrifugation is performed in 99% dimethyl sulphoxide (DMSO). To establish the gradient, two solutions are prepared: (i) 1 mM EDTA pH 7·0, 99% DMSO; (ii) 10%[4] sucrose, 1 mM EDTA pH 7·0 dissolved in 99% d_6-DMSO.[5] The linear gradient is established between (ii) (bottom of tube) and a mixture of (i)

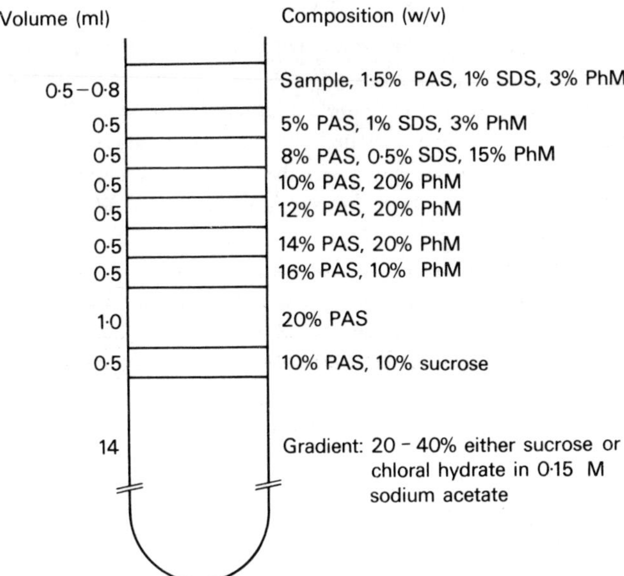

FIG 4.13   Discontinuous 'phenol gradient' of Hastings.

and (ii) (9:1 by vol.) at the meniscus. The RNA (in 5 μl) is mixed with 99% DMSO (90 μl) and formdimethylamide[6] (5 μl). The sedimentation position of zones can be readily related to the molecular weights of their constituents (p. 388).

The method of Sedat and others (1969) has only been succesfully applied so far to phage mRNA (p. 373) and bacterial mRNA. It is clearly rather expensive. An alternative method was originally devised by J. R. B. Hastings (unpublished work). The RNA is sedimented as its triethylammonium salt through a gradient of methanol to methoxyethanol. In a modification of the recipe that has been used in the author's laboratory by Marjorie Brown, glycerol is included in the gradient to enable the RNA to float over a density shelf at the start of the centrifugation. A linear gradient from 90% (v/v) methanol, 0·1 M triethylammonium acetate pH 7, 1% glycerol (w/v) to 90% (v/v) 2-methoxyethanol (same glycerol and buffer concentrations) is established in a polypropylene tube for an ultracentrifuge. Bacterial RNA is dissolved in water and methanol and triethylammonium acetate solution are added so that the solution will float on the 'methanol end' of the gradient and the triethylammonium acetate is 0·1 M. Using the 3 x 20 ml rotor of the MSE Superspeed 50 centrifuge, centrifugation for 15 h at 17°C and 25 000 rpm sediments (and resolves) 16 S and 23 S RNA; rapidly labelled RNA is distributed through the gradient.

DNA   The application of zonal centrifugation to DNA is limited by two considerations. For one thing the zones only separate on the basis of their sedimentation coefficients; most cellular DNA preparations contain heterodisperse material resulting from partial breakdown during the preparation. On the whole therefore a fractionation of this material on the basis of its molecular weight is unlikely to be of any great biological significance. Minor components of DNA are more likely to be separated by a method which exploits differences in their base ratios; separations which resolve molecules of different bases ratios are possible in a quite different type of centrifugation (p. 150). The other limitation is that, as we shall see in chapter 13, double stranded DNA has very anomalous hydrodynamic properties; one of these properties is that the sedimentation coefficients are extremely dependent on concentration, more so in fact than for any other class of macromolecule. When this is added to the fact that relatively strong solutions of DNA show a type of aggregation one has an explanation for the fact that if a significant amount of DNA is put on to a sucrose gradient (say 1 mg on a 30 ml gradient) the resulting distribution bears practically no relationship to the molecular species present. What is observed is a slow-moving peak with a trail of faster-moving material on its leading edge. Subsequent analysis of fractions from such a gradient indicates that the faster-moving molecules are not of higher molecular weight. They have merely broken away from the peak, and, finding themselves in an environment more dilute with respect to DNA, they sediment faster. Having said this much about the limitations it must be added that there are special circumstances in which sucrose gradient centrifugation of DNA is of value. The problem of anomalous sedimentation properties can be overcome in two ways, namely by using very small amounts of radioactive DNA or by using denatured DNA. The latter situation is achieved by using alkaline sucrose gradients.

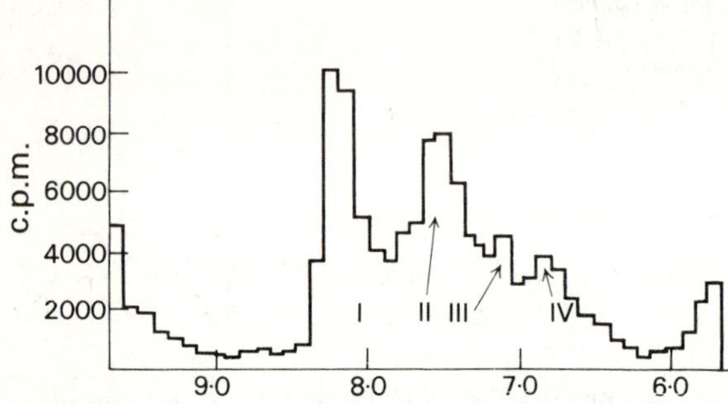

Radius (distance from centre of rotor in cm)

FIG 4.14   Demonstration of discontinuity in the strands of DNA in a bacteriophage *E. coli* H growing in a medium containing inorganic ^{32}P was infected with a suspension of T5st(O). Phage were purified from the resulting lysate and were deproteinized by suspending them in 0·1 M Na_3PO_4. This solution (which contains [^{32}P] DNA and unlabelled protein) was layered (0·1 ml sample) over 5 ml linear gradients of 5–20% sucrose in 0·9 M NaCl, 0·1 M NaOH in a tube for the SW 39L rotor for the Beckman model L untracentrifuge and was centrifuged for 3 h at 35 000 rpm at 20°C. The tube was punctured and drops were collected and the radioactivity determined. If the original duplex had consisted of two molecules of identical length we should expect there to be one homogeneous zone. In fact there are four zones (possibly five in some results). The authors calculate a chain length for these fragments. See p. 386. Reproduced from Abelson and Thomas (1966).

In subsequent chapters we shall consider examples of the use of alkaline sucrose gradients for detecting damage in cellular DNA (p. 257). We shall notice that it is often difficult to interpret the data unambiguously. In this chapter I have tried to choose clear-cut applications of the fractionation techniques and therefore I give only two examples of the use of sucrose gradients for DNA-fractionation. The first relates to bacteriophage T5, one of the T-odd phage infecting *E. coli* (p. 33). It is now recognized that the DNA of this phage is double stranded and contains single-strand breaks; that is to say the two molecules of the double helix each have a number of discontinuities in them. Obviously, provided that the double strands remain associated with one another in the duplex, and provided the breaks are not opposite one another, the duplex will have a molecular weight the same as that of a molecule in which the two component strands are unbroken. However if such a molecule is denatured such that the strands come apart the molecules which are produced will be less than half the molecular weight of the duplex. In Fig. 4.14 which is taken from the work of Abelson and Thomas (1966) we see that when such DNA is centrifuged through alkaline sucrose gradients the anomalously small single-strand fragments are identified.

Sucrose gradients can be employed to separate single- and double-stranded DNA. A very clever partial fractionation of transforming DNA (in *B. subtilis*, see p. 191) was achieved by Fried and Rapport (1970). DNA dissolved in 0·15 M sodium chloride, 0·015 M sodium citrate pH 7·0 was heated to 93·6°C for 15 min to denature all but the most stable (? high GC) regions. The DNA had been labelled *in vivo* with ^{32}P (see p. 175) and very little material (0·1 mg DNA) was layered over a 10-30% linear sucrose gradient (7·5 mM sodium phosphate buffer, 1 mM EDTA, pH 6·8) in 60 ml tubes (Beckman SW 25·2 rotor). Centrifugation was for 7 h at 25°C, 25 000 rpm. Two zones were found and the faster sedimenting (double stranded) one contained the histidine marker (see p. 191).

Caesium chloride gradients

In this chapter we shall only consider the application of buoyant density centrifugation to preparative fractionation. The very important application of these gradients to analytical methods is discussed in chapter 13. As far as preparative methods are concerned, the technique is restricted to DNA solution (see p. 133).

Caesium chloride gradients separate DNA molecules on the bases of differences in their density but not differences in their molecular weight.[7] Two factors can affect the density of DNA; base ratio and secondary structure. Although this method of separating native (double-stranded) from denatured DNA was used in the past there are rather simpler methods available these days (see p. 166) so we shall consider situations in which all the DNA is either single or double stranded throughout. Thus we have to look for applications of a technique separating molecules on the basis of their nucleotide content. The technique is an extremely sensitive one. For example in the Meselson and Stahl experiment (p. 389) *E. coli* DNA was fractionated in the analytical centrifuge on the basis of whether the bases contained ^{14}N or ^{15}N. In conventional (low-molecular-weight) chemistry such a mixture could only be analysed by a method using a mass spectrometer. I have grouped the use of such gradients under four headings with selective examples of each.

Preparation of gradients Caesium chloride solutions are usually referred to in terms of density of the final solution. Such solutions are most easily made up as w/w percentages. Thus if you have DNA in 1 ml of solution (= 1 g of aqueous solution) to make up an x% w/w solution you simply weigh out y g of CsCl and dissolve it where:

$$x = \frac{y}{1+y}$$

There is no need to make up to a volume as is needed to make up w/v solutions. To make up a solution of given density at 25°C (ρ^{25}) the requisite equation is:

$$x \ (\text{w/w} \ \%) = 137·48 - \frac{138·11}{\rho^{25}}$$

As with sucrose solutions (see p. 136) the refractive index is a useful parameter for checking concentration. Expressed in terms of ρ^{25} the relationship is

$$\rho^{25} = 10·8601 \ n_D^{25} - 13·4974$$

where n_D^{25} is the refractive index at 25°C at the wavelength of Na-D light. If, despite careful weighing, a check of refractive shows the density is too low by $\Delta\rho$ then you add ($\Delta\rho^{25} \times V \times$

1·32) g of solid CsCl to V ml of solution. If it is too high you add ($\Delta\rho^{25}$ x V x 1·52) ml water (or buffer).

All these relationships are accurate in the range $1·25 > \rho > 1·9$ g/cm^3 (Vinograd and Hearst, 1962).

Use of CsCl gradients for isolating DNA   Cairns's method for isolating intact bacterial chromosomes has been described (p. 108). An alternative method is due to Davern (1966). In this procedure the cells are lysed and the resulting suspension of protoplasts is inserted into the discontinuous gradient shown in Fig. 4.17. The gradient is left for 16 h (not in the centrifuge). During this period the pronase (a proteolytic enzyme) and detergent slowly diffuse into the protoplast layer and release the DNA. Pronase is effective in the presence of detergent. The gradient is then centrifuged and the DNA forms a band in the CsCl.

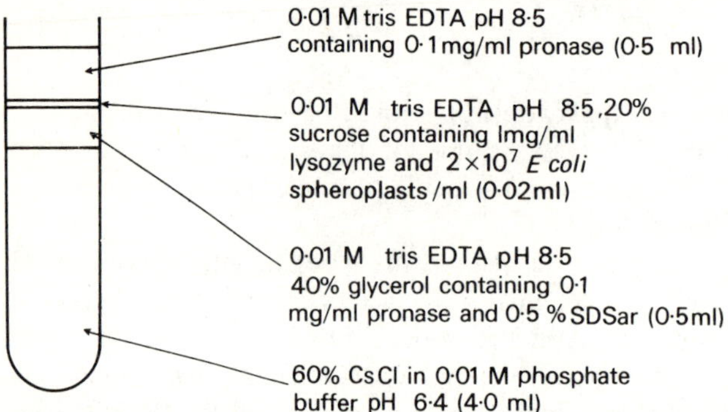

FIG 4.15   Davern's method (Davern, 1966) of isolating whole chromosomes from *E. coli*. The layers are made up in a tube for the Beckman SW 39 rotor. SDSar is sodium dodecyl sarcosinate, a detergent which, unlike SDS (sodium dodecyl sulphate) is not precipitated by strong salt. The '*E. coli* layer' is incubated at 37°C for $\frac{1}{2}$ h before being introduced. After standing for 16 h (see text) the tube is centrifuged for at least 50 h at 35 000 rpm at 15–20°C.

Minor components of DNA   Both prokaryotic and eukaryotic cells may contain minor DNA components. In the former case the small DNA molecules (plasmids) have been isolated by CsCl centrifugation in those fortunate cases where they are of different base composition to the chromosomal DNA. This method of isolation is now of largely historical interest as we shall see in chapter 9. Additionally the DNA of chloroplasts, mitochondria and other organelles from eukaryotic cells is normally isolated from the purified organelles and not by isolating DNA of anomalous density and then ascribing a function to it. Nevertheless there are DNA molecules of obscure physiological role which can only be isolated by this technique and also chromosomal DNA can be fractionated (in favourable circumstances) to yield cistrons with base compositions and hence densities, significantly different to the bulk material. Examples of some of these components are discussed below and on p. 278.

A very puzzling case which remains to be explained is the organism *Bacillus subtilis* strain Cbl-I whose spores contain an anomalous 'heavy' DNA molecule which appears to have the same base composition as normal *B. subtilis* DNA (Szulmajster, Arnaud and Yang, 1969). There may be a simple explanation in terms of different secondary structure and I recognize that this particular organism may be regarded as of restricted interest but it is worth bearing in mind, when devising schemes for handling the nucleic acids from 'new' organisms, that anomalous results may turn up. In chapter 9 there is a discussion of some of the pitfalls in using CsCl banding and other indirect methods for measuring the base ratios of DNA.

The classical case of a minor DNA component of grossly different base composition is the

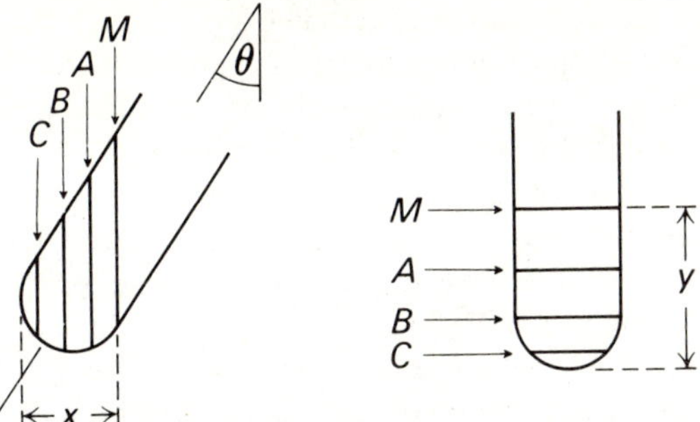

FIG 4.16   Effect of re-orientation in a fixed-angle centrifuge rotor. *M* is the meniscus; *A, B* and *C* can be regarded as very thin zones or as boundaries defining the width of zones. During the centrifugation (in this hypothetical case) when equilibrium is attained *MA*, *AB*, *BC* are all equal distances. In the sampling position (right-hand diagram) *MA* > *AB* > *BC*. In the well-known 40 and 50 Ti fixed-angle rotors for the Beckman centrifuges, angle θ is 26° and typically *x* and *y* would be respectively 1·6 and 2·6 cm.

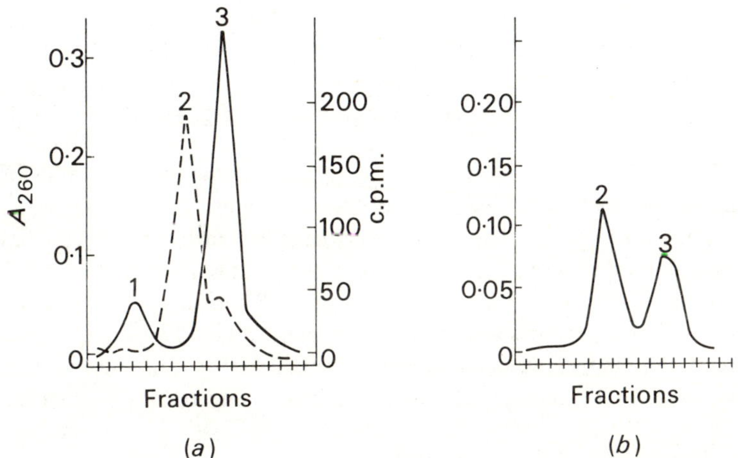

FIG 4.17   Isolation of the ribosomal cistrons from the toad *Xenopus laevis*. Total DNA was isolated by a conventional phenol procedure. In both gradients the dense end is to the left. (*a*) A preliminary experiment to identify the density of the ribosomal cistrons. *Xenopus* DNA was banded in CsCl (details below under (*b*)) together with a 'marker' (DNA of known density, in this case denatured DNA from *Pseudomonas aeruginosa*). There are two peaks of optical density (———) namely 1 (marker) and 3 (*Xenopus* DNA). Using radioactive RNA, the samples were assayed for complementarity with rRNA and another peak of density 1·723 (no. 2) was identified.

The preparative experiment consisted of dissolving sheared DNA (molecular weight around 20 million) in CsCl (mean density 1·715; 0·15 mg DNA/ml): the grandients were centrifuged in high capacity angle rotors (15 ml in each of tubes with a total capacity of 30 ml) in the MSE 2406 rotor. A total of 16 tubes (two rotors full) were centrifuged at 28 000 rpm at 20°C for 6 days in the MSE Superspeed 50 ultracentrifuge. Density was monitored by refractive index and the material of density 1·723 was collected and redistributed in the SW 39 rotor of the Beckman model L ultracentrifuge. The second centrifugation (*b*) was for 36 h at 33 000 rpm at 20°C. Fractions were taken and further RNA-hybridization tests (not shown here) confirmed the identities of peaks 2 and 3. Analytical ultracentrifugation established that peak 2 2 collected from gradient (*b*) was pure material of density 1·723 and hence purified ribosomal cistrons. Data from Birnstiel (1967).

molecule which is about 97% A and T found in the testis tissues of certain crabs. In these cases the two bands can be obtained in fixed angle rotors and the minor (light) component can be subsequently purified in a swinging bucket rotor. For references to work on the isolation of minor components of animal DNA by this method and an example of the technique as applied to the crabs *Cancer pagarus* and *Geocarcinus lateralis* see the paper by Skinner (1967). As this author points out in her paper the second of these two animals (a land crab) would appear to be especially useful for further work on these minor components as it can be handled readily as an experimental animal in the laboratory. The structure of this DNA is discussed on p. 278.

A technique of using fixed-angle rotors for CsCl gradients was devised by Flamm, Bord and Burr (1966). Obviously as the gradients are spun to equilibrium and one is not dealing with a sedimentation rate effect there is not the objection to using such rotors which obtains with sucrose gradients (p. 133). The only thing that does need to be born in mind is that the gradient is sampled *after* it has re-oriented. While it is spinning the density/radius slope is approximately linear. However, unlike the bucket of a swing-out rotor, the volume of 'slices' of the gradient of equal width are not independent of radius. Thus after the gradient has re-oriented to its 'dripping-out' position the density is *not* a linear function of tube length any more. The effect of the distortion is to squeeze up zones towards the bottom of the tube and spread out those near the top (Fig. 4.16).

An important example of the use of such gradients is the isolation of ribosomal cistrons (the sequences of DNA determining synthesis of rRNA). The work of M. L. Birnstiel and his colleagues on the identification of these cistrons is described on p. 368. As an example of the isolation of ribosomal cistrons, I have chosen the data on *Xenopus laevis* (Fig. 4.17).

Separating DNA strands   Chargaff's rules (p. 7) follow as requirements for the base ratios of any pair of complementary molecules in a double helix. It does not follow, of course, that the two component molecules need obey these rules. Thus it might happen that one molecule has the greater number of G and A residues in it and the other one the greater number of C and T. In this case the former (purine-rich) molecule would be the lighter. When this situation arises the two molecules will have different densities and should be separable in CsCl gradients. As we might expect the chances of this happening being greater in smaller DNA molecules than in larger molecules it would seem to be more likely that the base compositions might even themselves out.

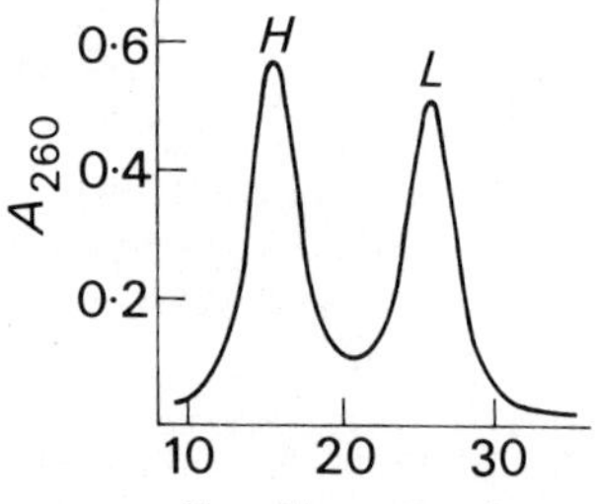

FIG 4.18   Separation of DNA strands. Mouse satellite DNA, a minor (light) component of mouse DNA was first isolated by a protocol similar to that given for *Xenopus* ribosomal DNA in Fig. 4.17. The satellite DNA (0·06 mg) was dissolved in 3·2 ml of 0·01 M tris HCl pH 8·5. This this was added 0·1 ml of 1% SDS, 0·1 ml of 1·0 N NaOH and 4·615 g CsCl. Centrifugation in the 10 x 10 ml fixed angle rotor of the MSE Superspeed 50 ultracentrifuge was for 24 h at 42 000 rpm at 25°C. The gradient was dripped and the optical density of the fractions at 260 nm was recorded. A complete separation of heavy (*H*) and light (*L*) strands was achieved. Data from Flamm, Birnstiel and Walker (1969).

This expectation is not altogether born out, as it has been possible (though not with CsCl gradients) to separate the molecules of DNA from *B. subtilis* (p. 165). Moreover it has been shown recently (Hershberger, Mickel and Rownd, 1971) that there is an asymmetric distribution of residues between the two strands of DNA from *E. coli, Proteus mirabilis* and *Serratia marcescens* and that the strands from these organisms (all Enterobacteriaceae) can be partially resolved on alkaline CsCl gradients. Nevertheless most success has been obtained with viral DNA and the minor DNA components of cells. I have delayed discussing the most remarkable case of strand separation and its exploitation, namely the experiments on DNA of bacteriophage λ until p. 373.

In certain cases, the densities are so different that it is possible to separate the two strands completely in an alkaline CsCl gradient. An example of this is the mouse satellite DNA (a minor component of uncertain function—see p. 278) in which the purine/pyrimidine ratio is 0·5 for the heavy (*H*-) strand and 1·8 for the light (*L*-) strand. The separation of these strands is illustrated in Fig. 4.18, taken from the review by Flamm, Birnstiel and Walker (1969).

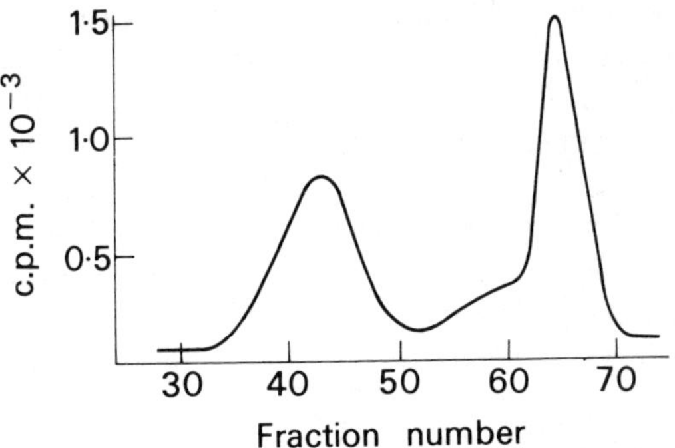

FIG 4.19   Strand separation in the DNA of a phage. *Bacillus subtilis* BS25 grown in a medium containing [^{32}P]phosphate was infected with phage SPPI. The resulting lysate was purified for phage which were themselves further purified by CsCl banding. Phage were deproteinized with phenol and phosphate buffer. [^{32}P]phage-DNA (0·05 mg) and rRNA (0·15 mg) were dissolved in 3·0 ml of 0·03 M NaCl, 0·003 M trisodium citrate. The solution was heated to 100°C for 10 min and chilled. The artefactual 'hybrids' (see text) were thus produced. CsCl and water were added to a final total volume of 13 ml and density 1·700. Centrifugation in fixed angle 40 rotor of the Beckman model L ultracentrifuge was for 36 h at 37 500 rpm and 20°C. Two-drop fractions were collected and the radioactivity was assayed. Only a portion of the gradient is shown above—120 fractions were taken altogether. Two peaks of densities 1·741 (on the left) and 1·713 (on the right) were obtained. RNA was removed from the pooled fractions by treatment with ribonuclease and subsequent dialysis. Data from the work of Riva, Polsinelli and Falaschi (1968).

In those cases where the difference in the base ratios of the two strands are less marked it is possible to 'amplify' the difference by partially re-annealing (see p. 357) the denatured DNA in the presence of an excess of some other nucleic acid (normally RNA). In such cases the new structures which are produced are artefactual. Nothing resembling a duplex with extended regions of base pairing is produced; rather the association of very short regions of RNA with a few bases in the DNA produces an open structure, largely denatured, and tied together at a few points by short base-paired regions. An example of a separation of this type is the two strands of DNA from *B. subtilis* phage SPPI (Fig. 4.19, Riva, Polsinelli and Faschi, 1968).

For many purposes, synthetic heteropolymers, such as poly(U, G) are employed for this type of fractionation (p. 373; Szybalski, Kubinski, Hradecna and Summers, 1971).

Phage replicative forms   Many applications of CsCl fractionation to identifying modes of DNA replication have been reported (see pp. 390 and 431). In particular, the work of R. L. Sinsheimer on the replication of DNA in the single-stranded circular DNA of coliphage φ X 174 has been one of the major achievements of studies on DNA synthesis. An example of his work is the proof that these are two double-stranded replicative forms, RF II and RF I.[8] Replication of the single-strand phage DNA (called the + strand) leads to a continuous closed duplex (RF II); a single-strand break (see p. 187) yields a molecule (RF I) which is still a continuous duplex but with a break in one strand (see p. 424). Knippers, Komano and Sinskeimer (1968) were able to show that RF I is in fact a mixture of two structures: the one in which the break is in the + strand and the other in which it is in the complementary (−) strand. Single-stranded phage DNA, RF I and RF II were separated by zonal centrifugation through a pre-formed CsCl gradient (ρ = 1·20–1·77) at pH 7. RF I was isolated and centrifuged to equilibrium in a CsCl gradient (ρ = 1·74–1·77) at pH 12·4. Three separate zones were found corresponding to single-strand continuous circles, linear single-strand + and linear single-strand −.

Density labelling   There are two methods of obtaining DNA of anomalous density. One is to grow the organism in a medium in which the nitrogen is present as ^{15}N; this is the basis of the Meselson and Stahl experiment (p. 390). The other is to incorporate a modified base into the structure. Both of these techniques are only feasible in microbiological systems. An example of the latter approach, is the experiment of Pettijohn and Hanawalt (1964). They used a strain of *E. coli* which is a thymine auxotroph. When the organism is suspended in a medium lacking thymine it will not grow;[9] however, if 5-bromouracil (br^{5}Ura) is present in the medium, it will grow and incorporate into its DNA this base in place of thymine. DNA in which brU is present in place of T is more dense. In Fig. 4.20 is shown a fractionation of molecules of normal

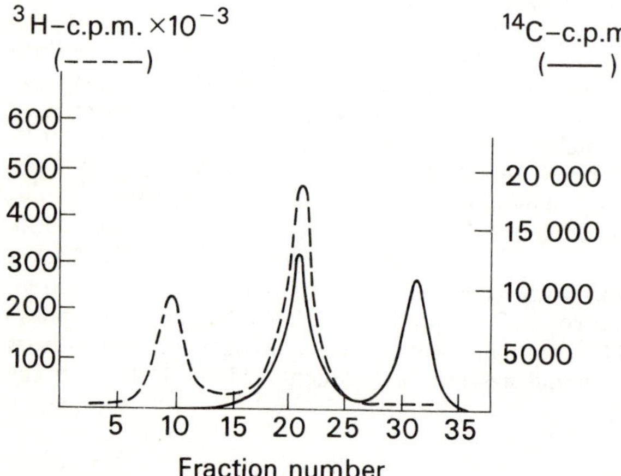

FIG 4.20   Fractionation of DNA containing bromouracil. *E. coli* TAU-bar was grown for four generations in a medium containing [^{14}C]Thy and was then transferred to a medium lacking Thy but containing [^{3}H]br^{5}Ura for a short time (less than a generation). DNA was isolated by the method of p. 107 and was dissolved in 0·1 M NaCl, 10 mM tris EDTA pH 8. CsCl (3·95 g) was dissolved in 3 ml of this solution and the preparation was centrifuged for 48 h at 37 000 rpm in the SW 39 rotor of the Beckman model L ultracentrifuge at 20°C. Fractions were obtained by tube-piercing and drop counting (p. 171) and the two isotopes were counted (p. 179). Three categories of DNA duplex are separated. That containing only ^{3}H contains [br^5]Ura-DNA in both strands; that containing only ^{14}C contains normal (Thy-) DNA in both strands; that containing both isotopes consists of a duplex with one strand of each. Data from Pettijohn and Hanawalt (1964).

(T-containing DNA from br^{5}Ura-DNA and hybrid (one strand of each). Detailed studies on br^{5}Ura-DNA by Hanawalt's group and others have proved extremely valuable in elucidating details of the mechanisms of synthesis and repair of DNA (see p. 255).

Disc electrophoresis

For the analytical fractionation of RNA, the best resolved method is disc electrophoresis. Although it has limitations as a preparative technique, as a method of seeing what sort of RNA there is in a mixture, the technique has everything. Resolution is unequalled by any other method, the apparatus is cheap to construct and the running time is very short. The gel material of choice is polyacrylamide and the techniques to follow are those of U. E. Loening who has revolutionized RNA research (and dealt ultracentrifuge manufacturers a blow from which they must still be smarting) by painstakingly perfecting the conditions for gel electrophoresis of RNA. A good review of his methods is Loening (1968b). The technical points that follow are those of Loening with a few notes from the author's own experience and the polyacrylamide lore which circulates among RNA laboratories.

Technical methods

Principles of gel electrophoresis   Charged molecules (polyanions in the case of RNA) migrate in electric fields. In the technique of gel electrophoresis the molecular species from tight zones or 'discs' in the field and migrate at characteristic rates through a gel matrix. In the

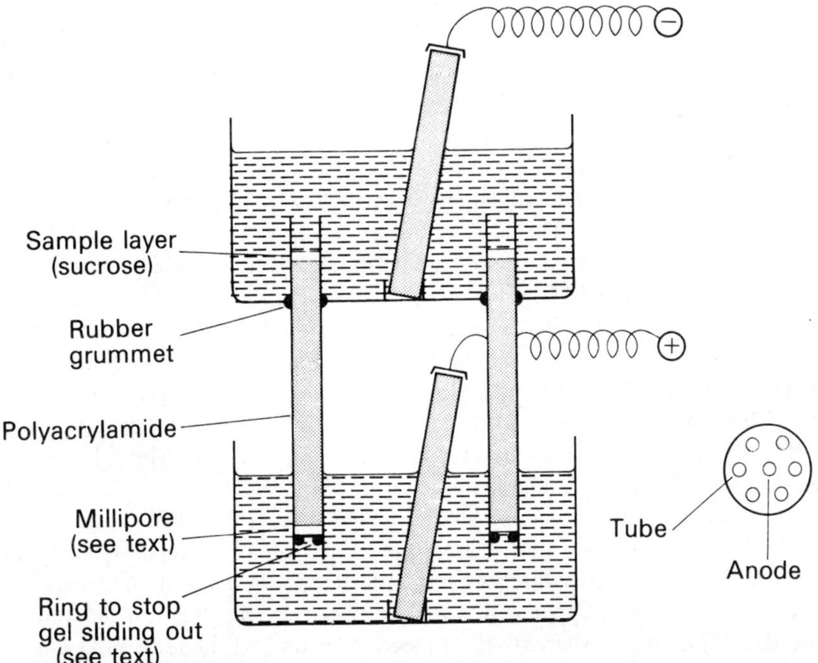

FIG 4.21   Apparatus for disc electrophoresis. The upper electrolyte chamber (top view shown in insert) is a plastic sandwich box with holes drilled in it. The lower chamber is either a second such box or beaker or other suitable container. The electrodes are either platinum (very expensive) or carbon rods taken out of scrapped dry batteries (cheap).

electrophoresis of proteins, tightening of the discs is achieved by allowing the proteins to pass through a dilute 'spacer gel' where the polyelectrolytes are concentrated by the distribution of the rapidly diffusible buffer ions (quantitated by consideration of Kohlrausch's regulating function). With RNA this spacer gel is omitted as the zones are self-sharpening anyway for the same reason that boundaries in the ultracentrifuge are self-sharpening (see p. 384). The apparatus that is required is shown in outline in Fig. 4.21. I have illustrated the simple device where the gel is present in tubes. If you are working in a laboratory where there is a slab gel apparatus, the methods of use should be apparent. Designs for slab electrophoresis systems are described by Gordon (1969). A simple method of making a slab gel is described on p. 311. In the tube apparatus, the RNA is layered over the top of the gel as a solution in sucrose.

Making up the gels   Polyacrylamide gels are made by co-polymerizing acrylamide (CH_2=CHCONH$_2$) and 'BIS' (methylenebisacrylamide, NH_2COCH=CH–CH_2–CH=CHCONH$_2$) in solution. The BIS is a cross-linking agent. The other compounds are buffer, an initiator ('TEMED' tetramethylethylenediamine, $Me_2NCH_2CH_2NMe_2$) and the catalyst for the polymerization (ammonium persulphate). Care is required for RNA electrophoresis as for certain purposes very dilute gels have to be made. These very dilute gels required ultra-pure acrylamide and BIS and in addition there is a problem in getting them out of the tubes. Polyacrylamide adheres to ordinary glass tubing and it is impossible to get the gels out without breaking them up. The tubing should be made of Perspex or alternatively glass tubing can be cleaned thoroughly with chromic acid, carefully rinsed and dried and then siliconized. The easiest way to siliconize it is to soak the tubing for some time (an hour or so) in a dilute solution (about 1%) of hexamethyldisilazane or trimethylsilyl chloride in light petroleum. (These reagents can be used over and over again.) The tubing is removed and baked in an oven at 110°C for several hours.

Acrylamide is purified by recrystallization from chloroform. A 7% solution is filtered at 50°C and slowly cooled to −20°C and the crystals are filtered off in the cold. BIS is dissolved in the minimum volume of acetone at 50°C and recrystallized in the same way. AnalaR solvents are essential for these purifications. Two stock solutions are then made up, *A* (15% w/v acrylamide, 0·75% w/v BIS) and *B* (2·42% w/v tris, 0·82% w/v anhydrous sodium acetate, 0·185% EDTA Na$_2$ salt and acetic acid to adjust the pH to 7·8). The gels are defined by the final concentrations of polyacrylamide. The stock solutions are mixed with water as follows; percentages refer to final gel concentrations. You start with 5 ml of *A* and add: 7·5 ml *B* and 24·7 ml water for 2·0%; 6·8 ml *B* and 22·0 ml water for 2·2%; 6·25 ml *B* and 19·7 ml water for 2·4%; 6·0 ml *B* and 18·7 ml water for 2·5%; 5·8 ml of *B* and 17·8 ml water for 2·6%; 5·0 ml *B* and 14·7 ml water for 3·0%; 3·0 ml *B* and 6·7 ml water for 5·0%. For stronger gels (>5%), solution *A* contains only 0·375% BIS; everything else is the same. As an example of a stronger gel: for 7·5% to the 5·0 ml of *A*, is added 2·0 ml *B* and 2·7 ml water. The gel tubes (typically 7–10 cm long and 7 mm internal diameter) are stoppered and clamped vertically. Everything must be ready because you have to work fast from now on. The mixture (*A*, *B* and water) is degassed under vacuum for half a minute or so. To it is added 0·025 ml TEMED and 0·25 ml freshly prepared 10% (w/v) ammonium persulphate and it is now pipetted into the tubes immediately (it sets quite quickly). After the gel has set the stopper in the bottom of the tube is removed. It is wise to replace the stopper with a little rubber ring that fits into the tube to stop the gel sliding out of the tube. Alternatively a piece of matchstick or something can be jammed into the end for the same reason. For the 2% or 2·2% gels it is a good idea to have a disc of Millipore filter under the gel; the gel is so dilute it can flow past the ring (or matchstick). The buffer for this type of electrophoresis is *B* diluted by 5. The electrophoresis is performed either in the cold or at room temperature. In the latter case it is possible to include a detergent as ribonuclease inhibitor in the buffer (either SDS or TIPNS about 0·2%). After the tubes are in place, make sure there is not a bubble of air trapped underneath the gel; such bubbles can be removed with a hypodermic syringe with a bent needle.

The gel is pre-run for 30 min at 8 V/cm. The current flow should be about 5 mA per tube. The current is switched off and the RNA (up to 0·1 mg) in running buffer (up to 0·05 ml) containing enough sucrose for it to form a tight band (about 10% sucrose) is layered over the gel. The gel is run for up to 4 h (examples are given later). After electrophoresis the rubber ring (or whatever) is removed from the bottom of the tube and the gel is gently slid out of the tube; it may need carefully blowing. If it will not come out easily, the tube was not clean enough or was insufficiently siliconized. If this is the case, there is nothing to be done except be more careful next time. These gels cannot be released with a wire in the same way as stiffer gels (used for protein electrophoresis) or they will break up.

Identification of RNA on the gels   The most direct measurement of the RNA is to scan the gel in a Chromoscan with ultra-violet optics. In the absence of such a device, the gel may be stained. The best stains are probably either 0·1% toluidine blue or 0·1% methylene blue dissolved in 40% (v/v) aqueous 2-ethoxyethanol. The gel should be left in the stain for about 2 h and then destained in the 40% ethoxyethanol overnight. The stained bands can again be recorded on a Chromoscan (with visible optics). All these procedures involve handling the gels outside the tube. With thin gels (less than 4%) it is very difficult to handle them with tweezers or by hand without breaking them. The simplest method is to have a siliconized glass tube (or a Perspex tube) of the same size as the one in which the gel was run with a teat on the end. With a little practice it is easy to suck the gel back into a tube without damaging it at all, and then it can be gently blown out into whatever vessel is used for the next part of its processing.

For many purposes it is necessary to slice the gel into segments (of 1 mm or so) for radioactive counting, recovery of sample and so on. Methods of radioactive counting are discussed on p. 180. The best way to slice the gels up is to freeze them (solid carbon dioxide) and slice them on a commercially available gel slicer. Alternatively some laboratory lash-up can be constructed. Probably the simplest is a set of razor blades bolted together with washers or other spacers between the blades. This device is dropped gently on to the frozen gel and the segments then adhere to the spaces between the blades and can be poked out with a spatula.

Recovery of RNA for subsequent experiments   Elution of RNA from gel slices is not an easy job. If the RNA zones are stained, the simplest method is to grind up the segments with a little phenol and buffer. The dye is extracted by the phenol. Otherwise the RNA can be soaked out in strong salt (1 or 2 M) containing 1% SDS (this method requires slices 1 mm thick or even thinner) or it can be electrophoresed out. The simplest way of doing this is to close the end of an electrophoresis tube (Perspex) by glueing a piece of dialysis tubing over the end, putting buffer containing say 30% sucrose in to a depth of around 5 mm and topping up with buffer. Gel pieces will float above the sucrose/buffer interface and the RNA can be electrophoresed into the sucrose layer. The trouble with all these methods is that the RNA is contaminated with a sticky substance (presumably linear polyacrylamide) with the same solubility properties as RNA. If it is merely to be used for base-ratio determinations or some other procedure where the volume of the solution is not critical, this is unimportant. However if very radioactive ^{32}P-labelled RNA is recovered from gels for studies on its primary structure (see p. 393), it is vital to get rid of the stuff. Either the solution can be made about 2 M with respect to sodium chloride and the RNA aggregates (there is no visible precipitate) can be spun out in the ultracentrifuge (this normally requires several repeated centrifugations) or alternatively it can be removed by running the RNA through a small column of Sephadex G-25 Superfine pre-equilibrated in water. The RNA is recovered from the eluate by freeze-drying.

Examples of the use of polyacrylamide-gel electrophoresis of RNA

High-molecular-weight RNA   DNA does not migrate in these gels at all so removal of DNA is not necessary in preparations to be analysed in this way. For RNA of the sort of molecular

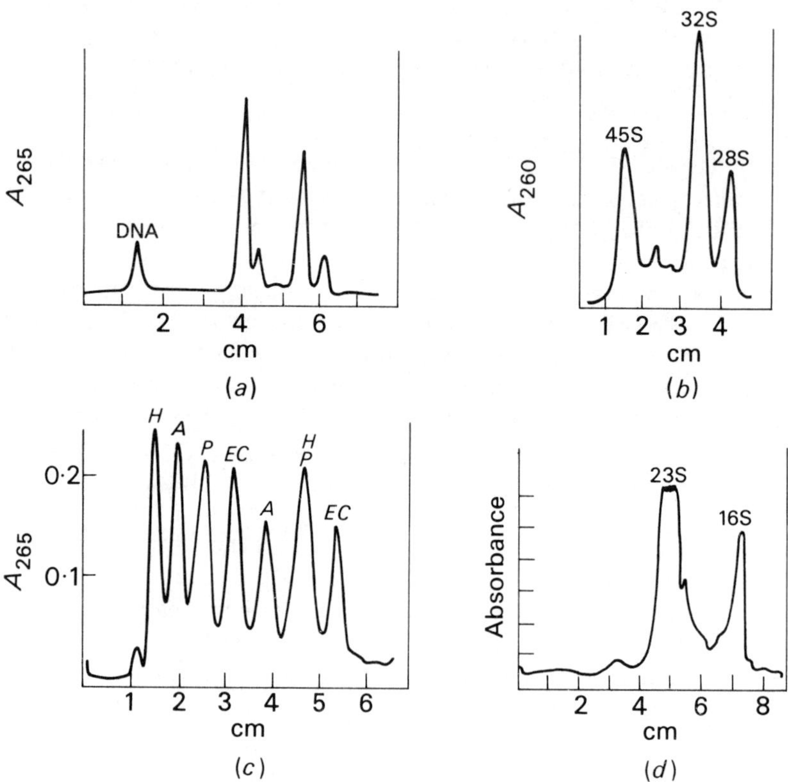

FIG 4.22   Examples of the fractionation of high-molecular-weight RNA by gel electrophoresis on polyacrylamide. All gels were 7–8 cm long and migration from left to right.

(*a*) RNA from *Phaseolus aureus* (mung bean) leaves. 2·2% gels electrophoresed at 50 V for 3·5 h. The gel was scanned directly at 260 nm. The two larger peaks are cytoplasmic rRNA; the two smaller ones are chloroplast rRNA. Data from Grierson, Rogers, Sartirana and Loening (1970).

(*b*) RNA extracted from HeLa nucleoli (p. 140). 2·4% gels electrophoresed at 50 V for 4 h. The gel was scanned directly at 265 nm. Numbers refer to S-values. Thus 18 S and 28 S are mature rRNA. The others are precursors of various types (see p. 182). Data from Weinberg, Loening, Willems and Penman (1967).

(*c*) Synthetic mixture of rRNA from HeLa (*H*), *E. coli* (*EC*), *Acanthamoeba* (*A*) and Pea (*P*). 2·4% gel electrophoresed for 3·5 h at 50 V. The gel was scanned directly at 260 nm. The larger molecule in each case is the peak to the left of the electrophoretogram; the smaller one is the peak to the right (18S RNA from pea and HeLa forms an unresolved peak). Data from Loening (1968a).

(*d*) 'Pure 23 S RNA' (*E. coli*) extracted from a sucrose gradient (such as Fig. 4.5(*a*), p. 138). The gel was intentionally overloaded to show up minor components. 2·4% gel electrophoresed for 4 h at 50 V. The gel was stained with toluidine blue (see text) and scanned with an orange filter. Data from Parish, Khairul Bashar, Brown and Brown (1971).

weight of bacterial 23 S and 16 S RNA, the gel should be 2·6%. For eukaryotic RNA (18 S, 28 S and larger) the gel should be 2·4 or 2·2%. An example of the superb resolution obtainable is shown in Fig. 4.22. The nature of the large RNA precursors will be discussed later (pp. 183 and 395) as will the logarithmic relationship between molecular weight and distance of migration. This particular piece of work was performed in a way which rigorously excluded the possibility that the very large molecules were not aggregated. However it is possible for rRNA aggregates, formed during the extraction of RNA, to form spurious slow moving zones. An analysis of this problem with respect to RNA from a fungus (*Blastocladiella emersonii*) has been made by Lovett and Leaver (1969).

Low-molecular-weight RNA   For separating tRNA (4 S) and 5 S rRNA, the best procedure is to use 7·5% gels. The separation (but with less resolution) can be achieved on 5% gels. These gels are also useful for resolving components of around 7 S. These RNA molecules are found in many cells. The 6 S RNA from *E. coli* is mentioned later (p. 309). In eukaryotic cells there is a 7 S rRNA[10] associated by hydrogen bonding with the 28 S molecule. In the nucleolus there are several species of RNA in this size range. A paper which brings much of the earlier evidence together is by Prestayko, Tonato and Busch (1970). An example of the gel electrophoresis of these substances is shown in Fig. 4.23. Unfortunately it is not clear from the text exactly what conditions were employed by the authors; however we can safely assume that such resolutions were obtained using the standard conditions described on p. 157 for 2 h running times on 5% gels. Notice that in this particular case, the ribosomal (cytoplasmic) 28 S RNA is associated with a 7 S molecule but that in the nucleolus (the site of synthesis of all rRNA apart from the 5

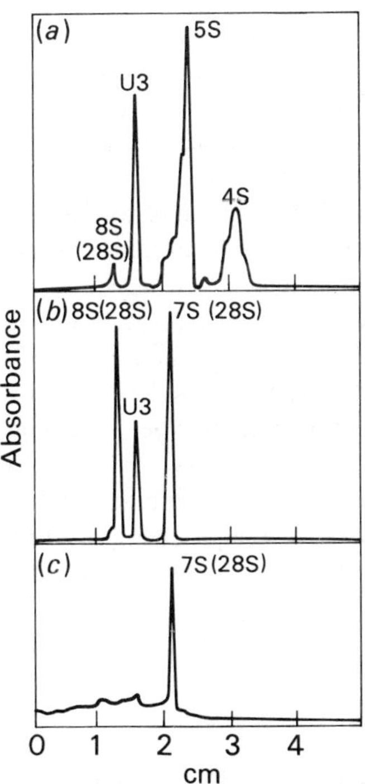

FIG 4.23   Fractionation by polyacrylamide-gel electrophoresis of low-molecular-weight mammalian RNA. (*a*) Total nucleolar RNA (low-molecular-weight components only); (*b*) low-molecular-weight RNA obtained following the heating of nucleolar 28 S RNA; (*c*) low-molecular-weight RNA obtained by heating cytoplasmic 28 S RNA. Data from Prestayko, Tonato and Busch (1970).

S) there is also an 8 S molecule apparently associated with 28 S RNA in a similar fashion. The RNA U3 (see Fig. 4.23) is a uridylic acid-rich species that is only found in nucleolar preparations and is of unknown function; however for a further discussion of this type of RNA see p. 372.

Other techniques Dingman and Peacock (1968) have done important experiments on the correlation of electrophoretic mobility with molecular weight and their work is discussed later (p. 394). A technical innovation of theirs is to mechanically strengthen the very dilute gels by incorporating agarose into the gel. The method is carefully described in their paper and is easy to follow. It is certainly of some advantage in making the gels a bit easier to handle, but with practice the gels can be handled as I described above and agarose is not really necessary. If you do use agarose, remember that the gels will run a bit more slowly and recovery of RNA from zones (at least in our hands) is more difficult.

A valuable contribution from A. C. Peacock's group (Dingman and Peacock, 1968; Dahlberg, Dingman and Peacock, 1969), which is outside the scope of this book, is the use of polyacrylamide gel electrophoresis for the analysis of ribosomes. The technique is a particularly elegant one as the ribosomes can be resolved on one gel; the zones can then be treated with SDS and re-run to reveal the nature of their RNA complement. The procedure is analogous to the use of SDS for deproteinization on sucrose gradients (p. 138).

Ion-exchange chromatography

Nucleic acids are polyanions and hence they will bind to insoluble cation matrices (anion-exchange resins). They can be eluted with solutions of salts whose anions displace the nucleic acid. For nucleic acids of varying molecular weight, obviously the larger molecules will have greater affinity for the resin than the smaller ones as there are more negatively charged (phosphodiester) groups per molecule and the number of binding sites will be greater. For nucleic acids of comparable molecular weight the molecules are selectively bound to the resin depending on their nucleotide composition and primary structure. The effects are extremely difficult to quantitate, largely because there are practically no systematic data on the elution of defined oligonucleotides and polynucleotides as model compounds from the commonly used resins. Qualitatively, one can envisage two effects. One is the small effect of the different charge distribution in the bases on the pK-values of individual phosphates. The other is the possibility of certain bases hydrogen bonding to the ionized groups of the resin. If the conditions employed for the ion-exchange chromatography of nucleic acids appear somewhat bizarre at first sight it is because a lot of the systems were devised in a highly empirical fashion and, when compared with other types of ion-exchange chromatography, these techniques are limited by the facts that nucleic acids bind very strongly to anion-exchange resins and can only be eluted in a relatively restricted pH range.

Ion-exchange celluloses and Sephadexes

DNA Several attempts have been made to elute DNA from DEAE-cellulose or ECTEOLA-cellulose (this latter substance is a weaker exchanger than DEAE; it is made by treating cellulose with epichlorhydrin and triethanolamine—the structure of the final resin is uncertain). Nucleic acids of high molecular weight are eluted at very high pH. Clearly this is of no use for RNA as it is hydrolysed. However denatured DNA can be eluted. The technique is not in favour these days, largely because the elution patterns are not necessarily at all significant in terms of differences in DNA structure. For a discussion of the method and a critical discussion of the artefacts produced, see the paper by Kit (1960). There are different ways of eluting big nucleic acids from DEAE, either by modifying the resin (below) or by using alkylammonium salts for elution (p. 163).

Fractionation of tRNA on DEAE   Transfer RNA is eluted from DEAE-cellulose by about 0·7 M salt. This is the basis of one of the methods of separating 4 S (tRNA) and 5 S RNA from the high-molecular-weight ribosomal RNA. The mixture is put on to a column of DEAE-cellulose and eluted with 0·8 M sodium chloride in buffer at pH 5. High-molecular-weight rRNA is retarded on the column (and cannot be recovered) but the low-molecular-weight RNA is eluted.

Using a gradient of salt concentration it is possible to achieve a fractionation of different tRNA species. Several factors can influence the elution characteristics of different tRNA molecules. Bock and Cherayil (1967) have exploited two of these, namely pH and the presence of urea to devise a set of chromatographic separations for the analysis of tRNA mixtures. Columns of 1 m length and 50 cm^2 cross-section resolve about 1 g of mixed tRNA. Proportionally smaller columns are used for smaller amounts. In all cases about 15 column volumes of gradient are pumped through slowly (elution time about two days). Examples of the use of this type of separation are discussed on p. 198, where the identification of specific tRNA acceptor activities are described. For the routine analysis of a mixture of tRNA molecules the most versatile system is probably the elution of a DEAE-cellulose column with a linear gradient of sodium chloride in buffer at pH 4·2 in 4 M urea. The nature of the salt gradient is determined by the nature of the source of tRNA and the grade of DEAE-cellulose. For *E. coli* tRNA and Watman DE23 DEAE-cellulose, the gradient is 0·50–0·61 M. For DEAE-cellulose with fewer ionizing groups (e.g. Calbiochem grade with 0·4 meq/g) the salt is lower by 0·08 M. For RNA from a non-bacterial source it is important to do preliminary experiments to establish the salt range which is optimal. Other systems of Bock which give different elution profiles are (with DE23): 0·41–0·48 M NaCl in 7 M urea pH 7·5; 2–5 M urea gradient (exponential) in 0·44 M NaCl pH 7·5; and 1–6 M urea gradient (exponential) in 0·48 M NaCl pH 4·2. And with DEAE-Sephadex A-50: 0·50–0·80 M NaCl pH 4·5; 0·55–0·75 M NaCl in 7 M urea pH 7·5. Other suggestions are presented in their paper and clearly many variations of this type of chromatography are possible. The nature of the buffer used to establish the pH is relatively unimportant. The characteristics of the column are greatly influenced by other cations however (such as magnesium). Thus it is wise either to exclude magnesium (as in the examples above) and incorporate a low concentration of a chelating agent in the buffer (such as EDTA) to ensure reproducibility, or to work out conditions with a defined low concentration (say 1 mM) of a magnesium salt.

Certain columns of this type are useful in addition for separating 4 S and 5 S RNA. The 5 S RNA emerges after the bulk of the tRNA on the DEAE-cellulose column eluted with the 2–5 M urea gradient under the conditions described above.

Modified DEAE-cellulose   The characteristics of DEAE-cellulose are enormously changed if a proportion of the hydroxyl groups in the cellulose matrix are substituted with benzoyl or naphthoyl groups. There are two effects of such substitution; one is to shield some of the ionized groups and hence weaken the affinity of RNA for the resin, the other is to introduce selective weak association between the bases and the aromatic groups introduced into the resin. The method was introduced by G. M. Tener and his group (Gillam *et al.*, 1967) and it is their procedure for making the resins which is usually employed. It is essential that all reagents are dried. DEAE-cellulose (0·9 meq/g) is dried *in vacuo* at 80°C overnight and pyridine is dried over calcium hybride.

Benzoylation of DEAE-cellulose is achieved by suspending the dry resin in 25 volumes of pyridine and benzoyl chloride (2·2 mole/mole of glucose in the resin) is added. The mixture is boiled under reflux for 15 min, cooled slightly, a further 1·1 mole of benzoyl chloride is added and heated again (30 min). The appearance of the reaction mixture at this stage is a viscous brown fairly homogeneous liquid (with an intolerable smell). Naphthoylation is carried out in the same way with naphtho-1-yl chloride in place of benzoyl chloride. A third type of resin is made using in the initial reaction a mixture of benzoyl chloride (2·0 mole/mole glucose) and naphthoyl chloride (0·3 mole/mole glucose) and adding for the second stage a further 1 mole (/mole glucose) of benzoyl chloride. In every case the work-up is identical. The viscous liquid in pyridine is allowed to cool a bit and (just before it solidifies) it is poured into a very large volume of water. As soon as it enters the water the mass forms a long disgusting-looking rope-like precipitate. It is allowed to set hard (several hours under water) and broken up in a Waring blendor and extensively washed with ethanol, 2 M NaCl in 25% ethanol and finally 2 M NaCl until no more optically absorbing material is eluted. The material is sieved (50 mesh) in the wet state. The saponification number can be determined by a standard back-titration method using sodium hydroxide as alkali. The resin can be dissolved in pyridine for this purpose and phenolphthalein is used as indicator. Saponification numbers corresponding to three benzoyl (or naphthoyl) groups per glucose can be obtained. The benzoylated product ('BD-cellulose') and the mixed benzoylated and naphthoylated cellulose ('BND-cellulose') are pale yellow or tan in colour. The naphthoylated product ('ND-cellulose') should be white.

The original intention of Gillam and others (1967) was to devise a system for the improved analysis and fractionation of tRNA. The best choice for the preparative fractionation of tRNA is probably BD-cellulose. A large column (3·2 x 110 cm) gives an excellent resolution of 5 g of tRNA. The sample is applied in 0·45 M sodium chloride, 0·01 M magnesium sulphate (500 ml). The column is eluted with 10 l of a gradient (0·45–1·0 M sodium chloride). It is not necessary to buffer the solution; indeed it is best to avoid buffers, many of which encourage the growth of micro-organisms during the prolonged elution process. The method for the identification of specific tRNA molecules in such an eluate is described on p. 198. There is often extremely good resolution of different tRNA species specific for the same amino acid. BD-Cellulose is the fractionation procedure which resolves tRNA$_F^{Met}$ from tRNA$_M^{Met}$ in *E. coli* (Seno, Kobayishi and Nishimura, 1968). In certain cases some tRNA is left on the column after the salt gradient. These species can usually be eluted with 1 M sodium chloride in 10% aqueous 2-ethoxyethanol (Gillam *et al.*, 1967). Further fractionation of tRNA eluted from BD-cellulose can be achieved on other chromatographic systems described in this section. Alternatively, a different pattern can be obtained from BD-cellulose with a salt gradient as above in 0·05 M sodium acetate buffer pH 4.

In addition to the 'hydrophobic interactions' between the aromatic groups in the resin and the bases in RNA, there are interactions between these aromatic groups and amino acids. Thus the fractionation may be performed with aminoacyl tRNA (i.e. tRNA that has been 'charged'—see p. 000) and the order of elution of the tRNA-species will be different. Thus the same column may be used for two different fractionations of the same mixture. Free tRNA mixtures that do not separate, may be resolved when the charged mixture is employed. Hydrophobic amino acids (such as phenylalanine) bind more strongly to the resin and consequently Phe-tRNA^{Phe} will be differently eluted to tRNA^{Phe}.

A different use of these modified DEAE-celluloses is to fractionate RNA of high molecular weight. Sedat, Lyon and Sinsheimer (1969) describe a remarkable system for the elution of pulse-labelled (? messenger see p. 194) RNA from BD-cellulose. *E. coli* was labelled for several generations with [^{3}H] uridine to ensure uniform labelling of tRNA and rRNA and then pulsed for short times (less than a minute) with [^{32}P] phosphate. Very small amounts of RNA were dissolved in about 2·5 ml 0·02 M tris HCl pH 7·5 and added to 10 volumes of the same buffer containing 0·3 M sodium chloride and 1 mM EDTA and applied to a very small column of BD-cellulose (2·5 x 5 cm) and washed through with this solution until ^{32}P (not RNA) was all washed off. The same buffer was then used in a gradient of 0·3–0·7 M sodium chloride. The same gradient contains increasing concentrations (0–30%) of dimethylsulphoxide. Transfer RNA, rRNA and some pulse-labelled RNA are eluted. Much pulse-labelled RNA was left on the column. It was eluted with 8 M urea containing 0·1 M acetic acid (pH gradient approaching 3·5). According to the authors it is essential to use ultra-pure urea to avoid degradation of high-molecular-weight RNA. This was not our experience (Parish and Hastings, 1966). The pulse-labelled RNA is eluted as quite a sharp peak. Some more ^{32}P is left bound to the column. The authors describe a way of getting it off; however it is not RNA.

As an alternative to using dimethylsulphoxide, rRNA and pulse-labelled RNA can be eluted (and separated) using a salt gradient (up to 0·9 M NaCl) in the presence of 0·02 M NaClO$_4$ (Maxwell, 1969).

Elution of high-molecular-weight RNA from DEAE-cellulose   The difficulty in using DEAE-cellulose for the chromatography of big RNA molecules, is that the conditions for elution require either such high concentrations of salt that RNA is precipitated or the pH is so high that it is hydrolysed. It occurred to me (Parish, 1968) that an alternative is to use high concentrations of alkylammonium salts. The system I originally advocated was a gradient of increasing concentrations of triethylammonium acetate in 15% *iso*propanol up to a mixture entirely lacking any water at all (nominally 4·35 M triethylammonium acetate, in effect essentially a mixture of covalent molecules). All nucleic acids are completely denatured in this system and they are all eventually eluted. There is no evidence that the system is of any value in fractionating tRNA or DNA. However 18 S and 28 S RNA from rat liver are separated and from *E. coli* there is a separation of rRNA and pulse-labelled RNA. We have not been able to significantly improve on the eluting systems although longer columns and slower flow rates do achieve much improved resolution and triethylammonium carbonate may be used in place of the acetate. Recent work in my laboratory by Miss M. Brown and H. A. Foster has established that DEAE-Sephadex or ECTEOLA-cellulose can be used in the same way, with no great effect on the patterns of elution except that DEAE-Sephadex has the minor advantage that less washing is required to get the optically absorbing material (eluted by these solvents) out of the resin and ECTEOLA has the disadvantage that this same problem is enormously greater. Just in case it has occurred to the reader that strong triethylammonium salts might be tried to elute BD-, BND- or ND-cellulose, a word of advice—don't. These resins dissolve in the eluting buffer.

Polypeptides

An alternative to using DEAE or some such synthetic resin for anion-exchange chromatography, one can use a polypeptide bound to an inert support. There are three types of column of this type that are commonly employed. These are methylated bovine serum albumin on kieselguhr (MAK), the same protein bound to silicic acid (MASA) and polylysine bound to kieselguhr (PLK). The methylation of the albumin prior to the construction of MAK or MASA columns is to esterify the carboxylic side chains of the glutamic and aspartic acid residues of the protein so the only ionizing groups of the protein are the basic side chains of lysine and arginine residues. We shall start by considering the recipes for making up these columns and then consider their applications.

Preparation of columns   MAK columns are made up according to the procedure of Mandell and Hershey (1960). Methanolic bovine serum albumin solution (5 g in 500 ml of methanol containing 4·2 ml of concentrated HCl) is allowed to stand in the dark for 3 days. The methylated protein precipitates during this time and the mixture should be occasionally shaken. The precipitate is collected, washed extensively with methanol to remove the HCl and is dried and stored over solid KOH. Well-washed kieselguhr (20 g) is suspended in 0·1 M NaCl, 0·05 M phosphate buffer pH 6·7 (100 ml); air is expelled by boiling and cooling the suspension. A solution of methylated albumin in the same buffer (1% solution, 5 ml) is added with gently swirling. This slurry is now MAK ready for packing into a column. The column must be well washed before the nucleic acid sample is applied. MASA is prepared in the same way except silicic acid of a grade suitable for chromatography is used in place of kieselguhr. The silicic acid is washed in distilled water before use to remove small particles that would slow up the chromatography. Also very much more methylated albumin is used (30 mg per g of silicic acid). PLK is made up in a very analogous fashion (Ayad and Blamire, 1968). In this case the buffer used in 0·4 sodium chloride in 0·02 M phosphate buffer pH 6·7. Kieselguhr (2 g) suspended in buffer and air is removed as earlier and polylysine hydrobromide of molecular weight in excess of 50 000 (solution of 10 mg/ml, 0·8 ml) is added.

It is the usual practice with all these columns to put the ion-exchange material as the middle layer of a sandwich. The original method for MAK described by Mandell and Hershey (1960) consists of packing a column (2 x 40 cm) as follows (using throughout the buffer described above). First a layer about 2 cm in depth of cellulose powder is packed. Then MAK freshly made as a slurry equivalent to 8 g of kieselguhr is packed and thoroughly washed to remove excess methylated albumin. A mixture of 10 ml of MAK slurry and an additional 6 g of kieselguhr suspended in 40 ml buffer is packed on top of the previous MAK layer; finally 1 g of kieselguhr suspended in 5 ml buffer is packed to form the top layer of the column. Similarly Ayad and Blamire (1968) advocate packing of PLK columns in a similar way using a cellulose powder layer at the bottom. For a column 1·2 x 25 cm, the total PLK preparation mentioned above is used. After it has been packed 1 ml of kieselguhr slurry is packed on top.

Applications   MAK chromatography has the distinction of being the only procedure whereby the major fractions of nucleic acid in the cell (tRNA, DNA and rRNA) can be adequately resolved in one go on a reasonably large scale. A column used for this purpose made up according to the procedure of Mandell and Hershey (above) will fractionate about 7 mg of *E. coli* nucleic acid. The nucleic acid is loaded in 0·1 M NaCl (buffered as described above) and washed on with this buffer if low-molecular-weight material (such as [^{32}P] phosphate—see p. 175) is to be removed. The chromatogram is developed at room temperature with a linear gradient of 0·1–1·2 M NaCl. The results of such a fractionation for a preparation which had been pulse labelled with [^{32}P] phosphate is shown in Fig. 4.24.

The technique may be applied to mammalian nucleic acids but with rather less success. The major peaks are eluted in the same concentrations of salts as their counterparts in the bacterial preparation. However the two peaks of high-molecular-weight rRNA (28 S and 18 S in this case) are less well resolved, presumably because they aggregate at this molarity (which is very close to the concentration at which they are insoluble.) Also there is a large amount of pulse-labelled material more strongly absorbed to the column than rRNA. On the whole MAK chromatography is not satisfactory for the separation of these big molecules (which include rRNA precursors, see p. 183).

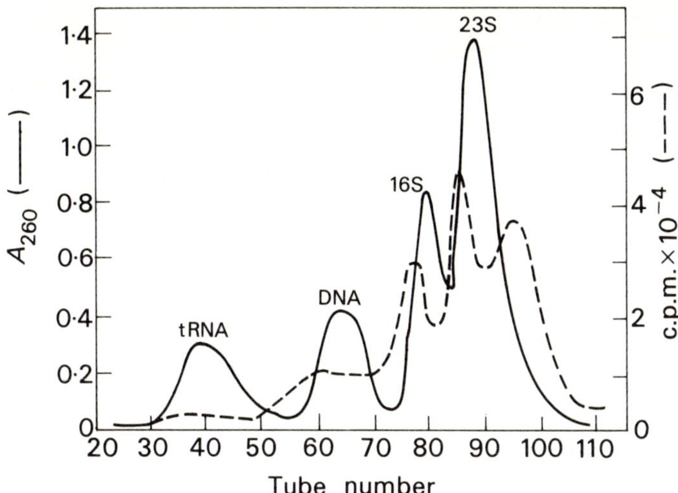

FIG 4.24   Fractionation of *E. coli* nucleic acids on MAK. The bacteria were labelled for 30 s with [^{32}P] phosphate. The radioactivity is thus associated with rapidly labelled nucleic acid (largely RNA). Conditions for elution are described in the text. Data from Otaka, Mitsui and Osawa (1962).

If shallower gradients, eluted more slowly, are employed the tRNA and 5 S RNA are easily resolved (Galibert, Larsen, Lelong and Boison, 1965). Moreover MAK and MASA may be used for analysing mixtures of aminoacyl tRNA (p. 162).

One of the most remarkable applications of MAK volumes is that of E. Chargaff and his collaborators to the separation of the complementary strands of denatured DNA. The evidence for this type of separation is discussed on p. 373. However here we may note that it is probably the most versatile method of separating DNA strands as it has been applied to the DNA of *Bacillus subtilis* (Rudner, Karkas and Chargaff, 1968) and to the DNA of various coliphage (Rudner, Karkas and Chargaff, 1969).

Ayad and Blamire (1969) have surveyed the use of PLK columns for the separation of high-molecular-weight nucleic acids. The technique is a relatively new one. My own impression of it is that it will probably prove of limited value in the separation of RNA molecules, but for DNA and possibly DNA/RNA hybrids (p. 164) the method could have many applications. DNA from *B. subtilis* is eluted with a gradient of 0·4–4·0 M sodium chloride (in the buffer used for making the column, p. 355). Evidence for the partial separation of transforming cistrons (p. 191) has been obtained. One attractive feature of PLK chromatography is that the defined nature of the ion-exchange material makes it possible to make precise and meaningful measurements of the binding of nucleic acids to the column (Ayad and Blamire, 1970).

Miscellaneous chromatographic procedures

The only common feature of these procedures is that they separate nucleic acids largely on the basis of their secondary structure. One of them (hydroxyapatite) is an extremely important technique for the separation of single- and double-stranded DNA and is the basis of a refinement of DNA/DNA hydridization studies (p. 365).

Silk fibroin   This is a simple and cheap method for the large-scale separation of native from denatured DNA. The physico-chemical basis of the procedure is obscure. The first job is to make the fibroin which may as well be done on a large scale as it is somewhat tedious. Once it has been prepared the method is very simple. The techniques are those of Huh and Helleiner (1967). Raw silk fibres are cut into lengths (less than 5 mm), boiled in water for an hour, drained, autoclaved for an hour (15 psi) rinsed and suspended in 0·5% aqueous solution of a non-ionic detergent. It is now autoclaved again for 3 h and thoroughly washed and dried. The crude fibroin is extracted for 2 days with methanol/benzene mixture (2:3 v/v; in a Soxhlet apparatus) and dried carefully. Before use for chromatography this product is oxidized (6% hydrogen peroxide in very dilute aqueous phosphoric acid pH 1·9, overnight at $37°C$). The fibroin is then washed extensively with phosphate buffer (pH of effluent to reach 3·8) and suspended in 0·1 M sodium acetate pH 3·8 and packed in a glass column under slight pressure to give an even bed. The packed column is washed with 0·2 M $MgCl_2$; DNA (about 1 mg per g of fibroin) is loaded on to the column in this magnesium chloride solution. Native double-stranded DNA is washed straight off again and denatured single-stranded DNA is adsorbed and can be eluted with 0·05 M potassium phosphate buffer pH 7·4.

Hydroxyapatite and hydromagnesite Hydroxyapatite is one of the many hydrated forms of calcium phosphate ($Ca_5(PO_4)_3OH$) and hydromagnesite is one of the hydrated forms of basic magnesium carbonate ($4MgCo_3Mg(OH)_24H_2O$). The discussion of the fractionation of DNA on hydroxyapatite (due to G. Bernardi) is delayed until p. 361 when this, nitrocellulose filtration and other methods of separating and analysing nucleic acid hybrids are discussed. However, hydroxyapatite can be used for the fractionation of tRNA (Pearson and Kelmers, 1966) and separating replicative forms of viral RNA (Pinck, Hirth and Bernardi, 1968). It can also be used for the separation of DNA and RNA. The most important point about hydroxyapatite is the preparation of the gel. On the whole commercial preparations are not satisfactory. The various recipes are all designed to overcome the fact that if hydroxyapatite is directly precipitated it forms such small crystals that it is virtually useless for chromatography as the flow rates are so slow. There are several recipes but the best is probably that of Levin (1962). The principle is to precipitate calcium phosphate as brushite ($CaHPO_4 \cdot 2H_2O$) which forms larger crystals, and convert this into hydroxyapatite by treatment with alkali. Disodium hydrogen phosphate (0·5 M, 2 l) and calcium chloride (0·5 M, 2 l) are carefully mixed. This is done by dripping them together at the same flow rate (15 ml/min or less) into a large, well stirred vessel. After completion of this, the precipitate (brushite) settles and the supernatant is poured off and discarded. It is essential that no centrifugation (which compacts and breaks up the gel) be used at this or any other stage in the procedure. The brushite is washed with about four changes of water (3 l at a time). It is now suspended in water (3 l) and 100 ml of 10 N sodium hydroxide (freshly and accurately made up) is added and the suspension is gently stirred and heated so that it reaches boiling in 45 min; it is then kept gently boiling for 1 h. Heat is removed and after the gel has settled for 5 min the supernatant (which contains much finely divided precipitate) is removed by suction and discarded. The alkali is washed out by adding 4 l water and decanting after 5 min; this washing is performed four times in all. It is now suspended in 4 l of 0·01 M phosphate buffer pH 6·8 (this corresponds to equimolar amounts of Na_2HPO_4 and NaH_2PO_4) and brought to the boil. After 5 min of actual boiling, heat is removed and after 5 min settling the supernatant is sucked off. This procedure is repeated two more times but with 15 min of boiling. The final sediment is hydroxyapatite suitable for chromatography and can be stored conveniently as a slurry under 1 mM phosphate buffer pH 6·8.

In the procedure of Pearson and Kelmers (1966) *E. coli* tRNA (100 mg) is loaded on to a column (2 x 68 cm) of hydroxyapatite in 0·1 M phosphate buffer pH 6·8 and eluted with a gradient of 0·100–0·200 M phosphate (2 l of eluting gradient, flow rate 0·2 ml/min at room temperature). A remarkable separation of acceptor activities is achieved.

As a method of separating RNA and DNA, hydroxyapatite is very useful if one is working on a very small scale so that precipitation methods are inappropriate. In their work on mitochondrial nucleic acids, Leffler, Creskoff, Lubowsky, McFarlane and Mora (1970) loaded radioactivity labelled mitochondrial total nucleic acids on to a small column (bed volume 5 ml, other dimensions are not given) of hydroxyapatite. Dilute (0·18 M) phosphate buffer pH 6·8 eluted the RNA (and low-molecular-weight nucleotides); DNA was eluted with 0·4 M buffer.

Lamb and Dukes (1964) used hydromagnesite to separate poliovirus RNA from cellular RNA. RNA from infected cells was loaded on to columns of commercial magnesite (obtained from Mallinckrodt). Cellular RNA was eluted with around 0·04 M phosphate buffer; viral RNA was eluted in the region of 0·11 M phosphate. The technique was successful when applied to the RNA molecules obtained from several other mammalian enteroviruses.

Molecular exclusion chromatography

Columns of Sephadex (such as G-25) are of course very useful for desalting or changing the salt concentration of nucleic acid (or protein) solutions. Here we are concerned with the relatively few fractionation problems of nucleic acid methodology which are suitable for molecular exclusion methods.

Teichoic acids   The teichoic acids are components of the cell walls of certain bacteria. With certain Gram positive organisms (such as *Bacillus subtilis*) they constitute about half of the mass of the wall. The teichoic acids bear a certain chemical resemblance to nucleic acids; they are linear macromolecules in which the components are held together with phosphodiester bonds. However the components are not nucleosides but polyhydric alcohols (either glycerol or ribitol). The structures are more complex than this, as some of the alcohol residues are substituted with amino acids (usually D-alanine) or glucose residues. Study of teichoic acids is a fascinating part of natural product research. From our point of view they must be regarded as an occasional menace. Although deproteinization and procedures for the removal of polysaccharides (p. 110) get rid of some teichoic acid, DNA from *B. subtilis* is almost inevitably contaminated to some extent with teichoic acids. The only satisfactory method of purifying this DNA is by chromatography on gels. In the method of Young and Jackson (1966) the column consists of a 4% agarose gel. The DNA is put on in 0·01 M tris HCl pH 7·1, DNA is eluted at the exclusion volume of the column and teichoic acid contaminants are considerably retarded. A column of 2 x 20 cm was suitable for the purification of 0·15 mg of DNA and was eluted at a flow rate of about 40 ml/h.

4 S and 5 S RNA   Sephadex G-100 has the correct exclusion range to separate 4 S and 5 S RNA. Galibert, Larsen, Lelong and Boison (1965) loaded 2 mg of RNA on a column (1·8 x 190 cm) and eluted with 0·01 M ammonium acetate pH 5·1 (presumably chosen to inhibit ribonuclease). High-molecular-weight RNA was eluted at the exclusion volume followed by 5 S rRNA and finally 4 S tRNA. The separation is complete and constitutes the best method for large-scale preparation of 5 S RNA.

Two-phase procedures

Mixtures of salts, organic solvents and water can be devised which produce (after equilibration) two-phase mixtures in which nucleic acids will distribute themselves between the phases. Typically one phase is more hydrophylic than the other and thus in the equilibrium mixture the nucleic acids in this phase tend to have more secondary structure than those in the more hydrophobic phase. Thus we have (at least in principle) a method of separating nucleic acids on the basis of their ease of denaturation. If the two phases contain mixed nucleic acids, those which denature more easily will partition into the hydrophobic phase more readily than those that denature less readily. Obviously both overall nucleotide composition and sequence can contribute to the ease of denaturation and the technique should be extremely versatile. In practice there are enormous problems associated with the general application of two-phase techniques to high-molecular-weight nucleic acids. I have pointed to one possible complication elsewhere (Parish, 1968). However as a method of separating tRNA species these methods are extremely useful although the actual basis of the fractionations is not certain. However, using samples which were in fact partly degraded rRNA and tRNA, Kirby (1962) demonstrated directly that the fractionation by counter-current distribution was determined by nucleotide composition. There are two ways of achieving the large number of equilibrations and re-partitions for the fractionation to be effective and these will be considered separately.

Counter-current distribution

Principle   Imagine we wish to repeatedly partition in a mixture of molecules between two liquid phases. The way in which it could be done would be to dissolve the mixture in one phase and shake it up with the other phase in a separating funnel. After settling has occurred, one phase (say the top one) is removed and shaken up with more lower phase. Meanwhile, the original lower phase (in its original funnel) is re-extracted with fresh upper phase. When both have settled, we make the second transfer, i.e. upper phase from funnel 2 is added to bottom phase in funnel 3, likewise upper phase from 2 is added to 2 and 1 is replenished with fresh upper phase, and so on. After N transfers a substance with a partition coefficient of 0·5 will form a Gaussian distribution among the funnels with its peak in funnel number $N/2$. Similarly components with other partition coefficients will form peaks of concentration in different tubes. This experiment is never done, except by a few hapless students doing practical chemistry or biochemistry in departments that believe in torture. A counter-current distribution apparatus is a device for performing these repeated extractions automatically. Assuming one wishes to make 200 transfers, about 200 tubes of the apparatus are filled with the correct volume of lower phase. The machine then pipettes an equal volume of upper phase from a reservoir into tube 1, shakes the tubes (for a pre-determined time) allows it to settle (again a pre-determined time chosen to allow any froth to disperse) and tips upper phase from each tube to the next one (replenishing upper phase in tube 1 from its reservoir). In practice it is wise to let it do a few transfers to ensure that neither phase is being depleted during equilibration (such depletion is a consequence of careless equilibration of the phases when the solvent mixture was made up) and then the sample is introduced; for the same reason it is wise to have excess lower phase in the reservoir and to introduce the sample, not into tube number 1, but rather a little way along (say tube 10). The nature of distributions that are performed depends on the equipment available. The number of transfers need not be limited to the number of tubes on the machine; most solvent systems do not move even the most fast-moving component to the upper phase front and one can thus just allow upper phase to spill out of the last tube if the instrument is set for more transfers than it has tubes. If you are afraid that there could be a miscalculation so that some RNA could be lost this way, the last two or three tubes can be made into an 'RNA trap', that is to say instead of containing lower phase they can consist of one empty tube (to catch any drops of lower phase carried through the machine) followed by two tubes containing (as lower phase) a strong salt solution from which RNA will not partition at all into upper phase. An absolute requirement for a counter-current machine is that it must be in a thermostatically controlled room as the composition of phases of many of the systems employed changes radically with temperature. Equally of course the solvent system must be made up in the same room.

tRNA   The advantage of counter-current distribution as a fractionation method for tRNA is that large amounts (of the order of a gramme) of tRNA can be fractionated in a modest apparatus that holds 10 ml of each phase per tube. Three systems are well established for this purpose; in all cases it is assumed that the machine will operate at $23°C$.

(1) 4 l of water is added to 550 g K_2HPO_4 and 850 g $NaH_2PO_4 . H_2O$; 300 ml of formamide and 1·3 l of isopropanol are then added; the phases produced are about equal in volume to one another (Apgar, Holley and Merrill, 1961). RNA is dissolved in the phosphate buffer and then the other components added in the same proportions to create the two phases in the sample tube.

(2) Ammonium sulphate (1·2 kg), glacial acetic acid (40 ml) and concentrated aqueous ammonia (8 ml) in water (final volume 4 l) are shaken with 160 ml formamide and 1·6 ml 2-ethoxyethanol. Again equal volumes of the two phases are obtained (Kirby, 1960). The sample is introduced as in method 2.

(3) In this case the cations are all alkylammonium. The mixture consists of 2 l *n*-butanol, 2·6 ml water, 200 ml tri-*n*-butylamine, 50 ml glacial acetic acid and 440 ml peroxide-free *n*-butyl ether. Again approximately equal volumes of the two phases are produced. Before introducing RNA into this system it must first be converted to the tributylammonium salt. With tRNA this can be done by precipitating it with 0·1 M hydrochloric acid at $0°C$; it is immediately centrifuged (only keep the RNA in acid for a few minutes), quickly washed with more dilute cold acid and dissolved in tributylammonium acetate (Zachau, 1965).

In these and any other procedures in which either ethoxyethanol or aliphatic amines are used as solvents for nucleic acids, the solvents must be purified to get rid of optically absorbing impurities (carbonyl compounds). The simplest procedure is to distil the compound (whether it be amine or alcohol) *in vacuo* over amidol (2, 4-diaminophenol).

DNA   Complete separation double- from single-stranded DNA can be achieved using a remarkable system devised by Albertsson (1962). The system is made up to yield final concentrations of 7·7% (w/w) of Dextran-500 (Pharmacia) and 4·4% (w/w) of polyethylene glycol (Carbowax-6000) in a appropriate buffer. This mixture forms two phases in which most of the polyethylene glycol is in the upper phase and most of the dextran is in the lower one. It is used at $0°C$. The partition coefficient (upper/lower) of native DNA is of the order of 30: that of denatured DNA is of the order of 0·05. It is thus a valuable analytical tool for estimating the degree of denaturation in a sample of DNA. As a fractionation procedure, the disadvantage of the system (the viscosity and consequent difficulty in separating the phases) are balanced by the fact that very few transfers are needed—it is even feasible to do it by hand with a pipette and about ten centrifuge tubes. Alternatively there are specially designed counter-current machines with very shallow flat distribution chambers suitable for this type of separation.

Reverse-phase chromatography

Reverse-phase chromatography is an extension of the idea of partition fractionation to column chromatography. In this case the stationary phase (lower phase in the counter-current systems above) is absorbed into an inert matrix in a column and the moving phase is used to elute the column. The advantages over counter-current distribution are that the apparatus and techniques are much simpler and one can use a gradient to elute the system (i.e. use varying partition coefficients to introduce another parameter into the experiment). As with all experiments on tRNA, the value of the fractionation is assessed by considering the resolution of amino-acid acceptor activities (p. 200). Here there is simply a description of the two most commonly used and versatile systems for reverse-phase chromatography of tRNA.

Use of Sephadex as support   In the system of Muench and Berg (1966), Sephadex G-25 fine grade is used. The two-phase systems consist of 6 volumes 1·25 M potassium phosphate buffer pH 6·88, 2 volumes of 2-ethoxyethanol, 1 volume of 2-butoxyethanol, x volume of triethylamine and 0·0005 volume of β-mercaptoethanol. This mixture is equilibrated at 23°C to yield two phases in which there is more than twice the volume in the upper phase as in the lower. The Sephadex is equilibrated in lower phase from a system in which x = 0·01 and packed without any added pressure in a column (174 cm x 4 cm^2). The column is washed with upper phase from the same system until no lower phase appears in the eluate. The elution rate is slow (14 ml/h). The sample (250 mg tRNA) is added to the column as a solution in water to which 5 volumes of upper phase have been added. It is washed on with a little more upper phase and the column is then developed with 1·8 l of a gradient of upper phase from the same x = 0·01 system through to upper phase from the x = 0·05 system. The same flow rate and temperature are maintained throughout. The authors describe a method of scaling the chromatography up to handle over 3 g tRNA.

The Freon/kieselguhr method   This technique was developed by Weiss and Kelmers (1967). The system is different to the preceding in that whereas Muench and Berg used potassium ions in the aqueous phase as stationary phase and eluted the column with an increasing concentration of organic amine, Weiss and Kelmers used an organic amine as stationary phase and eluted the column wih a gradient of sodium chloride. The two methods thus complement each other and produce quite different patterns of elution. The inert kieselguhr support consisted of Chromosorb W acid washed and treated with dimethyldichlorosilane to remove active groups which produce adsorption effects. Tricaprylammonium chloride was dissolved (5% v/v) in Freon 214 ($C_3Cl_4F_4$, a common refrigerant). This solution is extracted with 1 M NaOH, 1 M HCl and 0·5 M sodium chloride (2 volumes of each, successively) and dried over silica gel. The Freon solution (336 ml) is shaken with 600 g Chromasorb and allowed to stand for five days. It is then made into a slurry with 0·01 M $MgCl_2$ 0·01 M tris HCl pH 7·0 and packed into a jacketed column (1 x 240 cm) under 60 psi. A litre or so of magnesium chloride buffer is pumped through; up to 100 mg tRNA are loaded and the column is eluted with 3 l gradient of 0–0·3 M sodium chloride at a rate of about 1 ml/min. Various gradient shapes and alterations to this basic system are described in the paper. The column is run at either 25 or 37°C.

Recently, Pearson, Weiss and Kelmers (1971) have described a modification of this procedure in which the support material is a commercially available, synthetic plastic (replacing the kieselguhr of the above method). It seems that improved resolution is obtained with this system.

Notes to chapter 4

1   Another method which is still employed is to puncture the bottom of the tube by hand with a fine hypodermic needle and allow the gradient to drip out. The fractions are collected by a drop counter. Although this constitutes superb character training (one false move with the needle and the tube is split wide open and its precious contents are lost forever) it should be avoided as the surface tension, and consequently the drop-size, varies with sucrose-concentration.

2   There is a type of zonal rotor which will not be discussed in this book which fits large-capacity slow-speed centrifuges. It is intended for the large scale 'banding' of nuclei, microvilli and other organelles and membrane fractions.

3   In this section I have referred to the removable-seal rotors as B-XIV and B-XV as these are the designations used by centrifuge manufacturers. These are essentially the same as the B-X and B-XI rotors described in this reference.

4   Presumably w/v; the authors do not specify which.

5   Hexadeuterated DMSO, $(CD_3)_2SO$.

6    $HCONMe_2$, known as 'dimethylformamide'.

7   This is not to say that molecular weight has no effect on buoyant density centrifugations. Low-molecular-weight DNA produces a broader band than that formed by high-molecular-weight material (p. 389).

8   These two forms are numbered in the order of their discoveries. Metabolically they are the wrong way round: RF II is the precursor of RF I.

9   Microbiologists will recognize that there is more to it than that; it suffers the so-called 'thymineless death'. This phenomenon is outside the scope of this book. See p. 177 for an explanation of thymine auxotrophy.

10   Discovered by Knight, Pene and Darnell (1968).

Addendum

Dissociation of mammalian ribosomes and polysomes   The conditions for the isolation of functionally active mammalian ribosomal subunits have been described by A. K. Falvey and T. Staehelin (*J. molec. Biol.* 1970, **53**, 1) and G. Blobel and D. Sabatini (*Proc. Nat. Acad. Sci. U.S.* 1971, **108**, 390). Ribosomes lacking attached mRNA and polypeptidyl tRNA can be dissociated in 0·5 M KCl, 5 mM Mg^{2+}. Polysomes are not dissociated in this medium; however they can be dissociated if the polypeptidyl tRNA is removed by treatment with puromycin. These procedures are superior to those referred to on p. 141.

5   **Identification**

This chapter is concerned with methods for the assay of nucleic acids and the assignment of components of nucleic acid preparations to functional classes (DNA, mRNA, rRNA, tRNA, etc.). Detailed procedures for the physico-chemical characterization of nucleic acids (pp. 338 and 380) and use of hybridization (p. 355) are described elsewhere.

Optical methods, colour reactions and other methods

Here we are concerned with these procedures purely from the point of identifying nucleic acids, estimating their purity and determining the amount of nucleic acids in a preparation. Although some of these procedures are applicable to measuring the total RNA or DNA in tissue I have not considered such methods systematically.

Colour reactions and chemical methods

Nucleic acids can be assayed spectrophotometrically following a variety of colour reactions in which either the bases or the sugars react to form a chromophore. It is only necessary to describe two good methods (one for RNA, one for DNA) which are useful for measuring the efficiency of our isolation procedure. Both methods are based on the chemistry of the sugars.

The orcinol reaction for RNA   The reaction between acidified orcinol and aldopentoses to produce a green colour is one of the classical reactions of sugar chemistry (Bial, 1903). Its quantitation for RNA estimation is due to Dische and Schwartz (1937). This (and other applications of Bial's reaction) only detect the ribose moieties present as purine ribosides in RNA 9 (see pp. 55 and 203 for the reason); so they should be calibrated either with RNA of the same nucleotide composition as that of the sample, or expressed as purine riboside equivalents (in which case AMP is used as a standard). The reagent is made up by dissolving 100 mg of AnalaR $FeCl_3$. $6H_2O$ in 100 ml of AnalaR concentrated HCl and then adding 3·5 ml of orcinol dissolved in ethanol (6·0% w/v). The orcinol should be recrystallized from benzene. If it still looks a bit yellow it must be recrystallized again. The reagent itself is stable in the dark; it is bright yellow in colour. The solution containing RNA is diluted if necessary so that the concentration is in the range 10–100 μg/ml and two volumes of reagent are added to one volume of RNA solution. The mixture is heated to 90°C for half an hour and the tube is cooled under the tap and the extinction measured at 665 nm. Hexoses interfere with this reaction, producing a brown colour. If the hexose-contamination is not too great, a correction can be made by using glucose standards and measuring the extinction at 565 nm and 665 nm. By making the same two readings with the RNA sample, it is possible (by calculating the amount of the extinction at 665 nm due to hexose from the hexose absorption at 565 nm) to make a correction. However if there is a lot of hexose (or hexose glycoside) present the situation is hopeless. Thus the reaction cannot be applied to sucrose-gradient fractions unless the sucrose is thoroughly removed by dialysis.

The diphenylamine procedure for DNA   This procedure was originated by Dische (1930); the chemical reaction involved and the nature of the blue chromophore remain unknown to this day. The procedure that is actually employed these days is that of Burton (1956); it is both more sensitive and more accurate than Dische's original recipe. Diphenylamine (AnalaR) is dissolved in 100 ml AnalaR glacial acetic acid and 100 ml of concentrated sulphuric acid (preferably micronanalytical grade) is added. Just before use 0·1 ml of aqueous acetaldehyde (made up by pipetting—with a cold pipette!—1 ml of acetaldehyde into 50 ml water) is added to the reagent. If a batch of reagent is being made for the first time, try a blank reaction by adding half a volume of 0·5 M perchloric acid (see below) to a little of the reagent—if no colour develops all is well. If it does become coloured one reagent (probably the acetic acid) is no good. If it is the acetic acid it should be distilled over solid potassium dichromate. Similarly if success is achieved, the bottle of acetic acid should be kept in the dark and reserved for Burton assays. If the trouble lies with the diphenylamine, it can be easily recrystallized from light petroleum. The DNA solution or standard is made 0·5 M with respect to perchloric acid and is diluted so the DNA concentration is in the range 5–100 μg/ml. Two volumes of reagent are added to one volume of solution. They are left overnight at 25–30°C and the absorption is measured at 600 nm.

Nucleic acids in tissues and cells   Both these colour reactions can be conveniently applied to whole cells or tissues. Extraction of the tissue or cells (or an homogenate) at 90°C with 5% trichloracetic acid is probably the most popular method.

An alternative to colorimetric assay is to hydrolyse the nucleic acids (see p. 55) and to estimate the products of hydrolysis. A rapid and sensitive method for the assay of DNA, due to Zahn (1970), is based on such a hydrolysis and assay of the thymine by gas chromatography.

I have omitted any description of histochemical methods for nucleic acids in cells as I regard these as outside my scope. If the reader is familiar with these methods I hope that the methods and theories in this book will inspire him to isolate nucleic acids and study their biochemical and chemical properties.

Ultra-violet absorption spectroscopy

Absorption spectroscopy in the ultraviolet is a sensitive procedure for the assay of nucleic acids. The factors that contribute to the spectroscopic properties are discussed on pp. 83 and 338. However here we may note that for double-stranded DNA the specific absorption at 260 nm (the wavelength close to the maximum for all nucleic acids which is most usually employed for characterizing nucleic acids) per mean residue molecular weight is of the order of 6500. This mean residue molar absorptivity is usually given the symbol ϵ_P. It is the hypothetical absorbance (in a 1 cm pathlength) of a solution 1 M with respect to phosphate ester groups. The corresponding value for rRNA with the secondary structure found in preparations in which no denaturation has occurred is of the order of 6800. These values are obtained by making a solution of purified nucleic acid in a suitable buffer (lacking phosphate and containing only analytical grade salts) and accurately diluting a sample to measure A_{260} and accurately measuring the phosphate of a sample by a procedure such as that of Fiske and Subbarow (1925). A method of performing this assay, suited to spectrophotometry, is that of Griswold, Humoller and McIntyre (1951). Samples should contain of the order of 4 μg phosphorus in 0·2 ml, are added to 0·5 ml of 5 M sulphuric acid (this must be microanalytical grade) and heated in tubes inclined in a proper heater (such as is used for a microscale Kjehldahl estimation of nitrogen). Condensation is seen in the flask due to water being driven off. Heating is continued for 2 h after the further accumulation of this water has stopped. Water (3·8 ml) is added and the tube is heated to 100°C for 15 min (to hydrolyse any phospho-anhydrides formed during treatment with sulphuric acid). Aqueous ammonium molybdate (2·5% w/v, 5·5 ml) and 0·5 ml of 'reducing agent' are added and the sample is again heated to 100°C for 10 min. The reducing agent is made up by dissolving in 250 ml of warm water 5 g 1-amino-2-naphthol-4-sulphonic acid, 16 g of sodium metabisulphite and 10 g of anhydrous sodium sulphite. This reagent is stable at room temperature once it has been made up. After cooling, the sample is made up to 5·0 ml. The colour is dark blue and the extinction is read at 820 nm. Any phosphate is a suitable standard for this assay, however a nucleotide is particularly suitable as the exact concentration can be measured spectrophotometrically. For really accurate measurements of the extinction of RNA and DNA solutions, a correction should be made for light scattering effects (see p. 398). A simple method used by Burness (1970), in a paper which is a perfect example of the accurate quantitation of RNA assays, is to extrapolate the slightly positive base-line on the long wavelength side of the absorption envelope into the region under the envelope and to subtract this value from the extinctions under the peak. To construct this base-line, the line joining the extinction at 360 and 320 nm (which should be horizontal!) is used for the extrapolation.

Chemical criteria of purity of RNA and DNA

In general, it is never necessary to obtain nucleic acids absolutely dry. Indeed high-molecular-weight nucleic acids are very difficult to dissolve once they have dried out completely and frequently denature and form aggregates when kept in the dry state. If nucleic acids are precipitated with ethanol (p. 106) and are washed in ethanol, ether and air-dried they contain about 15% water. This is the assumption behind the approximate formulation that the absorbance at 260 nm of a 0·1% (i.e. 1 mg/ml) of nucleic acid is of the order of 15.

Otherwise it is largely a question of guessing what the likely contaminants might be and assaying for them. A method for assaying for protein is given on p. 111. A less sensitive and less satisfactory method of assaying for protein contamination is to perform a Lowry assay (Lowry, Rosebrough, Farr and Randall, 1951). On the other hand the orcinol (p. 172) and Lowry procedures performed on the sample provide an excellent way of measuring the RNA/protein ratio in ribonucleoprotein provided (and this is often neglected) that the standard for the Lowry reaction is the same protein as that present in the ribonucleoprotein, for the Lowry colour is dependent not only on protein concentration, but also on the amino-acid composition of the proteins being assayed. This point is well illustrated in the paper by Burness (1970) mentioned above.

As a check for possible contamination with polysaccharides, probably the best thing to do is to hydrolyse the preparation (6 M HCl at 100°C for 1 h) and to then chromatograph the hydrolysate on paper. Ribose is readily separated from hexoses on most of the acidic paper chromatographic systems designed for carbohydrates; details of these systems and methods of staining for them are to be found in any chromatographic bible (such as volume I of Ivor Smith's compilation—Smith, 1968).

Dyes   The binding of dyes to nucleic acids is discussed in chapter 14. Certain dyes are useful for the staining of nucleic acids on gels; suggested procedures are given on p. 157.

Radioactive labelling

No experimental molecular biologist is happy unless he has a laboratory full of radiochemicals and liquid scintillation counters churning out results on yards of paper. However before embarking on any radioactive labelling procedure it is extremely important to be quite sure what type of question is being asked and what precautions have to be taken to avoid misleading results. Basically, labelling of nucleic acids is undertaken to achieve one of two aims; one is simply to render detectable very small amounts of nucleic acids; the other is to enquire into the type of nucleic acid being synthesized or metabolized (either *in vivo* or *in vitro*) during a particular period of time. So let us briefly consider the rationale of labelling procedures and then consider applications to particular problems.

Labelling and counting methods

Pathways of incorporation   RNA is synthesized from the riboside 5′-triphosphates and DNA from the deoxyriboside 5′-triphosphates. Synthesis of nucleic acids *in vitro* is therefore followed by incorporating into the incubation mixture one (or more) of the triphosphates in a radioactive form. The triphosphates may be labelled either in the base with ^{3}H(tritium) or ^{14}C (see p. 58), or alternatively they can be labelled with ^{32}P with the α-phosphate including the label. Examples of this type of *in vitro* labelling are given on pp. 222, 225 and 271. This procedure is not appropriate for labelling *in vivo* as the nucleoside triphosphates (in common with all nucleotides) do not permeate cell walls. It is possible to obtain ^{32}P-labelled nucleic acids by inoculating the cells with inorganic ^{32}P. The procedure has certain advantages; the isotope is relatively cheap and obtainable carrier-free so that very high specific activities in the product can sometimes be obtained; only sometimes because there is a lot of inorganic phosphate in many cell-environments. For labelling whole animals, labelling with inorganic $[^{32}$P] phosphate is a difficult procedure as the $[^{32}$P] phosphate is considerably diluted by tissue phosphate and enormous amounts of radioactivity (like 5 mCi injected into some poor rat) have to be employed. With bacteria there is frequently an elegant way out of this problem. Many bacteria (such as *E. coli*) will grow in a medium in which all phosphorus is present as phosphate esters (β-glycerophosphate or serine phosphate in protein hydrolysates). Under these circumstances a phosphatase is induced to enable the organisms to have a constant supply of inorganic phosphate. However high levels of phosphate, never accumulate and if the culture is swamped with $[^{32}$P] phosphate, highly labelled nucleic acids can be obtained.

The other method is to label biosynthetic precursors of the nucleotides. One has to choose substances other than phosphate esters as only non-ionized molecules will get into the cells. In Fig. 5.1 is shown an outline of the synthesis of the pyrimidine nucleotides. Aspartic acid is not in practice used to label nucleic acids as such a technique would be inefficient, much of the label being 'lost' via transamination to Krebs cycle intermediates. Orotic acid is an efficient label in certain mammalian tissues for RNA and DNA; [5-^{3}H] orotic acid is a specific label for RNA as the labelled hydrogen is lost if subsequent metabolism produces thymine derivatives.

The major pathway for the biosynthesis of the RNA precursors (UTP and CTP) are from Oro via UMP and UDP to UTP and CTP. However there are several 'scavenging pathways' for the utilization of the free bases and nucleosides (derived from nucleic acid catabolism) so that Ura,

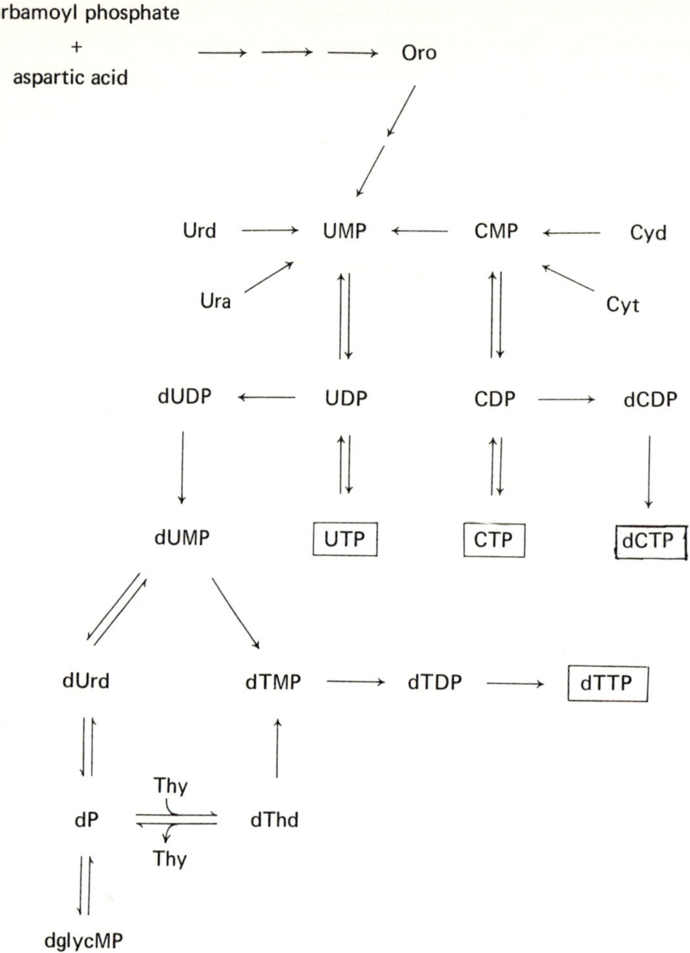

FIG 5.1   Pathways for the biosynthesis of pyrimidine nucleosides and nucleotides. dP = deoxyribose-1-phosphate; dglycMP = glycinamide deoxyribotide.

Urd, Cyt and Cyd can be used for labelling RNA (and DNA). If [5-^{3}H] Ura or Urd are used only the RNA will be labelled (see note above about [5-^{3}H] Oro). Also, labelled Ura or Urd will end up in both U and C residues in RNA (likewise labelled Cyt and Cyd). For many tissues and micro-organisims Ura or Urd are efficient labels. The relative efficiencies of labelling with Oro, Urd and Ura are determined by the pool sizes of these substances. With many bacteria, very high labelling with Ura can be achieved by using an organism which is a pyrimidine auxotroph.

The major routes for the synthesis of the pyrimidine precursors of DNA synthesis (dCTP and dTTP) are the reduction of the diphosphates (to dCDP and dUDP) and in the latter case the route through dUMP, dTMP, dTDP to dTTP.[1] However there are again scavenging routes for the synthesis of thymidine nucleotides. These pathways are employed for the specific labelling of DNA using dThd labelled with ^{3}H or ^{14}C. In principle it should be possible to use labelled Thy; in practice this technique does not work in general as the pool size of deoxyribose 1-phosphate is so low, that efficient incorporation is impossible. There is an important exception to this however, namely the thymine auxotrophs of bacteria. There are several categories of these organisms but they all require thymine to grow. At first sight the phenomenon of thymine auxotrophy is a puzzling one; a single mutation results in an organism which differs from the wild type in two respects, namely inability to grow without thymine and the ability (a necessity) to incorporate thymine into dTMP and dTTP. The explanation for the most straightforward type of Thy- mutant, one blocked at the step, dUMP to dTMP, is that to prevent the accumulation of dUMP, this substance is degraded to enhance the deoxyribose 1-phosphate pool and hence to open up the step which converts Thy into dThd.

Purine biosynthesis is much more complex than pyrimidine biosynthesis. The starting point for all the purine precursors of nucleic acid biosynthesis (ATP, GTP, dATP and dGTP) is IMP. The origin of the atoms in the purine skeleton of IMP is shown at the top of Fig. 5.2. This skeleton is built up on a ribose phosphate (starting with the nitrogen of the *N*-glycoside) and

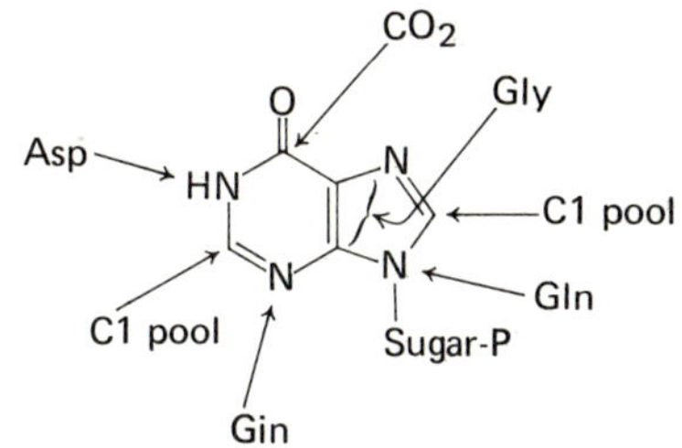

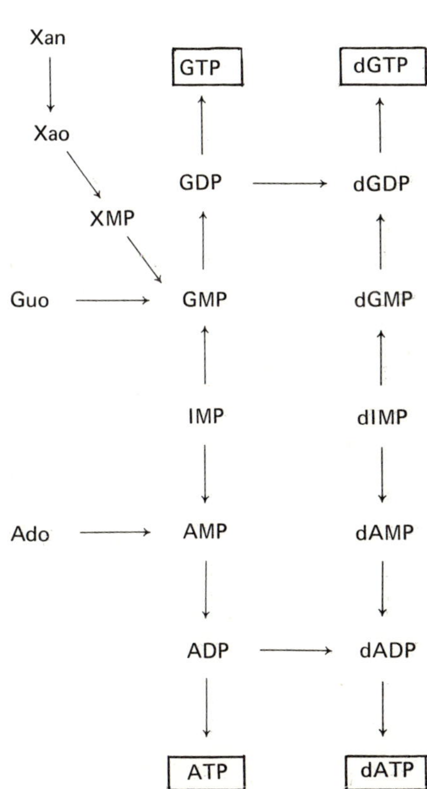

FIG 5.2   Biosynthesis of purines. Above: the skeleton of IMP and the metabolic origin of the ring atoms. Below: outline of the interconversions of purine nucleotides.

hence there is no uncharged precursor (like orotic acid) which can be used for labelling. In practice only the scavenging pathways are used for labelling purposes. Radioactive Ade is used for labelling the RNA of certain algae and radioactive Guo is efficient for labelling the RNA of prototrophic strains of *E. coli*. In this latter case, over 95% of the label is in G-residues (remainder in A).[2]

An important point to note from Figs. 5.1 and 5.2 is that certain amino acids are precursors of nucleotides (and hence nucleic acids). It is therefore very important, when labelling proteins *in vivo*, to either avoid the use of these amino acids (or other amino acids closely related to them metabolically) or to carefully purify the protein fractions if it is desired to follow protein synthesis.

Apart from the labelling of phosphate residues and the purine and pyrimidine nucleotide precursors of nucleic acids, it is often desirable to label specifically the minor bases of DNA, tRNA and rRNA. The only really feasible proposition is to label the residues that are methylated following synthesis of the nucleic acid chain. This is done by labelling the C1 pool using methyl-labelled (either ^{3}H or ^{14}C) methionine. Several examples of this technique are discussed later in this section. If the technique is employed, it is essential that the labelled sample be hydrolysed (see p. 203) and the resulting hydrolysate be analysed by a suitable chromatographic procedure to establish that all the label is in the base in question (say Thy) and that none of it is in methionyl residues of proteins which may be contaminating the nucleic acid preparation. A method of labelling thiated pyrimidines is described on p. 300.

Assay of radioactivity   Techniques of radioactive counting are largely dictated by the equipment available in the laboratory. The sad truth is that the average biochemist puts all his samples into a liquid scintillation counter with precious little idea about the basis of radioactive assay. For an informed and readable account of radioactive procedures consult the book by Fremlin (1964). The major techniques that are employed are as follows.

Autoradiography of chromatograms: This procedure is only suitable for ^{32}P, ^{14}C and ^{35}S. The paper is put against a sheet of X-ray film and left to expose in the dark. If ^{32}P is used, it is necessary to sandwich the paper and film between two sheets of lead, or other film in the vicinity will become fogged. The only difficulty with autoradiography is to guess the correct exposure time. If the amount of radioactivity is relatively low (say only a few hundred c.p.m. per spot) such that long exposures are needed, it is a good idea to put sheets of film on both sides of the paper and to just remove one of them after a few weeks and develop; if you then decide that more time is required, the remaining sheet of film has merely to be left in place. The same technique can be employed for the autoradiography of polyacrylamide gels provided there is a lot of ^{32}P on the gels; otherwise it is simpler to slice the gels up and count the pieces (see below). Remember that glass absorbs a lot of radiation so that a gel can only be autoradiographed while mounted in a glass tube if it is really very hot indeed (millions of c.p.m.).

Microscopic autoradiography: For this technique the only suitable isotope is ^{3}H; other 'harder' isotopes produce radiation with such long pathlengths that resolution is impaired. The sample is put on top of a piece of 'radioactive stripping film' which is used and developed according to the manufacturers' instructions. This is the technique employed in the Cairns procedure for studying bacterial DNA (see pp. 108 and 399). It can also be used to study the sites of DNA synthesis or RNA synthesis in whole cells. The organism is inoculated with [^{3}H] thymidine (for DNA labelling) or [^{3}H] uridine (for RNA synthesis) and then sections are cut and examined under the light or electron microscope, the section is then exposed to stripping film and the film then examined and the pictures superimposed. The techniques are outside the scope of this book; remember when reading papers in which such autoradiographs are presented, that under the electron micron microscope the individual decay patterns obtained from the disintegration of one tritium atom appear as little twisted strings each corresponding to the track made by the β-particle in the emulsion.

Cherenkov radiation: the simplest way to assay ^{32}P in samples is to make the total volume up to 5 ml (or thereabouts) with water and to count them in a liquid scintillation counter (without any added scintillant) using a low energy window setting (e.g. the 'tritium' window). The efficiency is lower than with scintillant but this disadvantage is outweighed by the simplicity of the procedure and the very low and consistent backgrounds.

Liquid scintillation counting: the principle of the liquid scintillation technique is that β-particles emitted by the radioactive nuclei interact with a chemical phosphor and the emitted photons are detected by banks of photomultiplier tubes. The instrument that is used for the assay is a spectrophotometer; as the wavelength of the emitted light is proportional to the intensity of the radiation it is possible to count two (or more) isotopes at once by recording counts in different 'windows' which can be selected on the spectrophotometer. Some overlap is to be expected between different isotopes, for example if ^{14}C and ^{3}H are being counted in the same sample a significant proportion of the ^{14}C counts will appear in the tritium channel. Careful calibration is required before calculating double-labelled counts, as the proportion of the counts that appear in different windows is a function of the quenching due to the scintillant. In modern scintillation spectrophotometers, it is possible to estimate the quenching (and efficiency of counting) of every sample with the external standard which has a radiation source with a broad emission spectrum. The samples are automatically counted on their own and also exposed to this source and simple calibration is possible. Liquid scintillation counting is not reliable if there are more than about 10^6 c.p.m. in a sample as coincidence errors are then significant. Many scintillation systems are recommended in the literature—some of them are extremely complicated. A useful selection of recipes is presented by Turner (1967). For general use,[3] let me propose two simple ones, both based on the use of diphenyloxazole (PPO) as primary phosphor and with *bis*phenyloxazolylbenzene (POPOP) as a secondary phosphor (to shift the wavelength to a region where the photomultipliers are most sensitive). The simplest one is made by dissolving 10 g of PPO and 0·25 g of POPOP in a Winchester (2·5 l) of sulphur-free toluene.[4] The second system, which is of lower efficiency but more convenient for counting aqueous solutions, consists of a solution in dioxan of naphthalene (10% w/v), PPO (0·5% w/v) and POPOP (0·0125% w/v). The naphthalene must be 'scintillation grade'. Dioxan of this grade is also commercially available but is incredibly expensive. Many technical grade preparations of dioxan are suitable, but some trial and error is inevitable as the impurities in dioxan which produce a poor scintillant (low efficiency and high background) are not identified.

Gas-flow counting: Gas-flow counters are Geiger-Müller tubes which have a constant flow of gas (an inert gas such as helium containing a small amount of a hydrocarbon) flowing through the tube. They are two types: those with an aluminium foil window at one end of the tube, and the windowless type. The former type counts dry samples prepared in planchettes; the latter is mainly useful in the radioactive chromatogram scanner which is a device in which a strip of chromatography paper is slowly fed past a gas-flow tube and the radioactivity present in the spots is drawn as peaks on a chart recorder. Planchette counting is not suitable for tritium as the efficiency is so low. Chromatograms containing very large amounts of tritium can be scanned in the second type of counter but on the whole these gas-flow counters are best used only for ^{32}P, ^{14}C or ^{35}S. As with any Geiger counter, the 'plateau region' has to be measured for the isotope being counted prior to feeding the samples through.

Preparation of samples for counting   Aqueous solutions of nucleic acids in dilute salt or buffer are readily counted in the liquid scintillation counter. In the case of ^{32}P-labelled material, Cherenkov counting (see p. 179) is the simplest method. Otherwise up to 1 ml can be counted by adding 10 ml of the dioxan-base system (above). If the solution contains sucrose, less can be added (say 0·2 up to 0·5 ml to 10 ml scintillant). The trouble is that sucrose crystallizes out and nucleic acids may co-precipitate; a certain amount of this is not important for ^{14}C-counting but with ^{3}H, whose β-particles have an extremely short pathlength, self-absorption in the precipitate of sucrose leads to low efficiencies and irreproducible results. An alternative is to extract the nucleic acid from the aqueous solution with a solution of an aliphatic ammonium acetate in a solvent such as butanol (Kirby, 1968). The important point to remember is that the technique relies upon the organic ions being in excess, so that the alkylammonium salt of the nucleic acid will partition into the organic phase. A suitable procedure for samples containing up to 0·2 mmole of salt (or buffer) is to make up an extracting solution by adding 23·4 ml of *bis* (4-ethylhexyl) amine and 4·0 ml of glacial acetic acid to 100 ml *n*-butanol. The nucleic acid solution is extracted with 1 ml of this solution and 0·5 ml of the upper phase is pipetted off and added to 10 ml of the dioxan-based scintillant. The method can be adapted to counting nucleoprotein in which the nucleic acid is labelled. To 0·8 ml fractions of RNP, is added 1 ml of EDTA disodium salt (2·5% w/v). The solution is then heated to 100°C for 10 min, cooled, 0·3 ml of *n*-butanol is added and then the nucleic acids are extracted and counted as above.

Alternatively precipitates of nucleic acid can be counted. Nucleic acids can be precipitated from aqueous solution with 5% trichloroacetic acid (or TCA-precipitates from tissues can be made—see p. 173) or CTAB (p. 135) may be used. If the nucleic acids are very dilute they can be bulked up by dissolving unlabelled nucleic acid (such as crude commercial 'RNA') prior to precipitation. There are two ways of counting these precipitates. They can be filtered off and the filter paper discs can be counted either in planchettes in a gas-flow counter, or by soaking them in scintillant in vials for a scintillation counter. Alternatively the TCA-precipitates are spun out in a bench centrifuge and dissolved in a suitable alkaline solution. Although various quaternary ammonium hydroxides can be used with varying degrees of success, the simplest thing to do these days is to invest in a bottle of jungle juice (such as Packard's 'Soluene') especially concocted for this purpose by scintillation counter manufacturers. Instructions for the use of these preparation is supplied with the bottles. For counting TCA-precipitates with Soluene, the procedure is to allow the precipitates to soak and dissolve in 0·2 ml Soluene overnight (preferably at 30–35°C) and to then add 10 ml of the toluene-based scintillant. The resulting sol is quite transparent.

Slices from polyacrylamide gels (see p. 155) can be counted in a variety of ways. A good reference for different procedures is the paper by Gressel and Wolowelsky (1968). If the gel contains ^{32}P, the simplest procedure is to put the slices in 10 ml of water in counting vials and count the Cherenkov radiation (p. 179). Otherwise they can be soaked overnight in Soluene as above and then toluene-based scintillant added as described above for TCA-precipitates. The slices contract in the Soluene and then swell in the toluene; nucleotides (nucleic acids are hydrolysed in Soluene) diffuse out of the gels under these circumstances.

Radioactive nucleotides on paper chromatograms are identified either by autoradiography (see p. 178), or scanning in a windowless counter, or by cutting spots or strips off the paper and putting them in vials for the scintillation counter and adding scintillant. For papers containing tritium, more reproducible results are obtained if Soluene is added to the pieces of paper; the paper does not dissolve completely but the nucleotides are eluted and counted in solution.

Applications of radioactive labelling

In this section there are miscellaneous applications of radioactive labelling to the identification of RNA or DNA synthesized *in vivo*. Labelling of nucleic acids *in vitro* is discussed on p. 175 and hybridization techniques on p. 360.

'Rapidly labelled' RNA   If cells are pulsed with RNA precursors, a large amount of labelling occurs in RNA fractions other than rRNA. Interpretations of these patterns of incorporation are very tenuous as methods for assaying mRNA are in general very unsatisfactory (see p. 194). Low-molecular-weight RNA labelled in this way must be carefully fractionated and separated from tRNA as the 3'-terminal sequence -C-C-A (see p. 13) is turning over *in vivo* (see p. 228). With increasingly long labelling times the labelling pattern more and more resembles that of high-molecular-weight rRNA. For rapid labelling of bacteria something of the order of 1 min labelling is suitable (see Fig. 5.7, p. 185). For eukaryotic cells the labelling times are much longer. For example in rat liver, there is extensive labelling of nuclear RNA within 15 min with very little labelling in the cytoplasm. In half an hour there is labelled RNA of a highly heterogeneous character in the cytoplasm. It is assumed that this is mRNA although there is no direct proof of the proposition.

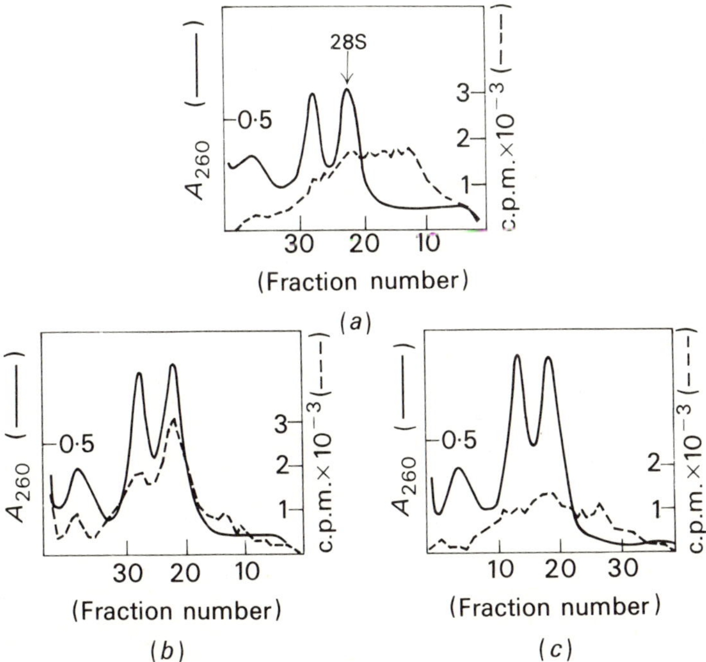

FIG 5.3   Labelling of duck erythroblast RNA. The results of sucrose-gradient analysis of RNA extracted from the labelled cells (sedimentation from left to right) are shown here. The gradients (15–30% sucrose in 0·05 M NaCl buffered to pH 7·0—see reference for details) were centrifuged for 19 h at 22 500 rpm in the SW 25 rotor of the Beckman model L ultracentrifuge for 19 h at 0°C. The erythroblasts were labelled with [3H]uridine for 30 min. In (*a*) they were harvested and RNA was extracted by a 'hot phenol method' (see p. 130). In (*b*) they were 'chased' with excess unlabelled Urd for 2·5 h; the conversion of much of the high-molecular-weight RNA into rRNA can be observed. In (*c*) the chase was performed in the presence of Actinomycin D. Data of Scherrer *et al.* (1966).

All labelling patterns for rapidly labelled RNA are confused by the presence of rRNA precursors (p. 183). There have been several attempts to sort out the muddle. One well known series of experiments (Scherrer, Marcaud, Zajdela, London and Gros, 1966) concerned the RNA synthesized in duck erythroblasts. Ducks (in common with all birds) have a system of red cell maturation different to that in mammals. The nucleus and ribosomes are not lost from the maturing cells; they are merely 'shut down'. In the erythroblast (which corresponds to the mammalian reticulocyte, see p. 125) the main protein being made is haemoglobin. Ribosomal RNA and presumably mRNA are being transcribed. Rapidly labelled cytoplasmic RNA is about 8 S; this is conveniently about the right size for haemoglobin message (see p. 398). The rapidly labelled nuclear RNA is enormous; from sucrose gradients it appears to be a polydisperse mixture with sedimentation coefficients between 30 S and 50 S. The effect of 'chasing' the label is for the rRNA precursors in this envelope to form rRNA. By comparing this pattern with that resulting from inhibiting RNA synthesis with actinomycin D (see p. 413) it is possible to deduce that a proportion of this 30–50 fraction is rapidly turning over RNA that apparently never leaves the nucleus and is of unknown function (Fig. 5.3).

If cytoplasmic or bacterial RNA is fractionated (either on sucrose gradients or poly-acrylamide gels) most of the rapidly labelled RNA (apart from the rRNA precursors in the latter case) forms a heterodisperse zone between about 8 S and around 25 S.

The characterization of particular components of this type of mixture is discussed elsewhere (pp. 194 and 376). Interesting examples of rapidly labelled RNA fractionation from isolated salivary-gland chromosomes (p. 116), and more especially from Balbiani rings[5] are described by Pelling (1970) and Daneholt, Edström, Egyházi, Lambert and Ringborg (1970).

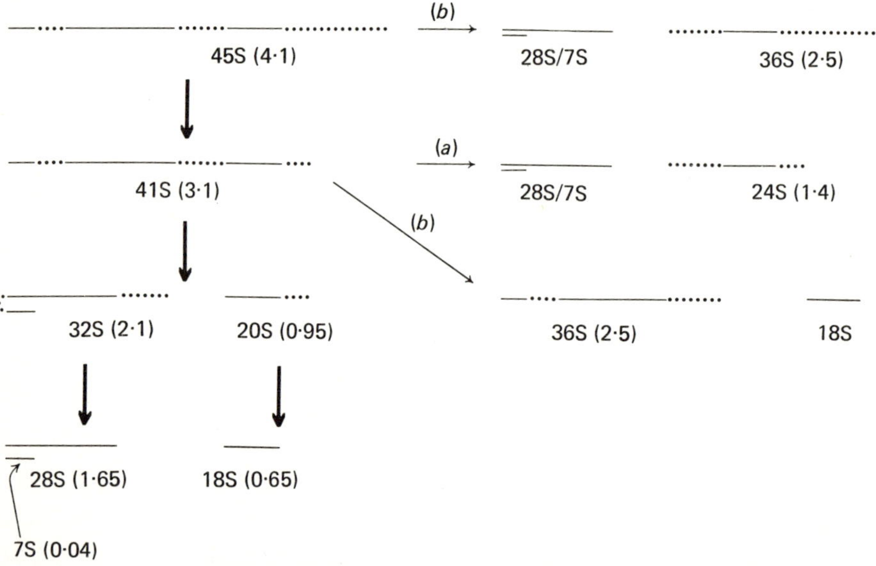

FIG 5.4   Proposed scheme for the processing of nucleolar RNA. The initial precursor (45 S, MW 4·1 x 10^6) contains the sequences for the final 28 S and 18 S molecules (continuous lines). The major route for the maturation of this precursor is shown by the heavy arrows. Note the suggestion (in the last stage) for the origin of 7 S RNA (see p. 159). Occasionally nucleolar 24 S RNA is found. Route (a) is designed to account for it; likewise routes (b) are designed to account for the occasional appearance of nucleolar 36 S RNA. The molecules are identified by their sedimentation coefficients and (in brackets) their molecular weights x 10^{-6}. Schemes from Penman *et al.* (1969).

Maturation of RNA   The processing of the immediate product of transcription to produce rRNA and tRNA is known as 'RNA maturation'. In the following paragraphs are illustrations of experiments which elucidate the stages in this process. I have not included any reference to mRNA maturation as it is not clear at the moment whether mRNA is a direct product of transcription or not. It is fairly clear (p. 189) that at least most, and probably all, *E. coli* mRNA is such a product. Whether some of the giant nuclear RNA species referred to in the previous section are mRNA precurosrs is still uncertain.

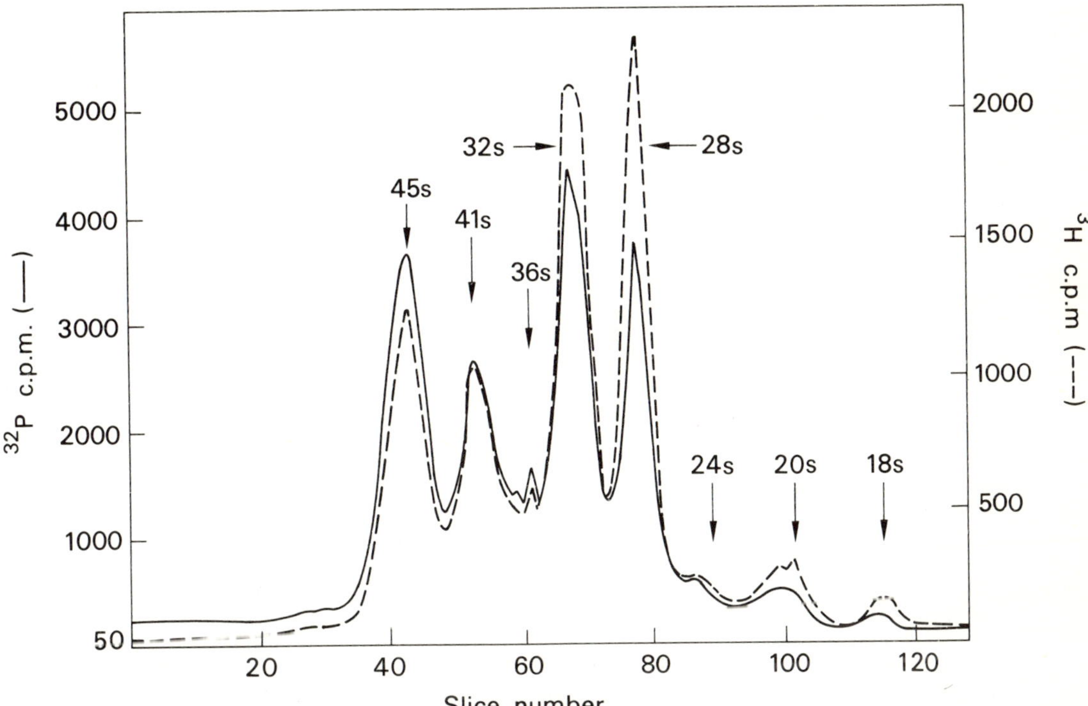

FIG 5.5   Methylation of HeLa rRNA and its precursors. Cells were labelled for 3 h with [^{32}P] phosphate and [methyl-^{3}H] methionine. They were then infected with poliovirus to hold up RNA maturation. After 30 min they were harvested and resuspended in the absence of virus for 3 h. Nucleolar RNA was extracted. Electrophoresis was performed on 2·6% polyacrylamide (p. 156) employing 5 mA/gel for 9 h. Data from Penman *et al.* (1969).

Much of the credit for establishing the correct maturation timetable for mammalian rRNA is due to S. Penman and his group. Their final scheme, based exclusively on following the labelling of different RNA fractions with either a general RNA label (such as inorganic phosphate) or radioactive methionine is shown in Fig. 5.4. As a single example of their procedure, Fig. 5.5 shows the gel electrophoresis of nucleolar RNA (see p. 114) isolated from HeLa cells (see p. 125) infected with poliovirus. The point about using infected cells is that this particular virus has the effect of slowing down the maturation processes so that rRNA precursors accumulate in the nucleolus. The data are taken from Penman, Vesco, Weinberg and Zylker (1969).

The literature on the rRNA precursors emphasizes the importance of isolating the RNA carefully, with the minimum manipulation, to avoid the degradation of the very large molecules such as the 45 S RNA. For this reason, the 'deproteinization gradient' of Hastings (p. 147) is particularly suited to the study of these molecules. An example of isolation of 45 S and 35 S RNA by this method is given in Fig. 5.6 (from the work of I. H. Maxwell).

The next phase of research into the maturation of eukaryotic rRNA will be concerned with

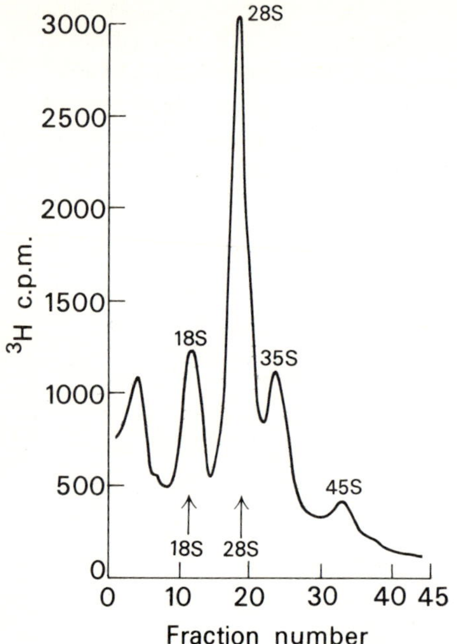

FIG 5.6   Isolation of rRNA precursors from BHK cells transformed with polyoma virus. The cells were labelled 2 h with [5-³H] Urd and were then lysed and layered over a gradient as described in Fig. 4.13, p. 000. The gradient contained the equivalent of 10^4 cells. The gradient was centrifuged in the MSE Superspeed 50 ultracentrifuge at 10°C at 1000 rpm for 1 h; the centrifuge was then accelerated and the sample centrifuged at 20 000 rpm for 14·5 h. The positions of rRNA (18 S and 28 S) were determined by running a parallel gradient containing the lysate of cells labelled with [¹⁴C] Urd for a generation. (BHK are baby hamster kidney cells in tissue culture). Unpublished data kindly provided by I. H. Maxwell.

an identification of the cleavage reactions summarized in schemes such as that of Fig. 5.4. For example, are the specificities of the reaction determined by specific sequences in the precursors or by conformational aspects of either 45 S RNA or the RNP of the nucleolus? Clearly the work of Shankar Narayan and Birnstiel (1969) referred to on p. 121, on the isolation of nucleolar RNP, represents the first phase of this research. The RNP they isolated contains heterodisperse particles with a peak around 60 S. These particles contained no 45 S RNA but did contain 18 S RNA and a molecule designated 31 S RNA. Whether these results reflect the fragmentation of rRNA precursors by enzymes in the particles themselves or in the 'nucleolar sap' remains uncertain.

In the case of bacterial rRNA, the precursors are not giant molecules (such as 45 S); rather there are two precursors, one for 16 S RNA and one for 23 S RNA, each slightly larger than the final product (Hecht and Woese, 1968; Fig. 5.7).

All the studies on rRNA maturation have been complemented by hydridization data (p. 368). Studies on the primary structure of the 16 S precursor in *E. coli* are referred to on p. 321.

The maturation of tRNA precursors has been extensively studied (in animal cells) by R. H. Burdon. In their admirably concise paper, Lal and Burdon (1967) explain a labelling schedule (see legend to Fig. 5.8) in which the RNA is labelled with [³H] uridine in the presence of excess unlabelled cytidine (to minimize labelling of the -C-C-A) and the methyl groups in RNA are labelled with [¹⁴C] methionine in the presence of excess unlabelled formate. This formate is present to dilute any of the label which enters the tetrahydrofolate C1 pool (and would hence be a precursor for purine synthesis—see Fig. 5.2, p. 177). The RNA was characterized by

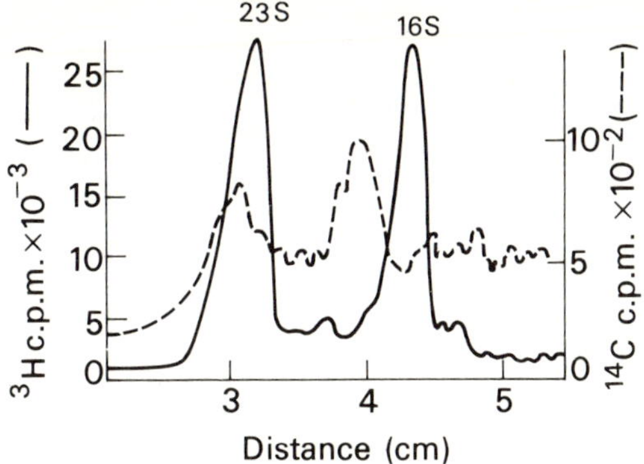

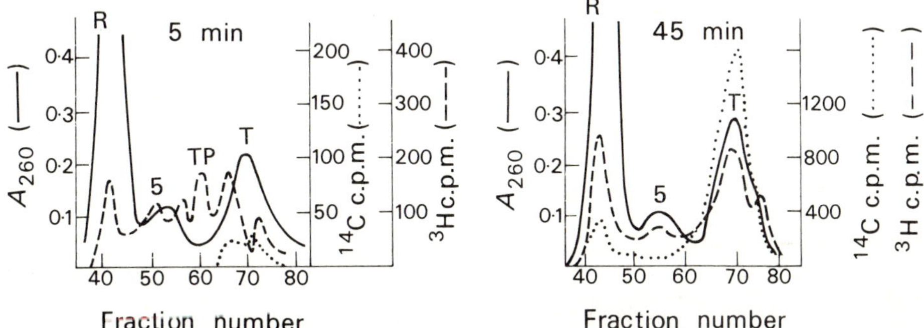

FIG 5.7   Precursors of rRNA in *B. subtilis*. Analysis by gel electrophoresis (2·4% polyacrylamide) of RNA extracted from double-labelled cells. The cells had been labelled for two generations with [³H] Urd and for 1 min with [¹⁴C] Urd. Thus the ³H-radioactivity is in the mature rRNA (16 S and 23 S) and the ¹⁴C-radioactivity in the precursors. Data from Hecht and Woese (1968).

FIG 5.8   Precursors of mammalian tRNA. Krebs II ascites tumour cells were labelled with [³H] Urd and [¹⁴C] methionine as described in the text. The RNA was extracted after the two labelling times indicated and analysed by column chromatography on Sephadex G-100 (see p. 167). The absorbance at 260 nm of each fraction was measured and then the fractions were incubated a high pH to remove amino acids from the tRNA (p. 125) and precipitated with trichloroacetic acid (p. 180) and counted. The peaks correspond to: (R) high-molecular-weight rRNA, (5) 5 S RNA, (T) 4 S tRNA and (TP) tRNA precursors. Note that ¹⁴C-(methyl) labelling is only found in the mature tRNA (as well as 'R'). Data from Lal and Burdon (1967).

Sephadex chromatography (see p. 167). Some of their data are shown in Fig. 5.8 which demonstrates that the tRNA precursors are broken down to form 4 S tRNA molecules with concomitant methylation.

DNA synthesis   Methods of studying DNA synthesis are discussed elsewhere (pp. 381 and 429). The experiment which established that newly synthesized DNA consists of small single-strand fragments was performed by Okazaki, Okazaki, Sakabe, Sugimoto and Sugino (1968). These newly synthesized pieces are now referred to as 'Okazaki fragments'. The experiment consisted of denaturing the DNA in alkali and analysing it on sucrose gradients (p. 148) following very short labelling times. The newly synthesized fragments were shown to be of low molecular weight (see p. 386). The internal molecular-weight marker was single-stranded δA phage DNA which could be identified in a spheroplast infectivity assay (p. 193). An illustration is shown in Fig. 5.9.

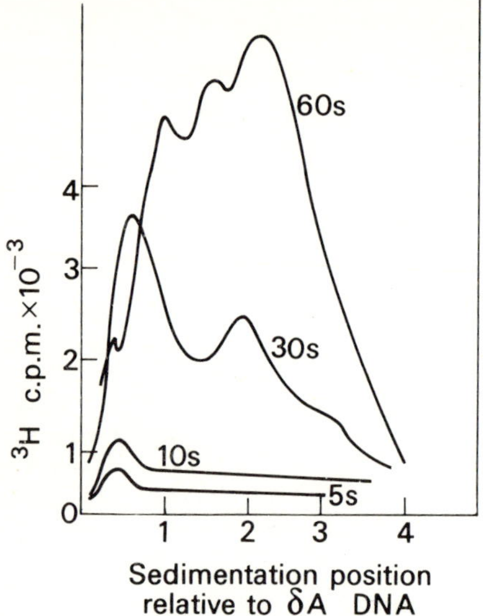

FIG 5.9 Okazaki fragments. *E. coli* B was labelled with [³H]dThd and for the times indicated. DNA was not purified from these cells. Rather they were incubated at 37°C for 20 min in 0·1 M NaOH, 0·01 M EDTA and the insoluble material was removed by slow centrifugation. The DNA was layered over a linear sucrose gradient (5–20% w/v) in the same 'buffer' and centrifuged for 10 h at 22 500 rpm in the Beckman SW 25 rotor at 4°C. Radioactivity was assayed by precipitating the samples with 5% trichloroacetic acid at 5°C. The precipitates were dissolved in the same acid at 90°C and re-precipitated in the cold. In employing such a procedure it is essential to establish the validity of the method (as was done in the present case) by doing some comparisons with phenol-purified DNA samples. The x-axis is calibrated in terms of the sedimentation position of infective δA DNA (see text). Sedimentation is from left to right. Data from Okazaki *et al*. (1968).

The rolling circle The rolling circle is one of the models for DNA replication (this and the alternative 'knife and fork model' are discussed further on pp. 430 and 432). At this stage let us just say that the rolling circle is a tenable model for phage DNA replication but not for cellular chromosomal replication. In this section and on p. 189 there are two sets of data which support the rolling circle model for the replication of φX 174 DNA both from the work of D. Dressler. This model is not universally supported for φX 174 replication (see for example Knippers, Whalley and Sinsheimer, 1969, and also p. 154). However attempts to prove the model for this system have led to extremely ingenious experiments.

DNA in φX 174 phage particles consists of a single-stranded closed loop. This molecule (see also p. 154) is called the + strand. The process of replication consists of two stages; the production of a − strand (complementary to the + strand) and the use of the − strand as template for the synthesis of more + strand. The rolling circle as applied to this particular system is illustrated schematically in Fig. 5.10. The results illustrated in Fig. 5.11 are designed to test two consequences of this model. The first of these is that during the stage at which replicative form (RF see p. 154. actually supercoils see p. 424) is occurring, the labelling of the two strands in the supercoil should be asymmetric. This is in fact found (Fig. 5.11(*a*)); at early stages in the synthesis the − strand is perferentially labelled as implied by the model. The second point is even more directly and dramatically established; when the circle is actually rolling according to the model there is one linear molecule longer than the total genome of the phage. The existence of these 'longer-than-φX' + strands is demonstrated in Fig. 5.11(*b*). The

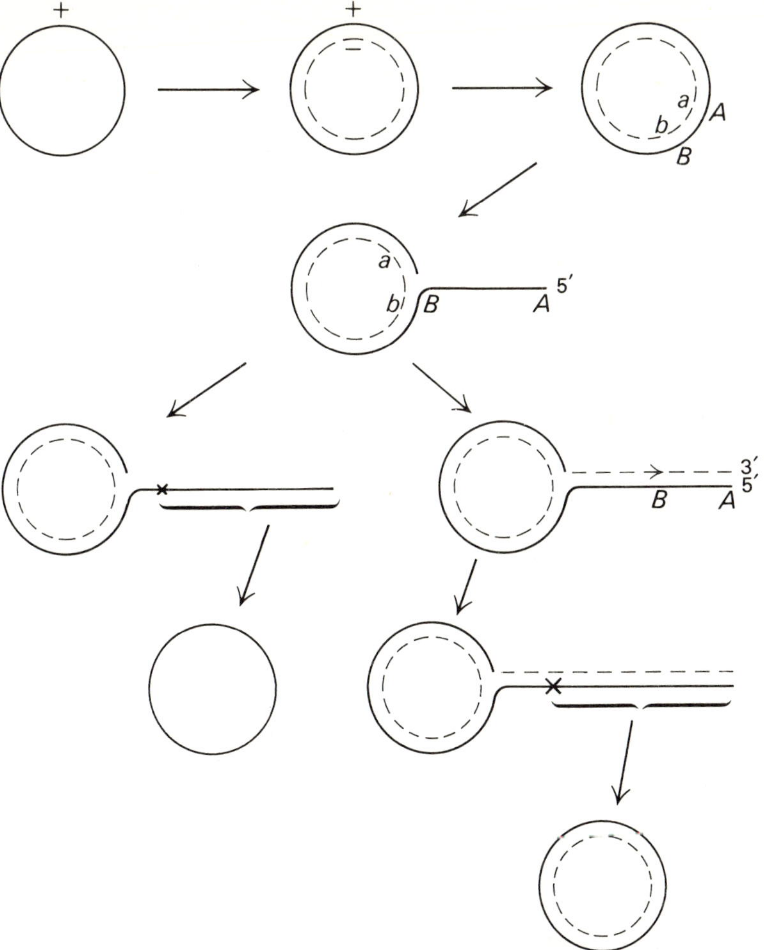

FIG 5.10   The rolling circle model for DNA synthesis as applied to the replication of a single-strand, closed-circle DNA (such as that of coliphage φX 174). The continuous line represents the + strand, the discontinuous one the − strand. The arrow indicates the 5′ to 3′ direction. *A* and *B* are merely points on the + strand chosen to illustrate the operation of the model; *a* and *b* are corresponding points in the complementary (−) strand. They are not of any structural or genetic significance. There are two alternatives represented here. Either the + strand produced as the circle rolls is used to produce a 'concatenate' (see p. 430) from which RF's can be cut off or alternatively, a long '+ tape' is read out, which is cut up to yield the (single-strand) mature phage genomes. Clearly, with a single-strand phage such as φX, the latter process predominates. Taken from Dressler (references in text and the caption to Fig. 5.11).

experiments on the asymmetric labelling of RF are by Dressler and Wolfson (1970); the long + strands were demonstrated by Dressler (1970).

    The experimental protocol was complicated and requires the expertise of φX-ologists. Briefly experiment (*b*) was as follows. A suitable *E. coli* strain (HF 4704 which is a thymine auxotroph) was infected in a system which was imperfect in three respects: (i) there was no Thy so DNA synthesis was inhibited, (ii) mytomycin C (see p. 251) was present to inhibit cellular DNA synthesis and (iii) the phage was φX *am*-3, a mutant incapable of lysing the cells (except in suppressor hosts of which HF 4704 is *not* one). After 15 min (to let the phage DNA into the cells), Thy was added. In the presence of the mitomycin this is tantamount to allowing

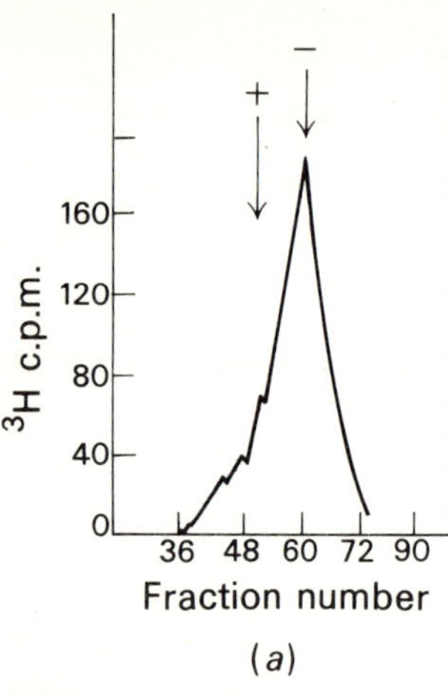
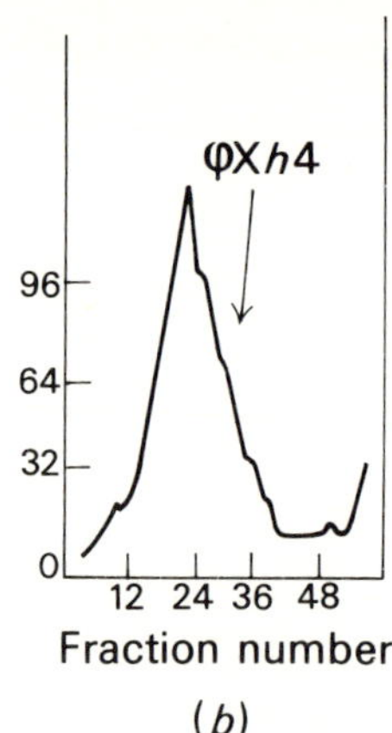

FIG 5.11   The rolling circle of φX 174. (*a*) is an alkaline CsCl gradient with the positions of single + and − strands as indicated. The asymmetric synthesis (predominatly −) in the early stages of synthesis is demonstrated. (*b*) is an alkaline sucrose gradient demonstrating the existence of + strand longer than one genome in the later stages of synthesis (sedimentation right to left).

φX synthesis to start but to inhibit *E. coli* DNA synthesis. After 50 min the cells were pulsed with [³H] Thy for 50 s and harvested. Some cells were chased with cold Thy and allowed to go on for a further 10 min. DNA was extracted from the cells and separated by neutral-sucrose-gradient centrifugation. The chased DNA was used as a marker (it should be RF). The pulsed-and-harvested DNA was isolated from various regions of the gradient and denatured by exposure to 0·25 M NaOH and neutralized with HCl (see p. 361). This DNA was mixed with φX*h*4 DNA as a marker (it was assayed by a spheroplast assay—see p. 193). The analysis was by centrifugation through a 10–30% sucrose gradient in 1 M NaCl at 64 000 rpm at 8°C in the Beckman SW 65 rotor. The position of the φX*h*4 is shown by the arrow. The larger molecule is denatured + strand from a rolling circle.

Experiment (*a*) was as follows. Cells (details as above) inhibited with mitomycin C, were infected with φX *am*-3 at low multiplicity. The phage replication was synchronized (Thy) and pulsed with [³H] Thy (for 40 s) after 6 min of phage development. RF was isolated from these cells by treatment with alkali at low temperatures (when the chains of the double-stranded continuous molecules are not separated). The RF was broken into single-strand linear fragments (shown to be half a φX chromosome in separate experiments) by heating to 100°C at pH 8·2 for 3 h. Centrifugation (in the presence of + and − strand markers treated in the same way) were centrifuged to equilibrium in a CsCl gradient at pH 12·5 (see p. 154). The result is as shown; most of the label is in the − strand. More label was found in the + strand with longer pulse times. Data from Dressler and Wolfson (1970) and Dressler (1970).

The electron microscope

The technique for the electron microcopy of nucleic acid molecules spread over a grid is of great value in determining molecular weight (p. 399) and tertiary structure (p. 427). The crucial part of the procedure is the spreading of the nucleic acid (as one component in a carefully constructed protein monolayer on top of an aqueous solution). The methodology has been carefully worked out by A. K. Kleinschmidt. Reference to the papers quoted here and elsewhere in the book, together with experience in operating the electron microscope and its ancillary equipment are essential before any attempt is made at this technique. In this section two procedures illustrative of the very best of nucleic-acid electron microscopy are given.

Examples

Shadowing of DNA An extremely elegant method (adapted from Kleinschmidt and Zahn, 1959) is that of Westmoreland, Szybalski and Ris (1969). The advantage of this particular method is that double- and single-stranded DNA are clearly distinguished and are of different widths when viewed under the electron microscope. All the solutions are filtered to remove dust (0·22 Millipore nitrocellulose filters for everything except the cytochrome for which 0·2 μ Elotronics (FM-13) membranes are used. DNA in 50% formamide (about 1 μg/ml; 0·1 ml) is mixed with 0·1% aqueous cytochrome a (0·01 ml) and carefully pipetted down an inclined glass ramp (ultra-clean) on to the surface of distilled water to form a protein monolayer with the DNA dispersed in it. The protein monolayer contains of the order of 1 mg protein/m^2 (see the paper by Kleinschmidt and Zahn for the method of measuring this value). The monolayer is picked up on Formvar-coated platinum grids and passed through ethanol, amyl acetate and isopentane (5 s in each). The DNA is shadowed using a rotating sample and a uranium-covered tungsten wire basket so that uranium is the shadowing metal. An illustration of the power of the instrument in distinguishing between single- and double-stranded DNA is shown in Frontispiece (a). The molecule being studied is a hybrid composed of two strands re-annealed (see p. 373). One strand was obtained from a wild-type phage, the other was obtained from a mutant in which one region (b2) was deleted so there is a significant length of DNA in the wild type absent from the mutant DNA (which is therefore shorter). Additionally there is a mutation in the b5 region which results in the two stretches of DNA in this region being non-complementary.

Another application of the same method by Dressler (1970) to the φX 174 rolling circle problem (see p. 186 for a discussion and additional evidence) resulted in the visualization of rolling circle itself (Frontispiece (b)).

RNA synthesis really happens Certain tissues are especially suited to the study of rRNA biosynthesis. One such is the oöcyte of many amphibians, in which there are large numbers (up to one thousand) extrachromosomal nucleoli in each nucleus. These nucleoli are active in the synthesis of rRNA (see pp. 370 and 378). In the experiment of Miller and Beatty (1969), the nucleolar nucleic acids from the oöcytes of *Triturus viridescens* (a species of newt) were visualized by disrupting the nucleus, dispersing the contents in water and spinning them through a solution of 0·1 M sucrose in 10% formalin on to a carbon-coated electron microscope grid. The grid was washed, stained for 1 min with 1% phosphotungstic acid in 50% aqueous ethanol, rinsed in 95% ethanol, 100% ethanol and finally isopentane. The remarkable result, a picture of the ribosomal cistrons with the rRNA precursors (see p. 183) being synthesized and getting longer along each cistron is shown in Frontispiece (c). Gaps between the cistrons (see p. 370 for other evidence of this) are clearly seen. For an extension of the technique to other organisms, see Miller, Beatty, Hamkalo and Thomas (1970). This latter paper contains direct photographic evidence for the coupled transcription and translation of *E. coli* genes (see p. 27).

DNA and DNA polymerase A quite different technique has been applied to the visualization of DNA polymerase bound to DNA by Griffith, Huberman and Kornberg (1971). The technique should prove of general application in studying the binding of proteins to nucleic acids (see p. 456). The principle is that carbon films are made on mica, floated off and picked up on grids previously treated with a solution of polybutene in xylene (to act as glue). These perfectly flat carbon films backed with grids are subjected to a glow discharge in a high vacuum and are then floated (face down) on tiny drops of DNA solution. The DNA sticks on to the carbon film (provided no great time has elapsed between the exposure to the discharge and contact with the DNA solution). After washing and drying, the DNA is shadowed with tungsten.

Biological and biochemical assays

This section is necessarily very superficial as its subject matter could be expanded into a discussion of a significant part of Molecular Biology (whatever that may be). The purpose is to survey the procedures for thebiological and biochemical assessment of the functional purity or heterogeneity or intactness of preparations of nucleic acids. One very important technique, hybridization, has been omitted altogether as this is the subject of a later chapter (p. 368). No mention of controversial or ill-established procedures (such as eukaryotic transformation) has been made as these techniques (even if they really work) are not viable, everyday assay procedures for nucleic acids. However, the recent report by Merrill, Geier and Petricciani (1971) that a bacterial gene (carried by a transducing phage) can be expressed in mammalian cells implies that we can look forward to a time when a variety of extremely versatile, heterologous assay systems involving eukaryotes may be available.

Transformation and transfection

Transformation is the name given to the process whereby DNA is isolated from one type of organism and is used to inoculate a different type of organism with a concomitant acquisition by a proportion of the second (recipient) organisms of some of the characteristics of the first. Transfection is an analogous process where DNA is isolated from a virus and this viral DNA is used to inoculate organisms with concomitant maturation of viral particles. Transformation has only been reliably demonstrated to date in certain species of bacteria; moreover it is usually (but not exclusively) the case that donor and recipient are different strains of the same species.[6] Similarly complete transfection (in which viral DNA replication as well as assembly of new viral particles occurs) has only been properly demonstrated with bacteriophage. In certain bacteria (such as *E. coli*) in which transformation and transfection cannot be directly demonstrated (see however p. 192), it is possible to infect spheroplasts (see p. 193) with phage DNA and obtain viruses from them.

Transformation in bacteria   Bacterial transformation was first demonstrated (using heat-killed donors) between pathogenic and non-pathogenic strains of *Diplococcus pneumoniae*[7] by F. Griffith in 1928. The classical analysis of this phenomenon by O. T. Avery and his group (Avery, MacLeod and McCarty, 1944) established that transforming principle was DNA and was the starting point for biochemical research on nucleic acids. Since then studies on the mechanism of transformation and its exploitation for the genetic analysis of transformable bacteria and studies on bacterial DNA have proved of enormous value in assessing the inter-relatedness of different bacteria and an understanding of the mechanisms of genetic recombination. Here we are only concerned with it as a method of characterizing DNA. For further details of the process read the two very authoritative reviews by Spizizen, Reilly and Evans (1966) and by Hotchkiss and Gabor (1970).

For bacterial transformation to work there are two requirements, that the species is transformable at all and that the recipients are 'competent' to receive the DNA from the donors. With respect to the first of these points, the number of transformable bacteria is growing all the time but still represents a very small proportion of the species used for experimental work. In addition to *Diplococcus*, transformation has been demonstrated in members of the genera *Bacillus*, *Haemophilus*, *Streptococcus*, *Staphylococcus* and (more recently) *Rhizobium* and *Neisseria*. Anyone researching on these organisms will presumably be familiar with the relevant literature. For physico-chemical studies on transforming DNA, probably the best organism to use is *Bacillus subtilis*. It is a saprophytic Gram positive organism, convenient to grow and handle, well mapped genetically and with many auxotrophs and other mutants readily available.

Competence is less easily dismissed. There is a genetic character ('competence factor') and some strains are invariably incompetent and cannot be used for transformation. A discussion of the genetics of this system is to be found in Spizizen, Reilly and Evans (1966). It is clear that genetic incompetence has something to do with the transformation process as such and not with genetic recombination in general as these incompetent strains can be transduced (see p. 30) with normal efficiency. However the character of competence is only expressed under specific growth conditions. In *B. subtilis* a culture is competent after the cells have entered stationary phase. The factors required for maximum competence in this organism have been carefully analysed by Wilson and Bott (1968). Their experiment is summarized here as it is a good example of how transformation is done. Transforming DNA was isolated from *B. subtilis* 168 T$^+$ (a protrophic strain very useful for biochemical work). The recipient was *B. subtilis* Mu8u5ul *leu$^-$ ilv$^-$ met$^-$*.[8] This organism is just 'the recipient' from now on. The recipient was grown overnight in a rich medium ('broth'). The cells were harvested, washed and transferred to 'transformation medium' so that the turbimetric extinction at 500 nm was about 0·1. The cells were grown until they had just reached stationary phase. They were kept stationary for several hours and then treated with DNA (0·6 μg/ml); after 30 min transformation was halted by adding deoxyribonuclease. The cells were then plated on defined agar media lacking one or two of the amino acids required by the untransformed recipient. They were also plated on a nutrient agar to establish the total number of viable cells. The transformation medium consisted of minimal glucose-salts medium supplemented with 50 μg/ml of each of leu, ile and met to which various other amino acids were added. Certain amino acids were found to be stimulatory for transformation, others inhibitory. It is thus better to use a defined mixture of amino acids for this purpose rather than a protein hydrolysate. The optimal conditions found by Wilson and Bott are 25 μg/ml of each of the following: his, trp, arg, val, lys, thr, gly, asp and met. It was also established for this medium at 37°C, that the optimal delay between the cells entering stationary phase and additio of transforming DNA is 4 h. Under these conditions the frequencies of transformation of the single markers (*ilv$^+$*, *leu$^+$*, *met$^+$*) were of the order of three in a hundred and for the linked markers the frequencies were for *ilv$^+$ met$^+$* 4·6 in 1000, for *ilv$^+$ leu$^+$* 0·5 in 100, and for *met$^+$ leu$^+$* 0·7 in 100. These last data are included to illustrate the simplest way in which transformation can be used for genetic analysis—by measuring the relative co-transformation frequencies of genes.

It has been emphasized that these experiments were discussed as an example not as a rule. There is no universal method of optimizing competence. For example in *Rhizobium*, competence occurs in early logarithmic phase (Balassa, 1960).

In order to assay competence, it is not necessary to measure transformation frequencies. If the DNA is radioactive, its uptake into the recipient cells can be measured directly. A typical procedure would be to add to the recipient culture [^{3}H]DNA (specific activity of 10^5 c.p.m./μg and about 0·1–0·5 g of DNA/ml of culture). After incubation (say 10 min) the DNA that has not been taken up is destroyed by adding deoxyribonuclease (0·1 mg/ml of pancreatic ribonuclease, p. 209). The preparation is incubated for a further 10 min, the cells are spun out and dispersed in buffer. The cells can be counted as a TCA-precipitate on filter paper using the technique described on p. 180. The assay as described above can be done on a very small scale (0·5 ml of a culture of 10^8 cells/ml for the incubation, resuspended in 0·5 ml buffer and 0·1 ml aliquots used for the counting).

This technique gives a measure of the uptake of DNA by the cells. It does not tell us anything about the fate of the DNA once it is in the cells. Assay of the integration of DNA into the recipient chromosome is far harder. It is not too difficult in *Haemophilus influenzae* as an excellent assay system has been worked out by Steinhart and Herriott (1968). In this case the radioactive labelling is done the other way round. The recipient cells are labelled *in vivo* prior to their transfer to medium for competence. They are then incubated with non-radioactive DNA. The release of radioactivity into the medium is a measure of the integration of transforming DNA into the genome.

Several examples of transformation assay are found in this book (pp. 149 and 246). Brief mention has been made of its application to genetic analysis. Another application is to study the evolutionary relatedness of transformable bacteria. In an interesting piece of work, Kloos (1969) isolated DNA from several members of the family Micrococcaceae and attempted transformation of an auxotrophic strain of *Micrococcus lysodeikticus*. The results confirmed some other taxonomic data on this family and the paper contains several references to other applications of transformation to taxonomy. The work also demonstrates the variation in conditions for competence—in *M. lysodeikticus* the cells are transformed while suspended in a buffer (not growth medium) containing glutamate and a strontium salt.

Transfection   It is probably generally true that DNA from a phage of a transformable bacterium will transfect the bacterium under conditions of competence. Some greater degree of flexibility is possible in transfection than in phage infection. The interaction between phage particle and cell wall is eliminated. As an example, in the work of Wilson and Bott (1968) mentioned earlier, use was also made of a transfection procedure. The recipient organisms (a strain of *Bacillus subtilis*) were inoculated with DNA from φ29, a phage for *B. amyloliquefaciens* Hw$^+$. After transfection (and removal of excess DNA with deoxyribonuclease as before) the cultures were incubated for 4 h to allow all transfectants to mature and emerge. The phage titres were assayed (as plaque forming units) in the normal way using *B. amyloliquefaciens* Hw$^+$ as indicator. The advantage of using a heterologous system of this type is of course that the results are never confused by phage maturing early and re-infecting the recipients. The experiments of Wilson and Bott established that the conditions for competence were essentially the same for transfection as for transformation. In practice transfection is a simpler procedure to use if conditions for competence are being explored. The SP family of phage (see p. 76) are transfectable so that biological assay of their DNA preparations is possible.

A special type of transfection is possible in *E. coli*; cells are competent for λ-phage DNA if they are pre-infected with the phage (see p. 30). The phenomenon can be demonstrated and explored using two different strains of the phage which can be variously defective. The phenomenon was discovered by Kaiser and Hogness (1960). It has been exploited in studies on the phenomenon of 'restriction' in phage λ; this is the observation that phage particles produced by lysis of certain lysogenic strains of *E. coli* (strains B and C) cannot infect strains of the K group successfully, as the λ-DNA is degraded in the new cell. However concomitant transfection with DNA from phage grown in the same host strain will 'rescue' genetic markers in the phage. There is evidence that host specific methylation of λ-DNA is the biochemical basis for restriction (Arber, 1968).

Spheroplasts and viral nucleic acids; plant virus RNA   Phage nucleic acid (either RNA or DNA) can frequently be assayed by treating spheroplasts with the preparation and then assaying the incubation mixture for plaque-forming units. The technology of these procedures is outside the scope of this book although certain specific examples will be given later. Additionally it is often possible to infect plants by treatment with viral RNA; the leaves are scratched with solutions of RNA. One example of this procedure with tobacco mosaic virus (TMV) has proved of enormous value (p. 447).

A similar system can be applied to the virulent phage T4. In this case it is important that there are no normally viable phage particles in the original incubation mixture as this would confuse the outcome of the cross. Thus the recipient phage are treated with strong urea to make them non-infective for *E. coli* cells but capable of infecting spheroplasts (see below). The procedure involves mixedly infecting spheroplasts with the urea-treated recipient phage and DNA from the donor. The technique and its applications is discussed by Veldhuisen and Goldberg (1968).

Spheroplast assays of phage nucleic acid   This technique is an extension of transfection applicable to the nucleic acids from the small single-strand phage of *E. coli* namely DNA from φX 174 (and related phage) and RNA from R17 (and related phage such as M12, MS2, f2 and fr). The procedure consists of making lysozyme spheroplasts and treating them with viral nucleic acid; virus particles produced are assayed as plaque-forming units by standard procedures for the phage in question. A good procedure for general use is that of Hofschneider and Delius (1968). Sterile techniques and solutions are used throughout. The host bacteria are grown in a rich ('broth') medium at $20°C$ until there are 4×10^8 cells/ml; 200 ml aliquots are harvested and the pellets are suspended in 5 ml of a medium made by mixing in the following proportions (by volume) of 0·5 M sucrose (7 parts), 0·1 M tris HCl pH 8·1 M tris HCl pH 8·1 (2 parts), 4% EDTA Na$_2$ salt pH 8·1 (0·4 part) and 0·4% crystalline lysozyme (0·2 part). After suspension in this medium the cells are left at room temperature until they are spherical (phase-contrast microscope). It should take about 10 min. The spheroplasts are now added to a 'dilution medium'. For the procedure of Hofschneider and Delius (above) the basic dilution medium is 1% nutrient broth, 1% casein amino acids, 0·1% glucose and 10% sucrose. After spheroplasts have been formed, the 5 ml suspension is added to 20 ml of this dilution buffer containing additionally 0·1% magnesium sulphate and 2% bovine serum albumin (this must be ribonuclease-free). After a further 20 min, 200 ml of medium lacking albumin (but with the magnesium salt) is added and the spheroplasts are harvested by centrifugation and resuspended in 10 ml of this medium (lacking albumin). This procedure can almost certainly be simplified by leaving out the incubation with albumin—the advantage of this method is that the spheroplasts are relatively stable and can be kept at $0–4°C$ for a week or so.

The assay is performed simply by incubating the spheroplast suspension with the phage RNA or DNA. This is achieved by diluting the suspension with three volumes of dilution medium (no albumin nor magnesium) then 0·1 ml of nucleic acid dissolved in a suitable buffer (3 mM EDTA according to the authors—I cannot understand this choice) is added and the mixture is incubated for 30 min at room temperature. There are sufficient viable cells in the spheroplasts for addition of indicator bacteria (for a plaque-forming assay) to be unnecessary. Thus all that is needed is to just warm them up to about 37°C, quickly add molten agar suitable for forming a lawn and then pour the whole lot on to an agar base in the usual way. The amount of nucleic acid to be used in the assay can be calculated from the approximate efficiencies of the process. With φX 174 DNA the efficiency is about one plaque per 10^3 phage equivalents of DNA. With the RNA phage it is more like one plaque per 10^6 equivalents.

An alternative recipe for these assays is that of Guthrie and Sinsheimer (1960); their method was originally designed for φX 174 DNA but works all right with the RNA phage; similarly the above method is primarily intended for RNA but works with φX 174 DNA.

Protein biosynthesis

It might seem perverse to relegate protein synthesis to a mere sub-section of one chapter. The truth is that as a method for identifying specific mRNA molecules, the technique is of limited application at the moment. An enormous amount of detailed understanding of the mechanism of protein synthesis and characterization of the elements of the system (see p. 25) has been accumulated over the past few years. However in general it is not possible to do the 'obvious' experiment in which RNA is isolated, fractionated and the fractions are assayed for their capacity to direct the synthesis of specific proteins *in vitro* using ribosomes and all the supernatant factors required. For this reason I have avoided a lot of references and recipes for doing protein synthesis *in vitro* using polysomes and other systems. For reference to these techniques, the reader must look elsewhere (such as the articles edited by Campbell and Sargent, 1967, 1969). In this section there are two techniques of reasonably general application to the study of mRNA activity—but for a new system, do not expect miracles.

The *E. coli* S-30 system   This method is virtually the same as that described by Matthaei and Nirenberg (1961). In its present form, the recipe is as follows. The cells are harvested in mid-log phase and are washed with TKM buffer (10 mM magnesium acetate, 50 mM potassium chloride, 10 mM tris HCl pH 7·4) and are then frozen. This frozen paste can be kept at −20°C for up to a month. From now on all the procedures are at 0–4°C (apart from the incubations). The frozen paste is put into a mortar and $2\frac{1}{2}$ times their weight of bacteriological grade alumina (such as Alcoa A-301) is ground in as described on p. 118. Now TM buffer (same as TKM but without the potassium chloride) is added ($1\frac{1}{2}$ times the weight of cell paste). This also has to be ground in a little at a time to produce a thick creamy liquid. This is left for 15 min, deoxyribonuclease (2 μg/ml) is added and the extract is centrifuged (15 000 g for 10 min—these figures are not critical). The supernatant is centrifuged for 30 000 g for 30 min (this is fairly important). About the top $\frac{3}{4}$ or $\frac{2}{3}$ is pipetted off—this is the S-30 fraction. It is made 50 mM with respect to KCl (by adding a 1 M solution) and 5 mM with respect to β-mercaptoethanol and incubated at 35° for 45 min and is then dialysed overnight against TKM containing 5 mM β-mercapto-ethanol. The incubation (known as the 'pre-incubation') is to allow protein synthesis to continue to run the ribosomes off fragments of endogenous mRNA; the dialysis is to remove the endogenous amino acids, mRNA fragments and other low-molecular-weight cellular components.

Protein synthesis is performed using a final total volume of between 0·03 and 1·0 ml. The final overall composition of the components in the system should be as follows (or thereabouts): 1·5 mM ATP, 25 mM phosphoenol pyruvate, 15 μg/ml of pyruvate kinase (from rabbit skeletal muscle), 0·05 mM GTP, 5 mM reduced glutathione, 0·25 mg/ml of stripped tRNA (from *E. coli*—see p. 125), 1×10^{-5} M of each of 19 amino acids, the radioactive amino acid (usually leucine), mRNA (between 10 and 500 g/ml), the components of TKM[9] buffer and pre-incubated S-30 diluted by a factor of 1:4 or 1:5. Protein synthesis occurs during incubation of this system; a plateau of incorporation is reached after about 40 min.

This recipe assumes that the initiation, elongation and termination factors (p. 24) are present in the preparation. For descriptions of the purification and assay of these factors see the articles edited by Moldave and Grossman (1971).

This type of experiment is terminated by placing the tubes in ice. The newly synthesized protein has now to be characterized: this, to say the least of it, is the tricky bit. If you are merely interested in total newly synthesized protein, the protein is precipitated with trichloroacetic acid and these precipitates are counted (after radioactive free amino acid(s) have been thoroughly washed off). Any interest which once attached to this experiment has largely evaporated (at least as far as the characterization of RNA is concerned). It is of some interest in assessing the total capacity of different fractions of RNA to direct the synthesis of polypeptides. The simplest procedure is that of Mans and Novelli (1961) which can be easily adapted to perform large numbers of assays at once. An aliquot (up to 0·03 ml) of the reaction mixture is spotted on to a disc (2·3 cm diameter) of thick filter paper (such as Whatman 3MM) and dried in a stream of air. The disc is then washed in the following solutions (TCA is trichloroacetic acid): 10% TCA at room temperature for 30 min; 5% TCA at 90°C for 15 min; 5% TCA at room temperature for 5 min; this last procedure is repeated three times; two washes with a 1:1 (v/v) mixture of ethanol and ether and once with ether. After it is dried the disc is counted in the toluene-based scintillant (p. 179). Large numbers of discs (previously numbered in pencil) can be handled by pinning them to a plastic sheet or by threading them on to a piece of wire.

If it is wished to identify specific mRNA activity, of course the protein must be characterized properly. There are very few cases in which this can be done. Basically one chooses between a biological assay for newly synthesized protein (in which case the radioactive amino acid may be omitted) or alternatively radioactive newly synthesized protein may be characterized properly. Specific examples are given below but it must be stressed that there is no general procedure which is universally applicable—on the contrary the RNA-directed synthesis of specific proteins in general does not work.

Viral RNA   RNA isolated from the coliphage f2, MS2 or R17 (which are all very similar and infect Hfr cells—see p. 31) can be added to a system of the above type. The assay can also be applied to RNA isolated from infected cells prior to lysis. These phage contain only three genes (see p. 313). Translation *in vitro* results largely in the synthesis of coat protein; the possible reasons why the other genes are less efficiently translated are discussed later (p. 315). Radioactive products can be identified by purifying the reaction mixture for phage proteins after adding non-radioactive phage proteins to the preparation following protein synthesis. The phage proteins (mainly coat protein) are identified by polyacrylamide gel electrophoresis (Webster, Engelhardt, Zinder and Konigsberg, 1967) or by mapping of the tryptic peptides (Nathans, 1965). Similar techniques (also using an S-30 system from *E. coli*) can be used to translate the coat protein cistrons of certain plant viruses (van Ravenswaay Claasen, van Leeuwen, Duijts and Bosch, 1967) or poliovirus (Rekosh, Lodish and Baltimore, 1969).

mRNA from bacteria and DNA phage   There are two problems about applying the S-30 system to the study of specific RNA synthesis by mRNA preparations from bacteria: one is that only very small amounts of protein can be made *in vitro*; the other is that many bacterial proteins are made up by a coupled transcriptional/translational system (see p. 27). Thus we need to find proteins which can be easily assayed in very small amounts and which are made in polysomes independent of DNA. The general strategy has been described on p. 195. Messenger RNA molecules which are synthesized on polysomes not associated with RNA polymerase and DNA certainly include those which are formed by the transcription of phage DNA. One of the best systems is the lysozyme mRNA from phage T4. RNA isolated from the cells, 30 min after infection with the phage, can be added to the S-30 system and enzyme activity (there is a very sensitive turbimetric assay for this enzyme using *E. coli* cell-wall preparations as substrate) increases with incubation time and follows the pattern of overall protein synthesis (measured by the TCA method, p. 195). Salser, Gesteland and Ricaud (1969) have carefully characterized the lysozyme synthesized *in vitro* and have partially purified the mRNA in question. The system is one of the most powerful ones available for studying details of protein biosynthesis in *E. coli*. As an example of its exploitation, Gesteland, Salser and Bolle (1970) have been able to suppress an amber mutation (see p. 23) *in vitro*. RNA was isolated from cells infected with a lysozyme-amber mutant of T4. When this RNA was used to programme an S-30 system obtained from wild-type (Su*am*⁻) cells no lysozyme was synthesized because at a point at what should be the middle of the message, there is a termination codon, UAA. However, with S-30 obtained from an amber suppressor strain (Su*am*⁺), lysozyme synthesis was complete. This is because the S-30 in this case contains tRNA which has anti-codon complementary to UAA and inserts an amino acid so that chain elongation can continue beyond this point.

More recently, Young (1970) has obtained a cell-free synthesis of a different T4-specific enzyme, glucosyl transferase.

For bacterial mRNA, the problem seems to be largely to find a suitable protein (such as an inducible enzyme) which is synthesized on free polysomes. There certainly are some—it is a question of finding them and characterizing the synthesis of the protein *in vitro* using polysomes. There is good evidence that penicillinase induced in *Bacillus* and *Staphylococcus* species is such an enzyme. The polysome system has been carefully studied by Davies (1969). Although it is slightly out of context in this book as the RNA/S-30 system is not described there, Davies' paper is worth reading as it describes another good method of checking whether protein synthesis is genuine, namely studying the rate of erosion of newly synthesized protein by carboxypeptidases. Possibly the gene products from episomes (see p. 27) are generally useful for such experiments.

On the basis of just one protein, I tentatively suggest that enzymes induced as the bacteria come out of log phase and enter stationary phase may be synthesized on free polysomes. I say this because we have been studying tryptophanase (an enzyme involved in indole metabolism in Enterobacteria) for the past two or three years. This enzyme is synthesized (i.e. chains are completed and active subunits are assembled) on polysomes isolated from *E. coli* induced for the production of tryptophanase, and RNA from these cells will direct the incorporation of amino acids into a protein with many of the properties of trypotphanase subunits with the correct N-terminal amino acid labelled and the same tryptic peptides but with apparently incomplete or inefficient chain termination (Parish, Khairul Bashar, Brown and Brown, 1971). The system is sufficiently complex to discourage anyone from such an assay method unless a very carefully thought-out question is being asked.

Eukaryotic 'S-30 systems'   The mRNA-programmed synthesis of proteins in eukaryotic systems is an extremely difficult experiment to do. However the following system is probably the best and it has been used (Weinstein, 1963) for an extremely distinguished series of experiments in which the universality of the genetic code was demonstrated by programming the system with artificial mRNA molecules consisting of defined repeating sequences (see p. 233). The procedure is here described for rat liver but can be applied to tumours (Ochoa and

Weinstein, 1964) or rabbit reticulocytes (see p. 121; Weinstein and Schechter, 1962) with only minor modifications. All the operations are carried out at 0–4°C. The buffer used throughout is 0·06 M potassium chloride, 0·005 M magnesium acetate, 0·01 M tris HCl pH 7·8, 6 mM β-mercaptoethanol. The livers are excised and immediately superficially washed in buffer containing 0·25 M sucrose, minced and added to an equal volume of the same solution (buffer plus sucrose) and homogenized (see p. 120). An equal volume of the same solution is added and the nuclei and debris are spun out (8000 g for 10 min). An S-30 fraction is made from the supernatant exactly as described on p. 120. This supernatant is now centrifuged at 122 250 g for 60 min. The top two-thirds are removed, the remaining supernatant is thrown away and the pellets are superficially washed with buffer and dispersed with a little hand homogenizer (about 1 ml per 10 ml of S-30 from which the pellets were obtained). This suspension is centrifuged (10 000 g for 10 min) to remove any aggregate or insoluble material). The resulting suspension is referred to by mammalian biochemists as 'light microsomes'. The 'heavy microsomes' (which constitute the greater part of the RNP) are discarded with the pellet formed during the preparation of S-30.

There are various ways of reconstituting this system for protein synthesis. The simplest is to just use the S-30, but the results tend to be unsatisfactory as the translation of endogenous mRNA is extensive and difficult to entirely eliminate by pre-incubation. Rather the light microsomes are used together with supernatant enzymes and other proteins which are first precipitated at low pH. The supernatant from the centrifugation at 122 000 g (the 'S-122 fraction') is diluted three-fold with buffer, 1 M acetic acid is added to pH 5 and the precipitate is spun out (3 000 g for 10 min). The precipitate is washed (by centrifugation) with water and is then dissolved in buffer (about 3 ml/20 ml of S-122 taken initially). The pH is usually lowered by the protein and is readjusted 7·5 with dilute potassium hydroxide. Insoluble material is spun out (10 000 g, 10 min); the supernatant is known as the pH 5 fraction. Weinstein's system for protein synthesis is to make up a 'reagent mix' by adding together 0·4 ml of 5 mM ATP, 0·25 mM GTP, pH 7; 0·1 ml of 0·1 M trisodium phosphoenol pyruvate; 0·1 ml of radioactive amino acid; 0·1 ml of a solution 1 mM with respect to the other nineteen amino acids; 0·01 ml of a suspension of rabbit skeletal muscle pyruvate kinase (10 mg/ml) and 0·1 ml of '8 × buffer'. The incubations are performed by adding standard buffer (to bring the final volume to 0·4 ml) to a test tube followed by 0·01–0·04 ml RNA (containing about 200 μg RNA), 0·1 ml reagent mix, 0·05 M pH 5 fraction (containing 1·5 A_{260} units) and 0·05 ml light microsomes (containing 3.0 A_{260} units). Incubation is for up to 45 min at 37°C and the reaction can be stopped with TCA, etc. (see p. 195).

The biosynthesis of globin   After a series of false starts, it now seems fairly clear what the requirements are for the cell-free synthesis of globin (the protein component of haemoglobin). A definitive experiment has been performed by Heywood (1970). Rabbit reticulocyte polysomes (see p. 121) are suspended in 20 mM sodium acetate, 40 mM tris acetate pH 7·6, 5 mM EDTA and 0·5% sodium dodecyl sulphate. Centrifugation through a sucrose gradient (in the same buffer but lacking the sodium acetate) yields (in addition to rRNA) a peak of 9–13 S (Scherrer and Marcaud, 1968). This is 'globin mRNA'. In the system of Heywood, this material is incubated in a mixture basically similar to the one above, except creatine phosphate and creatine kinase replace the pyruvate kinase system, the ribosomes are from muscle and the supernatant from reticulocytes. The labelled globin was identified by polyacrylamide gel electrophoresis of the labelled proteins. Success is dependent upon the addition to the mixture of the proteins (initiation factors—see p. 24) obtained by crushing reticulocyte ribosomes in 1 M KCl.

Animal viral RNA   In the light of the success which has been obtained with the translation *in vitro* of plant and bacterial viral RNA (p. 195), it might be expected that the translation of an animal viral RNA genome would constitute the best mRNA for optimizing mammalian cell-free systems for protein synthesis. One virus (encephalomyocarditis, EMC) is particularly suited to this experiment. Details of the properties of the virion are to be found in the paper by Burness mentioned on p. 174 and in Burness and Clothier (1970). The cell-free system that is used for the translation of this RNA (to yield viral polypeptides) has been perfected by Aviv, Boine and Leder (1971; their paper includes references to earlier work on this system).

Use of living cells for translation of exogenous mRNA   Gurdon, Lane, Woodland and Marbaix (1971) have described an elegant method for synthesizing rabbit haemoglobin by injecting non-differentiated amphibian cells (frogs' eggs and oöcytes) with RNA from reticulocytes. The product has been very thoroughly characterized. The result is an extremely important one in two respects. It offers a method for assaying mRNA in intact cells. Additionally by using chemically modified RNA, it should be possible to elucidate details of translational control *in vivo*, in eukaryotic cells.

DNA-directed protein synthesis   If coupled transcription and translation occurs *in vivo*, it should be possible to reconstruct such a system *in vitro*. There is one superb example, the cell-free synthesis of the β-galactosidase by Zubay and Chambers (1969). The principle is to use DNA from the transducing phage $\varphi 80_{dlac}$[10] in which the *lac* operon occupies a much larger proportion of the genome than it does in *E. coli* DNA (see p. 366). This DNA is incubated with S-30 which apparently contains sufficient RNA polymerase. The final components of the incubation system are as follows: tris acetate pH 8·2 (44 mM); dithioerythritol (1·37 mM); potassium acetate (55 mM); 20 amino acids (all 0·22 mM); CTP, UTP, GTP (all 0·55 mM); ATP (2·2 mM); trisodium phosphoenol pyruvate (21 mM); ammonium acetate (27 mM); 3′5′-cyclic AMP (needed for transcription—see p. 21, 1 mM); tRNA (0·1 mg/ml); pyridoxine hydrochloride, NADP, FAD, folinic acid (27 μg/ml of each); *p*-aminobenzoic acid (11 μg/ml); magnesium acetate (14·7 mM); calcium chloride (7·4 mM); $\varphi 80_{dlac}$ DNA (of the order of 0·1 mg/ml); S-30 (obtained from *E. coli* 514 a mutant in which the entire *lac* operon is deleted, 6·5 mg protein/ml). The paper describes the various tests which were carried out to establish that the enzyme activity which is produced during incubation is the result of *de novo* transcription and translation.

Transfer RNA

Aminoacyl tRNA synthetases   These enzymes are specific for each of the 20 amino acids which are incorporated during protein biosynthesis. The reactions catalysed by the enzymes are described on p. 14. For the purpose of identifying, say tRNA^{Leu} the procedure is to incubate the tRNA with radioactive leucine, ATP and the synthetase and to assay the incubation mixture for the production of radioactive leu-tRNA (necessarily leu-tRNA^{Leu} provided an endogenous system is being used, which does not produce any erroneous mismatched aminoacyl tRNA). The synthetase preparation can contain all the other 19 synthetases of course; the result will not be ambiguous provided only one amino acid is labelled. If the experiment is being performed to monitor the fractionation of tRNA molecules, the way in which it is done is to assay each fraction for several amino acid acceptor activities by performing several incubations, each involving the use of the same mixture of synthetase enzymes but each with a different labelled amino acid. Obviously the assays can be extremely time consuming unless a rapid assay method is employed.

A very rapid method, which is ideally suited to the assay of the large number of samples obtained by column chromatography, is that of Cherayil, Hamperl and Bock (1968). You start with a whole set of numbered thick filter paper discs (e.g. Whatman 3MM paper, 2·3 cm diameter) pinned either on a non-wettable surface (Teflon) or pinned on a wooden board so the paper is away from the surface. Aliquots of the fractions (about 0·1 ml) are dropped on to the discs. They are dried with a hair dryer and transferred to ice-cold, 0·03 M KCl in 75% aqueous ethanol. They are then washed free of urea, buffer, etc., with this solution. This can be done on a large number of discs at once in a Buchner funnel. They are now pinned back (as before) and 0·1 ml of 'reaction mixture' is added to each. The reaction mixture consists of 40 mM potassium maleate buffer pH 7·0, 4 mM $MgCl_2$, 0·5 mM EDTA, 1 mM ATP, 0·35 M CTP, 5 mM β-mercaptoethanol (or an equivalent, e.g. 2 mM dithioerythritol), 40 mM potassium chloride, 0·01 mM of 19 amino acids, the hot one (e.g. 2 μCi of [^{3}H] amino acid with a specific activity of around 0·5 Ci/mmole) and tenth of a volume of the 'enzyme mixture'. The discs are incubated at room temperature for 30 min. It is important not to let them dry out. The simplest procedure is to put the trays, or whatever is holding the pins, into a chromatography tube lying on its side and with water in the bottom. The lid is put on (the side) and the air is thus kept moist. After incubation, the discs are dried again and dropped into 10% trichloroacetic acid (TCA) containing some (about 1 mM) of the unlabelled amino acid (the same one that was labelled in the incubation). They are washed (about 50 at a time in a Buchner funnel) with the following solutions (all containing the amino acids except the last): 0·5 M sodium chloride in 66% aqueous ethanol, 10% TCA, 5% TCA and ethanol/ether (3:1). They are air-dried and suspended in a scintillant (toluene-based, see p. 000) and counted. For many organisms, the 'enzyme mixture' can be simply a fraction obtained by centrifuging an S-30 preparation (p. 194) for 3 h at 15 000 g.

In an actual fractionation procedure, the extinction of the solution eluted from the column is monitored to obtain a measure of the total RNA present: the result on total tRNA is typically a diffuse envelope with no clear peaks. Following the assay of acceptor activities it is possible to superimpose upon this the pattern of distribution of all those amino acids that have been assayed. As a single example of the results, see the distribution of acceptor activities for 15 amino acids in the fractionation of *E. coli* tRNA on one of reverse-phase systems of Weiss and Kelmers (1967—see p. 170) as shown in Fig. 5.12.

It is not necessary to perform the aminoacylation reaction after fractionation. In some cases it is more appropriate to do it before. As an example, Vold (1970) was looking for evidence that there might be a different distribution of tRNA species (with the same acceptor activities) in vegetative cells and spores from *Bacillus subtilis*. With any experiment of this sort it is essential that the preparations from the two sources be mixed and fractionated together. Clearly the only way to do this is to obtain tRNA from each source and label one with a [^{14}C] amino acid and the other with a [^{3}H] amino acid. In this particular work, the amino acid chosen was lysine. The radioactive lysyl tRNA (containing all the other RNA species as well of course) were deproteinized by extraction with aqueous phenol. The paper also draws attention to the fact that not all aminoacyl tRNA synthetases are as stable as those in *E. coli*. In particular some of them do not like tris. If work on a new organism is being undertaken it is worth trying a different buffer (such as cacodylic acid/sodium cacodylate).

In addition to assaying different acceptor activities in fractionated tRNA, the aminoacyl tRNA synthetase reaction may be used for the assay of tRNA in a total preparation. A remarkable discovery has been made by Pinck, Yot, Chapeville and Duranton (1970) regarding

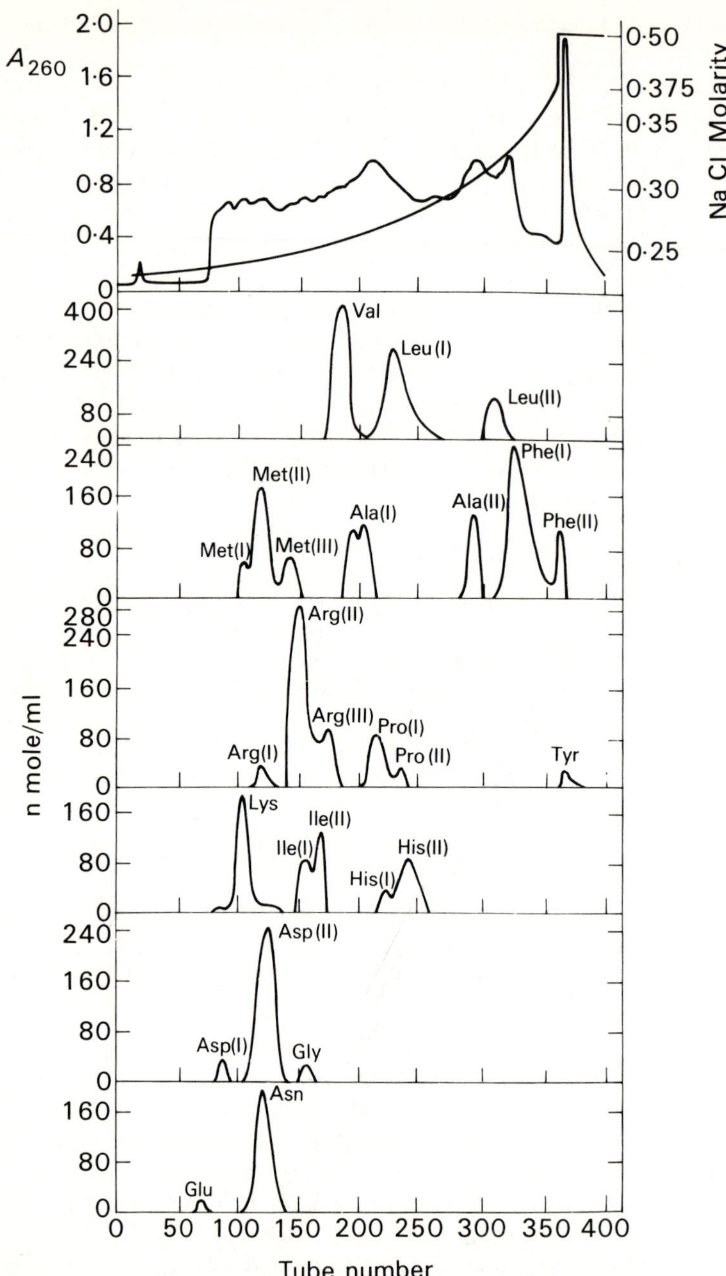

FIG 5.12   An example of the fractionation of tRNA into components of different acceptor activities. The fractionation illustrated here is the reverse-phase procedure described on p. 170. The upper panel shows the shape of the NaCl gradient and the absorbance profile due to the eluted tRNA. The lower panels illustrate various acceptor activities. In this experiment, 120 mg yeast tRNA was applied to a column (240 x 1 cm). Data from Weiss and Kelmers (1967).

a plant viral RNA. Valine-acceptor activity is present in the RNA from turnip yellow mosaic virus (TYMV). Thus the 3′-end of this molecule is apparently a tRNA sequence. The possible occurence of this feature in other viral RNA molecules and their biological significance should be of great interest and value in understanding the control, effected by viral tRNA, on the translational machinery of the host cell.

Formylmethionyl tRNA   From *E. coli* rRNA$_F^{Met}$ is isolated by BD-cellulose chromatography (p. 161). Total tRNA is first fractionated on DEAE-Sephadex (p. 162). The methionine tRNA species are eluted first from such a column; they are not resolved. They are resolved by re-chromatography on BD-cellulose, the tRNA$_F^{Met}$ being eluted first. The specificity of this molecule is easy to assay, as a crude aminoacyl tRNA synthetase preparation (p. 199) contains the transformylase. The assay has to be performed on aliquots of the fractions, separately from the methionine acceptor assay. The reaction mixture (Seno, Kobayishi and Nishimura, 1968) contains in a total volume of 0·1 ml, 10 μmole of cacodylate pH 7·4, 1 μmole of magnesium acetate, 1 μmole potassium chloride, 0·2 μmole of ATP, 5 μmole of L-methionine (unlabelled), 0·4 nmole of $[^{14}C]$formyltetrahedrofolate (about 30 mCi/mmole), the tRNA (about 0·01 ml or so) and 0·1–0·2 mg of *E. coli* synthetase preparation. After incubation (37°C for 10 min), the mixture is spotted on to filter paper discs and treated with trichloroacetic acid and counted.

Other assays for tRNA   Of course there are other ways of testing for the activity of tRNA. Its efficiency in protein synthesis (using a really well purified system with no endogenous tRNA) or the ribosome binding reaction (p. 213) are both possible. On the whole these are laborious and unsatisfactory methods. There is one striking case however in which they are essential. Bacterial cell walls contain both polysaccharide elements (forming linear backbones) and peptide chains (that cross-link the structure). The details are outside the scope of this book but basically one can envisage a section of the structure as follows. Every other monosaccharide unit in the polysaccharide is *N*-acetylglucosamine linked via a lactyl ether to a pentapeptide (which is assembled enzymically before the monosaccharides are polymerized). The middle amino acid of this pentapeptide is always basic (has an amino side chain) and to it is linked a short chain of glycine residues, which cross-link the structure to form a three-dimensional 'bacterium-shaped molecule'. The details of this structure and its biosynthesis has been largely due to J. L. Strominger and his group. The glycine residues are incorporated as glycyl tRNA. This is a different role for tRNA as no ribosomes or mRNA molecules are involved. The special tRNA responsible for this synthesis has been identified (Stewart, Roberts and Strominger, 1971). Large amounts (375 000 A_{260} units) of tRNA from *Staphylococcus epidermis* were fractionated by chromatography on a column (6 x 100 cm) of Whatman DE 52 microgranular DEAE-cellulose. The details of the chromatographic system were that the basic buffer contained 0·02 M tris HCl pH 7·4, 1 mM magnesium chloride, 5 mM sodium hyposulphite and 1 mM EDTA. The column was developed with a 16 l gradient of 0·25–0·45 M sodium choride and 20 ml fractions were taken. This Herculean undertaking proved worthwhile as (around fraction 500) just before the bulk of the optically absorbing material was eluted, molecule tRNA$_I^{Gly}$ was eluted. It was further purified by hydroxyapatite chromatography (p. 166). Its glycine acceptor activity was the same as that of the other tRNA^{Gly} species from *Staph. epidermis*. However the tRNA$_I^{Gly}$ was totally incapable of incorporating glycine into proteins when incubated (with the other tRNA molecules) with ribosomes and phage RNA nor was it bound to ribosomes in the presence of GGG, GGA, GGU and GGC (the glycine codons—see pp. 22 and 213). Thus there is in bacteria a class of glycine tRNA molecules which lack any functional anti-codon. Sequences of these molecules will prove of great interest. In the meantime it is known that the other sorts of tRNA$_I^{Gly}$ (at least in *Staph. epidermis*) are as good as tRNA$_I^{Gly}$ at synthesizing the glycine peptides in the wall—it is just that this particular molecule is defective in the mRNA-directed synthesis of peptide bonds.

Immunological methods; interferon

The antigenic behaviour of nucleic acids is of no real use in characterizing nucleic acids, although it is in an interesting problem in immunology. Sela, Ungar-Waron and Shechter (1964) showed that purines and pyrimidines are haptens when bound to a suitable protein (such as albumin). Nucleic acids also behave as macromolecular haptens; they are not covalently bound to proteins for this purpose—they are merely associated with a basic protein such as methylated albumin (see p. 164). For the antigenic properties of DNA to be expressed the molecules have to be denatured before binding to the protein. Techniques involved are described by Plescia (1968). The subject is of clinical importance as there is evidence that at least one disease, lupus erythematosus,[11] is due to antibodies to DNA in the patient's serum (Diecker, Holman and Kunkel, 1959).

Interferon   Interferon is a protein which is produced in animal cells in response to viral infection. It interferes with viral maturation. The existence of such a substance was inferred by Isaacs and Lindenmann (1957) from the observation that viral infection could be inhibited by a cell-free preparation isolated from the chorioallantoic membrane of fertilized chicken eggs that had been inoculated with heat-killed influenza virus. Interferon has been the subject of intensive research from this discovery to the present day. It has proved an elusive substance to those who have attempted its purification. Nevertheless, extremely sensitive assays for interferon are possible. The particular interest of interferon is that its induction can be produced, not only by viruses (and certain micro-organisms such as mycoplasma—see note 12 on p. 379) but also by double-stranded RNA. Interferon induction has been of particular value in identifying mycophage (fungal viruses) which contain double-stranded RNA (p. 396). In general, however, interferon is the province of specialists. For an informative review about interferon, which includes many references to RNA-induced production, see Colby (1971).

Notes to chapter 5

1   Biochemists will recognize that Fig. 5.1 is grossly oversimplified. In particular there is (in many tissues) a route to the thymidine nucleotides from dCMP.

2   Unpublished data of Miss M. Brown (in the author's laboratory). Labelling with Guo is useful in studies on the alkylation of RNA (p. 242).

3   The techniques described here for liquid scintillation counting are all suitable for liquid scintillation spectrophotometers that operate at ambient temperatures. Instruments that operate below ambient may require an 'anti-freeze' in the naphthalene/dioxan system. Additionally the problem of the crystallization of sucrose (mentioned later) is more serious at low temperatures.

4   'Scintillation grade' toluene is a pointless extravagance.

5   Balbiani rings are swollen regions seen in polytene (salivary gland) chromosomes from insects. These regions are known, from autoradiographic studies, to be active sites of RNA synthesis. These papers (especially the one by Daneholt and others) describe methods whereby clean preparations of Balbiani rings can be obtained by microdissection and used for studies on transcription *in vitro*.

6   Some care should be taken when referring to bacterial species. Bacterial taxonomy is confused by the fact that no single set of criteria for the definition of a taxon is appropriate to all the groups of bacteria. In this context we are thinking of two organisms isogenic except for defined phenotypic differences.

7   This organism is referred to as *Pneumococcus* in the earlier literature.

8   Valine and isoleucine have a common biosynthetic pathway. The genotype ilv^- has a requirement for both val and ile.

9   Or thereabouts. For many mRNA preparations, the Mg^{2+} concentration should be lower (8 mM).

10    $\varphi\, 80_{dlac}$: In the nomenclature for transducing phage, the subscript d indicates that the phage is defective and that successful infection requires a helper phage. The subscript p indicates a non-defective strain.

11   Lupus erythematosus: the commoner, cutaneous form of the disease results in red, slightly raised patches on the skin (usually of the face). There is an acute form (disseminated lupus erythematosus), which is mainly found in women, in which internal organs (including the heart) are involved.

6 Degradation

Nucleic acids are hydrolysed, either partially or completely, to determine features of their primary structure. The application of these techniques to sequence determinations is given in chapters 8 and 9. In this chapter the procedures for obtaining mixtures suitable for determining the nucleotide composition of a nucleic acid (or oligonucleotide) are described.

There are two methods of degrading nucleic acids—by chemical and enzymic reactions—and these are discussed separately.

Chemical methods

The hydrolysis of RNA and DNA occurs under quite different conditions and it is convenient to discuss them separately.

RNA

Acid Aqueous solutions of mineral acids release the purine bases from RNA; the pyrimidines are not released (p. 55). For the purpose of quantitatively analysing the nucleotide composition of RNA by acid hydrolysis the usual practice is to hydrolyse the RNA under conditions in which there is a quantitative production of Ade, Gua, UMP and CMP (Smith and Markham, 1950). For an actual recipe: make a constriction in a Pyrex glass ignition tube (about a third of the way down from the open end). Weigh out about 2 mg of RNA (ethanol-precipitated and by washing in ether and dried *in vacuo*) into the tube; add 0·05 ml of 1 M hydrochloric acid and seal the tube carefully (i.e. do not pull the tube out to a fine point of thin glass—melt the constriction in a tiny oxygen flame so the glass just drops in to form a really strong point). All you have to do now is to pop the tube into boiling water for 1 h. Then take it out, cool it in ice, break the top off the tube and pipette aliquots (0·01 ml) on to a piece of chromatography paper as shown in Fig. 6.1. The chromatogram can be developed in the system described on p. 48. After the chromatogram is dry, view it under a UV light. All the spots are marked out in pencil. A grid is then marked out as shown (Fig. 6.1) and the rectangles are cut out with ultra-clean scissors—using equally clean tweezers to handle the paper. The rectangles are cut up into strips and all the strips from one rectangle are put into a test tube and 5 ml of the eluting solution (see Fig. 6.1) is added. The tubes are stoppered, not shaken but allowed to stand in the dark overnight. Next morning, give each tube a limited shake-up; say a flick on the side or one inversion, pour the liquid out into spectrophotometer cells and read the extinctions at the maximum for the compound in question (see p. 85). It is now a simple arithmetic job to convert these readings into relative molar concentrations (extinction coefficients are given on p. 85). The results are normalized and standard errors of the mean are then calculated. Accurate and reproducible results are obtained; the only source of error is the slight degree of deamination of CMP (to yield UMP). This problem is discussed by Loring, Fairley, Bortner and Seagram (1951).

There is no satisfactory quantitative method of releasing the bases (Ade, Gua, Ura and Cyt) from RNA. However for certain radioactive labelling experiments it may be required to isolate the free pyrimidines from RNA. The most reliable method is to use 12 M perchloric acid at 100°C for an hour (Marshak and Vogel, 1951). The technique is similar to that described above except that it is essential at the end to remove the perchloric acid. The simplest method is to dilute the hydrolysate and to neutralize it with potassium hydroxide. Potassium perchlorate, which is insoluble, is removed in the bench centrifuge.

Alkali   RNA is easily hydrolysed by aqueous alkali (see p. 3) to produce mixtures of 2'- and 3'-nucleotides. For a consideration of different conditions and the quantitativity of the reaction, see the paper by Crosbie, Smellie and Davidson (1953). For the determination of nucleotide composition, the usual practice is to dissolve the RNA in 0·3 M KOH and incubate at 37°C for 18 h.

The advantage of employing these conditions is that 5'-nucleoside triphosphates are not hydrolysed. As all nascent RNA molecules (p. 226) and many, if not all, viral RNA molecules contain triphosphate at their 5'-end, this end can be identified in oligomers derived by the enzymic hydrolysis of RNA by digesting all the fractionated oligomers (p. 292) with alkali and analysing these digests for pppN.

Methods for analysing the products (GMP, CMP, AMP, UMP) are discussed on p. 47. The method is a valuable one for the determination of the nucleotide composition of [^{32}P] RNA.

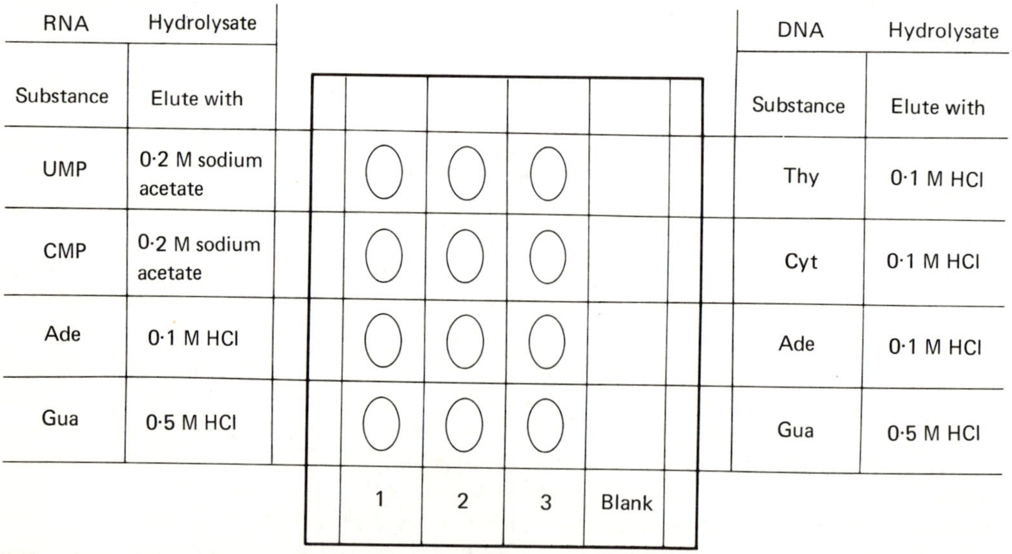

RNA	Hydrolysate					DNA	Hydrolysate
Substance	Elute with					Substance	Elute with
UMP	0·2 M sodium acetate					Thy	0·1 M HCl
CMP	0·2 M sodium acetate					Cyt	0·1 M HCl
Ade	0·1 M HCl					Ade	0·1 M HCl
Gua	0·5 M HCl					Gua	0·5 M HCl
		1	2	3	Blank		

FIG 6.1   Chromatography of acid hydrolysates of RNA and DNA for the purpose of measuring nucleotide composition. Conditions for hydrolysis are described in the text and the chromatographic systems on p. 48 (for RNA). Three identical aliquots (1, 2, and 3) are spotted. Space 4 is left blank. After chromatography and identification of the spots the grid is ruled out and the rectangles are cut out and eluted in the solvent given in the figure by the method described in the text. Each set of eluates are assayed spectrophotometrically (see text). The spectrophotometer is zeroed for each set of three spots, using as blank the eluate from corresponding blank rectangle (above position 4).

The main trouble is that it relies upon the even labelling of the four nucleotides, i.e. on equal pool sizes for the four precursors if the labelling has been for a short time. If possible, the specific activities of the nucleotides should be determined. In practice this is often impossible as the amounts of the labelled RNA available are very small.

There is one type of phosphodiester link in RNA which is not readily hydrolysed by alkali (nor by any ribonucleases). This is the bond following a $2'$-methyl nucleoside (see p. 4). Thus in the sequence below, mild alkaline hydrolysis would yield the four common $2'/3'$-nucleotides and Gm-U-.

$5'$ . . . -A-G-Gm-U-C-U- . . .$3'$

Production of nucleosides   The conversion of RNA to a mixture of the nucleosides (Urd, Ado, Cyd and Guo) is of great historical interest but of little practical importance to-day as the reaction is not quantitative. The conditions of Levene and Jacobs (1910) were to heat the RNA in a sealed tube with concentrated aqueous ammonia to $180°C$ for 3.5 h. Nowadays, it is usual to use enzymes (p. 55).

DNA

Acid   It is possible to hydrolyse DNA quantitatively with dilute acid in such a way that only the purines are released and the remaining 'apurinic acid' is left as a polymeric product. Typical conditions are 0.1 M sulphuric acid at $100°C$ for 35 min (Spencer and Chargaff, 1963). This reaction and its application to determining partial DNA sequences is described in detail on p. 275. For the quantitative determination of nucleotide composition, a satisfactory procedure is the following one based on the original method of Vischer and Chargaff (1948). In the version which I propose, which is derived from various methods found in the literature, DNA is dried out and introduced into a constricted ignition tube as described above for RNA. Formic acid (90% v/v, 0.5 ml) is added, the tube is sealed and put into a flask containing 2-butoxyethanol. The temperature is increased so that the solvent is boiling in 20 min and it is then refluxed for 40 min. The solvent is poured off (after it has just come off the boil). The tubes are cooled in ice, broken open and the contents (brownish in colour if there is any polysaccharide present) are reduced to dryness. The residue is extracted several times with 1 M HCl; the acid is then reduced to dryness, redissolved in 0.05 ml of the same acid and chromatographed (see p. 48 for a suitable system) and the results obtained as described for RNA (see Fig. 6.1). Remember that the spectrophotometric properties of Cyt and Thy are different to those of the UMP and CMP isolated from the RNA hydrolysis (p. 275).

Alkali   Alkaline hydrolysis of DNA requires conditions that are so strong that quantitative recovery of the deoxyribonucleotides is difficult, and enzymic methods are normally employed (p. 209). Mild alkali is a useful procedure for the denaturation of DNA (p. 361).

As the conditions described for the alkaline hydrolysis of RNA do not result in hydrolysis of DNA, they may be used for selectively destroying RNA in a mixture. A procedure for counting radioactive DNA in the presence of radioactive RNA can easily be devised. If the solution is made say 0.1 M with respect to sodium hydroxide and heated to $100°C$ for 10 min or so, the DNA (but not the ribonucleotides derived from the RNA) can be precipitated with trichloroacetic acid (p. 180).

Enzymic hydrolysis

Classification

Nucleases can be classified according to the type of specificity they show. The sorts of specificity that are encountered are as follows:

(1) Specificity for the sugar. Nucleases for which the only substrates contain ribose are called ribonucleases, those for which the only substrates contain deoxyribose are called deoxyribonucleases, those which lack specificity of these types are called non-specific nucleases or phosphodiesterases.

(2) Endonucleases and exonucleases. These groups are respectively enzymes that hydrolyse phosphodiester groups anywhere in a chain (subject to the restrictions imposed by their other specificities) and those enzymes that erode the chain from one end. Obviously there are two classes of exonucleases, those that erode the chain from the $3'$-end (the products from a hydrolysis are $5'$-nucleotides) or from the $5'$-end (the products in this case are usually $3'$-nucleotides but there are exceptions—see later).

(3) Secondary structure. Probably all nucleases exhibit some sort of specificity in this regard; after all a double helix is a very different substrate to a single polynucleotide strand. However there are certain nucleases with a very high degree of specificity in this respect. Obviously there are different sorts of double-strand nucleases, those that hydrolyse the two strands at the same point and those that only hydrolyse a phosphodiester bond in one strand at a time.

(4) Base specificity. So far, only one deoxyribonuclease has been shown to have base specificity, and the recognition of this specificity is so recent that no extensive use of the enzyme for sequence studies has been reported (see p. 211). On the other hand, several base-specific ribonucleases have been used for several years and have enabled extended RNA sequences to be determined (p. 291).

Ribonucleases

Bovine pancreatic ribonuclease A   This enzyme is commercially available in either a crude or purified form. The purified enzyme should always be used if it is required to hydrolyse RNA (and leave the DNA alone) or for sequence studies. It is an endoribonuclease, has specificity for single-stranded RNA and has an absolute specificity for pyrimidine nucleotides. Thus all the oligonucleotides derived from complete digestion with the enzyme end in a pyrimidine nucleotide moiety (with a $3'$-phosphate—it is the $5'$-phosphoester that is hydrolysed). An illustration of the hydrolysis of a hypothetical oligonucleotide by this and other ribonucleases is shown in Fig. 6.2. The pH-optimum for the enzyme is around 8; no divalent cations are required.

The takadiastase enzymes   The enzymes in this group and the next one have been isolated and identified by F. Egami. Without the painstaking work of Egami and his colleagues, which led to the characterization and purification of these enzymes, the work on RNA primary structure, which has proved one of the most exciting areas of biochemistry over the past few years, would have been largely impossible. Takadiastase is the name given to a crude extract from *Aspergillus oryzae*, a fungal rice pathogen. It contains several degradative enzymes. Two of these (T_1 and

T_2) are nucleases. Ribonuclease T_2 is not much used for experiments with RNA. It contains carbohydrate. As an enzyme it is of a non-specific acid ribonuclease (pH-optimum 4·5). Full details regarding its purification and properties are given by Uchida and Egami (1967).

Ribonuclease T_1 on the other hand is one of the most useful tools for studying RNA structure. The enzyme is commercially available in a pure form. It is an endonuclease with an absolute specificity for guanylic acid residues (see Fig. 6.2). Unlike recipes for ribonuclease A which usually quote amounts of enzyme as a weight of protein, recipes for T_1 usually quote units. The specific activity is measured (Uchida and Egami, 1967) by incubating the enzyme in a mixture composed of 0·2 M tris HCl pH 7·5 (0·25 ml), 20 mM EDTA (0·1 ml), enzyme in water (0·1 ml) and 12 mg/ml RNA (0·25 ml). After incubation at $37°C$ for 15 min, the reaction is stopped (they recommend 0·25 ml of 0·75% uranyl acetate in 25% perchloric acid). The precipitate is removed, the supernatant diluted fivefold with water and the extinction at 260

—G—G—A—C—G—C—U—U—A—G—C—A—G—A

U_4, T_2, OH^-	A	T_1, U_1	U_2, U_3
—G—	—G—G—A—C—	—G—	—G—
4 x G—	G—C—	G—	3 x G—
3 x A—	2 x U—	A—C—G—	A—
A	A—G—C—	C—U—U—A—G—	C—G—
3 x C—	A—G—A	C—A—G—	C—U—U—A—
2 x U—		A	C—A—
			A

FIG 6.2   Effects of various ribonucleases (and alkali) on a hypothetical oligoribonucleotide to illustrate the different base specificities of the reagents.

nm is measured and compared with that from a blank incubation lacking the enzyme. The hydrolysis results in the production of acid-soluble nucleotides. A unit of enzyme is defined as the amount that results in an increase of one unit of extinction under the conditions of the assay.

In a buffer such as the above, T_1 has little specificity for secondary structure; however if the enzyme is inhibited by divalent cations it has considerable specificity for single-stranded RNA (see p. 299).

The Ustilago enzymes   A more recent development from Egami's laboratory is the analysis of the nucleases from *Ustilago sphaerogena* (a lower basidiomycete, pathogenic in plants but capable of being grown in defined media in liquid culture). Four enzymes (U_1, U_2, U_3 and U_4) have been fractionated and characterized (Ariana, Uchida and Egami, 1968). Enzyme U_1 seems to have the same specificities as T_1; U_4 appears to be a non-specific ribonuclease similar to T_2. U_2 and U_3 however complement the other base specific nucleases very well. They both have the same properties, namely an absolute requirement for purine nucleotides and single-stranded RNA. They are inhibited by divalent cations and have a pH optimum around 4·5. Ribonuclease U_2 has been successfully used in sequencing methods (see p. 312).

Non-specific nucleases

Snake venom phosphodiesterase   The usual source of this enzyme is the venom of *Crotalus adamanteus* (the diamond-backed rattlesnake). It is a non-specific exonuclease with a pH optimum of around 8–9. Both the lyophilized venom (which is suitable for some purposes) and the purified enzyme are commercially available. It erodes nucleic acids from the 3′-end to yield the 5′-nucleotides. The enzyme has two experimental uses; by limiting its rate of reaction it is useful for the sequencing of oligonucleotides (see p. 299) and it is useful for the complete degradation of RNA. In fact when used in conjunction with a phosphomonoesterase, it is used in the only satisfactory method of converting RNA into a mixture of nucleosides. In this method for the fractionation of the minor nucleosides from tRNA (see p. 52), Hall (1967) obtains the nucleoside mixture by dissolving 5 g of tRNA in 500 ml of 5 mM $MgCl_2$. To this is added 0·5 g of lyophilized venom and 15 g of bacterial alkaline phosphatase (see p. 55). A drop of toluene is added to stop bugs growing during the incubation. The mixture is gently stirred at $37°C$ for 7 h. During this time dilute alkali is added to maintain the pH at 8·6. After the 7 h, more $MgCl_2$ (0·5 ml of a 1 M solution), phosphatase (5 mg) and venom (0·2 g) are added and incubation is continued until no more alkali is consumed (overnight). The proteins are removed by heating to $60°C$ for 30 min, cooling to $4°C$ for 4 h and centrifuging (to remove the precipitate). The nucleosides are in the supernatant which may be lyophilized. This technique can be easily scaled down. For the procedure to work quantitatively with large RNA molecules (e.g. high-molecular-weight rRNA) it is a good idea to break the RNA up a bit by giving it a preliminary bashing with an endonuclease (such as ribonuclease A). For small amounts of RNA, the high pH may be maintained with a suitable buffer, and addition of alkali is unnecessary.

Spleen exonuclease   This enzyme complements the foregoing venom enzyme; like it, it is a commercially available non-specific exonuclease. However it liberates, not the 5′-nucleotides but the 3′-nucleotides. It works from the 3′-end and it is essential that the substrate should have a 3′-phosphate on the end. The enzyme is thus suitable for degradation of the fragments obtained from digestion of RNA with ribonuclease A or ribonuclease T_1 or digestion of DNA with the spleen deoxyribonuclease (see p. 210). The pH-optimum for the enzyme is low (around 5·0). Assay conditions, using as substrate PNP-T- (where PNP is *p*-nitrophenyl) are described by Razzell and Khorana (1961).

Neurospora endonuclease   This enzyme has no base-specificity and limited sugar-specificity but is highly selective for the hydrolysis of denatured or single-stranded nucleic acids. It is isolated and purified from *N. crassa* by the method of Linn and Lehman (1965). Their preparation is apparently contaminated with a small amount of another nuclease with activity for the hydrolysis of double-stranded nucleic acids. This contaminant can be inactivated (and hence the selectivity of the preparation can be improved) by warming the purified preparation to $55°C$ for 15 min. The enzyme has Co^{2+} as a prosthetic group; thus it is inhibited by EDTA and activated by Co^{2+} salts (about 10 M). The pH optimum is about 8·2. The effect of divalent cations (other than Co^{2+}) is to alter the sugar-specificity. Magnesium salts (10 mM) increase the rate of degradation of denatured DNA about 2·5 fold; RNA hydrolysis is inhibited (40%) by these conditions. An example of its use is given on p. 368.

Staphylococcal nuclease   This enzyme is an exonuclease that erodes either RNA or DNA from the 5′-end to yield the 3′-nucleotides. It is the same enzyme as 'deoxyribonuclease from *Micrococcus pyogenes*'. It is not a specific deoxyribonuclease and '*M. pyogenes*' is merely a group of strains of *Staph. aureus*. The enzyme is commercially available. It is of great practical importance as it is employed in the determination of nearest neighbour frequencies (p. 271). It has an absolute requirement for Ca^{2+}. For assay purposes, the reaction mixture consists of 1%

nucleic acid in 0·1 M sodium borate buffer pH 8·8 (0·5 ml), 0·1 M calcium chloride (0·1 ml) and 0·1 ml of enzyme. After incubation at 37°C, 2 M hydrochloric acid (0·33 ml) is added and the precipitate (protein and undegraded substrate) is removed by centrifugation. The supernatant is diluted with water and its extinction at 260 nm is measured (Reddi, 1959).

Deoxyribonucleases

The endonucleases of *Escherichia coli*   There are several endonucleases in *E. coli*. Their metabolic role is largely unknown. Endonuclease I is an enzyme with limited specificity for double-stranded DNA (rate of hydrolysis about seven times that for single-stranded DNA). It never completely degrades DNA (compare the pancreatic enzyme). The oligonucleotide products (on average they are heptanucleotides) have a 5′-phosphate residue. The enzyme (like alkaline phosphatase—see p. 55) is located in the periplasmic space. It has absolute specificity for DNA but RNA is an inhibitor (Lehman, Roussos and Platt, 1962). When DNA is degraded, the breaks are predominantly in both strands (see however Bernardi and Cordonnier, 1965).

Endonuclease II is an interesting enzyme as the most rapidly hydrolysed substrate is alkylated, double-stranded DNA; it has therefore been proposed that it plays a role in repair processes (see p. 257 and Howard-Flanders, 1968).

The exonucleases of *Escherichia coli*   These enzymes are all deoxyribonucleases. Their experimental value derives from the different specificities for secondary structure. They are conventionally referred to by roman numerals.

Exonuclease II is in fact DNA polymerase; as a nuclease it has limited specificity (although it catalyses a faster hydrolysis with double-stranded substrate—see p. 223).

Exonuclease I and exonuclease III complement one another. The former hydrolyses single-stranded DNA; it erodes the chain from the 3′-end (to yield 5′-nucleotides—see p. 206). The enzyme has been carefully characterized by Lehman and Nussbaum (1964). Exonuclease III catalyses a similar reaction except double-stranded DNA is the substrate. Thus if the substrate is a DNA molecule with complete duplex structure right up to each end, enzyme molecules will start nibbling along from each end (but along a different strand in each case). Unlike exonuclease I which will completely degrade a chain until it has got down to the last dinucleotide, exonuclease III stops after a time, apparently inhibited by the single strands produced by the reaction (Richardson, Lehamn and Kornberg, 1964).

Exonucleases IVA and IVB are two enzymes with similar properties but separable by ion-exchange chromatography (Jorgensen and Koerner, 1966). They are similar to exonuclease I in their mode of action except that their preferred substrates are the oligonucleotides produced by endonucleolysis of DNA; thus they will quantitatively convert the 'pancreatic fragments' (p. 210) to 5′-nucleotides.

The activities of these and other deoxyribonucleases are summarized in Fig. 6.3.

Bovine pancreatic deoxyribonuclease and calf spleen deoxyribonuclease   Pancreatic deoxyribonuclease is commercially available and is the best enzyme to use for selectively degrading DNA (and leaving RNA unharmed). It is important that if it is to be used for this purpose the highly purified 'RNAse-free' grade must be purchased (regardless of cost). The enzyme is an endonuclease with little specificity for secondary structure. The pH-optimum is broad (around 8); it has a requirement for magnesium. If it is to be used to degrade DNA in an RNA solution, all that is needed is to add about 5–10 μg of the enzyme/ml to the solution (pH about 7–8)

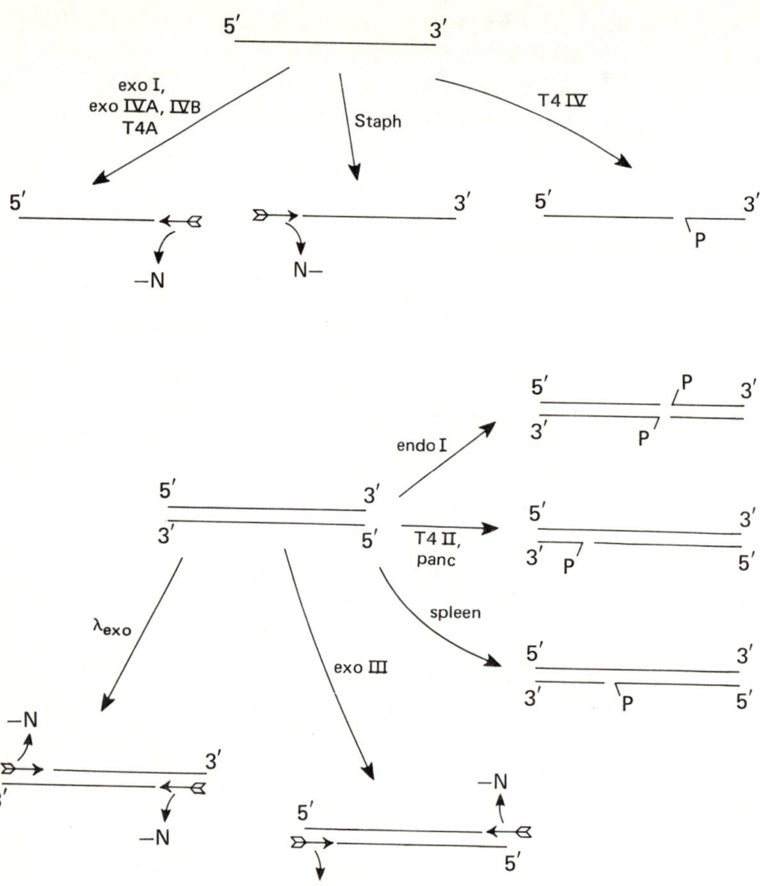

FIG 6.3   Summary of the activities of several nucleases on DNA substrates. Above, single-stranded DNA, below, a duplex. The direction of hydrolysis by the exonucleases is indicated thus

Abbreviations:

exo I, II, IVA, IVB: *E. coli* exonucleases	(p. 209)
endo I; *E. coli* endonuclease	(p. 209)
T4 II, IV; phage T4 endonucleases	(p. 211)
T4 A; exonuclease A	(p. 211)
λ exo; exonuclease of phage λ	(p. 211)
Staph; staphylococcal nuclease	(p. 208)
spleen; calf-spleen deoxyribonuclease	(p. 209)
panc; bovine pancreatic deoxyribonuclease	(p. 209)

and a magnesium salt to a final concentration of about 1 mM. The incubation can be at room temperature. The enzyme never degrades DNA completely. If it is required to obtain the mononucleotides, the oligonucleotides must be further degraded with exonucleases. Unfortunately there is no clear evidence that the incompleteness of this reaction is due to any characteristic sequences in the DNA.

The spleen enzyme (also commercially available) is another endonuclease similar to the above, except that it leaves the phosphate groups on the 3′-position. It is used (in conjunction with Staphylococcal nuclease) in nearest-neighbour-frequency analysis (p. 271).

Phage nucleases   Many DNA phage contain genes for deoxyribonucleases. Several have been well characterized.

One of these enzymes can be regarded as a routine analytical tool for studies on DNA; it is λ-exonuclease. Lysogeny is induced in an *E. coli* λ-lysogen and the enzyme is extracted from the cells. The highest yield of enzyme is obtained if the phage is a mutant of a class called TII (Radding and Shreffer, 1966). This paper describes a purification of the enzyme although a more complete recipe is given by Little (1967). The enzyme is magnesium-dependent and has a high pH-optimum (9·4). It has a preference for double-stranded DNA (rate of reaction for native DNA is about 50 times that for denatured DNA). The enzyme erodes a chain (of the double helix) from the 5′-end; the products are 5′-nucleotides (not 3′-nucleotides). Thus it yields the same products as exonuclease III (above) but works down the chain from the other end (Little, 1967).

Several deoxyribonucleases present in *E. coli* infected with either T2 or T4 have been characterized. Obviously phage-infected bacteria are a good place to look for nucleases as host cell DNA is degraded during infection (p. 30). Exonuclease A (Short and Koerner, 1969) is an enzyme with very similar properties to exonucleases IVA and IVB (see p. 209); that is the preferred substrate is an oligodeoxyribonucleotide and 5′-nucleotides are produced. Two endonucleases from T4-infected cells (II and IV) have been purified; it is almost certain that both these enzymes are involved in degradation of host-cell DNA. T4 endonuclease II introduces a limited number of single-strand breaks (5′-phosphate) in double-stranded DNA. (Sadowski and Hurwitz, 1969); endonuclease IV (same reference) has a similar reaction but with single-stranded DNA as substrate. Ling (1971) has recently re-examined the reaction catalysed by endonuclease IV and has shown it to be C-specific. Thus there is a fairly readily purified specific deoxyribonuclease available which should revolutionize work on the primary structure of DNA in the future. Several other enzymes have been partially characterized from cells infected with T-even phage. One that remains to be satisfactorily purified is the RNA-inhibited, phage-specific deoxyribonuclease for which indirect evidence was obtained by Stone and Burton (1962).

From the above it must be clear that, as with so much microbial biochemistry, the bulk of the work on phage nucleases has been done using the *E. coli* system. However, for the search for new deoxyribonucleases which could prove of value in sequencing studies perhaps we should be looking elsewhere. The most interesting deoxyribonuclease to have been studied recently is the enzyme produced in *B. subtilis* infected with phage SP3. The enzyme achieves the remarkable feat of lopping off two nucleotides at a time. It is an exonuclease; it erodes chains from the 5′-end to yield fragments of the type -N-N. The enzyme has been carefully characterized by H. V. Aposhian and his group. A recent paper, which includes their earlier references is by Aposhian, Friedman, Nishihara, Heimer and Nussbaum (1970).

7   Synthesis

This chapter is about the experimental methods required for two separate types of investigation. On the one hand, synthetic oligo- and polynucleotides are required as model compounds for structural studies on nucleic acids and for biochemical studies on the metabolism of nucleic acids and the genetic code. On the other hand, an understanding of the synthesis of nucleic acids *in vitro* is a prerequisite for the study of these processes *in vivo*.

There are two general methods of achieving the synthesis of nucleic acids, chemical or enzymic procedures. The former method is only relevant to the problem of synthesizing the model compounds referred to above as, at the moment, there are no non-enzymic methods of synthesis which throw any light on the enzymic processes. As a method of synthesizing specific oligo- or polynucleotides, chemical methods are more versatile in principle. There are no absolute restrictions on the type of nucleosides to be incorporated and the chain can be extended in either direction. Technically however, chemical synthesis of oligonucleotides is extremely difficult. The problems derive from the fact that nucleotides contain many functional groups and selective blocking (and unblocking) of particular groups is required. This is true of all organic synthesis. The particular problem with the oligonucleotides is that the purification of the products of these reactions, and their recovery in high yield, require particularly exacting expertise. For this reason, chemical synthesis is outside the capacity of the average biochemistry laboratory and my account of these techniques is limited to illustrating the more important procedures so the literature relating to these methods can be critically evaluated.

Enzymic synthesis on the other hand is within the capacity of the average biochemist and I have included reference to the isolation of the important enzymes. As a synthetic tool, enzymic reactions are limited by the facts that the chain-elongation enzymes only work in one direction ($5'$ to $3'$), have considerable substrate-specificity (so that the incorporation of unnatural residues is largely a matter of luck) and, in general, are difficult to stop at specific points in the chain. Having said this, it is nevertheless the case that the most ambitious syntheses of nucleic acids (see p. 235) rely absolutely on enzymic steps for all the major joining reactions.

The chapter is divided into sections describing chemical, enzymic and combined methods of synthesis.

Chemical synthesis

Oligoribonucleotides

On the whole oligoribonucleotides are harder to synthesize than deoxyoligomers of comparable length because of the additional functional group ($2'$-OH) in RNA and the difficulty in obtaining uniquely 3′5′-phosphodiesters (with no 2′,5′-product). Nevertheless the synthesis of all 64 possible trinucleoside diphosphates (N-N-N) was one of the first great achievements of synthetic nucleotide chemistry. The interest in this set of compounds derived from their use by M. Nirenberg and his group for elucidating the bulk of the genetic code. The method used by Nirenberg was based on the ability of these compounds to mimic one of the reactions of a codon in mRNA, namely to catalyse the binding of the appropriate aminoacyl-tRNA to ribosomes. The procedure consists of incubating a trinucleoside diphosphate (let us say G-C-A) with *E. coli* ribosomes and each of 20 radioactive aminoacyl-tRNA preparations. It is not necessary to fractionate the tRNA species for this experiment. Unfractionated tRNA is incubated with the aminoacyl-tRNA synthetases (also unfractionated) with 19 unlabelled amino acids and one (say arg) radioactive. This preparation is then incubated with the ribosomes and the G-C-A; the ribosomes are separated by binding to cellulose nitrate membranes (see p. 361) and counted. If GCA is a codon for arg (it is), appreciable binding of radioactivity (present as arg-tRNA^{Arg}) to the ribosomes will be obtained. With the other 19 possible radioactive preparations no binding above background should be observed. For a review of Nirenberg's work on the genetic code using this technique see Nirenberg *et al.* (1966). The dinucleoside triphosphates were synthesized independently by the groups of Nirenberg and H. G. Khorana. I have chosen Khorana's synthesis of G-C-A to illustrate the use of protecting groups in the strategy for the synthesis of such compounds (and incidentally the simpler dinucleotides).[1]

Synthesis of G-C-A   This synthesis is from Lohrmann, Söll, Hayatsu, Ohtsoka and Khorana (1966), a paper which describes the synthesis of all 64 codons. The strategy is to start with a derivative of 3′-CMP in which 5′-OH is blocked with an acid-labile protecting group and 2′-OH and the reactive group in the pyrimidine (4-NH_2) are protected by alkali-labile groups. This phosphate is condensed with an alcohol, namely the 5′-OH of a derivative of Ado in which 2′-OH, 3′-OH, 6-NH and N-1 are also protected by alkali-labile groups. When the dinucleoside monophosphate product (a derivative of C-A) is treated with acid the 5′-OH on the Cyd is exposed but the remaining reactive groups are still protected. This compound then supplies the alcohol group for another condensation with a suitably protected derivative of 3′-GMP to form the other phosphodiester link; finally treatment with alkali removes all the protecting groups to yield G-C-A. The actual reagents used are given in Fig. 7.1. Note that the acid-labile group is a simple derivative of the trityl group (see p. 70).

Alternative procedures   In the synthesis described above, the crucial phosphorylation reactions are condensations of a 3′-phosphomonoester with an alcohol using dicyclohexylcarbodiimide (DCC) as condensing agent. There are several alternatives to the use of DCC; among these are tri-*iso*propylbenzenesulphonyl chloride (Lohrmann *et al.*, 1966), trichloroacetonitrile (Cramer, Rittersdorf and Bohm, 1962), picryl chloride (Cramer, Wittmann, Daneck and Weimann, 1963) and $Me_2N{=}CHClCl^-$ ('dimethylformamide chloride'; Jacob and Khorana, 1964). All these compounds form phosphodiester intermediates which react as follows. If N and N′ are two nucleosides and X–Cl is the chloride reagent, a 5′-nucleotide, $O_3^{2-}PO{-}N$, reacts with X–Cl to form $X{-}O{-}PO_2{-}O{-}N$, which then reacts with N′ to form $N'{-}O{-}PO_2{-}O{-}N$ (N′–N in the shorthand nomenclature) and X–OH. In the case of the dimethylformamide chloride, X–OH ($Me_2N{=}CH{-}OH$) is simply the imide tautomer of the ionized form of dimethylformamide, $Me_2NCH{=}O$.

It is equally possible to employ a 5′-nucleotide as the phosphorylating species and to employ as 'alcohol' the 3′-OH of an otherwise protected nucleoside. According to C. B. Reese, who has

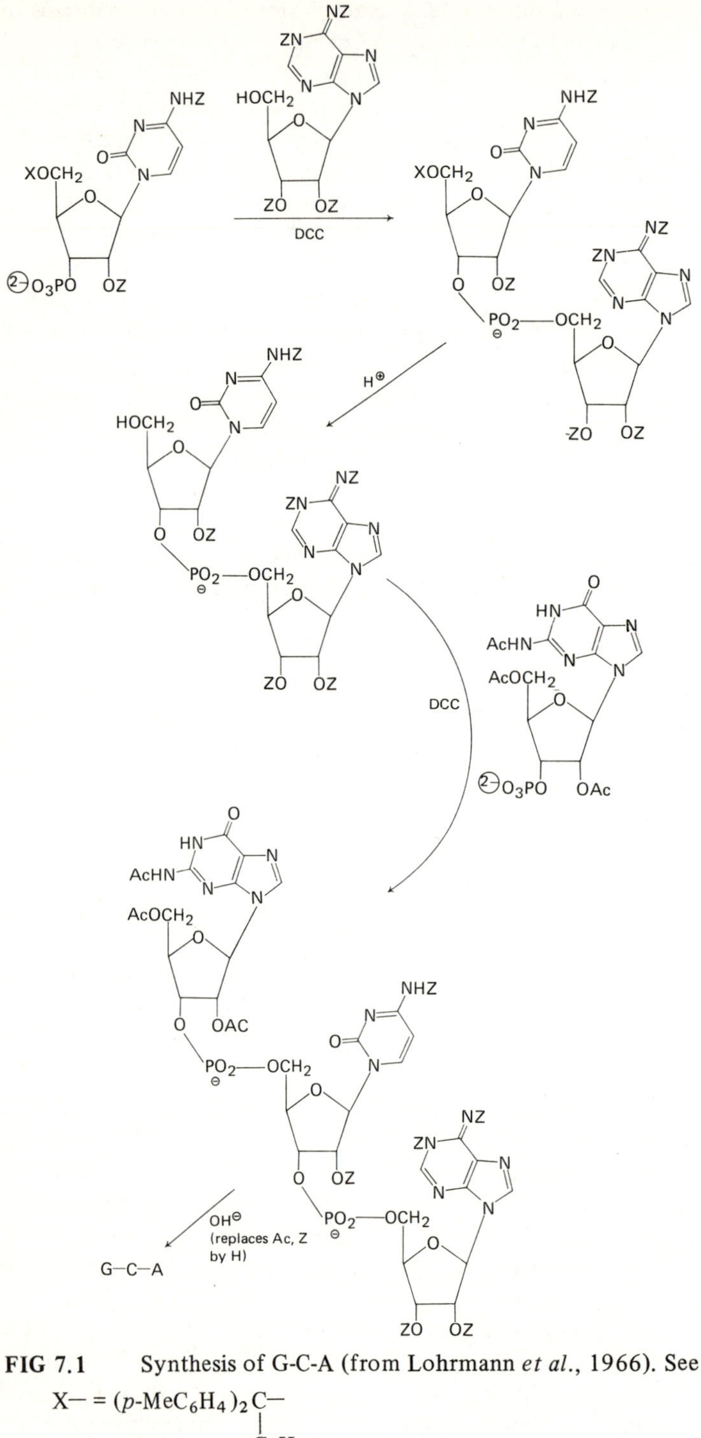

FIG 7.1   Synthesis of G-C-A (from Lohrmann *et al.*, 1966). See text for discussion.

X— = $(p\text{-MeC}_6\text{H}_4)_2\text{C}$—
         |
         C_6H_5

Z— = $\text{C}_6\text{H}_5\text{CO}$—,   Ac— = CH_3CO—

DCC = dicyclohexyl carbodiimide $(\text{C}_6\text{H}_{11}-\text{N}{=}\text{C}{=}\text{N}-\text{C}_6\text{H}_{11})$

made an extensive study of the reactions involved in this procedure, the technique has certain advantages from the point of view of stereospecificity and retention of the correct 3′-5′-phosphodiester structural isomers (Griffin, Jarman and Reese, 1968).

Synthesis of oligoribotides can be extended to making somewhat longer chains than the trinucleotide diphosphates. For example the substance U-U-U-U-U has been synthesized by a stepwise procedure (Smrt and Šorm, 1964). However in general the sequential chemical synthesis of longer defined oligoribotides is extremely difficult.

A different extension of these reactions is the synthesis of dinucleoside triphosphates containing aberrant chemical features (such as 'odd bases'). One of the most interesting classes to have been studied are those in which a phosphodiester group is replaced by the thiolated group $-O-POS-O-$. The nucleoside thiophosphates and their derivatives have been carefully studied (Lisý, Eckstein and Škoda, 1968; see also p. 225).

Oligo- and polydeoxyribonucleotides

For the reasons stated above, the synthesis of oligodeoxyribotides is technically easier than the synthesis of the ribotides. Three types of chemical synthesis have been used: polymerization of nucleotides to yield polymeric products of undefined length (which can be subsequently fractionated to yield homogeneous species); stepwise synthesis; and block condensation (the linking up of pre-formed lengths). In addition to protecting groups of the types discussed so far, these processes require phosphate-protecting groups as well. Probably the most widely used group of this type is cyanoethyl. The cyanoethyl derivative itself is a phosphate ester of β-cyanoethyl alcohol (so that the derivative of a nucleotide would have the general formula $NC-CH_2-CH_2-O-PO_2-O-$ nucleoside). Aqueous alkali effects the β-elimination of such a compound to yield $NC-CH=CH_2$ (acrylonitrile) and the free nucleotide. However several other groups have been surveyed (Söle and Khorana, 1965). I have chosen examples of each of the three approaches to the synthesis of the oligodeoxyribonucleotides to illustrate the chemical methods involved. Most are chosen from the work of Khorana, as later in the chapter we shall see how he has used these synthetic products in conjunction with enzyme reactions to make RNA molecules of defined, repeating sequence and synthetic DNA molecules of great complexity.

Polymerization methods   Poly dT (Tener, Khorana, Markham and Pol, 1958), poly dA (Ralph and Khorana, 1961), poly dC (Khorana, Turner and Vizolyi, 1961) and poly dG (Ralph, Connors, Schaller and Khorana, 1963) can all be synthesized by treatment of the appropriate 5′-deoxyribotide with a suitable condensing agent (such as DCC). The products can be fractionated by suitable chromatographic procedures. For example, Cramer (1966) quotes an elegant analysis of the poly dT products by G. Hoffarth (working in Cramer's laboratory). The oligomers up to $(-T)_{14}$ were clearly resolved by column chromatography on a DEAE-cellulose column eluted with a gradient (0·1–0·75 M) of triethylammonium bicarbonate. The same review by Cramer shows an analysis of such a mixture by thin-layer chromatography using polyethylene imine/cellulose (see p. 49). In this case the data were those of S. Rittner and G. Weimann.

Stepwise synthesis   The use of the cyanoethyl 5′-nucleotides for the synthesis of oligodeoxyribotides was first proposed by Cramer, Rittersdorf and Böhm (1962). As an example of the use of this reagent, consider the synthesis of the tetranucleotide d-A-T-C-G (Jacob, Narang and Khorana, 1967) as each of the four nucleotides is involved. Throughout this synthesis the bases are protected by benzoyl (or anisoyl p-MeOC$_6$H$_4$CO—) groups which are alkali-labile (see p. 213). The synthesis is summarized in Fig. 7.2. The main features of the synthesis are the repeated condensation (with either DCC or a sulphonyl chloride as condensing agent) of a 5'-nucleotide with a blocked 3'-hydroxyl and a nucleotide (or oligomer) with a blocked 5'-phosphate. The 3'-blocking group (acetyl) must be removed (with alkali) after each condensation reaction in order to generate the 3'-hydroxyl necessary for the next round of phosphorylation. These conditions avoid removal of the bases' protecting groups but they do remove the cyanoethyl from the 5'-phosphate, so that a condensation with cyanoethyl alcohol must be inserted at each stage. The limitations to this type of progressive synthesis are the difficulty of purifying the oligomers at each stage (this becomes progressively harder as the chain grows) and the loss of yield at each stage. This latter problem can be partly overcome by using the single-nucleotide reagent in excess.

Oligomers longer than the tetranucleotide described above can be made by these methods. However the synthesis by the stepwise method of significantly longer oligomers requires a breakthrough in the technology of the process. This breakthrough will probably consist of some sort of polymer support technique (similar to the Merrifield method for the chemical synthesis of proteins (Merrifield, 1965). So far the methods that have been tried are less successful than the conventional procedures exemplified above. However developments in this field can be reasonably expected in the future.

In one method, the nucleotide is attached to an insoluble polymeric support so that the

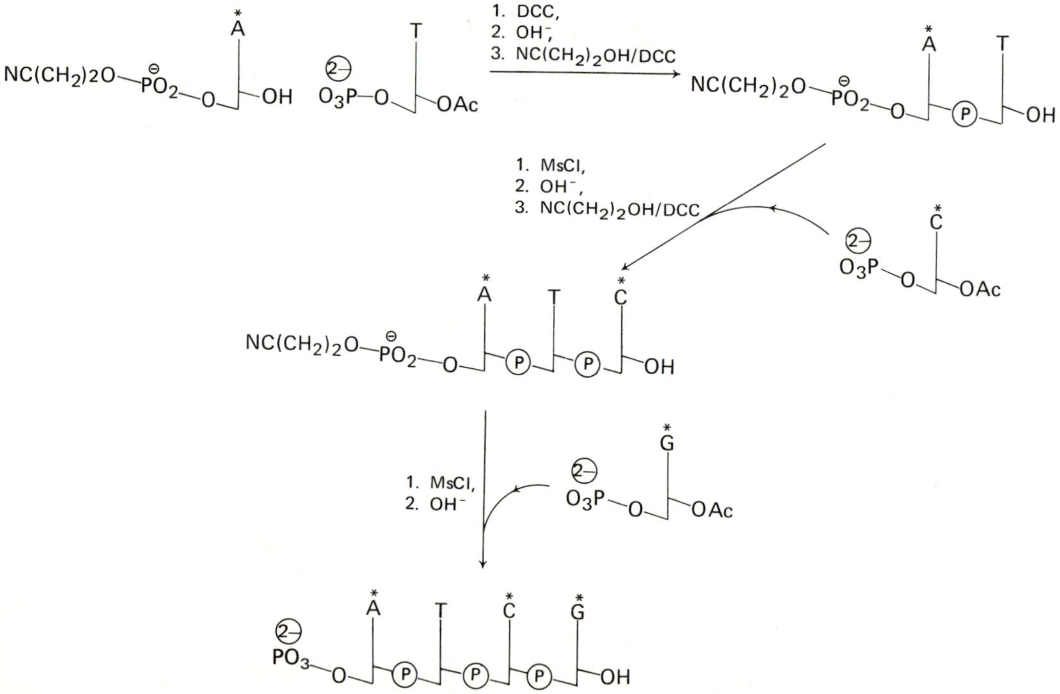

FIG 7.2   Synthesis of d-A-T-C-G (from Jacob *et al.*, 1967). See text for discussion.

DCC = dicyclohexylcarbodiimide
MsCl = mesitylene sulphonyl chloride
A*   = N^6-benzoyl-Ade
C*   = N^4-amisoyl-Cyt
G*   = N^2-amisoyl-Gua.

reagents are washed on and off the column (Blackburn, Brown and Harris, 1967). Alternatively the polymer support can be linear and the products from each stage are precipitated as a polymer matrix. At the moment this latter approach is probably the more hopeful (Hayatsu and Khorana, 1967; Cramer, Helbig, Hettler, Scheit and Seliger, 1966).

Block condensations   Block condensation is the process whereby two or more oligonucleotides are joined together chemically. The reactions are outlined in Fig. 7.3. The simplest example of block condensation is the polymerization of a pre-formed oligomer with a sulphonyl chloride as condensing agent (Fig. 7.3(a)). The products from such dinucleotides contain up to eight dinucleotide units (i.e. up to 16 residues, Ohtsuka, Moon and Khorana, 1965). If a proportion of the oligomers have a blocked 3′-end (Fig. 7.3(b)) then the size of the product is limited by the ratio of blocked to unblocked oligomers. Thus in this case the protected tetranucleotide from Fig. 7.2 is condensed with a 3′-blocked derivative with a condensing agent to yield a 16-residues oligomer. The protecting groups (which were not removed by the alkaline conditions used in the earlier synthesis of the tetranucleotide—Fig. 7.2) are removed with ammonium hydroxide (Jacob, Narang and Khorana, 1967). This type of synthesis can be achieved one step at a time (Fig. 7.3(c)). The first stage in this sequence is the condensation of a trinucleotide with a blocked 5′-end and a free 3′-hydroxyl and another trinucleotide with a free 5′-phosphate and a blocked 3′-hydroxyl. The product is a 6-nucleotide oligomer; its 3′-protecting group is alkali-labile (the 5′- one is not; see p. 213) so that it can be converted into

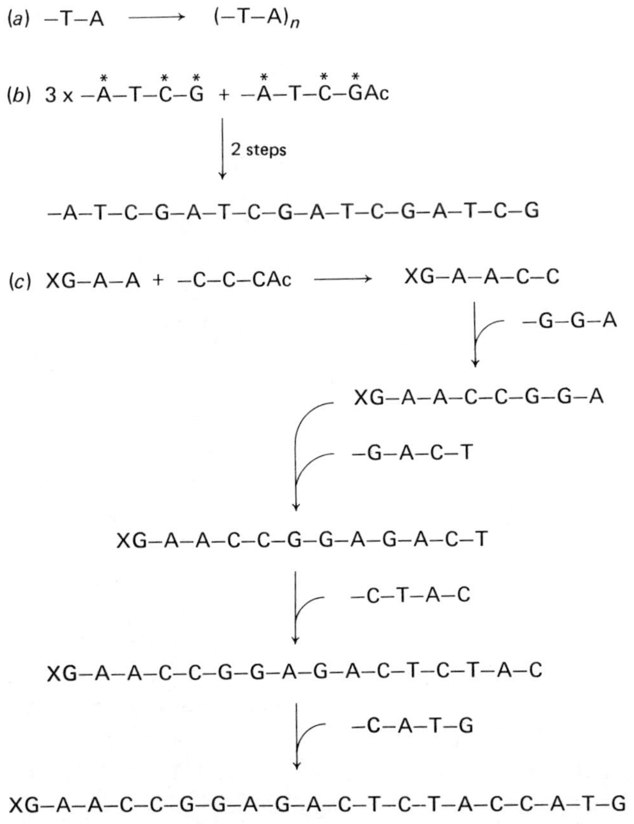

(a) −T−A $\longrightarrow$ (−T−A)$_n$

(b) 3 × −$\overset{*}{\text{A}}$−T−$\overset{*}{\text{C}}$−$\overset{*}{\text{G}}$ + −$\overset{*}{\text{A}}$−T−$\overset{*}{\text{C}}$−$\overset{*}{\text{G}}$Ac

$\quad\quad\quad$ ↓ 2 steps

$\quad$ −A−T−C−G−A−T−C−G−A−T−C−G−A−T−C−G

(c) XG−A−A + −C−C−CAc $\longrightarrow$ XG−A−A−C−C

$\quad\quad\quad\quad\quad\quad\quad\quad\quad$ ↓ −G−G−A

$\quad\quad\quad\quad$ XG−A−A−C−C−G−G−A

$\quad\quad\quad\quad\quad\quad$ −G−A−C−T

$\quad\quad$ XG−A−A−C−C−G−G−A−G−A−C−T

$\quad\quad\quad\quad\quad\quad$ −C−T−A−C

$\quad$ XG−A−A−C−C−G−G−A−G−A−C−T−C−T−A−C

$\quad\quad\quad\quad\quad\quad$ −C−A−T−G

XG−A−A−C−C−G−G−A−G−A−C−T−C−T−A−C−C−A−T−G

FIG 7.3   Examples of block condensation. See text for discussion and references. Symbols have the same meanings as in Figs. 7.1 and 7.2 (pp. 214 and 216).

a reactant for a further condensation. The sequence of reactions shown in the rest of Fig. 7.3(*c*) omits the protection and removal of the 3′-ends to make the picture less cumbersome. The 20-nucleotide oligomer product will appear again later in this chapter (p. 234; Gupta *et al.*, (1968).

Enzymic synthesis

Polynucleotide synthesis independent of template

Polynucleotide phosphorylase This enzyme catalyses the equilibrium between riboside diphosphates and polyribotide and inorganic phosphate:

$$\ldots\text{-N-N-N-N} + ppN' \rightleftharpoons \ldots\text{-N-N-N-N-N}' + p_i$$

Under suitable conditions the enzyme can be used either to synthesize the polynucleotide (left to right reaction above) or as a nuclease (right to left reaction above). The physiological role of these enzymes remains obscure.

It is probably true to say that these enzymes are present in all bacteria. As the enzymes are easy to isolate and assay and purify, it is probably true to say that anyone might as well use whatever bacteria happen to be available in the laboratory at the time. However, for largely historical reasons, the preparations that are used for polynucleotide synthesis are obtained from *Micrococcus sp.* strain ATCC 4968,[2] *Azotobacter vinelandii* or *E. coli* strain B. The *Micrococcus* enzyme is commercially available although it is hardly worth buying as the organism is very easy to grow on a large scale and the enzyme is easy to isolate. Recipes for the isolation of the enzymes are to be found in Steiner and Beers (1961) for the enzymes from *Micrococcus* and *A. vinelandii* (the latter recipe being due to S. Ochoa); the isolation of the *E. coli* is described by Williams and Grunberg-Manago (1964).

There are many assay procedures for the enzyme. The polymerization reaction can be assayed either by measuring the release of inorganic phosphate or by measuring the accumulation of polynucleotide. Alternatively the phosphorylysis reaction can be assayed by measuring the release of nucleoside diphosphate from a polynucleotide substrate or the equilibration rate can be measured by assaying phosphate exchange (between inorganic phosphate and nucleoside diphosphate). For the assay of the enzyme during a purification scheme it is wise to use adenylate substrates as the crude fractions obtained early in the procedure are contaminated with pyrimidine-specific ribonucleases.

Probably the simplest procedure is to measure the rate of formation of poly A from ADP. The polymerization reaction is favoured by high pH. In the procedure of Beers (1957) which was described for the enzyme from *Micrococcus*, the reaction volume is 2·0 ml and contains enzyme, 1·3 mM $MgCl_2$, 0·2 M NaCl, 20 mM tris HCl pH 9·9 and ADP (a suitable concentration up to 2 mM). After incubation at $37°C$ for up to 20 min, the poly A (together with protein) is precipitated with 1 ml of 10% $HClO_4$. This precipitate is spun out, washed (twice with 10% $HClO_4$ and twice with 95% aqueous ethanol) and suspended in 3 ml 0·1 M tris HCl pH 8·0. The suspension is shaken for 15 min, the insoluble material (protein) is removed by centrifugation and the extinction at 257 nm (due to poly A, ϵ_p 9800) is measured. The blank value is obtained by adding the same reagents but putting the ADP in after the perchloric acid (zero-time control). High blank values are due either to poor experimental technique (such as inadequate washing of the perchloric-acid precipitate so that ADP is left behind) or alternatively can be due to a poor preparation of enzyme, contaminated with nucleic acids. If highly purified enzyme is used, the amount of enzyme-protein employed in the assay is very small; in order to obtain a perchloric acid precipitate under these circumstances, a co-precipitant in the form of a protein (such as bovine serum albumin) must be added to a final concentration of about 1·1 mg/ml. The procedure of Grunberg-Manago, Ortiz and Ochoa (1965) for the *A vinelandii* enzyme is similar but is run at pH 8·1; in it and in an alternative method described by Beers (1957), phosphate-release is assayed by estimating inorganic phosphate colorimetrically.

Assay procedures for the *E. coli* enzyme using either [^{14}C]ADP as a substrate for the polymerization reaction or [^{32}P]phosphate for measuring phosphate-exchange are given by Littauer and Kornberg (1957). All these and several other methods are described in detail in Steiner and Beers (1961).

Use of polynucleotide phosphorylase   For the preparation of homopolymers, the procedure is to incubate the appropriate nucleoside diphosphate(s) with the enzyme in a buffer of high pH in the presence of divalent cations. At the end of the incubation, which can be prolonged (overnight), the preparation is deproteinized with phenol and the polymers are precipitated with ethanol. The only problem really is the requirement of some of the enzymes for 'primer', a short oligonucleotide chain with a free 3'-end to start work on. The primer requirements for the *Micrococcus* enzyme has been extensively studied by Moses and Singer (1970). If the purified enzyme is made by the procedure of Klee and Singer (1968) it is independent of primer (Form-1). For preparative purposes it is probably best to follow exactly the recipe for the preparation of the enzyme used for a given polymerization and not to fiddle about with the composition of the incubation system.

Homopolymers of high molecular weight (about 10 S) are synthesized. There is really no difficulty about the synthesis of poly A, poly U, and poly C with any of the polynucleotide phosphorylases. Conditions for the reaction with the *Micrococcus* enzyme are described by Steiner and Beers (1961). Analysis of the optimal conditions for the *A. vinelandii* enzyme has been made by F. Cramer and his group (see Cramer, 1966 and references therein). Many modified nucleotides may be polymerized by the enzyme to yield (for example poly I and poly Ψ (Pochon, Michelson, Grunberg-Manago, Cohn and Dondon, 1964). The real problem comes with attempts to make poly G. However there is one reliable method using the *E. coli* enzyme in the presence of manganese (rather than magnesium) ions at 60°C (Thang, Graffe and Grunberg-Manago, 1965). Other examples of the use of the enzyme for the synthesis of homopolymers are given on p. 335.

If the enzyme has as substrates a mixture containing more than one nucleoside diphosphate, the product will be a random polymer with a nucleotide composition determined by the ratio of diphosphates in the reaction mixture. A particular application of this type of synthesis is to make polymers of uniform nucleotide type with a different nucleotide at the 3'-end. Let us take a classical example, the substance -A-A-A-A-A- . . . . -A-A-A-C. This homopolymer acts as a synthetic mRNA in a system for protein synthesis *in vitro*. The triplet AAA is the codon for lys; the 3'-terminal triplet (AAC) is asn. The fact that the polypeptide products from such an incubation consisted of polylysine with arginine present exclusively as a C-terminal amino acid and the complete absence of any histidine (CAA is his) established unequivocally the 5' to 3' reading direction of mRNA *in vitro* (see p. 16; Salas, Smith, Stanley, Wahba and Ochoa, 1965). The poly A with 3'-terminal C was made by polymerizing ADP in the presence of a small amount of CDP. The product of course is a mixed polymer consisting (on average) of extended runs of . . . A-A-A-A- . . . with an occasional . . . -C- . . . . When this product is hydrolysed with ribonuclease A (see p. 206) the chains are cleaved following every C to yield the correct polymer type. The same technique an be used to synthesize . . . A-A-A-U, . . . I-I-I-C and . . . I-I-I-U. If an appropriate reaction mixture for the polynucleotide phosphorylase is incubated and the product is treated with ribonuclease T_1, it is possible to make ... U-U-U-G, ... C-C-C-G and ... A-A-A-G. The products of these reactions have been analysed by H. Küntzel; his data are reported by Cramer (1966).

The enzyme can be used to synthesize homopolymers using dinucleotides as primers for the polymerization. For example if A-U is the primer dinucleotide and UDP is substrate, the product, A-U-U-U-U- ..., is poly U with a different initial triplet (AUU). The conditions for the synthesis of several such polymers using the *Micrococcus* enzyme are described by Küntzel/Cramer (same reference as above).

An alternative method of using polynucleotide phosphorylase to synthesize oligo- (rather than poly-) nucleotides is to slow it down by performing the reaction under excessively non-optimal conditions. A detailed description of a method for making oligomers containing six to nine nucleotides, in which sodium chloride is employed to limit the reaction, has been given by Thach and Doty (1965).

All these examples illustrate the use of nucleoside diphosphates as substrates.

The *Micrococcus* enzyme will accept as substrates a mixture of UDP and the thiophosphate analogue PO_3^{2-}-O-POS-Urd (see p. 215) to synthesize a mixed polymer containing both uridylic acid and uridine thiophosphate residues (Eckstein and Grindl, 1970).

'Poly A enzymes'   There are several enzymes that synthesize poly A or oligo A fragments. The physiological role of these enzymes is obscure; however poly A is a naturally occurring substance. The enzymes could prove useful for synthetic work but have not been exploited to date. They all catalyse the same reaction as RNA polymerase (pp. 12 and 225) but without template direction of the synthesis. In other words the 3'-OH of the growing chain reacts with ATP to produce pyrophosphate and the chain extended by A-residues. All the enzymes require an RNA 'primer'. Two classes are recognized. One type is typified by enzymes from *E. coli* (August, Ortiz and Hurwitz, 1962) and thymus nuclei (Edmonds and Abrams, 1962) in which a specific RNA primer appears to be associated with the enzyme. Whether or not fragments of primer sequence are present in the product is not certain but this product is essentially poly A. The other type extends the chain of specific cellular RNA molecules from the 3'-end with a relatively short oligo-A chain. Such enzymes have been isolated from rat liver (Klemperer, 1963), *E. coli* (Gottesman, Canellakis and Canellakis, 1962) and maize seedlings (Walter and Mans, 1970). The first two use as primer rRNA; the last uses tRNA. This enzyme extends the chains of both maize and heterologous tRNA but will not react with rRNA.

'Terminal addition enzyme'   This enzyme, which is more properly called non-replicative (or template-independent) deoxyribonucleotidyl transferase, catalyses the same reaction as does DNA polymerase but without a template. It is thus analogous to the poly A enzymes above but has no base-specificity. It has an absolute requirement for a primer, which is either denatured DNA or a synthetic oligodeoxyribonucleotide with a free 3'-end. The reaction is as follows (N and N' are two deoxyribosides, either the same or different; -N-N-N is the primer):

$$\text{-N-N-N} + n \text{ dN'TP} \rightarrow \text{-N-N-N(-N')}_n + n \text{ pp}$$

The enzyme is found in certain mammalian tissues, of which calf thymus seems to be the most convenient source. A description for its isolation, separation from DNA polymerase and purification on a really massive scale (starting with 30 kg of thymus) has been described by Yoneda and Bollum (1965). The enzyme is magnesium-dependent and has a neutral pH-optimum. The optimal conditions for preparation of oligodeoxyribonucleotides are described by Kato, Goncalves, Houts and Bollum (1967). An example of the use of the enzyme is given on p. 224.

Ribonucleases The reaction catalysed by ribonucleases is reversible. The immediate product of the hydrolysis of the bond between phosphate and the oxygen of C-5$'$ by an enzyme such as ribonuclease A (p. 206) or ribonuclease T_1 (p. 207) is the 2$'$,3$'$-cyclic phosphate (see p. 3). It is the subsequent hydrolysis of the cyclic phosphates that yields the mixture of 2$'$- and 3$'$-nucleotides, characteristic of the products of such hydrolyses. Therefore we can represent the equilibrium established by such an enzyme as follows:

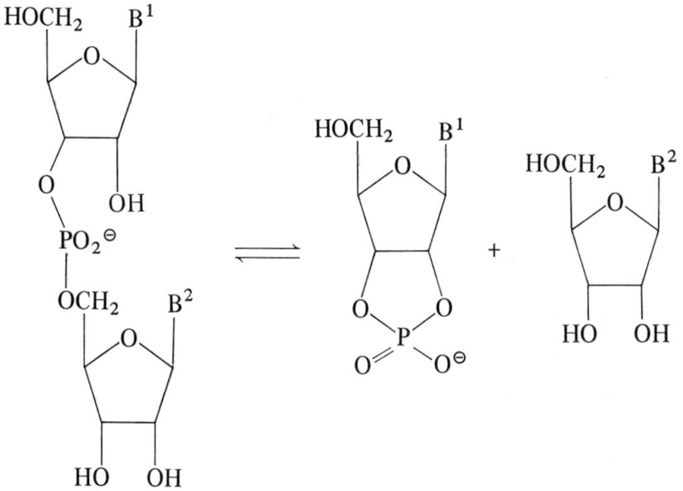

$$\text{(structures of a dinucleoside phosphate)} \;\rightleftharpoons\; \text{(2}',3'\text{-cyclic phosphate)} \;+\; \text{(nucleoside)}$$

In this case the substrate for the hydrolytic reaction is simply a dinucleoside phosphate. If the ribosides of B^1 and B^2 are N^1 and N^2, the shorthand version of this equilibrium is:

$$N^1\text{-}N^2 \rightleftharpoons N^1! + N^2$$

The important point about this reaction is that it is catalysed by an enzyme specific for 3$'$-5$'$-phosphodiesters, so the product of the reverse reaction is uniquely the natural isomer, and the major problem in the chemical synthesis of oligoribotides (contamination of the product of each phosphorylation step with the 2$'$-5$'$-isomer) is overcome. The conditions for reversal of the ribonuclease reaction are high concentrations of the cyclic phosphates (greater than 0·1 M). As the reaction is an equilibrium, it is not possible (by incubating cyclic phosphates with ribonuclease) to synthesize polymers. Heppel, Whitfield and Markham (1955), who pioneered the method with ribonuclease A, were unable to find any oligomers larger than C-C-C- when C! was used as substrate. The major product from this particular reaction is C-C! As a method of adding nucleotides to a chain, the reaction extends the chain in the 3$'$ to 5$'$ direction. The reaction can be used with ribonuclease T_1 in which case G! is the cyclic phosphate substrate. The application of this reaction to make trinucleoside diphosphates of the type G-N-N is described by Scheit and Cramer (1964).

Template-dependent polymerases

All template-dependent polymerases identified to date have their substrates the 5$'$-nucleoside triphosphates and extend the growing nucleic acid chain from the 3$'$-end. There are four classes of these enzymes: DNA-dependent DNA polymerase, RNA-dependent RNA polymerase and RNA-dependent DNA polymerase. In normal cells (from whatever type of organism) the only classes represented are the first two. In cells infected with RNA viruses there are molecules of RNA-dependent RNA polymerase (often known as RNA replicase) and in animal cells infected with certain oncogenic RNA viruses, there is an RNA-dependent DNA polymerase. Examples of enzymes from each of the two classes are discussed below.

The DNA polymerases of *E. coli* Two separate DNA polymerases have been discovered in *E. coli*. One is a cytoplasmic soluble enzyme which has been known and used extensively for several years and is known variously as DNA polymerase I or the Kornberg enzyme (after its discoverer, A. Kornberg). The properties of this enzyme are discussed in the next paragraph. However, in all probability this enzyme is not involved in chromosome replication. This fact was established with almost complete certainty when de Lucia and Cairns (1969) succeeded in isolating mutants (known as *pol A*$^-$) which lacked any detectable amount of Kornberg enzyme, but still grew normally. Subsequently it was shown by Smith, Schaller and Bonhoeffer (1970) and by Knippers and Strätling (1970) using different techniques that these mutants (as well as wild type *pol A*$^+$ strains) did contain a membrane-bound polymerase (DNA polymerase II) with similar properties to the Kornberg enzyme. As an experimental tool for the synthesis of DNA *in vitro*, and also as a model for studying the details of the reactions catalysed by DNA polymerases, the Kornberg enzyme is used exclusively. However with the report by Knippers (1970) that he has succeeded in solubilizing DNA polymerase II, we can expect to see extensive studies on the characterization of this enzyme in the future.

The assay of Kornberg enzyme (or DNA polymerases in general) consists of incubating the enzyme in 7 mM $MgCl_2$, 1 mM β-mercaptoethanol, buffer (such as 70 mM glycine) of pH 9·2, the four deoxyriboside triphosphates, one of which (usually dATP) is labelled with $[^{32}P]$, and a DNA template. After incubation, perchloric acid is added to a final concentration of 0·6 M and the precipitate is filtered off, washed and counted. The purified enzyme is inefficient at copying native DNA. The usual practice is to employ partly degraded DNA (treated with pancreatic deoxyribonuclease, see p. 209) as template for the enzyme in early stages of purification and synthetic poly d AT as template for the more highly purified preparations. The unit of polymerase is defined as the amount of enzyme that will convert 10 nmole of total dNTP into acid-insoluble product in 30 min at 37°C.

For preparation of the enzyme on a scale suitable for the average biochemist's laboratory, one can start with about 500 g of cell paste of *E. coli* (any fast-growing prototrophic strain such as B is suitable). This paste can be stored frozen at −20°C without losing activity so several batches can be accumulated by anyone who has not the luxury of a large fermentor. The isolation procedure for a preparation suitable for the applications of the enzyme described in this book, together with full details of the assay conditions are given by Richardson, Schildkraut, Aposhian and Kornberg (1964). The enzyme should be purified to stage VII in their scheme.

The detailed properties of the enzyme have been studied by Kornberg's group. Using an enormous amount of *E. coli* (90 kg of paste) they obtain 600 mg of purified enzyme as a homogeneous protein. There is a very well-written review of this work (Kornberg, 1969). Several important facts can be summarized as follows. The enzyme contains only one polypeptide chain (molecular weight 109 000). The reaction after which the enzyme is named is the following equilibrium, in which . . . -N-N-N-N is the growing DNA chain:

$$\ldots \text{N-N-N-N} + \text{pppN}' \rightleftharpoons \ldots \text{N-N-N-N-N}' + \text{pp}$$

This equilibrium reaction can be assayed in three ways. In the presence of DNA template (either single- or double-stranded DNA) chain elongation can be measured. This is 'the polymerase reaction'—see the assay method above. In the absence of dNTP molecules, and in the presence of pyrophosphate, the reverse reaction (right to left in the above equation) occurs.

This is the pyrophosphorylase reaction. The third method is based on the fact that (in the presence of DNA) the enzyme will equilibrate phosphate groups between pyrophosphate and the triphosphates. This reaction (pyrophosphate–triphosphate exchange) can be measured by using labelled pyrophosphate (or triphosphates) and monitoring the appearance of the ^{32}P in the triphosphates (or pyrophosphate).

The enzyme is also a nuclease; this is not a reference to the pyrophosphorylase reaction described above: it really is a 'genuine' exodeoxyribonuclease. Denatured DNA (and synthetic polydeoxyribotides) are degraded from the 3'-end. Under suitable conditions chains of DNA present in a double helix are eroded by the polymerase in both directions. One of these reactions, namely the 5'- to 3'-nucleolysis, is probably of physiological importance as it is known that DNA polymerase never seems to leave the 5'-triphosphate on the end of its newly synthesized chain (this is in contradistinction to the calf thymus enzyme and the RNA polymerizing enzymes—see pp. 224 and 225). This is presumably due to the exonucleolytic reaction. Kornberg was able to test this directly by using as substrate a synthetic double-stranded molecule in which the polymer pppT-T-T-T- . . . has been annealed to poly dA (. . . A-A-A-A-A-A- . . .). During the first few seconds of incubation of this substance with DNA polymerase the 5'-triphosphate was removed as the dinucleoside tetraphosphate pppTpT.

The review by Kornberg presents a plausible and elegant model to account for the surprising fact that a single-subunit enzyme can accommodate so many and varied functions. It also summarizes data that support the view that when acting as a polymerase, it binds preferentially to a 3'-end of a polynucleotide chain present at a discontinuity in a duplex template. The use of the enzyme and the nature of DNA molecules synthesized by it are discussed further on pp. 232 and 271.

A property of the enzyme that has been known for some time is its ability to make DNA strands either without template or with an RNA template. The former reaction (Schachman, Adler, Radding, Lehman and Kornberg, 1960) is a surprising one. The enzyme is incubated with dATP and dTTP (whether or not dGTP and dCTP are included is of no consequence—they have no effect on this reaction). After a lag period of several hours, poly dAT is made. The molecule is of high molecular weight (23 S) and double stranded; each chain has the sequence . . . A-T-A-T-A-T . . . . This substance is a natural product (pp. 152 and 278). It acts as a template for DNA polymerase in the normal way (no lag period). RNA templates are transcribed inefficiently on the whole. However a mixture of poly U and poly A (themselves both easily made—p. 219) will act as template. The product is poly dA:dT (Lee Hung and Cavalieri, 1963).

Other DNA polymerases   The DNA polymerase from calf thymus was purified in the same experiment in which the 'terminal addition enzyme' was prepared (Yoneda and Bollum, 1965, see p. 220). The enzyme behaves *in vitro* in a different fashion to the Kornberg enzyme. The template must be denatured or single-strand DNA. Also, when synthetic substrates are used, the 5'-triphosphate is left attached to the chain.

Several phage-specific DNA polymerases have been isolated and characterized, notably those induced in *E. coli* following infection with phages T2 (Aposhian and Kornberg, 1962), T4 (Goulian, Lucas and Kornberg, 1968) and T5 (Steuart, Anand and Bessman, 1968). These enzymes differ (in their reactions *in vitro*) from *E. coli* polymerase I in two important respects. One is their inability to use the 3'-OH discontinuities in DNA double helices as points to start synthesis; in other words they cannot use double-stranded templates at all (this may be the reason why the calf enzyme required single-strand templates). The other difference is that they lack the 5'-to-3' exonuclease action (although they have the 3'-to-5' nuclease activity described above for DNA polymerase I). Kornberg proposes (1969) that these facts are probably related; that is that the only reason that DNA polymerase I can replicate double-stranded DNA is that its concomitant 5′-to3-3′ nuclease activity clears a path for the enzyme ahead of the replication point. This is precisely what is expected of a repair polymerase (see p. 257). Meanwhile let us just summarize the template-directed synthesis catalysed by the polymerases and the nature of their products *in vitro* (Fig. 7.4).

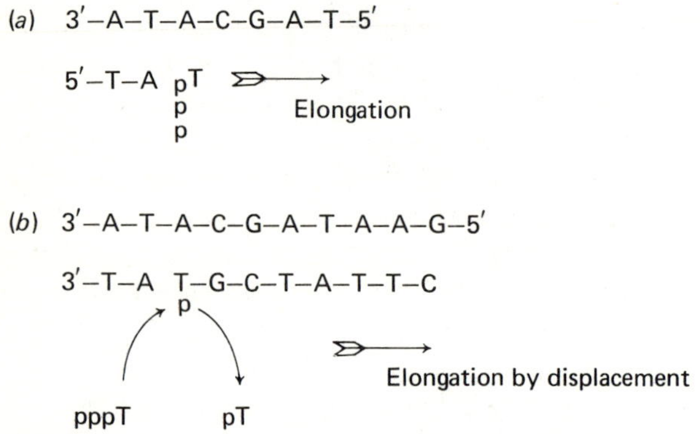

(*a*) 3′–A–T–A–C–G–A–T–5′

5′–T–A pT ⇒ ⟶
   p   Elongation
   p

(*b*) 3′–A–T–A–C–G–A–T–A–A–G–5′

3′–T–A T–G–C–T–A–T–T–C
    p

⇒ ⟶
      Elongation by displacement

pppT   pT

FIG 7.4 Hypothetical substrates and products in reactions catalysed by DNA polymerases. In (*a*), the template is single-stranded DNA; this reaction is catalysed by all DNA polymerases. In (*b*) the template is double-stranded DNA with a discontinuity. This reaction is catalysed by *E. coli* DNA polymerase I but not by the T-phage polymerases.

Evidence for the fact that the calf thymus polymerase does not hydrolyse the 5′-triphosphate is one of the consequences of the synthesis by Hansbury *et al.* (1970) of the internally hydrogen-bonded oligodeoxyribotide shown below.

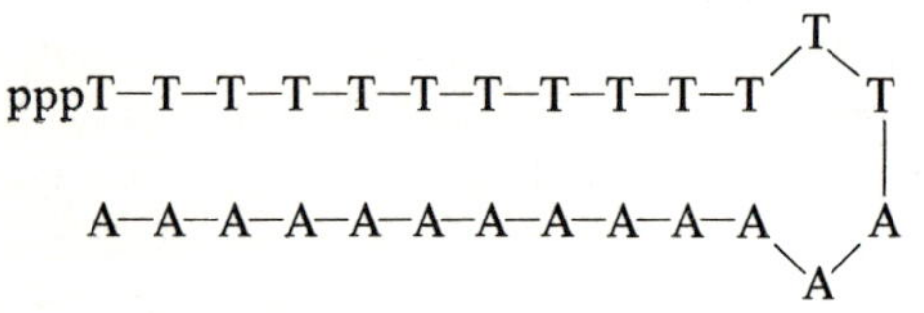

pppT–T–T–T–T–T–T–T–T–T–T T
A–A–A–A–A–A–A–A–A–A–A A

This substance was made by first synthesizing -T-T-T-T-T-T, converting this to pppT-T-T-T-T (see p. 58 for the method) and then extending the chain to a total of 13 thymidylate residues (average value) by limited use of terminal addition enzyme (p. 220). When this material is used as primer and template and dATP as substrate for the polymerase, the above product is made. If the final step is performed with end-addition enzyme a much larger product (with the polydeoxyadenylate section extending far beyond the oligodeoxythymidylate region) is produced.

RNA-dependent DNA polymerase The discovery of these remarkable enzymes (for which the rather stupid name 'reverse transcriptases' has been proposed) by Temin and Mizutani (1970) has been one of the most surprising discoveries about nucleic acid metabolism in recent years. The enzymes are found in mammalian cells infected with certain oncogenic RNA viruses. The reaction which they catalyse is synthesis of DNA from deoxyriboside triphosphates (in the same way as other DNA polymerases) using as template not DNA but viral RNA. The importance of the role of these enzymes in viral carcinogenesis and their significance for cancer research can hardly be overestimated. We shall be further considering the DNA synthesized by them *in vivo* later (p. 373). Here there is rather little to say about them; they have not been employed for synthetic purposes *in vitro*, nor would seem to be likely that they will have any general application in this respect, as the difficulties involved in RNA synthesis are in general much greater than those involved in DNA synthesis (see p. 212) and synthetically it is the DNA-dependent RNA polymerases that are so important.

For assay of the enzymes, the preferred templates are double-stranded RNA or DNA/RNA hybrids (Spiegelman *et al.*, 1970). A good assay method using poly r (A:U) as template is described by Scolnick, Aaronson and Todaro (1970). Of the four triphosphate substrates, one (usually dTTP) is labelled and the assay consists of measuring incorporation of radioactivity into ribonuclease-insensitive acid-precipitable polynucleotide. The same group (Scolnick, Rands, Aaronson and Todaro, 1970) have made a survey of the divalent-cation requirements for a variety of the enzymes and describe a procedure which is obviously very suitable for the extension of such surveys to other sources of the enzymes.

A recent discovery regarding these enzymes is that they are present in normal mammalian tissue (mouse BALB/3T3 cell line and human fibroblasts) as well as in many cancerous cells including lymphocytes from patients with acute leukaemia (Scolnick, Aaronson, Todaro and Parks, 1971). The exact significance of this finding remains to be evaluated; it is possible, as the authors suggest, that their presence indicates the presence of latent genomes of oncogenic 'RNA viruses' in the cells. This suggestion has enormous implications for cancer researchers—and for all of us.

DNA-dependent RNA polymerase from *E. coli*   A great deal is now known about the subunit structure of this enzyme and the role of the σ-factors in the initiation of RNA synthesis. However for a preparation that is required just to transcribe DNA (or a polydeoxyribotide) the classical older preparative procedure of Chamberlin and Berg (1962) is simple to make and use. The frozen *E. coli* paste, with which the preparation starts, can be stored in a deep freeze for several weeks if necessary without losing much of its activity. The assay conditions for RNA polymerase (which also apply to the use of the enzyme for transcription *in vitro*) consist of incubation of the enzyme 30 mM tris HCl pH 7·9, 0·13 M NH_4Cl, 0·03 M magnesium acetate with the template and triphosphates, together with an ATP-generating system (see p. 194). In the Chamberlin and Berg procedure these are present as 1 mM GTP, CTP and UTP and 0·5 mM [^{14}C] ATP with 10 mM phosphenol pyruvate and 20 μg/ml of rabbit skeletal muscle pyruvate kinase and 0·1 mg/ml of DNA. One unit is defined as the amount of enzyme which catalyses the incorporation of 1 mole of ATP into RNA in 10 min at 37°C. In the absence of factors, the enzyme will use either single- or double-stranded DNA as template. In the latter case, both strands are transcribed.

The enzyme is restrictive with respect to the bases and sugars in the template and triphosphate substrates. Not so over the phosphate however. Poly thiophosphato uridine (represented as $_sU_sU_sU_s$ . . .) is made if poly dA is used as template and if the triphosphate is $O_3\bar{P}$-O-$P\bar{O}_2$-O-$P\bar{O}S$-Urd (see p. 215) is employed as substrate (Eckstein and Grindl, 1970).

A discussion of the nature of the RNA synthesized in the presence of particular σ and ρ factors and with particular DNA templates is delayed until hybridization has been considered (see p. 372). For a review of the subunit data on the enzyme see papers by Burgess (1969). Characterization of σ factor in *E. coli* is described by Burgess, Travers, Dunn and Bautz (1969).

The role of the σ-factor in the establishing of the transcriptional initiation complex is summarized by Zillig *et al.* (1970) as follows. Roman numerals designate the four complexes that are formed. In these equations, E is the 'core enzyme' ($\alpha_2\beta\beta'$—see p. 21) and E* is the active form (? allosterically modified) in transcription. Four equilibria are recognized:

$$E \quad + DNA \quad \rightleftharpoons \quad (E\text{-}DNA)_I$$

$$(E\text{-}DNA)_I \quad + \sigma \quad \rightleftharpoons \quad (E\sigma\text{-}DNA)_{II}$$

$$(E\sigma\text{-}DNA)_{II} \quad \underset{12^\circ C}{\overset{20^\circ C}{\rightleftharpoons}} \quad (E^*\sigma\text{-}DNA)_{III}$$

$$(E^*\sigma DNA)_{III} \quad + pppPu \quad \longrightarrow \quad (pppPuE^*\sigma\text{-}DNA)_{IV}$$

Bautz and Bautz (1970) describe one of the characteristics of the σ-factor in detail, namely its ability to convert the DNA-bound enzyme into a form resistant to the antibiotic ricampicin. Although this area is really outside the scope of this book, the result is an important one; rifampicin (see note 11, p. 416) inhibits transcription by binding to the polymerase; thus it does not prevent initiation of RNA synthesis but 'freezes' it once it has commenced. It is therefore of great experimental value in studying transcription *in vivo*. Phage-infected cells contain phage-specific σ-factors. The factors specified by T4 (Travers, 1969) and T7 (Summers and Siegel, 1969) have been isolated.

Another antibiotic which may be of great value in the control of reactions catalysed by RNA polymerase is streptolydigin. This substance considerably slows down the transcription rate but does not affect the fidelity of transcription (Cassini, Burgess, Goodman and Gold, 1971).

In addition to σ-factors, there are other classes of factors involved in the control of RNA polymerase action. One of these is the CAP factor. It is a substance that has been fully characterized (Zubay, Schwartz and Beckwith, 1970). The assay system for it is the coupled transcriptional and translation synthesis of β-galactosidase *in vitro* (p. 198). The CAP factor, in the presence of σ, seems to be required for the transcription of the operons containing structural genes for inducible enzymes. It requires $3',5'$-cyclic AMP for its binding to the enzyme—DNA complex. A representative of another class is the Ψ_r factor, that is possibly required for the transcription of the rRNA cistrons (Travers, Kamen and Schleif, 1970—see also p. 372). The operation of Ψ_r is inhibited by the nucleoside tetraphosphate ppGpp. It is probable that this nucleotide inhibits transcription *in vivo*, as Cashel (1970) has shown that it is accumulated during the stringent response to amino-acid starvation (p. 229).

Another possible method of control involves the phosphorylation of RNA polymerase. Evidence comes from the observation of Martelo, Woo, Reimann and Davie (1970) that phosphorylation of the polymerase (with a mammalian protein kinase) greatly enhances the efficiency of transcription (*in vitro*).

Other microbial RNA polymerases In normal *E. coli*, a common σ-factor is probably required for the binding of the enzyme to the initiation sites for transcription on the transcribing strand of DNA. CAP- and Ψ-factors presumably effect a greater degree of selectivity between various classes of operons. During radical changes in the transcription process, such as phage infection and the requirement for the transcription of phage genes, a phage-specific σ-factor is produced (see p. 30). A yet more radical change occurs late in the infection with at least two of the T-odd phage, T3 and T7. This consists of the synthesis of a phage-specific polymerase. The T7 enzyme was the first to be identified. The work followed well-established data on the control of the production of the 'late enzymes' (phage-specified proteins that are synthesized late in infection) was controlled by a genetic locus designated gene-1 (Summers and Siegel, 1969). Chamberlin, McGrath and Waskell (1970) identified this gene product as a new RNA polymerase. The enzyme is very different to the *E. coli* polymerase. It contains only **one**

polypeptide chain and has very specific template requirements. A comparison of the T7 and T3 polymerases has been made by Dunn, Bautz and Bautz (1971).

Another example of a radical change in transcription pattern accompanies the process of sporulation (see note 3, p. 125) in bacteria. In this instance the process is associated with a modification of the 'core' of the polymerase (the subunits other than the σ-factor (Losick, Sonnenshein and Shorenstein, 1970). Another example of an alteration in the core enzyme is found during the development of phage λ (Walter, Seifert and Zillig, 1968).

Other RNA polymerases   The RNA polymerases of eukaryotic cells are less fully characterized than the bacterial and viral enzymes. The generalization that seems to be emerging (at least for animal cells) is that separate enzymes are present in the nucleolus and the nucleoplasm (part of the nucleus outside the nucleolus). The enzymes have different cation requirements and other differences are being established. No eukaryotic σ-factors have been identified so far. For a short paper on the differences between the two enzymes in sea urchin embryos and also in rat liver, together with several references, see Roeder and Rutter (1969).

RNA-dependent RNA polymerases   These enzymes are the replicases produced by RNA viruses. One of them has been used more extensively than others and (in an unpurified form) resulted in one of the landmarks for nucleic acid synthesis *in vitro* (p. 231). I therefore restrict this section to a description of this particular enzyme.

Of the RNA phage in *E. coli*, most belong to a very closely related group R17, f2, MS2, etc. These phage all cross-react serologically and in fact their coat proteins are of almost identical amino-acid sequence. The odd one out, which is in no way related to these other phage is $Q\beta$. The importance of $Q\beta$ lies in the fact that the total synthesis of infectious RNA (detected by a spheroplast assay—see p. 193) can be achieved *in vitro*.

There are two methods described in the literature for the purification of the polymerase, that of Eoyang and August (1968) relies on fairly conventional protein purification procedures, ammonium-sulphate cuts, ion-exchange chromatography, gel-exclusion chromatography and zonal centrifugation. The other (Kamen, 1970) is an interesting approach using a two-phase partition fractionation employing polyethylene glycol and dextran (see p. 169 for the use of such systems for nucleic acids). A complete description of experimental details has not yet appeared. The following points emerge about the enzyme, from the work of Franze de Fernandez, Eoyang and August (1968), (1970) and Kondo, Gallerani and Weissman (1970).

The enzyme consists of four dissimilar polypeptide chains (α, β, γ and δ). Of these only β is phage-directed. The others are all *E. coli* proteins. These other polypeptide chains are not subunits of the *E. coli* polymerase although one of them is apparently identical with the Ψ_r factor (p. 226). The enzyme has astonishing specificity. In the presence of two other factors, F_I and F_{II} (also not phage-specific proteins) it employs phage RNA (+ strand) as template and makes its complementary copy (− strand). In the absence of these factors it uses − strand as template and synthesises + strand. The only other natural RNA that can be used as a template is 6 S RNA (see p. 309), a substance of totally unknown function but known sequence, which is present in *E. coli*. The enzyme can also use poly C (or certain other synthetic polymers rich in C) as templates. However it cannot then employ the product of the synthesis (poly G in the former case) as template for subsequent synthesis of the starting polymer ('+ strand').

Enzymic modifications of nucleic acids

The enzymes described in this section do not polymerize nucleotides but effect various modifications, either to the bases or to the terminal nucleoside or nucleotide. The collection is very incomplete; I have concentrated on enzymes which are well characterized and are either used in synthetic studies or could readily be used if a problem requiring them were to arise.

Polynucleotide kinase   This enzyme catalyses the following reaction at the 5′-end of a chain,

N-N-N-N- . . . + ATP → -N-N-N-N- . . . + ADP

The enzyme is found in *E. coli* infected with phage T4. In practice an amber mutant (T4 *am* XF 1) is employed. The phage are maintained using a suppressor host and are then used to infect a wild-type strain. The infected cells are then harvested (after 2 h of infection) and disrupted. Enzyme can be purified from the lysate by the method of Richardson (1965); some technical improvements to his procedure are described by Hänggi, Streeck, Voigt and Zachau (1970). The enzyme is a remarkable one for the variety of substrates that are phosphorylated by it. Thus N-N-N-N- . . . in the equation above can be either RNA, DNA or short oligomers (either ribo- or deoxyribo-) provided of course that the 5′-end is not phosphorylated. The enzyme is assayed at pH 7·6 in the presence of 10 mM magnesium salt. A suitable substrate is the mixture of oligomers obtained by the digestion of DNA with pancreatic deoxyribonuclease and the unit is defined as the amount of enzyme that will phosphorylate 1 nmole of DNA substrate in 30 min under the conditions described by Richardson (1965). The transfer of radioactivity from $[^{32}P]$ ATP to acid-insoluble oligomer or nucleic acid is measured. RNA (or its oligomers) are phosphorylated rather less readily than DNA fragments. The paper by Zachau's group (above) describes the reactions with oligoribotides and tRNA (from which the 5′-phosphate had been removed with alkaline phosphatase) and is referred to later (p. 421).

tRNA . . . C-C-A pyrophosphorylase   This enzyme is of great practical importance whenever protein biosynthesis *in vitro* or the synthesis of aminoacyl tRNA is achieved (in the assay of acceptor activities—see p. 201). The necessary activity is present in the supernatant fraction from most cells and tissues; however if one of these processes is not working in a new experimental system, this enzyme should be checked out as one of the possible causes of trouble. The enzyme performs a patching up job on uncharged tRNA molecules. In the cell the final C-C-A sequence, common to all tRNA molecules, turns over much more rapidly than the tRNA molecules themselves. There is no apparent physiological significance for this turnover. It is generally assumed to result from accidental degradation of the end of the chain by single-strand exonucleases, with the consequent evolution of a system for repairing the molecules. This is 'cheaper' for the cell than the transcription of new tRNA molecules.

The assay system consists of incubating the enzyme preparation with incubated tRNA preparations in which at least a proportion of the molecules have lost their C-C-A termini. A procedure for making such a preparation is described by Hecht, Zamecnik, Stephenson and Scott (1958). The incorporation of $[^{32}P]$ CTP and ATP into the molecules is assayed. The *E. coli* enzyme has been characterized by Preiss, Dieckmann and Berg (1961) and the rat liver enzyme by Daniel and Littauer (1963). There is a difference between these enzymes in that the latter (but not the former) can incorporate U into the end of the tRNA molecule (provided the reaction mixture contains UTP but no CTP).

tRNA methylases   Of the many enzymes involved in the maturation of rRNA and tRNA (see p. 183) the tRNA methylases are the most extensively studied. The reason is that there is a convenient way of making methyl-deficient tRNA, which is a suitable substrate for them. As these methyl-deficient molecules are of considerable importance in studying RNA metabolism and as the basic techniques involved are employed extensively in several other areas of RNA metabolism, it is worth explaining in outline how it is done.

Amino-acid auxotrophs of *E. coli* will not grow in the absence of the required amino acid. Of these auxotrophs there are two classes 'stringent' (genetic nomenclature RC^{str}) and 'relaxed' (RC^{rel}). The RC^{str} genotype is wild type. RC^{rel} involves a separate mutation (apart from the auxotrophy). When a stringent auxotroph is starved of its required amino acid, protein synthesis stops (necessarily) and RNA synthesis is stopped as well. In a relaxed mutant this control of RNA synthesis by protein synthesis does not occur (it is this *control* that is 'relaxed'). Therefore the cells accumulate RNA. If the mutant is *met*$^-$ RC^{rel}, this uncontrolled synthesis of RNA produces RNA molecules which cannot be methylated, as methionine is (indirectly) the methylating agent. The first relaxed mutant to be isolated was actually a *met*$^-$ strain of *E. coli* K12 W6 (Hoagland, Stephenson, Scott, Hecht and Zaemcnik, 1958) but it is now recognized that the RC gene is not linked to any particular amino-acid requirement. However this particular organism is a very good source of methyl-deficient RNA when it is starved of methionine (Mandel and Borek, 1963).

The methylation reaction is shown below for the methylation of a C-residue in tRNA.

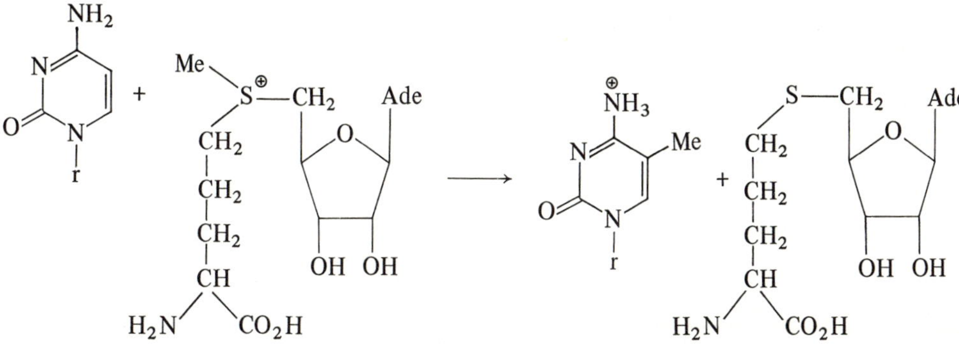

The methylating agent is *S*-adenosylmethionine. Similar reactions provide the other methylated bases of RNA (see p. 40). The assay system consists of incubating methyl-deficient tRNA with the enzyme and [Me-^{14}C]*S*-adenosylmethionine and measuring the incorporation of radioactivity into acid-insoluble polynucleotides. A survey of these enzymes was made by Hurwitz, Gold and Anders (1964a). Specific enzymes are involved in the methylation of G, C, U and A.

DNA methylases These enzymes are formally similar to the tRNA methylases; the same co-factor is employed. The main sites of methylation in DNA are C and A (see p. 271). The enzymes have been studied by Hurwitz, Gold and Anders (1964b). The C-specific enzymes from several mammalian sources have been studied by Scheid, Srinivasan and Borek (1968). They employed salmon-testis DNA as substrate. Their more important conclusions are that the enzyme is species-specific, it is a nuclear enzyme (present in the so-called nuclear insoluble fraction) and the amounts of the enzyme differ in different tissues of the animal. Thus in the rat, there is more in kidney than liver, more in adult liver than embryonic liver and very little indeed in brain and spleen. The significance of these data is not clear; for example is the methylation controlling replication or transcription or both? As we are a long way from being able to sequence animal DNA (see p. 266) it looks as though the study of this reaction is one of the few areas where clues regarding the control of the template activity of DNA (both for replication and transcription) are going to be found.

Glucosylation of DNA The DNA of the T-even phage of *E. coli* contains in place of cytosine, 5-hydroxymethylcytosine. This substance is incorporated into the DNA molecule as the corresponding deoxyribonucleoside triphosphate. However in the finished product, phage DNA, a proportion of these hydroxymethyl groups are substituted with carbohydrate side-chains. The enzymes involved have been characterized by Kornberg, Zimmerman, Kornberg and Josse (1961). The glucosylating agent is UDP-glucose and the assay consists of incubation of the

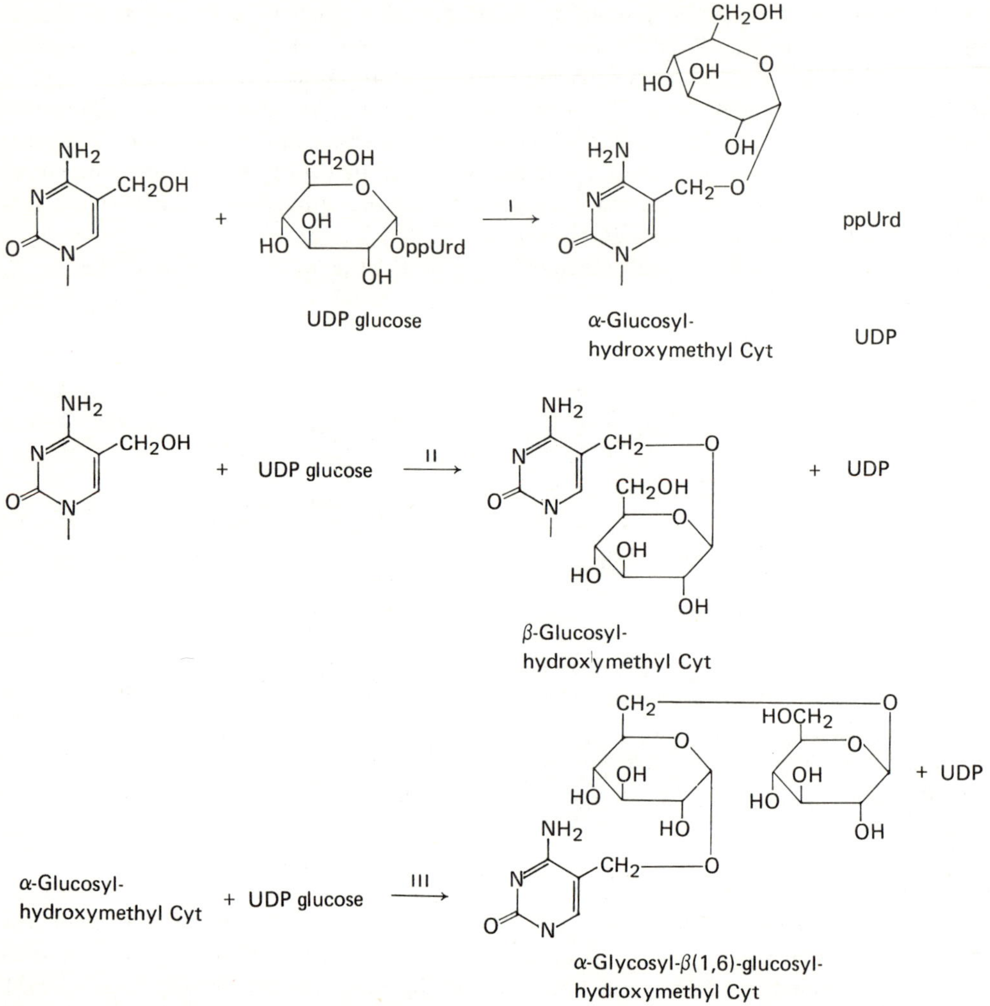

FIG 7.5   DNA-glucosylation reactions. Enzymes catalysing reaction I are found in *E. coli* infected with phage T2, T4 or T6. An enzyme catalysing II is found in cells infected with T4 and one catalysing III in cells infected with T6. From Kornberg, Zimmerman and Kornberg (1961).

enzyme with DNA containing hydroxy-methylcytosine with [^{14}C]UDP-glucose. Incorporation of ^{14}C into acid-insoluble DNA is measured. Three categories of enzyme have been identified (Fig. 7.5); these activities correlate well with the known frequencies of these side chains in the DNA of the different phage (p. 270).

DNA ligase   The reaction catalysed by this enzyme is illustrated on p. 18. The *E. coli* enzyme, which is the one employed for most synthetic work has as its high-energy

addition of Qβ RNA. They demonstrated the time-dependent incorporation of radioactive nucleoside triphosphates into Qβ RNA and the concomitant synthesis of units of infectivity as assayed by the spheroplast system. This experiment was the first demonstration of the synthesis of a biologically active and complete genome *in vitro*. Since then the Qβ system for complete RNA synthesis *in vitro* has been exploited in a most imaginative way for the partial sequencing of this RNA (see p. 316).

φX 174 DNA   The first complete DNA genome to be synthesized *in vitro* is that of phage φX 174 (Goulian, Kornberg and Sinsheimer, 1967). The authors established unambiguously the synthesis of both − strands and + strands separately (see p. 154). The DNA of φX 174 is a single-stranded continuous molecule. The experiment was performed using the DNA polymerase and the DNA ligase of *E. coli*. Both + strands and − strands can be assayed in the spheroplast system. The plan of the experiment is summarized in Fig. 7.6. For simplicity the + strand is shown inside the circle. [^{3}H] plus strand is initially used as a template for the polymerase, the substrates are [^{32}P] triphosphates dATP, dCTP, dGTP and d br^{5}UTP. There is no dTTP; the 5-bromouracil replaces thymine in DNA (see p. 000). The new − strand is shown as a heavy line in the figure, it is labelled with ^{32}P and is denser than the normal thymine-DNA (see p. 154). The new − circle is joined with ligase. The double-stranded product is now treated with very limited amounts of nuclease so that some molecules are unaffected (*a*), others have a break in the − strand (*b*) and others have a break in the + strand (*c*). The mixture is heated to 90°C so that the broken strands are peeled off the structures to produce single-stranded, denatured DNA molecules. When the mixture is centrifuged in a caesium chloride gradient (p. 154) the denatured molecules (which have no infectivity), the closed duplices and the closed single circles form separate bands. Moreover as the newly synthesized − strands are denser then the original + strands they too form a separate zone. This zone, which is used for the next stage, is the most dense of all; it is labelled with ^{32}P but not ^{3}H and it is infective. This isolated, newly

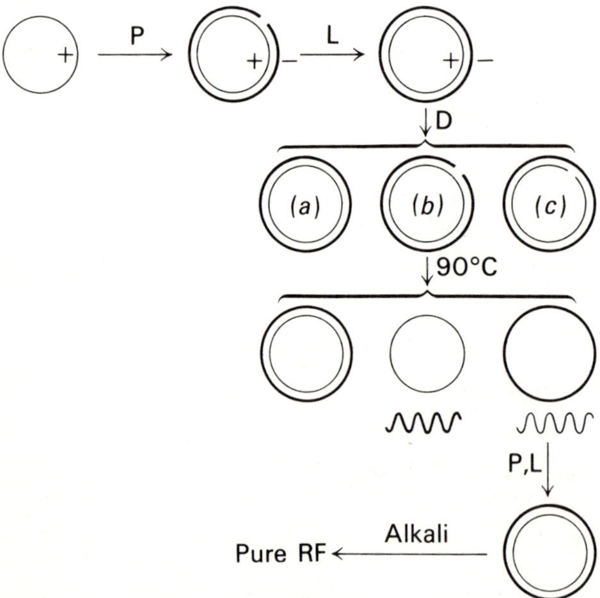

FIG 7.6   Replication of φX 174 DNA *in vitro*. See text for explanation. P = polymerase, L = ligase and D = pancreatic deoxyribonuclease. (From Goulian, Kornberg and Sinsheimer, 1967.)

synthesized — strand is now used as template for the next round of replication. [^{3}H] deoxyriboside triphosphates are employed this time and the product is treated with alkali (to denature any incomplete fragments and so on) and centrifuged through sucrose gradients; purified, infectious RF labelled with ^{32}P and brUra and ^{3}H is isolated from the gradient.

Mixed synthesis

Finally there are methods of synthesis which employ both chemical and enzymic procedures. These procedures have been used (notably by Khorana) to make defined repeating polynucleotides and to make defined non-repeating synthetic DNA.

Defined repeating polynucleotides One of the first applications of *E. coli* RNA polymerase was the synthesis of poly AU (alternating) using poly dAT (see p. 223) as template (Chamberlin, Baldwin and Berg, 1963).

The method has been extended to synthesize several synthetic mRNA molecules using as templates synthetic DNA produced by using short oligodeoxyribotides as templates for DNA polymerase. As a simple example, consider the synthesis by Khorana of poly (UG) and poly (AC) (Jones, Nishimura and Khorana, 1966). d(TG)$_6$ and d(AC)$_6$ were mixed and used as template for DNA polymerase. The enzyme does not stop when it gets to the end of such a template; the product is poly d(TG:AC). The mechanism for this type of reaction is discussed on p. 336. If this polymer is used as template for RNA polymerase, the products are the two complementary polyribotides. To avoid the fag of separating them, the enzyme can be restricted. If only UTP and GTP are supplied, just the poly d(AC) strand is transcribed (to yield poly (UG)); if only ATP and CTP are supplied the product is poly (AC). About the same time the same technique was employed by Cramer to prepare from poly d(TC:AG), poly (AC) and poly (GU) (Cramer *et al.*, 1966).

Many more examples of defined polymers have been studied by Khorana. A particularly interesting one is poly AUG (made by the transcription of poly d ATC:GAT; Ghosh, Söll and Khorana, 1967). When this substance is used as a synthetic mRNA, there are clearly three possible polypeptide products depending on the 'reading frame' chosen by the ribosome. The efficient choice of one particular frame, AUG, AUG, AUG, . . . (= Met Met Met) was greatly enhanced by the presence in the incubation mixture of fMet-tRNA$_F^{Met}$ (p. 24) and initiation factors. The initiating role of tRNA$_F^{Met}$ was thus directly demonstrated *in vitro*.

Synthesis of the yeast ala tRNA gene The greatest achievement to date in polynucleotide synthesis is Khorana's successful preparation of the DNA double helix, one of whose component molecules is the transcribing strand for yeast tRNA^{Ala} (see p. 24). The strategy of this synthesis is described by Agarwal *et al.* (1970). The plan consisted of annealing (see p. 355) preformed oligodeoxyribotides with overlapping ends and joining the segments together with DNA ligase (p. 231). Fifteen oligomers were made by the combination of stepwise synthesis and block condensation as illustrated earlier in the chapter; in fact the oligomer illustrated at the very bottom of Fig. 7.3, p. 217, is oligonucleotide 6 in the following scheme. The oligomers were all purified (both before and again after removal of the protecting groups) by chromatography on DEAE-cellulose using as eluants buffers containing strong urea (see p. 292).

The overall scheme for the synthesis is illustrated in Fig. 7.7. The details of the modified bases are not given in the tRNA sequence; the complete sequence of this molecule is given in Fig. 1.17, p. 24. The modified adenosines (xA) in Fig. 7.7 include I and mI as I in tRNA is produced by deamination of A. Likewise the residues xU include hU, Ψ and T. I have

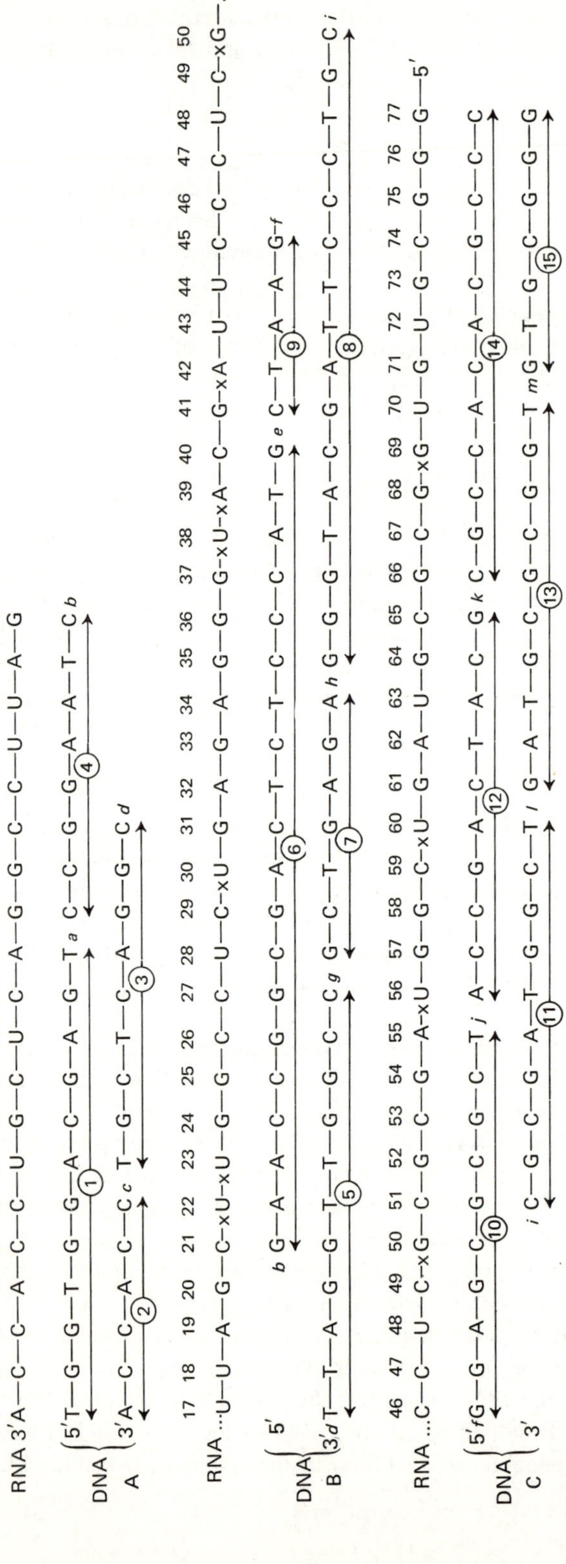

```
           1   2   3   4   5   6   7   8   9  10  11  12  13  14  15  16  17  18  19  20
RNA 3'A—C—C—A—C—C—U—G—C—U—C—A—G—G—C—C—U—U—A—G

DNA  ⎧5'T—G—G—T—G—G—A—C—G—A—G—T a  C—C—G—G—A—A—T—C b
     ⎨            ①                              ④
A    ⎩3'A—C—C—A—C—C c  T—G—C—T—C—A—G—G—C d
                  ②                    ③

        17  18  19  20  21  22  23  24  25  26  27  28  29  30  31  32  33  34  35  36  37  38  39  40  41  42  43  44  45  46  47  48  49  50
RNA ...U—U—A—G—C—xU—xU—G—G—C—C—U—C—xU—G—A—G—A—G—G—G—xU—xA—C—G—xA—U—U—C—C—C—U—C—xG—...

DNA  ⎧5'        b G—A—A—C—C—G—G—C—G—A—C—T—C—T—C—C—C—A—T—G e  C—T—A—A—G f
     ⎨                              ⑥                              ⑨
B    ⎩3'd T—T—A—G—G—T—T—G—G—C—C g  G—C—T—G—A—G—A h  G—G—G—T—A—C—G—A—T—T—C—C—C—T—G—C i
                  ⑤                    ⑦                              ⑧

        46  47  48  49  50  51  52  53  54  55  56  57  58  59  60  61  62  63  64  65  66  67  68  69  70  71  72  73  74  75  76  77
RNA ...C—C—U—C—xG—C—G—C—G—A—xU—G—G—C—xU—G—A—U—G—C—G—C—G—xG—U—G—U—G—C—G—G—G—5'

DNA  ⎧5'f G—G—A—G—C—G—C—G—C—T j  A—C—C—G—A—C—T—A—C—G k  C—G—C—C—A—C—A—C—G—C—C—C
     ⎨                  ⑩                    ⑫                              ⑭
C    ⎩3'        i C—G—C—G—A—T—G—G—C—T l  G—A—T—G—C—G—C—G—T m G—T—G—C—G—G—G
                          ⑪                    ⑬                    ⑮
```

FIG 7.7   Khorana's synthesis of the gene for $tRNA^{Ala}_{yeast}$. See text for discussion.

employed the same terminology as Khorana, including the numbering of the residues (1 to 77 in the 3′ to 5′ direction of the tRNA sequence; that is the 5′-terminal -G- is residue 77 and the 3′-terminal -C-C-A are residues 3, 2 and 1 respectively). The linkage positions for the ligase reactions (*a, b, c, d,* etc.) are my own nomenclature and are for reference to the rest of this section.

The synthesis was achieved by making the three large segments A, B and C, annealing them and joining up the breaks with ligase. The same overall process was employed for building the segments A, B and C themselves from the chemically synthesized oligonucleotides 1 to 15. The most straightforward synthesis was actually that of the longest segment, B. For the first stage, oligomers 6, 7, 8 and 9 were 5′-phosphorylated using the T4 polynucleotide kinase (see p. 228). They were then annealed (p. 355) and DNA ligase (p. 231) was employed to seal discontinuities *e* and *h*. The remaining oligomer (5) was 5′-phosphorylated and added to the segment and the break at *g* was joined with ligase. Segment B was purified (separating from non-annealed oligomers) by gel exclusion chromatography using Agarose.

Segment A was synthesized by a more complicated method designed to overcome the problem that oligomer 3 contains at its 5′-end residues (13 to 16) which can (and in fact do) form a duplex with another molecule of itself so that when oligomers 1, 2, 3 and 4 are annealed the product is not A but free oligomer 4 and a dimer of segments 1, 2 and 3. It is easy to envisage (or sketch) the structure of this dimer. Imagine two segments A without oligomer 4, one drawn as in Fig. 7.7 the other one 'upside down'. The two fit together perfectly to produce a segment containing 28 base pairs. Starting from one end of the structure there are base pairs corresponding to positions 1 to 12 followed by the complementary overlap in oligomer 3 (covering positions 13, 14 and 15 and 16 in one chain and 16, 15, 14, 13 in the other) and then positions 12 to 1. This type of self-complementarity in a DNA molecule is known as 'sticky-endedness'. In large DNA molecules it is a type of binding that produces closed molecules (see p. 363). In this case the consequences of the effect have to be overcome. It was done by first annealing 5′-phosphorylated oligomers 1, 4 and 3. The single break (*a*) was joined (ligase); the structure was then melted, the strands separated and oligomer 2 was annealed to form the double-stranded structure in positions 1 to 5. When 3 is added to this, it anneals faithfully and break *c* is joined with ligase. In this way no double-stranded segment with the sticky end of 3 exposed is ever given a chance to form. A was purified in the same way as B.

Segment C was made in a similar fashion (again using 5′-phosphorylated oligomers and ligase) except in this case trouble due to the formation of stable but mis-matched complexes was encountered. Segment C is very rich in G and C so that it can form stable conformations in which short sections of double-stranded structure link denatured (and non-complementary) sections. The problem was overcome by first annealing oligomers 14 and 15 carefully and using this as the matrix to build up the rest of the segment except for oligomer 10 which was left out. So oligomers 11, 12 and 13 were added and breaks *k, l* and *m* were joined.

The complete gene was made in two ways; either by joining A and B and then adding C or by joining B and C and then adding A. In the latter case, B, C and oligomer 10 were annealed and the breaks (*f, i* and *j*) were joined. The product was purified by Agarose chromatography. Segment A was added and following the annealing and joining of breaks *b* and *d*, the product was finally chromatographed on Agarose. The evidence for the reactions involved in the annealing and joining of the oligomers was carefully assessed by following the kinetics of the annealing, the use of ^{32}P (or ^{33}P) to label the 5′-phosphate groups and to show that following ligase, these groups are insensitive to phosphomonoesterase (see p. 231) and by degrading the sections with exonucleases.

What is the future of this type of work and what is its status? Is it no more than an enormous chemical *tour de force* or does it herald a new era in biological research? The future of this particular project must depend on the feasibility of efficiently replicating the structure *in vitro* to ensure future supplies. In principle the transcription of this structure and of extended synthetic genes (with the structural gene surrounded by possible signals for the starting and stopping of RNA polymerase and or protein synthesis) should enable experiments on the control of these processes by DNA sequences and RNA sequences and conformations (see p. 19) to be performed in a direct and unambiguous way. Nevertheless it must be admitted that the amount of time, money and chemical expertise required to follow through a project of this type is so enormous that the chemical synthesis of genes must remain the territory of a small elite. On the other hand the great value of these complex programmes is that if they are carefully and exactly described, technical methods that were devised to overcome problems in such an undertaking can be of use in other fields. In this respect, Khorana cannot be faulted. He has kept the informed readership of the technical literature completely up-to-date with a series of elegantly written papers full of technical detail. The most important aspect of the techniques devised in the present programme must surely be the annealing of DNA fragments with only quite short overlapping regions (such as the four base pairs (numbers 17 to 20) that join segments A and B).[3] This could in principle be adapted to join either isolated genes (see p. 366) or synthetic genes or other DNA fragments on to such DNA molecules as transforming or transfecting DNA. This really is a vision of a new realm of molecular genetics.

Notes to chapter 7

[1]   This is an example of a 'difficult one'. G-residues are particularly hard to protect selectively in this type of synthesis.

[2]   This is the reference number of the American Type Culture Collection. This strain is known to biochemists as *M. lysodeikticus* but recent taxonomic work places it in the species *M. luteus* (ATCC Catalogue, 1968).

[3]   See Gupta, Ohtsuka, Weber, Chang and Khorana (1968) for details of this step.

8 Chemical reactions

The chemical reactions of nucleic acids are used in structural studies and references to these applications are given in the later chapters. The biological consequences of certain reactions are extremely important, partly because of the use of chemicals and radiation for the production of mutations in micro-organisms and partly because the reactions with nucleic acid are an important feature of the cytotoxic and carcinogenic activities of certain chemicals.

The only processes that are described here are reactions in which covalent bonds are formed and/or broken. The association of molecules with nucleic acids by other forms of bonding is deferred until chapters 14 and 16.

Chemical reactions 'in vitro'

Reactions of sugars and phosphates

Methods of hydrolysing N-glycoside or phosphate ester bonds are discussed in chapter 5. The other characteristic reactions of these groups are of little practical importance with the following exceptions.

Derivatives of the phosphate groups Diazomethane, which reacts predominantly with the bases (p. 60) also methylates the phosphate esters to some extent to produce methyl derivatives which are phosphotriesters. Unsymmetrically substituted phosphotriesters are more easily hydrolysed than are phosphodiesters. The reactions have been studied by Holy and Scheit (1967).

Primary phosphates have characteristic reactions; one of these, the amidation of terminal phosphate in nucleic acids has been examined by RajBhandary, Young and Khorana (1964) but has not caught on as a method for end-group determinations (p. 289).

Reactions of the sugar moieties The 2′-hydroxyl groups in RNA can be acetylated; Stuart and Khorana (1964) describe the conditions. The reaction is not a particularly useful one. Not surprisingly the double helical structure is destabilized (Knorre and Shamovsky, 1967).

The *cis*-glycol group of ribose is present in the 3′-terminal nucleoside of RNA. The periodate ring-opening reaction (p. 72) can therefore be applied to RNA. Applications of this reaction to determining the identity of this nucleoside are discussed on p. 287. The oxidation of the 5′-CH_2OH group of dThd (p. 74) can be applied to the 5′-end of oligodeoxyribotides (p. 281).

Reactions of the bases

A systematic account of the reactions of the bases has been given in chapter 2. The reactions of bases in nucleic acids are broadly the same as the reactions of the low-molecular-weight compounds. In this chapter the reactions are grouped in a way which is chemically less systematic but is of more relevance to the biological consequences of the reactions, which are discussed in the later sections of the chapter. For completion however, I have summarized first some reactions, which although of little biological significance (as the reagents are so non-specific) are employed in structural studies on nucleic acids and will be referred to later.

Miscellaneous reactions   The reactions of halogens and HOBr (see p. 59) are of such low specificity that the reaction with nucleic acids is of virtually no interest. The hydrolysis of N-glycoside bonds and phosphodiesters has been discussed on p. 55. The selectivity of the acid hydrolysis for purine nucleosides is of use in partial degradative studies in DNA (p. 275).

Nitrous acid and hydroxylamine   The reaction of nitrous acid with nucleic acids is a seemingly much more complex process than one might expect. It is probable that the reactions of p. 66 do not account for all the processes that occur in nucleic acids. Moreover the reactions are limited by conformational and other factors. Thus although, of the common nucleosides, the most reactive one is Guo, the G-residues in tobacco mosaic virus (TMV) are untouched by the reagent (Schuster and Wilhelm, 1963). When the experiment was performed with isolated TMV RNA, a significant portion of the G-residues were converted to an unidentified derivative. Perhaps the reaction would be re-examined in the light of the proposal (p. 67) that a diazonium ion is an intermediate in the reaction. The relationship of this process to mutagenesis is discussed on p. 353.

The reactions of hydroxylamine and methoxyamine with pyrimidines has been discussed (p. 65). In nucleic acids the reactions both with C- and U- (or T-) residues can be studied (pp 290 and 353). However there is every reason to suppose that the conditions used for mutagenesis are ones in which the reaction with U (or T) is of no significance. As these substances are potent mutagens the nature of the products obtained from the somewhat complex sequence of reactions between these reagents and Cyd in a polynucleotide is of great interest. Phillips, Brown and Grossman (1966) have analysed the reaction between methoxyamine and poly C in detail. If the reaction is allowed to continue for a long time (3 h) and the yield of modified C-residues is assayed (in aliquots removed at intervals) it is found that the total percentage of modified C-residues increases linearly with time but that of the two types of modified residue (see below), I reaches a steady state (about 1%) after 2 h and that II lags and then increases linearly, implying that I is precursor for II. This is in agreement with the assumptions made about the reaction between methoxyamine and Cyd (p. 65).

I

II

Alkylation reactions   All the alkylating agents described on p. 60 react with DNA and RNA. The reactions are measured by incubating the nucleic acid with the reagent and then hydrolysing the product and measuring the proportions of alkylated bases. For reagents that effect considerable levels of alkylation, the products can be identified chemically. Examples of such reagents are the sulphur- and nitrogen-mustards, methyl methane sulphomate and dimethyl sulphate. The main products are the 7-alkylguanines (Brookes and Lawley, 1961). Otherwise the reaction is performed with radioactive alkylating agents and the products are identified on the basis of their co-chromatography with authentic samples. Thus if DNA has been incubated with a radioactive methylating agent, the product is digested with acid and radioactivity is measured in the material that co-chromatographs with 7-methylguanine. If the alkylating agent is of the nitrosamide type (like NMG p. 60 or methyl-N-nitroso urea) the alkylated bases will include O^6-methylguanine (see p. 68). As this substance is acid-labile, the product is digested enzymically to nucleosides (p. 205) and the O^6-methylguanosine is identified (Lawley and Thatcher, 1970).

Not that guanine is the only target for alkylation in nucleic acids. In adenine the nitrogen atom with the highest negative charge (and hence the one most likely to react with electrophilic agents) is N-1 (see Fig. 2.16, p. 76). This prediction is confirmed when studying the alkylation products of adenine (p. 60) or the adenine residues in RNA or poly A (Ludlum, 1965). However in a Watson-Crick base-pair N-1 is involved in H-bonding (see Fig. 1.5, p. 8). Thus in native DNA or in poly A:U the major alkylated adenine is 3-methyladenine (Lawley and Brookes, 1963). In methylated DNA there is always some 7-methyladenine (Lawley and Brookes, 1964). The reactive site for alkylation in the pyrimidines is N-3 (see the electronic distributions in Fig. 2). With alkylating agents at neutral pH, the only reactive site is the N-3 of C-residues. As this atom is involved in H-bonding in the Watson-Crick structure (as with N-1 of adenine above), there is very little pyrimidine-alkylation in DNA. The product can be identified in enzymic digests alkylated RNA, as can 3-methyluridine; this latter product is obtained as consequence of the base-catalysed deamination of 3-methyl-C-residues (Lawley and Brookes, 1963; Brookes and Lawley, 1964; Kriek and Emmelot, 1964).

Cross-linking reactions   It has been pointed out on p. 60 that bi-functional alkylating agents, such as the mustards can react with two nucleosides (such as Guo) to link them together. Let us take as a simple case the two sulphur mustards full mustard ($Cl(CH_2)_2S(CH_2)_2Cl$) and half mustard ($Cl(CH_2)_2S(CH_2)_2OH$).[1]

If DNA or RNA is treated with half mustard in buffer at pH 6-8 and acid-hydrolysed (p. 205) the 7-alkylated product is obtained. Provided an excess of mustard is employed, it is not necessary to employ [^{35}S] half mustard to detect the substance. If the experiment is performed with full mustard then both the above product and a smaller amount of *bis* guaninylethyl sulphide are obtained. These products are shown in Fig. 2.8, p. 61.

These bases are easily resolved by paper chromatography using the acid system described for DNA bases on p. 48 as the *bis* guaninyl-mustard has an R_f-value of zero in this system whereas the guaninyl-mustard has an R_f of 0·4. Some care is needed in interpreting such a chromatogram as minor sulphur-containing bases are also present in many of the acid digests. These compounds are produced by the oxidation of the sulphide group and are presumed to be sulphoxides and sulphones (Brookes and Lawley, 1960). The interesting question is how are two G-residues which react with the same mustard molecule related to one another in the original structure. By an analysis of the three-dimensional structure of the B-form of DNA (see p. 9), it is apparent that there are two types of environment in which the distance between the N-7 atoms of two guanine moieties is about right for the two atoms to be spanned by a mustard molecule without any significant distortion of the geometry of the helix. These positions are shown diagrammatically below where -m- represents a portion of the mustard molecule (full mustard would be represented as Cl-m-Cl).

$$5'-\text{N}-\text{N}-\text{G}-\text{C}-\text{N}-\text{N}-\text{N}-\overset{\overset{\displaystyle m}{\diagup\;\diagdown}}{\text{G}\quad\text{G}}-\text{N}-\text{N}-\text{N}-\text{C}-\text{G}-\text{N}-\text{N}-3'$$

$$3'-\text{N}-\text{N}-\text{C}-\underset{\underset{\displaystyle m}{\diagdown}}{\text{G}}-\text{N}-\text{N}-\text{N}-\text{C}-\text{C}-\text{N}-\text{N}-\text{N}-\text{G}-\text{C}-\text{N}-\text{N}-5'$$

(1)   (2)   (3)

Thus there are two possibilities in which either the two G-residues are present in -G-C-sequences in opposite chains (inter-strand cross-linking, 1 above) or are present as a -G-C- sequence (intra-strand cross-linking, 2 above). If the model is studied it is found that G-residues present in opposite strands in -C-G- sequences (position 3 above) are too remote from one another and interstrand cross-linking is not possible.

What is the evidence for inter-strand cross-linking (as in 1 above)? The criterion that is used is the effect of such links on the denaturation of the nucleic acid (see p. 356). We shall see on p. 268 that when DNA in solution is heated, the denaturation is a co-operative phenomenon and that over a small temperature range, the molecules change from a duplex to a random-coil conformation; this effect is monitored as hyperchromic shift at 260 nm. When DNA solutions are cooled down, the absorbance at 260 nm approached the original (hypochromic) value. However in general, the value that is achieved is still higher than that of the original solution (see p. 363). The reason is that a metastable form is produced in which some base-pairing has been produced as a result of the artefactual 'annealing' of regions of partial complementarity. However if the molecules are cross-linked, the denatured structures are still joined together in the original duplex pairs and subsequent annealing is faithful. Consequently the reversal of denaturation results in a return to the original conformation and hence the original absorbance. The transitions are illustrated schematically (for very short DNA fragments) in Fig. 8.1.

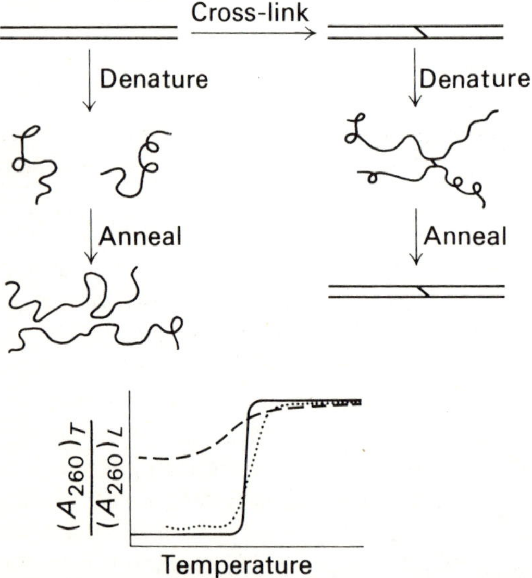

FIG 8.1   Schematic representation of the effects of inter-strand cross-linking on the denaturation and renaturation of DNA. The mustard linkers are shown as — (see text). In the upper part of the figure is the denaturation and renaturation of control and cross-linked molecules and in the lower half is the way in which these changes are reflected in the optical properties of the solution. In this graph the ratio of the absorbance at temperature T° $(A_{260})_T$ to the absorbance at low temperature $(A_{260})_L$ is plotted as a function of T: ——— during the heating of either sample, - - - - - - during the cooling of the control sample and . . . . during the cooling of the cross-linked sample.

The two types of experiment (identification of alkylated bases in hydrolysates and annealing studies) have been compared by Lawley and Brookes (1967). They employed [35S] full mustard and alkylated preparations of DNA of known weight-average molecular weight (p. 387). Using paper chromatography to analyse mixtures of the [35S]-labelled bases, following treatment of salmon-sperm DNA in aqueous buffer and subsequent acid hydrolysis, they found that of the total alkylated bases, 22% was the *bis* guaninyl compound and the remainder were single-alkylated products (substituent was therefore $-(CH_2)_2S(CH_2)_2OH$) as follows: 7-substituted Gua 59%, 1-substituted Ade 4%, and 3-substituted Ade 15% SCH. If the buffer was replaced with water containing methanol to produce denatured DNA (see p. 10) then the amount of the *bis* guaninyl compound was reduced and the relative amount of the 1-substituted Ade was increased (the reason for this is given on p. 239). In separate experiments an attempt was made to quantitate the relationship between inter-strand cross-linking and the effect on re-annealing. With *E. coli* DNA they found that an average of one *bis* guaninyl derivative per molecule was sufficient to produce reversible denaturation implying that (at least with these low levels of alkylation) intra-strand cross-linking was quantitatively of little importance. The levels of alkylation required to effect the reversibility of denaturation of salmon-sperm or rat-liver DNA was about ten times this figure. This is not taken to mean that intra-strand cross-linking is quantitatively more important in these cases however. There is far more partial complementarity between different stretches of eukaryotic DNA than is the case with bacterial DNA (see p. 365). In other words the chances are that the sort of mis-pairing that is illustrated in Fig. 8.7 can occur within regions of the same helix in a single DNA molecule.

An interesting technical point that arises in Lawley and Brookes (1967) is that prolonged heating of DNA cross-linked produces a non-renaturable structure, apparently because of a type of depurination reaction as illustrated below (d = deoxyribose).

$5'-N-N-N-G-C-N-N-3'$ $5'-N-N-N-d-C-N-N-3'$
$\quad\quad\quad\quad\quad \backslash$
$\quad\quad\quad\quad\quad\quad m$ $\longrightarrow$ $+$
$\quad\quad\quad\quad\quad\quad \backslash$
$3'-N-N-N-C-G-N-N-5'$ $3'-N-N-N-C-G-N-N-5'$
                                               $m-Gua$

Do all bi-functional alkylating agents cross-link DNA? Provided the molecular parameters are about right, they apparently do. The substance $Cl(CH_2)_2NMe(CH_2)_2Cl$ ('nitrogen mustard' or HN2) has exactly the same distance between its two outside carbon atoms as has sulphur mustard. This reagent has the same effect on the reversibility of DNA-denaturation as has sulphur mustard. Historically it was the first reagent for which this effect was demonstrated (Geiduschek, 1961). The reaction differs in certain respects from that with sulphur mustard. A higher proportion of *bis* guaninyl derivatives is obtained and these derivatives are even formed with denatured DNA (Doskočil and Šormová, 1965). HN2 is a less reactive substance than sulphur mustard and the intermediate in the reaction, viz. $Cl(CH_2)_2NMe(CH_2)_2-G$ presumably has a significantly longer half-life and has time to react with other G-residues before it is hydrolysed. Lawley and Brookes (1967) showed that the effects of sulphur mustard on the reversible denaturation of DNA could be mimicked by two other substances, butadiene di-epoxide and triethylenemelamine. The structures of these two compounds are given below.

Both of these compounds raise interesting problems. It is easy to see how the former substance might act as a bi-functional alkylating agent to link G-residues together. The *bis* guaninyl derivative has the structure G–CH$_2$–CHOH–CHOH–CH$_2$–G. However the 'bridge' between the two N-7 atoms in this case involves only four atoms. Study of the three-dimensional model of DNA implies that some distortion of the helix is required to effect this reaction. Triethylenemelamine raises problems of a different sort. There is no satisfactory account of the reaction of the compound with nucleotides. However Lawley and Brookes construe that the reaction with DNA involves G-residues. This being the case there is no steric objection to the reaction. The reaction envisaged is one in which two of the ethylene imine groups have reacted to produce structures G–CH$_2$–CH$_2$–NH–X in which X represents the heterocyclic part of the triethylenemelamine residue. Although there are nine atoms spanning the two N-7 positions of the G-residues, there is no stereochemical difficulty as the free rotation about the single bonds in the –CH$_2$–CH$_2$–NH– part of the structure allows conformations in which the N-7 to N-7 distance is the same as that in the double helix.

And RNA? The cross-linking of RNA is less well studied. Preliminary work of Brookes and Lawley (1961) demonstrated that cross-linking does occur when sulphur mustard reacts with RNA. Our own results (Malbon and Parish, 1971) have established that at least in the case of the highly alkylated RNA, that the effect of alkylation is to restrict denaturation so that in the case of mustard-treated 16 S rRNA from *E. coli* there is no apparent hyperchromicity on heating. In this case we assume that the cross-linking is between G-residues in the 'stalks' of the structure (see p. 10). It is difficult to perform the reversible denaturation experiments on rRNA as the molecule renatures readily anyway.

There is evidence of cross-linking of a type not yet chemically identified when DNA is treated with various other reagents, notably nitrous acid (Becker, Zimmerman and Geiduschek, 1964).

A special type of cross-linking is produced in RNA containing thiolated residues. Chemically the simplest is the oxidation of these residues to form disulphides as below (for 4-thioU-residues). Clearly this reaction can only occur if there are two thioU-residues in the same

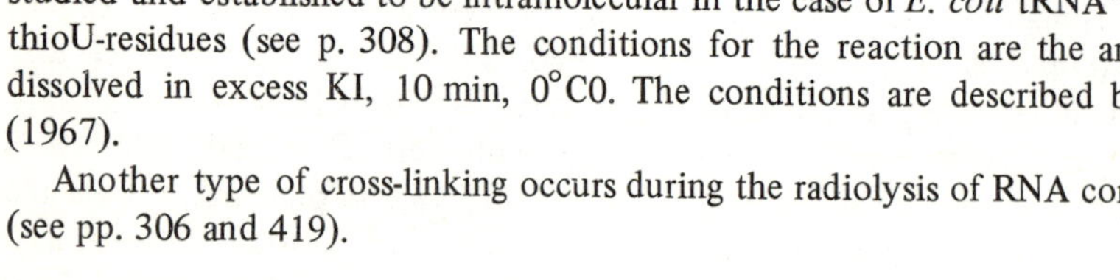

molecule (unless intermolecular cross-linking of this sort is possible). The reaction has been studied and established to be intramolecular in the case of *E. coli* tRNA^{Tyr} which contains two thioU-residues (see p. 308). The conditions for the reaction are the anion I$_3{}^-$ (1 mM iodine dissolved in excess KI, 10 min, 0°C0. The conditions are described by Lipsett and Doctor (1967).

Another type of cross-linking occurs during the radiolysis of RNA containing thioU-residues (see pp. 306 and 419).

Ultra-violet radiation and polynucleotides

Luminescence and electron spin resonance   The luminescence and electron spin resonance spectra of low-molecular-weight components of nucleic acids have been discussed in chapter 2 (pp. 87 and 99). These spectral properties of polynucleotides has been used to answer two questions: what are the major molecular contributions to the triplet state of DNA? and can the properties of excimers be exploited to probe the secondary structure of nucleic acids? This second question is deferred until chapter 11 when the secondary structures of nucleic acids are considered. The former question is relevant to the present discussion as the interaction of radiation with DNA has profound chemical (and biological) consequences and we must enquire into the nature of the initial molecular perturbations that give rise to these reactions.

Two lines of evidence that the UV-irradiated DNA contains thymine free radicals (with the unpaired electron on C-5, see p. 99). The most direct evidence comes from a study of the ESR spectrum. The first derivative spectrum resembles (in a rather rough-and-ready way) that of thymine (see Fig. 2.27, p. 100). We saw on p. 100 that the interpretation of the thymine spectrum could be simplified by using methyl-deuterated Thy so that much of the hyperfine splitting is eliminated and a simple triplet is produced. The same triplet is observed in the ESR spectrum of irradiated DNA in which all the Thy-methyls are deuterated (Eisinger and Shulman, 1963).[2]

The second line of evidence is the phosphorescence spectrum of DNA. The spectrum shows a broad envelope with a maximum around 440 nm. This does not resemble the spectrum of any of the nucleotides but the Thy and its derivatives do not have any phosphorescence except in alkali (see p. 88). However solutions of (say) TMP can be sensitized by a substance such as acetone. The effect of such a 'sensitizer' is to populate the triplet state of the TMP by triplet-triplet energy transfer (with the triplet state of acetone acting as an intermediate donor). This type of sensitization is only effective if a triplet state with lower energy than the first excited state (see p. 87) really does exist. Normally the rate at which singlet states change to triplets is very low. In this context however, the important thing is that the phosphorescence emission spectrum of sensitized TMP is very similar to that of DNA (Lamola, Guéron, Yamane, Eisinger and Shulman, 1965). Hence it appears the triplet state of Thy in DNA is similar to that in TMP.

Photocatalysed reactions of DNA and RNA   In chapter 2 we noted two UV-catalysed reactions of pyrimidines: the self-addition of thymines to yield the thymine dimers (p. 64) and the photohydration of cytosine (p. 63). The first of these reactions is of great importance in the photochemistry of DNA. The reaction occurs between two thymines adjacent to one another in a strand of double helical DNA. The reaction is a UV-catalysed equilibrium, usually represented thus:

$$\cdots -T-T- \cdots \quad \underset{}{\overset{h\nu}{\rightleftharpoons}} \quad \cdots -\overline{T-T}- \cdots$$

The reaction is not the only one that occurs in UV-irradiated DNA. The other types of pyrimidine dimers are produced as well; sequences containing them are represented as $\ldots -\overline{C-T}- \ldots$, $\ldots -\overline{T-C}- \ldots$ and $\ldots -\overline{C-C}- \ldots$. These features can be analysed by acid hydrolysis of the DNA and chromatography.

The resulting cyclobutane bases are frequently abbreviated as $\overline{TT}$, etc. This is clearly inappropriate in the context of the present rules for abbreviated structures (see p. 5) and I depict them as Thy□Thy, etc. Chemically these materials are rather badly characterized. It is certain that Thy□Thy isolated from irradiated DNA is the *cis syn* compound (see Fig. 2.9g, p. 64) as the structure has been unambiguously established (Blackburn and Davies, 1966). Presumably the others have the same configuration. However Cyt□Thy and Thy□Cyt (which are not the same compound) have not been resolved so far.

Cyt□Cyt and Cyt□Thy are acid-labile. This is generally true of 5,6-saturated cytosine compounds (see p. 68). Therefore the best procedure is to convert the cytosine components of dimers to the corresponding uracil derivatives prior to hydrolysis. In the procedure of Setlow and Carrier (1966) the DNA solution is irradiated and then heated to 100°C for 5 min. This converts the cytosine dimers to Ura□Ura and Ura□Thy. The DNA is hydrolysed with formic acid (p. 205) and analysed by paper chromatography, developed with a mixture of n-butanol, acetic acid and water (80:12:30 by volume). The order of elution (from the start towards the solvent front) is Ura□Ura, Ura□Thy, Thy□Thy/Cyt (unresolved), Ura and Thy.

Alternatively prpyrimidine-dimer formation can be followed spectrophotometrically. Thymine dimer has no absorption at the maximum for Thy at 270 nm and so dimer formation can be followed as drop in extinction at this wavelength. The spectroscopic properties of saturated derivatives of Cyt are more complex; thus in 5,6-dihydrocytidine there is no peak at the Cyd-maximum (280 nm) although there is extinction at that wavelength and there is a peak (present only as a shoulder in Cyd) at 240 nm (Setlow, 1966).

The position of the equilibrium between normal sequence and dimers (illustrated for thymine dimers on p. 243) depends on the wavelength of irradiation. However at constant wavelength a dynamic equilibrium is established. At wavelengths in the region of the first $\pi \to \pi^*$ (260-270 nm) transition of pyrimidines (see p. 84) the equilibrium favours dimer formation, as the dimers, which are saturated compounds, do not absorb UV in this region. At lower wavelengths (around 235 nm) there is absorption by the dimers (characteristic of their carbonyl groups) and the equilibrium favours monomerization. Additionally the yield of dimer in the forward reaction or the extent of monomerization from dimers (the back reaction) is dependent on the UV flux (usually measured in erg/nm^2). The reactions have been carefully analysed by Wulff (1962). If $E.\ coli$ DNA is irradiated at 275 nm, up to 20% of the thymines produce dimers (the values plateau at around 20 erg/nm^2). At 235 nm the plateau (only of the order of 1% of the thymines) is achieved at 4 erg/nm^2. Moreover if the highly irradiated DNA (20% thymines dimerized) is irradiated at 235 nm then the product is monomerized to produce the 1% level. These processes do genuinely reflect the equilibration of these forms; this was confirmed by giving native DNA a very short dose (2 erg/nm^2) of radiation at 275 nm so that equilibrium was not reached and only about 0·5% of the thymines were dimerized; when this material was irradiated at 235 nm, the dimerization actually increased until the plateau level (about 1%) for the '235-equilibrium' was reached.

Setlow and Carrier (1966) surveyed the effect on the dimerization reaction (1) of DNA nucleotide composition and (2) of the wavelength of irradiation. Their results (Table 8.1) demonstrate that both these factors considerably affect the distribution of products in the dimer mixture.

TABLE 8.1   Pyrimidine-dimer formation in DNA. Data taken from Setlow and Carrier (1966). See text for discussions.

Source of DNA	AT-content (%)	Wave-length (nm)	Flux (erg/nm^2)	Proportion of dimers (%)	Proportion dimer types (%)		
					$\overline{C\text{-}C}$	$\overline{C\text{-}T}$ / $\overline{T\text{-}C}$	$\overline{T\text{-}T}$
Haemophilus	62	265	2×10^3	0·27	5	24	71
influenzae		280	2×10^4	2·06	3	19	78
E. coli	50	265	2×10^3	0·20	7	34	59
		280	2×10^4	1·34	6	26	68
M. lysodeikticus	30	265	2×10^3	0·14	26	55	19
		280	4×10^4	0·69	23	50	27

phosphate reactant NAD^+. The overall reaction is as below, where ... N-N-N-N ... and the other molecule of the same form are the two strands of a DNA double helix.

$$5'...N-N-N-N \quad N-N-N-N-... 3' \qquad 5'...N-N-N-N-N-N-N-N-... 3'$$

$$+ NAD^{\oplus} \longrightarrow$$

$$3'...N-N-N-N-N-N-N-N-... 5' \qquad 3'...N-N-N-N-N-N-N-N-... 5'$$

with a P joining above the first strand, plus

$$NMN^{\oplus} + AMP$$

There are several methods of assaying the enzyme (see p. 232 for a method employing phage DNA). The simplest reaction is to obtain a substrate of the above type in which the free 5'-phosphate is radioactive and to demonstrate that the ^{32}P originally present as a phosphomonoester (and consequently sensitive to alkaline phosphatase) is converted into a phosphodiester (and consequently becomes insensitive to this enzyme). Suitable substrates can be synthesized by introducing a few single-strand breaks into DNA with pancreatic deoxyribonuclease and then labelling the 5'-phosphomonoesters (by removing these groups with alkaline phosphatase and then putting them back again with $[^{32}P]$ATP and poly-nucleotide kinase, p. 228). A more satisfactory method is to use a synthetic substrate made by annealing stretches of one homopolymer with longer stretches of its complementary polymer. This was the method employed by Olivera and Lehman (1967) in the work on the enzyme that established that NAD is the co-factor. Poly dT was partially hydrolysed with Staphyloccal nuclease (p. 208); the progress of the hydrolysis was assessed by phosphorylation with polynucleotide kinase and lengths of around 250 nucleotides were obtained. Their 5'-ends were dephosphorylated (alkaline phosphatase) and re-phosphorylated ($[^{32}P]$ATP) and were then annealed to poly dA. The assay mixture consisted of 10 mM tris HCl pH 8·0, 2 mM $MgCl_2$, 1 mM EDTA 1·6 mM (mean residue concentration) of poly dA and 1·6 mM of poly dT prepared as above; thus the mixture was about 10^{-8} M with respect to $[^{32}P]$ 5'-termini. The mixture was incubated with enzyme preparation plus either NAD^+ (or boiled *E. coli* extract before the nature of the co-factor was established). After 30 min incubation at $37°C$ the preparation was treated with 5 units of alkaline phosphatase. One unit of ligase is defined as the amount of enzyme that converts one pmole of ^{32}P into a phosphatase-stable form under the conditions of the assay.

DNA ligase is found in all tissues and cells and several phage-specific ligases have been studied. An important recent discovery (Mizutani, Temin, Fodana and Wells, 1971) is that a ligase (together with a virus-specific exonuclease) is found in Rous sarcoma virus particles. This is one of the oncogenic RNA viruses which produces an RNA-dependent DNA polymerase (p. 224). Presumably the nuclease and ligase are involved in the integration of DNA complementary to the viral RNA into the chromosomes of the cell.

Total synthesis *in vitro* of viral nucleic acid

As there is a spheroplast assay available for both the RNA from RNA phage and the DNA from DNA phage (p. 193) it has been possible to demonstrate the synthesis *in vitro* of two complete viral genomes (one of each type). As these experiments are landmarks in the story of research into nucleic synthesis, they are briefly described here.

Qβ RNA   Some time before the Qβ replicase enzyme and its subunits had been characterized (p. 227), Haruna and Spiegelman (1965) demonstrated that a cell-free system with *E. coli* infected by the phage could synthesize Qβ RNA. The system was absolutely dependent on the

The same effect can be demonstrated using transforming DNA. We shall see shortly that transformation has been successfully employed to clarify various features of the interaction of X-irradiation of DNA. In the present context the experiment consisted of correlating transforming activity of DNA irradiated *in vitro* with the other parameters of inactivation. As the monomerization of dimers in the sample was followed spectrophotometrically it followed that high levels of dimerization and consequently large amounts of irradiation were required. For these reasons it was essential to choose a radiation-insensitive marker for the transformation assay. The marker that was used was resistance to the antibiotic cathomycin in *Haemophilus influenzae*. Transforming DNA from a cathomycin-resistant strain was irradiated according to various schedules and assayed for transforming activity. The results resembled those of Wulff (above). Irradiation at 280 nm produced a considerable reduction in transforming activity. Following irradiation at 280 nm for up to 5×10^4 erg/nm^2 (transforming activity reduced to 0·03%) the DNA was exposed to radiation at 239 nm. Partial recovery of transforming activity was achieved. The kinetics of this recovery exactly followed the kinetics of the increase in absorbance at 273 nm (no monitor production of thymine from dimers). Following initial irradiation at 239 nm, inactivation of transforming activity formed a pleateau at about 1% of the control level. Subsequent irradiation at 280 nm produced further reduction in the transforming activity. Quantitation of the data implied that about one-half of the biological inactivation of transforming activity could be ascribed to pyrimidine dimerization (Setlow and Setlow, 1962, 1965).

Several additional factors affect the pyrimidine dimerization reaction. One of these (the effect of acridine dyes on the equilibrium) is discussed on p. 411. Dimer-formation occurs most readily in double-stranded DNA as the adjacent pyrimidines are in the correct position to react. In other polynucleotides the situation is more complicated. Thus in poly U, dimer-formation is co-operative, that is when a dimer has formed, the production of another dimer in the chain is facilitated (see p. 334). The reaction is extremely difficult to analyse as there is another complication; the formation of photohydrates (see Fig. 2·9f, p. 64 for the structure of the cytosine photohydrate) competes with dimer-formation and effectively removes pyrimidines from the reaction (Swenson and Setlow, 1963).

Photocatalysed reactions other than dimerization   UV-irradiation of RNA in the presence of NaBH$_4$ results in conversion of U-residues into hU-residues and consequent ring-opening (Cerutti and Miller, 1967; p. 69). Additionally UV-irradiation cross-links ^{4}sU- and C-residues; this reaction has been studied in detail in the case of tRNA^{Val} from *E. coli* (see p. 421). Another reaction (that also produces cross-linking) is the photocatalysed reaction between DNA and psoralen; this reaction is discussed on p. 420.

X-radiation and DNA

The effect of X-rays on DNA in solution is understood in less chemical detail than is effect of UV-irradiation of DNA. Additionally it is uncertain to what extent the effects of X-rays on DNA *in vitro* can be regarded as analogous to the reaction *in vivo*.

The available data have been collated by Szybalski (1967) who has proposed a unified model for the reactions.

Three types of process are envisaged. (1) is the direct interaction of X-irradiation with DNA. The effect of this is to produce double-strand breaks in the sugar-phosphate backbone thus:

```
. . . -N-N-N-N-N-N-N- . . .   . . . -N-N-N-   N-N-N-N- . . .
. . . -N-N-N-N-N-N-N- . . .   . . . -N-N-N   -N-N-N-N- . . .
```

(2) is the reaction of a radical produced in the water which then reacts with DNA to produce either double-strand breaks (as above) or single-strand breaks (to produce a structure similar to that shown on p. 231). (3) is the reaction of a radical with a base to effect a base-modification.

The radical-reactions (2) and (3) differ in that the (3) is O_2-dependent. This does not necessarily mean that the radicals (OH or something of the sort) are actually different in these two situations. It may mean that the oxygen is required to promote the reaction between radical and base. Be this as it may, reaction (3) is inhibited in strictly anaerobic conditions or in the presence of a mercaptan. Reaction (2) is inhibited by 'radical scavengers' either mercaptans or histidine (and hence any histidine-containing substance such as broth[3]).

In solution in distilled water or buffer without scavengers, the major reaction is double-strand breakage via process (2). In histidine-scavenged solution reactions (1) and (3) occur with roughly equal frequency. Freifelder (1966b) examined the effect of such radiation on the infectivity of phage T7. The result was that 40% of the lethal events are due to double-strand breaks (presumably via process 1) and 60% to base-modification. Of these base-modifications, the major reactions appear to involve pyrimidine bases. The evidence is discussed by Weiss (1964).

Methods of measuring DNA-breakdown during X-irradiation are perfectly illustrated by Randolph and Setlow (1971). Their work employed the transformation system (see p. 192) for *Haemophilus influenzae*. The transformation assays were performed using only one marker, resistance to the antibiotic cathomycin. This marker was chosen as previous work had established that it is radiation-insensitive (p. 245) and presumably the complications arising from reaction (3) are minimized. DNA was extracted from a cathomycin-resistant strain. The assay was based on the transformation of cathomycin-sensitive bacteria. The transformation rate could be measured directly by counting the colonies which grew on agar containing the drug. The number-average ($\overline{MW}_n$) and weight-average ($\overline{MW}_w$) molecular weights were measured from sucrose gradients (see p. 384) in neutral or alkaline sucrose gradients (to calculate the double-stranded and single-stranded molecular weight distributions respectively). The DNA was irradiated with X-rays for a given number of kilorads.[4]

The samples were employed either for sedimentation studies (above) or for transformation assays or alternatively for measuring their uptake into competent cells. *Haemophilus* is particularly suitable for this type of work as a method for establishing the extent of integration of transforming DNA into the recipient chromosome has been carefully quantitated (p. 192).

In summary the results are as follows. The effects of radiation on transformation and DNA uptake are exactly parallel establishing that the damaged DNA will not transform because it cannot enter the cell. Irradiation was performed either in phosphate buffer or in 10% yeast extract (to act as histidine source—see above). The presence of the yeast extract had an enormous effect on the dosages required to obtain comparable effects. Thus to obtain inactivation of the transformation assay that the number of transformants was reduced to 1% of control, 20 krad were required if the DNA dissolved in phosphate and 1000 krad were required if the DNA was dissolved in the yeast extract. It is therefore assumed that in buffer, process of type (1) (direct interaction of X-rays with DNA to produce breaks) constitutes no more than 1/50 th of the total inactivating events. The values of $\overline{MW}_n$ and $\overline{MW}_w$ for native or denatured DNA were obtained from various irradiated DNA samples. The number of double-strand breaks were calculated from the equations:

$$\text{Number of breaks/strand} = \frac{(\overline{MW}_n)_{\text{con}}}{(\overline{MW}_n)_{\text{irrad}}} - 1$$

$$= \frac{(\overline{MW}_w)_{con}}{(\overline{MW}_w)_{irrad}} - 1$$

$$= \frac{2 \times (\overline{MW}_n)_{con}}{(\overline{MW}_w)_{irrad}} - 1$$

The measurements were made by including control [^{14}C]DNA ('con') in the same gradients as the [^{3}H]DNA (irradiated, 'irrad'). These values were plotted against dose to calculate the dose required for either one single-strand or one double-strand break. The data established that (1) transformation is slightly more sensitive to radiation than integration (so some damaged DNA is integrated into the recipient genome) that (ii) uptake is about 50 times less sensitive than either transformation or integration and (iii) the inhibition of transformation (and integration) correlate with the double-strand breaks but not with the single-strand breaks. Thus in the direct effect (process 1, in the presence of yeast extract) 42 krad is required to produce (on average) one single-strand break per molecule (of $\overline{MW}_w$ 58 x 10^6) whereas 104 krad is required to produce one double-strand break.

Chemical reactions *in vivo* and their biological consequences

When cells are treated with chemicals that react with nucleic acids or are irradiated, chemical changes similar to those identified *in vitro* occur in the cellular nucleic acids. The first part of this section deals with the methods of identifying these changes. There are two types of consequence of such reactions. The cell may repair the damage to DNA or the changes themselves may result in temporary or permanent changes in the cell's genetic or metabolic properties. The later parts of the section discuss methods of identifying repair processes and some thoughts about the other biological consequences.

Reactions of nucleic acids *in vivo*

Evidence for alkylation, arylation and other similar reactions *in vivo*.   The procedure is simple enough. You take some radioactive chemical and inoculate bacteria with it or inject it into a rat (or some such unfortunate animal). You then harvest the cells (or chop out various organs) and isolate DNA and RNA and measure the radioactivity 'bound' to the macromolecules. If this experiment is being done with RNA, the procedure then is to separate the tRNA and species of rRNA by sucrose-gradient centrifugation or gel electrophoresis. With DNA, it is usual to do rather little with it except count it. A more worthwhile experiment is to subject the sample to buoyant density centrifugation to establish that the 'bound radioactivity' is genuinely associated with material of the same density as DNA. After all this has been done, the nucleic acids can be hydrolysed and the modified nucleotides, nucleosides or bases containing the radioactivity can be characterized. If the cells have been treated with (for example) a well-established methylating agent, the acid-digests can be analysed for 7-methylguanine. With some of the reagents, that undoubtedly do react with nucleic acids *in vivo*, it is not as simple as that as the nature of the modified residues is unknown. In this case what is needed is inspired guesswork, so that the chromatographic properties of the labelled base can be compared with those of some likely candidates run as standards. If this does not work, you are left with a very difficult chemical problem. The extent of reaction of many of these substances with DNA is very small and so you can never hope to work on anything more than a radioactive scale. With all these difficulties in mind, what is known about the chemical reactions *in vivo?* In fact a great deal, mainly as a result of the work of P. Brookes and P. D. Lawley.

Alkylation of DNA (and RNA) *in vivo* can be achieved with the nitrosamides and NMG (see p. 239) and the mustards. It is also effected by a rather surprising group of substances, the nitroso secondary amines, such as nitroso dimethylamine.[5]

All the straight chain 'nitrosamines' alkylate DNA and RNA apparently always yielding 7-methylguanine residues. The reactions require metabolic oxidation of the compound. The alkylation of animal DNA *in vivo* by these compounds has been extensively studied by P. N. Magee and his colleagues (see for example Lee, Lijinsky and Magee, 1964; Craddock and Magee, 1965). It is however important to recognize that C- and A-residues are also alkylation targets (Brookes, 1968).

Other carcinogens bind to nucleic acids *in vivo*. In many cases the nature of the modification to the nucleotide is unknown. The subject forms an active part of cancer research. The principle

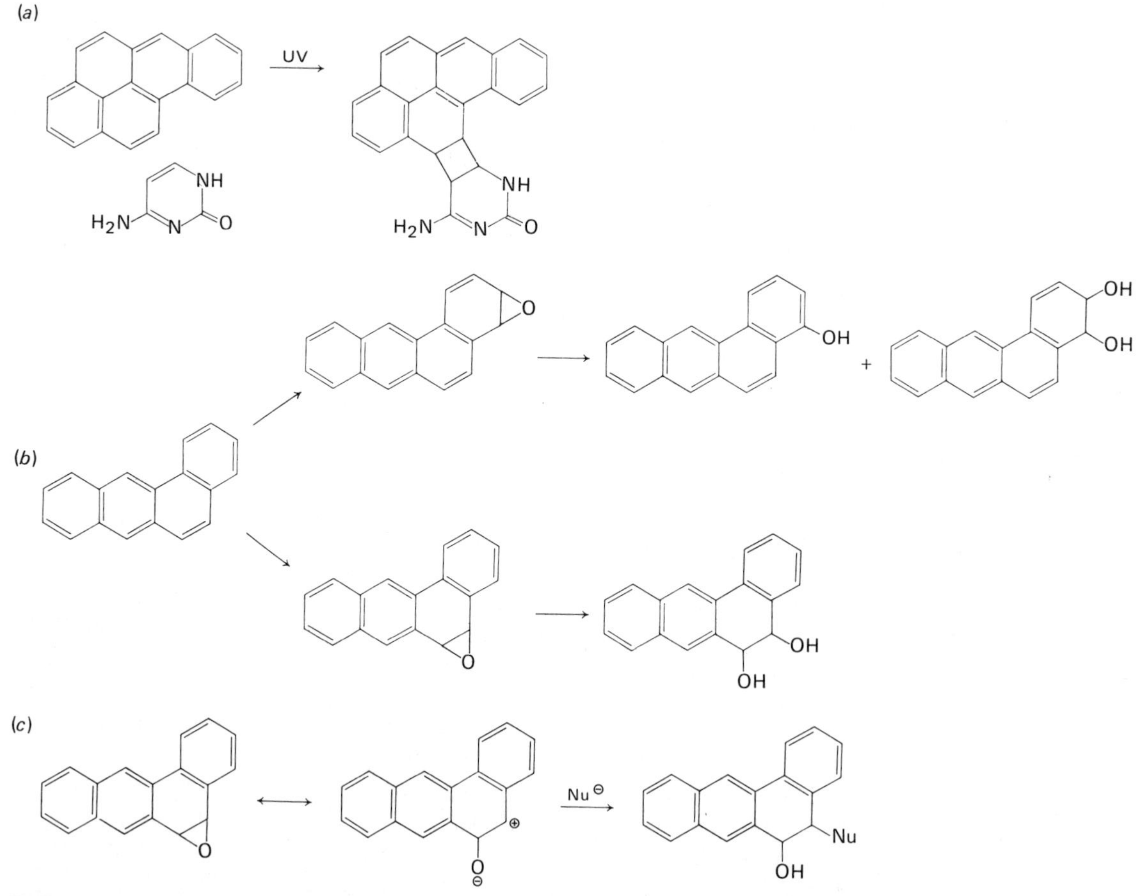

FIG 8.2   Interactions between polynuclear aromatic hydrocarbons and DNA. (*a*) Proposed photocatalysed addition of benzpyrene to Cyt. (*b*) Metabolites produced during the oxidative metabolism of benzanthracene. (*c*) Possible arylation reaction between metabolite of benzanthracene and a nucleophile (such as, perhaps, N-7 of Gua). See the accompanying text and Rice (1964), Boyland (1964) and Dipple, Lawley and Brookes (1968).

classes of compounds involved are the polycyclic aromatic hydrocarbons, the arylamines and azo dyes.

Several years ago Brookes and Lawley (1964) reviewed a number of these compounds and found that they bound to mouse-skin DNA. The nature of the reaction remains extremely obscure. There seem to be two possibilities. Either the hydrocarbons add on to the bases or alternatively they are metabolized to arylating reagents. The former possibility appears to be remote, even though Rice (1964) was able to demonstrate the UV-catalysed addition of benzpyrene to Cyt (see Fig. 8.2 (a)). The reaction requires a detergent to solubilize the hydrocarbon and alkali. Ts'o and Lu (1964) demonstrated the UV-catalysed binding of the same compound to DNA. The second possibility is based on real metabolic processes. It is well established that mammalian cells (expecially the parenchymal cells of liver) contain $NADPH/O_2$-dependent hydroxylases which hydroxylate these hydrocarbons to form hydroxyl compounds with epoxide intermediates (see the review by Boyland, 1964 and Fig. 8.2 (b)). The proposal of Dipple, Lawley and Brookes (1968) is that such intermediates might function in the arylation of bases in DNA (Fig. 8.2 (c)). The hydrocarbon derivatives of nucleotides are difficult to characterize as they contain large aromatic groups that make them insoluble in polar solvents and the ionized phosphate group and the hydrophilic sugar residues that make them insoluble in non-polar solvents. There is no doubt that the electrophoretic methods of Brookes

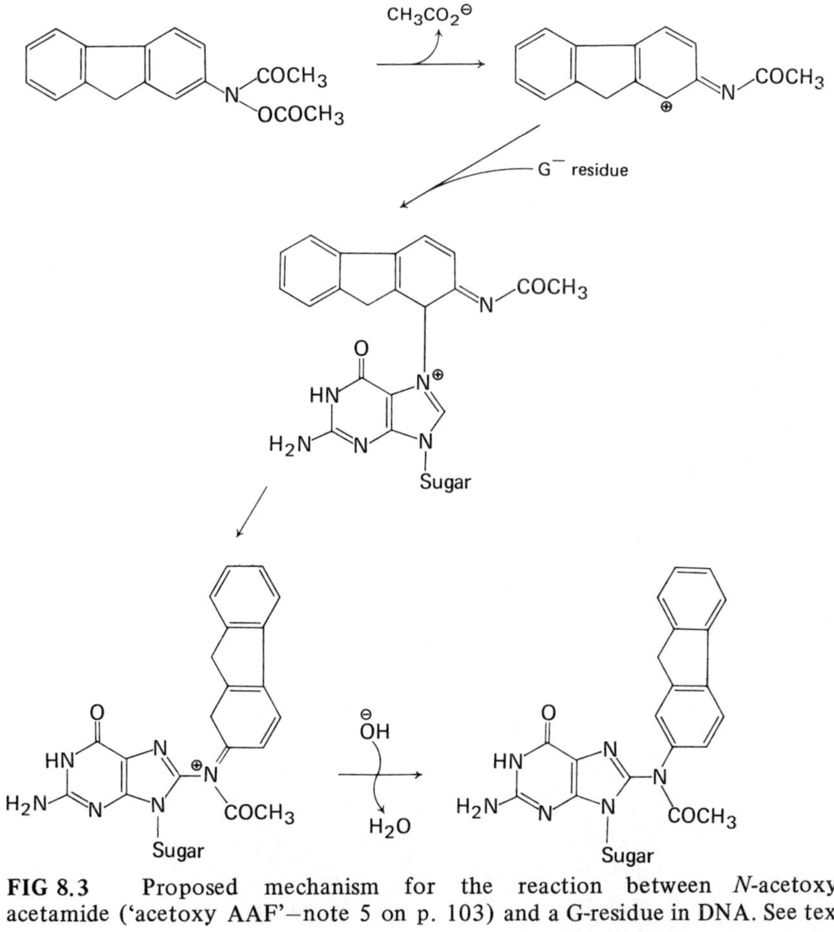

FIG 8.3   Proposed mechanism for the reaction between N-acetoxy-N-(2-fluorenyl) acetamide ('acetoxy AAF'—note 5 on p. 103) and a G-residue in DNA. See text for discussion.

and Heidelberger (1969) are essential for any further characterization of these products. The possibility of a physical interaction between DNA and the aromatic hydrocarbons is discussed on p. 415.

The reaction of the aromatic amines and amides with DNA is presumed to require their hydroxylation. In a classical piece of work, Miller, Miller and Hartman (1961) showed that *N*-acetoxy-*N*-(2-fluorenyl) acetamide was more carcinogenic than *N*-(2-fluorenyl) acetamide. The *N*-acetoxy compound is the acetate of the hydroxylated derivative. The reaction between *N*-acetoxy-*N*-(2-fluorenyl) acetamide and DNA results in substitution of G-residues at C-8 (see Fig. 2.8 (*f*), p. 62). Kriek *et al.* (1967) propose a mechanism involving direct attack on C-8. I regard an alternative mechanism in which the initial attack is on N-7 as more plausible (Fig. 8.3). This mechanism involves the role of N-7 as the nucleophile and is thus consistent with other substitutions of guanine derivatives; the subsequent rearrangement is similar to a mechanism we have proposed to account for an 'ortho effect' in the arylation of benzenoid nucleophiles by $C_6H_5NH^+$ (derived from phenyl hydroxylamine; Parish and Whiting, 1964).

The remaining class (for aromatic carcinogens) are the azo-dyes, of which the best known is 4-dimethylaminoazobenzene (p-Me_2N–C_6H_4–N–N–C_6H_5, 'butter yellow'[6]). The nature of the compounds formed when these dyes bind to DNA and RNA are unknown. However the reaction has been extensively studied by J. J. Roberts and G. P. Warwick. They have two lines

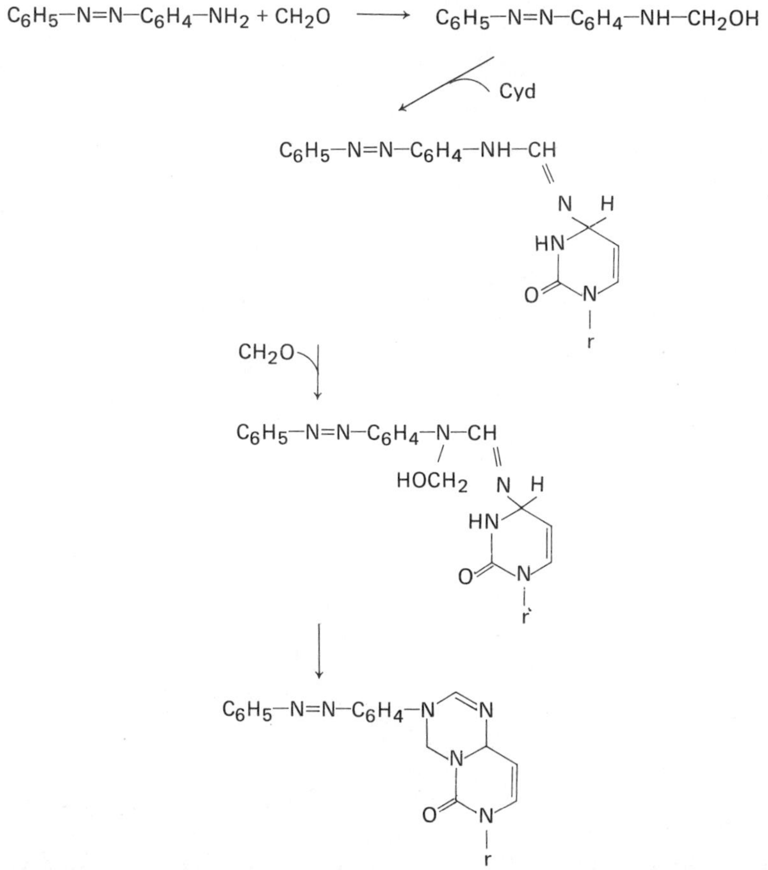

$$C_6H_5-N{=}N-C_6H_4-NH_2 + CH_2O \longrightarrow C_6H_5-N{=}N-C_6H_4-NH-CH_2OH$$

FIG 8.4   Reaction between 4-aminoazobenzene, formaldehyde and Cyd (see text for discussion; Roberts and Warwick, 1966a).

of evidence. On the one hand they have demonstrated that in the presence of formaldehyde, 4-aminoazobenzene reacts with Cyd or CMP as shown in Fig. 8.4 (Roberts and Warwick, 1966a).

It has not been established that such a reaction occurs between butter yellow and the C-residues of nucleic acids *in vivo* although clearly the enzymic methyl-hydroxylation would produce the necessary intermediates for the above (Mannich) condensation. The binding of butter yellow to nucleic acids *in vivo* was studied by feeding [^{3}H] butter yellow to rats and isolating liver nucleic acids. It was established that the reaction is primarily to RNA but that the activity is lost from RNA as the various RNA species turn over and that very low, but persistent binding of the dye to DNA was observed (Roberts and Warwick, 1966b).

Cross-linking of DNA *in vivo* There is every reason to suppose that the bi-functional alkylating agents yield inter-strand cross-links *in vivo*. Several studies of the biological consequences of the action of these agents in living cells have been made. The cross-linked products can be identified by methods described already (see p. 240). In addition to the bi-functional agents described already it seems that the *bis* methane sulphonyl butan-1,4-diol ($MeSO_2O(CH_4)_4OSO_2Me$, 'Myleran' an anti-cancer drug) cross-links DNA (see p. 253). This is the example of a reagent which like butadiene diepoxide (see p. 241) must bridge the strands with a four-carbon link.

A group of important antibiotics cross-link DNA *in vivo*. These are the mitomycins and porphyromycins. The evidence for this reaction is very clear-cut. If *E. coli* is treated with mitomycin$_a$ C (10 mg/ml for 15 min) DNA isolated from the cells is reversibly renaturable (p. 240). If tne experiment is performed with *B. subtilis* and the DNA is denatured and cooled, transforming activity is regained under conditons where control DNA loses virtually all its transforming activity. Furthermore if this DNA is centrifuged to equilibrium in Cs_2SO_4 (p. 389) the DNA from the mitomycin-treated cells bands at the density characteristic of native DNA whereas the control DNA bands at the density characteristic of denatured DNA (Iyer and Szybalski, 1963; Szybalski and Iyer, 1964).

Now let us consider the structure of these compounds (Fig. 8.5(*a*)); they do not look much like bi-functional alkylating agents. If you are not a chemist they may look simply like nothing on earth. They are not obvious alkylating agents because an alkylating agent requires some sort of stabilization of the carbonium-ion intermediate for alkylation (such as the three-membered ring in the mustard intermediate, Fig. 2.8(*d*) p. 62). In this case the attempt is to obtain carbonium ions centred on C-1 and C-10. Such carbonium ions (especially a primary one such as the one on C-10) are very unstable and are unlikely intermediates unless their charge can be delocalized in some way. You may not be convinced but the facts are that these antibiotics do not themselves cross-link DNA *in vitro*. However the antibiotics can be 'activated' to cross-link DNA solutions by incubating them in the presence of a cell-free bacterial extract, or in the presence of diaphorase[7] or in the absence of enzymes but in the presence of sodiumborohydride. It was thus assumed that reduction of the molecule produced the reactive species. The species is an unstable one; the borohydride-reduced product could not be kept in a bottle; either the reagents had to be all there together or the DNA had to be added immediately the borohydride reaction (easily followed by a colour change—purple to yellow—in the mitomycin chromophore) was complete (Szybalski and Iyer, 1964). The proposal is that the quinone ring is reduced to yield an indole system, which following loss of the oxygen group at C-9a, will ionize nicely at C-1 or C-10 to form stable, delocalized intermediates (see Fig. 8.5(*b*)).

The remaining question is for the nature of the nucleophiles in Fig. 8.5. Only four atoms span C-1 to C-10. There is every reason to suppose that such bridges can be formed (see p. 241).

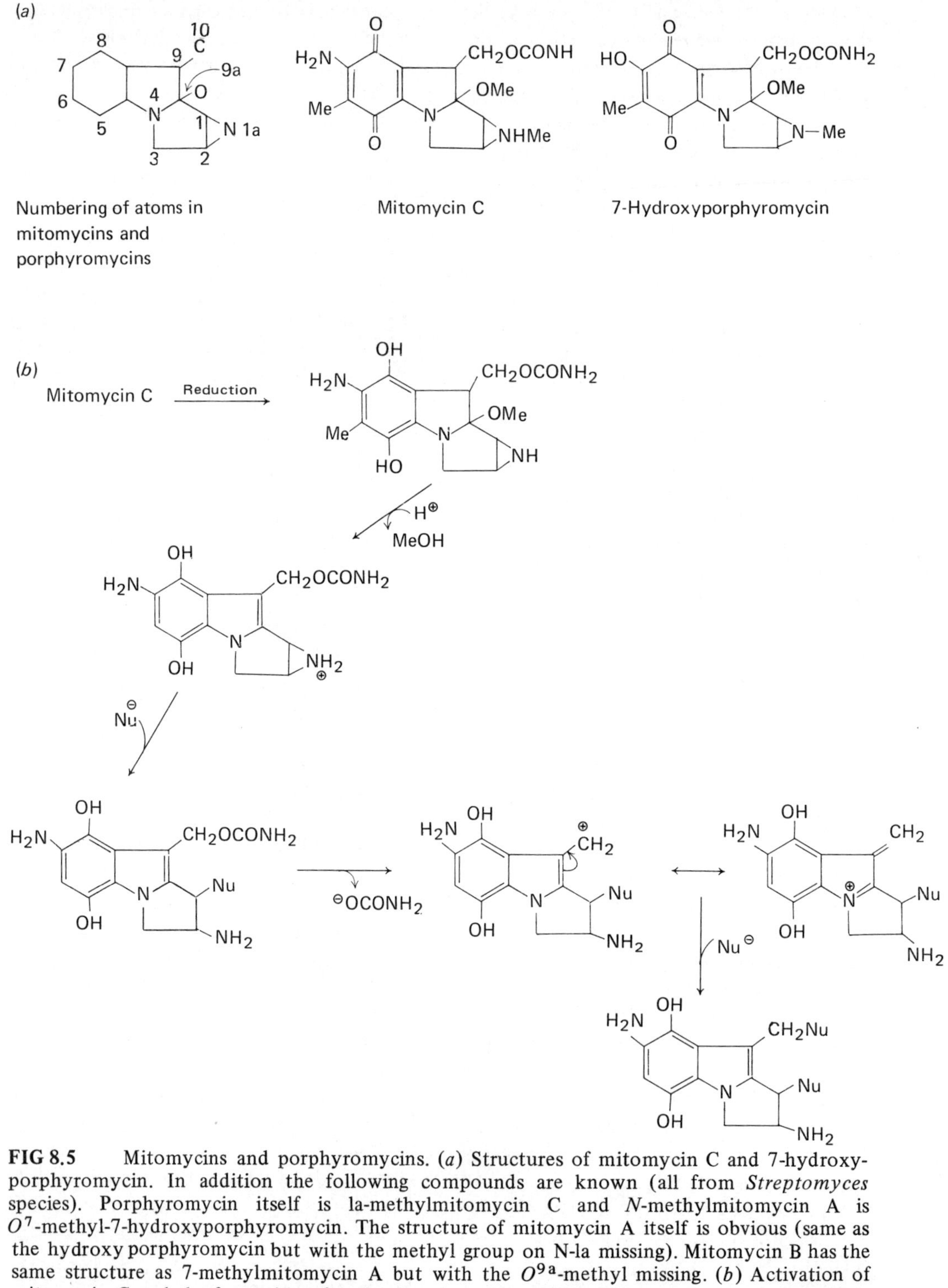

(a)

(b)

FIG 8.5   Mitomycins and porphyromycins. (a) Structures of mitomycin C and 7-hydroxy-porphyromycin. In addition the following compounds are known (all from *Streptomyces* species). Porphyromycin itself is la-methylmitomycin C and *N*-methylmitomycin A is O^7-methyl-7-hydroxyporphyromycin. The structure of mitomycin A itself is obvious (same as the hydroxyporphyromycin but with the methyl group on N-la missing). Mitomycin B has the same structure as 7-methylmitomycin A but with the O^{9a}-methyl missing. (b) Activation of mitomycin C and the formation of carbonium ions (hypothetical intermediates in alkylation). See text for discussion. Nu⁻ is a nucleophile.

Szybalski and Iyer (1964) inspected space-filling models of the DNA duplex and concluded that the least distortion of the helix was required if the links were between O^6-groups in G-residues. They obtained some data in support of the fact that this may be the case by observing that the extents of cross-linking of phage DNA molecules *in vitro* (using activated mitomycins) was about the same for molecules containing a variety of substituted or modified pyrimidines (see below) and confirmed that (on the basis of measurements with the model that such changes as glucosylation of 5-hydroxymethylcytosine should not interfere with this particular reaction.

It is well established that the cytotoxic effects of many bifunctional alkylating agents is more than ten times greater than that of their mono-functional analogues. Does this mean that all these compounds cross-link DNA? Almost certainly not in my view. DNA is only one of the alkylation-targets in cells. However there are reasons to suppose that several of these compounds (such as Myleran, p. 251) do alkylate and possibly cross-link DNA. As an example of the sort of indirect evidence that is useful, see the work on the mutagenic effect of Myleran in *Chalmydomonas reinhardii*, a unicellular alga (Verly, Dewandre, Moutschen-Damen and Moutschen-Damen, 1962). A consequence of cross-linking is that DNA polymerase gets stuck at this 'knot' in the helix so the lesion is lethal unless the damaged region is either repaired (see p. 260) or alternatively deleted. In their work Verly *et al.* (1962) established that deletions were produced in *Chlamydomonas* by Myleran and that the extent of the deletions was consistent with levels of alkylation.

Effects of radiation on DNA *in vivo*   The effects of UV-irradiation on DNA *in vivo* are essentially the same as the effects *in vitro*. The way in which pyrimidine-dimer formation can be assessed *in vivo* are discussed later in the context of DNA-repair (see p. 256).

With X-irradiation (or the related γ-irradiation) the situation is more complex. There is every reason to believe that the lethal consequences of X-irradiation are due to the reaction with DNA. Evidence includes: (1) the sensitivity of bacteria to the lethal consequences of X-irradiation is a linear function of the GC-content of their DNA (Kaplan and Zavarine, 1962); (2) an analogous effect is observed in the inactivation of genetic markers in *B. subtilis* transforming DNA of different base compositions (see p. 149); (3) following doses of X-irradiation that kill yeast, DNA synthesis is closed down but RNA- and protein-synthesis continue (Eckstein, Paduch and Hilz, 1967); (4) lethality is correlated with mutagenesis and in synchronized cultures of *Paramoecium*[8] the number of mutations is a function of the closeness of the time of irradiation to the time of DNA-synthesis (Kimball, 1966); (5) the sensitivity of synchronized cultures of mammalian cells to X-irradiation is a similar function of the period of the cell-cycle in which they are irradiated—they are most sensitive during the S-phase (Brent, Butler and Crathorn, 1966); and (6) the phenomenon of 'radiosensitization'.

This last process, X-ray sensitization is the most direct evidence for DNA-involvement in the effects of X-irradiation. It also a powerful experimental tool. Bacteria are more sensitive to X-irradiation if they are grown under conditions in which either they are starved of thymine (see p. 154) or in which a base analogue is incorporated into DNA (Kaplan, Earle and Howsden, 1964). The base analogues that are employed are 5-bromo- and 5-iodo-uracil (both analogues for Thy, incorporated by growing Thy⁻ mutants in the presence of the deoxyribotides—see p. 177) or 6-thioguanine. In the case of a *B. subtilis* phage that contains Ura in place of Thy (p. 270) sensitization is achieved by the uracil analogue 5-flurouracil (Lozeron and Szybalski, 1968).[9]

Precisely why these events lead to radiosensitization is not altogether clear. One chemical consequence of the UV-irradiation of DNA containing them has been recognized for some time, that is the photocatalysed dehalogenation of bases to produce U-residues (Wacker, Mennigmann and Szybalski, 1962). It is not clear why this should be a lethal event. Two further observations are relevant to the process. DNA-repair mutants (*uvr⁻*, p. 258) of *E. coli* are more sensitive to irradiation than are wild-type cells. However their sensitivity is not increased by sensitizing agents (Hotz and Reuschel, 1967). Thus it appears that sensitization is due to a non-repairable change. It seems likely that the observation is related to the facts that sensitized DNA is more susceptible to radiation-induced cross-linking and double-strand breaking (see p. 245) although there is no clear chemical rationale of the reactions involved (Szybalski and Opara-Kubinska, 1965; Kaplan, 1966).

Furocoumarins   The furocoumarins are a group of oxygen heterocyclic natural products (and their semi-synthetic derivatives) produced by certain higher plants, notably members of the genera *Ficus* and *Angelica*. Structures of three of the naturally occurring compounds are given in Fig. 8.6 (*a*). These compounds are employed clinically to treat patients who suffer from excess skin-sensitivity to sunlight or from the disease vitiligo (the ocurrence of very pale patches of skin). It is known that these substances have a photo-sensitizing effect on phage, bacteria and tissue culture cells. It is also established that they bind reversibly to DNA

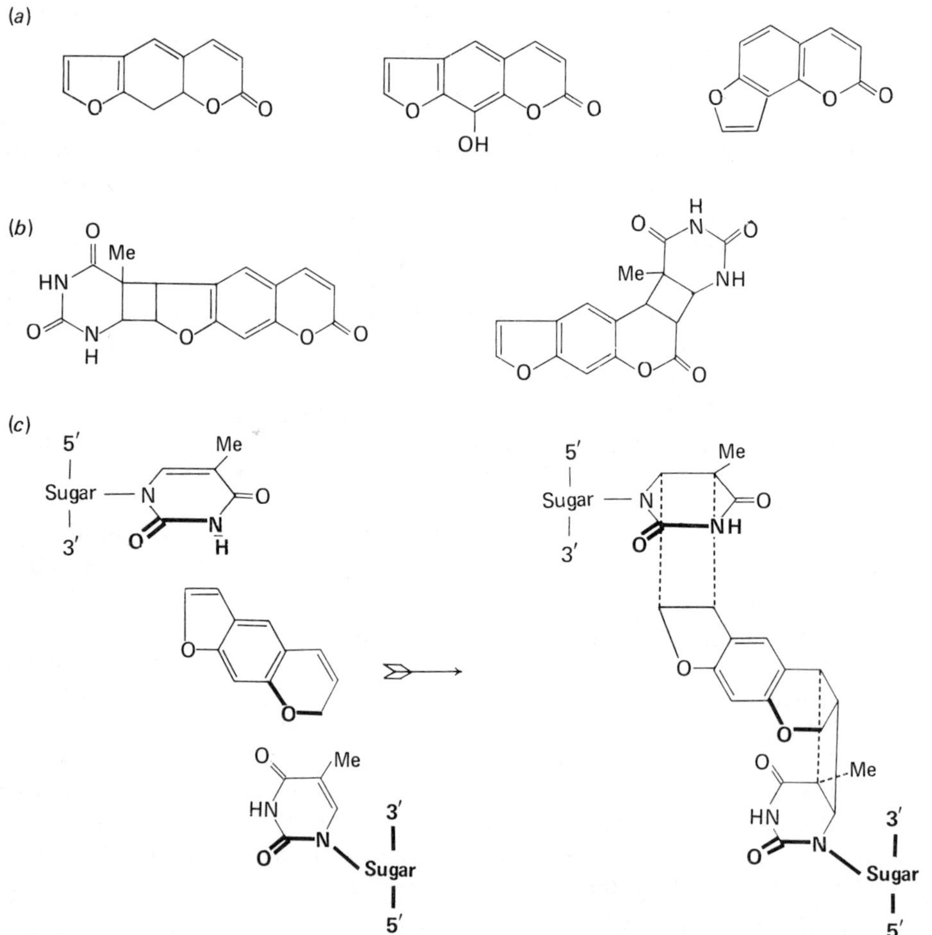

FIG 8.6   Furocoumarins. (*a*) Structures (from left to right) of psoralen, xanthotoxal and angelicin. (*b*) Addition products of psoralen and an unsaturated ketone (Thy). (*c*) Suggested reaction between two T-residues and intercalated psoralen (see text).

(apparently by intercalation—see p. 410); however if the DNA solutions or cells treated with the furocoumarin are irradiated with long-wavelength UV (about 350 nm) the binding is irreversible. The correlation between the DNA-reaction (and a less efficient rRNA reaction) and the biological activity has been made by Rodighiero *et al.* (1970). They employed [3H] furocoumarins and studied the binding of radioactivity to the macromolecules. A more detailed analysis of the reaction with DNA has been made by Cole (1970). He showed that if DNA was mixed with psoralen and irradiated, the DNA was cross-linked. The same observation was made when DNA was isolated from λ phage, *E. coli* or mammalian tissue cultures, which had been treated with psoralen and irradiated. The cross-linking was demonstrated by the reversibility of denaturation (p. 240) and by using either hydroxyapatite chromatography (p. 365) or alkaline sucrose gradients (p. 257) to identify the cross-linked structures in a denatured DNA preparation. He proposes a very plausible model of the psoralen reaction. It is known (see Cole's paper for the references) that UV-catalysed addition of psoralen to α,β-unsaturated ketones (including pyrimidines) can occur in one of two ways (see Fig. 8.6(*b*)) involving either the pyrone or furan double bond. The suggestion is that 'intercalated' psoralen[10] can react with two T-residues in opposite DNA chains as shown in Fig. 8.6(*c*).

Repair of DNA

DNA is unique among macromolecules (with the possible exception of cell wall) in being capable of repair when it has been damaged. There are two types of repair process in DNA; the second of these is the more widespread (and in general the easier to study) and is the process usually referred to as 'repair of DNA'.

Light repair   Light repair is the process of enzymic photoreactivation of cells previously irradiated with UV light. The reaction that is catalysed is the following one:

$$\ldots \text{N-N-N-}\overline{\text{T-T}}\text{-N-N-N} \ldots \to \ldots \text{N-N-N-T-T-N-N-N} \ldots$$

It is thus the reverse of the pyrimidine dimerization reaction (p. 243). This process is distinct from the observation *in vitro* (p. 244) that the dimer/monomer equilibrium can be shifted back in the direction of monomer formation at low wavelengths (235 nm). The photoreactivation is observed if cells exposed to UV radiation are subsequently exposed to light of between 300 and 400 nm. The experimental systems available for the study of this process in bacteria are very well defined (see the next paragraph). However if reactivation is studied in a hitherto uncharacterized system, great care must be taken over intepreting the results. The mere fact that UV-irradiated cells can be made to recover by subsequent irradiation at (say) 350 nm is not unequivocal evidence for enzymic monomerization of pyrimidine dimers in DNA. The problems of interpretation derive from the facts that what is being observed may have nothing to do with DNA (these 'other' processes are lumped together and referred to as indirect photoreactivation) and also recovery from radiation is usually scored after several hours (or days) when clones of cells have grown up. In this case of course there is great difficulty in pinning down the reactivating event to the period of irradiation.

   Photoreactivation in *E. coli* has been characterized both genetically and enzymically. The genetic locus is *phr*; mutants that are *phr⁻* can be isolated. The locus is closely linked to the *gal* gene in *E. coli* K12 (van de Putte, Zwenk and Rorsch, 1966). Enzymically, the process is characterized by assaying an enzyme that performs the reaction in cell-free extracts. It is difficult to assay the enzyme directly. In principle it should be fairly easy as the enzyme catalyses the formation of a structure (. . . -T-T- . . .) which has an absorption maximum at

267 nm from one that has no such maximum (. . . -T-T- . . .). However the residues involved form such a small proportion of the total residues contributing to the UV absorption in this region that accurate measurements are difficult. There are two ways round the difficulty. One is to use transforming DNA that has been irradiated *in vitro* as substrate. The assay then consists of irradiating (at 350 nm or thereabouts) a mixture of this DNA and the extract and then assaying the restoration of transforming activity. The alternative method is to assay directly the monomerization of dimers in irradiated poly d I:C (Setlow, Carrier and Bollum, 1965). When this material is irradiated, the poly dI chain is unaffected but some of the residues of the poly dC chain are dimerited to produce structures such as . . . -C-C̅-C̅-C- . . .. This product is heated to convert any dimerized structures as above into . . . -C-U̅-U̅-C- . . . (see p. 244), which is the substrate for the assay. After incubation, the polymer is acid-hydrolysed and the bases are separated (p. 244) and the amounts of Cyt, Ura□Ura and Ura are measured. The extent of the conversion of Ura□Ura into Ura is a measure of the activity of the enzyme.

Using the transformation assay, the enzyme has been identified in extracts from many bacterial, fungal and animal (but not mammalian) sources (see for example Terry and Setlow, 1967; Cook and McGrath, 1967). The enzyme requires as substrate a reasonable stretch of DNA (nine nucleotides) so that substances such as T̅-T̅ or -T̅-T̅-, are not suitable substrates (Setlow and Bollum, 1968).

The fact that this enzyme has been identified *in vitro* does not prove that the reaction is of importance *in vivo*. Fewer examples of the direct demonstration of monomerization of dimers *in vivo* are available but such evidence is available for bacteria (Boyce and Howard-Flanders, 1964), *Paramoecium* (Sutherland, Carrier and Setlow, 1967) and in amphibian tissue-culture cells (Regan and Cook, 1967).

Polymerase-dependent repair–general features   This is the process referred to either as 'dark repair' (to distinguish it from light repair, above) or more simply just as DNA repair. The reactions that are involved are summarized in Fig. 8.7. I have illustrated repair in the special case where the damage (or 'lesion') in the DNA consists of a pyrimidine dimer as this is the system which has proved of most use in analysing the problem.

Little is known about processes (1) and (2); questions such as 'Are they achieved by the same enzyme?' remain largely unanswered. Excision is assayed by the following procedure. Cells are labelled *in vivo* with [³H] thymidine or [³H] thymine[11] and are then transferred to non-radioactive medium and are irradiated. The process of excision is followed by measuring the amount of radioactive pyrimidine dimers (see p. 244) in acid-hydrolysates of (*a*) acid-insoluble cellular material (i.e. DNA–see p. 180), (*b*) acid-soluble cellular material, and (*c*) the medium. It is essential to measure the amounts of [³H] Thy in (*b*) and (*c*) as the results can be confused by random degradation of DNA. Such experiments do not establish that the excised portion is a dinucleotide (such as the -T-T that has been excised in the scheme of Fig. 8.7.) but merely that the pyrimidine dimers are excised in short, acid-soluble, oligomers. Experiments of this type on bacteria have established that the rate of dimer-excision is approximately the same in 'normal' and in radiation-resistant strains of *E. coli* and also in the extremely radiation resistant organism *Micrococcus radiodurans*. Moreover the rate of excision in a thymine auxotroph (*E. coli* 15 T̅) is independent of the presence or absence of thymine and the process is hence independent of DNA-synthesis (Setlow and Carrier, 1964; Setlow, 1968). In certain radiation-sensitive strains of *E. coli* (see p. 258) no dimer excision can be detected. An interesting point about *M. radiodurans* is that the dimers are very efficiently removed from the cells (i.e. most of the Thy□Thy is found in the hydrolysates obtained from the medium, Boling and Setlow (1966)).

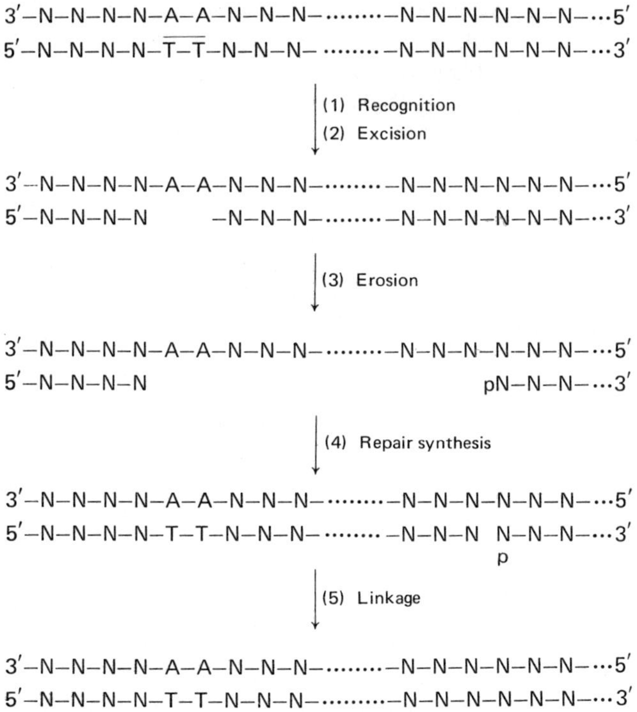

```
3'—N—N—N—N—A—A—N—N—N—·········—N—N—N—N—N—N—···5'
5'—N—N—N—N—T̅—T̅—N—N—N— ········ —N—N—N—N—N—N—···3'
```

|(1) Recognition
|(2) Excision
↓

```
3'—N—N—N—N—A—A—N—N—N—········ —N—N—N—N—N—N—···5'
5'—N—N—N—N        —N—N—N—········ —N—N—N—N—N—N—···3'
```

|(3) Erosion
↓

```
3'—N—N—N—N—A—A—N—N—N—·········—N—N—N—N—N—N—···5'
5'—N—N—N—N                              pN—N—N—···3'
```

|(4) Repair synthesis
↓

```
3'—N—N—N—N—A—A—N—N—N—········ —N—N—N—N—N—N—···5'
5'—N—N—N—N—T—T—N—N—N—········ —N—N—N  N—N—N—···3'
                                          p
```

|(5) Linkage
↓

```
3'—N—N—N—N—A—A—N—N—N—·········—N—N—N—N—N—N—···5'
5'—N—N—N—N—T—T—N—N—N—··········—N—N—N—N—N—N—···3'
```

FIG 8.7   Scheme of DNA repair, following the production of a thymine dimer in DNA (by UV irradiation).

Returning to our scheme (Fig. 8.7), the remaining processes involve enzymes of types discussed elsewhere in this book. In fact it is not necessary to imagine that processes (3) and (4) are separated in time and space as we have seen that DNA polymerase I (p. 224) has the necessary activities for both the erosion and repair synthesis.

A consequence of processes (2) and (3) is that during repair the molecular weight of the single strands of DNA is reduced. This effect is difficult to demonstrate directly as DNA is isolated as fragments of the genome anyway (p. 105). However the average molecular weight of denatured DNA should be reduced if it is isolated from cells undergoing DNA-repair, provided of course the fragments that are isolated from the control cells are significantly larger than the fragments produced *in vivo* by the repair process. The method that is employed (McGrath and Williams, 1966) is to use cells whose DNA has been labelled *in vivo* with tritium. The cells are lysed gently and centrifuged through alkaline-sucrose gradients (p. 186). The sedimentation of the zone of radioactivity is taken as a measure of the sedimentation properties of the single-strand fragments of DNA. Using these data as a parameter for assessing single-strand breaks, McGrath and Williams were able to demonstrate the production of such breaks during the recovery of *E. coli* B/r (see p. 258) from X-irradiation.

Genetic aspects or repair in *E. coli* and its phage   The experimental methods that are employed in this work require some explanation. First of all consider the inactivation of phage by UV-irradiation. Such an inactivated phage can be re-activated by DNA repair. The repair process can either be effected by gene products of the chromosome of the host cell or alternatively by gene products of the phage genome. The latter situation obtains in the T-even phage in which case 'host cell reactivation' is impossible. It is therefore possible to study

UV-resistance or sensitivity in the phage. The system has been analysed in phage T4. Two genes are involved (v and x). Although x remains something of a mystery, it is established that v is responsible for dimer excision (Setlow, Carrier and Setlow, 1969). Moreover if v^+ and v^- strains are irradiated, mixed and used to infect the same host cells, not only do the v^+ phage recover and replicate but they 'rescue' the v^- phage as well. In this case it seems certain that the v-gene product is an excision enzyme specific for T4 DNA.

The former possibility (that host cell reactivation effects the rescue of UV-irradiated phage) is tested by establishing whether the cells can propagate the irradiated phage. Examples of phage that can be used for the assay (and hence lack the gene(s) for excision themselves) are T1, T3 and λ The process cannot be performed with the single-stranded DNA phage φX 174. However it is possible to achieve phage titres from irradiated φX RF DNA (p. 154) with the spheroplast assay (p. 193).

E. coli strains capable of supporting the propagation of UV-inactivated phage are said to be Hcr$^+$ (host cell reactivation). Those incapable of such reactivation are said to be Hcr$^-$. These Hcr$^-$ strains are incapable of excising pyrimidine dimers (Setlow and Carrier, 1964) and if their spheroplasts are employed in assay with irradiated φX RF (see above) a lower phage titre is obtained than with Hcr$^+$ spheroplasts (Yarus and Sinsheimer, 1967). The Hcr phenotype is controlled by three genetic loci *uvr*A, *uvr*B and *uvr*C. These three loci are widely spaced on the *E. coli* chromosome (Howard-Flanders, Simpson and Thériot, 1966). The different gene products involved have not been identified.

Before leaving this consideration of the *uvr* loci, we must just comment on the two strains that are probably most frequently employed for comparative studies on repair in *E. coli* (see p. 260). These are the resistant strain B/r and the sensitive strain B_{s-1}. The second of these is of great historical importance as it was the first Hcr$^-$ strain to be obtained. It was isolated by Hill (1958) and I do not wish for a moment to play down the importance of her work on this organism as it really formed the foundation of the subsequent studies by Howard-Flanders and others on this very important aspect of bacterial physiology. The only point of warning is that if data on B/r and B_{s-1} are being compared it is essential to remember that the two organisms are not isogenic for all loci except a *uvr* locus. In addition to being *uvr*B$^-$, B_{s-1} has a mutation in a locus known as *exr* which is in some way related to UV-sensitivity (Greenberg, 1967). Another difference between the strains is in filament-forming locus *fil*. Thus there may be more to B_{s-1} than its inability to excise dimers, and indeed it does differ from other UV-sensitive strains in the absolute inability of the organisms to incorporate any exogenous dThd into DNA. Recent data of Werner (1971) suggest that the DNA-synthesis for repair and replication use different pools of thymidylate precursors (Thy for synthesis and dThd for repair). The implications of this work are not easy to evaluate as it is difficult immediately to reconcile with the commonplace findings about the use of these substances for DNA-labelling (p. 176). However they have some bearing on the properties of B_{s-1}.

The genetic analysis of the *uvr* loci in *B. subtilis* is almost as complete as that in *E. coli*. Details of system are to be found in Reiter and Strauss (1965).

Sensitivity to radiation and other genetic defects in *E. coli*   There are other classes of mutants (apart from those with defects in *uvr* and *exr*) which are sensitive to radiation. Of these one class is completely understood in terms of the gene product, namely the *pol*A$^-$ mutants of Cairns (see p. 222). These strains are UV-sensitive. Coukell and Yanofsky (1970) studied the spontaneous deletion frequency in a particular region of the chromosome (at the end of the tryptophan operan overlapping into the adjacent gene *ton* B which determines sensitivity to bacteriophage T1) in a variety of UV-sensitive strains. The deletion frequency (the rate at which these gene functions are concomitantly lost indicating a deletion) was of the order of 1 in 10^7 cells for all the strains except polA$^-$ in which it is of the order of 1 in 10^6. So the consequences of the lack of Kornberg's enzyme (p. 222) for a cell are two-fold. UV-damage cannot be repaired and spontaneous deletions occur more easily. Presumably we are merely looking at different consequences of the same fact. When there is a lot of damage (UV-irradiated cells) some of the damage is certain to be lethal unless it is repaired. The rare accidental damage that occurs in normal cells is normally adequately repaired; when the polymerase cannot effect this repair synthesis, DNA is deleted and the deletions can be scored if (like Yanofsky) one chooses a region in which the deleted genes are not essential for growth in certain media.

So far we have discussed mutations involved in the processes of excision and repair synthesis (2 and 4 in Fig. 8.7). However there are other mutations which can result in UV-sensitivity in $E.$ $coli$. The complexity of this sytem implies that there is more to the picture than the simple representation of Fig. 8.7 implies. We can consider the additional processes as accompanying either excision (2) or following repair (i.e. in between 4 and 5).

Mutations in genes identified as uvrD and ras result in organisms that are UV-sensitive, that can excise pyrimidine dimers and are apparently defective in a process subsequent to excision but prior to repair synthesis (Ogawa, Shimida and Tomizawa, 1968; Walker, 1969; Coukell and Yanofsky, 1970).

The other group are those in which joining of DNA single strands is incomplete. These organisms degrade structures such as the penultimate one in Fig. 8.7. These are not ligase$^-$ mutants, as the DNA is replicated normally and DNA replication requires the action of the ligase (see p. 17). The phenotypes are referred to as Rec^+ and Rec^-. Rec stands for recombination and although Rec^- mutants are in fact UV-sensitive, they were identified in a different way. When such a mutation is present in an F^- organism (p. 31) the outcome of conjugation is (to pursue the analogy) abortive. Thus there is a process (or there are processes) involved in the healing of structures containing a single-strand break which occurs in cells that have attempted to repair UV-damage and also occurs in conjugation-recipients which does (do) not occur during chromosome replication. There are three Rec-genes, recA, recB and recC. The precise way in which their functions are interrelated is not known although it is certain that the activity of the revA gene-product precedes in time the activities of the other gene products (van de Putte and others, 1966). Rec^- mutations do not have high spontaneous deletion frequencies (Coukell and Yanofsky, 1970). Rec^- mutants are further discussed on p. 133.

Repair following X-irradiation   In general the process of repair following X-irradiation is similar to that following UV-irradiation with the important difference that double-strand breaks induced by X-rays are lethal and non-repairable.

The situation is slightly more complicated in another respect. If X-ray sensitive strains of $E.$ $coli$ are scored, mutations are picked up in a gene designated lex; lex^- mutants are four times more sensitive to X-rays than are wild type whereas uvr^- mutants are only about twice as sensitive. Moreover lex^- cells are Hcr$^+$ (see p. 258); thus lex is not involved in excision although, intriguingly, lex is very close to uvrA in the chromosome (Howard-Flanders and Boyce, 1966).

Base-excision during repair of alkylated DNA   We have already seen that bi-functional alkylating agents are more toxic to cells than are their mono-functional analogues. These reagents are therefore useful for the study of repair following chemical alkylation. Lawley and Brookes (1965, 1968) were able to show that the uvr^- strain $E.$ $coli$ B$_{s-1}$ (see p. 258) survives treatment with sulphur mustard (p. 251) much less well than the resistant strain B/r. It was therefore concluded that recovery from mustard treatment involves excision and repair-synthesis (p. 257). Unfortunately there are difficulties in interpreting this comparison as these two strains are not isogenic (see p. 258). However Brookes and Lawley obtained chemical evidence in support of the excision stage in the repair process (p. 257). The cells were treated with [^{35}S] sulphur mustard and cellular DNA and acid-soluble material from the cells and the medium were separately hydrolysed with acid (see p. 239). In this way it was possible to show that Gua-mustard-Gua moieties were excised from B/r DNA with much higher efficiency than with B_{s-1} and that the excision of Gua-mustard by B/r was of much lower efficiency than that of this *bis* guaninyl compound. The conclusion therefore is that cross links are excised and the lesions are repaired, but that single-alkylated nucleotides are largely left alone. The presumed mode of cross-link excision is illustrated in Fig. 8.8.

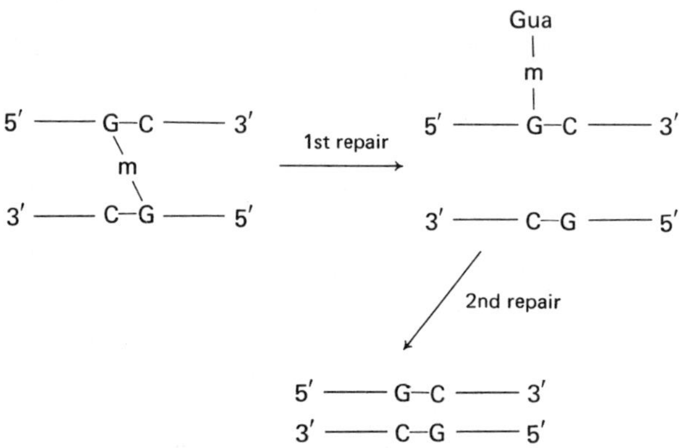

FIG 8.8   Excision of mustard-induced cross-links from the DNA of *E. coli* B/r (m = mustard 'bridge'—see p. 240).

Excision of modified bases following exposure of cells to mono-functional alkylating agents has been rather little studied although the relevance of this process to the experimental induction of mutations with such agents as NMG (p. 60) is enormous. An important study of NMG-alkylated DNA has been made by Lawley and Orr (1970). *E. coli* (either strain B/r or B_{s-1}) was treated with [Me-^{3}H] NMG (between 0·6 and 6 nM final concentration for 20 min). Cells were then transferred to growth medium and harvested after various time intervals and DNA was extracted and hydrolysed with 0·1 M HCl (70°C for 20 min). These mild conditions were employed as one of the products (O^6-methylguanine—see p. 68) is destroyed in strong acid. The hydrolysates were analysed by two-dimensional paper chromatography employing as solvents (1) a mixture of isopropanol, conc. NH$_3$ and water (7:1:2 by volume) and (2) *n*-butanol, conc. NH$_3$ and water (86:2:12 by volume). A certain proportion of radioactivity remained at the origin. It presumably consisted of oligomers not digested by the acid. This material is perhaps worth further study. From the conditions of hydrolysis, I guess it contains pyrimidine oligomers (p. 275) and hence possibly methylated C-residues. The results can be summarized as follows.

B/r treated with 0·74 nM NMG: The extent of methylation (nmole of ^{3}H$_3$C-/mole of DNA phosphate) was 1·9 at time 0 and dropped to 0·26 over 3 h during which time the cells had grown (about 3 generations). The percentage of 7-methylguanine in the total methylated bases increased over this time. 3-Methyladenine and O^6-methylguanine dropped from 2 and 11 respectively to zero. 'Origin material' dropped from 10 to 3%. The data implied that 7-methylguanine is not excised but that the other alkylated bases are.

B_{s-1} treated with even 0·62 mM NMG did not grow. However over a $1\frac{1}{2}$ h period O^6-methylguanine was fairly efficiently excised (10 to 2% of total alkylated bases). At 6 nM NMG, there was less excision of this base but some excision (7·3%) of 3-methyladenine. (3-Methyladenine could not be detected in DNA from cells treated with 0·62 nM NMG.)

Although these data do not really clarify the biochemical reason(s) for the differing resistance to NMG in these two strains, they do demonstrate that (at least to a first approximation) 7-methylguanine is not excised from *E. coli* DNA but that O^6-methylguanine is totally excised (at least in B/r).

Evidence for repair synthesis   The degradation of DNA produced by ionizing radiation is due to two processes. One is the initial single-strand or double-strand breaks induced by the radiation (p. 253). The other is due to the erosion process. The effects of radiation-induced breakdown and the kinetics of the process have been studied by Kaplan (1966). It is known that in *B. subtilis*[12] the degradation is increased by actinomycin D (Grady and Pollard, 1967). This antibiotic inhibits RNA synthesis (p. 413) and the synergism between actinomycin and irradiation reflects the requirement for transcription and translation in repair synthesis. However under these synergistic conditions, DNA-degradation was continued after irradiation had ceased. Thus it appears that erosion is achieved by constitutive nuclease(s) and it is the synthetic step which requires the *de novo* synthesis of a protein.

Further evidence of this type of breakdown of DNA resulting from enzymic degradation following irradiation (and presumably excision) is that DNA-breakdown is much reduced in most Hcr⁻ (*uvr⁻*) strains (Emmerson and Howard-Flanders, 1965). This is not the case with *E. coli* B_{s-1} (and other 'reckless' strains, p. 434) in which extensive DNA-degradation follows UV-irradiation (Suzuki, Moriguchi and Hori, 1966).

Repair synthesis is more simply demonstrated. The technique is based on a density-labelling technique. Let us consider DNA-synthesis in a system in which two interchangeable nucleosides (X and Y) of different density can be incorporated into DNA. The cells are grown in a medium in which X is incorporated into the molecule thus:

3′. . . N-N-N-X-N-X-X-N-N-N-N-N-X-N- . . . 5′
5′. . . N-X-N-N-X-N-N-X-X-N-X-X-N-N- . . . 3′

It is now irradiated or alkylated to produce a lesion (1)

3′. . . -N-N-N-X-N-X-X-N-N-N-N-N-X-N- . . . 5′
5′. . . -N-X-N-N-X-N-N-X-X-N-X-X-N-N- . . . 3′
        (1)   (2)

The cells are now transferred to medium which contains radioactive Y (Ẏ). We are trying to establish whether normal, semi-conservative DNA-replication or excision, erosion and limited repair synthesis (say between 1 and 2 above) occurs. In the former case (replication), the result will be that Ẏ becomes incorporated into extended duplices of hybrid density (see p. 389). In the latter (repair) the product will be:

3′. . . -N-N-N-X-N-X-X-N-N-N-N-N-X-N- . . . 5′
5′. . . -N-X-N-N-Ẏ-N-N-Ẏ-Ẏ-N-Ẏ-X-N-N- . . . 3′

In this structure the Ẏ residues are present in a short stretch of a chain containing essentially X-residues. Thus if the DNA is isolated and banded by buoyant density centrifugation in CsCl (p. 154) the radioactivity will be associated with molecules of approximately the same density as the original DNA. Similarly if the DNA is denatured, it will still be associated with single-strand DNA with the same density as that in the initial cell. There are two ways of doing the experiment: either X can be dThd and $\overset{*}{Y}$ [^{3}H] 5-Br-dUrd or alternatively X can be 5-Br-dUrd and $\overset{*}{Y}$ can be [^{3}H] dThd.

The first convincing demonstration of repair synthesis by this technique was achieved by Pettijohn and Hanawalt (1963, 1964). They employed the method in which thymine auxotrophs of *E. coli* were incubated (following UV-irradiation) in a medium lacking Thy or dThd but in the presence of [^{3}H] 5-Br-dUrd. The radioactivity was found to be associated with 'light' duplex DNA. Denatured DNA (alkaline CsC1 gradients—see p. 152) also had its radioactivity associated with light DNA. When DNA was sonicated and denatured so that fragments ($\overline{MW}_w$) of the order of 10^6) were obtained, again the radioactivity was found to be associated with the light DNA. The conclusion that can be reached from this is that the length of eroded and repaired DNA is small compared to 10^6 Daltons.

A recent very thorough study of repair synthesis following treatment of tissue-culture cells with alkylating agents has been made by J. J. Roberts, A. R. Crathorn and their colleagues. The cells are most sensitive to the agents during the DNA-synthetic phase of the cell cycle (Crathorn and Roberts, 1966). The experiments are summarized by Roberts, Brent and Crathorn (1968). Roberts, Crathorn and Brent (1968) perfected two applications of the assay for repair synthesis for the purpose of studying the process in tissue-culture cells. They employed the method of Pettijohn and Hanawalt (above) and also performed the experiment the other way round so that the cells were grown in 5-Br-dUrd (so that X above is br^{5}U) and then studied repair in the presence of [^{3}H] dThd (so that $\overset{*}{Y}$ is [^{3}H] T). The experimental system that was employed was one in which two cell lines (HeLa, see p. 124 and V79-379A obtained from Chinese hamsters) were treated with either NMG, or methyl methane sulphonate (see p. 60). Previous work had established that the two cell lines had a very different relative sensitivity to the lethal consequences of alkylation with these agents; HeLa cells show exceptionally high sensitivity to the two nitrosamide derivatives (i.e. the ones that produce O^6-methylguanine p. 60; Roberts,

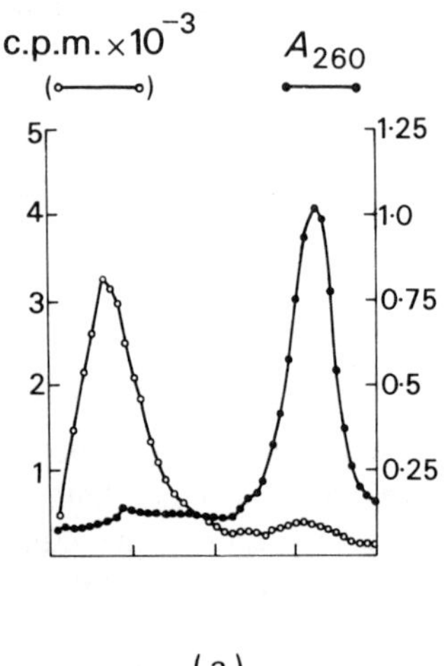

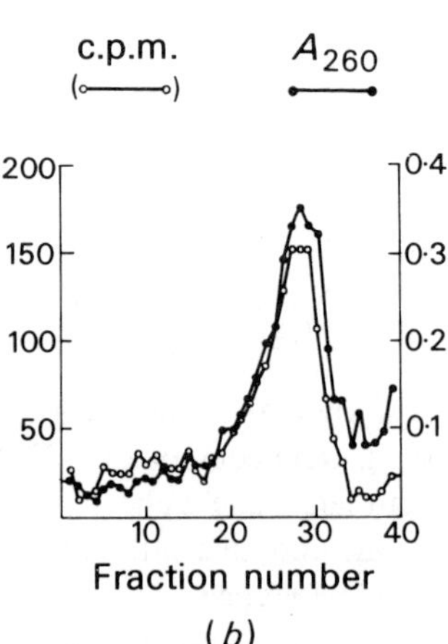

FIG 8.9   Repair synthesis in Chinese hamster cells treated with NMG. Cells were transferred to 5-Br-dUrd and after 2 h were treated with 50 nM NMG for 1 h. They were transferred to medium lacking NMG but containing 5-Br-dUrd and [^{3}H] dThd. DNA was isolated after 3 h and fractionated by buoyant density centrifugation. The graphs are distributions in CsCl gradients at pH 12·5 (p. 188). (a) total DNA; (b) peak I from (a) re-run under identical conditions. Data of Roberts, Pascoe, Smith and Crathorn (1971).

Pasco, Plant, Sturrock and Crathorn, 1971). The work on repair was performed as indicated above. An example of the experimental methodology is given in Fig. 8.9.

The experiment illustrates one difference in technique which can only be applied to a new system if the necessary controls are done (as they were here). Rather than using [³H] 5-Br-dUrd, the labelling was done with [³H] dThd in the presence of a vast molar excess of cold 5-Br-dUrd. The reason for transfer of the cells to 5-Br-dUrd before treatment with NMG was to ensure that no 'light' semi-conservative, replicative DNA synthesis was going on utilizing pre-existing intracellular pools of dThd (or other T-residue precursor). Note that the amount of *de novo* light synthesis during the repair phase is very small (Fig. 8.9(*b*)). In order to ensure that the counts under the light DNA peak are genuine, and not merely the tail from the heavy DNA counts (reflecting chromosome-replication in the 5-Br-dUrd) this peak was re-run (Fig. 8.9(*c*)) and the genuine identity of the counts was confirmed.

The result of the survey (Roberts, Pascoe, Smith and Crathorn, 1971) was to show that the greater sensitivity of HeLa cells to the nitrosamides is apparently unrelated to the repair process as there was comparable repair synthesis in both cell types. A similar conclusion (Ball and Roberts, 1970) was reached in a study of the effects of sulphur mustard on two cell lines of the Yoshida sarcoma,[13] one of which was resistant and the other sensitive to this agent. Both cell types excised alkylated residues (as assessed by the loss of ^{35}S-counts from cellular DNA) and achieved repair synthesis with comparable efficiency. Probably there is a nonrepairable lethal reaction with DNA, forming a small proportion of the total 'lesions', that accounts for the great sensitivity of these cells.

Other biological consequences of chemical reactions with DNA

Apart from the consequences of repair and the lethal effect (by rendering chromosome replication impossible) of such processes as double-strand breakage, unrepaired cross-linking and so on, what are the other biological consequences of the modification of base residues in DNA? I mention briefly three types of effect that can be considered; however it must be emphasized that this book is not the place to consider these processes systematically. Most chemicals that react with DNA are either very reactive electrophiles (or are metabolized to such compounds) such as NMG, sulphur mustard, etc., which alkylate proteins as well as nucleic acids, or are highly 'hydrophobic' compounds such as the aromatic hydrocarbons and aromatic amines which can presumably have a dramatic effect on membrane processes. Therefore, with the exception of mutagenesis, the greatest care must be taken in correlating any biological change with the reaction involving nucleic acids.

Transcription   If X-irradiated DNA is employed as template for RNA polymerase the efficiency of nucleoside triphosphate incorporation is impaired and the extent of the inhibition is proportional to the radiation dose (Zimmerman, Kröger, Hagen and Keck, 1964). In a similar experiment, transcription of homopolymers treated in various ways has been studied. For these experiments RNA polymerase from *Micrococcus lysodeikticus* (see footnote to p. 236) was employed. Although this is strictly a DNA-dependent polymerase (p. 225), it is possible for it to operate as an RNA-dependent polymerase if molecules such as poly C are used as template (Fox, Robinson, Haselkorn and Weiss, 1964). Ono, Wilson and Grossman (1965) demonstrated that UV-irradiation of poly C reduced its template-activity for this reaction. If poly C is treated with hydroxylamine, its template efficiency for the incorporation of GTP into polynucleotide is reduced. This inactivation can be partly recovered by using a mixture of ATP and GTP. The A-residues introduced in this way were shown by nearest-neighbour frequency analysis to be flanked by G-residues (Phillips, Brown, Adman and Grossman, 1965; Wilson and Caicuts, 1966). It is thus possible that hydroxylamine results in direct mis-reading by the polymerase.

Treatment of cells with alkylating agents or the irradiation of cells both result in changes in RNA metabolism. There is evidence that RNA synthesis is inhibited (sometimes this inhibition is followed by a stimulation) and that (on the basis of both the nucleotide composition of the newly synthesized RNA and on the evidence of the shape of the 'rapidly labelled RNA' distribution) the conclusion is reached that different fractions of RNA are synthesized following the treatment. The same sort of evidence can be obtained from experiments involving whole animals. I am not going to discuss these data as it would involve straying too far from the central purpose of this book which is to illustrate experiments which lead to a fairly unambiguous interpretation of processes involving nucleic acids. There are so many possible consequences of irradiation and alkylation (on ribosomes, on the endoplasmic reticulum at which so much eukaryotic protein synthesis occurs, on precursor pool sizes, on the nuclear envelope, on repressors and so on) that at the moment it is impossible to identify the nature of events which produce these transcriptional changes. It is my hope that any reader who is involved in experiments of this sort will find in the rest of the book useful indications of the properties of nucleic acids which might help him to unravel results in this difficult field.

Mutagenesis   Mutagenesis arises during the unfaithful replication of DNA (or RNA in an RNA virus). Clearly the presence of a modified base in the template might lead to erroneous or ambiguous selection of the incoming nucleoside triphosphate so that mutations can arise. Compounds such as NMG are in everyday use as bacterial mutagens. However a study of the process of mutagenesis itself can throw much light on the nature of the changes in the template which are originally produced. The nature of the mis-pairing that can occur and its relation to mutagenesis is discussed in the context of the secondary structure of nucleic acids (p. 351).

And cancer?   'Carcinogens are substances which react with DNA to form a chain with a modified base in it. When this structure is replicated the modified base produces mis-pairing and a mutant daughter chromosome is produced. The cell that arises from this process is therefore of different genotype. There is (are) a (some) genetic locus (loci) which, when mutated in this fashion, produce as their phenotype a cancer cell.; This is the so-called 'somatic mutation theory of cancer'. It is not a popular view to hold these days. At least not in that form.

There are certainly some tough problems raised by it. For example: some tumours are produced by viruses; not all carcinogens react with DNA; there is reason to suppose that carcinogenesis is not a one-step process at all but rather consists of 'induction' followed by stages lumped together and called 'promotion'; for those carcinogens that do react with DNA, the correlation between extent of reaction and carcinogenic potency is not terribly impressive; X-irradiation (which can perhaps be regarded as a carcinogen which reacts with DNA) has some strange effects, for example irradiation of one organ may produce tumours in a different organ.

Is DNA concerned in carcinogenesis at all? Certainly it is quite possible to defend the position that it is not. For myself I think the simplest and most plausible position is that it must be involved in some way. After all, most carcinogens do react with DNA and oncogenic viruses are all either DNA viruses or RNA viruses that can put a copy of their genome into a DNA form (p. 224). So how can the somatic mutation theory be patched up? Something like this: Reaction with DNA is involved in the process of tumour initiation but not necessarily promotion. The carcinogen itself is not necessarily a reactant with DNA but is metabolized to a 'proximal carcinogen' (like the aromatic amines, the nitrosamines and the aromatic hydro-carbons—see p. 248). The effect of the reaction is not necessarily to 'mutate a cancer gene'. It could be that the repair processes that follow reaction with DNA result in either the release of a 'lysogenic' DNA virus or the transcription of an RNA virus's complementary genome. Alternatively the repair machinery could encourage recombination between the alleles of the chromosome pairs (cancer is a disease of diploid organisms). We must wait and see. However there is every reason for Cancer Research to be concerned with nucleic acids for some time yet.

Notes to chapter 8

1 In many ways these two compounds are the most convenient mustards to use for experimental purpose. Both can be obtained commercially as ^{35}S-labelled compounds. However they are extremely toxic (full mustard is of course mustard gas). They can be converted into innocuous compounds by treating with strong alkali, boiling in water or (and this is the most convenient way of decontaminating apparatus) by treating them with a freshly made slurry of 'bleaching powder' and water. Dry bleaching powder must not be used—it ignites mustard gas. Mustard gas is commercially available. Half mustard is not (at least not as far as I know except for the radioactive preparations). However it is very easy to make half mustard in the laboratory by the reaction between $HO(CH_2)_2SNa$ and $Cl(CH_2)_2Cl$. The reaction is conducted in the fume-cupboard as, apart from the toxic nature of the product, the smell is atrocious. Sodium hydride (11 g) is carefully dissolved in a mixture of β-mercaptoethanol (36·8 ml) and methanol (300 ml). Ethylene chloride (300 ml) is added and the whole lot is kept at 0°C for a few days (3–7). A heavy precipitate (NaCl) is deposited. The liquid is decanted, the solvents are removed *in vacuo* (water pump), the residue is dissolved in ether (300 ml) extracted with ice-water (to remove NaCl), dried (Na_2SO_4) and the ether is removed *in vacuo*. The product (refractive index, $n_D^{27} = 1·5110$) is suitable for alkylation studies without further purification or distillation. (Recipe of J.H.P. based on Rutenberg, Persky and Friedman, 1952.)

2 Such DNA is obtained by growing a thymine auxotroph of *E. coli* in a medium containing methyl-deuterated Thy.

3 Broth. Crude partial proteolytic digests of proteins such as casein, meat extract and so on, used for growing micro-organisms. It is not clear why this particular usage (rather than soup or gravy) has caught on amongst microbiologists.

4 The total radiation can be calculated from the radiation flux from the X-ray machine and the exposure time. In practice, the flux is measured in an ionization chamber or alternatively (for approximate work) the total radiation can be measured in a radiation dosimeter.

5 This substance is known as dimethyl nitrosamine. It is not a very happy name as the 'parent compound', nitrosamine does not exist. Nevertheless these compounds have become known as the nitrosamines and it looks as though the name is here to stay. Interest in them originated when it was recognized that nitrosodimethylamine ($Me_2N–NO$), which was used as a solvent for an industrial process, is an extremely potent and dangerous carcinogen (Magee and Barnes, 1956).

6 So-called because it was used for making margarine yellow (and hence to look like butter) before its carcinogenic properties were discovered.

7 Diaphorase: an NADPH-dependent quinone reductase; the commercial source of the enzyme is *Clostridium Kluyveri*. There is a problem however; the preparations tend to be contaminated with nucleases (Szybalski and Iyer, 1964).

8 Although animals such as *Paramoecium* are in many ways very difficult to use for genetic analysis on account of their extensive gene-amplification (p. 363), they do have one advantage over bacteria. During the cell cycle, DNA synthesis occurs only during one period of time (the S-phase—S standing for 'synthetic'). This phase can be studied if the cells are in synchrony. Methods of synchronizing such cells (or tissue culture cells—see p. 117) are outside the scope of this book but let it be said that there are several methods of arresting the mitotic cycle of eukaryotic cells at particular points. One method has been described already in a different context—p. 116.

9 In Thy, the substituent on C-5 ($–CH_3$) has a van der Waals radius of 2·0 Å. Thus those derivatives of Ura with substituents at C-5 with a radius of this order of magnitude ($–Cl$ 1·8 Å, $–Br$ 1·95 Å or $–I$ 2·15 Å) behave as analogues of Thy in DNA. In Ura the substituent at C-5 ($–H$) has a radius of 1·2 Å. The radius of $–F$ is 1·35 Å and so fluorouracil is an analogue of Ura rather than Thy.

10 Intercalation is described elsewhere (p. 410) in detail. Essentially the intercalated substance is sandwiched between adjacent base pairs in the duplex.

11 This requires thymine auxotrophs—see p. 176.

12 Any experiment which involves the use of actinomycin D to inhibit transcription *in vivo* must be performed on a Gram positive bacterium. Although the drug is a universal inhibitor of RNA-synthesis it has no effect on most Gram negative bacteria (such as *E. coli*) if it is added to a culture, as it does not enter the cells.

13 An experimental tumour in rats.

9   Primary structure of DNA

The determination of the primary structure of DNA is beset with enormous technical difficulties. For one thing DNA molecules are extremely large (the smallest, φX 174 DNA, contains 5000 nucleotides). For another there are no base-specific deoxyribonucleases known. Thus the only extended DNA structure of complex defined sequence that is of known primary structure is Khorana's tRNA^{Ala} gene (p. 233). So briefly what do we know about the primary structure of DNA and what is the future?

For the present, we can establish the base ratios of DNA samples and the 'nearest-neighbour frequency'. This latter phrase covers the relative proportion of the sixteen possible pairs of consecutive nucleotides. We can also obtain the DNA fragments containing runs of pyrimidine nucleotides or runs of purine nucleotides. We can enquire into the occurrence of minor bases in the DNA chains. We can also determine the basic structure of non-informational DNA (such as the polydAT described on p. 152).

For the future, there is no doubt that endonuclease IV (p. 211) must play a large role in DNA sequencing. Ling (1971) has characterized the oligomers obtained by the digestion of a small single-stranded phage DNA with this enzyme. Two other inroads into the primary structure of DNA can be envisaged. One type of structure to probe would be an isolated gene either for tRNA (see p. 372) or for proteins (see p. 366). The other possibility would be to employ alkylating agents to put 'handles' into the molecule and sequence round the alkylation sites (see p. 239).

The following sections describe techniques employed at present for studies on DNA primary structure together with some of the more important results.

Overall nucleotide composition

Base ratios

There are several methods of measuring base ratios of DNA. In this section, I have chosen only those three techniques which are commonly employed. Between them these methods cover practically any problem that is likely to be encountered.

Hydrolysis In general, if 2 mg or so of DNA is available the simplest thing to do is to hydrolyse DNA with acid and separate the bases and quantitatively measure their molar proportions (see p. 205). This is not possible with very small amounts of DNA. One of the two indirect methods can be employed in this case. Alternatively, if the DNA can be obtained uniformly labelled with ^{32}P, it is possible to hydrolyse it to the 3'-nucleotides; these substances can be fractionated and counted. The method for the hydrolysis is to treat the DNA with a suitable nuclease. In practice it is best to employ successively the Staphylococcal nuclease and the calf spleen nuclease. Details of this procedure and a method for establishing that the hydrolysis is complete are described on p. 272 in connexion with the nearest-neighbour method. A survey of DNA base ratios that contains da covering a wide range of GC-contents is by Belozersky and Spirin (1960). Samples suitable for the calibration of the following indirect methods can be chosen from their list.

Correlation with density   The GC-content of double-stranded DNA is related to its density, (see p. 127) at 25°C by the following equation (Schildkraut, Marmur and Doty, 1962),

$$GC = \frac{\rho - 1\cdot660}{0\cdot098} \times 100$$

The density of DNA may be measured by buoyant-density centrifugation. If the DNA is obtainable in optically dense quantities, the experiment is most conveniently performed in the analytical ultracentrifuge (see p. 381). Alternatively the preparative centrifuge (using a swing-out rotor) may be used; the position of the DNA zone is established by fractionating the gradient and measuring extinction, radioactivity or (for certain phage and bacterial DNA preparations) infectivity or transformation.

There are several methods of calculating the buoyant density of DNA from the position of its zone. The subject is discussed in more detail on p. 389. For general purposes, the simplest thing to do is to compare the position of the zone with that of a zone of marker DNA included in the same gradient. Methods of making up caesium chloride gradients are described on p. 149. The initial solution is made up containing the two DNA samples. If ρ_m is the known density of the marker, the density of the unknown DNA is given by the equation:

$$= \rho_m + \frac{\omega^2}{2\beta_0} (r_s{}^2 - r_m{}^2)$$

Where ω is the angular velocity (see p. 131), r_s is the distance from the rotor-centre to the maximum of the DNA zone and r_m is the distance from the rotor-centre to the maximum of the marker DNA zone and β_0 is a function which can be read off from the graph (Fig. 9.1).

If for some reason it is inconvenient to include a DNA marker in the solution, the same equation can be applied with the initial density of the solution substituted for ρ_m and the

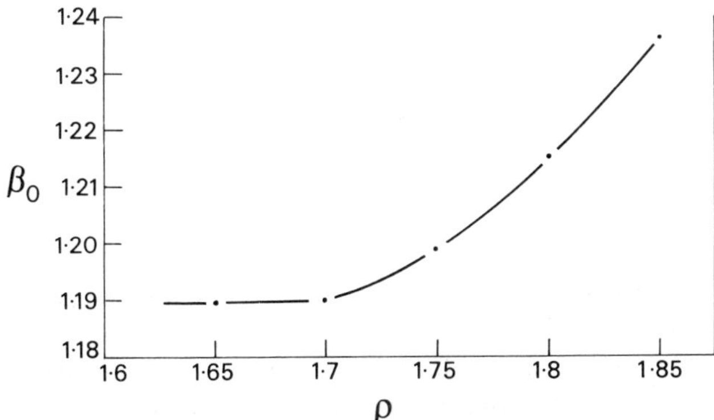

FIG 9.1   Values of β_0 as a function of density of the initial solution, plotted from data of Ifft, Voet and Virograd (1961).

'isoconcentration distance' r_i substituted for r_m. The value of r_i can be determined from the following equation,

$$r_i = [\tfrac{1}{2}(r_b{}^2 + r_t{}^2)]^{\tfrac{1}{2}}$$

where r_b is the distance from the rotor-centre to the bottom of the tube and r_t is the distance from the rotor-centre to the top of the gradient (meniscus). This equation is taken from the superb review of buoyant density centrifugation by Vinograd and Hearst (1962).

Melting temperature   When aqueous solutions of DNA are slowly heated, there is no change in secondary structure until a temperature is reached when co-operative denaturation occurs and the molecules denature ('melt') over a temperature range of a few degrees. The progress of the melting can be followed by measuring the extinction at 260 nm as denaturation produces a hypochromic shift of around 40%. The idealized appearance of a plot of extinction (or more usually the ratio of extinction at temperature T to extinction at 25°C) versus temperature is shown in Fig. 9.2. This type of plot is referred to as the melting profile. The melting temperature T_m is defined as the temperature at which the hypochromic effect is 50% of its maximum value.

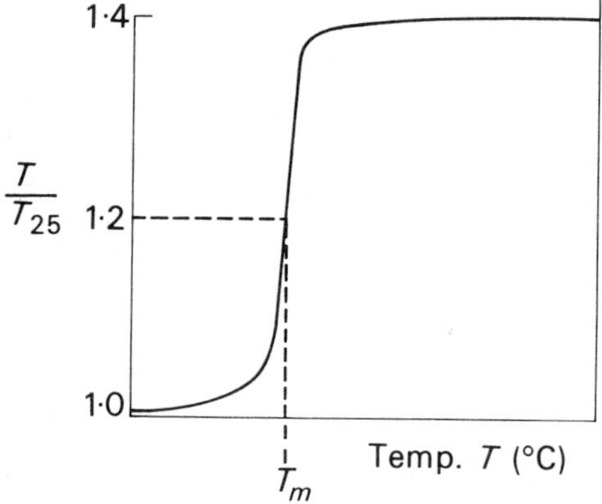

FIG 9.2   Melting profile of DNA. T_m is the melting temperature.

The melting of DNA is discussed further on p. 338 where an alternative method measuring base ratios is described. Here we merely need to point out that the value of T_m is a function of the ionic strength of the buffer and the nucleotide composition of the DNA. For solutions of different DNA samples in a standard buffer, the higher the GC-content, the higher the value of Tm. It is thus possible to deduce the nucleotide composition from these values. The relationship that is usually used is that of Marmur and Doty (1962) which refers to solutions of DNA in 'SSC buffer' (0·15 M sodium chloride, 0·015 M trisodium citrate). The equation (which gives GC as a mole percentage) is:

$$GC = (T_m - 69\cdot3) \times 2\cdot44$$

If the GC-content is to be measured this way, it is essential that certain precautions be taken. The DNA is thoroughly dialysed against SSC and introduced into well-stoppered spectrophotometer cells with a temperature-controlled cell holder. Water (or glycol) is circulated through the cell holder from a heating bath. The temperature of the cell holder is monitored with a miniature thermocouple. Methods of modifying a spectrophotometer for this purpose are described by Inman and Baldwin (1962) and by Szybalski and Menningman (1962). The temperature must be increased slowly (1 deg steps in the region of the 'melt' and equilibrated for 5–10 min at each temperature before a reading is taken). A correction must be made for the fact that water expands on heating (so the solution becomes more dilute). A graph for reading off this dilution factor is given (Fig. 9.3). Alternatively the instrument can do the compensation itself, if, in a separate cuvette, in the same holder, a solution of adenine with the same extinction as the DNA at 25°C is read in parallel during the melting. This suggestion is made by Szybalski and Menningham (above).

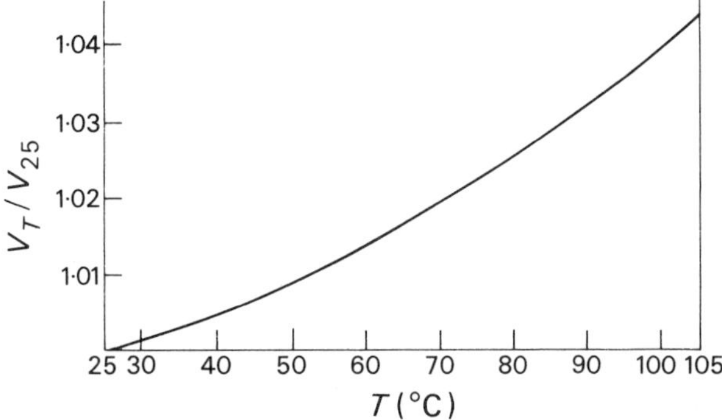

FIG 9.3   Effect of temperature on the expansion of water. V_T/T_{25} is the ratio of volume at temperature T to the volume of the same mass of water at 25°C, plotted from data of Szybalski and Menningman (1962).

Unusual bases

There have been no reports so far of any sugar other than deoxyribose in the sugar-phosphate backbone of DNA. The modifications in the 'odd' nucleosides in DNA (unlike those in RNA—see p. 205) consist solely of modifications in the bases. The types of modified base are also more limited than the wide range found in tRNA. This last conclusion may be largely an artefact of our very incomplete knowledge of the structure of DNA. As Chargaff (1968) has quite rightly pointed out, the identification of a very small proportion of minor components in a giant molecule of several million nucleotides is a job which will tax analytical chemical procedures to their limit. Hence we cannot pretend that a survey of the known modified bases in DNA is anything like a final catalogue of these substances.

The unusual bases in DNA fall into two classes. First of all there are bases which entirely replace one of the four normal bases. In these cases the odd base is incorporated into the DNA during DNA replication. An artificial example of this is the incorporation of 5-bromouracil into the DNA of thymine auxotrophs of *E. coli* (see p. 154). The only cases of DNA naturally synthesized in this way (with one of the pyrimidines replaced by an analogue) are found in the DNA of certain phage. The second class is the occurrence of a small proportion of modified bases that arise from the enzymic modification of DNA following its synthesis. Examples of these groups are given in the following sections.

Phage DNA   The T-even phage of *E. coli* contain no cytosine; in its place is 5-hydroxy-methylcytosine which is incorporated as the corresponding deoxyriboside triphosphate. These residues are glucosylated to varying extents (see p. 229). The nature of the side-chains and the proportion of the different glycosylated bases are summarized in Fig. 9.4. The data are from Kornberg, Zimmerman, Kornberg and Josse (1959). The figures correlate well with the relative activities of the different glucosylating enzymes present in cells infected with these three phage (p. 230).

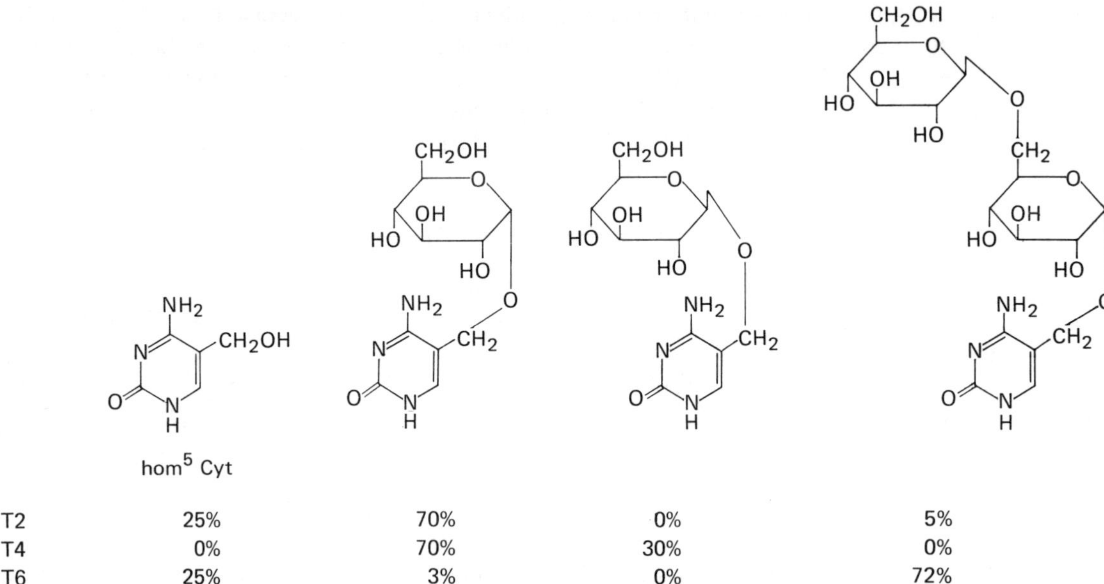

T2	25%	70%	0%	5%
T4	0%	70%	30%	0%
T6	25%	3%	0%	72%

FIG 9.4   Occurrence of hydroxymethyl cytosine and its derivatives in the DNA of T-even phage.

Other examples of DNA with different bases are in *B. subtilis* phage. In SP8 5-hydroxymethyluracil replaces thymine (Kallen, Simon and Marmur, 1962) and in phages PBS1 and PBS2, uracil replaces thymine (Takahashi and Marmur, 1962). In *Xanthomonas oryzae* phage XP12 5-methylcytosine replace cytosine (Kuo, Huang and Teng, 1968).

Modified bases   The occurrence of methylated bases in DNA from a variety of sources has been curveyed by A. N. Belozersky and his colleagues. The methylated bases consist largely of 5-methylcytosine and N^6-methyladenine. Belozersky's group identified these bases by paper chromatography. A DNA hydrolysate was spotted on to the paper and the chromatogram was developed with a mixture of *n*-butanol, water and concentrated aqueous ammonia (60:10:0·1 by volume) to isolate 5-methylcytosine, or with *n*-butanol and water in the presence of ammonia (43:7 by volume; a beaker of concentrated aqueous ammonia is placed in the bottom of the tank to saturate the air with ammonia vapour) to isolate N^6-methyladenine.

Most bacterial DNA samples contain both of these bases (Vanyushin, Belozersky, Kokurina and Kadirova, 1968). Enterobacteriaceae contain quite large amounts (0·5% of total bases) of N^6-methyladenine. *Alcaligenes faecalis* is another organism with a high proportion of this base (0·54%). Most *Bacilli* are of intermediate content (0·3% or so). Of the other species surveyed, the amounts were between 0 and 0·1% for the majority. The amounts of 5-methylcytosine were in general much smaller and several species had no detectable amounts at all. An interesting point about 5-methylcytosine is that the amounts differ from strain to strain. For example *E. coli* C contains 0·25%, whereas *E. coli* B has no detectable amounts; similarly *Bacillus brevis* S contains 0·45% whereas *B. brevis* R contains 0·06% *Alcaligenes faecalis* contains the largest amounts of methylated bases (0·54% of N^6-methyladenine and 0·62% of 5-methylcytosine) of any bacterium studied. Of the bacteria they studied, two (*Micrococcus lysodeikticus*, *Streptobacterium plantarum*) contain no detectable amounts of either methylated base. The significance of these data is unknown. The subject is extremely tantalizing; presumably the methylations have some role in the control of replication or (perhaps more plausibly) transcription. A systematic study of the extents of methylation of the DNA of (say) *A. faecalis* growing in different media and at different doubling times might give some clue.

N^6-Methyladenine is also found in phage DNA.

In animal DNA, the major modified base is 5-methylcytosine (Vanyushin, Tkacheva and Belozersky, 1970). The largest amounts are found in fish. An extremely fascinating aspect of the story is that the amounts of 5-methylcytosine vary from organ to organ in the same animal. Examples of this difference in DNA from different organs of a fish and a mammal are shown in Table 9.1.

TABLE 9.1   Occurrence of 5-methylcytosine in DNA from various organs of *O. gorbuscha* (species of salmon) and sheep taken from the data of Vanyushin, Tkacheva and Belozersky (1970). The values are molar percentages of total DNA bases with standard errors of the mean.

	Onorhynchus gorbuscha	Sheep
Liver	2·13 ± 0·06	1·13 ± 0·09
Kidney	1·98 ± 0·03	1·15 ± 0·05
Spleen	2·05 ± 0·06	1·07 ± 0·06
Muscle	1·80 ± 0·03	Not determined
Sperm	1·64 ± 0·04	0·76 ± 0·02

Fragmentary sequences

Nearest-neighbour-frequency analysis

After his isolation of DNA polymerase I (see p. 222), Kornberg exploited the enzyme in the determination of nearest-neighbour frequency. The basis of the method is extremely simple. In the reaction catalysed by the polymerase, the chain is extended from the 3′-end and the incoming nucleotide is the 5′-triphosphate (p. 12). On the other hand certain nucleases (such as the Staphylococcal nuclease, p. 208, or the calf spleen enzyme, p. 210) hydrolyse DNA to produce the 3′-nucleotides. In other words, looked at from the point of view of an α-phosphate group on the triphosphate, if it is employed as a substrate for DNA polymerase and the newly synthesized product is then digested with a nuclease, this phosphate enters the sequence attached to one nucleoside, and leaves it attached to a different one. This is shown diagrammatically below, in which the phosphate in question is represented in heavy type.

$$\cdots N{-}N{-}N{-}N{-}N{-}N \; + \; pp\mathbf{p}N' \xrightarrow{\text{polymerase}} \cdots N{-}N{-}N{-}N{-}N{-}N\mathbf{p}N'$$

further incorporation

$$N'\mathbf{p} \xleftarrow{\text{nuclease}} \cdots N{-}N{-}N{-}N\mathbf{p}N'{-}N{-}N{-}\cdots$$

(and other 3′— nucleotides)

Clearly, if the polymerase reaction is conducted with the four deoxynucleoside triphosphates ppG, pppC, pppT and pppA, in which the **p** is identified as a radioactive phosphate group (i.e. [α-^{32}P] dATP), then the radioactivity will end up in all four 3′-nucleotides. Moreover the ratio of the radioactivity in these four compounds will be a measure of the frequencies with which dATP has been incorporated into the growing chain following each of the four nucleotides; this is the frequence with which these nearest neighbours occur in the newly synthesized strand. The same experiment can be repeated with each of the other three triphosphates labelled in the same fashion and the frequencies of all 16 nearest-neighbour arrangements can be assigned. In order to illustrate the experimental method in more detail and explain the calculations, I describe the classical experiment by Josse, Kaiser and Kornberg (1961) for a particular microbial DNA.

Nearest-neighbour-frequency analysis of DNA from *Micrococcus phlei*   The incubations with DNA polymerase were performed on a very small scale using an enzyme preparation with a specific activity of 500 units/mg (see p. 222 for the reference to the method of isolation and definition of the unit of activity). The reaction mixtures (four in all) contained in a total volume of 0·3 ml, 66 mM glycine buffer pH 9·2, 6·6 mM $MgCl_2$, 1 mM β-mercaptoethanol, *Micrococcus* DNA (a total of 20 nmole of DNA-phosphate), the four triphosphates (a total of 5 nmole of each in the 0·3 ml with one of them labelled with ^{32}P in the α-phosphate and of very high specific activity, at least 50 Ci/mole) and about 1 unit of enzyme. After incubation (37°C, 30 min), the reaction was stopped by cooling to 0°C. Carrier DNA (5 mmole DNA-phosphate) and perchloric acid to a final concentration of 3·5% are added to precipitate the DNA. The nucleotides were removed from the preparation by successively dissolving (0·2 M alkali) and re-precipitating (perchloric acid) the DNA. The DNA was hydrolysed to the 3′-nucleotides by successive treatments with Staphylococcal nuclease and spleen nuclease. The DNA was dissolved in 0·5 ml of 4 mM tris HCl pH 8·6, 2 mM $CaCl_2$ and Staphylococcal nuclease (180 units) was added to make a final volume of 1 ml. After incubation (37°C, 2 h) the pH was reduced to 7·0 (hydrochloric acid) for digestion with the spleen enzyme. The total incubation time for this stage was 3 h. Enzyme (0·2 unit) was added at 0·1 and 2 h. The completeness of digestion was assessed (95% or more of the ^{32}P should be sensitive to alkaline phosphatase).

Such mixtures of 3′-nucleotides are separated electrophoretically (see p. 46). The four nucleotide spots are identified under UV light (p. 46), cut out, eluted and counted. The data obtained in the experiment are summarized in Table 9.2.

TABLE 9.2   Experimental data from a nearest-neighbour determination with *M. phlei* DNA. The numbers are c.p.m. in the nucleotides isolated from incubations with polymerase and nucleases as described in the text. Data from Josse, Kaiser and Kornberg (1961)

3′-nucleotide	Initial reaction with labelled			
	dATP	dTTP	dGTP	dCTP
Tp	873	1665	3490	4130
Ap	1710	2065	2500	4300
Cp	4430	2980	7730	6070
Gp	4690	3945	4960	8200

The next job is to normalize these four columns of figures to give the relative frequencies of the four classes of doubles obtained from DNA synthesized using each of the four labelled triphosphates. These normalized data are shown in Table 9.3.

Finally these four sets of nearest-neighbour data have to be weighted according to the nucleotide composition for the newly synthesized DNA. If the two strands of the original template have been copied with equal fidelity, these values (the 'incorporation ratios') should be the same as the base ratios of the template. In fact it is not necessary to make this assumption and it is better to calculate the nucleotide composition from the nearest-neighbour data and compare these with the known base ratio for this DNA as a test of the efficiency and

TABLE 9.3   Normalized values obtained from the data of Table 9.2. See text for discussion

T-A	0·075	T-T	0·157	T-G	0·197	T-C	0·182
A-A	0·146	A-T	0·194	A-G	0·134	A-C	0·189
C-A	0·378	C-T	0·279	C-G	0·414	C-C	0·268
G-A	0·401	G-T	0·370	G-G	0·265	G-C	0·361
Totals	1·000		1·000		1·000		1·000

fidelity of the synthesis *in vitro*. The way in which it is done is as follows. Let us call the incorporation ratios a, t, g and c so that:

$$a + t + g + c = 1·000$$

Now using the normalized values from Table 9.3 we can write four equations which are mathematical descriptions of the fact that the total amount of (for example) dGuo incorporated as pppG must be recovered as a mixture of 3′-nucleotides or:

$$\text{T-G} + \text{A-G} + \text{C-G} + \text{G-G} = \text{G-A} + \text{G-T} + \text{G-G} + \text{G-C}$$

Re-writing this in terms of a, t, g and c, we obtain:

$$g = 0·401a + 0·370t + 0·265g + 0·361c$$

Similarly it is possible to write down the corresponding equations for a, t and c. If these five simultaneous equations (the definition that a, t, g and c total unity is the fifth one) are solved, we find the values $a = 0·164$, $t = 0·162$, $g = 0·337$ and $c = 0·337$. When these figures are compared with the base ratio for the primer DNA (A = 0·162, T = 0·165, G = 0·338 and C = 0·335) it can be seen that the copying has apparently been faithful. The final job is to weight the values in order to take account of the incorporation ratios. The results then turn out as in Table 9.4.

TABLE 9.4   Normalized values for nearest-neighbour frequencies for DNA from *M. phlei*. The 16 numbers total 1·000 and the horizontal lines add up to the incorporation ratios (thus the top line adds up to 0·162 (= t).

T-A	0·012	T-T	0·026	T-G	0·063	T-C	0·061
A-A	0·024	A-T	0·031	A-G	0·045	A-C	0·064
C-A	0·063	C-T	0·045	C-G	0·139	C-C	0·090
G-A	0·065	G-T	0·060	G-G	0·090	G-C	0·122

Several applications of the method are discussed in the next section. However these data themselves illustrate one important result which can be obtained from such measurements, namely a proof that the two strands of the double helix are anti-parallel.

This proof can be seen by considering two hypothetical strands of DNA in the syn-parallel and anti-parallel configurations (see below—the structure on the left is syn-parallel, that on the right is anti-parallel).

5′-N-N-N-N-N-A-C-N-N-N-T-A-N-N-N-3′   3′-N-N-N-N-N-A-C-N-N-N-T-A-N-N-N-5′

5′-N-N-N-N-N-T-G-N-N-N-A-T-N-N-N-3′   5′-N-N-N-N-N-T-G-N-N-N-A-T-N-N-N-3′

In the syn-parallel structure there is a T-A in one chain opposite an A-T in the other. Clearly this will obtain where even either of these doublets appears in a strand so for the complete duplex, T-A = A-T. This is not the case for the anti-parallel structure as the A-T in one strand is opposite another A-T in the other. However the other pair of nucleotides shown do determine one another in this case; in other words everywhere there is a T-G in one strand there is a C-A in the other so T-G = C-A (in the complete structure). These particular examples were chosen arbitrarily; if all sixteen doublets are considered in this way, both models will imply that the frequencies of G-G and C-C, and A-A and T-T are identical. The syn-parallel model implies in addition that the following pairs of frequencies are identical: T-A and A-T, C-A and G-T, G-A and C-T, T-G and A-C, T-C and A-G, C-G and G-G. The anti-parallel model implies the identity of the following pairs of frequencies: C-A and T-G, G-A and T-C, C-T and A-G, G-T and A-C. Inspection of the data of Table 9.4 gives support to the anti-parallel model.

If we accept that any double-stranded DNA is anti-parallel, the argument above leads to an additional conclusion. Most of the doublet frequencies are each identical to the frequency of another doublet. However four doublet frequencies A-T, T-A, C-G and G-C (those in which a nucleotide is followed in sequence by its complementary equivalent) are completely independent of any other doublet. This is obvious as the only requirement of the structure for A-T is that there should be another A-T in the complementary DNA strand.

Other applications of nearest-neighbour-frequency analysis One technically important application of the method is a test for the fidelity of copying of DNA by a polymerase preparation. The same type of experiment can be carried out with RNA polymerase. In this latter case, one is enquiring whether the nearest-neighbour frequencies in the newly synthesized RNA are complementary to the frequencies in the template DNA strand. The experiment is technically rather simpler as the RNA can be hydrolysed to the 3'- (and 2'-)nucleotides with alkali (see p. 204) and nucleases are not required.

Nearest-neighbour-frequency analysis can be employed to confirm the structure of polynucleotides of alternating sequence, such as poly dAT (see p. 278). In this case incubation of the polymerase reaction with poly dAT template and [^{32}P] dATP should yield (following degradation) ^{32}P present uniquely in dTMP. The structure of some naturally occurring DNA molecules of simple repeating sequence has been established in this way.

The procedure can also be employed to confirm the single-strandedness of certain viral DNA molecules. Swartz, Trautner and Kornberg (1962) established that φX 174 DNA is single stranded by comparing the nearest-neighbour frequencies obtained following incubation with the polymerase for periods of time corresponding to the replication of only a fifth of the genome or corresponding to six times the genome. In the former case, the incorporation ratios differed from the base ratios and the expected relationships between the pairs of doublets were not obtained. In the second case, the enzyme has used its own synthetic product as template to make more + strand (see p. 232) and the resulting incorporation ratios and doublet frequencies were in good agreement with those predicted from the former data.

An application of this technique which is of great biological interest is a comparison of nearest-neighbour-frequency patterns in DNA from different organisms. The assumption is that the extent of the relatedness of nearest-neighbour patterns is a measure of the evolutionary relatedness of the organisms. It is outside my scope (and knowledge) to evaluate this type of correlation in any detail. The most valuable enquires of this type are probably directed to the rather more accessible question of whether viral DNA is similar to the DNA of the host cell. If it is, it supports the view that the virus in question contains a bit of DNA derived from the organism in something like its present form. Several coliphage were analysed in this way by Swartz, Trautner and Kornberg (1962). Their conclusion was that DNA from the temperate phage, λ is very similar to host cell DNA but that that of the T-phage was dissimilar and apparently of exogenous origin. A similar analysis of mammalian viruses has been made by

Subak-Sharpe *et al.* (1966). In this case it appears that the picodna viruses (polyoma and Shope papilloma) are derived from mammalian DNA (which has a fairly constant nearest-neighbour pattern irrespective of source). The large DNA viruses (vaccinia, herpes, pseudorabies and equine rhinopneumonitis) have very different patterns and are presumed to be of a different origin.

A recent result concerns the very interesting cauliflower mosaic virus. This particle, unlike most plant viruses (see p. 32), is a DNA virus. As with other small DNA viruses, it has been shown that the nearest-neighbour-frequency pattern resembles that of the host-cell DNA (cauliflower curd in this case—Russell, Follett, Subak-Sharpe and Harrison, 1971).

Pyrimidine and purine oligomers

The pyrimidine oligomers   *N*-Glycoside bonds are more easily hydrolysed if the base attached to the sugar is a purine than if it is a pyrimidine. In RNA, where the bonds are less labile in acid anyway, it is almost impossible to hydrolyse the pyrimidine *N*-glycoside bonds at all (p. 55). With DNA it is possible to choose conditions in which the sugar phosphate backbone is undamaged (essentially at least) and the purines are removed to produce a molecule known as apurinic acid.[1] Subsequent treatment of the apurinic acid with either alkali or stronger acid yields the pyrimidine oligomers. The two reactions are illustrated below for a hypothetical DNA molecule. In the sequence for the apurinic acid, d stands for deoxyribose.

...C–C–T–A–T–A–G–C–A–T–T–T–C–G–G–A–T–G–A–.....

$$\downarrow$$

...–C–C–C–d–T–d–d–C–d–T–T–T–C–d–d–d–T–d–d–....

$$\downarrow$$

...–C–C–C– –T–   –C– –T–T–T–C–   –T–

Note that in this case the oligomers are flanked by phosphates on both their 5'- and 3'-ends. The hydrolysis of apurinic acid involves β-elimination reactions of the deoxyribose phosphate moieties. The mechanism is discussed by Dekker, Michelson and Todd (1953).

The methods chosen depend on the purpose of the experiment. If it is simply required to prepare the pyrimidine oligomers (so that dupurination and hydrolysis of apurinic acid can occur concurrently) the conditions are simple. A 2% solution of DNA in 0·1 M sulphuric acid is heated to 100°C for 35 min (Shapiro and Chargaff, 1957).

If on the other hand, it is desired to make apurinic acid in as undegraded form as possible, it is preferable to use formic acid for the depurination step. The procedure of Peterson and Burton (1964) gives preparations of average molecular weight 57 000. The important thing is to keep the acidified solutions of apurinic acid (free acid form) cold throughout (except for the initial hydrolysis). DNA solution is completely desalted (dialysis) and incubated with two volumes of 98% formic acid at 30°C overnight; formic acid is removed by repeated extraction with ether and the preparation is added to a column of Dowex 50 (hydrogen form) and eluted with water; the apurinic acid comes off and the purines remain on the column. The column is operated at 0°C and the volume of solution is reduced by freeze drying (if it is completely dried, the preparation warms up and is degraded). The sodium salt of the acid is obtained by neutralizing this solution with sodium hydroxide, this salt is sufficiently stable to be completely freeze-dried. The same authors, describe a modification of this method to obtain the pyrimidine oligomers. It is based on the reaction of diphenylamine (see p. 173) with the deoxyribose moieties which have no base attached. The incubation conditions are exactly as above except that the 98% formic acid contains 3% (w/v) diphenylamine. The work-up procedure is just the same except that the product from the column (a mixture of pyrimidine oligomers) is more stable and can be freeze-dried in the acid-form—this is a useful way of removing the last traces of formic acid. According to Peterson and Burton, their method is preferable to the sulphuric-acid procedure and leads to less degradation of the oligomers.

The purine oligomers   The preparation of these fragments of DNA requires the selective removal of the pyrimidine bases. This requires a more complicated (and extremely ingenious) reaction sequence and was developed by Chargaff. The method is based on the reaction between pyrimidine nucleosides and hydrazine (see p. 69 for details of this reaction). In addition to fragments of the base, the product is the sugar hydrazone. The hydrazine can be removed from this hydrazone with excess benzaldehyde to form the benzalazine (see below: RCH=O is the sugar).

$$2C_6H_5CH{=}O + R{-}CH{=}N{-}NH_2 \rightarrow C_6H_5CH{=}N{-}N{=}CHC_6H_5 + R{-}CH{=}O$$

If these reactions are performed on DNA, the products are benzalazine and apyrimidinic acid, which can be hydrolysed to yield the purine oligomers with alkali. The reaction is the same type of β-elimination mentioned on p. 275 for apurinic acid; obviously acid hydrolysis cannot be employed in the present case. The details of the procedure (Temperli, Türler, Rüst, Danon and Chargaff, 1964) are as follows. Freshly distilled anhydrous hydrazine (1 ml) and 0·1 g of thoroughly dried DNA are sealed in an ampoule and heated to 60°C for 4 h. The hydrazine is removed *in vacuo* and the residue is dissolved in 50 ml of 0·2 M borate buffer pH 6·9 and stirred with excess (3 ml) benzaldehyde in the cold overnight. Benzaldehyde and benzalazine are removed by repeated extraction with ether. The product is extensively dialysed and freeze-dried. It is hydrolysed to form the purine oligomers with 0·3 M KOH (100°C for 35 min). This solution is suitable for fractionation of the oligomers. As with the pyrimidine oligomers, the products are flanked at both 5′- and 3′-ends by phosphomonoesters. Alternatively the apyrimidinic acid may be hydrolysed by incubation in 3% aqueous aniline (pH 5·0) at 37°C for 5 h (Vanyushin, Buryanov and Belozersky, 1971).

Separation of oligomers   In principle, the methods described in the following chapter for the separation of oligoribonucleotides are suitable for the fractionation of these oligodeoxyribo-nucleotides. The oligomers can be either fractionated in their original state with the two terminal phosphates still in place or they can be dephosphorylated with alkaline phosphatase. In the procedures of Chargaff, the oligomers (either runs of purines or runs of pyrimidines) are first sorted out into groups with the same number of residues per oligomer. Chargaff calls these groups 'isostiches'. Thus purine isostich 2 is the mixture of -G-G-, -G-A-, -A-A- and -A-G-. The isostiches can be separated by preparing the sample in 0·1 M lithium acetate and applying it to a column of DEAE-cellulose. Free bases are washed off with the same buffer; the isostiches (in the order 1, 2, 3, 4, . . .) are eluted with a linear gradient of 0–0·32 M lithium chloride in the same buffer (Rudner, Shapiro and Chargaff, 1966). In their work on the isostiche 2, Chargaff, Buchowicz, Türler and Shapiro (1965) dephosphorylated the isostiches (after the fractions had been desalted[2]) as above and separated the products by paper chromatography using as solvent a mixture of isobutyric acid (5 volumes) and 0·5 M aqueous ammonia (3 volumes).

This same solvent is a suitable one for separating oligomers (not grouped into isostiches).

Shapiro and Chargaff (1963) have listed the positions of many oligomers using a two-dimensional system in which this solvent is used for developing one dimension and 70% (v/v) aqueous isopropanol containing concentrated aqueous ammonia (0·35 ml per 1 of tank volume).

The two-dimensional electrophoretic method devised for oligoribotides by Sanger is suitable for fractionating the oligodeoxyribotides (p. 293). A method used by Southern (1970) is similar to that of Sanger (p. 278). Pyrimidine oligomers were fractionated on 2·3% polyacrylamide gels in glass tubes, using as running buffer 1 M pyridinium acetate pH 3·5. The gel was then laid along a piece of DEAE-paper (parallel to and a little way in from one edge). The gel was allowed to dry out. It formed a thin film on the paper and the nucleotides were absorbed by and bound to the DEAE groups. The paper was then wetted with 7% formic acid and electrophoresed under the conditions described on p. 295. Alternatively, the method of Southern and Mitchell (1971; p. 296) is suitable for the oligodeoxyribotides.

Applications of the technique involving purine and pyrimidine runs   It has been possible to establish the anti-parallel configuration of the helix using the isostich 2 fractions in a manner analogous to the 'nearest neighbour' proof but by a technique which does not employ the use of polymerase and hence makes no assumptions about the fidelity of DNA replication *in vitro*. The argument can be seen by just considering four types of double structure (all isostiches) in a hypothetical stretch of DNA in a syn-parallel (structure on left) or anti-parallel (structure on right) configuration.

5′-N-N-G-A-N-N-A-G-N-N-3′   3′-N-N-G-A-N-N-A-G-N-N-5′

5′-N-N-C-T-N-N-T-C-N-N-3′   5′-N-N-C-T-N-N-T-C-N-N-3′

The former structure implies that [-A-G-] = [-T-C-] and [-G-A-] = [-C-T-]; the latter implies that [-G-A-] = [-T-C-] and [-A-G-] = [-C-T-]. It is not possible to directly compare these amounts of concentrations as the purine and pyrimidine isostiches must be isolated by separate experiments with different efficiencies of release. However the ratio of the two sorts of pyrimidine isostich can be accurately measured; so can the ratio of the two sorts of purine isostich. If we take the ratio [-A-G-]/[-G-A-], the former (syn-parallel) case then predicts that this ratio is equal to [-T-C-]/[-C-T-]; the latter (anti-parallel) model predicts it is equal to [-C-T-]/[-T-C-]. The experimental finding (Chargaff, Buchowicz, Türler and Shapiro, 1964) was that both [-A-G-]/[-G-A-] and [-C-T-]/[-T-C-] were equal to 1·3 thus confirming the anti-parallel configuration.

Chargaff's group have analysed the frequency of the occurrence of different pyrimidine isostiches in DNA from several sources. In a random DNA sequence one would expect a characteristic frequency of different isostich frequency (the moles of isostich/DNA unit length say 100 moles of DNA pyrimidines—should be a logarithmic function of isostich length). This relationship is found among bacterial DNA molecules. These data and their possible evolutionary significance are discussed by Chargaff (1968).

The pyrimidine isostiches (or 'isopliths' in Sinsheimer's papers) of single-stranded DNA coli phage have been catalogued. Hall and Sinsheimer (1963) studied the isostich distribution in φX 174 DNA. More recently Darby, Dumas and Sinsheimer (1970) have studied the pyrimidine isostiches of the minus strand of a mutant (φX*am*3) and Černý, Černá and Spencer (1969) have studied the isostiches from the minus strand of phage S13*su*N15. These minus strands were synthesized *in vitro*. The paper by Darby and others describes useful additional techniques for the isolation and fractionation of isostiches. The isostiches were obtained by fractionation of the initial mixture of pyrimidine oligomers on DEAE-cellulose in the presence of urea (see p. 292) and the isostiches were fractionated by chromatography on DEAE-Sephadex (see p. 161).

The work establishes that there is a great similarity between the patterns from the minus strands from the two phage implying that the phage from which these mutants were obtained (φX 174 and S13) have a common ancestor. One other application of the isostich method is described below.

In phage λ and phage T7, it is possible to separate the two DNA strands. Mushynski and Spencer (1970) have prepared complete catalogues of the pyrimidine isostiches in each of the two strands in these two phage. Both of these phage show an asymmetric distribution of the isostiches between their two complementary strands.

Miscellaneous results

In this section I have summarized results in which particular features of DNA molecules have been exploited to obtain more information about sequences than are typically available.

Satellite DNA   Satellite DNA is of different density to the main DNA of the cell which consequently can be isolated as a 'satellite zone' in buoyant density centrifugation. An up-to-date review of satellite DNA is by Coudray, Quetier and Guille (1970). Satellite DNA falls into four groups. There is the DNA from organelles such as chloroplasts, mitochondria, kinetoplasts and so on; there are bacterial plasmids and episomes; there are the ribosomal cistrons of certain organisms which can be isolated from the bulk of the DNA on account of the high GC-content of rRNA (p. 152); finally there is the hard core of mystery molecules which form satellites in many animal DNA preparations. This section is about the last group. Although buoyant density centrifugation may be a convenient method of isolating the other sorts of DNA, they are discussed elsewhere in this book as there is no special feature of their primary structure that facilitates their sequencing.

The poly dAT satellite of crab DNA has been mentioned elsewhere (p. 152). The use of nearest-neighbour-frequency analysis for confirming this structure is described by Sueoka (1961).

A more complex repeating sequence is that of guinea pig α-satellite DNA. The two strands of the satellite are easily separated by centrifugation in caesium chloride gradients of denatured preparations (see p. 152). Southern (1970) prepared the pyrimidine oligomers from the heavy strand by the diphenylamine procedure (p. 276) and separated them by two-dimensional electrophoresis (p. 277). Only three main spots were identified. These oligomers were sequenced by removing the terminal phosphomonoesters (alkaline phosphatase) and phosphorylating the $5'$-end (polynucleotide kinase) followed by digestion with the spleen nuclease (p. 208). The result of the experiments is to establish unambiguously that basically the DNA consists of the following repeating sequence,

$$5'\text{-(C-C-C-T-A-A-)}_n \ldots 3'$$

$$3' \ldots \text{(-G-G-G-A-T-T)}_n\text{-}5'$$

An analysis of the minor oligomers from the same sample showed that certain (apparently non-random) base substitutions have occurred as a result of evolution of the basic structure.

What is the role of this non-informational, genetically inactive DNA and where is it located in the nucleus? Really we have no idea about an answer to the former question. The two major possibilities would seem to be that either it represents a store of deoxyribonucleotides for use in synthesis or repair of chromosomal DNA or alternatively that its synthesis represents some kind of dynamic control of the activity of DNA polymerase, channelling its activity away from gene duplication. As for the second question, it is experimentally accessible with present techniques for the sub-fractionation of deoxyribonucleoprotein. Mattoccia and Comings (1971) have produced what should be a most valuable demonstration of how such experiments should be performed. Nuclei from mouse liver were sonicated to release nucleoli (p. 114). The initial centrifugation of sonicate (after a slow spin to remove unbroken nuclei) yields a supernatant that contains the 'euchromatin' (this is the bulk of the non-nucleolar chromatin); the pellet contains the nucleoli and the 'heterochromatin', darkly staining material associated with the centromeres during mitosis and apparently replicated late in the DNA-synthetic phase of the cell cycle. Analysis of DNA from various fractions established that the satellite DNA (which is lighter than the bulk of mouse DNA) is associated with heterochromatin and also apparently with the nucleoli. The experiments established that there is no simple relationship between the fractions, such as saying that satellite is the only sort of DNA in heterochromatin; indeed a specific function is associated with insect heterochromatin (p. 370).

The sequences round modified bases   Vanyushin, Buryanov and Belozersky (1971) have extended their studies on the occurrence of N^6-methyladenine in bacterial and phage DNA to examining the sequences in which this base occurs. Purine oligomers were prepared from DNA from *E. coli* and also phage T2. The results that follow apply to both sorts of DNA. The oligomers were sorted into their isostiches (see p. 276) and the occurrence of N^6-methyladenine in the classes of isostiches was compared with the frequency of the isostiches. Clearly if the methylation of Ade had been random, there would be a linear correlation between the two. In fact the results were strikingly different to this. The N^6-methyladenine was found to be concentrated in the small isostiches. Over a third of the total complement of N^6-methyladenine was found in the isostich-2 fraction. This fraction was isolated from bacteria labelled *in vivo* with $[^{14}C]$ methionine to label the methylated bases. Fractionation of the isostiches established that all the radioactivity (present in isostich 2) was in the (A,G) group. Following treatment of this mixture (of -A-G- and -G-A-) with (i) alkaline phosphatase (to yield A-G and G-A) and (ii) with 'snake venom' enzyme (see p. 208—this yields 5'-nucleotides), the radioactivity was found to be exclusively associated with the nucleoside fraction so that the product containing the radioactivity must have been m^{6}A-G. In addition to the isostiches 2, about 10% of the N^6-methyladenine was found to be present as free -m^{6}A-. As these sequences must be flanked by pyrimidines, a total of over 40% of the methylated base must be part of a sequence Py-m^{6}A-N, where N is either a pyrimidine or a G.

The argument can be taken a little further, Falaschi and Kornberg (1965) had already examined the activity of *E. coli* DNA methylase towards different synthetic adenine-containing oligomers. Their conclusion was that the nucleotide to the right of the A had no effect on the specificity of the reaction but that the trinucleotides A-T-A and A-C-A are not methylated. If this is really a valid restriction of the possible sequences for m^{6}A, then the total possible sequences are as shown below. The triplets shown are the transcription products of the triplets in DNA starting with m^{6}A (remembering that transcription is an anti-parallel complementary copying).

T-T-m^{6}A-N	UAA
T-C-m^{6}A-N	UGA
C-T-m^{6}A-N	UAG
C-C-m^{6}A-N	UGG
G-T-m^{6}A-N	UAC
G-C-m^{6}A-N	UGC

Before commenting on the very striking feature of these results, let us just be quite clear what assumptions have been made. We have neglected the 60% of m^6A which is present in sequences other than Py-m^6A-N, we are assuming that Falaschi and Kornberg's data on the specificity of the methylase can be extrapolated to its activity *in vivo*; and we have only considered the transcription product in which RNA polymerase is reading m^6A as the first nucleotide of a codon.

Having said all this, it is remarkable that of the six possible triplets in the RNA product, three of them (UAA, UGA and UAG) are the three termination signals in protein synthesis. Of undoubted significance is the similarity of the environments of m^6A in *E. coli* and T2. An extension of this work both in the direction of characterizing the larger isostiches containing m^6A and in surveying more bacteria and viruses could be very exciting.

Sticky ends of phage DNA   The DNA of phage λ can occur in either a straight or continuous form. The straight form has the following structure shown in Fig. 9.5. The two single-stranded ends (X and X′) are complementary in sequence and hence facilitate the joining of the molecule up to produce a continuous molecule as illustrated in the figure. These two strands are known as the 'sticky ends'.

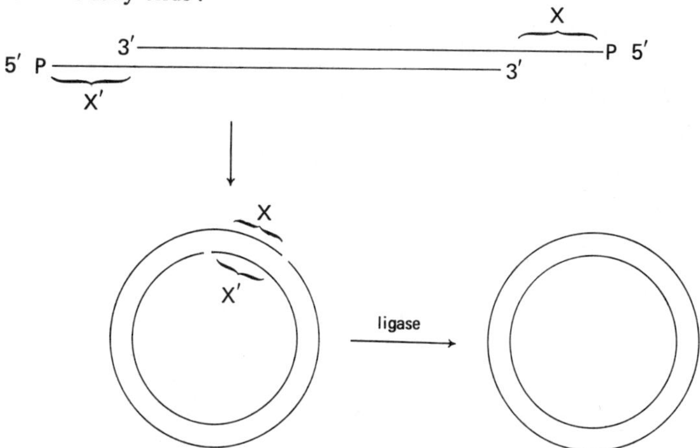

FIG 9.5   Schematic structure of DNA from phage λ. For explanation see text.

The sequences of X and X′ have been established by Wu and Taylor (1971). The experiment consisted of using the linear DNA as a substrate for DNA polymerase. In other words the enzyme was given the job of 'repairing' the ends. Radioactive triphosphatases were employed. The product molecule was subject to limited digestion with either the snake venom of spleen phosphodiesterases and the radioactive products were analysed. In this way it has been possible to tentatively assign the sequence -A-G-G-T-C-G-C-C-G-C-C-C . . . to X and the complementary sequence (read it backwards to see the complementarity) -G-G-G-C-G-G-C-G-A-C-C-T- . . . to X′.

This approach is capable of extension to other types of DNA structure if one chain of a double helical structure is eroded with an enzyme from the 3′-ends (such as *E. coli* exonuclease III—see p. 209). The product from such limited digestion is a molecule that looks like λDNA (except of course the ends are not sticky).

End groups   Naturally occurring DNA molecules are so large that chemical assays of end groups are of relatively little interest. However it it should become possible to prepare the fractionate specific DNA fragments a chemical procedure for identifying an end of the molecule would be of great value. The oxidation of dThd with hydrogen peroxide by the procedure of Vizsolyi and Tener (1962; p. 174) is easily adapted to this purpose. The product of the reaction is deoxyriburonyl thymine. If the thymidine is at the 5'-end of a chain the reduction still occurs and can be used to remove the terminal thymidine quantitatively. The procedure is outlined in Fig. 9.6. The oligomer must have a free 5'-OH (if it has not, the terminal phosphate can be removed with alkaline phosphatase). The propylamide derivative of the uronic acid is prepared and decomposed in alkali. The base, furfurylic acid propylamide and an oligomer one residue shorter are produced. The terminal 5'-phosphate of this oligomer can be removed with phosphatase and the sequence can be repeated. Vizsolyi and Tener studied the reaction with oligothymidylates; it is not quite certain what modifications to their conditions may be required for quantitative removal of different terminal nucleosides.

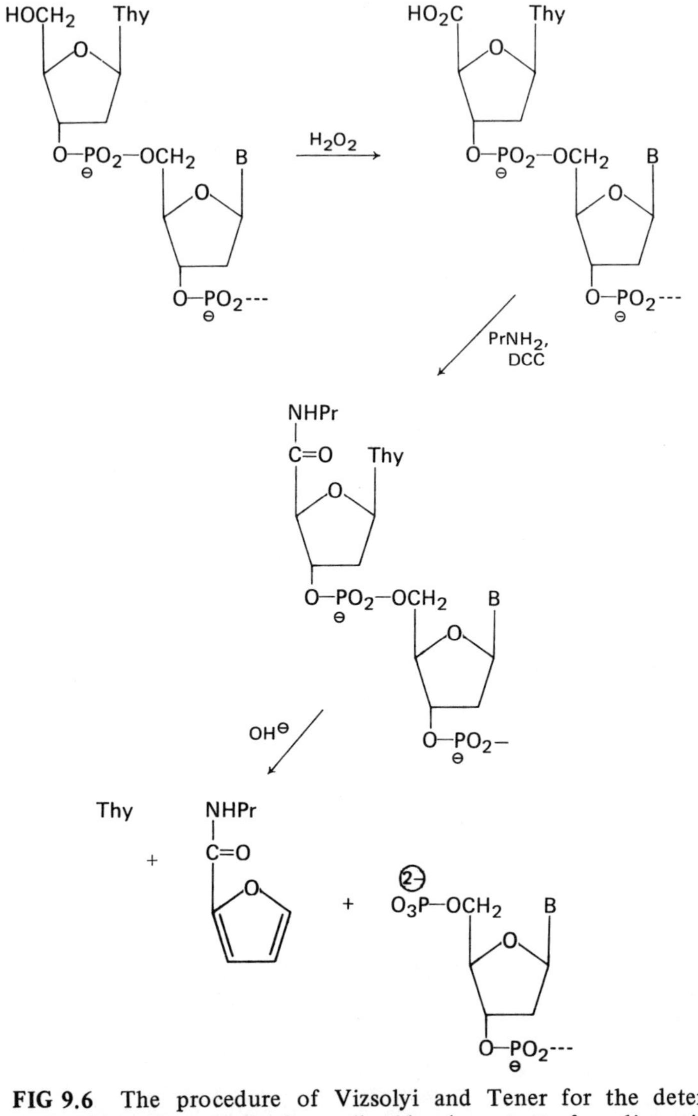

FIG 9.6   The procedure of Vizsolyi and Tener for the determination of the 5'-terminal nucleoside of an oligodeoxyribotide (see text for discussion). Pr = propyl, DCC = dicyclohexycarbodiimide.

Notes to chapter 9

[1] An alternative method of making apurinic acid is to treat DNA with a mercaptan in the presence of a catalyst of $ZnCl_2$. The purines are removed and leave thioacetal derivatives of deoxyribose.

The mercapto-groups are removed (to yield the apurinic acid) with aqueous alkali. The reaction has been extensively studied (Jones, Letham and Stacey, 1956) but is not used these days. It must be a pretty smelly procedure.

[2] There are several methods of desalting such mixtures. The simplest one is to dilute the mixture so that it will bind to DEAE and to pour it through a very short, wide 'column' of DEAE-cellulose, wash the lithium salts out with water and then elute the nucleotides with a volatile buffer (such as ammonium bicarbonate—2 M in this case). The desalted nucleotides can be obtained from this eluate by freeze-drying.

10 Primary structure of RNA

There are three types of macromolecules consisting of a defined sequence of chemically similar units; proteins, RNA and DNA. Techniques for the sequence determination of RNA and DNA are a long way behind those for the sequence determination of proteins. The subject is historically behind protein sequencing and the nucleic acids are in general larger molecules so the problem is technically much more difficult. An additional technical difficulty results from the fact that there are fewer categories of units (20 amino acids in proteins; 4 major nucleotides in a nucleic acid) so that the selective fragmentation of nucleic acids is more difficult, and the fractionation of the fragments so formed is harder, as there are fewer differences in chemical and ionic properties between different oligonucleotides than there are between the peptides of the mixture which forms a partial digest of a protein. We have seen (chapter 9) that two of these problems (molecular size and lack of specific methods of fragmentation) have proved so great in the case of DNA that no extended sequences of fragments of natural DNA molecules have been determined. With RNA the problems are not quite so great. There are base-specific endoribonucleases available (p. 206) and there is at least one class of RNA species (tRNA) in which molecular weights are such that the prospect of sequencing them seems less daunting. The consequence is that there is no technical objection to successfully sequencing any tRNA molecule provided it is available as a pure preparation and that the technical resources of the laboratory are up to it. RNA sequencing methods have been extended to studying the sequence of large fragments from other RNA molecules of high molecular weight.

The first part of this chapter deals with the determination of the simpler features of RNA primary structure and a survey of some of the chemical properties which have been (or could be) exploited in extensions of sequencing methods. In the next section there is an account of the techniques involved in sequencing RNA molecules containing of the order of 100 residues (including tRNA) with some illustrative examples. Finally I have a section which deals with approaches to the sequencing of large RNA molecules and a survey of some of the results that have been obtained to date.

Chemical degradation and reactions and primary structure

Total hydrolysis

Methods for the hydrolysis of RNA and their application to determining the nucleotide composition have been described on p. 203. It is merely necessary here to highlight one or two results that have been obtained in this way.

Base ratios In addition to the direct methods of measuring base ratios (pp. 203 and 204) there is one indirect method. Following hydrolysis in 0·5 H HClO$_4$ at 70°C for 20 min, the absorbance is read at 260, 280 and 290 nm. From the ratios $A_{290}{:}A_{260}$ and $A_{280}{:}A_{260}$ it is possible to calculate (from a calibrated series of data) the GC-content. The method (and the calibration) are described by Eisenstadt and Brawerman (1964). In general there seems to be no particular value in doing this as it requires optically dense amounts of RNA and only a slightly larger scale of preparation would yield enough for a proper analysis to be performed. The experiment can be useful for a rapid assay of GC-content in certain circumstances. Other indirect methods of measuring the GC-content which are useful for DNA (see p. 267 and 268) are not suitable for RNA for reasons that are apparent when these properties (hyperchromic effect on heating and buoyant density centrifugation) of RNA are considered in any detail (pp. 133 and 340).

TABLE 10.1 Normalized percentage nucleotide composition of high-molecular-weight rRNA from various sources. U* includes Ψ (see p. 286).

Source	Molecule[1]	G	A	C	U*	G + C	$\dfrac{G+U}{A+C}$	$\dfrac{G+A}{U+C}$	Ref.
B. subtilis	large	30·0	26·4	22·8	20·7	52·8	1·03	1·30	
	small	30·8	24·8	23·4	21·1	54·2	1·08	1·25	a
E. coli	large	32·5	25·5	21·0	21·0	53·5	1·15	1·38	
	small	32·1	34·2	22·3	21·3	54·4	1·15	1·29	b
Pseudomonas aeruginiosa	large	30·0	26·3	22·1	21·5	52·1	1·07	1·29	
	small	30·3	26·1	22·5	21·1	52·8	1·06	1·29	c
Yeast[2] cytoplasm	large	28·4	26·4	19·2	26·0	47·6	1·19	1·21	
	small	26·1	26·6	19·1	28·1	45·2	1·19	1·12	d
Yeast[2] mitochondria	large	14·0	40·3	11·0	34·6	25·0	1·19	1·19	
	small	16·1	38·4	11·0	34·5	27·1	1·02	1·20	d
Neurospora[3] cytoplasm	large	29·4	24·8	21·9	23·9	51·3	1·14	1·18	
	small	27·7	25·3	21·6	25·4	49·3	1·13	1·13	e
Neurospora[3] mitochondria	large	19·1	33·9	15·0	31·9	34·1	1·04	1·13	
	small	20·4	31·8	16·0	31·7	36·4	1·09	1·09	e
Psalliota[4] cytoplasm	large	27·8	27·7	20·1	24·3	47·9	1·09	1·25	
	small	30·1	24·9	18·0	27·1	48·1	1·33	1·22	f
Pea cytoplasm	large	32·1	23·6	22·7	21·5	54·8	1·16	1·26	
	small	31·1	23·7	20·1	25·2	51·2	1·13	1·21	g
Pea chloroplast	mixed	29·8	24·2	24·8	21·2	54·6	1·04	1·17	h
Broad bean cytoplasm	mixed	29·4	25·1	24·4	21·2	53·8	1·02	1·19	h
Broad bean chloroplasts	mixed	29·2	25·6	24·3	20·8	53·5	1·00	1·21	h
Lamium[5] cytoplasm	mixed	28·9	24·5	24·5	22·0	54·4	1·04	1·15	h
Lamium[5] chloroplasts	mixed	28·2	24·8	25·3	20·7	53·5	0·96	1·15	h

TABLE 10.1—continued

Source	Molecule[1]	G	A	C	U*	G + C	$\dfrac{G+U}{A+C}$	$\dfrac{G+A}{U+C}$	Ref.
Euglena[6] cytoplasm	mixed	29·5	22·7	27·1	20·7	56·6	1·01	1·09	i
Euglena[6] chloroplasts	mixed	27·0	30·6	17·0	25·0	44·0	1·09	1·37	i
Ascaris[7]	large	29·6	26·8	22·3	21·2	51·9	1·04	1·27	j
	small	30·0	26·8	21·6	21·5	51·6	1·06	1·32	
Drosophila[8]	large	22·5	30·8	19·6	27·1	42·1	0·98	1·14	k
	small	23·5	28·8	20·3	27·4	43·8	1·04	1·10	
Arbacia[9]	large	34·9	19·9	26·0	19·3	60·9	1·18	1·21	l
	small	29·9	23·1	23·8	23·2	53·7	1·13	1·13	
Xenopus[10]	large	35·0	19·7	27·9	17·4	62·9	1·10	1·21	m
	small	28·9	24·1	24·1	22·9	53·0	1·08	1·13	
Duck red cells[11]	large	37·3	17·8	27·7	17·2	65·0	1·20	1·23	n
	small	33·0	21·4	24·1	24·1	57·1	1·20	1·20	
Rat liver	large	32·9	18·3	29·8	19·0	62·7	1·08	1·05	o
	small	30·2	22·4	27·8	19·6	58·0	0·99	1·11	
HeLa[12]	large	35·1	15·9	32·2	16·8	67·3	1·08	1·04	p
	small	30·8	20·0	27·6	21·6	58·4	1·10	1·02	

Notes:
1. Large: the rRNA of the larger ribosomal subunit; small: that of the smaller. Thus for bacteria, large = 23 S, small = 16 S, etc.
2. *Saccharomyces cerivisiae*
3. *N. crassa* ascomycete fungus
4. *P. compestris* basidiomycete fungus
5. The weed 'dead nettle'
6. *E. gracilis*, a photosynthetic flagellate
7. *A. lumbriciodes*, a nematode worm. In this and all subsequent animal sources, the data refer to cytoplasmic rRNA
8. *A. punctulata*, a sea-urchin
9. *D. melanogaster*, the fruit fly
10. *X. laevis*, South American toad, see p. 152
11. See p. 182
12. Human cancer cells in tissue culture (see p. 117).

References:
a. Doi and Igarashi (1964)
b. Stanley and Bock (1965)
c. Spiegelman (1961)
d. Fauman, Rabinowik and Getz (1969)
e. Rifkin, Wood and Luck (1967)
f. Pollard (1964)
g. Click and Hackett (1966)
h. Odinstova, Golubeva and Sissakian (1964)
i. Eisenstadt and Brawerman (1967)
j. Grummt and Bielka (1968)
k. Hastings and Kirby (1966)
l. Slater and Spiegelman (1966)
m. Birnstiel, Speirs, Purdom, Jones and Loening (1968)
n. Attardi, Parnas, Hwang and Attardi (1966)
o. Hirsch (1966)
p. Amaldi and Attardi (1968).

The results of studies on the nucleotide composition of rRNA contains quite a high proportion of modified nucleosides and is of high GC-content. Now that so many sequence data on tRNA are available for comparative study (p. 305), there is no interest in the small differences in overall nucleotide composition between different tRNA species.

With high-molecular-weight rRNA however the position is rather different. For one thing we know rather little about the sequence of rRNA in general and the task of elucidating the sequence of significant tracts of the molecules is of sufficient complexity (p. 319) that I think we can expect to rely on base-composition data and a knowledge of molecular weight (p. 395) for some time in any comparison of rRNA from a variety of organisms. Moreover there are extremely intriguing differences in these properties of rRNA and a tabulation of the properties is worthwhile. A summary of the base ratios from a variety of rRNA species is summarized in Table 10.1.

Several generalizations emerge from the data of Table 10.1. There is no apparent correlation with the DNA base composition (compare *B. subtilis* and *Ps. aeruginosa* whose DNA compositions are 42% (G + C) and 67% (G + C) respectively; Belozersky and Spirin, 1960). A possible exception to this statement is the mitochondrial rRNA which is of extremely anomalous base composition (very low G + C) possibly related to the fact that the GC-constant of mitochondrial DNA is very low (for example yeast mitochondrial DNA contains 25% (G + C); Wood and Luck, 1969). As far as GC-content is concerned, *Drosophila* and (to a lesser extent) *Ascaris* stand out as having anomalously low GC-content. The rRNA of other invertebrates should prove of interest for comparison. Notice that the rRNA from the chloroplasts of higher plants is of rather similar base composition to that of the cytoplasmic rRNA from the same plant. In *Euglena* there is a striking difference between the two values.

In vertebrates the GC-content is consistently high and there is a striking difference between the overall base composition of the large (higher GC-content) and small (lower GC-content).

The last two columns of Table 10.1 are 'tests' of Chargaff's rules (see p. 7). In truly complementary molecules (DNA) these ratios should both be unity. Notice that they do not deviate grossly from unity for rRNA, even in cases (yeast mitochondria) in which the base composition is extremely typical. Thus rRNA has at least the overall composition required for considerable intramolecular nucleotide complementarity. Sequence studies and assessment of secondary structure in rRNA (p. 320 and 351) confirm that there is considerable intermolecular base-pairing in these molecules. On the whole, we can say that deviations from the 'rule', G + U = A + C, are less great than those from, G + A = U + C, and that if this latter test is used (last column of Table 10.1), it can be seen that bacterial rRNA is relatively rich in purines.

Minor nucleosides   Transfer RNA, high-molecular-weight rRNA but not 5 S rRNA, nor (as far as we know) mRNA contain minor nucleosides. Their position in tRNA is discussed on p. 305. It is worth considering their occurrence in rRNA although their role in *E. coli* rRNA is discussed further on p. 321.

There seem to be differences between rRNA from bacteria and eukaryotic cells in this respect. Briefly the differences are as follows. In bacteria (i.e. *E. coli*) Ψ-residues account for of the order of 0·1% of the total U-residues with a higher proportion in 23 S molecule than in the 16 S molecule (Dubin and Günalp, 1967). In animal-cell RNA the proportion is of the order of 1% and in this case the higher proportion is in the smaller (18 S) molecule (Amaldi and Attardi, 1968). In all rRNA molecules examined so far the methylated nucleosides form a somewhat higher percentage of the total residues in the smaller (16 S or 18 S) molecule than they do in the larger (23 S or 28 S). However the overall proportion of methylated nucleosides in bacterial rRNA (of the order of 0·8%—Fellner and Sanger, 1968) is lower than the comparable figure for animal rRNA (of the order of 1·5%—Iwanimi and Brown, 1968). Moreover a lower proportion of the methyl groups in the bacterial RNA are attached to the 2′-residues of ribose (as opposed to the bases) than is the case for eukaryotic RNA. However this last generalization may be an oversimplification, as many methylated bases are unstable (see p. 70); some of the figures which imply that there is a high proportion (among the total methylated nucleosides) or 2′-methyl nucleosides may be due to the fact that the degradation conditions employed in some of the earlier studies on animal rRNA have destroyed methylated base residues and hence artificially created the impression that, of the methylated nucleosides, a high proportion have the sugar methylated. For a survey of the available data and a critical appraisal of the earlier results see the reference by Iwanami and Brown (above).

Nearest-neighbour-frequency analysis   Clearly a nearest-neighbour analysis of RNA can be done for RNA synthesized *in vitro* in a manner exactly analogous to the experiment described for DNA (p. 271). In this case the template DNA is incubated with RNA polymerase and (i) [α-^{32}P] ATP with unlabelled CTP, UTP and GTP, (ii) [α-^{32}P] CTP and unlabelled ATP, GTP and UTP and so on. It is not necessary to use nucleases to digest the products (to produce the mixtures of 3′-nucleotides) as alkaline hydrolysis is sufficient (p. 204). The technique is useful for assessing the fidelity of transcription *in vitro*.

End groups

Oxidation of RNA with periodate   The 3′-end of RNA contains a *cis* glycol (see p. 13), absent from the rest of the molecule. A characteristic reaction of *cis* glycols, that can be easily exploited in aqueous solution, is that they are oxidized by periodate to yield a dialdehyde. The product of such an oxidation, together with some of the reactions of the product are summarized in Fig. 10.1. The periodate reaction is performed by dissolving the RNA in 0·15 M sodium acetate/acetic acid buffer pH 5·3 containing a large excess of sodium metaperiodate ($NaIO_4$). The reaction is easy to work up if the volumes are kept fairly small. Assume we have a solution containing 5 mg/ml of RNA of a molecular weight around a million. If the $NaIO_4$ concentration is around 1 mM, this is a good large excess suitable for the experiment. The reaction mixture should be kept cold (ice bath) and dark to prevent any other oxidative degradation of the chains. After about 2 h, the reaction is stopped by removing either the periodate or the RNA from the solution. The periodate can be removed by adding potassium chloride (KIO_4, is insoluble) and centrifugation or the RNA can be precipitated with ethanol (p. 106). If this latter course is adopted, the precipitate must be re-dissolved and re-precipitated several times until the solution is really free of periodate (assayed with starch-iodide paper), as some $NaIO_4$ co-precipitates with the RNA. If very small (radioactive) amounts of RNA are being oxidized it is obviously necessary to use either the former method (precipitation of KIO_4) or to co-precipitate the RNA with added carrier RNA. In this case, if you are having difficulty in removing the last traces of $NaIO_4$ from solution it can always be mopped up with a drop of a *cis* glycol (such as ethylene glycol or glycerol). After the periodate has been removed, the RNA (Fig. 10.1(*b*)) is quite stable and can be run (for example) on sucrose gradients, to check its molecular integrity.[1]

   Having oxidized the RNA there are one of two things to do with it; either the 3′-end can be converted into a derivative or the 3′-terminus can be selectively degraded chemically. The former experiment can be done with more-or-less any aldehyde reagent. If a radioactive reagent is employed it is possible to use the method for estimating molecular weight (p. 400) or for

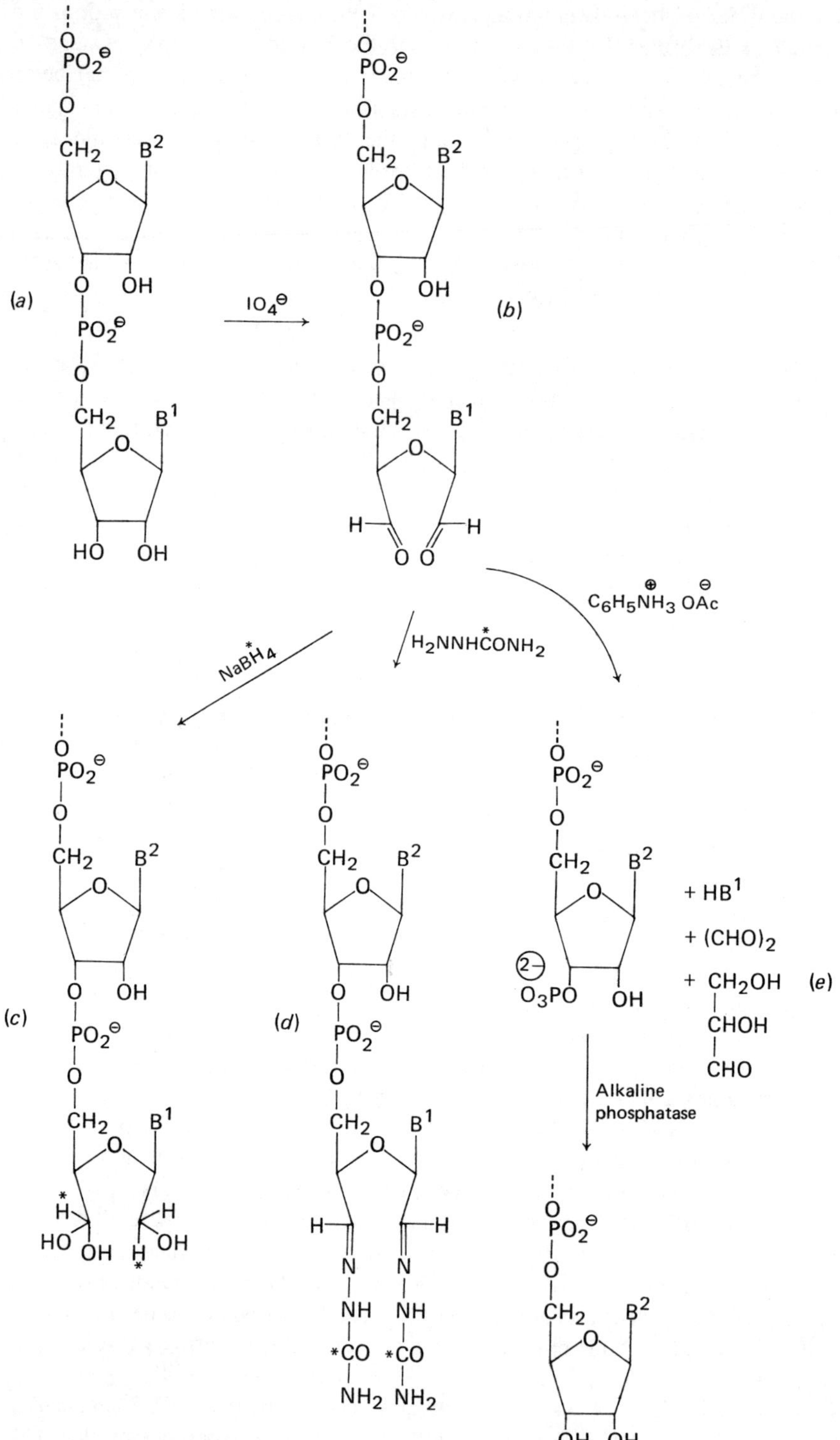

FIG 10.1   Reaction of RNA (*a*) with $NaIO_4$ (details in text) to yield the aldehyde product (*b*) and subsequent reactions of this substance (see text for discussion).

characterizing the terminal nucleoside. Several aldehyde reagents are suitable. Undoubtedly the simplest is sodium borohydride ($NaBH_4$ or the labelled sodium borotritiide NaB^3H_4). In the procedure of Leppla, Bjoraker and Bock (1968) the oxidized RNA is treated with about a 500-fold excess of NaB^3H_4 in solution at pH 8. After the reaction (1 h in the dark—this is important, see p. 245 for the reason) the excess NaB^3H_4 is destroyed (fume cupboard—$^3H_2$ is evolved!) by adding acetic acid to adjust the pH to around 4. The product (Fig. 10.1(c)) is RNA with a labelled triol at the 3′-end. Alkaline hydrolysis (p. 204) yields nucleotides plus the triol derivative of the terminal nucleoside. These triol nucleosides can be fractionated by any system that separates the nucleosides (p. 48) although the R_f-values are a little higher than those for the corresponding nucleosides. Other, less satisfactory, methods that have been employed are to make the isonicotinic acid hydrazide derivatives (p. 74) or the semicarbazide (Steinschneider & Fraenkal-Conrat, 1966). In this latter case the conditions are to treat the RNA in sodium acetate/acetic acid buffer pH 5·3 with a 1000-fold excess of $[^{14}C]$ semicarbazide hydrochloride at 25°C for 1 h. The RNA is precipitated with alcohol. The structure of the product is shown in Fig. 10.1(d).

The alternative procedure (Steinschneider and Fraenkal-Conrat, above) is to hydrolyse the 3′-dialdehyde either with very dilute alkali or (preferably) with the more selective reagent 0·3 M anilinium acetate pH 5·3 (adjusted with HCl). After 3 h at 25°C, the RNA is precipitated with alcohol and the supernatant is analysed for bases (p. 48). The method used by the authors is fairly complicated. The problem is that excess aniline is present. They used two separate paper chromatograms, the first one on heavy chromatography paper. It would seem to be an ideal problem for solution by a gas chromatographic method (see p. 53). If the residual RNA is treated with alkaline phosphatase, the sequence of reactions can be repeated to identify the penultimate base and so on (Fig. 10.1(e)).

Recognition of the 3′- and 5′-ends in mixtures of oligomers   We shall soon see that when endonuclease digests of large RNA molecules are analysed it is useful to be able to identify the end sequences readily. If the 3′-end is labelled by one of the methods given above the oligomer carrying the label can be identified. An elegant alternative method of identifying this end is to make use of the fact that the 3′-terminal oligomer has no phosphomonoester groups (see Table 10.2, p. 301). Sometimes this fact can be immediately recognized (see p. 301). Alternatively samples of all the separated oligomers can be treated with alkaline phosphatase. The 3′-end oligomer is the only one that is not affected and hence the only one whose electrophoretic mobility is the same before and after phosphatase-treatment (Dahlberg, 1968).

The 5′-end oligomer differs from the others in that it is the only one containing two phosphomonoester groups. There are several ways of identifying it. One is to treat the undigested RNA first with alkaline phosphatase and second with $[^{32}P]$ ATP and polynucleotide kinase (Suzuki and Haselkorn, 1968). The labelled oligomer is the 5′-terminus. As so many sequence studies these days are done with $[^{32}P]$ RNA anyway (p. 291) it would seem that ^{33}P-labelling should be a useful alternative (see p. 235). Viral RNA is a special case (see p. 318) in that the 5′-end carried a triphosphate. In this case the 5′-oligomer is identified as the one that yields a triphosphate (apparently invariably GTP) on alkaline hydrolysis (see p. 204).

Chemical reactions and internal degradation

In protein sequencing one amino-acid-specific chemical reagent (cyanogen bromide that fragments the chains at methionine residues) complements specific enzymic proteolysis in structural studies. There are several chemical reactions of the bases in nucleic acids which can in principle be applied to the selective degradation of nucleic acids. With DNA, these methods are of importance because there is rather little else that can be attempted (see p. 266). With RNA these methods are of less importance as the enzymic procedures (to which the remainder of this chapter is devoted) are so versatile. However there are several reagents which should be born in mind as potentially useful in sequencing particularly recalcitrant oligomers. One of these methods has been exploited in one of the most important pieces of sequencing work to have been performed so far (p. 312).

Reactions of uracil residues   Uracil residues in RNA can be reduced to the dihydroderivative by $NaBH_4$ in the presence of UV light (p. 245). As dihydrouridine is easily hydrolysed (p. 68) this is a method of introducing weak points in the molecule at every U-residue.

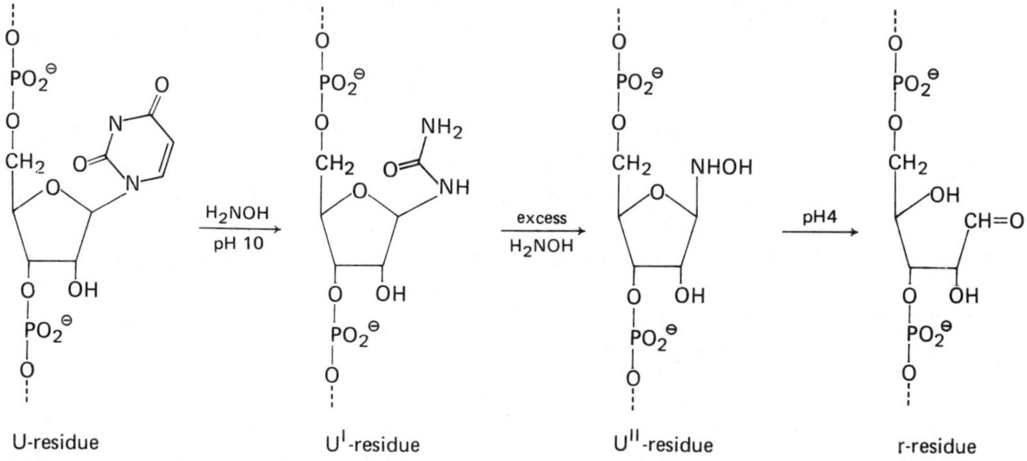

U-residue $\xrightarrow[\text{pH 10}]{H_2NOH}$ U^I-residue $\xrightarrow[\text{H}_2\text{NOH}]{\text{excess}}$ U^{II}-residue $\xrightarrow{\text{pH4}}$ r-residue

FIG 10.2   Reactions of a U-residue in RNA with hydroxylamine. See text for discussion.

A more versatile sequence of reactions involving the U-residues is that with hydroxylamine at high pH (see p. 70). This reaction has been very carefully studied as a possible tool in sequence analysis by N. K. Kotchetkov and E. I. Budowsky and their colleagues. Many of the original data are in the Russian literature. In my references, I have assumed that the reader is capable of understanding scientific papers in English, French and German but (like me) is confounded by Russian, Chinese and Japanese. Fortunately, in the present case, Kotchekov and Budowsky (1969) have written a review article in English which includes an excellent summary of their own work on the hydroxylamine reaction with many useful technical details in the text and legends to the figures.

The reactions of hydroxylamine with a U-residue in RNA are summarized in Fig. 10.2.

The reaction at pH 10 is to produce ribosylurea (U^I) residues. In the presence of KCl it is possible to stop the reaction at this point. The concentration of H_2NOH is usually 10 M (methanol is needed to keep it in solution). In the absence of KCl the reaction goes further to yield ribosylhydroxylamine residues (U^{II}). U^{II} decomposes at pH 4 to produce free ribose residues. These are now weak points in the chain and amines will cleave them by β-elimination (see p. 275). RNA chains containing U^I or U^{II} are stable, not easily hydrolysed and the U^I and U^{II} residues are not attacked by ribonuclease A (p. 206).

The potential application of these techniques to sequencing should be fairly obvious. Consider a hypothetical large T_1-fragment (p. 207),

A-A-U-C-A-C-C-A-C-U-U-C-C-A-G-

First of all let us imagine that it be treated with H_2NOH:

A-A-U^I-C-A-C-C-A-C-U^I-U^I-C-C-A-G-

If this product is treated with ribonuclease A, the chain will just be hydrolysed at the C-residues, so the specificity of the enzyme has been increased. Alternatively if it is converted via the U^{II}-containing molecule into the ribose-containing molecule, the product will be:

A-A-r-C-A-C-C-A-C-r-r-C-C-A-G-

Treatment with an amine (anisidine, pH 5) yields the compounds A-A-, r-C-A-C-C-A-C-, r-, r-C-C-A-G-.

Acylation reactions   The reaction of glyoxal and kethoxal with Gua has been described on p. 65. Acylated residues are resistant to nuclease digestion. Thus these reagents should effectively produce an RNA molecule which is hydrolysed specifically at the A-residues by an enzyme much as ribonuclease U_2 (p. 207). According to Shapiro, Cohen and Clagett (1970) who have recently re-evaluated these reactions, kethoxal is a genuine G-specific reagent and should prove useful in sequencing studies. Glyoxal shows much less specificity and is presumably of less value.

Water-soluble carbodiimides   These reagents react with the residues, G, C, Ψ and I (see p. 63); the reactions are readily reversible. They can be employed to alter the specificity of nucleases readily. Thus if an RNA molecule (or oligomer) is treated with CMCT (p. 63) the U- and G-residues will be modified. If the product is digested with ribonuclease U_2, it will be hydrolysed at the A-residues only; these products can then be fractionated and digested again with the enzyme when they will be cleaved at the G-residues. Alternatively, if the CMCT-treated molecule is digested with ribonuclease A, the molecule will be cleaved specifically at the C-residues. Once again, treatment with dilute alkali followed by more nuclease A will hydrolyse the oligomers at the U-residues. The conditions required for the reaction have been studied by Ho, Uchida, Egami and Gilham (1969); an application of the use of the reagent is to be found on p. 304.

The sequencing of small RNA molecules

Basically the procedure is as follows. (i) The RNA is digested with a base-specific ribonuclease (usually T_1); (ii) the resulting oligomers are fractionated; (iii) they are sequenced; (iv) the RNA is digested with a second base-specific ribonuclease (usually A); (v) these oligomers are fractionated; (vi) ... and sequenced; (vii) overlapping regions are sought and extended sequences are deduced; (viii) the RNA is digested with an enzyme (usually T_1 again) under non-optimal conditions so that larger oligomers are produced; (ix) the large oligomers are fractionated; (x) these oligomers are digested and the smaller oligomers produced by complete digestion of them are characterized and identified as components in the earlier complete digests.

The crucial stages in this procedure are the fractionation steps (ii, v). There are two approaches to the fractionation of the oligomers; one is to use column chromatography of one sort or another; the other is to employ a two-dimensional separation in one of a variety of systems developed by F. Sanger. These methods have revolutionized sequence analysis of RNA but they are limited by the fact that the first dimension of all the systems is electrophoresis on a cellulose acetate strip. These strips have a very limited capacity for oligomers (0·1 mg is a maximum). Therefore the procedures are all employed on a very small scale and extremely highly labelled [^{32}P] RNA is required. However if the RNA can be obtained in this very radioactive form, so many of the other procedures, notably the oligomer sequencing (iii and vi) and the fractionation of the big oligomers (ix) are speeded up and simplified enormously. If one scans the literature for the latest RNA sequences, it is easy to reach the conclusion that the macro-scale techniques are largely out of date. This is not the case; the large-scale fractionation and characterization of oligomers obtained by the digestion of RNA are important in several respects. These methods are essential if a previously unknown (or unsuspected) minor nucleoside is present. Equally, partial digestion is of great importance in studies on the secondary structure and on the structure-and-function relationships of tRNA.

Thus I have subdivided this section into discussions of the various techniques in groups and then illustrate the applications of the methods to a few special examples of actual sequence determinations.

Enzymes and oligomers

Enzymic digestion   Digestion with either ribonuclease T_1 or ribonuclease A is performed using high levels of enzyme for short incubation times; in this way the specificity of the enzymes is greatest (Sanger, Brownlee and Barrell, 1965). A good general recipe is to use an enzyme:RNA ratio (on a weight-for-weight basis) of 1:20. The buffer (for either enzyme) is 2 mM EDTA, 20 mM tris HCl pH 7·4 and the incubation time 30 min at 37°C. For digestion of highly labelled [^{32}P] RNA, the reaction is conveniently performed in a small capillary tube (or micro-pepette) containing of the order of 10 μg of RNA (pecific activity of 1 mCi/mg or greater) in 5 μl of buffer. Partial digestion is most usefully achieved with T_1 as this enzyme has considerable specificity for single-stranded RNA under appropriate conditions (see p. 300). A good buffer for this purpose is 0·2 M NaCl, 20 mM magnesium acetate, 50 mM tris HCl pH 7·5. Enzyme:substrate ratios of between 1:100 to 1:1000 should be surveyed by analysis of the fragments (see p. 311) for optimal conditions. Incubation times of 1 h at 0°C are suitable.

Fractionation of oligomers: columns   One of the most versatile methods for the fractionation of oligomers is ion-exchange chromatography in the presence of urea to reduce intermolecular association on DEAE-cellulose or DEAE-Sephadex (see p. 161). The methods have been described by Tomlinson and Tener (1962, 1963); typical procedures are described below. In all cases simple linear gradient is employed to elute the column.

If DEAE-cellulose is employed, it is packed in 0·5 M ammonium carbonate under 5 psi pressure and washed with 2 M NaCl to remove optically absorbing impurities. It is then washed with 5 mM tris HCl pH 7·8. DEAE-Sephadex (A-25) is packed in the presence of 7 M urea. The column must be very carefully loaded with digest to prevent the surface being disturbed in any way, and then the gradient (including the urea) is applied. For a 2 x 100 cm column, the total volume of eluate is about 7 l and the pumping rate about 1·5 ml/min. Over 500 fractions should be collected. In the following examples there is 7 M urea in the gradient throughout. The gradient volume and pumping rates are adjusted according to the dimensions of the column. If the digest is produced with ribonuclease A, the gradient is 0·02–0·3 M NaCl in 5 mM tris HCl pH 7·8 and the column is of DEAE-cellulose. A column 2 x 100 cm will fractionate up to 0·5 g of digest on the basis of oligomer length (peaks corresponding to oligomers up to nonanucleatodes are obtained). A similar fractionation of a T_1-digest employs DEAE-Sephadex eluted with 0–0·28 M NaCl in 20 mM tris HCl pH 7·6 (and 7 M urea). Such a column (4 x 50 cm) will again fractionate 0·5 g of digest. Subfractionation of peaks from these columns (corresponding to say the pentamers or hexamers) is achieved on columns of approximately the same dimensions but with much less (10 mg or so) of material and eluted with a smaller gradient (2 l) under more acid conditions. For example DEAE-Sephadex eluted with 0–2·0 M

NaCl pH adjusted to 2·7 by the addition of HCl (and in 7 M urea) is suitable. Alternatively DEAE-cellulose can be eluted with ammonium acetate (0·4 M pH 4·0) containing a very shallow gradient (0·1–0·12 M) NaCl. So the first fractionation (higher pH) is on the basis of molecular weight; the second one (low pH) is on the basis of nucleotide composition.

An alternative to the use of urea is to employ a somewhat higher pH. In the systems of Zachau, Dütting and Feldman (1967), the DEAE-column (6·5 meq/g) is eluted with ammonium carbonate buffer at 8·6. As an example of their procedure, for the fractionation of 40 mg of T_1-digest (treated with alkaline phosphatase to remove the 3′-terminal phosphate) of tRNA a column (1 x 40 cm) was eluted with successive gradients of ammonium carbonate: 0·01–0·06 M (900 ml), 0·05–0·32 M (1·9 l), 0·25–0·6 M (1·6 l). Any components that were poorly resolved were fractionated by a urea procedure (0·05–0·12 M NaCl in 0·1 M formic acid and 7 M urea).

Fractionation of oligomers: cellulose acetate and DEAE   Studies on RNA primary structure have been revolutionized by these techniques that were designed and perfected by F. Sanger. There are several techniques involved but they are all based on the same preliminary technique. The oligomers are fractioned in one dimension on cellulose-acetate paper strips and the electrophoretogram is then transferred to DEAE (either paper or a TLC plate—see p. 295). The transfer from the cellulose-acetate strip to the DEAE is the crucial stage and it is effected by placing the strip on top of the DEAE and then layer on top of the cellulose acetate strips of wet chromatography paper. Water seeps through the cellulose acetate on to the DEAE. The oligomers are eluted from the cellulose acetate by the water but when they reach the DEAE they are bound (by ionic interactions). The wet paper and the cellulose acetate are now peeled off to leave the oligomers 'printed' on to the DEAE. The nucleotides are now subjected to the second dimension of the fractionation which is either further electrophoresis or a very novel type of chromatography to be described below.

The cellulose-acetate electrophoresis is performed at pH 3·5 (Sanger *et al.*, 1965; Brownlee, Sanger and Barrell, 1968). This is the optimal pH for nucleotide separations—see p. 47. As pointed out before, this is the stage that is easily overloaded with excess oligomers (p. 292). The digestion is performed in a capillary tube and transferred to the cellulose acetate directly. As this and certain subsequent stages in these procedures involves the quantitative transfer of extremely small volumes of liquid, it is essential to employ siliconized test tubes for dissolving the RNA.[2]

In what follows reference is made to the positions of certain coloured markers. The dye consists of a mixture of equal volumes of xylene cyanol F.F., orange G and acid fuchsin (all 1% solutions in water). The buffer employed is either 0·5% pyridine, 5% acetic acid, pH 3·5 or the same containing 7 M urea. The pH of these buffers should always be checked before use. Buffer left lying in electrophoresis tanks tends to become less acid with time. A strip of cellulose acetate[3] 3 cm wide is floated on buffer and when the buffer has soaked in[4] it is rapidly blotted to remove excess buffer. From now on you have to work fast and avoid the cellulose acetate drying out. The small amount of digest (p. 292) is applied to the middle of the strip about 10 cm from one end; it is flanked (see Fig. 10.3) by the dye markers on either side and is then electrophoresed. The electrophoresis is performed in a tank under white spirit with platinum electrodes and a high-voltage power pack. Suitable apparatus is described by Sanger and Brownlee (1967). The electrophoresis is performed with the spot at the cathode end of the strip. Either 60 cm strips are employed or longer (95 cm on strips) may be used. Assume we are using the 60 cm size. Electrophoresis is performed so that the length of strip between the buffer compartments is about 50 cm and at 3000 V until the yellow marker has almost reached the anode (2 h or so—longer of course if your power pack is not capable of reaching 3000 V).

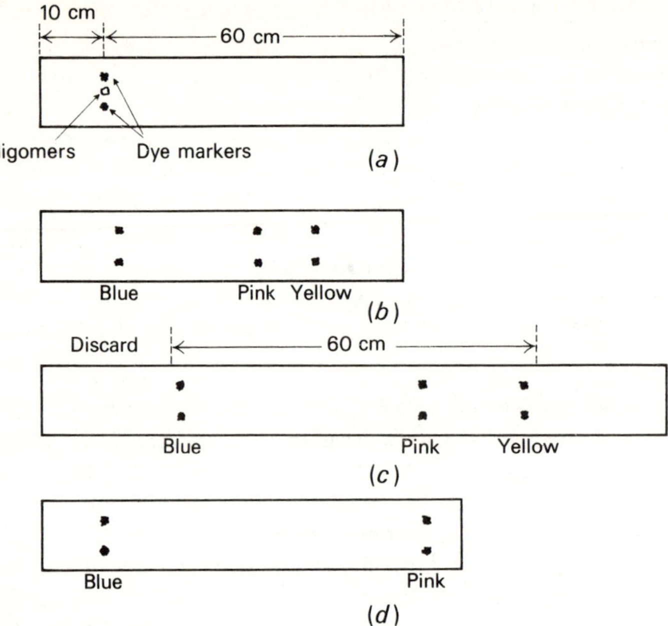

FIG 10.3 The cellulose-acetate dimension of Sanger's separations of oligomers (X = region of oligomers on the strip).
(a) The 60-cm strip before electrophoresis
(b) The 60-cm strip after electrophoresis
(c) A 90-cm strip after the first electrophoresis
(d) The same strip (as in c) after a further electrophoresis.

For the longer strips, which are useful for the larger nucleotides, the electrophoresis does not need to be done in a bigger tank (with an even higher voltage). You start off with excess cellulose acetate immersed in the anode compartment. At the end of the run as above, the strip is removed, the excess buffer on the end that had been in the anode compartment is blotted off, the region behind the blue marker may be cut off (see Fig. 10.3) and the strip electrophoresed again until the pink marker had approached the anode. This process is repeated until the blue–pink distance is 40 cm. Whichever method is used, the oligomers (which can be identified with a hand-held radioactive monitor) are between the blue and pink markers. It is easy to check that the electrophoresis has gone appropriately as the pairs of markers (on either side of the oligomers) should be opposite one another on the paper. If they have skewed it is because, either the strip was incorrectly located in the tank or because the strip had dried out too much in the processing.

There are five alternatives for the second dimension of the separation, (i) electrophoresis on DEAE-paper at pH 1·9 (2·5% formic acid, 8·7% acetic acid), (ii) electrophoresis in 10% formic acid, (iii) 'homochromatography' (see below) on DEAE-paper, (iv) homochromatography on DEAE-TLC and (v) TLC on polyethyleneimine.

For methods (i), (ii) and (iii) the nucleotides are 'printed' on to the DEAE by the following method. The cellulose acetate strip is hung up to allow all the white spirit to drip off. Before the buffer has dried out, it is laid on top of a large sheet (see below) of DEAE-paper 10 cm from one edge of the DEAE and on top of the cellulose acetate four 2-cm-wide strips of heavy (Whatman 3MM) paper previously soaked in water. A glass plate is pressed on top of the wet sandwich of paper and the nucleotides are quantitatively (about 90% efficiency) transferred to the DEAE-paper. This paper is now dried and is ready for use. For method (iv) the same procedure is used but (Brownlee and Sanger, 1969) greater care is required and the cellulose acetate should still contain a little white spirit. The TLC plates are prepared beforehand by blending a mixture of 1·5 g DEAE-cellulose (TLC grade), 15 g cellulose (TLC grade) and 102 ml water for three min, degassing the slurry and spreading (with a TLC spreader) glass plates with a 0·25 mm layer. The plates are allowed to dry at room temperature (see p. 49). The oligomers are 'printed' through from the cellulose acetate 3 cm from one edge. For method (v) the nucleotides may be transferred to polyethyleneimine plates (made according to procedure of Randerath and Randerath, p. 49) by the former technique (Southern and Mitchell, 1971).

Having transferred the oligomers to DEAE or polyethylene imine the alternative procedures are as follows.

(i) and (ii). The oligomers are 10 cm in from a short edge of a huge DEAE sheet 46 x 57 cm or (preferably for most purposes) 46 x 90 cm. The paper is draped over a frame designed to fit an electrophoresis tank under white spirit (see Sanger and Brownlee, 1967, for a simple design). The paper is wetted with one of the two acid-buffer systems by squirting the buffer on with a plastic wash bottle. It is impossible to handle wet DEAE-paper as the sheet must be properly positioned for the electrophoresis while it is dry. The wetting is done by squirting away but not giving the line of oligomers a direct hit. The sheet is wetted on either side and the buffer is allowed to seep up to the oligomers from either side. The frame is now put into the tank and electrophoresed until the blue marker has reached about the half way mark (that is up to the top of the point where the paper wraps over the top of the frame). With the long papers 1500 V overnight is about right.

(iii) and (iv). The technique described above is a unique type of electrophoresis in which the electrophoretic drift (from cathode to anode) is retarded by the ionic binding of the nucleotides to DEAE. Sanger's other invention in this connection, 'homochromatography' is an even more novel procedure. In essence it is a special type of frontal analysis chromatography. For the moment forget about the radioactive oligomers. Imagine a sheet of DEAE-paper (or a DEAE-TLC plate) dipped in a strong solution of mixed oligomers of varying chain length. The mononucleotides N^- saturate the binding sites in the DEAE. As the solvent rises this saturation will proceed up the paper but the less mobile dinucleotides $N\text{-}N^{2-}$ will displace the mononucleotides as they are of higher valency (using the old-fashioned meaning of the word). As the elution proceeds they will be displaced by $N\text{-}N\text{-}N^{3-}$ and so on. Thus fronts of the oligomer classes will be produced. Now imagine the DEAE with a band of radioactive oligomers on it. These radioactive nucleotides form a very small proportion of the total nucleotides in the eluting system (and do not saturate the DEAE binding sites themselves). Thus the fronts will have radioactive nucleotides present. Examples of the nucleotide mixtures that can be used are as follows. For homochromatography on DEAE-paper, crude commercial RNA is dissolved (10 g in 200 ml of 9 M urea, pH adjusted to 7·5 with 10 M KOH). If smaller oligomers are to be separated this way it is a good idea to degrade the 'RNA' even further. For this purpose, 10 g RNA in 100 ml water are made alkaline by the addition of 10 ml of 10 M KOH. After 30 min at room temperature the pH adjusted to 6·5 with 26% $HClO_4$, the $KClO_4$ is allowed to coagulate (1 h at 0°C) and removed by centrifugation. The solution is then saturated with urea at 0° before use (at room temperature). The former system is suitable for TLC homochromatography although 7 M urea is employed in this case. For separating larger oligomers in this system it is wise to remove some of the smaller 'RNA molecules' from the commercial preparation by dialysing the solution against 7 M urea before use. A more defined mixture for TLC homochromatography is prepared by hydrolysing an RNA solution as before but for only

15 min; the solution is neutralized with concentrated HCl (to pH 7·5) dialysed against distilled water for 4 h. Urea (84 g) and water (to final total volume of 200 ml) is added. This system (see p. 312) is used to elute large T_1-oligomers from viral RNA. The TLC plates used with this system differ slightly from those described above. Instead of a 1:10 mixture of DEAE: cellulose, a 1:7·5 mixture is employed. Homochromatography on DEAE-paper is usually performed on 46 x 57 cm sheets and the papers take about 24 h to develop. On TLC, the plates are either 20 x 40 cm and are developed in an oven at 60°C. The longer plates take 5 h, the shorter ones 2 h. In the case of the TLC system the plates are equilibrated at 60°C (outside the tank) before chromatography. Excess urea (transferred during the 'printing' process) is removed from the hot plates by lightly spraying the region of the oligomers with water.

(v) TLC plates (polyethyleneimine) are developed with a suitable buffer. The technique has been developed by Southern and Mitchell (1971). The procedure appears to be a more versatile one than the above procedures of Sanger, and it should prove to be extremely valuable. In general 20 x 40 cm plates are convenient. In all cases, the plate is developed with water so that the water runs 1 cm past the origin prior to development proper. In this way, the concentration of salts at the solvent front is avoided. The buffers used for developing the chromatogram consist of mixtures of pyridine and formic acid. In the following recipe, the molarity refers to the formic acid; the pH is then adjusted by adding pyridine. A superb resolution is achieved by developing the plate sequentially with (a) 1·5 M buffer, pH 3·5 for 4 h; (b) 1.5 M buffer, pH 3·7 for 2 h; (c) sprayed with dilute formic acid pH 3·2 followed by development in 3 M buffer pH 3·6 and finally (d) sprayed as above and developed with 1·6 M buffer pH 3·6 for 20 h. Southern and Mitchell describe many other applications of this type of TLC to the separation of oligomers, including a two-dimensional TLC method in which the plate is developed in 0·8 M LiCl, the plate is dried and washed in anhydrous methanol[5] and then developed (in the second dimension) with one of the pyridine–formate systems. Applications of these methods to both oligoribotides (T_1-digests) and DNA isostiches (p. 277) are described.

The DEAE-papers or TLC plates are dried and autoradiographed with large sheets of X-ray film. No X-ray film is large enough to cope with the giant DEAE sheets so the chromatogram must be cut up and autoradiographed in three sections. The position of the paper is located accurately either by cutting notches in the film and paper or by drawing guide lines on the chromatogram with radioactive ink. The length of exposure depends on the amount of radioactivity present. For 1 μCi *in toto* a day is enough. Proportionally longer periods are required for smaller amounts. Once the spots have been identified, they are cut out and eluted. The eluting system devised by Sanger, consists of a triethylammonium carbonate solution made by bubbling CO_2 through a mixture of Et_3N and water (30 ml redistilled amine plus 70 ml water) until the liquids form one phase; excess Et_3N is added to bring the pH to 10 and nucleotides (and alkaline digest of RNA—see p. 000) are added (0·4 mg/100 ml). The method of elution differs slightly according to the nature of the two-dimensional chromatogram. In the case of the electrophoresis methods, the spots are cut out and connected via little wicks to a trough of eluting solvent. Very small capillary tubes are propped up against the paper. The whole thing is enclosed (in an empty chromatography tank) to prevent evaporation. The solvent flows through spot and up into the tube. Typically in 30 min about 0·05 ml will have collected and essentially all the radioactivity will be eluted. Homochromatograms on DEAE-paper are eluted in the same way except the pieces of paper must be washed in 95% ethanol (to remove urea) and dried prior to elution. TLC spots are sucked off the plate and washed on a glass sinter (no. 1 or 2) with ethanol and eluted by draining a drop of eluting system through the sinter and collected in a siliconized test tube. Brownlee and Sanger (1969) describes an elegant little device for the removal of these spots from the plates. The triethylammonium carbonate is removed under partial vacuum. If it is present in capillaries it is first blown into siliconized test tubes or (more simply) on to a sheet of polythene. The dried samples are suspended in a drop of water and the drying is repeated about three times to ensure complete removal of the triethylammonium carbonate.

Sequencing of oligomers

Complete hydrolysis and digestion with endonuclease   The nucleotide composition of an oligomer is determined by alkaline hydrolysis and separation of the nucleotides (see p. 204). Alternatively it can be deduced from the electrophoretic properties of the oligomers (see the following section). For a dinucleotide, this information is sufficient to establish the sequence. Thus if a T_1-dimer contains G and A it must be A-G-. With larger oligomers it is largely a matter of luck. Consider tetranucleotides obtained from a T_1-digest. If such a nucleotide has the nucleotide composition A_3G there is only one possible structure (A-A-A-G-). If it is A_2UG there are three possibilities (A-A-U-G-, U-A-A-G- and A-U-A-G-). The nucleotide can be digested completely with snake venom phosphodiesterase (see p. 207). In this way the 5′-end can be identified as it is the only nucleoside released as such (the others all appear as their 3′-phosphates). If the free nucleoside is Urd, the sequence must be U-A-A-G-. If it is Ado, there are two possibilities that can be sorted out by treating the oligomer with ribonuclease A. If the products are two dinucleotides (A-U- and A-G-) then the sequence was A-U-A-G-. If these products are GMP and a trinucleotide, the sequence must have been A-A-U-G-. For the sequencing of oligomers obtained from a digest with ribonuclease A, T_1 is used for the second endonucleolysis.

Using these techniques it is possible to sequence unambiguously 90% of the possible tetranucleotides obtained from digestion with the two nucleases and 60% of the possible pentanucleotides (Holley, 1968).

To establish the sequences of oligomers which were still ambiguous according to these procedures, it is possible to use different exonucleases which do not go right up to the end. For example Staphylococcal nuclease (p. 208) degrades chains from the 3′-end but leaves the final phosphodiester bond untouched. Thus it may be used to identify the 5′-terminal dinucleotide residue of an oligomer (Zamir, Holley and Marquisee, 1964). The use of this technique confirmed for the first time the sequence of T-Ψ-C-G-, a common feature of most tRNA molecules (p. 305; Madison, Everett and Kung, 1967). A similar technique (Madison, Holley, Poucher and Connett, 1967) employs the use of polynucleotide phosphorylase (p. 000). As a degradative enzyme, it erodes the chain from the 3′-end but (unlike the Staphylococcal enzyme that removes the nucleoside residues as 3′-nucleotides) this enzyme yields the 5′-diphosphates. Moreover it will stop at the 5′-trinucleotide under appropriate conditions. Consider a particularly tough T_1-oligomer C-U-C-U-C-G-. Complete digestion with Staphylococcal nuclease yields C-U- plus Gp, 2 x Cp and Up. Removal of the 3′-phosphate with alkaline phosphatase[6] and treatment with polynucleotide phosphorylase in phosphate yields ppG, ppC, ppU and C-U-C.

Nucleotide composition and electrophoretic properties   The remarkable two-dimensional electrophoretic separation of Sanger (p. 295) yields a pattern (or 'fingerprint') in which the oligomers are distributed according to their nucleotide composition. Let us have a look at the effect of such an electrophoresis and see the what is meant. In Fig. 10.4 there are tracings of the spots (identified by autoradiography) of two such fingerprints of T_1-digests.

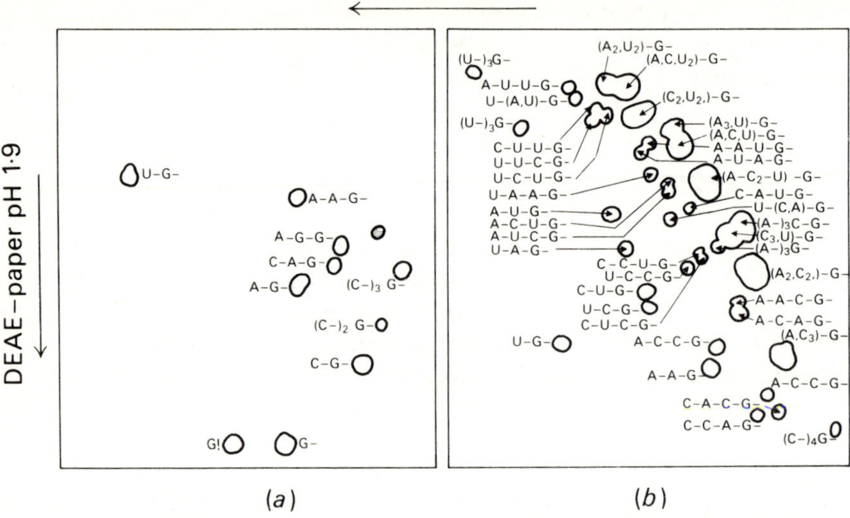

FIG 10.4   Two-dimensional separations of a mixture of T_1-oligomers. The same digest has been run in both cases but in (b) the longer DEAE-run (p. 295) has been used and the bottom of the chromatogram is not shown. The shaded spot is the blue marker. From Sanger, Brownlee and Barrell (1965). Only selected oligomers are included in this figure.

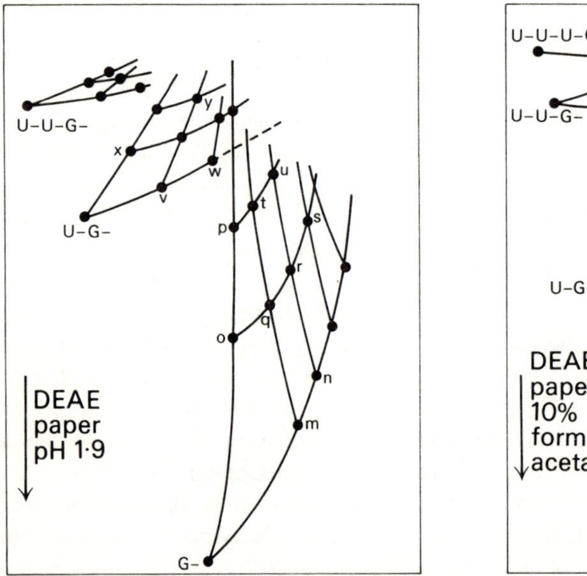

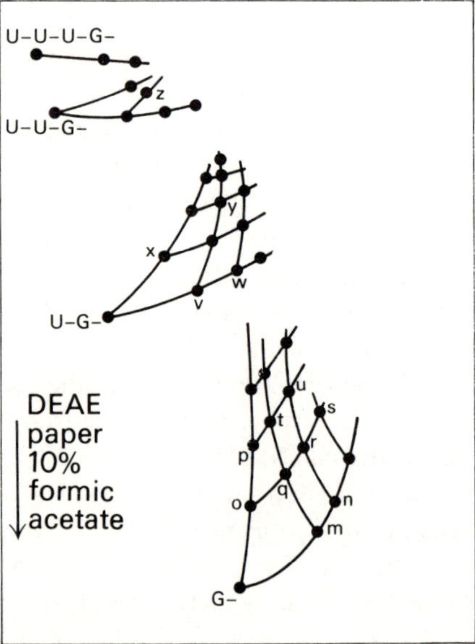

FIG 10.5   Electrophoretic 'graticules' in the system of Sanger, G, C, A and U have their usual meanings and the numbers are the numbers of residues in oligomers. The small letters are for identification purposes only. Oligonucleotide M has the composition CG, n is G_2G (and so on up that line of the graticule). Oligomer o is AG, p is A_2G; q is ACG, r is AC_2G, s is AC_3G and so on; t is A_2CG, u is A_2C_2G. Note that the UG and UUG graticules are better resolved in the 10% formic acid system. The same numbering applies. Oligomer v is CIG, w is C_2UG, x is AUG, y is A_2CUG and z is ACU_2G. (From Sanger, Brownlee and Barell, 1965.)

A glance at these pictures should reveal one or two obvious patterns. Obviously G- runs fastest in the DEAE dimension. Now look (Fig. 10.4(*a*)) for G-, A-G-, A-A-G-, ... and now for G-, C-G-, C-C-G-, C-C-C-G-. These families lie on well defined lines. Several other lines can be identified in Fig. 10.4(*b*). For example look for the oligomers with the base compositions UG, UCG, UC_2G, UC_3G. The patterns can be rationalized in terms of graticules defining the positions of the spots. Such graticules for the pH 1·9 system are illustrated by a real electrophoretogram in Fig. 10.4 and also for the 10% formic acid system in Fig. 10.5.

We have already commented that all these techniques require minute amounts of very highly labelled [^{32}P] RNA. Székely and Sanger (1969) describe a method in which non-radioactive oligomers may be fractionated in the same way by treating the digest with (1) alkaline phosphates and (2) polynucleotide kinase and [^{32}P] ATP. The method has found only limited application in RNA sequencing; as the authors point out, it is equally applicable to the fingerprinting of pyrimidine stretches from DNA (see p. 277).

Partial digestion with exonucleases   As the exonucleases progressively erode on oligomer chain from one end, their controlled, partial action can be employed for sequencing studies. The techniques were pioneered by R. W. Holley whose group achieved the first sequence determination of a naturally occurring nucleic acid (yeast tRNA^{Ala}). He employed the snake venom enzyme (p. 208) which erodes the molecule from the 3'-end. As an example of his method, consider the sequencing of the oligomer A-U-U-C-C-G-.[7] Partial digestion with the enzyme yields (in addition to the mononucleotides) the oligomers A-U-U-C-C, A-U-U-C, A-U-U, A-U.

These oligomers were sorted out according to chain length by chromatography on DEAE-cellulose in the presence of 7 M urea (see p. 292) and their nucleotide compositions were determined. This yields sufficient information for the sequence to be written down unambiguously (Holley, Madison and Zamir, 1964).

In the method of Sanger, Brownlee and Barrell (1965) the spleen enzyme (p. 208), that erodes the chain from the 5'-end was employed. The methodology is designed for the sequencing of several [^{32}P] oligomers (see p. 291) at once. The enzyme is dissolved in 0·1 M ammonium acetate pH 5·7, 2 mM EDTA, 0·05% Tween 80 (a non-ionic detergent). For T_1-digests the enzyme concentration is 0·2 mg/ml; for pancreatic digests it is 0·1 mg/ml. The sample is treated with 10 µl of this mixture for 30 min (60 min for T_1-oligomers containing two or more C-residues which are less easily hydrolysed by the enzyme) at 37°C in a capillary tube. They are then transferred direct to DEAE-paper (together with an undigested sample of the same oligomer and the dye markers). The electrophoresis (1500 V for 4 h or so) is contained until the blue marker is at the top of the paper. Now what is observed (when these papers are autoradiographed) are bands corresponding to the undigested oligomer and the products of its degradation from the 5'-end. Just consider the unchanged oligomer and its first degradation product (say N'-N''-N-G- and N''-N-G-. The difference in electrophoretic mobility between these two compounds is determined by the nature of the nucleotide (N'-) that has been lost. A value M is measured. M is the ratio of the distance between the nucleotides divided by the distance from the origin to N'-N''-N-G-. The values of M for all oligomers surveyed by Sanger are as follows:

If N'- = C-, then M is 0·05–0·3
if N'- = A-, then M is 0·5–1·0
and   if N'- = U-, then M is 1·5–2·5.

If several bands are produced it is possible to deduce the sequence along from the 5'-end. For example let us assume that we know that N'-N''-N-G- has the composition CAUG. If

N′-N″-N-G- and the next oligomer up the paper (N″-N-G-) are compared and M is found to be (for example) 2·3, then N′- must be U-. Let us assume that there is a third band (apart from the nucleotides right at the top) beyond N″-N-G-. This must be N-G-. Assume that its M-value (obtained this time by dividing the distance between N-G- and N″-N-G- by the distance from N″-N-G- to the origin and ignoring the position of the undigested N′-N″-N-G-) is 0·9. The N″- must be A- and the sequence of the original oligomer must be U-A-C-G-.

Minor nucleosides It is difficult to generalize about methods for the identification of minor nucleosides. Examples of their recognition are discussed in the subsequent sections of this chapter. However there are one or two useful points to be borne in mind.

The presence of a minor nucleotide in the alkaline digest of an oligomer is detected on the basis of its different electrophoretic or chromatographic properties to those of any of the four major nucleotides. If the experiment has been done on a large scale, the odd nucleotides can be converted to the corresponding nucleosides (see p. 55) and their spectral and chromatographic properties can be compared with those of known standards. In the event of a previously unknown nucleoside cropping up in this way, of course there is a purely chemical task to be undertaken to determine its structure on the basis of its spectral and chemical properties. As the amounts of nucleoside that are available are small, this can prove a fairly formidable task and it may take several years (as was the case with the Y-base, p. 101). If the RNA is only available in small amounts and has been analysed by the ^{32}P-methods of Sanger, the nucleotides must not be digested with alkaline phosphatase or the only recognizable feature of the structure will disappear. In this case, the electrophoretic mobility of the nucleotide is the only direct clue about its identity. However the nature of the modification can be guessed by a different method. For example, if *E. coli* tRNA is being studied, the cells can be labelled with very high specific activity [Me-^{14}C] methionine. RNA can be isolated, fractionated, digested and electrophoresed from these cells. The autoradiograph from the T_1-two-dimensional fractionation will consist only of those nucleotides which contain methylated nucleosides and a comparison of this picture with the ^{32}P-labelled autoradiograph will identify the starting point for a search for the methylated nucleotides. Similarly, the cells can be labelled with [^{35}S] sulphate and an analogous procedure will identify those stretches that contain such residues as s^{4}U.

A particularly difficult problem is posed by hU-residues. For one thing hUMP has no UV-spectrum (p. 51), for another it is alkali-labile (p. 68). Holley's identification of this nucleoside as a component of tRNA has been mentioned (p. 51). An ingenious method of identifying it in the ^{32}P-labelled digests is due to Sanger (Sanger, Brownlee and Barrell, 1965). They identified an odd nucleotide in an alkaline digest of ^{32}P-labelled tRNA. They measured its electrophoretic mobility on paper (expressed in their paper as an R_u value, i.e. ratio of migration distance to that of UMP) as a function of pH. They found that the R_u-value increased sharply between pH 4 and 5 and in fact the plot resembled the titration curve for a group with a pK of 4·1. This implies the presence of carboxylic acid group. It was therefore suggested that the unknown compound might be the alkaline hydrolysis product of hUMP, β-ureidopropionic acid ribotide (p. 68). The electrophoretic and paper chromatographic properties of the substance were then shown to be identical to those of an authentic sample of β-ureidopropionic acid ribotide and hence the method for locating hU-residues was devised.

Partial digestion with ribonucleases Both ribonucleases (T_1 and A) can be used for the incomplete digestion of RNA. They are limited either by using higher substrate:enzyme ratios or in the presence of magnesium salts (or both). Examples of the methods (Brownlee, Sanger and Barrell, 1968) are to use either enzyme in 10 mM tris HCl pH 7·5, 20 mM $MgCl_2$ with enzyme:substrate ratios of 1:100 (for T_1 or 1:1000 for A). According to the same authors, replacement of the $MgCl_2$ in the above by 10 mM EDTA and an enzyme:substrate ratio of 1:1000 are even milder conditions for T_1. All these digestions are for 30 min at 30°C.

For the separation of these partial digest fragments from small molecules either one of the homochromatography systems or chromatography on DEAE-Sephadex is employed. This is a step for which the techniques of Sanger are of great value. The chromatographic separation of large oligomers on columns is an extremely tedious method. Very long columns of DEAE-Sephadex are eluted with the 7 M urea systems (p. 292) using NaCl gradients up to about 0·6 M (see the various references to Zachau's work).

Results of the sequencing of small RNA molecules

I have chosen three examples of sequencing experiments to illustrate the way in which sequences of small RNA molecules have been built up. One is a tRNA sequence deduced from a 'macroscale' analysis. The other two are done to applications of the ^{32}P-labelling methods of Sanger, one a tRNA, the other 5 S RNA (which contains no minor nucleosides). Finally I have summarized further sequencing data that are available.

Yeast phenylalanine tRNA   Yeast tRNA^{Phe} was sequenced by Khorana (RajBhandary *et al.* 1967) using tRNA fractionated by counter current distribution (see p. 168).

Complete digestion of the molecule with T_1 or ribonuclease A produced a series of oligomers, which were sequenced by combinations of the methods given on p. 297. Additionally the octanucleotide G-G-G-A-G-A-G-C- was sequenced by a chemical 3′-end oxidation method (see p. 287). The total oligomers derived from the molecule are shown in Table 10.2.

Several points emerge from the data at this stage. The 5′-end of the molecule (-G-C-...) can be identified (oligomer (4) in Table 10.2). Likewise the 3′-end (1 and 15). Clearly the 3′-terminal A (common to all tRNA sequences—see p. 13) has been lost during the isolation. Note also that of the minor pyrimidine nucleosides, Cm is not a substrate for ribonuclease A

TABLE 10.2   Fragments obtained in the sequencing of yeast tRNA^{Phe}. Abbreviations used are defined on p. 6. See p. 39 for an explanation of Y. Numbers in parentheses are for reference to the text. From RajBhandary *et al.* (1967).

Ribonuclease A fragments		T_1-fragments	
Cyd (1)	G-U-	4 x G-	T-Ψ-C-G- (16)
hU! (2)	-G-C- (4)	-G-	C-C-A-G-
7 x C-	A-m$_2$G-C- (5)	C-m$_2^2$G- (13)	C-U-C-A-G-
m^{2}C-	G-m^{6}A-U- (6)	C-G-	A-U-U-U-A-m^{2}G! (17)
6 x U-	A-G-hU- (7)	U-G-	m^{7}G-U-C-m^{2}C-U-G- (18)
Ψ-	G-G-A-U- (8)	3 x A-G-	A-A-U-U-C-G- (19)
2 x A-C-	A-G-A-A-U- (9)	hU-hU-G- (14)	m^{6}A-U-C-C-A-C-A-G- (2)
m$_2^2$G-C-	A-G-A-Cm-U- (10)	C-A-C-C- (15)	
G-C-	Gm-A-A-Y-A-Ψ- (11)	A-Cm-U-Gm-A-A-Y-A-Ψ-m^{2}C-U-G- (21)	
G-T- (3)	G-G-A-G-m^{7}G-U- (12)		
G-G-G-A-G-A-G-C-			

(10), and that the others are, but that hU remains as the cyclic phosphate (2, 3 and 11). Similarly Gm is not a substrate for T_1 (21) but the base-methylated guanosines are (13 and 17). From the overlapping regions in these fragments certain longer fragments can be put together. Thus the only T in the sequence can be used to bridge the region covered by (3) and (16) and we can write down Py-G-T-Ψ-C-G-. The Py can be put in because (3) is an A-fragment and must have been produced by the hydrolysis of a Py-G bond. In a similar way (7) and (14) can be put together (Py-G-hU-hU-G-); so can (9) and (19) (Py-A-G-A-A-U-U-C-G-); (8), (17) and (5) (Py-G-G-A-U-U-U-A-m^2G-C-); (12) and (18) (Py-G-G-A-G-m^7G-U-C-m^2C-U-G-); (6) and (2) (Py-G-m^6A-U-C-C-A-C-A-G-); and finally (10) and (21) (Py-A-G-A-Cm-U-Gm-A-A-Y-A-Ψ-m^2C-U-G-). This last stretch contains the anti-codon. The codons for Phe (Table 1.16, p. 22) are UUU and UUC. The anticodons (reading it backwards 3′-5′- so that it lines up with the codon) is AAGm so that Gm is the nucleoside in the wobble position.

The rest of structure was put together by analysis of partial digests of the molecule so that the more overlaps could be recognized. The final structure is illustrated in Fig. 10.6, in which the products of the partial digests are indicated.

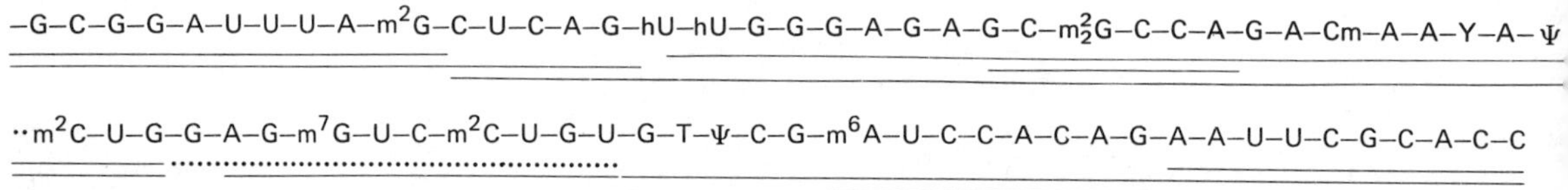

$$-G-C-G-G-A-U-U-U-A-m^2G-C-U-C-A-G-hU-hU-G-G-G-A-G-A-G-C-m_2^2G-C-C-A-G-A-Cm-A-A-Y-A-Ψ$$

$$\cdot\cdot m^2C-U-G-G-A-G-m^7G-U-C-m^2C-U-G-U-G-T-Ψ-C-G-m^6A-U-C-C-A-C-A-G-A-A-U-U-C-G-C-A-C-C$$

FIG 10.6   Sequence of yeast tRNA^{Phe}. The lines indicate the stretches isolated either as partial T_1-oligomers (———) or as a partial A-oligomer (..........) and sequenced. The oligomers characterized in this way furnish sufficient unambiguous overlaps for the complete molecule to be assembled as shown. The 3′-terminal C is discussed in the text.

The tyrosine tRNA molecules from *E. coli* CA strains   Many tRNA sequences, determined by the Sanger techniques, have appeared in the literature. It is unfortunate that in several cases only very fragmentary experimental details have been described. A notable exception to this is the work on the tyrosine tRNA species of *E. coli* by J. D. Smith and his colleagues (Goodman, Abelson, Landy, Zadražil and Smith, 1970). Their work employed several CA[8] (K12-derived) strains of *E. coli* and also defective transducing phage carrying a tRNA gene. Without discussing the genetic aspects of the subject in any details it is important to understand exactly what molecules were studied. The tRNA designated su_{III} tRNA is a molecule present in certain suppressor strains of *E. coli* which can suppress the amber nonsense codon UAG by inserting tyr in the chain. Tyrosine tRNA molecules (there are two in *E. coli*) recognize the two tyrosine codons UAU and UAC.

The experimental procedure that was adopted was first of all to analyse (by the two-dimensional electrophoretic methods) the complete T_1- and A-digests and then to study the partial T_1-digests, separated by two-dimensional methods, in which the second dimension was homochromatography on DEAE-paper (p. 295). These partial-T_1-fragments were analysed by complete digestion with T_1 or A. The partial T_1-fragments for su_{III} tRNA are given in Table 10.3.

Not all the sequences in these fragments could be deduced from the complete digests of T_1 and A. For example in (*a*) there is a nasty long complete-T_1-fragment covering the bulk of the molecule (look for the G's and you will spot it). In this region there is a substantial A-fragment (A-A-A-Ψ-) which does not overlap the end of the T_1-fragment at all. The problem was overcome by isolating a partial A-fragment C-U-A-ms^2i^6A-A-Ψ-C-U-G-C-C-G-U-C-.[9] A similar problem occurs in (*c*) but is solved by a knowledge of the structure of (*a*). In (*d*) a different T_1-fragment, a bit 'less partial' than this oligomer (see p. 292) had to be identified. The fragment running from T-Ψ-C-...up to ....-C-C-A was isolated and sequenced.

TABLE 10.3   Structures of partial T_1-fragments from su_{III} tRNA. From Goodman *et al.* (1970).

(*a*) A-G-C-A-G-A-C-U-C-U-A-ms^{2}i^{6}A-A-Ψ-C-U-G-C-C-G-

(*b*) -G-G-U-G-G-G-G-s^{4}U-U-C-C-C-G-

(*c*) A-G-C-A-C-A-C-U-C-U-A-ms^{2}i^{6}A-A-Ψ-C-U-G-C-C-G-U-C-A-U-C-G-

(*d*) U-C-A-U-C-G-A-C-U-U-C-G-A-A-G-G-T-Ψ-C-G-A-A-U-C-C-U-U . . .

    . . . -C-C-C-C-C-A-C-C-A-C-C-A

(*e*) U-G-G-G-G-s^{4}U-U-C-C-C-G-A-G-C-Gm-G-C-C-A-A-A-G-

From these data the whole sequence can be put together. It, together with tRNA$_I^{Tyr}$ and tRNA$_{II}^{Tyr}$ appear later (Fig. 10.9). Note that there are very few modified nucleosides in this structure (no hU-residues at all). Those that there are were identified by methods discussed on p. 300. Thus [Me^{3}H] methionine labelled both Gm and ms^{2}i^{6}A; [^{25}S] sulphate labelled s^{4}U and also the modified A. The former of these was identified by comparing the electrophoretic properties of its 3′- (and 2′-) phosphate with those of authentic 4-thiouridylic acid. The ms^{2}i^{6}A was identified separately in a digest from a large amount of the tRNA^{Tyr} by Harada *et al.* (1968). When the tRNA^{Tyr} sequences are considered (Fig. 10.9) it will be seen that they contain an odd nucleoside G̊. The phosphate of this substance has been obtained on an optically dense scale but its chemical nature is unknown. Its spectrum is similar to that of GMP but it probably contains two $-NH_2$ groups. It is known to be derived from G metabolically but it is not labelled by radioactive methionine or sulphate.

E. coli **5 S RNA**   The determination of the sequence of *E. coli* 5 S RNA by Brownlee, Sanger and Barrell (1968) was of great importance as the experiments involved the introduction of several novel techniques in sequence studies. Some of these have been mentioned already (p. 295). Some of the other techniques are mentioned below. The molecule contains 120 nucleotides, none of which is 'minor'. Complete digestion with ribonuclease T_1 gives 19 different oligomers, none longer than ten nucleotides; complete digestion with ribonuclease A gives 21 oligomers, none longer than six nucleotides. Clearly the amount of information available from these data is insufficient to identify significantly longer stretches of sequence. Similarly partial digests with T_1 or ribonuclease A yielded insufficient information as there are significant stretches of degenerate (repeated) sequence in the molecule. In order to condense a very large amount of experimental information, it is simplest to look at the complete sequence that was finally elucidated and pick out some of the essential points. In Fig. 10.7 is the

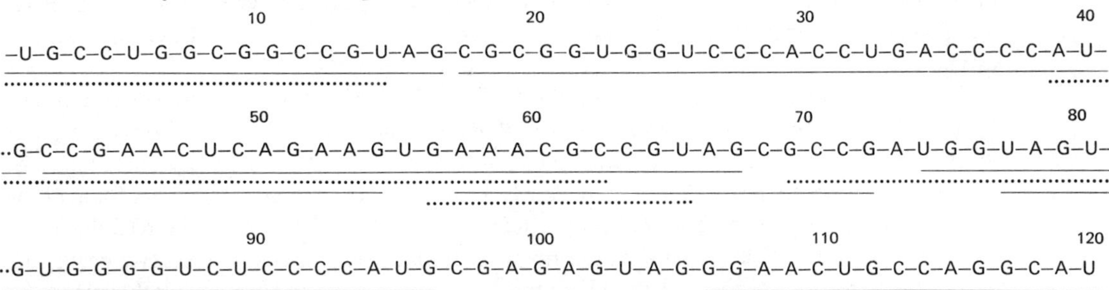

FIG 10.7   Sequence of *E. coli* 5 S RNA. Stretches underlined with a continuous line are partial T_1-fragments; those underlined in dots are partial A-fragments. Many more partial fragments were isolated and sequenced by Brownlee, Sanger and Barrell (1968). In this summary those which are merely internal fragments of larger oligomers are not illustrated except for one by way of illustration. This is the partial T_1-fragment 42–54 which is the 5′-portion of the larger 42–67.

sequence of *E. coli* 5 S RNA written in such a way that it is easy to identify particular nucleotides in the following discussion with the partial T_1- and A-digests shown.

The details of the structure that were required to put the fragments together unambiguously were deduced by using chemically modified 5 S RNA. Two reactions were employed. Brief technical details are given followed by an example of the way in which they were used.

The first method was the use of CMCT (see p. 63). Although the reaction occurs with both U and G, the major reaction in an RNA chain is with U. As the reagent was used by Sanger to modify the substrate for ribonuclease A, the minor reaction (with G) is neglected in the following discussion. In the 'partial reaction', a solution containing about $1\,\mu$Ci of $[^{32}P]$ RNA (i.e. about $1\,\mu$g) in between 0.01 and 0.05 ml of 20 mM tris HCl pH 8.9, 0.4–0.75% CMCT (freshly dissolved). Incubation was for 16 h at 37°C. For 'complete reaction' the concentration of CMCT was 10% but the conditions were otherwise identical. RNA was isolated from the reaction mixtures by adding NaCl to a final concentration of 0.1 M and two volumes of ethanol. The RNA was digested with ribonuclease under the conditions of p. 292. In the case of the products following partial reaction with CMCT, the oligomers (still bearing their CMCT-groups—see Fig. 2.9e, p. 64) were fractionated by the method of p. 294 (employing the pH 1.9 system for the DEAE-dimension). Spots present in the chromatogram, but absent in the control A-fingerprint, were eluted (p. 296) and treated with 0.01 ml of 0.2 M ammonia at 37°C for 6h, to remove the blocking groups (see p. 63). These oligomers were sequenced by conventional methods, employing ribonuclease A or T_1. The oligomers obtained from the digest following complete reaction with CMCT, had their blocking groups removed prior to electrophoresis. Obviously this cannot be done until the ribonuclease is inactivated as new reactive sites in the oligomers are exposed as the CMCT-groups come off. Therefore the solution of oligomers (plus ribonuclease) was dried down on polythene (see p. 297) and taken up in a mixture of $5\,\mu$l β-mercaptoethanol and $5\,\mu$l dimethylformamide. This solution was incubated (in a capillary tube) at 37°C for 8 h to reduce and denature the nuclease. After again evaporating the solution it was digested with 0.01 ml of pronase solution (0.5 mg/ml) at 37°C for 16 h to destroy the ribonuclease. The solution was mixed with an equal volume of concentrated NG_3 and left at room temperature for 2 h to remove the blocking groups. The resulting mixture of oligomers was fractionated as above.

The alternative chemical modification employed dimethyl sulphate. This substance methylates G-residues in nucleic acids (p. 60). A mixture of $2\mu1\,Me_2SO_4$ and 0.08 ml of RNA dissolved in 5% sodium acetate (pH adjusted to 5.4 with acetic acid) was incubated at room temperature for 20 min. The experiment was done with about 1–$10\,\mu$Ci of $[^{32}P]$ RNA with unlabelled RNA added to give a total amount of RNA of 0.02 mg. After the reaction the RNA was precipitated (ethanol) and the degree of alkylation (about 30–50% of the G-residues) was estimated.[10]

The ionic properties of m^7G-residues differ markedly from those of G-residues (see p. 60) Consequently the electrophoretic properties of the oligomers containing m^7G-residues differ greatly from their analogues containing G. In general, in the two-dimensional map, the methylated oligomers move significantly slowly in the cellulose-acetate dimension (behind the blue marker largely, see p. 294) and a little faster in the DEAE-dimension. The technique that was employed was to fractionate the mixture on a very long strip of cellulose-acetate paper, and to cut the paper in two through the blue marker. The two sections were separately printed (p. 294) on to two sheets of DEAE-paper and were fractionated either in the pH 1.9 system or the 7% formic acid system (p. 294). The sequences of these oligomers was determined by digestion with ribonuclease A and fractionation of the oligomers by paper electrophoresis at pH 3.5.

These two reagents were used (for example) in identifying the $5'$-sequence of the molecule.

One of the methylated T_1-oligomers had the composition -U(C, U, M-C)-G-. One of the partial CMCT products was -U-G-C- and hence the sequence must be -U-G-C-C-U-G- . . . .. This gives one a good start in the subsequent sequencing of the first really large partial T_1-oligomer (1–16).

It is obvious from Fig. 10.7 that there are not enough overlaps in the sequence as shown and additionally, there was not quite enough information for the unambiguous sequencing of the larger oligomers. For this reason two additional enzymes were employed. With the advent of ribonuclease U_2 (see pp. 207 and 312) these two non-specific nucleases will probably be used less in future. However they do illustrate the way in which ribonuclease can be tried as a method of obtaining oligomers by employing it in non-optimal conditions. One was the spleen-acid ribonuclease; it was employed as a 0·2% solution in 0·1 M acetate buffer pH 5·0. RNA (10 μg) were incubated in 2 μl of this solution at 37°C for 30 min. *B. subtilis* ribonuclease was employed as a 0·01% solution in tris HCl pH 8·5. Again 10 μg of RNA was incubated with 2 μl of enzyme solution; in this case the conditions are 0°C for 30 min.

Transfer RNA sequences   All tRNA sequences determined so far contain intramolecular complementarity of sequence so that they can be arranged in the clover-leaf conformation (see p. 23). The general appearance of this clover leaf is shown in Fig. 10.8.

All the stems of this structure are H-bonded although there are occasionally mis-paired bases in the 'stalk' of the leaf. The names of the loops are already archaic as it is now recognized that in the tRNA^{Tyr} species from *E. coli* (and their suppressor *su*$_{III}$—see p. 303) there is no dihydroU in the sequence at all. The other loops are still appropriately named. The anti-codon always appears in the same place (flanked by a very characteristic pattern of nucleosides as shown) and the T-Ψ-C sequence in the 'T-pseudoU-C loop' is apparently universal.

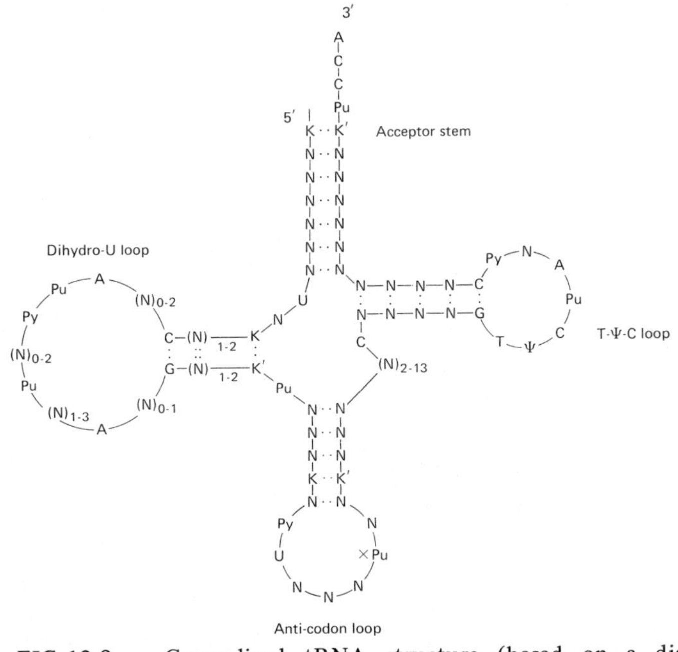

FIG 10.8   Generalized tRNA structure (based on a diagram of Waring, 1968). The conventions used here are those employed throughout the rest of the book. Additionally K, K' are a G:C base pair (i.e. either K = G and K' = C or vice versa) and xPu is a purine, modified in most molecules. The symbol $(N)_{1-3}$ refers to a stretch of nucleotides between 1 and 3 residues long (similarly other symbols of the same form).

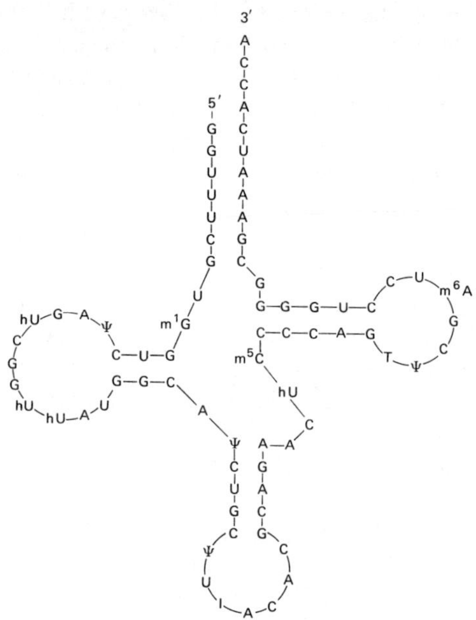

(a) tRNA$_1^{Val}$ (*E. coli*)

(b) tRNA^{Val} (*Saccharomyces* yeast)

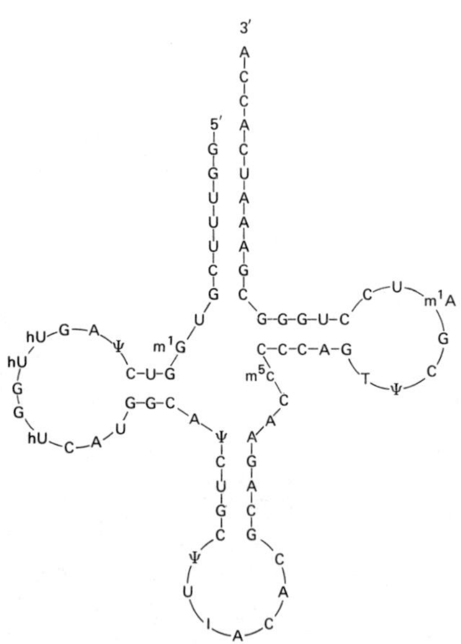

(c) tRNA$_1^{Val}$ (*Torulopsis* yeast)

FIG 10.9   Sequences of various tRNA molecules.
References: (*a*) Yaniv, Favre and Barrell (1969)
(*b*) Bayev *et al.* (1967)
(*c*) Takemura, Mizutani and Miyazawa (1968)

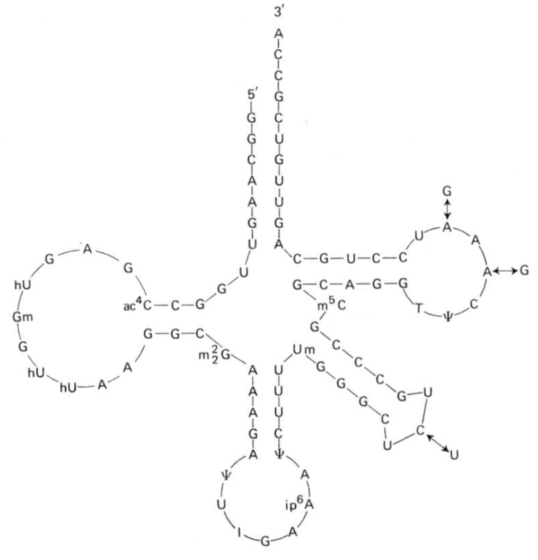

(*d*) tRNA$_1^{Ser}$ (yeast)
tRNA$_2^{Ser}$ differs in the
three residues shown

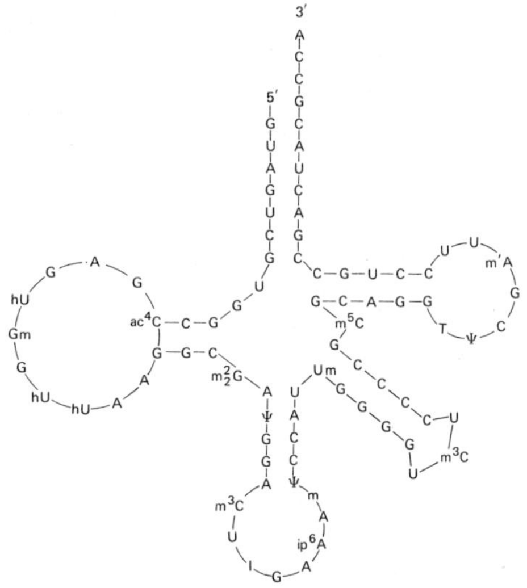

(*e*) tRNA^{Ser} (rat liver)

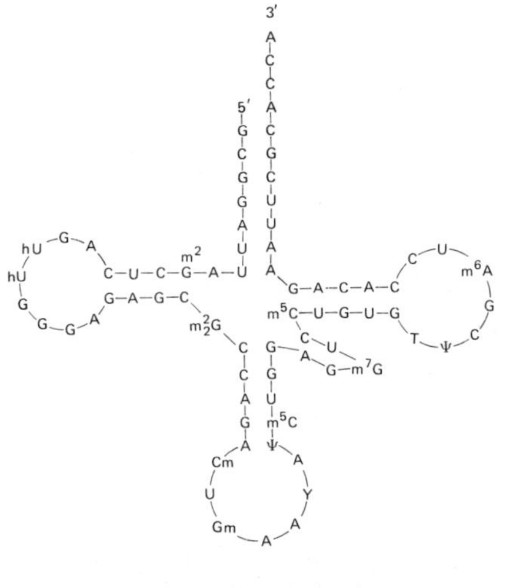

(*f*) tRNA^{Phe} (yeast)

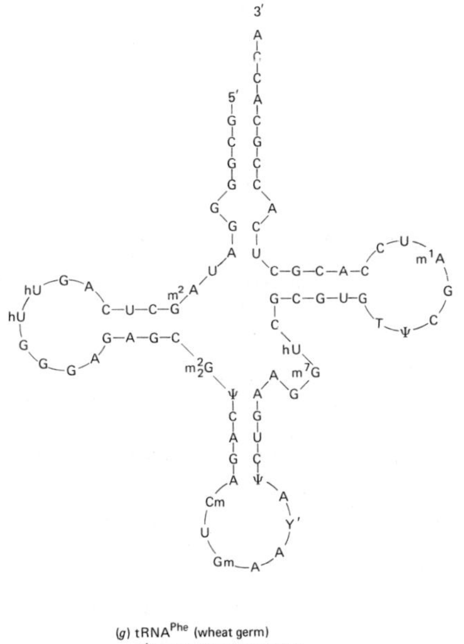

(*g*) tRNA^{Phe} (wheat germ)
Y′ is an unidentified residue
with similar spectroscopic properties
to those of Y

(*d*) Zachau, Dütting and Feldmann (1966)
(*e*) Staehelin, Rogg, Baguley, Ginsberg and Wehrli (1968)
(*f*) RajBhandary *et al.* (1967)
(*g*) Dudock, Katz, Taylor and Holley (1969)

(*h*) tRNA$_\text{M}^\text{Met}$ (*E. coli*)

*
N is a modified N

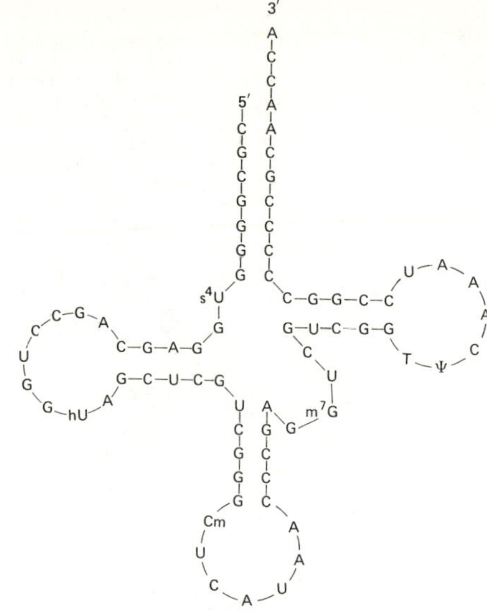

(*i*) tRNA$_\text{F}^\text{Met}$ (*E. coli*)

(*j*) tRNA$_\text{I}^\text{Tyr}$ (*E. coli*)
Modifications 'II' occur in tRNA$_\text{II}^\text{Tyr}$
Modification '*su*' occurs in tRNA *su* III

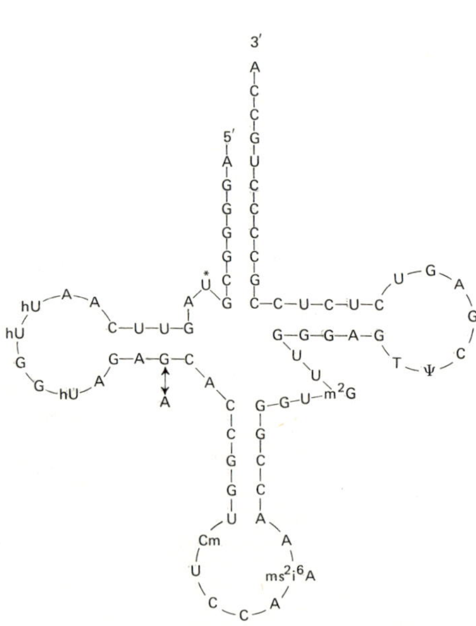

(*k*) tRNA$^\text{Trp}$ (*E. coli*)
Modification occurs
in tRNA su_UGA^+

(*h*) Cory, Marcker, Dube and Clark (1968)
(*i*) Dube, Marcker, Clark and Cory (1968)
(*j*) Goodman *et al.* (1970)
(*k*) Hirsch (1971)

The sequences of several tRNA species are shown in Fig. 10.9. This is not a complete record of all the sequences that are known. They have been chosen to illustrate certain remarkable features of the structure, especially the ways in which various structural features have been conserved. Note that the sequences of the tRNA^{Val} from the two species of yeast are almost identical but that the sequence from *E. coli* is quite different. These three structures also emphasize that bacterial tRNA contains thiolated residues which have not been found in eukaryotic tRNA so far. The *E. coli* tRNA^{Val} is further discussed on p. 421. A more remarkable similarity exists between the tRNA^{Ser} structures from a yeast and from rat liver. There has been a lot of evolution between yeasts and mammals but note that the fourth loop (in what is the most variable region of tRNA) is present in both these molecules. The yeast tRNA^{Ser} is one of the largest tRNA structures, it contains many odd nucleosides and yet it was one of the first tRNA sequences to be determined. Zachau's achievement with this structure is the more remarkable because it is in fact a mixture of the two structures as illustrated and methods of resolving the mixture were not available at the start of the work. The two tRNA^{Phe} structures shown are of interest too. Note that there are several similarities here (including the Y-base in the xPu position—see Fig. 10.8). The two tRNA^{Met} molecules from *E. coli* are of special interest as in this case the different metabolic roles of these substances are well defined (p. 24).

Of particular interest are the two examples of suppressor tRNA sequences. The tyrosine-suppressor carries a mutation in the anti-codon so that $\overset{*}{G}$-U-A (complementary to the tyr codons UAPy) becomes C-U-A (complementary to UAG, 'amber'). In contrast, the anti-codons in tRNA^{Trp} and the 'umber' suppressor are identical (C-C-A). In the normal molecule, the first C of this anti-codon reads the final nucleotide (G) in the trp codon (UGG) faithfully. Following a rather inocuous looking mutation in the dihydroU stem, the conformation of the anti-codon loop is changed so that the first C (of the anti-codon) can read its 'wobble alternative' of A (p. 23) and the molecule will incorporate trp when the umber (UGA) codon appears in mRNA.

Other aspects of function/structure relationships in tRNA are discussed on p. 417 and 448. It will be seen that the sequences throw considerable light on the specificity of tRNA in the synthetase reaction. We are still waiting for information which will enable us to correlate primary structure with other functions of tRNA (such as its binding to ribosomes). It will also be interesting to learn something of the sequence of those tRNA species that have other functions. One of these functions is mentioned on p. 201 (cell-wall synthesis); another interesting example is the tRNA^{Trp} of *Drosophila* which inhibits one of the enzymes of tryptophan catabolism (Jacobson, 1971b).

5 S RNA   The sequences of 5 S RNA from *E. coli* (p. 303), *Pseudomonas fluorescens* (DuBoy and Weissman, 1971) and human tissue culture cell-line KB (Forget and Weissman, 1968, 1969) have been determined. There is considerable homology between the two bacterial sequences, very little between the bacterial and human sequences.

The role of 5 S RNA in the ribosome is unknown. It is unfortunately not the case that a knowledge of the sequences suggests a role for them. Moreover no very obvious secondary structure can be devised which will accommodate all three sequences. The question of the possible secondary structures for the molecules is briefly considered in the next chapter (p. 350).

6 S RNA   The sequence of 6S RNA from *E. coli* has been determined (Fig. 10.10) and a preliminary report of the experiments has appeared. 6 S RNA is even more of a puzzle than 5 S RNA. It is a soluble cytoplasmic substance. The structure can be arranged in a H-bonded

...5′–A–U–U–U–C–U–C–U–G–A–G–A–U–G–U–U–C–G–C–A–A–G–C-...

...G–G–G–C–C–A–G–U–C–C–C–C–U–G–A–G–C–C–G–A–U–A–U–U-...

...U–C–A–U–A–C–C–A–C–A–A–G–A–A–U–G–U–G–G–C–G-...

...C–U–C–C–G–C–G–G–U–U–G–G–U–G–A–G–C–A–U–G–C-...

...U–C–G–G–U–C–C–G–U–C–C–G–A–G–A–A–G–C–C–U–U–A-...

...A–A–A–C–U–G–C–G–A–C–G–A–C–A–C–A–U–U–C–A–G–C–U–...

...U–G–A–A–C–C–A–A–G–G–G–U–U–C–A–A–G–G–G–U–U-...

...A–C–A–G–C–C–U–G–C–G–G–C–G–G–C–A–U–C–U–C–G-...

...G–C–A–U–C–U–C–G–G–A–G–A–U–U–C–C–3′
                              ↕
                              A

FIG 10.10   Sequence of *E. coli* 6 S RNA (Brownlee, 1971).

conformation (Brownlee, 1971) but no possible function is suggested by this shape. Nor is this likely to be the end of the story. Both bacterial and animal cells contain several cytoplasmic RNA molecules, apparently homogeneous and not related to mRNA. Quite an amount of preliminary sequencing work has been done on these molecules and is surveyed by Larsen, Lebowitz, Weissman and DuBoy (1970).

tRNA precursors   Altman and Smith (1971) have established that the 5′-terminus of the precursor for tRNA^{Tyr} is the following (where $\underline{G}$ is the 5′-terminal G of the mature molecule (Fig. 10.9(*j*)). Several secondary structures are possible for this precursor; some involve the tRNA sequence in a structure different to the clover-leaf.

5′pppG-C-U-U-C-C-G-G-G-G-A-G-C-A-G-G-C-C-A-G-U-A-A-A-A-G-C-A-U-U-A-C-C-C-G-U-$\underline{G}$-...3′

The sequencing of large RNA molecules

Partial digestion

Partial digestion has been described before (p. 292). With large RNA molecules containing several thousand nucleotides, the methods that are available, are either to use the secondary structure of the RNA to limit the nuclease reaction or to limit by some biochemical interaction the RNA exposed to nuclease reaction. The central technical problem in both of these approaches is the fractionation of large oligomers. Once this has been achieved, the problem with the first approach is to identify which part of the RNA molecule has given rise to the fragment due to be studied. There are very few general rules that can be used to guide the approach to a particular problem and having dealt with the fractionation problem, I have grouped the various techniques under the headings of particular groups of molecules.

Fractionation of large oligomers   Two-dimensional electrophoresis (p. 294) is really only useful for the separation of oligomers up to about six or so nucleotides long. Homochromatography takes one up to about 15 nucleotide oligomers (perhaps a bit bigger in the TLC system of p. 295). For large oligomers one is left with either column chromatography or gel electrophoresis in polyacrylamide. There is no doubt that, provided very highly labelled [^{32}P] RNA is available, gel electrophoresis has everything to be said for it. The separations are quick, the resolution is remarkable and (provided the experiment is performed in slabs) they can be rapidly autoradiographed.

Probably the most versatile system is that of Adams, Jeppeson, Sanger and Barrell (1969). The electrophoresis is operated along the longer dimension of a gel trapped between two glass plates 20 x 40 cm. The plates must be scrupulously clean. The plates are separated by two thin strips of 'Perspex' along the two longer edges (see Fig. 10.11). The surfaces of the Perspex are smeared with petroleum jelly before use. The bottom of the device is sealed with 'Plasticine'. The whole assembly is held together with very strong clips (of the type designed for clipping sheafs of paper to wooden boards). A solution of 12·5% acrylamide, 0·4% BIS (both recrystallized as described on p. 156) in 40 mM tris acetate pH 8·3 (250 ml) is degassed and 0·6 ml of 10% ammonium persulphate and 0·25 ml of TEMED are added and the little plastic 'slotter' is pushed on to the top (see Fig. 10.11). The gel sets in about 5 min. The slotter and Plasticine are removed, the result is a slab of gel with slots (for the samples) in the top. The bottom of it is put into a tank of 2 l of buffer (40 mM tris acetate pH 8·3). The top part of the gel compartment is filled with buffer. The digest in 0·025 ml of buffer containing 10% sucrose and a marker of bromophenol blue is added to a slot (the other slots are filled up with other digests or markers—it is a good idea to run tRNA as one marker). The buffer in the top of the gel is connected with a thick chromatography-paper wick to another 2 l reservoir. The electrophoresis is run until the bromophenol blue has run to the bottom (overnight at 400 V and 40 mA in the cold room).[11]

In the case of T_1-digests to be analysed by this method, the digestion is stopped by extraction with phenol at 0°C (in the capillary tubes). The phenol is removed by three extractions with ether.

After the electrophoresis is over, the clips are taken off, the slab is laid on its side and the top plate is slid off. The gel surface is marked with radioactive ink, covered with cellophane and autoradiographed. In the experiment on R17 RNA (see p. 312) 30 μCi (!) of [^{32}P]RNA was used for each digest. The autoradiography took 30 min. The problem of eluting the oligomers is fairly difficult. The bands that are required are cut out with a razor blade. The method of elution recommended by Adams *et al.* (1969) is to pop the bits of gel into a little cylindrical glass tube covered at the bottom with a disc of DEAE-paper (supported with cellophane or dialysis membrane held in place with a rubber band). The tubes are then subjected to electrophoresis (in the same buffer). The oligomers are trapped on the disc of

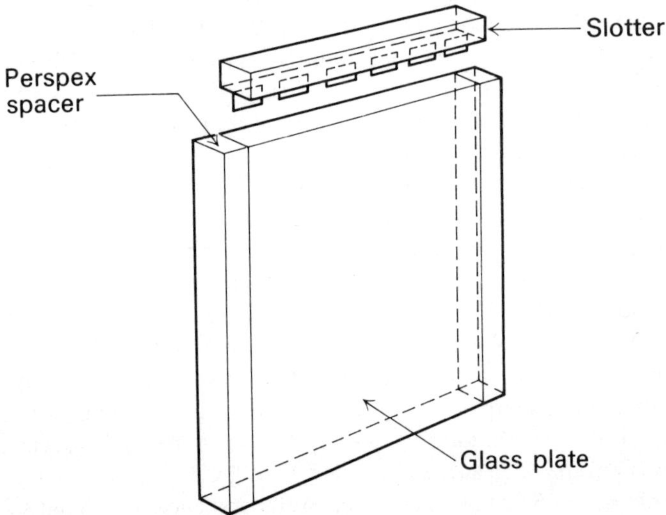

FIG 10.11   Design of apparatus for slab gel electrophoresis (see text for discussion).

DEAE-paper and these bits of paper are dried and the oligomers eluted as described on p. 296.[12]

According to de Wachter, Vandeberghe, Merregaert, Contreras and Fiers (1971), further purification of bands isolated in this way is very satisfactorily achieved on 10% polyacrylamide gels in 25 mM citric acid, 6 M urea. Full experimental details are described in de Wachter and Fiers (1971).

Examples of sequencing in large RNA molecules

Phage R17   Adams, Jeppeson, Sanger and Barrell (1969) wrote what must be one of the classical papers of molecular biology as it provided the first direct chemical proof of the reality of the genetic code *in vivo*. A total T_1-digest of R17 RNA was fractionated by the two-dimensional system in which the second dimension was TLC homochromatography (p. 295). One particularly large oligomer (oligonucleotide A) was isolated and its nucleotide composition determined (p. 204). The result (calling G = 1·00) was $U_{8·6}$, $A_{6·5}$, $C_{3·9}$, G. This

-G̲-A-A-U-U-A-A-C-U-A-U-U-C-C-A-A-U-U-U-U-C-G-

(1)		asn	ochre	leu	phe	gln	phe	ser
(2)		ile	asn	tyr	ser	asn	phe	arg
(3)	glu	leu	thr	ile	pro	ile	phe	

FIG 10.12   Sequence of oligonucleotide A and three possible translations of it in terms of an amino-acid sequence. G̲ is a residue not present in oligonucleotide A itself but its presence can be inferred from the fact that oligonucleotide A is itself a T_1-fragment.

oligomer was digested with ribonuclease A and the products A-A-C-, 2 x A-A-U-, A-U-, 3 x C-, 6 x U- and G- were identified. Oligonucleotide A was then treated with CMCT, followed by ribonuclease A (see p. 206). By analysing the products of this digestion, the following fragments were identified: C-, (A-A-U-, U)-A-A-C-, G-, (A-A-U-, U_3)-C- and (A-U-, U-U-)C-. Finally oligonucleotide A as incubated with ribonuclease U_2 (see p. 207; the conditions were 0·01 ml of a solution containing 2 mM EDTA, 50 mM sodium acetate pH 4·5, 0·1 U/ml U_2 and 0·1 mg/ml of albumin plus oligonucleotide A). The products identified from this reaction were A-, A-A-, (C,U)-A-, U-U-A-, U-U-A-A-, (U-U-, C-C-)A-, (U-U-, C-C-)A-A- and (U-U-U-U-, C-)G-. There is enough information there to write down an unambiguous sequence for oligonucleotide B (see Fig. 10.12). Now let us assume that this is part of a structural gene in the phage RNA. We do not know of course what the 'reading frame' is and so there are three possible amino-acid sequences that can be written down (Fig. 10.12).

Amino-acid sequence (3) is the same as the sequence of amino acids 89–95 in the coat protein of the phage. R17 RNA was next partially digested with T_1 and the fragments were isolated (by gel electrophoresis as on p. 311). These giant oligomers were completely digested with T_1 and analysed by the same homochromatography system as the one originally employed for the isolation of oligonucleotide A. One 57-nucleotide fragment was identified which contained oligonucleotide A and another large oligomer (B). The sequence of oligonucleotide B was determined in an exactly analogous manner to oligonucleotide A. It was G̲-U-A-C-U-U-A-A-A-U-A-U-G-. One of the three readings of this sequence is tyr-leu-asn-met which are amino acids 85–88 in the coat protein. The implication is that the situation in the 57-oligomer is B-G̲-A (i.e. the G̲ inferred at the 5'-end of A is not the same as the 3'-terminal G of B). With these clues, it was possible to sequence the whole of the 57-nucleotide fragment; its sequence was found to code exactly for amino acids 81–100 in the coat protein.

The work has since been considerably extended (Jeppeson, Nichols, Sanger and Barrell, 1970). There is a very remarkable conclusion from their work but this cannot be fully discussed until another method of fishing out fragments of viral RNA has been considered.

The genetic map of R17 is rather simple—there are only three genes (the total genome length is 33 000 nucleotides). It looks like this:

5′........(A)-A protein....(C)-coat protein-...(S)-synthetase..3′.

The words describe the structural genes, the letters in parentheses are the ribosomal binding sites for the translation of the genes (A protein is required for phage maturation) and the dots are the untranslated regions. The dots between 5′ and (A) are known as the leader sequence.

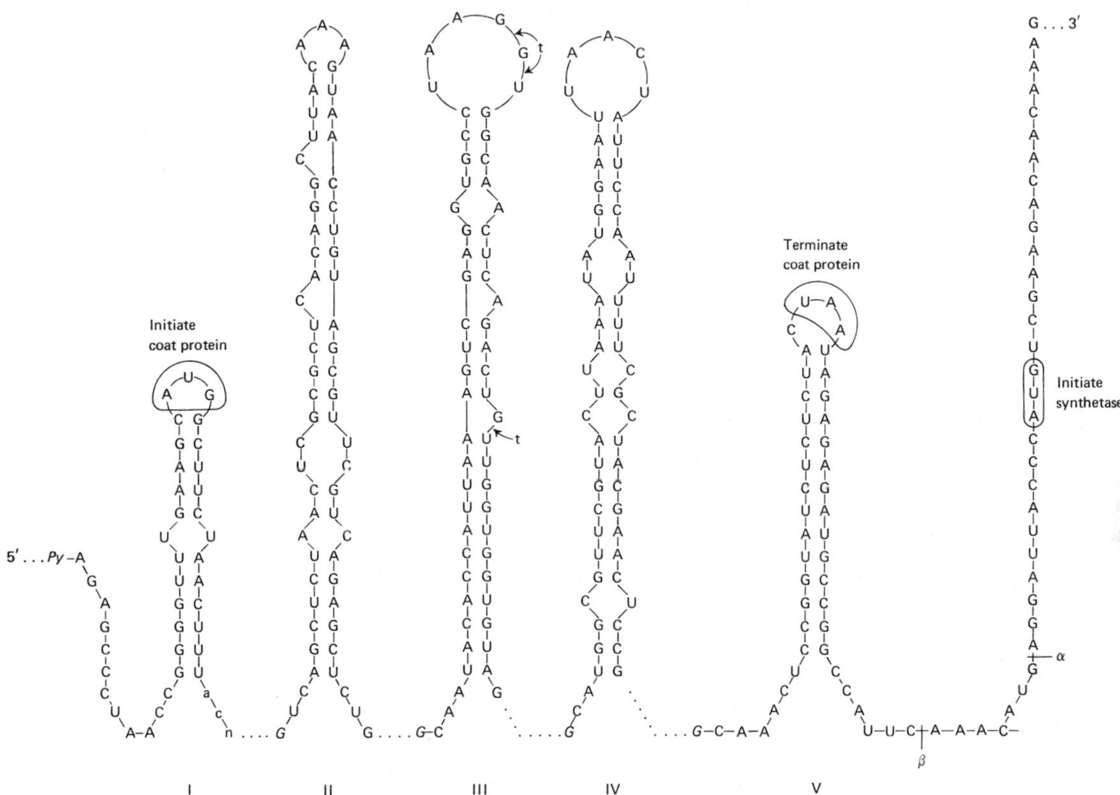

FIG 10.13   Extended sequences for the region of the R17 genome covering the coat protein gene. Oligomer I was isolated by Steitz from the ribosome-binding experiment. II, III and IV were isolated from partial T_1-digests by Jeppesen and others (see text for references; IV is the fragment discussed on p. 312). V is a composite structure. From the 5′-end to position α it is a partial T_1-fragment; from β to the 3′-end it is the synthetase initiation fragment isolated by Steitz. There is sufficient overlap (sequence β to α) to join them up unambiguously. The underlined nucleosides (G and Py) were not present in the oligomers; their positions in the chain are inferred by the nature of the enzyme involved in the production of the fragment following (see p. 312). The amino acids are those known to occur in either the coat protein sequence or the synthetase N-terminus. The sequence a-c-n is purely hypothetical; it is based on the known codons for thr (see Fig. 1.16, p. 22) which is known to be the sixth amino acid of the coat protein. References to the RNA sequencing are given in the text. The coat protein sequence was determined by Weber 1967; that of the N-terminus of the synthetase is deduced from the sequence. The positions 't' are the only positions in fragments II, III and IV that are hydrolysed readily by T_1 (i.e. under the partial conditions of p. 292).

The sequences around the ribosome binding sites have been determined by binding ribosomes (without the factors necessary for actual translation) and digesting the complex with ribonuclease. The fragments that are isolated are the bits of RNA protected by ribosomes at the point of initiation of protein synthesis. In this way the three binding sites have been sequenced (for a review including all the relevant results see Jeppesen, Steitz, Gesteland and Spahr, 1970).

In her work, Steitz (1969) employed the following technique. A reaction mixture (total volume of 0·05 ml) contained in 50 mM NH_4Cl, 5 mM magnesium acetate, 0·2 mM GTP, 0·1 M tris HCl pH 7·4, 0·1 mg of crude initiation factors, 6 units[13] of *E. coli* MRE 600[14] ribosomes, 2·2 units of fMet-tRNA and R17 RNA. The R17 RNA consisted of several microcuries of very highly labelled [32P] RNA plus extra unlabelled RNA to bring the total amount to 1·5 units. The R17 RNA was incubated at 37°C in 10 μl of mM EDTA for 8 min before addition to the reaction mixture. The mixture was incubated at 37°C for 12 min and then cooled to 22°C; ribonuclease A was added (0·025 mg) and after a further 12 min (at 22°C) the preparation was centrifuged through a sucrose gradient containing 5 mM Mg^{2+} and the 70 S ribosome peak (p. 22) was collected. A peak of radioactivity was associated with this. The ribosomes were denatured in 8 M urea (p. 461) and the R17 fragments separated from the bulk of the rRNA by DEAE-cellulose column chromatography (p. 161). The large oligomers were separated in the TLC homochromatography system (p. 295) and were sequenced by digestion with ribonuclease A or T_1. The result that was found was that when freshly prepared [32P] RNA was employed, the main initiator sequence isolated was that for the coat protein. The other initiator fragments

	Second letter				
First letter	U	C	A	G	Third letter
U	Phe(1) Phe(1) Leu(2) Leu(0)	Ser(3) Ser(2) Ser(1) Ser(2)	Tyr(0) Tyr(4) Stop(1) Stop(0)	Cys(1) Cys(0) Stop(0) Trp(1)	U C A G
C	Leu(0) Leu(0) Leu(0) Leu(0)	Pro(1) Pro(0) Pro(1) Pro(0)	His(0) His(0) Gln(0) Gln(4)	Arg(2) Arg(1) Arg(0) Arg(0)	U C A G
A	Ile(3) Ile(2) Ile(0) Met(1)	Thr(4) Thr(2) Thr(0) Thr(1)	Asn(1) Asn(4) Lys(3) Lys(1)	Ser(0) Ser(3) Arg(0) Arg(0)	U C A G
G	Val(2) Val(1) Val(2) Val(2)	Ala(3) Ala(0) Ala(3) Ala(0)	Asp(0) Asp(0) Glu(1) Glu(1)	Gly(3) Gly(0) Gly(0) Gly(0)	U C A G

FIG 10.14   The genetic code as it operated in the regions of R17 RNA sequenced so far. A number in brackets indicates the number of times the codon in question is employed for the amino acid (or full stop) in question.

were obtained using old and therefore degraded[15] RNA. This was confirmed as it is known that exogenous ribosomes from *Bacillus stearothermophilus* only translate the A-protein of R17 RNA (Lodish and Robertson, 1969) and when such ribosomes were employed in this experiment only the A-protein initiation oligomers were obtained.

In Fig. 10.13, the data on fragments of the sequence of R17 RNA in the region of the coat protein structural gene and the surrounding sequences are put together.[16] Several points emerge from these data. One is that a high degree of secondary structure is at least possible in the sequences. It is very likely that this secondary structure occurs in the RNA as isolated from the nature of the partial T_1-digests (see caption to the figure). It is interesting to see in this proposed conformation, that one of the initiating AUG sequences is at the apex of a loop. This feature is not apparently present in the initiation region of the synthetase (nor of the A-protein which is not shown here) but has been found in an initiation region in Qβ RNA isolated in the same way (Hindley and Staples, 1969).

In addition to confirming the genetic code directly and implying a secondary structure for R17 RNA this work demonstrates two features of the genetic code as it operates *in vivo*. One is the double full stop (UAUUGA, i.e. ochre-amber) at the end of the cost protein gene. It is now supposed that this feature is fairly widespread.[17] It does of course explain how an organism carrying an su^+ gene (see p. 23) ever manages to terminate one of its own proteins. The significance of the fact that the ochre is at the end of a loop is uncertain as yet. The other point is that we can begin to assess the frequency with which the various alternative codons for the various amino acids occur *in vivo*. The results as they occur here are summarized in Fig. 10.14. As yet there are insufficient data to make any generalizations. However an interesting possibility is that one of the factors that determines the optimal codon for an amino acid is that the secondary structure of loops such as those in Fig. 10.12 is maintained and stabilized (see also p. 438). If this is the case, it might turn out that mutations involved in the third position of a codon such as CCC (all four codons of the type CCN code for the same amino acid, pro) are not necessarily 'neutral' with no biological consequences if the secondary structure of the mRNA and the effectiveness of this structure in controlling translation are altered.

The leader and termination fragments of phage RNA   Phage RNA, labelled with ^{32}P, can be partially digested with a nuclease and the oligomers containing the two ends (pppPu- at 5'- and the 3'-OH) can be screened for. The screening for the 5'-end is described on p. 289. The screening for the 3'-end is more difficult; it is based on first of all deducing the terminal few nucleotides in the RNA using the method described on p. 288. Then the oligomers are digested completely with one of the nucleases and mapped (p. 293) and the ones containing the characteristic 3'-terminal short oligomer are used for further study. Using these procedures, Cory, Spahr and Adams (1970) have sequenced the first 74 nucleotides and the last 40 nucleotides of R17 RNA and de Wachter *et al.* (1971) have sequenced the first 123 nucleotides of the very closely related phage MS2. Their results are summarized in Fig. 10.15.

Apart from the opportunity of comparing the relatively minor differences in this region of the RNA of two very closely related phage (the coat proteins are almost identical and it is supposed that they have diverged quite recently from common stock), the work of de Wachter *et al.* on MS2 is of particular interest as they have managed to get further into the molecule than the R17 workers so far; so by analogy (see Fig. 10.15) the sequence overlaps into the first of the initiation regions of the genome. Thus we can say that the molecule has a 129-nucleotide leader before the first gene is reached. The other point (to arise from the data of Cory, Spahr and Adams) is that the two ends of both + and − strands (see p. 30) can be compared. If the whole of the 3'-end of the − strand (corresponding to the whole of the 74-nucleotide fragment at the 5'-end of the + strand) is constructed for R17 RNA, it is found that only the region

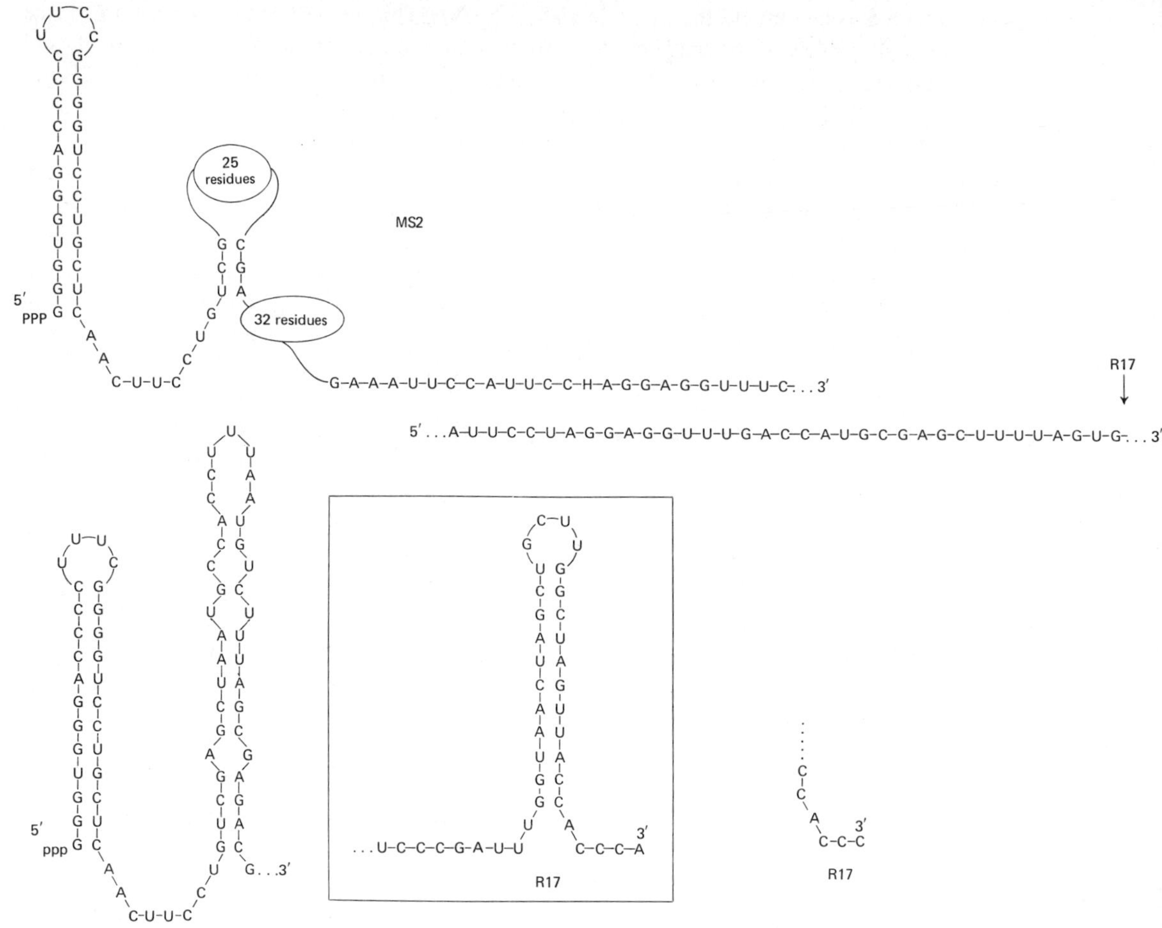

FIG 10.15   Nucleotide sequences from fragments of R17 and MS2 RNA. The 5'- and 3'-sequences were determined by the groups referred to in the accompanying text. The other fragment of R17 RNA (the A-protein initiation site) is taken from the work of Steitz (see p. 314). The secondary structure for the MS2 fragment is my design; the rest of the molecules are drawn as by the groups who determined their structures. The sequence in the box is the supposed structure for the 3'-end of the R17-strand (complementary to the 5'-end of the + strand).

illustrated demonstrates any significant homology. Moreover the two implied -strand ends cannot be assembled into the secondary structures shown for the + strand. The implication is that the secondary structure of the ends of strands of this type might be of significance in helping the synthetase to decide between the + and − strands during replication (see p. 227).

With Qβ RNA, the first 175 nucleotides have been sequenced by an even more ingenious experiment (Billeter, Dahlberg, Goodman, Hindley and Weissman, 1969). Their method is based on the fact that the RNA can be faithfully replicated *in vitro* (see p. 232). Briefly the method consisted of the following. Mixed Qβ + and − strands (which are difficult to separate completely) were incubated with Qβ polymerase in the absence of the host-specific proteins; in this way only the − strand is copied (i.e. only the + strand is synthesized, see p. 227). The rate of synthesis was synchronized and slowed down (for a reason shortly to be apparent). The synchrony was achieved by incubating the mixture in the presence of ATP and GTP, but without UPT and CTP. The synthetase puts in the first five nucleotides (pppG-G-G-G-A). When the other triphosphates are added they all start off together using the 3′-OH in this pentanucleotide as primer. The incubation was performed at 20°C using [^{32}P] triphosphates so that the nucleotides were incorporated at the rate of about 5 nucleotides/s (compared with 35/s at 37°C). The newly synthesized + strands were purified by annealing the product with excess Qβ RNA (i.e. more complete + strands). The material was analysed by sucrose gradient centrifugation, the annealed double-stranded product contained some radioactivity (due presumably to some synthesis of − strand). In the initial part of their work, Billeter *et al.* showed that with progressive incubation times, the length of the product increased as expected (2·5 S after 5 s up to 5·6 S after 30 s). They subsequently used the gradients for the routine purification of the fragments for sequencing. When very short incubation times were employed, the usual practice was to chase the label with excess unlabelled triphosphates so that the labelled portion was tacked on to the 5′-end of a larger fragment. The advantage of this procedure was that it made the removal of the excess triphosphates (by gel filtration) considerably easier. The procedure then was to take all the labelled fragments (after different

5′pppG–G–G–G–A–C–C–C–C–C–U–U–U–A–G–G–G–G–G–U–C–A–C–A–C–C–U–C–A–G–C–A–G–U–A–C–. .

. . . U–U–C–A–C–U–G–A–G–U–A–U–A–A–G–A–G–G–A–C–A–U–A–U–G–C–C–U–G–C–U–A–A–A–U–U–. .

. . . A–C–C–G–C–G–U–G–G–U–C–U–G–C–G–U–U–U–C–G–G–A–G–C–C–G–A–U–A–A–G–A–A–A–U–U–. .

. . . C–U–U–A–A–U–G–A–U–U–U–U–C–A–G–G–A–G–C–U–C–U–G–G–U–U–U–C–C–A–G–A–C–C–U–U–. .

. . . U–C–U–A–U–C–G–A–A–U–C–U–U–C–C–G–A–C–A–C–G–C–A–U–C–C–G–U–G–G . . . 3′

FIG 10.16   The first 175 nucleotides of the RNA of phage Qβ.

incubation times) and to sequence them by making the T_1-digests and separating them by the system of p. 295. The two features of this method that are so ingenious are that (i) it is possible to order the T_1-oligomers from the 5′-end (they are in the same order as the time-scale in which they appear) and (ii) it is possible to do a nearest-neighbour analysis on the oligomers which greatly facilitates their sequencing and does away with the need for much exonuclease analysis. The method is best illustrated with the example they chose in their papers (Billeter *et al.*, 1969). In this discussion, $\overset{*}{p}$ and $\overset{*}{}$ are symbols for [^{32}P] phosphate groups. One of the T_1-oligomers, was uniformly labelled with all four triphosphates. Alkaline hydrolysis gave the nucleotide composition ($U_{2·7}$, A, G, $C_{1·8}$). Pancreatic ribonuclease A gave only A-G-, UMP and CMP. Therefore the structure ends . . . A-G-. The experiment was repeated with ppp̌A (and the three other unlabelled triphosphates). The labelled nucleotides following alkaline hydrolysis were C_P^* and G_P^* and G_P^* in the proportion 1·8:1. Pancreatic digestion yielded just A-G- and C- as labelled products. Therefore the original sequence from which the T_1-digest was made must have ended up . . . C^*-A-G^*-A. Labelling with ppp̌U yielded only labelled UMP with either alkali or ribonuclease A. Therefore all the U-residues are in a row and the sequence must have been G-U-U-U-C-C-A-G-A. Incubations with either ppp̌C or ppp̌G provided the predicted results and confirmed the sequence. The final structure that was elucidated required some partial T_1- and A-digests to provide some of the links. The result is shown in Fig. 10.16.

Although there is evidence from the partial T_1 reaction for secondary structure towards the 5′-end, there is no clear evidence for secondary structure in the rest of the sequence. Moreover the initiation point for the first gene (coat protein) is not present in this structure. Although there are two AUG sequences in this sequence, neither fits this function. The N-terminal sequence of the coat protein is fMet-ala-. This would require (see Fig. 1.16, p. 22) AUGGCN (or possibly GUGGCN). Not only is there no sequence GGC in this structure but a pancreatic digest of the newly synthesized RNA corresponding to about the first 300 nucleotides shows no evidence for G-G-C-. It is therefore concluded that the leader sequence in Qβ RNA is very much longer than that in the R17/MS2 group of phage (see p. 227).

More fragmentary sequences from the ends of several phage RNA molecules are known. They are tabulated by Cory, Spahr and Adams (1970). Three of the R17 group have the 3′-terminal sequence ...G-U-U-A-C-C-A-C-C-A in common. The 3′-end of Qβ is different ...G-C-C-C-U-C-U-C-U-C-C-U-C-C-C-A. The ...C-C-A however seems to be universal to all viral RNA molecules (including those from plant viruses (Glity, Bradley and Fraenkel-Conrat, 1968)). The pppG-G-G-... sequence at the 5′-end seems to be true of all phage RNA molecules.

Cellular mRNA   It is likely to be some time before the sequences of cellular mRNA are studied in any detail as we hardly know how to isolate and assay the stuff (p. 194). Can it be done by genetics? On the whole it is outside the scope of this book to discuss fine-structure genetics but there are cases in which the methods of bacterial genetics can be used to make considerable deductions about nucleic acid sequences. As an example, it is worth just considering one extremely well worked-over genetic region which has yielded exact data regarding mRNA sequence. This is the *hisD* region in *Salmonella typhimurium*. A little background knowledge is required. The *his* operon in *Salmonella* has the structure *oGDCBHAFIE*. The meaning of *o* is discussed on p. 19. The capital letters are structural genes for enzymes in the biosynthetic pathway for histidine. The gene product for *D* is histidinol dehydrogenenase. All the work involved the use of frameshift mutations in gene *D*. A frameshift mutation is one in which a base pair in the DNA (and hence a nucleotide in the mRNA) is either inserted or deleted. The consequence is that every amino acid encoded beyond the shift will be wrong. The only 'cure' for a frameshift is another frameshift. The second frameshift does not need to be in exactly the same place as the original provided the amino acid coded in the region between the shifts are not essential for biological activity. The way in which the shifts are employed for deducing codon assignments in mRNA is illustrated by the work of Yourno (1970). A particular amino-acid sequence from wild-type histidinol dehydrogenase was shown to be as follows. The codons were deduced from work of the type being discussed.

N ... val   thr   ala   leu   arg   val   thr   pro   glu ... C

5′...GUN   ACA   GCG   PyUA   CGC   GUC   ACC   CCU   CAPu ... 3′
           1   2   3

The sequence of the corresponding region from the same enzyme isolated from a double frameshift mutant was determined. The mutant consisted of a spontaneous revertant (R5) of a frameshift mutant *hisD3018*. The organism was selected on the basis of the fact that it was D^+ (so the region involved is one where the amino acids are not all essential for enzyme activity). The sequence in *hisD3018(R5)* is as follows,

N ... val   lys   arg   tyr   ala   ser   pro   pro ... C

The explanation in this case is that the *hisD3018* mutation consists of the insertion of a C in the string of C-residues between 2 and 3 above and that R5 consists of the deletion of the C no. 1. The argument is of course presented backwards here. The paper by Yourno establishes the nature of the mutations genetically (or more correctly refers to earlier genetic work on *hisD3018*) and determines the peptide sequences and hence deduces the codons as above.

The most intriguing application of the *hisD* gene is by Rechler and Martin (1970). Here again I must refer to the paper (which is one of the clearest expositions of detailed bacterial genetics I have read) for references to much of the background genetics which was largely the achievement of M. J. Voll. The organisms all contained a mutation *his01242* which is an operator mutation with the result that the histidine operon is permanently depressed (such an organism is said to be constitutive). Constitutive mutants for this operon produce massive amounts of the enzymes in the histidine pathway and for reasons I shall not discuss can be directly recognized as they have an aberrant colony morphology on agar. Spontaneous revertants (to wild-type colonies) were selected. One of these was shown by transduction (into a group of recipients with overlapping established frameshift recipients) to be a frameshift in the distal[18] region of the *D* gene. The organism was dubbed *hisO1242 hisD2352*.

This particular strain was discovered to have very unusual properties. When compared with the parent *hisO1242*, the relative activities of the enzymes were G-enzyme 100%, D-enzyme 65%, C-enzyme 2% and B- and A-enzymes 10%. The fall-off of enzymic activity along an operator is referred to as 'polarity' (see p. 21). However the polarity from *D* to *C* is exceptional (especially in this particular operon). Moreover what is the explanation for the smaller effect of *hisD2352* on the genes (*B* and *A*) more remote from the operator than the one immediately following *D?* The C-terminal tryptic peptide of histidinol dehydrogenase from *hisO1242* is glu-gln-ala. This peptide is completely missing from the mutant enzyme. In its place is ala-ser-leu-thr. The conclusion is that the frameshift in *hisD2352* is either a base insertion (or deletion of two bases which is its equivalent). Considering overlapping sequences that will account for these data (and taking into account the fact that the codon after the one for the C-terminal amino acid must be one of the three nonsense codons), there is a unique sequence that can be written down. Before writing it down, we can take into account the fact that the N-terminal amino acid for the C-gene product (transaminase) is ser, and that in all probability the newly synthesized protein starts off fMet-ser- . . .. Taking this into account the only possible sequence for this region is the following:

5′. . . -G-A-G-C-A-A-G-C-Py-U-Pu-A-C-N-U-Pu-Pu-N-N- . . . 3′
    1 2 3 4 5 6 7 8 9 10 11 12 13 14 15 16 17 18 19

In *hisO1242* you start reading at 1 (GAG = glu) until you get to 7-8-9 (GCPy = ala) followed by 10-11-12 (full stop) then there is a choice (still undecided) about where the C-gene starts. But it is probably close because of the very strong polar effect of *hisD2352* on C-gene. Most probably it is 14-15-16 (=AUG). In *hisD2352*, you start reading at 3 (GCA = ala) and keep going until you get to 12-13-14 (ACN = thr) followed by 15-16-17 (full stop). The implications of the work are as follows. An unambiguous sequence for the end of the mRNA coding for the end of *D*-gene can be written down. In the wild type there is only one full stop (not two in a row as in the R17 coat protein—see p. 315). There is an intercistronic space but in all probability a very short one (not the long stretch characteristic of R17—see Fig. 10.13, p. 313). The two points of difference are separate issues. On p. 315, I said that the existence of a single amber full stop would lead to difficulties in a suppressor strain and amber suppressors are readily available mutations. In the case of this particular full stop in *Salmonella* it is clear from the frameshift sequence that the full stop cannot be amber (UAG) although either of the UPuA stops (ochre and umber) are possible. The other point is that R17 is not an operon; that is, it is not designed for coordinate translation, whereas the *his* operon is specifically designed for such translation. Thus the large intercistronic gap in R17 probably fulfils a role in a type of translational control that does not operate in a bacterial operon.

16 S and 23 S rRNA from *E. coli*   The primary structure of high-molecular-weight rRNA from *E. coli* is currently being extensively studied by P. Fellner and J. B. Ebel and their colleagues. Although the molecules are extremely large (1700 nucleotides in 16 S and about twice that number in 23 S), there are various methods of sequencing available for these molecules which are not available for phage RNA. For one thing they contain several methylated nucleosides which may be used as points of reference in sequencing studies (see p. 300). For another there

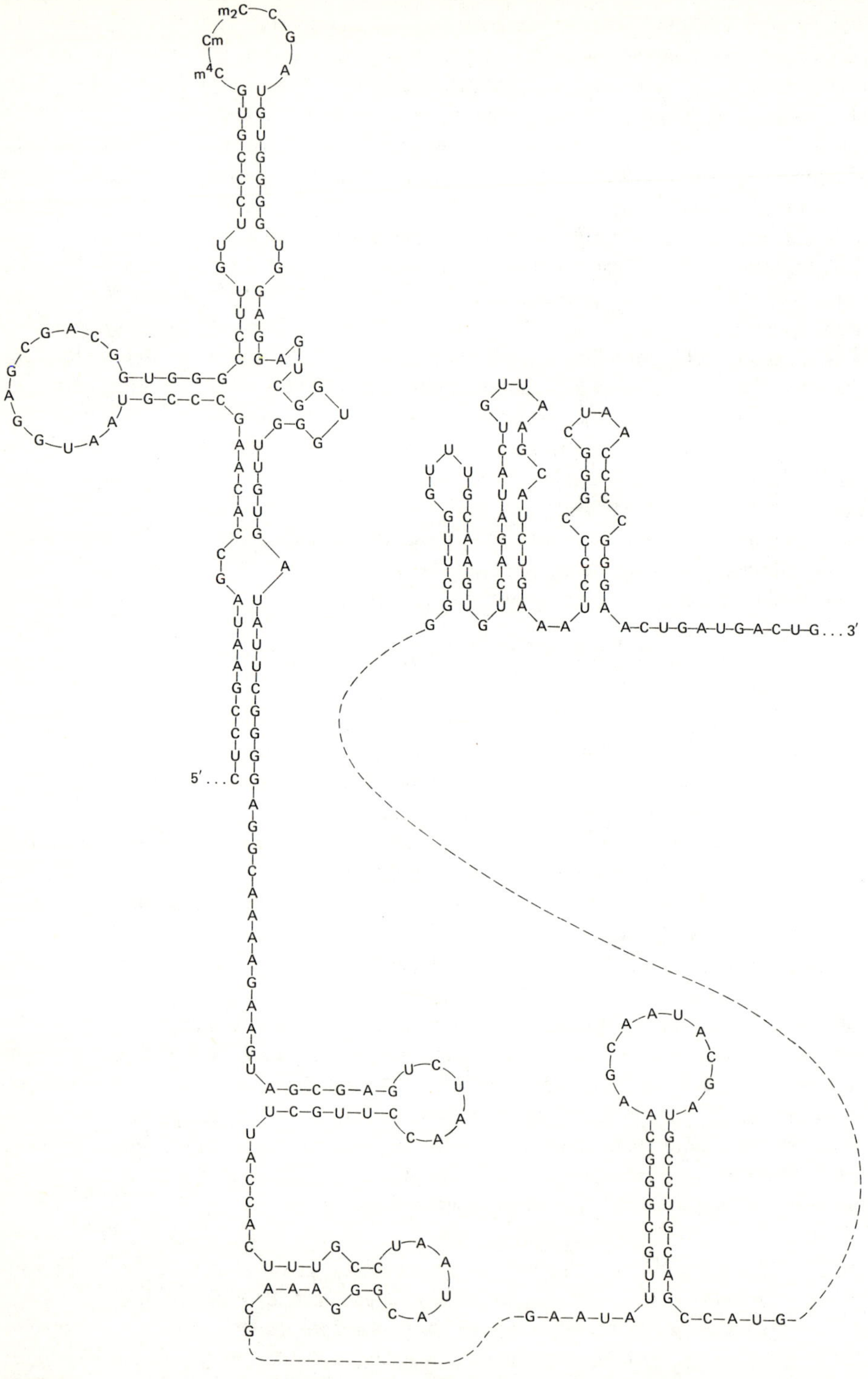

FIG 10.17   Sequences from the 16 S rRNA of *E. coli* (from Fellner, Ehresmann and Ebel, 1970).

are two separate ways in which ribonuclease digestion may be limited. As rRNA has a considerable degree of secondary structure (p. 339) partial digestion with ribonuclease T_1 will yield several large fragments. Additionally the molecules occur in a nucleoprotein matrix (the ribosome) and treatment of ribosomal subunits with ribonuclease results in the selective hydrolysis of those parts of the RNA that are exposed to the enzyme in the ribosome conformation.

The data at present available on the 16 S molecule are very well summarized by Fellner, Ehresmann and Ebel (1970). Some of their conclusions are summarized in Fig. 10.17. The large fragments shown were obtained from partial T_1-digests. Their relative positions in the 16 S molecule were deduced from the occurrence of certain characteristic sequences also identified as oligomers following the total T_1-digestion of much larger (but as yet incompletely sequenced) fragments obtained by the digestion of 30 S subunits with ribonuclease A. Most of the large oligomers were fractionated by the method described on p. 311. A modification in which a discontinuous gel (with a higher concentration of polyacrylamide at the bottom of the gel tube) is described in their paper.

There is every reason for maximizing the secondary structure within these fragments as is shown in the figure—see p. 251. An interesting feature of the sequence is that there is one of the rather rare clusters of methylated nucleosides of the 16 S RNA molecule at the top of a complex loop of largely double-stranded structure. It seems very likely that the role of these methylated C-residues is to form a sort of denatured (or base-stacked—see p. 330) hinge for the formation of this loop.

Fewer data are available about the 23 S molecule. On the basis of the occurrence of the methylated nucleosides in the oligomers from a complete T_1-digest of 23 S RNA (identified by labelling the cells with [14C]methionine—see p. 300), Fellner and Sanger (1968) concluded that most of these oligomers occurred twice in the molecule and therefore suggested that the 23 S molecule might consist of two halves of almost identical sequence. The truth, as it is now emerging, is very much more complex. Fellner and Ebel (1970) have catalogued a large number of complete T_1-oligomers from 23 S RNA and have concluded that most of them must occur only once in the structure. Therefore 23 S RNA is not a 'double molecule', with two almost identical halves, but consists of a defined linear sequence like any other RNA molecule but that those sequences which contain the clusters of methylated nucleosides (and possibly of course some other sequence not yet identified) are repeated.

Several groups are currently engaged in studies on the primary structure of the rRNA precursors. The 16 S precursor in *E. coli* was apparently of burning interest in July of 1971 as four consecutive short communications to *Nature New Biology* appeared, covering a surprisingly wide range of submission dates (Sogin, Pace, Pace and Woese; Brownlee and Cartwright; Lowry and Dahlberg; Hayes, Hayes, Fellner and Ehresmann; 1971). Several oligomers absent from the mature 16 S RNA are present in these sequences. One of them can be joined on to the 3'-end of the 'mature' sequence with reasonable certainty and oligomers corresponding to 16 S sequences but lacking base-modifications have been recognized. It is also clear that the 16 S precursor does not contain the sequences of 5 S or 6 S RNA.

Poly A   It is a well established fact that at least some mammalian cells contain significant amounts of cytoplasmic poly A. This 'extremely A-rich RNA' was discovered in rat liver by Hadjivasilou and Brawerman (1966). The problem has been examined recently by Kates (1970). There are two methods of identifying poly A (or poly A segments in RNA molecules). Either the RNA can be digested with ribonucleases A and T_1 (p. 206), in which case the only surviving polymer will be poly A, or alternatively (and this is the technique devised by Kates) poly U is bound to cellulose powder by treatment with UV light (see p. 63) and the RNA solution chromatographed. The poly A is retarded on the column as poly U:poly A hybrids are produced.

Kates own work has been largely concerned with RNA transcribed *in vitro* by vaccinia virus. Vaccinia is a large animal DNA-virus with a complex structure including membrane fragments from the host cell. If the virus particles are treated with non-ionic detergents and β-mercaptoethanol the particle is partially uncoated leaving the vaccinia core (a DNP particle), Kates and Beeson (1970). These cores contain an RNA polymerase, effective only for the synthesis of 'early virus RNA' (mRNA species required for the synthesis of the vaccinia-specific proteins produced in the period immediately following the entry of the particle into the host cell).

Using RNA synthesized *in vitro* and labelled with $[^3\text{H}]$NTP (p. 175) it was established that the poly A formed sequences attached to larger RNA molecules. The experiment consisted of isolating on the poly U column the 'extremely A-rich RNA' and estimating its sedimentation coefficient (sucrose gradients—see p. 385) before and after digestion with the nucleases. In this way it was shown that the poly A is covalently linked to 'normal' RNA sequences. The undigested material was then studied to see whether the poly A sequence was at one end of the RNA chain or was somewhere in the middle. Consider a molecule thus:

$$5'\text{-pppN}^1\text{-N}^1\text{-N}^1\text{-N}^1\text{-}\ldots\text{N}^1\text{-N}^1\text{-N}^2\text{-N}^2\text{-N}^2\text{-}\ldots\text{N}^2\text{-N}^2\text{-N}^3\text{-N}^3\text{-N}^3\text{-}\ldots\text{N}^3\text{-N}^3\ 3'$$

We are enquiring which of three types of sequence $(\ldots\text{-N}^1\text{-N}^1\text{-N}^1\ldots, \ldots\text{-N}^2\text{-N}^2\text{-N}^2 \ldots$ or $\ldots\text{-N}^3\text{-N}^3\text{-N}^3 \ldots)$ consists of $\ldots$ A-A-A $\ldots$. The method was to digest the RNA with ribonucleases as above, remove the nucleotides by dialysis (to leave the poly A sequences intact) and then to digest the product with 0·3 M KOH at 37°C (see p. 204). The digest was then analysed for Ado, AMP and ATP by paper chromatography. It was found that the digest consisted of Ado and AMP in the ratio of about 1:100. The implication is that in the above picture $N^3 = A$; i.e. that at the end of those RNA molecules containing poly A sequences, there is a run of about 100 A-residues at the 3'-end of the molecule. (If N^1 had been A, ATP would have been found in the alkaline digest. If N^2 had been A, only AMP would have been found.) In these experiments synthetic poly A containing a pppA-5'-terminus was used as control; the control must yield on alkaline hydrolysis, in addition to AMP, small amounts of ATP and Ado in equal proportions.

In a similar way, Burr and Lingrel (1971) have demonstrated a poly A sequence at the 3'-end of globin mRNA.

Notes to chapter 10

[1] I mention sucrose gradients because I know from experience that this experiment works. Whether RNA carrying the 3'-dialdehyde can be fractionated by polyacrylamide gel electrophoresis or not is something I just do not know. Although it is a stable molecule, it does have a very reactive group at one end and any aldehyde reagent is going to bind to it. Possibly polyacrylamide contains such residues.

[2] Siliconization of glassware is easily achieved by the method described on p. 156 for the preparation of tubes for gel electrophoresis.

[3] 'Oxoid' cellulose acetate for electrophoresis.

[4] It is important not to get bubbles of air between the strip and the buffer. For long strips the easiest method is to have the buffer in a little trough and to hold the paper so one end is on top of the buffer and then to gently pull the whole length of the paper across the surface of the buffer by hand. A little practice is required!

[5] LiCl is soluble in 100% MeOH.

[6] The 3'-terminal phosphate must be removed in order to convert the oligomer into an acceptable substrate for polynucleotide phosphorylase.

[7]   The 3′-terminal phosphate of this T_1-fragment had been removed with alkaline phosphatase—see note 6. The T_1-fragment itself is one of the oligomers from yeast $rRNA^{Ala}$. If you look at the structure of this molecule (Fig. 1.17, p. 24) the sequence can be seen towards the 3′-(CCA) end of the molecule.

[8]   This is not just microbiological obscurantism. *E. coli* B $tRNA^{Tyr}$ differs significantly from any of the three sequences described here (Doctor, Loebel, Sodd and Winter, 1970).

[9]   Although the paper by Goodman *et al.* (1970) is excellent in most respects, there is an unfortunate mis-print in one of the versions of this sequence; ms^2i^6A (which they represent as $\overset{*}{A}$ elsewhere) appears simply as 'A'.

[10]   There is a bit of a problem here. 7-Methyl-G-compounds are decomposed in alkali (see p. 70). In general the simplest solution to this sort of problem is to employ ^{14}C-labelled alkylating agent, to digest with acid (p. 205) and to analyse the mixture for $[^{14}C]$ 7-methylguanine chromatographically (see p. 239). In fact Sanger employed alkaline hydrolysis which converted the m^7G-residues into the 3′/2′-phosphate of the nucleoside shown in Fig. 2.13c, p. 71, together with another minor component. These substances were identified on the basis of their electrophoretic mobilities (R_u-values—see p. 300—of 0·82 and 1·2 respectively) at pH 3·5.

[11]   A cold room with plenty of dire warnings in it. This apparatus could be absolutely lethal. High-voltage electrophoresis equipment is usually protected with safety switches and so on. It is easy to be careless with this sort of medium-voltage equipment but the current that these little power packs can put out is quite high and the efficiency of the electrocution process (that would ensue if anyone touched the liquid in the two reservoirs) is very high.

[12]   This method does not work with high-molecular-weight RNA (see p. 157). Although the RNA binds to the DEAE easily it cannot be satisfactorily eluted with triethylammonium carbonate by this method.

[13]   1 ml of a solution of A_{260} of 1·00 is said to contain 1 unit.

[14]   MRE 600 is a strain of *E. coli* C containing little or no ribonuclease I and is hence very suitable for experiments on RNP fractionation.

[15]   Degraded RNA is simply a stored sample in this context. Every time a ^{32}P atom disintegrates (to form ^{32}S) the RNA (or DNA) chain is broken. With very highly labelled RNA such as this the events occur at a significant rate in a large molecule such as R17 RNA.

[16]   The secondary structure presented in this figure is not the most stable form. A method of re-evaluating these structures is discussed on p. 349.

[17]   But not universal (p. 319).

[18]   Distal: Points in a gene remote from a fixed point (the *his* operator in this case) are said to be distal; those closer to the fixed point are said to be proximal. Here of course it means that *hisD2352* is a mutation towards the 3′-end of the structural gene.

Addendum

The coat-protein gene of MS2   A sequencing marathon has recently been completed by W. Min Jou, G. Haegeman, M. Ysebaert and W. Fiers (*Nature*, 1972, **237**, 82). They have sequenced the entire gene for the coat protein in phage MS2 (see p. 315). They propose an ingenious secondary structure in which the duplex regions radiate from a central area like the petals of a flower. Their paper discusses the possible biological implications of the model. The model should be tested by the method of p. 348.

11   Secondary structure

The most detailed knowledge of the secondary structure of nucleic acids comes from an analysis of the X-ray diffraction data obtained from fibres of duplex DNA or RNA. As with any discussion of X-ray diffraction data, there is always the question, do these structures occur in solution? . . . and if it is broadly true to say that they do, what degrees of freedom are present in solution which are not available in the solid state? With nucleic acid structures there are the additional complications that several secondary structures are found in the solid state and it is not clear where the equilibrium between them lies in solution. Moreover it is important to be able to measure the transition from the duplex to non-duplex structures in solution and to assess the degree of secondary structure in molecules that cannot be obtained in a fibrous form (and in which the secondary structure only exists in particular fragments of the molecule). The contributions of the various possible base paired arrangements (G:C, G:U, A:U, A:C, etc. to the stability of the secondary structure and the effects of chemical modifications to the nucleosides on their base pairing potentialities are of importance in two regards; one is that pairs other than the Watson–Crick combinations of G with C and A with U (or T) do apparently occur in RNA; the other is that the formation of such 'mistaken' pairs during DNA replication is one cause of mutagenesis.

Before discussing some of these questions and the experimental methodology involved in some of the work, perhaps I should define the terms of reference of the chapter. Arbitrarily, I have defined secondary structure as the occurrence of random-coil, base-stacked or duplex conformations in molecules. Such questions as the supercoiling of DNA and the folding up of tRNA into a three-dimensional structure, I have called 'tertiary structure' which is discussed in chapter 15.

Results of X-ray analysis of nucleic acid fibres

The methods by which the structures of polymers are deduced from diffraction studies are outside the scope of this book. For a description of the three methods available for nucleic acids and a critical appraisal of their relative worth see Arnott (1970). All I shall do here is to present some of the more important facts that have emerged.

Description of three-dimensional structures in detail

Models of nucleic acids are difficult to visualize from flat photographs. The simplest thing to do is to define and then tabulate some of the parameters that define the structure. The conventions here are basically those used by Arnott (1970) with the trivial exceptions that I

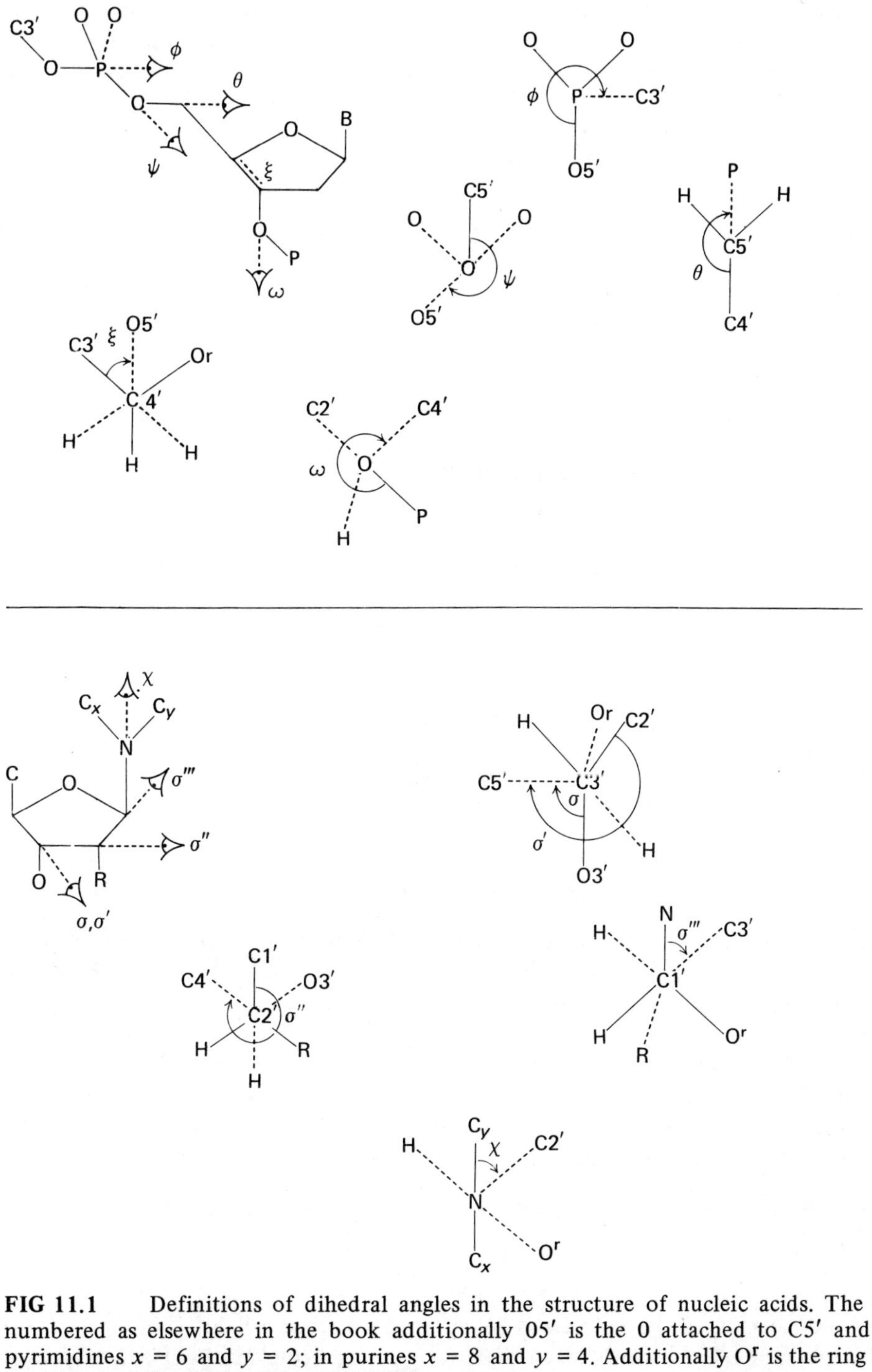

FIG 11.1   Definitions of dihedral angles in the structure of nucleic acids. The atoms are numbered as elsewhere in the book additionally $O5'$ is the O attached to $C5'$ and so on. In pyrimidines $x = 6$ and $y = 2$; in purines $x = 8$ and $y = 4$. Additionally O^r is the ring oxygen in the sugar and R = OH in RNA, H in DNA.

The eyes are looking along bonds in thoses senses required to 'see' the dihedral angle in question. For example, eye 'ϕ' is looking along a P—O bond. It 'sees' the P atom, surrounded by 0, 0, $O5'$. Behind the P (and not seen) is another O attached to $C3'$. This $O—C3'$ bond is drawn thus - - -.

prefer 'dihedral angle' to his 'conformational angle' (as 'dihedral angle' defines a unique type of angularity whereas presumably any angle in a conformation is a 'conformational angle') and I used the system for numbering the ring atoms of p. 4. (Arnott's paper employs an older system of numbering.)

The general symbol for bond length is l, that for bond angle is τ and that for dihedral angle ζ. The various sorts of ζ are given different greek letters and these symbols are defined in Fig. 11.1. In the following description, the values of σ are not quoted. In a five-membered ring, the values of $\sigma, \sigma', \sigma''$ and σ''' are related to one another. In particular $\sigma = \sigma' - 125 \cdot 4$.

The parameters described so far define the conformations of the atom in the nucleotide structure with respect to one another. We also need to define some parameters which will determine the positions of the components of the nucleotides with respect to one another and with respect to the coordinates of the helix geometry. There are two ways of doing this. One is to make the helix axis the core of a frame of reference for cylindrical polar coordinates and to define reference points in the nucleotides (such as the phosphorus atoms, the sugar carbons the glycoside nitrogen in the base and so on in terms of these coordinates). An alternative approach is to define a set of 'base parameters' which locate the bases unambiguously with respect to one another and with respect to the helix axis. To understand these parameters, let us look at the actual geometry of the nucleoside pairs A:U (or T) and C:G (or I). Some of the distances between the atoms in these pairs in planar configuration are shown in Fig. 11.2. Before

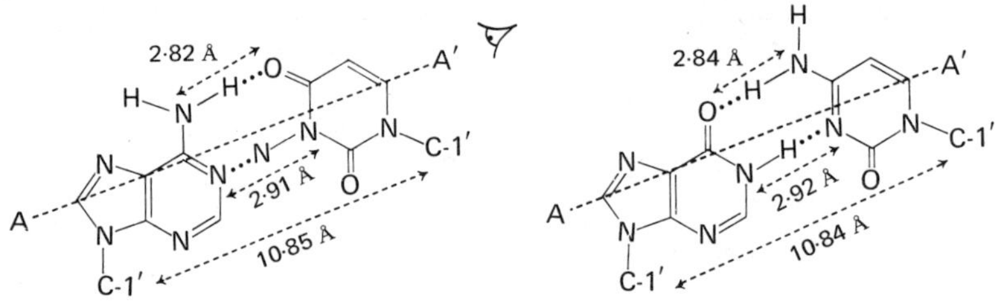

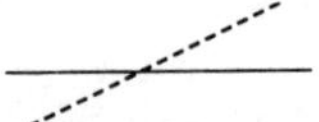

FIG 11.2   Geometry of planar Watson–Crick pairs. A:U and I:C are shown. The alterations required to arrive at A:T and G:C can be seen by comparing the picture with Fig. 1.5, p. 8. The reference line A A' and the little 'eye' are referred to in the text.

considering the parameters for defining the actual orientations of these pairs in the helix, note that the C-1' to C-1' distance in each of these two Watson–Crick pairs is identical (10·85 Å). This fact is sometimes referred to descriptively by saying that these two pairs are 'isometric'. We shall return to this point later.

The reference line A A' is employed to define the orientation of the base pair with respect to the helix axis and also to define the distortion of the pair from co-planarity. Let us take this second point first. Let us imagine the little eye looking down the reference line. If the pair is co-planar, what it will see is simply this:

If on the other hand the bases are twisted with respect to one another it will see this:

where the continuous line is the end view of the pyrimidine as shown and the dotted line is the end view of the purine. The angle between these two (0° for a co-planar pair) is referred to as the twist angle (symbol α in this book). The angle between A A' and its projection on the plane at right angles to the helix axis is referred to as tilt (β in this book). In the idealized B-DNA structure (see p. 9) it is 0°. The distance from the 'centre of gravity' of the pair (which lies along A A') to the helix axis is called D. Once again, in the 'ideal' B-DNA structure, $D = 0$ Å.

The parameters in duplex structures   The major parameters in several duplex structures are summarized in Table 11.1. The features of some of the structures are discussed in the following paragraphs.

TABLE 11.1   Values of conformational parameters of nucleotides in various duplex structures in fibres. All figures, except those for D are in degrees. The symbols are defined in Fig. 11.1 or in the text.

	B-DNA	A-DNA	A-RNA	A'-RNA	poly (rI:dC)
	(Ref. *a*)	(Ref. *a*)	(Ref. *b*)	(Ref. *c*)	(Ref. *d*)
ϕ	258·1	297·2	292·9	301·7	295·3
ψ	319·0	219·6	293·5	301·7	282·9
θ	194·0	191·4	186·4	194·7	183·5
ξ	33·4	49·5	49·5	50·0	56·3
σ	145·1	79·2	79·2	79·2	77·7
ω	170·5	199·8	201·8	193·2	194·0
χ	142·6	82·0	75·0	79·3	75·0
σ'	270·6	204·7	204·7	204·7	195·8
σ''	322·6	37·3	37·3	37·3	45·8
σ'''	153·5	94·4	94·4	94·4	90·6
D (Å)	−0·16	4·25	4·41	4·64	4·88
α	−2·1	−8	−4·5	−4·2	
β	4·9	20	13·6	9·2	9·5
Helix type	10-fold	11-fold	11-fold	12-fold	12-fold

References and notes: *a*. Arnott, Dover and Wonacott (1969).
    b. Reovirus RNA (same Ref. as *a*).
    c. poly (A:U), Arnott, Dover, Fuller and Hodgson (see Arnott, 1970).
    d. same Ref. as *c*.

Base pairing   Several of these structures are interconvertible in fibrous preparations by changing either the relative humidity of the environment or the nature (or amount) of the cations present in the preparation. The transitions are sufficiently facile that it is unlikely that the geometry of the in-plane (or nearly in-plane interactions in 'twisted' pairs) are radically altered. Thus it seems that the nature of the H-bonds between C and G (or I) and A and U (or T) is the same in all these duplices. The Watson–Crick configurations are not the only stereochemically possible structures. On p. 79, we commented on three others (anti-Watson–Crick, Hoogsteen and anti-Hoogsteen). In fact there are many other stereochemically feasible structures. Donahue and Trueblood (1960) produced 29 possible base pairs involving the four DNA-nucleosides. Recently, Donahue (1969) has attacked the original proposal of Watson and Crick (1953) on the grounds that their model does not, in detail, fit the observed diffraction pattern any better than certain alternatives in which Watson–Crick base pairs are not formed. The discussion which followed Donahue's paper is published as a consecutive series of articles (Wilkins *et al.*, 1970). There is no doubt that Watson and Crick's model for B-DNA needed some refining (see the data in Table 11.1). However all the subsequent X-ray data on DNA have confirmed that in essence the Watson and Crick model is correct but that certain refinements are needed in the detailed geometry, involving techniques of analysing X-ray data not available to Watson and Crick in 1953. It is unfortunately the case that no X-ray data have been obtained involving non-Watson–Crick pairs (with the exception of the protonated homopolymers mentioned briefly on p. 329), although there is every reason to suppose these pairs are of great importance in biological systems (see p. 434). Notice that the most significant consequence of the recent refinements to B-DNA is that the base pairs are somewhat inclined to the plane normal to the helix axis (i.e. α, $\beta \neq 0\cdot0$, see Table 11.1).

Dihedral angles in the nucleotides The ζ-values (Table 11.1) are all of much the same order of magnitude in different nucleic acids, except for the values of χ, σ, σ', σ'' and σ''' in B-DNA which are significantly different in this structure as opposed to all the rest. The qualitative descriptions of these differences are to say that in all structures in nucleic acid fibres, except B-DNA, the sugar ring is C3$'$ *endo* C2$'$ *exo* (see p. 98) whereas in B-DNA it is C3$'$ *exo* C2$'$ *endo* The value of χ is the quantitative description of whether the *N*-glycoside conformation is *syn* or *anti* (see p. 92). For a pure *anti* conformation, χ is about 90°. Thus we can see from Table 11.1 that they nearly all contain pretty good *anti* conformations except B-DNA in which the bases are swung round considerably away from this configuration.

DNA In addition to the A and B forms of DNA, there is a third form (C-DNA). It is essentially very similar to B-DNA but the duplices are packed differently in the fibre and moreover the number of base pairs per full turn of the helix is not an integer (Marvin, Spencer, Wilkins and Hamilton, 1961). The point that distinguishes DNA from RNA is the possibility of assuming the B-structure (Watson and Crick's original in fact). It can assume a structure (A) which is similar in form to RNA and indeed RNA/DNA hybrids of A-structure can form (see below). It is tempting to believe that this may be the biophysical essence behind why DNA is the genetic material, namely that is can assume discrete conformations appropriate for replication on the one hand and transcrition on the other. However this is no more than speculation as we just do not know what conformation is adopted by nucleic acids involved in metabolism (especially the intermediate in transcription—see p. 436).

RNA There are three forms of RNA (A, A$'$ and A$''$), which are broadly similar but differ in the following respects.

The A-structure occurs in two forms (α-A and β-A) which differ in the way in which the duplices are stacked together in the fibre. The differences in helix geometry between the two are very small. The A-RNA structure has been studied in the naturally occurring RNA duplex from reovirus and also in such synthetic substances poly A:U and poly I:C. The question of whether it is a 10-fold helix is still somewhat controversial. The reference given in Table 11.1 and also Arnott (1970) both come down clearly in favour of the 11-fold alternative and discuss the earlier results and evidence for the 10-fold model. A$'$-RNA is a 12-fold helix. It is apparently favoured by higher ionic strength and therefore is thought to be the stable form, analogous to B-DNA. A$''$-RNA (Arnott, Fuller, Hodgson and Prutton, 1968) has been less fully characterized but is related to A-RNA and has a non-integral helix (like C-DNA). In this case the number of base pairs per full turn is approximately 11·5.

Note that there is no evidence of RNA (nor of RNA/DNA hybrids—see below) adopting a B-configuration. Nor is there any evidence that the hydroxyl group at C2$'$ contributes to the intramolecular H-bonded stabilization of any RNA structure. On the contrary, in the fibres this extra OH-group is sticking away from the helix (see Fig. 1.6, p. 9) but it can contribute to the intermolecular association of the duplices in the fibre structure (Arnott, 1970).

RNA/DNA hybrids Fibrous RNA/DNA hybrids demonstrate one of the A-type conformations. The naturally occurring hybrids resemble A-DNA (Milman, Chamberlin, Langridge,

1967). The synthetic hybrids (for which more detailed data are available) differ from this structure in certain respects. Poly rI:dC (Table 11.1 and O'Brien and McEwan, 1970) resembles A′-RNA: poly dA:rU has two forms analogous to the A′- and A″-RNA structures (Arnott, 1970).

Homopolymers   Two duplex homopolymers have been studied in the fibrous state by X-ray diffraction. Although they are of no direct biological interest they do throw light on the protonation of two nucleoside residues (A and C) in polynucleotide chains. At low pH, it is possible to obtain a poly C duplex with half the residues protonated. The helix is six-fold (Langridge and Rich, 1963). In the poly A duplex (again at low pH) all the residues are protonated and an eight-fold helix is produced (Rich, Davies, Crick and Watson, 1961). The structures of the base pairs predicted for these structures are shown in Fig. 11.3.

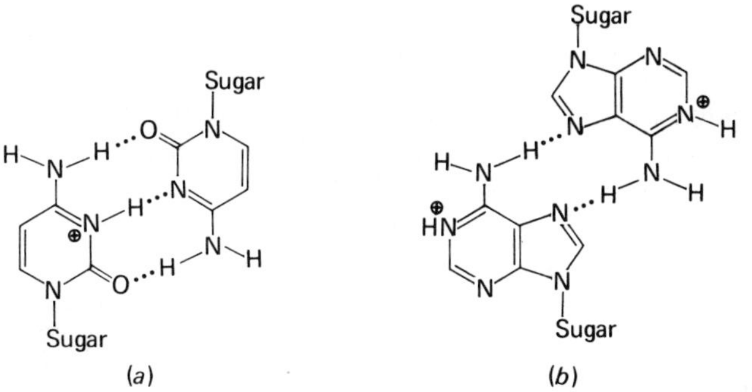

(a)   (b)

FIG 11.3   Structures of base pairs between (*a*) C and CH^+ and (*b*) AH^+ and AH^+ both found in homopolymer duplices at low pH (see text).

Conformation of nucleic acids in solution

In the discussion of the properties of nucleosides and their derivatives (chapter 2) the properties of the compounds were discussed specifically in the context of the physical methods available for their study. Having now reached the properties of oligomers and their derivatives, it is more useful to group the topics in a structural rather than their analytical context. However it is worth reviewing the methods that are available for the elucidation of the secondary structure of nucleic acid in solution as certain of these techniques are also suitable for application to the study of the tertiary structures of nucleic acids, which are discussed in chapter 15.

The conformations that are encountered in solutions of polynucleotides are base stacking (see p. 330) and the association of chains to form double- or triple-stranded structures. The most familiar technique for studying secondary structure is UV spectroscopy. The techniques rely on the fact that the absorption maximum for nucleotides in the region of 250–270 nm (see p. 84 for the transitions that produce this band in pyrimidines and p. 86 for the transition that produces it in purines) is affected by the environment of the chromophore. In particular the stacking of bases (either in a 'free stack' or in a duplex or triplex) produces a marked hypochromic effect (p. 338). When such an ordered structure is denatured, the absorbance increases to produce the familiar hyperchromic effect on 'melting' (see p. 268).

A more sensitive and more easily analysed effect is observed when the ORD or CD spectra (p. 88) are considered. In this case the effect of ordered structure is not solely due to the change in the electronic transition of the chromophore but additionally (and quantitatively this the more important change) the asymmetry of the molecule is increased in the oriented form.

Although the theoretical analysis of such spectra is semi-empirical, the optical properties of oligo- and polynucleotides provide the most sensitive test for order, especially in short-range ordering (see p. 333). IR spectroscopy of nucleic acids has been relatively little-used, it suffers from the same technical drawbacks as that of the IR spectra of nucleosides (p. 93). The related technique of Raman spectroscopy is probably one of the most important techniques for the study of nucleic acid conformation in the future (see p. 342). Polarized IR spectroscopy (ORD in the infra-red) is a technique which, provided certain technical limitations can be overcome (see p. 342), promises to be of use in determining one of the most intractable parameters in nucleic acid structure, namely the nature and stability of the hydration shell (p. 345). Another important technique that has been relatively little exploited so far is microcalorimetry (see p. 347). Hydrodynamic methods of analysis are delayed until chapter 13.

Base stacking

The optical properties of simple oligomers The evidence for base stacking in solution originated with the discovery by Holcomb and Tinoco (1965) that poly A assumed an ordered structure in solution. This was based on the observation that the ORD of poly A in solution at pH 5·4 was characteristic of a single-strand helix. At low temperature the curve (with a much larger Cotton effect) was that characteristic of a double-strand helix (presumably similar to that described on p. 341). They were also able to demonstrate the existence of an ordered structure in A-A. At about the same time, van Holde, Brahms and Michelson (1965) employed CD to investigate the same effects and analysed the 'melting' of A-A more systematically.

The ORD of oligomers has been discussed by Yang and Samejima (1969). Their treatment is summarized below. At a suitable wavelength the value of $[M]$ (see p. 89 for definition) is measured. The value of $[M]$ for an oligomer (such as N-N′) is expressed as a sum of the values of $[M]$ for the nucleoside components plus base interaction contributions (I or J). Thus we say that for N-N′:

$$[M_{\text{N-N'}}] = [M_{\text{N}}] + [M_{\text{N'}}] + I_{\text{NN'}}$$

and for N-N′-N″

$$[M_{\text{N-N'-N''}}] = [M_{\text{N}}] + [M_{\text{N'}}] + [M_{\text{N''}}] + [J_{\text{NN'}}] + [J_{\text{N'N''}}]$$

If we assume that only nearest-neighbour base interaction contributions are significant (i.e. $J_{\text{NN''}} = 0$) and that $J_{\text{NN'}} = I_{\text{NN'}}$ and $J_{\text{N'N''}} = I_{\text{N'N''}}$ (obtained from the equation—not shown—the value of $[M]$ for N′-N″), then we can re-write $[M_{\text{N-N'-N''}}]$ in terms of the values of $[M]$ for the dinucleoside phosphates thus:

$$[M_{\text{N-N'-N''}}] = [M_{\text{N-N'}}] + [M_{\text{N'-N''}}] - [M_{\text{N'}}]$$

This equation can be re-written in terms of mean residue rotations ($[m]$) thus:

$$[m_{\text{N-N'-N''}}] = \tfrac{1}{3}\{2[m_{\text{N-N'}}] + 2[m_{\text{N'-N''}}] - [m_{\text{N'}}]\}$$

An immediate consequence of this relationship is that it provides a method of sequencing trinucleotides. The relationship does in fact hold good provided that only trinucleoside diphosphates are employed: trinucleotides (N-N′-N″-) do not obey this relationship, apparently because of some sort of end-effect of the terminal phosphate on the conformation of the preceding nucleoside (Inoue, Agoyagi and Nakanishi, 1967).

If the treatment is extended to longer 'n-mers' the result is the following equation, provided that 'end effects' are neglected.

$$[m_{\text{polymer}}] = \sum_{N=1}^{4} \sum_{N'=1}^{4} 2\chi_{NN'} [m_{NN'}] - \sum_{N=1}^{4} \chi_N [m_N]$$

In this equation, it is assumed that there are four types of nucleoside and that the phosphodiester groups are omitted from such nearest-neighbour sequences as N-N$'$. The symbols χ represent mole fractions. If the co-polymer is random, $\chi_{NN'} = \chi_N \times \chi_{N'}$, so that the equation above is simplified to:

$$[m_{\text{polymer}}] = \sum_{N=1}^{4} \chi_N \left\{ 2 \sum_{N'=1}^{4} \chi_N [m_{NN'}] - [m_N] \right\}$$

In a homopolymer, the relationship is further simplified to:

$$[m_{\text{polymer}}] = 2[m_{NN}] - [m_N]$$

An implication of this equation is that the value of $[m_N]$ is about twice that of $[m_{NN}]$. This prediction has been born out in the case where N is A by Michelson, Ulbricht, Emerson and Swan (1966). The importance of this result is that it confirms that the most important assumption of the theory, namely that only nearest-neighbour contributions are significant is completely justified. In other words, ORD can only 'see' the effect of the base adjacent to another base in a stack.

Base stacking in mononucleotides   Molecules of GMP stack in aqueous solution. No sophisticated spectrophometric equipment is required to convince yourself of this fact. If the pH of a reasonably strong solution 5$'$-GMP is reduced to between 4 and 5, the solution becomes very viscous (rather like DNA in fact). The GMP eventually comes out of solution as a sort of gel. The reason is that the nucleotides form long 'polymeric' stacks. The effect has been studied quantitatively (using ORD) by Sarker and Yang (1963). Some such effect could be important in considering models for prebiotic evolution. After purines had been synthesized (see p. 36) and converted into sugar N-glycoside phosphates in some way, the formation of such stacks could form a matrix for nucleotide polymerization (to yield a primitive nucleic acid). Of less academic interest but of practical importance, when intramolecular base stacking in short oligomers is being studied, it is important to either work at low concentration or alternatively to make the measurements at a number of concentrations and to extrapolate to infinite dilution, to eliminate effects due to intermolecular stacks.

Base stacking in dinucleoside monophosphates   We have already seen (above) that ORD is useful for demonstrating the existence of an interaction parameter of some type between nucleotides in a stack. For an analysis of the thermodynamic parameters, CD measurements are more valuable. In the theoretical analysis of CD results, the 'rotational strength' of a CD band is useful. The symbol for rotational strength is R and it is an integral of the total envelope for a CD peak expressed in molecular units:

$$R = \frac{3hc}{8\pi 3N} \int_0^{\infty} [\Psi(\lambda)/l\lambda] \, \mathrm{d}\lambda,$$

where h is Planck's constant, N is Avogadro's number and the other symbols have the same significance as on pp. 88 and 89. Note that we write Ψ as a function of λ for the integration. For a method of applying this equation to a calculation of R for a CD band for a nucleotide, see Brahms (1963). In the work of Brahms and his colleagues on the dinucleoside phosphates, CD spectra were measured at a variety of temperatures. Examples of the spectra obtained are given in Fig. 11.4.

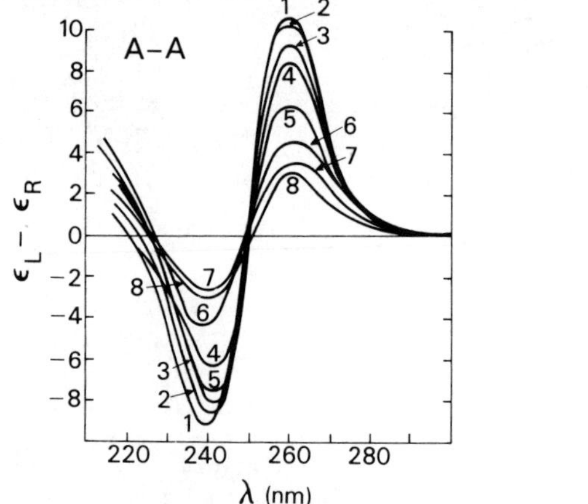

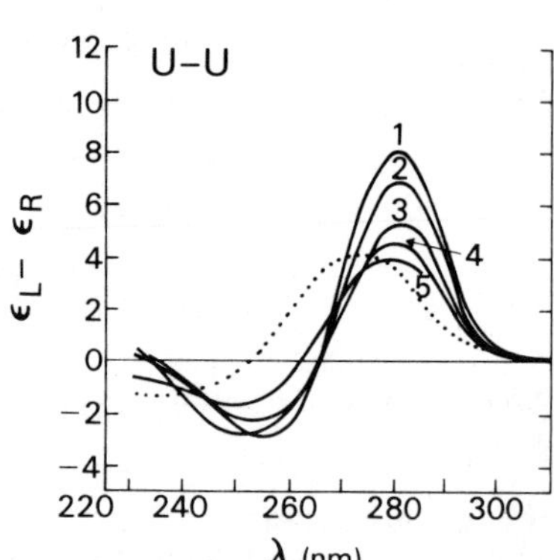

FIG 11.4   Circular dichroic melting of A-A and U-U. The A-A spectra were measured in 0·1 M NaCl, 0·1 M tris HCl pH 7·4 at (1) −1·5°C, (2) 3·5°C, (3) 5·5°C, (4) 15·5°C, (5) 25°C, (6) 35°C, (7) 6 46·5°C, (8) 58°C (from van Holde, Brahms and Michelson, 1965). The U-U spectra were measured in 4·7 M KF, 10 mM tris pH 7 at (1) −18°C, (2) 2°C, (3) 26°C, (4) 41°C, and (5) 67°C (from Brahms, Maurizot and Michelson, 1967).

The analysis consists of considering an equilibrium between stacked and unstacked forms. If the concentrations of these forms are c_s and c_u the value of the equilibrium constant at temperature $T°$C can be calculated from:

$$K_T = \frac{c_u}{c_s} \frac{R_s - R_T}{R_T},$$

where R_T is the value of R at $T°$C and R_s is the value of R for the purely stacked form. Clearly, if it is possible to measure R_s, the value of K at a variety of temperatures can be computed. It is not quite as simple to determine the optical parameters for the ordered form in a stacked structure; the process of denaturation is not co-operative as it is in a duplex structure and a continuum of semi-ordered structures is produced during the heating of a sample. Obviously one method is to extrapolate to very low temperatures. (This is why the solvent for the U-U data in Fig. 11.4 is one that is still liquid at −20°C). However the method that is usually employed is to plot R as a function of T. The result is a sigmoidal curve (Fig. 11.5). For a simple equilibrium of the type involved here, the value of R is at the point of inflexion as $R_s/2$.

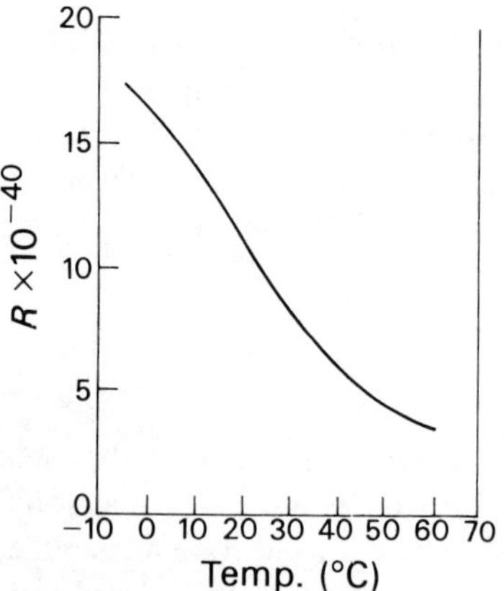

FIG 11.5   Plot of R as a function of temperature for the unstacking of A-A. The values are computed from the positive dichroic band for A-A shown in Fig. 11.4. (From van Holde, Brahms and Michelson, 1965.)

Once R_s is known and K is hence determined for a number of temperatures, it is possible to determine the thermodynamic parameters graphically.[1]

Using this procedure, Brahms, Maurizot and Michelson (1967) found that the thermodynamic parameters for destaking several dinucleoside phosphates were about the same (ΔH° approx. 6 kcal/mole; ΔS° approx. 22 kcal/mole/deg and ΔG° (at 0°C) between + 0·2 and + 0·7 kcal/mole). In all these values, a 'mole' is a mole of nucleoside residue. The value of T_m varied considerably (A-A 25°C, A-C 25°C, C-A 15°C, G-A 9°C, C-U 6°C).

Although the thermodynamic parameters computed from the CD curves are similar, the curves themselves differed significantly; for example, compare the curves for GA and G-C (Fig. 11.6). The differences are not simply due to the different contributions from the nucleosides; these dichroic maxima in the low temperature (stacked) structures are very much greater than the maxima for the nucleosides. They reflect different conformations in the stacked structure. For a tentative analysis of the significance of these spectra and their fit to theories of the dichroic spectra see the paper by Brahms, Maurizot and Michelson (1967).

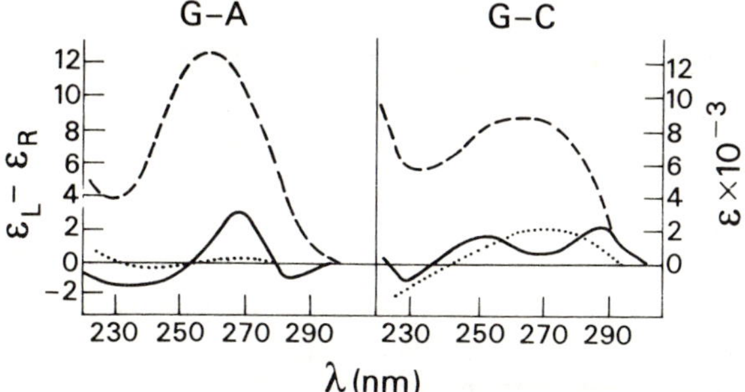

FIG 11.6   CD curves for G-A and G-C. The CD spectrum for the dinucleoside phosphate (————), the CD spectrum for the equivalent mixture (G-A, etc., ...) and the UV spectra (- - - - - - -) are shown. Data of Brahms, Maurizot and Michelson (1967).

An additional method for probing the secondary structure of oligomers is to employ PMR spectroscopy. As an example, the resonance due to protons H-8 and H-2 in A-A has been studied by Davis and Tinoco (1968). The bands are split because of non-equivalence of these protons in the two adenosine moieties in the nucleotide (see p. 97). Using this criterion, they were able to demonstrate that some secondary structure is still present in A-A at 90°C (in 25% LiCl). The measurements were made in D_2O solution in order to eliminate the signals due to the (exchangeable) hydroxyl protons in the sugar residues (p. 97).

Base stacking in longer oligomers   The main difficulty about studying the base stacking of polynucleotides is that duplex structures are so easily formed (see p. 343 and the following section). However it is established that poly A at pH greater than 7 is a single-strand base stacked helix. The ORD measurements were made by Holcomb and Tinoco (1965) The fact that a duplex was not involved at these pH-values was confirmed by van Holde, Brahms and Michelson (1965) who showed that similar optical properties were obtained with poly N^6-hydroxyethyl-A, in which duplex formation is impossible on steric grounds.[2] The stacking in poly A is apparently short range (p. 330).

Multiple-helix structures

Homopolymers   The multiple-helix structures formed by the polyribotides are summarized in Table 11.2.

TABLE 11.2   Summary of multiple helices formed by polynucleotides. A '+' indicates that a multiple helix can be formed between the two polymers in question: a '−' indicates it cannot. Thus a multiplex between molecules of poly A and poly U or poly U and itself do form: ones between poly U and poly C do not. H indicates that the multiplex only forms at acid pH. The data summarized here are taken from the review by Michelson, Massoulié and Guschlbauer (1967).

	A	C	U	G	I	X
X	+	−	+	−	+	+
I	+	+	−	−	+	
G	−	+	−	+		
U	+	−	+			
C	−	H				
A	H					

The significance of these results is probably not as great as seems at first sight. You might think that the base pairing of A with U and G with C are naturals and the 'reason' why these are the four common nucleosides is that they achieve less ambiguous pairings than the other two purines shown in the table. However all the polymers form complexes with themselves and moreover the stiochiometry of the various complexes is not universally 1:1 (duplex). The stoichiometry of a multiplex in solution is determined by mixing the components of a heterologous multiplex together and observing the proportion(s) at which minima in the absorbance is produced. With homopolymer multiplices, less direct deductions from optical properties must be used. Comments on some of the multiplices that have been characterized follow.

The secondary structure of poly G is unknown. It is extremely stable ($T_m > 100°$C) even at low pH at which the bases are protonated (Michelson and Monny, 1967). Poly I also forms a very stable secondary structure believed by Rich (1958) to be a triplex. Poly U forms a very low-melting structure which is probably a triplex (Millar and Mackenzie, 1970; see also p. 335). Poly A and poly C form duplices at low pH (see p. 329).

Two types of intermolecular complex occur between poly A and poly U. In the duplex structure (poly A:poly U) the base pairs have the same structure as the Watson−Crick A:U pair (see Fig. 1.5 p. 8). In the triplex (poly A:2 poly U) the bases are held together as shown in Fig. 11.7. The kinetics of associations in the poly U/poly A system have been analysed by Blake and Fresco (1966).

A similar situation occurs between poly G and poly C but the H-bonding is not so clear and the thermodynamics of the system are clearly different. With really long poly C and Poly G, the result is that only the duplex poly G:poly C is formed. However if either polymer is fragmented

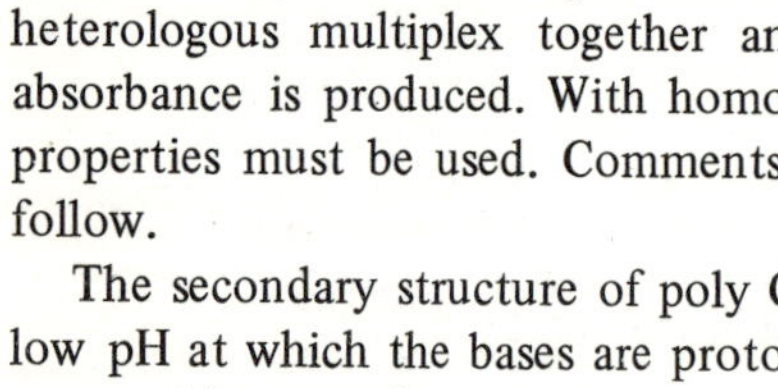

FIG 11.7   Hydrogen bonding between one A- and two U-residues.

(i.e. in a mixture of either poly C with oligo G or oligo C with poly G, a triplex is formed (poly (or oligo) C : 2 oligo (or poly) G). In the structues formed with one chain fragmented, there is sufficient flexibility in the duplex for the second molecule of poly G to bind. Presumably some distortion of the base-pair conformation in the duplex is required for this reaction (Pochon and Michelson, 1965).

So far we have commented on some of the multiplex arrangements shown in Table 11.2. What are the properties of other polynucleotides? The effects of substituents in U-residues are fairly well-studied. The secondary structure of poly Ψ is more stable than that of poly U. It forms similar multiplices with poly A (poly Ψ : poly A and 2 poly Ψ : poly A) to poly U (Pochon, Michelson, Grunberg-Manago, Cohn and Dondon, 1964; Massoulié, Michelson and Pochon, 1966). Poly br^5U does not form a multiplex in solution on its own. It (like poly cl^5U but not poly io^5U) forms a duplex with poly U which is more stable than the multiplex which poly U forms with itself (Millar and Mackenzie, 1970).

As an example of a detailed study on a modified purine polyribotide, Pochon and Michelson (1969) examined the properties of poly m_2^2G. This polymer does not form a multi-chain complex with poly C but it does form a duplex with itself. This duplex is acid-stable but alkali labile, thus confirming (see p. 94) the structure of the ionized forms of guanosine derivatives (see Fig. 11.8).

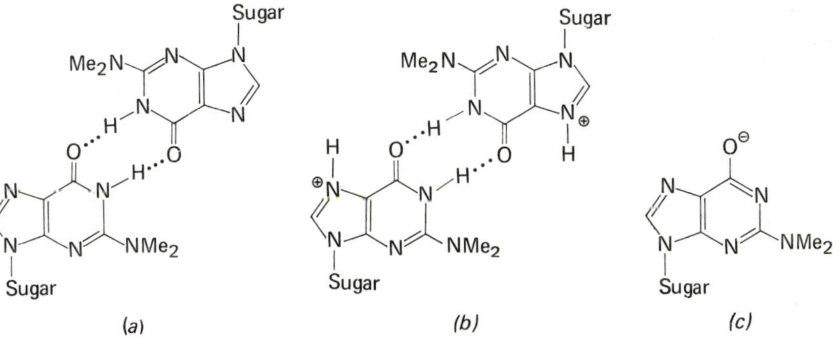

FIG 11.8   Structures of N^2-dimethylguanylate residues. (a) Base pairing at neutral pH, (b) base pairing at acid pH and (c) a residue at alkaline pH (from Pochon and Michelson, 1969).

Duplex formation in synthetic heteropolymers   Heteroligomers have proved of great value in establishing models for the base pairing that occurs in short loops of RNA structure. These experiments are discussed on p. 343. Here I draw attention to three experiments on heteropolymers which illustrate points of reasonably general interest.

Thiouridine residues occur in bacterial tRNA (p. 242). K. H. Scheit has shown that s^4U has similar H-bonding properties to U-residues. The hypochromism of complexes between poly A and poly U containing s^4U-residues was studied. The presence of s^4U-residue had no effect on the thermal stability of the helix. Moreover a shift in the pK of the s^4U-residues to 9·45 (compare with the value below 8, p. 51, for the free nucleotide) confirmed that the hydrogen at N-3 was involved in the H-bonding and so presumably the s^4U:A pair has the same H-bonding as an A:U pair (Scheit and Gaertner, 1969).

A curiosity that recently has come to light is the duplex formed by poly d(I-C) and poly d(C-G). In linear representation the complementarity of this structure can be drawn thus:

5' . . . - I-C- I-C- I-C- I-C- I-C- I- . . . 3'

3' . . . -C-G-C-G-C-G-C-G-C-G-C- . . . 5'

The CD data for this complex are consistent with it being an eight-fold left-handed helix (Mitsui *et al.*, 1970). Whether such helices are ever produced in naturally occurring polynucleotides is another matter, but is perhaps worth bearing in mind when complex tertiary structures of nucleic acids are being considered (p. 428).

The other heteropolymer is poly d(A-T). This molecule is of more than academic interest. For one thing it is a naturally occurring substance (see p. 152); for another its synthesis by DNA polymerase is of considerable interest. The enzyme itself will synthesize the polymer without template after a long lag period (p. 223). However, given a short primer of this sort (or other self-complementary oligomers—see p. 233) the enzyme synthesizes long polymer strands. A possible mechanism for this replication was proposed by Kornberg, Bertsch, Jackson and Khorana (1964). The suggestion is that the chain is extended as the result of the formation of an intramolecular duplex ('hairpin loop') which slips to expose a new unpaired region to act as template for the enzyme and bring the duplex back into register. Further slippage is necessary before further synthesis can occur; see Fig. 11.9.

Do branched structures of the type shown at the bottom of Fig. 11.9 really occur; if so what is the kinetics of their formation? The answer is that they undoubtedly do and that there are optical methods of following their formation. The melting and re-annealing of AT-oligomers (7,

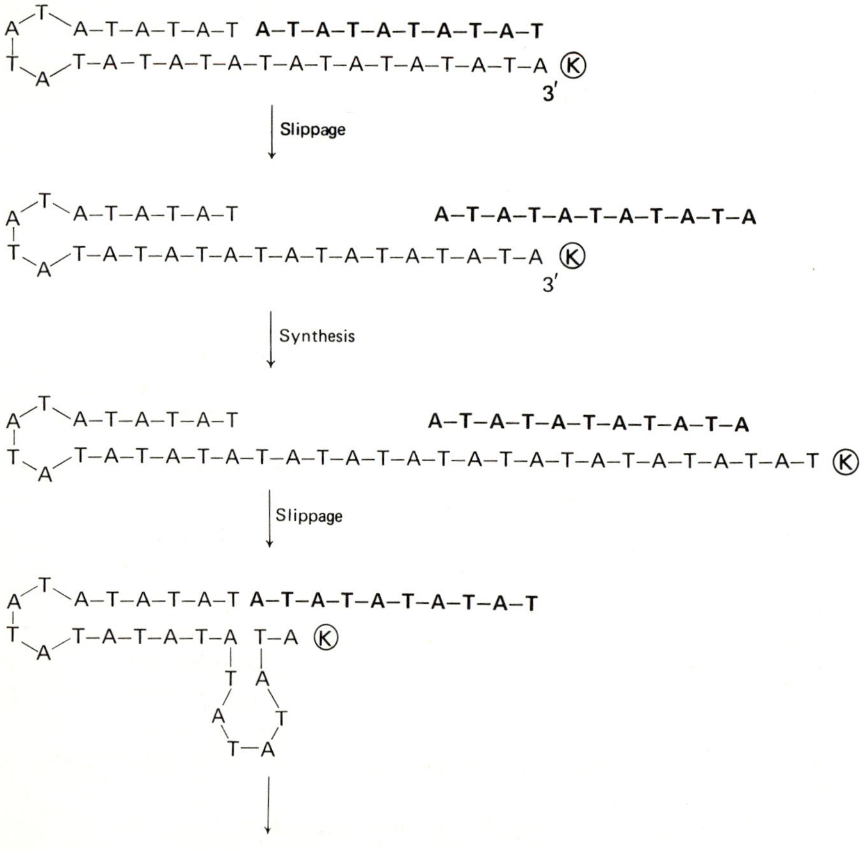

FIG 11.9   Slippage during the synthesis of poly d(AT) from an oligomeric template (from Kornberg *et al.*, 1964). The initial template is shown in heavy characters; K is Kornberg's polymerase (see p. 222).

9 and 11 nucleotides long) has been carefully analysed by I. E. Scheffler (reported in Baldwin, 1968). The appearance of such a melting curve is shown in Fig. 11.10. Note the slight inflexion at temperature below the melting temperature. The cooling of this zone shows hysteresis. The

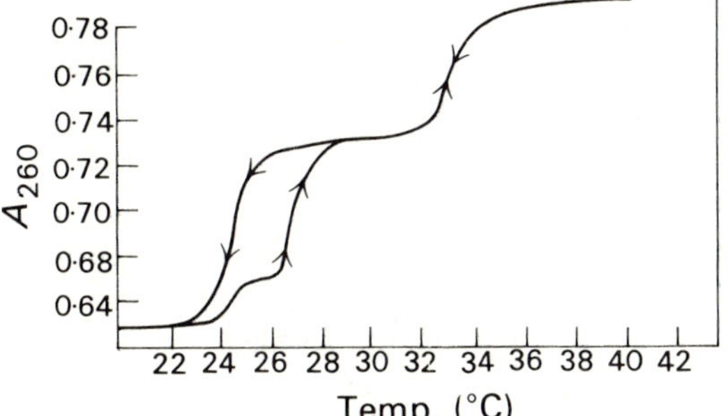

FIG 11.10   Melting and cooling profile for an AT-oligomer (based on pictures obtained by Scheffler—see Baldwin (1968)). The arrows pointing 'uphill' represent the melting profile; those pointing 'downhill' represent the cooling profile.

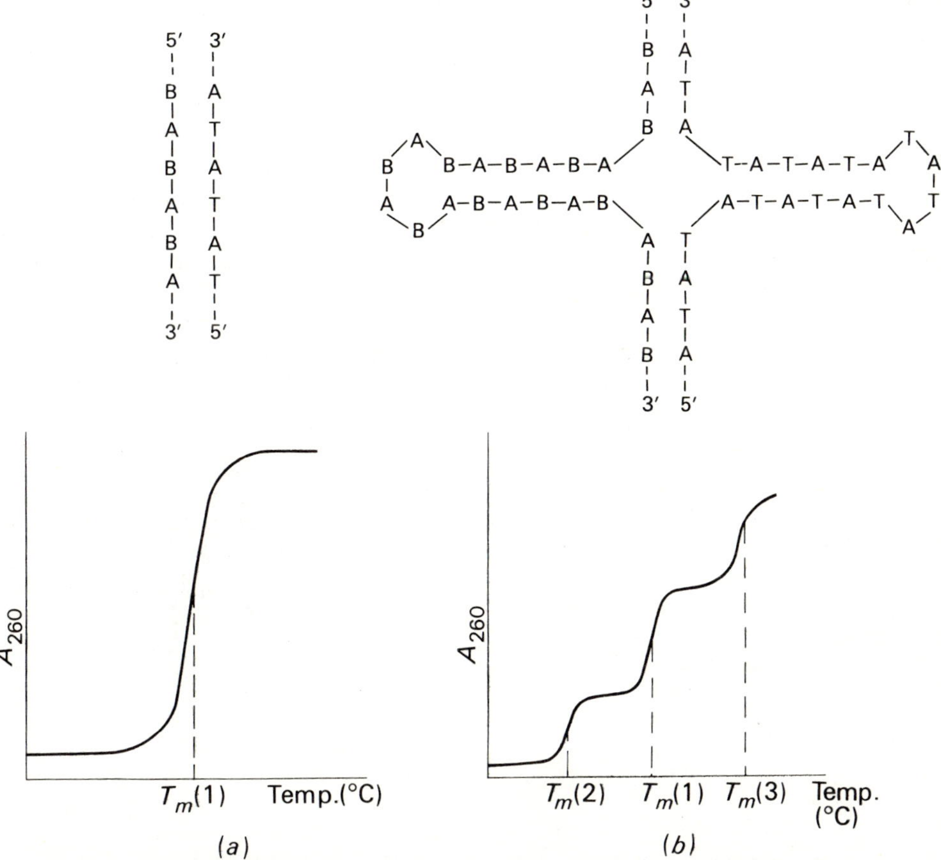

FIG 11.11   Structures for a poly dAT : poly dA br^{5}U co-polymer. Structure (*a*) shows a simple melting profile; the extent of looping in (*b*) can be analysed by comparing the contributions of the different helix types. Melting temperatures are for poly dAT : poly dA br^{5}U ($T_m(1)$), poly dAT : poly dAT ($T_m(2)$) and poly dA br^{5}U : poly dA br^{5}U ($T_m(3)$). B = br^{5}U.

simplest explanation of the effect is that during re-annealing, looped structures (which necessarily have bases which were paired in the 'native' conformation in an unpaired configuration) are produced. The results have been analysed from this point of view (Baldwin, 1968).

The difficulty in analysing data of this type is that it is hard to evaluate the relative contribution of intramolecular structures to 'concatamers'—see p. 343.

An extremely ingenious method of observing the formation of loops is due to Inman and Baldwin (1962). It is based on the fact that duplices of poly dAT : poly dAT, poly dA $br^5 U$: poly dA $br^5 U$ and poly dAT : poly dA $br^5 U$ have different melting temperatures. In this way the formation of loops can be detected by following annealing process (see Fig. 11.11).

In addition to demonstrating the existence and production of loops in a self-complementary oligomer, these results can be applied to an analysis of the biochemical problem of the kinetics of poly dAT synthesis by Kornberg's enzyme. For a discussion of this application see the review by Baldwin (1968).

Spectroscopic and optical properties of nucleic acids

Hypochromism of DNA   We have already pointed out that the hyperchromic effect upon the co-operative denaturation ('melting') of DNA may be correlated in an empirical manner with the GC-content of the sample. A more rigorous mathematical treatment of the hyperchromic effect in DNA-melting has been made by G. Felsenfield. His methods make the GC-determination possible with fewer data than those necessary for a full melting profile. However, the great advantage of his method is that it enables us to determine the degree of secondary structure in a partly denatured sample (such as the product of imperfect annealing—see p. 240). For references to the theory and practice of the following methods see Felsenfield and Hirschmann (1965), Hirschmann and Felsenfield (1966) and Felsenfield (1968).

For a partial or complete denaturation (helix-coli transition) the increase in absorbance at i nm (ΔA_i) can be factorized as follows:

$$\Delta A_i = C\{\phi a_i + (1 - \phi)g_i + \phi(\phi - 1)[(1 - 2k_i)(a_i + g_i)]\}$$

where C is the nucleotide-concentration, ϕ is the mole-fraction of AT-base pairs (both in the denatured region only) and a_i, g_i and k_i are three spectrophotometric parameters whose physical significance we can neglect for this discussion. The equation is only strictly valid for a random sequence. For real DNA, the deviation from randomness is accommodated by putting in a fudge factor, δ, so that the function $\phi(\phi - 1)$ is replaced by $\phi(\phi - 1 + \delta)$. Now the equation can be re-written with new parameters defined as follows:

$$\alpha_i = (1 - 2k_i)(a_i - g_i)$$

$$\beta_i = 2[k_i(a_i + g_i) - g_i]$$

and $$X = C(\phi^2 + \delta\phi)$$

whence,

$$\Delta A_i = \alpha_i X + \beta_i(C\phi) + g_i C$$

The methods of Felsenfield consist of making measurements of absorbance in four wavelengths, 250, 260, 270 and 280 nm. In what follows, the absorbance at 250 nm is called A_5 and so on to avoid cluttering the equations up with large subscripts. The treatments are based on a semi-empirical solution of the equation, assuming that the function $\alpha_i X$ can be neglected. In other words we are solving equations of the form:

$$\Delta A_i = \beta_i(\phi C) + g_i C.$$

for the unknowns C and $C\phi$ using observed values of ΔA_i and empirical values of β_i and g_i.

To determine the base composition (AT-content) of an unknown DNA of native structure, dissolve the DNA in a dilute buffer and accurately measure A_5, A_6, A_7 and A_8 at a low (native) and high (denatured) temperature. Precautions must be taken to correct for solvent-expansion (see p. 268). The simplest treatment is to solve the semi-empirical equation:

$$C = L_1\Sigma\Delta A_i\beta_i + L_2\Sigma\Delta A_i g_i$$

and

$$C\phi = L_3\Sigma\Delta A_i\beta_i + L_1\Sigma\Delta A_i g_i$$

where

$$\Sigma\Delta A_i\beta_i = \Delta A_5\beta_5 + \Delta A_6\beta_6 + \Delta A_7\beta_7 + \Delta A_8\beta_8$$

and so on. Values of L, β and g are given in Table 11.3.

To determine the degree of duplex structure in a partially native preparation, the same readings are taken and C and ϕ are calculated as before but additionally the values of A_i (at the lower temperature) are put into the equations:

$$C' = L_1'\Sigma A_i\beta_i + L_2'\Sigma A_i g_i'$$

and

$$C'\phi' = L_3'\Sigma A_i\beta_i' + L_1'\Sigma A_i g_i'$$

The values of L', β' and g' differ slightly from the corresponding values of L, β and g (see Table 11.3). Clearly C/C' is a measure of the proportion of duplex structure in the partly native sample. Moreover a comparison of ϕ and ϕ' indicates whether A:T or G:C pairs predominate in the duplex regions in this preparation.

TABLE 11.3   Values of parameters for the solution of equations for C, C', $C\phi$ and $C'\phi'$ (see text for explanation). Values taken from Felsenfield (1968).

Wavelength (nm)	i	β	β'	g	g'
250	5	43	−2669	2264	9703
260	6	1388	−282	1981	9397
270	7	−629	−1066	2892	8387
280	8	−2422	−2887	2920	6615

L_1 $3{\cdot}4359 \times 10^{-8}$ L_1' $3{\cdot}2466 \times 10^{-8}$
L_2 $4{\cdot}6556 \times 10^{-8}$ L_2' $9{\cdot}568 \times 10^{-9}$
L_3 $1{\cdot}47455 \times 10^{-7}$ L_3' $1{\cdot}70129 \times 10^{-7}$

. . . and RNA?   It has been recognized for a long time that RNA shows a melting profile similar to, but not identical with, that of DNA. The increase in absorbance typically occurs over a wider temperature-range than is true of DNA. However it is reasonable to suppose that at least the major contribution to the melting is the denaturation of loops of Watson−Crick structure. Apart from the properties of model oligomers (p. 343) and the self-complementarity of many RNA sequences (p. 10), native RNA has many chemical reactions suggesting that a proportion of the bases are in duplex structures (p. 419). Additionally we have shown that RNA cross-linked with mustard gas has limited hypochromic melting and indeed that in highly cross-linked 16 S rRNA from *E. coli* melting is apparently completely suppressed (see p. 242). Examples of the melting profiles of RNA are shown in Fig. 11.12.

How long are the loops? and how far is it possible to draw the secondary structure from a knowledge of the primary structure? and what about mis-paired bases in such structures? and are the bases in a tilted (A-type—see p. 9) conformation in this helices? The answers to these questions are all known with almost total certainty. However it is not possible to answer them from simple UV spectra and therefore it is necessary to enquire into the other spectroscopic properties of nucleic scids in more detail.

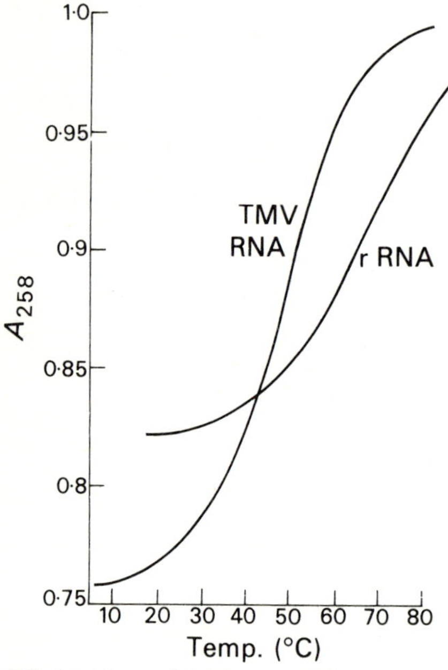

FIG 11.12   Melting profiles of tobacco mosaic virus RNA (TMV) and ribosomal RNA (rRNA-'microsomal RNA' in the publication) from bovine liver. Data taken from Doty *et al.* (1959).

ORD and CD of DNA and RNA   ORD and CD are methods of studying short-range stacking of bases (see p. 331). Duplex structures and stacks are not really different categories. The bases in a helix are stacked. In fact they form a perfect stack. Thus ORD and CD are very sensitive techniques for the determination of duplex structure. Rather than presenting a lot of optical data, let us just consider the general shapes for DNA and RNA (Fig. 11.13).

First of all let us look at the DNA curves as we probably feel to be on rather firmer ground in interpreting the spectra. The ORD spectra are similar for denatured DNA but the Cotton effects are of weaker amplitude. If some suitable wavelength is monitored (such as $[\alpha]_{290}$) melting curves, essentially super-imposable with the absorbance melting curve can be drawn. Moreover the size of this effect can be used to measure the GC-content. This is probably the simplest method to measure base composition (provided you have a DNA solution of accurately known concentration and an ORD machine). For native DNA the empirical relationship is as follows (GC = mole percentage of G + C):

$$[\alpha]_{290} = 26 \cdot 5 \times GC + 850$$

for denatured α (or single-strand) DNA, the relationship is:

$$[\alpha]_{290} = 15 \cdot 4 \times GC + 220$$

These relationships are taken from the work of Samejima and Yang (1965).

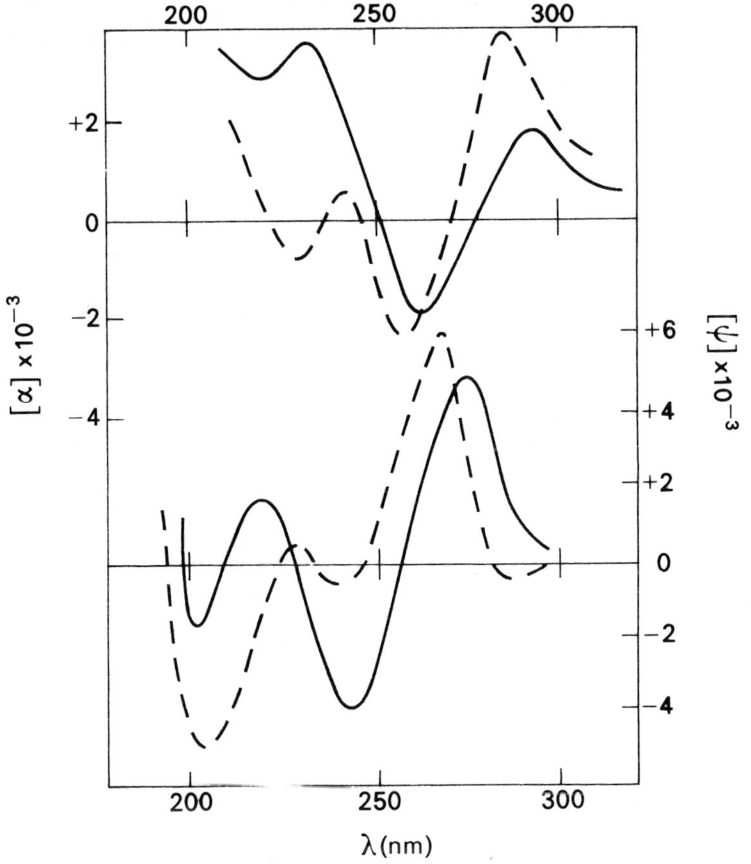

FIG 11.13   Idealized ORD (upper) and CD (lower) spectra of DNA (———————) and RNA (- - - - - - -).

Now let us have a look at the RNA picture in Fig. 11.13. It is qualitatively obviously different to the DNA picture. This difference is not due to the double-stranded nature of DNA as all DNA stacks (either duplex DNA or such stacked, single-strand structures as φX 174 DNA at low temperatures) resemble the DNA picture. Likewise all RNA structures (whether double-stranded such as reovirus RNA, or molecules with internal secondary structure such as TMV RNA) have the 'RNA shape'. Moreover the melting of RNA can be accurately monitored from these spectra. Upon the denaturation of RNA, the CD positive maximum of RNA ('right-hand' CD band in Fig. 11.13) is reduced (with a concomitant reduction in the magnitude of the corresponding Cotton effect's peak and trough—see Fig. 2.19(c), p. 90); additionally there is a small blue shift of the CD peak upon denaturation. The simplest explanation of the blue shift and other features of the spectra is that ordered RNA structures in solution contain bases tilted to the plane perpendicular to the helix axis (i.e. they have A-structure—see p. 328). For a detailed analysis, see Samejima, Hashizume, Imahori, Fujii and Miura (1968).

Conformation of N-glycoside bond in DNA nucleosides   It will undoubtedly be possible in time to refine the mathematical analysis of CD curves to measure the value of χ (see p. 325) in nucleic acids in solution. The problem is certainly capable of solution for nucleosides themselves (see p. 92). The prediction from the X-ray data on DNA (p. 328) is that if DNA is essentially in the B-form in solution, it should differ from RNA in that the bases are skewed away from the *anti*-conformation. The fact that the guanines of G-residues are in a different orientation to those in RNA (or in denatured DNA) has been established in an extremely ingenious manner by Kapuler and Michelson (1971). They analysed carefully the reaction of *N*-acetoxy-*N*-2-fluorenylacetamide with DNA and various polynucleotides. This substance reacts with C-8 of Gua-moieties (p. 61). The reaction with DNA is particularly facile, implying that at least the G-residues are not locked in an *anti*-conformation (in the strict *anti*-conformation the reaction is impossible on stereochemical grounds).

Other spectroscopic techniques   So far UV, ORD and CD have really furnished all the important information regarding the structures of nucleic acids in solution. Nevertheless IR spectroscopy of nucleic acids (either as solutions in D_2O, see p. 93, or as thin films) can be used to demonstrate intermolecular interactions and indeed DNA-melting curves can be obtained from suitably chosen IR bands. For a review of the IR properties of polynucleotides in general and the methods used for assignments of bands, see Shimanouchi, Tsuboi and Kyogoku (1964). It is very probable that polarized IR spectroscopy should be of value in determining features of base pairing and nucleoside conformation. The method has been largely held up by technical limitations; however an elegant method of obtaining the polarized IR spectra of DNA has been described by Rupprecht (1970) and the literature in this field should be well worth following over the next few years.

A potentially even more exciting technique (not only for nucleic acids but the whole of macromolecular research) is Raman spectroscopy. The traditional limitation to Raman spectroscopy is the low yield of measurable radiation (p. 95). However the use of laser beams for Raman sources has resulted in the potential of the technique suddenly to appear almost limitless. The great advantage of Raman over the other methods of measuring bond stretching and vibrational frequencies (IR) is that it can be employed in aqueous media. One's appetite for this application to nucleic acids is whetted by the paper by Thomas (1970), in which the Raman spectra of RNA are described. One of the many fascinating preliminary conclusions of Thomas is that the O-P-O stretching frequency (assigned to a band at 814 cm^{-1}) is particularly sensitive to ionic strength. In Raman spectroscopy we may have a technique for looking at the 'taking up of slack' in the chains of a duplex, a process central to the extremely important phenomenon of intercalation (p. 413).

The structure of RNA

Secondary structures for RNA molecules are written down in the literature with a considerable degree of confidence (glance through the structures in the figures for chapter 10). We are now able to evaluate the rationale behind this procedure with some confidence, so let us just ask a sort of examination question: what is the evidence in favour of the hairpin models for the secondary structures of RNA molecules? The answer must contain the following points. RNA has some sort of secondary structure from its hypochromism (see Fig. 11.12), its 'melting behaviour' and its anomalous ORD and CD spectra (see p. 340). Except for certain viral RNA molecules, it lacks the two complementarity strands of DNA, therefore if the secondary structure is helical, it must consist of loop structures of some kind. All RNA sequences determined so far contain considerable apparent intramolecular complementarity but all these schemes require some accommodation for 'non-Watson–Crick' pairs, notably G:U. All tRNA sequences discovered so far can be folded up in clover-leaf arrangements so that the common or analogous sequences appear in comparable conformations (see Fig. 10.8, p. 305) so it seems certain that this secondary structure is probably genuine. So what are we left with? Are short RNA duplices stable? What are the thermodynamic parameters of base pairing, odd-base pairing and looped structures? It is possible to quantitate the establishment of the 'most stable secondary structure' for a given sequence? Do the optical properties of RNA exactly correlate with proposed secondary structure? Is it possible to evaluate the helical content of RNA molecules without knowing the sequence? Each of the next five sections is an answer of sorts to each of the last five questions.

Multiplex formation in defined oligomers This section is the answer to 'are short RNA duplices stable?'. At first sight it might seem rather simple to use mixtures of short homo-oligomers. In practice such data are often difficult to evaluate. As Gennis and Cantor (1970) point out, the mixture of A-A-A-A and U-U-U-U, which forms stable secondary structures, probably contains 'concatamers' of the type shown below (reproduced using different symbols from Gennis and Cantor),

5′ . . . U-U-U-U U-U-U-U U-U-U-U U-U-U-U . . . 3′

3′ . . .   A-A-A-A A-A-A-A A-A-A-A   . . . 5′

5′ . . . U-U-U-U U-U-U-U U-U-U-U U-U-U-U . . . 3′

Gennis and Cantor synthesized oligomers of the type $(A-)_n C$ and $(U-)_n G$ (see p. 219 for the explanation of their mode of synthesis) and $C(-A)_n$ and $G(-U)_n$ (made with the primer-dependent polynucleotide phosphorylase preparations—see p. 220). The advantage these oligomers have is that concatamer-formation is rendered extremely unlikely as it involves mis-paired bases. For example the analogue of the concatamer shown above would require A:G pairs for which there is no evidence at all. Their work established that short oligomer helices are formed and are stable; they also developed an ingenious CD-test for strandedness. One of their most important discoveries was that, for short oligomers, the anti-parallel structures (both duplex and triplex) are more stable than the syn-parallel ones. The way in which it was done is best illustrated with a real case. Consider 1:1 mixtures of A-A-A-A-A-A-C with (i) G-U-U-U-U-U-U and (ii) U-U-U-U-U-U-G. If the configuration is anti-parallel, (i) will be more stable than (ii) (because the G and C will be paired in the former case but not the latter). In this particular case the T_m of (i) was found to be 19°C and that of (ii) to be 13·5°C.

Another very ingenious use of such defined oligomers is the study of Uhlenbeck, Martin and Doty (1971) on various pairs of oligomers such as A-A-A-A-G-U-U-U-U-U and A-A-A-A-U-G-U-U-U-U. If these compounds form intermolecular complexes (the conditions for this can be established—see p. 388), structures can be written down in which either G and U are paired or are looped out. If G:U pairs genuinely exist in helices, the secondary structures of the dimers of these two oligomers are very similar, viz.

5′ A-A-A-A-G-U-U-U-U-U 3′   5′ A-A-A-A-U-G-U-U-U-U-3′

3′ U-U-U-U-U-G-A-A-A-A 5′ and   3′ U-U-U-U-G-U-A-A-A-A 5′

If on the other hand, the G-residues are looped out, the structures are different to one another in geometry and also have one less base pair per dimer than the above structures:

<pre>
            G   G
           / \   / \
5′ A—A—A—A U—U—U—U—U 3′   5′ A—A—A—A—U U—U—U—U 3′
3′ U—U—U—U—U A—A—A—A 5′   3′ U—U—U—U U—A—A—A—A 5′
           \ /   \ /
            G   G
</pre>

The melting data of these (and other similar pairs) were found to be consistent with the former (G:U-paired) configuration.

Correlation of optical properties with proposed secondary structures for RNA of known sequence   Provided one has sufficient RNA to perform ORD measurements, it is possible to check the plausibility of a proposed secondary structure experimentally. This check has been made (Cantor, Jaskunas and Tinoco, 1966) for rRNA and the optical data have been found to be consistent with the clover-leaf structure for yeast tRNA^{Ala} (see Fig. 1.7, p. 24). The equation they applied was the following factorization of the value of $[m_{RNA}]$, the mean-residue molar rotation (p. 330) for the RNA molecule:

$$[m_{RNA}] = 2 \sum \sum \chi_{NN'} [m_{NN'}^D] - \sum \chi_N [m_N^D]$$

where $[m_N^D]$ is the molar rotation for a nucleoside residue in a 'double-stranded monomer' and $[m_{NN}^D]$ is the mean-residue molar rotation for a 'double-stranded dimer', i.e. the structure:

5′ . . . N -N′ . . . 3′

3′ . . . N′-N . . . 5′

As it stands, the equation is incapable of solution. It can be simpliefied however, if the following assumptions are made: the nearest-neighbour frequency will be regarded as random; the two strands of a helical region will be of identical nucleotide composition; it is legitimate to separate single-strand and double-strand interactions in the analysis. The equation may now be re-written as:

$$[m_{RNA}] \doteq \tfrac{1}{2}[m_1] + \tfrac{1}{2}[m_2] + \chi_{AU}^2 + 2\Delta m_{\sim AU} + \chi_{GC}^2 + 2\Delta m_{\sim GC} + (\chi_{AU}\chi_{GC} \times 2\Delta m_{\sim AU/GC})$$

where m_1 and m_2 are the mean residue rotations for the individual strands for the helical region (actually the composite sum of three helical regions). The '$2\Delta m$-values' are the mean residue rotations for the sequential A-U and G-C pairs and the average interactions ($2\Delta m_{\sim AU/GC}$) of all the remaining possible base-pair sequences. Note that this treatment assumes that only nearest-neighbour interactions are to be taken into account for calculating the ORD of stacks (see p. 330). χ_{AU} is the mole fraction of A:U base pairs. Then additionally:

$2\Delta m_{\sim AU}$ is assumed to equal $[m_{AU}] - \tfrac{1}{2}([m_A] + [m_U])$

where $[m_A]$ is the observed mean-residue rotation for poly A (likewise $[m_U]$) and $[m_{AU}]$ is the observed mean-residue rotation for poly A:poly U. The value of $2\Delta m_{GC}$ is computed in the same way and $2\Delta m_{\sim AU/GC}$ is calculated by a statistical treatment of the same data. In this way a theoretical value for $[m_{RNA}]$ was calculated for the clover-leaf conformation and was in good agreement with the experimental value.

Thermodynamic parameters of nucleotide associations   Is it possible to predict the stability of nucleic acid interactions directly from the predicted stability of different types of base-pairing given in Table 2.7 (p. 79)? The answer is 'No'. For one thing, RNA structures undoubtedly contain loops and the wave mechanical calculations of the stability of loops is difficult for, although we can estimate the stability of stacks of defined sequence (Table 2.8, p. 81), we must take into account that the nucleotides in a loop are constrained additionally by the restricted geometry of the two terminal nucleotides (those attached to the end of a duplex segment). There are however methods of evaluating the stability of loops of different size (see below). A bigger problem which has not been resolved so far is the role of hydration in the stabilization of nucleic acid structures. We have seen that the 'hydrophobic interactions' in nucleic acid structures consist partly of the out-of-plane base:base interactions (p. 77). The other class of forces which contribute to the 'hydrophobic interactions' does genuinely involve the water molecules, although not quite in the way implied by 'hydrophobic'. Water contains considerable structure itself. The molecules are H-bonded to produce short-range ordering of the same type that occurs (over long ranges) in ice. For a molecule, such as a heterocyclic base, to exist in such an environment, the ordering of the water lattice must be locally disrupted. The stacking of several such molecules requires less disorder in the water (to create a hole big enough to accommodate the stack) than would the individual unstacked molecules (which require a number of smaller holes). In other words one contribution to the stability of the stack is that the entropy of dislocation of the water lattice is smaller for the stacked than for the unstacked structure. The theory, for the quantitation of this phenomenon in polypeptide configurations has been worked out in detail by Némethy and Scheraga (1962). For a preliminary discussion of its application to nucleic acid structures see Sinanoğlu (1968). A quite different effect of water is the actual binding of 'shell' of water molecules to the nucleic acid structure. Although the contribution of this type of binding to the stability of duplices is unknown, the fact that hydration of nucleic acids involves the stable and relatively long-term association of water molecule with the phosphodiester backbone can be demonstrated in two separate ways. Lubos and Wilczok (1970) have described a method of following the magnetic relaxation times of DNA solutions in a spin-echo device. Their conclusion is that bound water dipoles do contribute to the secondary (and/or tertiary) structures of native DNA. An alternative technique (which most of us would feel more at home with) is to follow the kinetics of hydrogen-exchange between these 'exchangable hydrogens' of the nucleic acid structure ($-NH_2$ and $-OH$ groups) and the aqueous environment. The technique involves dissolving the nucleic acid in ^3H_2O and then following exchange by measuring the rate of loss of H^3-radioactivity, either by rapid microdialysis or with Sephadex columns. The methods are described by Englander (1968). At $4°C$, he finds that both RNA- and DNA-bound, exchangeable hydrogen has a half-life of 5–15 min. Further quantitation of this type of experiment in future should make it possible to make deductions regarding the half-life of molecules of water bound to the hydration shell (through which the H_2O molecules that have picked up our exchangable hydrogen have to pass).

The thermodynamic parameters for the formation of a base pair in a duplex of an oligomer or polymer may be calculated in a number of ways. One method is to employ the following relationship elucidated by Applequist and Damle (1965):

$$\frac{1}{T_m} = \frac{1}{T_m^\infty} + \frac{R}{\Delta H} \ln \beta c$$

where T_m is the melting temperature for a defined oligomer, T_m^∞ is the melting temperature for a polymer of equivalent nucleotide composition, ΔH is the heat-content change for base-pair formation, c is concentration and β is the equilibrium constant for the formation of the first pair. Hence ΔH can be calculated graphically from data in which T_m is measured as a function of concentration. An alternative method (Applequist and Damle, 1966) does not require any knowledge of T_m. In this case we merely plot f, the proportion of bases that exist in base pairs, as a function of temperature, when, at the melting temperature:

$$\left(\frac{df}{dt}\right)_m = \frac{N\Delta H}{6RT_m^2}$$

where N is the chain length. Clearly this sum requires a method of calculating f from a melting curve. If the denaturation of an oligomer is followed spectrophotometrically it is possible to analyse the profile in terms of the co-operative melting of duplex structures superimposed on two graphs; one is the non-cooperative denaturation of single-strand stacks; the other is the

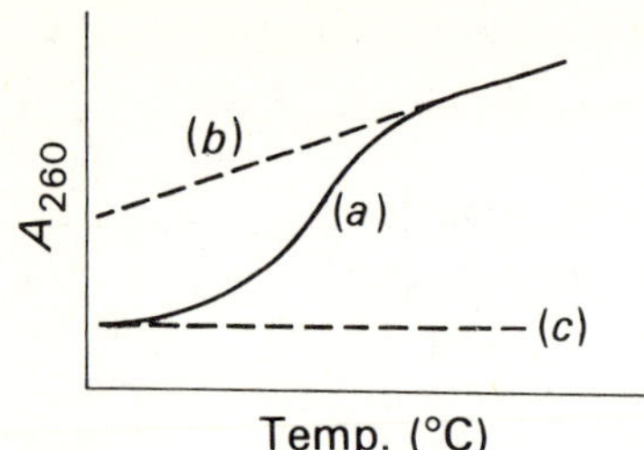

FIG 11.14   Analysis of a hypothetical melting curve for an oligomer (from Martin, Uhlenbeck and Doty, 1971). (*a*) is the observed melting curve, (*b*) is the unstacking contribution and (*c*) is the duplex contribution (see text for discussion).

variation in absorbance of native duplex with temperature. These two curves are deduced by extrapolation (although in practice the latter 'variation' is assumed to be zero). The method is illustrated diagrammatically in Fig. 11.14.

The observed absorbance at $T°C$ is $A(T)$, thus curve (*a*) in Fig. 11.14, is a plot of $A(T)$ against T; the other functions are defined as follows. Curve (*b*) is a plot of $A_s(T)$ against T and (*c*) is a plot of $A_d(T)$ against T. The T_m is the temperature at which there is a point of inflexion in curve (*a*). From these curves, f can be calculated from the equation following (Applequist and Damle, 1966):

$$f(T) = \frac{A_s(T) - A(T)}{A_s(T) - A_d(T)}$$

Similarly the percentage hypochromicity is calculated from such a curve:

$$\% \text{ hypochromicity} = 100 \times \frac{A_s(T) - A_d(T)}{A(T)}$$

In work on the optical properties and thermodynamics of oligomers, it is important to know what sort of secondary structure is being studied. The presence of triplices (as opposed to duplices) can be detected either in the ultracentrifuge (p. 388) or by a spectrophotometric method (see Blake, Massoulié and Fresco, 1967). The other question is whether the duplices are inter- or intra molecular. As an example, consider the oligomer

A-A-A-A-A-A-A-U-U-U-U-U-U-U

There are two types of secondary structure (apart from concentrates and triplices—see p. 343):

5' A—A—A—A—A—A—A—U—U—U—U—U—U—U 3'

3' U—U—U—U—U—U—U—A—A—A—A—A—A—A 5'

5' A—A—A—A—A $\nearrow$ A $\searrow$ U

3' U—U—U—U—U $\searrow$ A $\nearrow$ U

The melting temperature of the former will be concentration dependent (see p. 345); that of the latter will not. The formation of the former (intermolecular) duplex is favoured by high salt concentrations; that of the latter (intramolecular) is favoured by low salt concentrations (Martin, Uhlenbeck and Doty, 1971).

The conclusion from the data is that the change in heat content for the formation of a base pair in a helical segment is about −8 kcal/mole (for either A:U or G:C; the values are the same within the limits of experimental error). To form a helix of N-bases, the total heat-content change is given as:

$$\Delta H = (N - 1)(-8 \pm 0.7)$$

Note that $(N - 1)$ rather than N is employed as there is assumed to be no change in ΔH for the 'initiation reaction' (formation of the first base pair) as there is no stacking interaction in this case (Tinoco, Uhlenbeck and Levine, 1971).

Having assigned values for ΔH for the formation of base pairs, the next question is the assignment of free energy changes (ΔG-values). The method is based on work by Kallenbach (1968) in which the reciprocal of the melting temperature of an RNA molecule (or synthetic polynucleotide duplex) is expressed as a linear function of GC-content (expressed here as χ_{GC} or mole fraction of G + C; i.e. $\chi_{GC} = \frac{1}{100} \times$ GC-content of p. 8). The equation is:

$$\frac{1}{T_m} = \frac{1}{T_{AU}} + \chi_{GC}\left(\frac{1}{T_{GC}} - \frac{1}{T_{AU}}\right)$$

where T_{AU} and T_{GC} are respectively the melting temperatures (in the same buffer) of polymers containing only G:C or A:U base pairs. Uhlenbeck, Martin and Doty (1971) find that in 1 M NaCl, pH 7, T_{AU} is 78°C and T_{GC} is 152°C and hence ΔG for the formation of an A:U pair is -1.2 kcal/mole and for the formation of a G:C pair it is -2.4 kcal/mole. Their work on oligomers which form duplices containing G:U pairs (Martin, Uhlenbeck and Doty, 1971; see also p. 343) implied that ΔG for the formation of such a pair is about 0 kcal/mole.

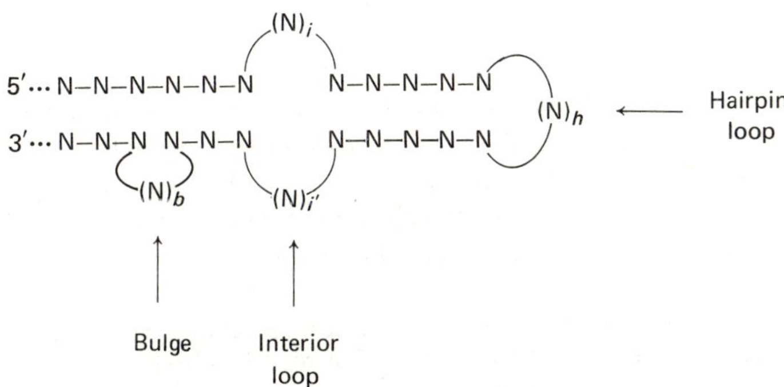

FIG 11.15   Explanation of the terms 'hairpin loop', 'interior loop' and 'bulge' in an RNA duplex (see text for explanation). Base pairing (G:C, G:U and A:U) is represented as N:N: h, i (and i') and b are simply integers defining the number of nucleotides in the three categories of non-helical sequence.

Finally there is the problem of assigning free-energy changes associated with the formation of non-helical regions in a mainly duplex stretch of an RNA structure. The three types of region that are theoretically possible are defined in Fig. 11.15. The nomenclature (hairpin loops, interior loops and bulges) is that of Tinoco, Uhlenbeck and Levine (1971).

From a study of space-filling models, it is possible to rule out certain structures as sterically impossible and conclude that (using the integers defined in Fig. 11.15) that h must be greater than 2^3 and i (and i') must be greater than 1 (of course i does not need to equal i'). There are no steric restrictions on bulges (b can equal 1, 2, 3, . . ., etc.).

Values for ΔG for formation of loops and bulges have been estimated by Tinoco *et al.* (1971); all the values are positive, as looping and bulging are unfavourable processes (they represent considerable loss of entropy when compared with the unrestricted, random-coil conformations). The free energy of initiation of loop formation can be factorized as:

$$\Delta G = -RT[B' - 1.5 \ln h]$$

where h has the meaning given in Fig. 11.15, and B' is constant, (Poland and Scheraga, 1970).

If this is re-written to eliminate the natural logarithm:

$$\Delta G = -2 \cdot 303\, RT[B - 1 \cdot 5 \log_{10} h]$$

where B is a new constant. On the basis of their own data (unpublished) on oligomers of the type A-A-A-A-A-A(-C)$_n$-U-U-U-U-U-U, Tinoco and others propose -3 to be a good value for B for hairpins in which $h = 4$. For $h = 3$, they find the loop is more stable by $1 \cdot 2$ kcal than predicted and that this figure should therefore be subtracted from the (positive) prediction of ΔG. The above equation stands for interior loops ($i + i'$ replacing h) but that for small interior loops with $i, i' = 2$ a blanket value of $+5$ kcal/mole is more useful. The free-energy change associated with the formation of small bulges is about $+2$ kcal/mole from work on the poly U : poly (A$_m$, U$_n$) systems (Steiner, 1960). For larger bulges the above equation (b replacing h) is applicable (Tinoco *et al.*, 1971).

With all this information, we can put together a general equation for the formation of an RNA secondary structure:

$$\Delta G = N_{AU}\bar{H}\left(1 - \frac{T}{T_{AU}}\right) + N_{GC}\Delta H\left(1 - \frac{T}{T_{GC}}\right) - \Delta H\left(1 - \frac{T}{T_m^\infty}\right) + \sum f(h, i, b)$$

where $\Delta\bar{H}$ is taken as -8 and the third term takes into account the '($N - 1$)' used for the calculation of ΔH for each helical section. Tinoco *et al.* (1971) take $1 \cdot 8$ kcal/mole to be a good general estimate for it. The final term is the sum of the values derived for the loops and bulges.

Estimation of the relative stabilities of conformations for RNA molecules or fragments of known sequence   The final equation of the preceding section may be used to compute the value of ΔG for the formation of any proposed secondary structure for an RNA molecule. The structure with the largest negative value for ΔG is the most stable. For the more straightforward evaluation of stabilities, Tinoco, Uhlenbeck and Levine (1971) propose that the stability of

TABLE 11.4   Stability units for structural features of RNA conformations. See text for discussion and Fig. 11.15 (p. 347) for definitions. Data from Tinoco, Uhlenbeck and Levine (1971).

	Stability units
Each A:U base pair	+1
Each G:C base pair	+2
Each G:U base pair	0
Hairpin loop $\begin{cases} h = 3 \\ h = 4\text{–}7 \\ h = 8 \text{ or above} \end{cases}$	$\begin{matrix} -5 \\ -6 \\ -7 \end{matrix}$
Interior loop $\begin{cases} i + i' = 3 \\ i + i' = 4\text{–}7 \\ i + i' = 8 \text{ or more} \end{cases}$	$\begin{matrix} -5 \\ -6 \\ -7 \end{matrix}$
Bulge $\begin{cases} b = 1 \text{ or } 2 \\ b = 3 \\ b = 4\text{–}15 \\ b = 16 \text{ or more} \end{cases}$	$\begin{matrix} 2 \\ -3 \\ -5 \\ -6 \end{matrix}$

various segments of the structure be evaluated in terms of 'A:U base-pairing units'. The stability constant for the formation of an A:U pair in a duplex is defined as 1. As ΔG for this process is $-1\cdot2$ kcal/mole (p. 347), obviously we transform ΔG into these new units by dividing by $-1\cdot2$. The most stable structure is the one that gives the largest positive value for the total stability units. The values for stability units for different structural features are given in Table 11.4. (h, i, i and b are defined in Fig. 11.15, p. 347):

Note that these values (Table 11.4) emphasize very clearly a feature of the thermodynamic estimates of the preceding section, namely that it is less expensive for non-complementary nucleotides to form a bulge than a loop. The molecular reason is that the stacking interactions are continuous in one strand (opposite the bulge) whereas a loop necessitates total discontinuity of stacking. This important point has been neglected in several proposals for RNA secondary structure.

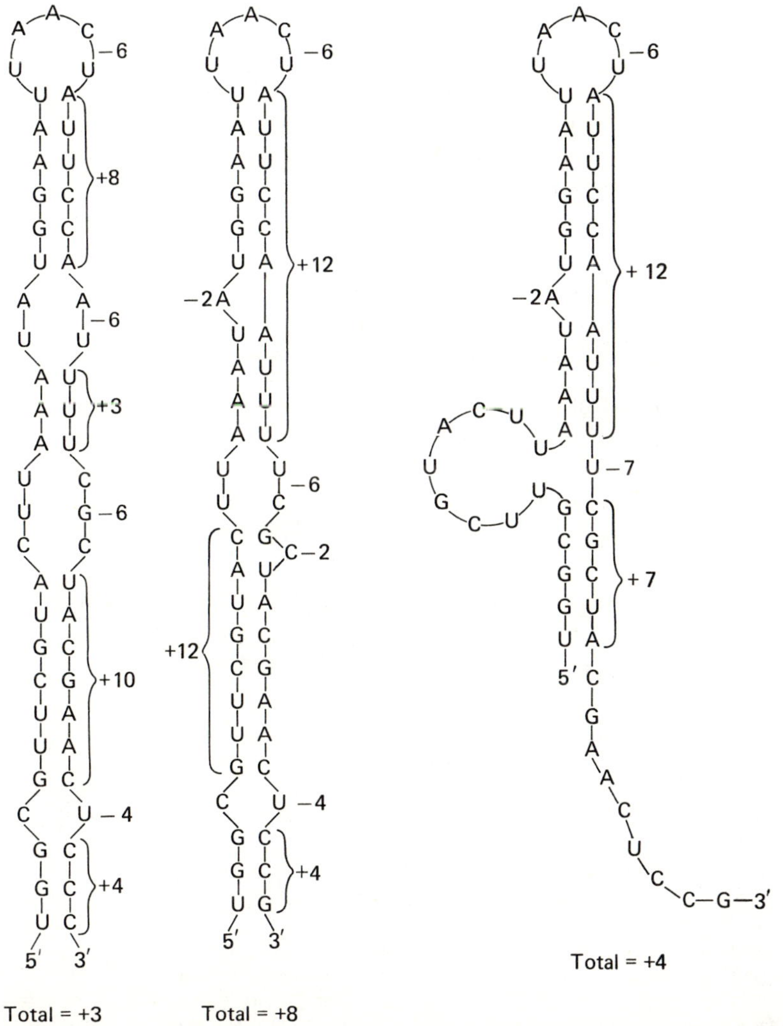

Total = +3   Total = +8

Total = +4

FIG 11.16   Three possible conformations for a sequence from R17 RNA with a calculation of the total stability units (see Table 11.4, p. 348) in each case. The sequence was determined by Adams, Jeppeson, Sanger and Barrell (1969); the present figure is from Tinoco, Uhlenbeck and Levine (1971). See text for discussion.

Before briefly considering the application of the values in Table 11.4 to real sequences, there remains one question. How far is it legitimate to construct a secondary structure for a fragment of an RNA molecule? These fragments are isolated by limited digestion with a nuclease which has preference for single-strand substrates (p. 207). It therefore seems almost certain that the large T_1-fragments from (say) R17 RNA or 16 S rRNA are genuinely isolated in their native conformation. Of course it may turn out that more extensive future knowledge of these sequences will suggest additional tertiary interactions in the intact molecules but we can safely assume that such interactions are of a weak character and are superimposed on the pre-formed secondary structure. For a discussion of this question as it applies to tRNA, see p. 417.

In Fig. 11.16 are computed the total stability values for one of the R17 fragments. This example is taken from Tinoco, Uhlenbeck and Levine (1971) and is of particular interest as the structure with maximal base pairing (which is the one that was proposed by Sanger and his colleagues who elucidated the sequence (see Fig. 10.3 (IV), p. 313)) is not the most stable.

The paper by Tinoco *et al.* (1971) describes a simple graphical method of searching for such structures in a known sequence of this type.

A second important application of the approach has been to analyse the many proposed secondary structures for 5 S RNA. In this case the structure of Cantor (1967; Fig. 11.17) with a total of 30 stability units is the most stable. For a comparison with other proposals for this molecule (and references to these proposals) see Tinoco, Uhlenbeck and Levine (1971).

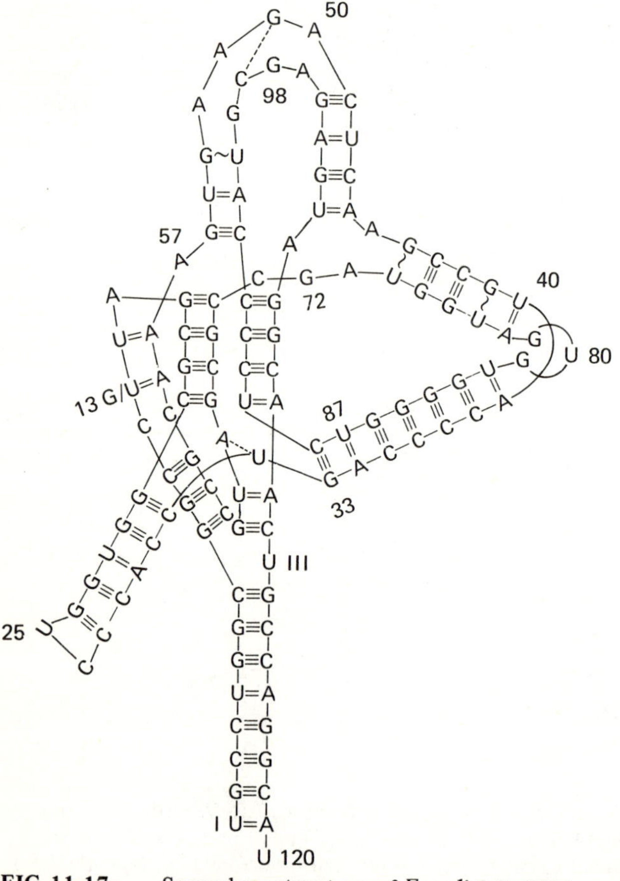

FIG 11.17   Secondary structure of *E. coli* 5 S rRNA according to Cantor (1967).

Estimation of the secondary structure in RNA molecules of unknown sequence Methods of estimating the amount of secondary structure in RNA samples have been developed by R. A. Cox. An extensive study was made (Cox and Kanagalingham, 1967) of total unfractionated (28 S and 18 S) rRNA from reticulocytes. The mathematical treatment of the data is relatively complex as the authors established the distribution of duplex/loop segments of different lengths without any assumption being made in the first instance that the RNA contained such loops at all. However their paper is a thorough and complete account of both the experimental methodology and the necessary mathematics and is easy to follow. Basically they hydrolysed the RNA to oligomers, either with limited treatment with ribonuclease A (p. 206) or with dilute alkali. The molecular weights of the fragments was measured in the ultracentrifuge (p. 387) and the fragments were analysed for the presence of 'hidden breaks' (discontinuities in a duplex/loop structure) by analysing them by electrophosesis[4] before and after denaturation (85°C). Clearly an oligomer containing a hidden break will produce two zones after denaturation.

The melting curves and especially the differences spectra for the oligomers were measured and fitted to theoretical spectra derived from model compounds. The question of difference spectra will be considered in the next paragraph. The data were consistent with the model that many duplex/loop structures (containing on average between 10 and 20 residues) were present in the molecule.

More recent work (Cox, 1970; Cox, Kanagalingham and Sutherland, 1970) has been directed to establish that an RNA sample is a genuine duplex molecule. Their work has concerned a very interesting double-stranded RNA found in *Penicillium chrysogenum*. The nature of the particles containing this RNA is still somewhat obscure but in all probability they are virus. Fungal viruses or mycophage are almost unknown and this duplex RNA is of great potential biological interest. Two RNA species are obtained from the 'viruses' (see p. 292). Both are now known to be duplices containing two complementary strands. There are two types of difference spectrum for a nucleic acid. One is obtained by subtracting the spectrum (plotted as ϵ_p—see p. 000—as a function of λ) from the spectrum for the hydrolysed sample (i.e. the mixture of the component nucleotides); the other is obtained by subtracting the native spectrum from that of the denatured nucleic acid. For RNA, these two types of difference spectrum are not superimposable on account of the residual stacking (that does not melt co-operatively) in the strands following the melting of the duplex (see Fig. 11.12, p. 340). The former (hydrolysis-) difference spectra have been determined for mixed homopolymer duplices by Haselkorn and Fox (1965). The difference spectrum of poly A : poly U has a maximum at 260 nm ($\Delta\epsilon$ at mas = 6000); that for poly G : poly C has a minimum at 260 nm ($\Delta\epsilon$ = 2000) and maxima at 245 nm ($\Delta\epsilon$ = 3000) and 280 nm ($\Delta\epsilon$ = 4500). From a knowledge of the nucleotide composition of the two species of this RNA, Cox (1970) calculated a hydrolysis difference spectrum from the above data of Haselkorn and Fox. The difference spectra of the RNA species were found to be superimposable with the calculated ones. Moreover the denaturation-difference spectrum was found to be superimposable too provided it had been multiplied by 1·33 to counteract the hypochromicity due to base stacking in the denatured form.

Mis-pairing of DNA in duplices and mutagenesis

The theory behind mutagenesis is really outside the scope of this book as it requires considerable familiarity with genetics to understand the analysis of the results. However an important facet of the study of nucleic acid structure is the insight that is gained into the molecular processes involved in mutagenesis. There are two sorts of mutagenesis, spontaneous and induced. For the purposes of the present discussion, I am going to assume that mutations that arise spontaneously are the result of a rare event that has a real chance of occurring in the absence of any external influence. Obviously we cannot be certain that a particular 'spontaneous' mutation did not arise as the result of a rare event involving (say) a chemical impurity in the environment of the organism or UV-radiation in sunlight.

Spontaneous mutation   The model for spontaneous mutagenesis is that it arises from non-Watson–Crick base pairs. The only plausible structures are those in which the bases are in the rare tautomeric form. In the discussion of these forms I follow the convention of Pullman and Pullman (1969) and use an asterisk to represent the minor (imino or lactim) form. The base pairs that could be formed involving these forms are shown in Fig. 11.18. Notice that the pairs involving two minor forms show Watson–Crick stoicheiometry (G*:C* and A*:T*). However the pairs with one minor tautomer and one major tautomer are the other Pu:Py pairs.

The factor expected to determine the chance of such mispairings occurring is the relative stabilities of the major and minor tautomers. This is quantitated in terms of resonance energies (p. 81). From the values of these energies (Table 2.9, p. 82) it is predicted that the 'most common' rare tautomer should be C*. The result of C* occurring in the template for DNA polymerase is to incorporate A in the newly synthesized strand. Further replication will yield an A:T pair in place of the original G:C. There is considerable evidence that this change (G:C to A:T) is a more frequent spontaneous mutagenic event than the converse (Drake, 1966). So far I

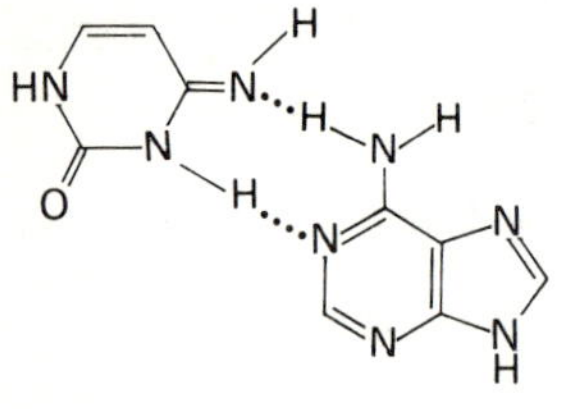

C*:A 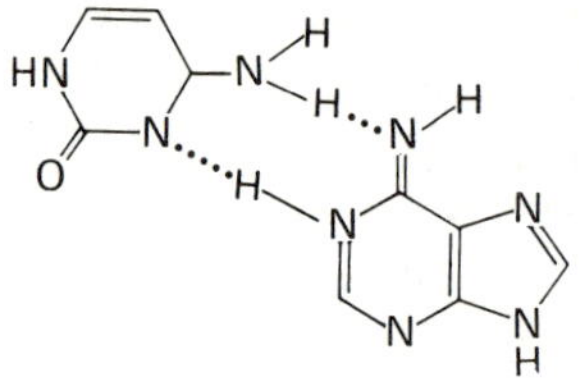   C:A*

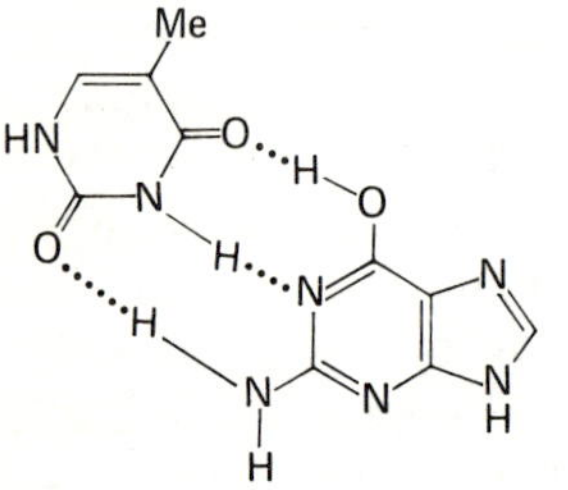

G*:T 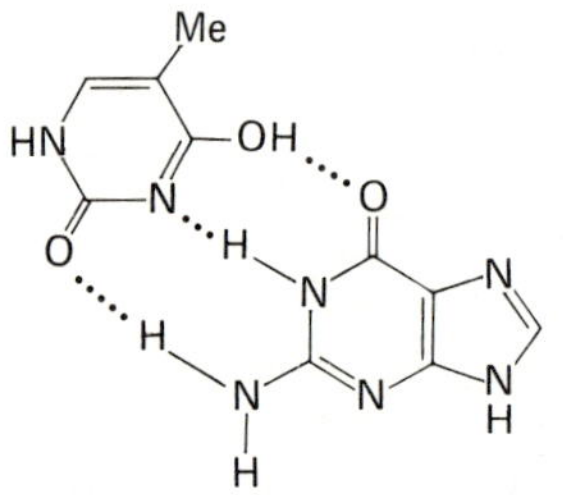   G:T*

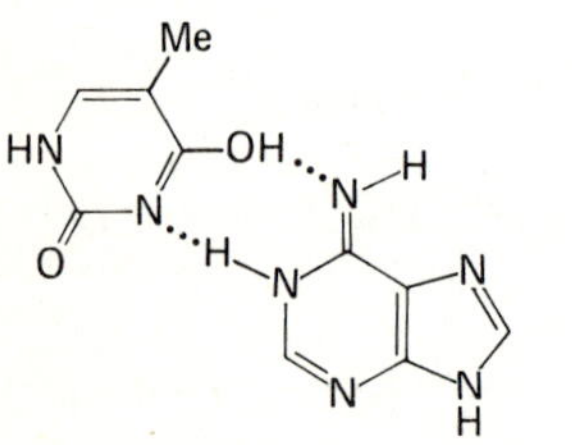

T*:A* 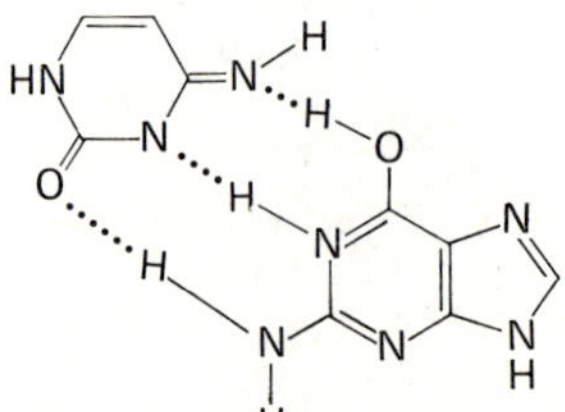   C*:G*

FIG 11.18   Base pairing schemes involving the rare tautomers of the bases.

have not commented on the interaction energies for the structures shown in Fig. 11.18. In fact they turn out to be as stable as the Watson–Crick structures although the relative stabilities are different. When the values of E_M (see p. 77) are calculated for the pairs involving one major and one minor tautomer it turns out that A*:C (E_M = −19 kcal/mole) and T*:G = −17 kcal/mole) are much more stable than the other two (E_M-values of −8 and −9 kcal/mole). These data and other details are to be found in Pullman and Pullman (1969).

An alternative suggestion is due to Löwdin (1965) who regards the initial mutagenic event being a type of electron-tunnelling. The overall effect of Löwdin-tunnelling is to convert A:T into A*:T* and G:C into G*:C*. The implication is that a stable pair of minor tautomers occurs in the parent duplex but that DNA polymerase catalyses the formation of anomalously paired structures so that the mutagenic lesion is carried in both the daughter molecules.

Ultra-violet radiation   There are several models for UV-induced mutagenesis involving deamination of bases and other similar reactions. A more straightforward explanation is due to Pullman and Pullman (1969), who point out that the difference in resonance energies (ΔR—see p. 82) are much smaller for two of the bases, namely the amino-imino equilibrium in Cyt and Gua, in the first excited state than in the ground state. For the first excited state of Cyt ΔR = −0·01 β-units; for the first excited state of Ade, ΔR = −0·14 β-units (compare with the ground-state values on p. 82). Thus the production of excited states in the bases of DNA by radiation would be expected to significantly increase the mutation frequency as a result of the greater tendency for the structures of Fig. 11.18 to form.

Chemical mutagenesis   Chemicals that react with the bases of DNA cause mutations—it sounds so simple. In fact, the actual molecular event that causes mutation (as opposed to death) is difficult to identify with certainty with many reagents. The position as it stands with some well-known mutagens is summarized below.

Nitrous acid: nitrous acid is a deaminating reagent. Thus it can convert Cyt to Ura, Ade to Hyp and Gua to Xan. The first two events are undoubtedly mutagenic (the I-residues formed in the second case are equivalent to G-residues for H-bonding). The last event is probably inactivating but not mutagenic. X-Residues do not readily base-pair with anything (see for example Table 11.2, p. 334).

Alkylation: Of the various alkylated products obtained (pp. 60 and 239) from nucleic acids and their constituents, it is generally assumed that all events are inactivating except for those involving G-residues. Both 7-alkyl-G-residues (Fig. 2.8a, p. 61) and O^6-methyl-G-residues (Fig. 2.11f, p. 68) are essentially G*-type structures so both may be assumed to pair with T (Fig. 11.18, p. 352).

Hydroxylamine: There is some controversy regarding which of the reaction products of hydroxylamine and C-residues in nucleic acids is responsible for mutagenesis as opposed to inactivation. For a discussion of this problem, see the reviews by Phillips and Brown (1967) and Singer and Fraenkel-Conrat (1969). This latter paper reviews a great deal of material relating to chemical mutagenesis and is very informative. It is unfortunate that it was written before the significance of O^6-methyl-guanosine was recognized.

Notes to chapter 11

[1]   By plotting $\ln K$ against $1/T$ a straight line is produced in accordance with Van't Hoff's isochore:

$$\ln K = \text{const.} - \frac{\Delta H}{RT}$$

Additionally, $\Delta G° = -RT \ln K$ and $\Delta G = \Delta H - T\Delta S$.

2   This polymer was made with polynucleotide phosphorylase using as substrate N^6-hydroxethyl-ADP (see p. 219).

3   This may be invalid in the special circumstances that arise when one of the nucleosides in the loop is a pyrimidine. Seeman, Sussman, Berman and Kim (1971) have studied the crystal structure of U-A by X-ray diffraction. The structure of the crystal is rather complicated and two conformations of the substance occur in the unit cell. However in one of these conformations, the bases are twisted round in such a way that it would be possible to continue a duplex 'beneath them'. I am grateful to Dr. W. Fuller for drawing my attention to this consequence of an extremely important paper (U-A is the first 'natural', i.e. 3'-5'-dinucleoside phosphate to be analysed by X-ray diffraction).

4   If you read this paper, I advise you not to employ the electrophoretic techniques described in it (which were not Cox's own methods). The methods of casting gels, and especially of staining them are much inferior to the procedures of Loening, Sanger and others that are described elsewhere in this book (pp. 155 and 311).

12 Hybridization

General and theoretical

Introduction

What is hybridization?   Hybridization is the experiment in which a mixture of denatured nucleic acids are re-annealed and hybrid duplices are isolated. From the yield of hybrids it is possible to deduce that various complementary regions in the various components of the mixture. The procedures involved, consist of incubating the denatured mixture at a temperature (typically about 25 deg. below T_m) and identifying double-strand products by a chromatographic or other procedure. As an example, consider one of the most familiar experiments involving hybridization. Radioactive rRNA (of known specific activity) is annealed with a known amount of denatured DNA from the same organism. The duplices are isolated (for example by cellulose-nitrate absorption—see p. 361). Radioactivity in such duplex preparations is due to hybrids consisting of rRNA molecules annealed to transcribing strands of ribosomal cistrons. By quantitating the assay it is possible to calculate the number of moles rRNA nucleotides bound per mole of DNA nucleotides and (from a knowledge of the molecular weight of the rRNA and the size of the genome) one can calculate the number of rRNA cistrons per genome.

And what are the problems?   The most cursory glance at the hybridization literature will convince anyone that it is not quite as simple as that. To take the experiment described above (which is a relatively simple application of hybridization) it has been applied to several species with widely disparate results. For example, estimates of the number of rRNA cistrons per rat-liver genome have varied from 280 to 1100. The following are some of the points to be borne in mind when hybridization data are evaluated.

Hybridization is not the reverse of melting. Many faulty encounters between non-complementary molecules will (on average) precede the formation of a faithful duplex. Thus hybridization has very slow kinetics; moreover much 'artefactual' annealing may occur and lead to the production of structures in which there is partial hybrid structure with much looped out non-complementary sequence. The production of these artefacts is eliminated partly by performing the annealing at a fairly high temperature (above the melting temperatures of many of the artefacts); they are also minimized by digesting the hybrids with single-strand-specific nuclease. The presence of artefacts in the hybrid may be detected by measuring the melting of the hybrid. This is difficult in many of the experimental systems that are employed.

Another important point is that the counting of gene copies of RNA:DNA hybridization requires assumptions to be made about the 'saturation' of the sites on the DNA by the RNA. Perhaps the saturation value is approached asymptotically as the ratio of RNA/DNA is increased? Well it is of course, but you never reach the limiting value experimentally as the hybridization process is an equilibrium. There is a simple method of performing this extrapolation (p. 368) but it is not always employed in the literature.

Then there is the problem of minor components in the RNA. If there are minor labelled components in (for example) our rRNA preparation to be used for the hybridization experiment considered above, a false conclusion will be reached. As increased amounts of RNA are employed, the hybridization of minor components with their complementary sequences will complicate the data: they might rise and appear above the background in the middle of a set of comparative data (using different RNA/DNA ratios). The problem of minor components may be even more serious in the case of a 'competition experiment'. (This is an experiment in which the genuineness of RNA/DNA hybrids is checked by examining the various effects of unlabelled homologous or heterologous RNA on the reaction.)

With these points in mind, let us consider some of the theoretical and practical aspects of hybridization in general and then consider the application of these techniques to particular hybridization experiments.

Kinetics

Denaturation of a duplex   What is the driving force for the denaturation of a duplex? The simplest model to consider would be one in which the two strands separate through diffusion. As the denaturation involves an uncoiling of the duplex, the ends of the chains would be seen to perform 'drunken walks' with an eventual unravelling of the structure. At the melting point, the implication for such a model is that the time of unwinding would be $20 \text{ min}/10^4$ base pairs. The experimental value is close to 0.02 s. Thus there must be some unwinding force predominating in the process. The problem has been considered by Longuet-Higgins and Zimm (1960) and Fixman (1963) who conclude that the explanation must be found in the entropy-release which follows the dissociation of the specific base pairs. In the initial duplex one particular residue (say a G) is specifically bound to a specific C-residue. Although the H-bonds unique to GC are not a major contributory factor in the stability of the duplex (p. 77) the release of the structure from this restriction results in a large positive entropy change; it is thus thermodynamically an extremely favourable process. The suggestion (put in qualitative terms) of Longuet-Higgins, Zimm and Fixman is that this entropy release is coupled to drive another unfavourable process—the restriction of the drunken walks into a helical precession to unwind the duplex.

Kinetics of duplex formation   There is no theoretical consideration of the basis of renaturation kinetics. It is an incredibly difficult problem. Presumably the favourable process of stacking the bases drives the coiling-up of the helix. How the processes are coupled to base-base recognition is almost impossible to contemplate.[1]

However, there is no objection to using simple second-order kinetics to analyse the kinetics of the hybridization (or re-association reaction) at constant temperature. If the nucleotide residues in one chain are N, and those in the other M, the equilibrium is:

$$M + N \underset{k_2}{\overset{k_1}{\rightleftharpoons}} MN$$

If m and n are the initial reactant-concentrations and x is the concentration of duplex (at time t):

$$\frac{dx}{dt} = k_1(m - x)(n - x) - k_2 x$$

In the special case of the renaturation of DNA, there are equal amounts of the two strands, then $m = n$; moreover k_1 is much greater than k_2. Putting these conditions into the equation above, we obtain:

$$\frac{dx}{dt} = k_1(m - x)^2$$

Integration (and use of the relationship that $x = o$ when $t = o$) and re-arrangement gives:

$$\frac{x}{m} = mk_1 t/(1 + mk_1 t)$$

This equation is used in the so-called 'Cot method' for determining the extent of redundancy in DNA sequence (p. 364). The basis of the method is to regard (mt) as an independent variable and to plot (x/m) as the dependent function and to determine the value of (mt) for which the complex is half-formed. Note that the half-life of the reaction is given by the relationship:

$$t_{1/2} = 1/k_1 m$$

A different method of approximating the fundamental rate equation, is to consider the condition when one of the components (say n) is limiting. In this case, the equation becomes that for a first-order process:

$$\frac{dx}{dt} = k_1 n(m - x) - k_2 x$$

integration and re-arrangement as before gives us:

$$k_1 b + k_2 = \frac{1}{t} \ln \frac{r_n}{r_n - x}$$

where r_n is the amount of N bound at equilibrium.

The half-life (value of t when $x = \frac{1}{2}r_n$) is given by:

$$t_{1/2} = \ln\ [2/(k_1 m + k_2)]$$

This is the equation that determines the rate of duplex-formation when one strand is in great excess. If the additional approximation is made (see above) that k_2 is very small compared with k_1 (i.e. that the helix, once formed, does not open again) then: $dx/dt = K_1 m(n - x)$, which on integration and rearrangement gives:

$$t = \frac{1}{k_1 m} \ln \frac{n}{n - x}$$

and

$$t_{1/2} = \ln\ (2/k_1 m)$$

If we assume that m and n are of the same order of magnitude (and make the assumption that k_2 is negligible), the corresponding equation is (for t)

$$t = \frac{1}{k_1(n - m)} \ln \frac{k(n - x)}{n(m - 1)}$$

In the simplifying case (see above) of $m = n$, this becomes:

$$t = \frac{x}{m(m - x)}$$

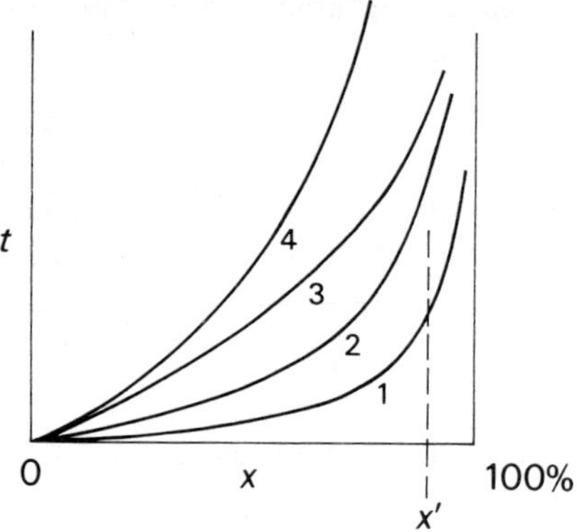

FIG 12.1   Plots of x (in percentages) as a function of time. In all these plots, it is assumed that temperature, m and k_1 are the same (and k_2 is assumed to be 0) and n is varied.

Curve *1:* a first-order plot for the situation in which n is vanishingly small so that

$$t = \frac{1}{k_1 m} \ln \frac{j}{n - x}$$

Curves *2* and *3:* second-order plots

$$t = \frac{1}{k_1(n - m)} \ln \frac{m(n - x)}{n(m - x)}$$

the ratio m/n is greater in (*2*) than in (*3*).

Curve *4:* the special case of $m = n$;

$$t = \frac{x}{m(m - x)}$$

The implications of these equations are shown in the plots in Fig. 12.1. For a more complete (and quantitative) discussion see Kennell (1971), whose article is essential reading for anyone who wishes to take hybridization seriously.

We consider the kinetics that obtain when RNA:DNA hybridization are being studied. If n is the initial concentration of DNA and m that of RNA; and we regard some value of x (x' in Fig. 12.1–typically x' might be 97%) as a value close enough to 'complete binding' for experimental evaluation, then it is clear that the more the DNA is in excess (with curve *1* as the limiting case) the faster will x' be achieved.

The equilibrium

In the previous sections, we have discussed the consequences of certain assumptions upon the kinetics of the processes of renaturation (or hybridization) and denaturation. The assumption has been made that at the temperatures (below T_m) at which hybridization studies are undertaken, the equilibrium lies essentially in 100% duplex position. Although this assumption is justified (as we see below) for the purpose of simplifying the rate equations, the equilibrium constant is still of such a magnitude that, with very low input-concentrations of one of the reactants, the available sites may remain unfilled. The problem of course is that if mixed RNA populations are used for RNA:DNA hybridizations, the concentration of some of the RNA species may be extremely low. Therefore it is necessary to consider this equilibrium in some detail. This discussion follows that of Kennell (1971). For simplicity let us consider RNA:DNA hybridization exclusively.

Concentration-dependence of RNA:DNA hybridization   Let R be the concentration of an RNA species and D the concentration of its corresponding 'site' (complementary sequence) in DNA. If y moles are bound at equilibrium:

$$K = \frac{(R - y)(D - y)}{y}$$

This is a quadratic equation and can be rearranged as such; a more usual rearrangement is as follows:

$$y/R = D/K - yD/KR - y/K + y^2/KR$$

as y/R is the fractional yield of hybridized RNA. In the limiting case of vanishingly little hybridization ($R \gg y \ll D$), y/R approaches D/K. Curves corresponding to plots of this equation can be obtained by plotting y/R as a function of R or D. The usual practice is to plot it as a function of R and to obtain a family of curves for different values of R/D. In general, the RNA

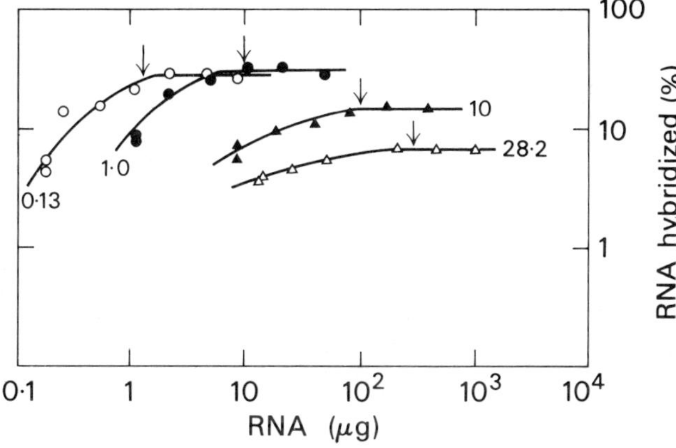

FIG 12.2   An example of RNA:DNA hybridization data. Pulse-labelled RNA (between 21 and 24 min following infection) isolated from T4-infected *E. coli* incubated with T4 DNA in 0·9 M NaCl, 0·09 M sodium citrate for 20 h at 66°C. The numbers against the curves (0·13, 1, 10 and 28·2) are values of R/D (weight/weight). The little arrows point to positions corresponding to 10 µg DNA. They illustrate the point that the amount of DNA required for this is independent of input ratio for such a mixed population of RNA (see text). Data of Kennel and Kotoulas (1968).

will consist of a mixture of molecules. If we let Σy be the sum of all the bound RNA species (this is the figure deduced when the hybrid is harvested and counted) then a plot of $\Sigma y/R$ against R will be a composite curve made up of a number of curves; one for each species; the species varying in their values of R/D. It will always be the case that the most abundant RNA species will have a value of y/R close to its maximum, even when the DNA input is so low that $y/R \approx D/K$ (see above). The most important conclusion from Kennell's analysis of these data is that if the RNA consists of a mixture of species with a roughly continuous distribution of number per species, the amount of DNA required to give a maximum yield of hybrids is independent of R/D. That this is the case is illustrated for T4 RNA as shown in Fig. 12.2. This figure is a good illustration of the appearance of these curves and the log/log method of plotting them which is most convenient (in the light of the extreme range of values adopted by the variables).

Alternatively the equilibrium may be studied by following the release of RNA from the complex by a melting (Kennell and Kotoulas, 1968—see also p. 361).

The magnitude of the equilibrium constant Several estimates of the magnitude of K have been made. The values are compared by Kennell (1971). As an example of the methodology, Kennell and Kotoulas (1968) hybridized RNA from *E. coli* induced for the *lac* operon (p. 20) with DNA from λ_{dlac} (p. 367) at a variety of R/D inputs. The proportion of the phage genome that is the *lac* operon is known to be 11·3%.[2] Thus the input DNA was expressed in terms of operon-DNA. From the definition of K (p. 359), it follows that when y is half-maximum, $R - y = y$ and $D - y = K$. From one set of their data, when $R/D = 2$, for $\frac{1}{2}$-maximum-y, $D = 2\mu g$ equivalent to $2 \times 0·113 \times 0·5\,\mu g$ of single-strand *lac*-operon DNA. Since the mRNA is 3% of the RNA mass and contains 50% of the pulse-label,[3] the *lac*-mRNA input was calculable (0·0072 μg in the reaction volume of 0·05 ml; y is therefore 0·0036 μg). Thus $D - y$ was 0·1094 μg/0·5 ml. To convert this to the conventional units of K, we must convert to moles/l (MW of *lac* DNA = $1·8 \times 10^6$) and obtain $1·2 \times 10^{-10}$. This may be taken as a realistic general value for the RNA/DNA equilibrium constant at 66°C.

From this value it is possible to calculate the free-energy change associated with hybrid-dissociation:

$$\Delta G° = -RT \ln K$$
$$= 2 \times 339°K \ln (1·2 \times 10^{-10}) = +15\ 600\ \text{cal}^4$$

Thus 16 kcal/mole is driving the hybridization reaction. As explained on p. 347 this value is composed (presumably) of at least two terms; one is the free-energy of the nucleation event, the other is the stacking interaction (see Kennell, 1971 and p. 330). If indeed these assumptions are valid, the value can be factorized thus:

$$\Delta G° = -RT \ln \sigma - RT \ln s^n$$

where σ is the equilibrium constant for the nucleation event. This is presumed to be of the order of 10^{-4} (p. 347). The other term describes the stacking equilibrium constant(s) for n base pairs. Then:

$$\Delta G° = -RT \ln (\sigma s^n)$$

We can estimate ΔH (about -7 kcal/mole—see p. 333). The equilibrium constant is related ΔH by the differential equation:

$$\frac{d(\ln s)}{dT} = \frac{\Delta H}{RT^2}$$

Now at T_m, $s = 0$. Kennell (1971) has measured the T_m for the *lac* mRNA/*lac* DNA hybrid to the 90°C. If the differential equation is integrated between the limits of the normal hybridization temperature (66°C; 339°K) and T_m (363°K) it transpires that at 66°C, $s = 1·95$ and $n = 49$. This last figure is a really striking one. Remember n is the number of nucleotides in the hybrid duplex. Now *lac* mRNA should contain molecules up to 5500 nucleotides long. This is admittedly an upper level; much (or all) *lac* mRNA as isolated undoubtedly consists of fragments as a consequence of the coupled transcription–translation and nucleolysis (p. 000). However it means that in the duplex (at 66°C) only about 2% of the nucleotides are actually duplex and the 'hybrid' is in fact in a dynamic state with only a small fraction of the RNA actually H-bonded to the DNA at any instant.

Technical methods

The methods of identifying hybrids are extensions of the fractionation procedures for nucleic acids as applied to a very special problem (the separation of duplex from single-strand structures). In particular there is a need for analytical studies to devise a method for the rapid analysis of large numbers of samples. The simplest and most quantitative procedure is to make use of the fact that duplices bind to cellulose nitrate. The methods involve either performing the hybridization in solution and subsequent 'filtration' of the hybrids through cellulose nitrate or alternatively denatured DNA may be immobilized in the filter and incubated with RNA. In this case, the nonhybridized RNA is degraded with ribonuclease and the fragments are washed off. Details of this and reference to other technical methods follow.

Cellulose-nitrate procedures   As was implied by the foregoing these procedures are suitable for RNA/DNA hybridization. The procedure was devised by Gillespie and Spiegelman (1965) and is usually employed in their original form with little modification. In these discussions, SSC is the name of a buffer 0·15 M NaCl, 0·015 M trisodium citrate; 2 x SSC is twice as strong (0·3 M and 0·03 M) and so on.

DNA in solution is denatured by either heating to 100°C and rapidly cooling (by plunging the tube into freezing mixture) to 0°C or by adding NaOH (pH 12) followed by HCl (to pH 7). Salts are added to make the solution 2 x SSC. The solution is passed through a cellulose-nitrate disc (radius 25 mm; pore size 0·45 μ).[5] Such a disc is conveniently loaded with between 2 and 100 μg DNA. The discs are washed with 10 ml 2 x SSC, dried (room temperature overnight followed by 4 h at 80°C *in vacuo* over solid KOH) and are stored at 4°C (*in vacuo* over solid KOH).

The discs are incubated in scintillation vials with 0·5 ml 2 x SSC containing the RNA at 66°C for 20 h. The discs are then washed under standard conditions (10 ml 2 x SSC on each side under suction) and are incubated with 25 μg ribonuclease in 1 ml 2 x SSC at 37°C for 1 h. After these incubations, the discs are washed with 2 x SSC (60 ml on each side as before).

Fry and Artman (1969) have modified this procedure to overcome the degradation of RNA that accompanies these prolonged incubations at high temperatures. DNA (50 μg) is loaded on to a disc as above in 0·33 M KCl, 10 mM tris HCl, pH 7·4 (TK). They used rRNA (0·25 mg) and employed an incubation of 3 h at 33°C in 50% (v/v) formamide in TK. After incubation the disc was washed (100 ml of TK on each side) and the RNA was eluted with 3 ml formamide (25°C for 30 min). Carrier RNA (0·2 mg), NaCl (1/10 vol. of 10% aqueous solution) and ice-cold ethanol (2 volumes) were added and the precipitate collected. The RNA was shown (by sucrose-gradient centrifugation) to be essentially undegraded. The method would appear to be of considerable value for the isolation of RNA from hybrids.

Spadari, Sgaramella, Mazza and Falaschi (1971) have modified the procedure to isolate, not intact RNA but the duplex RNA:DNA hybrid. In this case, rRNA:DNA hybrids were digested with ribonuclease and deoxyribonuclease (a single-strand-specific enzyme—see p. 209). In this case the hybrid was filtered off on a cellulose-nitrate column.

Several references have been made to the melting profiles of RNA:DNA hybrids (p. 360). These profiles may be determined using cellulose-nitrate discs containing the hybrid. The discs are incubated for 15 min in SSC at those temperatures required for points in the plot. After this time, they are cooled, treated with ribonuclease, washed and counted. The melting profile is obtained as a loss of bound RNA-radioactivity. The conditions have been optimized (to minimize loss of DNA from the discs) by Kennell and Kotoulas (1968). An example discussed on p. 360, the melting of the *lac* mRNA:*lac* DNA hybrid is illustrated in Fig. 12.3.

Hydroxyapatite   Of the various large-scale chromatographic procedures for the fractionation of double-strand and single-strand nucleic acids, this is the most versatile. The general aspects of hydroxyapatite chromatography are discussed on p. 166. The use of the technique is exemplified below. The sort of experiment that can be performed is to anneal DNA at (say)

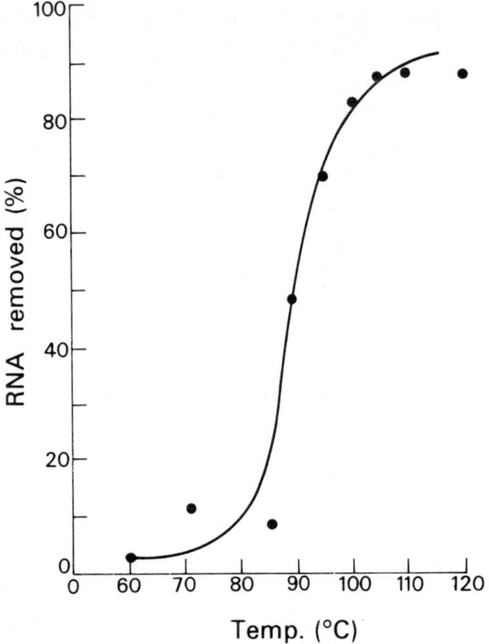

FIG 12.3   Melting profile of *lac* mRNA hybridized with λ_{dlac}-DNA. The profile was determined on cellulose-nitrate discs (see text). Data of Kennell and Kotoulas (1968).

60°C. The mixture is then passed through a water-jacketed column at 60°C. The double-stranded DNA is retained on the column. The column is then eluted at a variety of (elevated) temperatures and DNA is obtained in fractions on the basis of their different T_m-values.

DNA:DNA hybridization

Analytical experiments

The procedures of DNA:DNA hybridization are useful for detecting homologies between DNA sequences. In particular repetitious DNA sequences may be detected. I have chosen two examples from very different biological problems.

Terminal repetition in double-stranded DNA phage   Many (if not all) DNA phage have repetitious sequences at their termini. Let us first of all define the three types of repetition that occur and then consider how a new phage may be allocated to one of the groups. In the following examples, a, b, c, d, etc., are points along the phage chromosome (they could be genetic markers).

The simplest case is the unique, terminally redundant, continuous-duplex molecule (examples are coliphage T3, T5 and T7). The DNA from such a phage is as follows:

$5'$ a   b   c   d   e   f   g   h   a   b $3'$
<hr>
$3'$ a   b   c   d   e   f   g   h   a   b $5'$

The next case is a special case of the unique, terminally redundant molecules that are not continuous duplices at the end. These are the 'sticky-ended phage' of which λ (and related specialized transducing phage) is the classical example. The DNA in this case is of the form:

$$5'\ a\quad b\quad c\quad d\quad e\quad f\quad g\quad h\ 3'$$
$$3'\quad c\quad d\quad e\quad f\quad g\quad h\quad a\quad b\ 5'$$

Finally there are the circularly permuted, terminally redundant (continuous-duplex) molecules. Examples are T2, T4 and P22.[7] In this case the phage DNA (as isolated) consists of a mixture of molecules such as those shown below. In other words, not all T4 phage are strictly identical. The primary structures of their DNA molecules differ in this special sense.[8]

$$5'\ a\quad b\quad c\quad d\quad e\quad f\quad g\quad h\quad a\quad b\ 3'$$
$$3'\ a\quad b\quad c\quad d\quad e\quad f\quad g\quad h\quad a\quad b\ 5'$$

$$5'\ b\quad c\quad d\quad e\quad f\quad g\quad h\quad a\quad b\quad c\ 3'$$
$$3'\ b\quad c\quad d\quad e\quad f\quad g\quad h\quad a\quad b\quad c\ 5'$$

$$5'\ c\quad d\quad e\quad f\quad g\quad h\quad a\quad b\quad c\quad d\ 3'$$
$$3'\ c\quad d\quad e\quad f\quad g\quad h\quad a\quad b\quad c\quad d\ 5'$$

The evidence for sticky-endedness comes from the fact that the molecules can form continuous ('circular') duplices in solution. Methods of identifying these structures are discussed in chapter 15; a method particularly suitable for recognizing such molecules in DNA from λ phage is described on p. 425. Accepting this, the other types of terminally redundant phage can be converted into sticky-ended molecules by limited digestion with an exonuclease (see p. 280). Thus when these phage are digested with exonuclease and annealed, rings are formed (Ritchie, Thomas, MacHattie and Wensink, 1967). In the special case of the circularly permuted molecules, a special type of sticky-endedness is created by simply denaturing and re-annealing the molecules. Partial hybrids occur which will form rings in solution (MacHattie, Ritchie, Thomas and Richardson, 1967). This type of ring formation is illustrated schematically in Fig. 12.4.

Repetitive sequences in eukaryotic DNA   The procedure to be introduced here is due to Britten and Kohne (1968) and is commonly referred to as the 'Cot method'. The method consists of following the renaturation kinetics of denatured DNA at constant temperature. Mammalian (or other eukaryotic) DNA is fragmented before being employed in this assay. It is broken by forcing through a needle valve (in a French pressure cell or an equivalent device) at about 50 000 psi. The result is that essentially denatured pieces about 500 nucleotides long (i.e. MW-range around 2×10^5, measured by the method of p. 387) are obtained. Long DNA molecules cannot be used for this experiment as extensive partial mis-matching confuses the results.

We have briefly considered the renaturation kinetics (p. 357) and deduced that:

$$\frac{x}{m} = \frac{mk_1 t}{1 + mk_1 t}$$

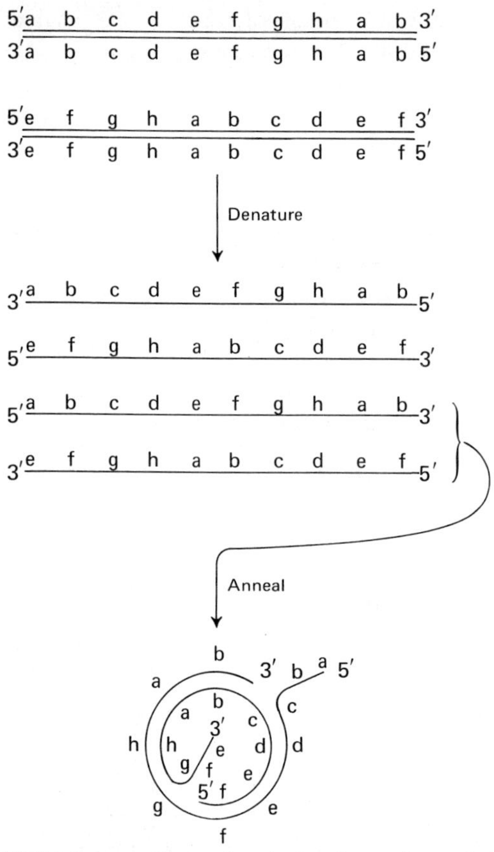

FIG 12.4   Example of the formation of rings by the denaturation and annealing of DNA from a circularly permuted, terminally redundant phage. Just two of the permutations are chosen as an example. The conventions are explained in the text.

where x is the concentration of DNA-nucleotides in the duplex. For the present discussion, it is more convenient to express the results in terms of the concentration of nucleotides in single-strand DNA. As $x = m - y$; the equation can be re-written:

$$\frac{x}{m} = \frac{1}{1 + mk_1 t}$$

The nomenclature of Britten and Kohne calls y, C and m, C_0; $mk_1 t$ thus becomes $kC_0 t$. Britten and Kohne proposed a method of following the renaturation by using standard conditions (60°C in 0·12 M phosphate buffer pH 7·4) and plotting y/m as a function of time, to be precise mt (or $C_0 t$). As the mt-axis covers an enormous range (as we shall see in a moment) it is convenient to use a logarithmic scale. The results of such a plot for a variety of nucleic acids are shown in Fig. 12.5. Note that the values of mt for $y/m = 50\%$ $[(mt)_{1/2}]$ [9] are a function of molecular weight. The longer the molecule, the more abortive collisions occur before complementary molecules find one another and form a duplex, and the larger the value of $(mt)_{1/2}$.

In addition to this procedure for measuring molecular weight, Britten and Kohne (1968) extended their studies to examine the redundancy in eukaryotic DNA. These experiments are extremely important as they are the only method of studying this redundancy in general (as opposed to the special cases of the ribosomal cistrons and so on—see p. 368) and also an

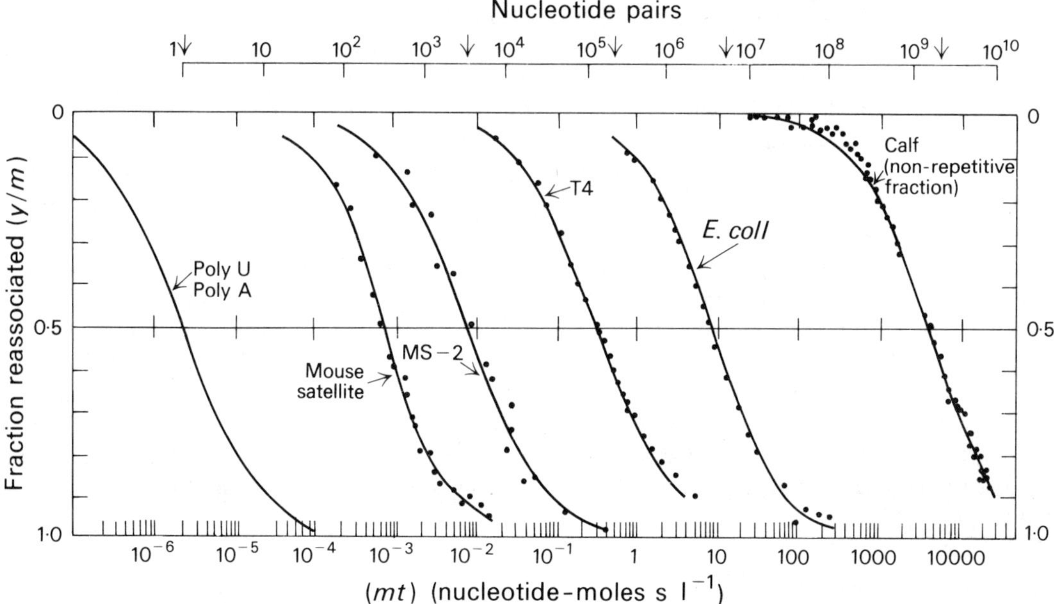

FIG 12.5   Plot of y/m against mt for a variety of nucleic acids. The values for MS2 RF were obtained by following the increased susceptibility of the RNA to ribonuclease. The others were obtained by monitoring hyperchromicity as the DNA renatured (p. 240). The standard conditions of the renaturation were 0·12 M phosphate buffer pH 7·4 at 60°C. Data of Britten and Kohne (1968).

understanding of this redundancy is essential for any real studies on hybridization between mixed RNA populations ('mRNA') and eukaryotic DNA. Several methods of studying redundancy are available. One is to examine curves such as those shown in Fig. 12.5 and analyse them for evidence of several points of inflexion. A more powerful method is to renature for a time and at an initial DNA concentration so that mt reaches an intermediate value (see p. 352 for a specific example) and to pass the DNA through a column of hydroxyapatite (see p. 166). The non-repetitive DNA is eluted. The repetitive DNA is then eluted by raising the temperature. The advantage is that not only is the proportion of repetitive DNA more accurately measured, but there is a fractionation of DNA fragments on the basis of the repetitiveness of the sequences from which they were derived. Such repetition is extensive. In mouse DNA, about 10% of the genome is extremely repetitive (about 10^6 copies per gene); about 20% is 'fairly' repetitive (10^3–10^5 copies per gene) and 70% is non-repetitive (within experimental error only one copy or so per gene).

Preparative applications

There has been one extremely ingenious isolation of a specific gene by a hybridization method. The principle was to obtain two molecules which were of largely complementary sequence with the exception of one section which was of reversed polarity in one molecule. The two equivalent strands of denatured DNA from the two sources were isolated and re-annealed. Only the piece of opposite polarity in the two strands could form a duplex and this duplex region could be isolated by chewing off the single-strand pieces with a single-strand-specific nuclease. The problem is to obtain two molecules in which one gene is of reversed polarity and in which the two strands may be separated. However it has been done once and presumably it could be extended to other genes if the necessary genetic manipulation could be satisfactorily effected.

The isolation of the *lac* operon   The essential feature of this experiment is the isolation of transposition mutants of *E. coli* in which the *lac* operon is present in opposite senses. The method (Beckwith, Singer and Epstein, 1966) involves obtaining from an Hfr *lac*$^+$ strain an episome containing the F-factor and the *lac* operon. This episome is maintained in a *lac*$^-$ strain. Following mutagenesis, a temperature-sensitive (TS) episome is obtained. The properties of a strain containing $F_{TS}lac$ are that the lower (permissive) temperature the episome is replicated and daughter cells contain episomes; at higher (restrictive) temperatures the episome is not replicated and the episomes (and their *lac* operons) are lost in subsequent gene rations. Integration of the episome into the chromosome can be demonstrated by scoring *lac*$^+$ colonies after bacteria (grown at the restrictive temperature) are plated out. If the chromosomal *lac*$^-$ genotype is due to a point mutation, the frequency of incorporation is quite high (1 in 10^2) due to homology between the *lac* operon in the episome and the defective *lac* locus in the chromosome. If on the other hand the *lac*$^-$ genotype is due to a deletion of the operon in the

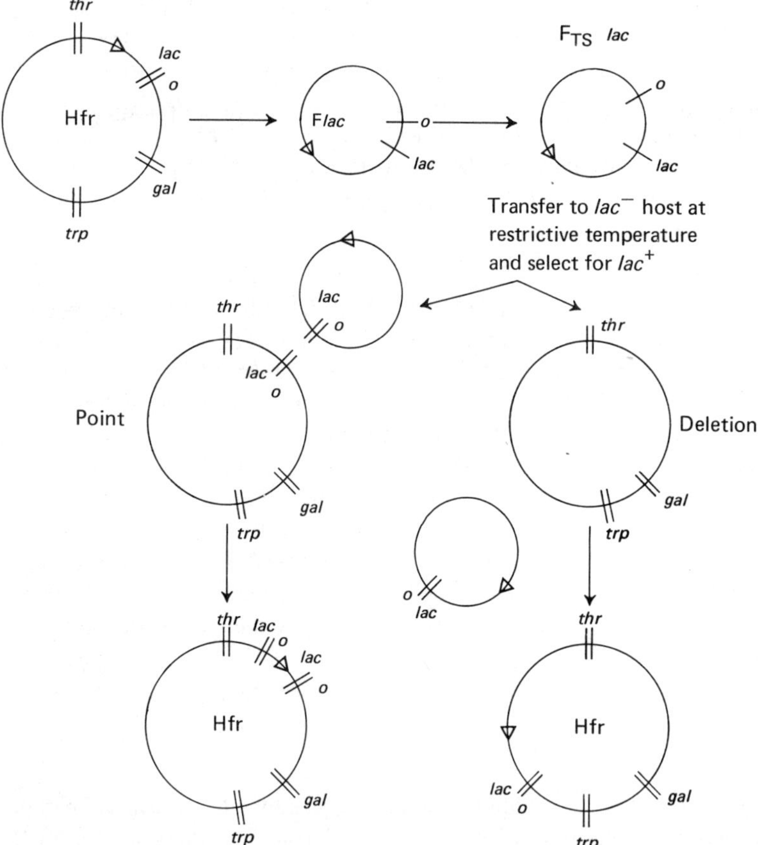

FIG 12.6   Integration of the F_{TS} *lac* episome into the chromosome of a *lac*$^-$ organism. The isolation of F_{TS} *lac* is illustrated at the top. On the left is an example of the integration into a chromosome carrying a point mutation in the *lac* locus. On the right is the integration into a chromosome with *lac* deleted. See text for additional discussion.

Genetic markers: *thr*, threonine; *lac*, β-galactosidase, permease and transacetylase; *o*, the *lac* operator; trp, tryptophane; small triangle, F.

chromosome, the rarer incorporations (1 in 10^4) is produced by (accidental) homology between a region of the episome with a position on the chromosome. The two types of incorporation are illustrated schematically in Fig. 12.6.

The organism in the bottom right-hand corner of Fig. 12.6 is a 'transposition mutant'; the *lac* operon has been transposed to a position in the chromosome remote from its normal locus. As this position is determined by a chance homology, there is no reason why the operon should be of the same polarity as it is in 'normal' *E. coli*. In the figure, I have shown a case where the polarity is in fact reversed.[10]

For the work to be described, two such mutants were isolated; in both cases the operon was inserted in the region of the attachment site of either phage λ or the related temperate phage φ80.

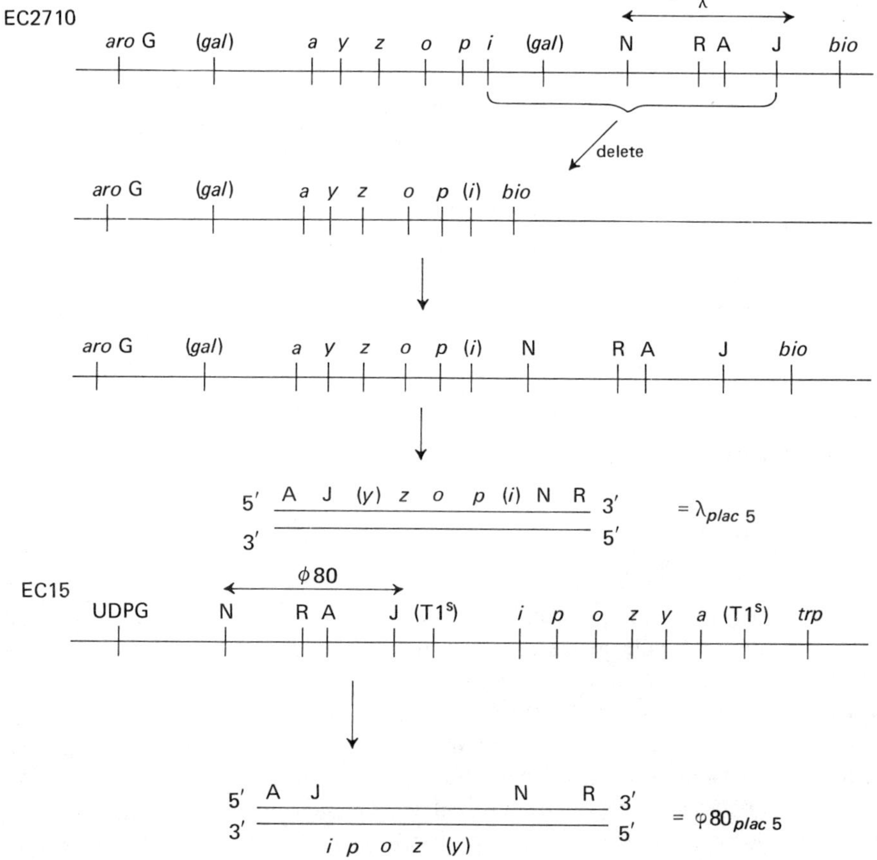

FIG 12.7 The *lac* operon saga continued (see text for discussion). The *lac* operon consists of i-gene (*i*), operator (*o*), promoter (*p*) and three structural genes (*z, y, a*). N, R, A and J are markers in λ (and φ80). The order of these markers is different in the integrated and non-integrated forms of the phage genome (see p. 374). Markers in parentheses have been split by episomal insertion of some kind or are partly deleted. At the top is shown the *gal* region of EC2710 (*gal*, capacity to metabolize galactose; *bio*, biotin requirement). Deletion and re-infection and lysosegenization as shown produces $\lambda_{plac\,5}$. The phage is shown with both strands of the DNA. The priming is included just to indicate the different complementarities of the markers.

Lower down is the *trp* region of EC15. T1^s is the phage-T1-sensitivity marker. Excision of φ80 produced spontaneously the necessary partial deletion of the operon to yield $\varphi 80_{plac\,5}$.

These two lysogenic strains (EC2710 and EC15) and the isolation of transducing particles from them are shown diagrammatically in Fig. 12.7.

With the isolation of the two *plac*5 phage, the bulk of the project was over. These phage genomes are related to one another in the way described in general terms on p. 366. Moreover the two strands of λ DNA can be separated in CsCl in the presence of poly GC (see p. 153). The density of the strands is dictated by the phage markers. In other words, with reference to Fig. 12.7, let us say that the two primed strands are the 'heavy' ones. Shapiro *et al.* (1969) isolated these two heavy strands and annealed them. The only complementary region is a portion[11] of the operon. This can be readily seen by writing down the complementary heavy (primed) strands again:

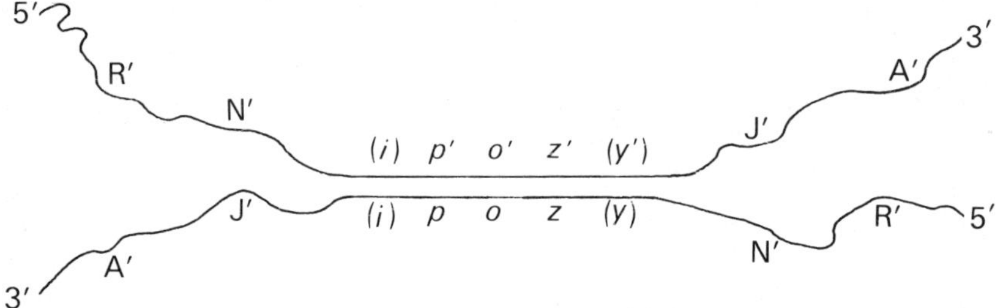

This annealed product was trimmed down to produce pure duplex DNA of the operon by incubation with *Neurospora* nuclease (p. 208). The identity of molecules of the type illustrated above and the pure duplex operons was confirmed by electron microscopy (p. 188). The identity of the operon itself was established by hybridization with *lac* operon mRNA (p. 360).

RNA:DNA hybridization

Ribosomal and transfer RNA

Ribosomal cistrons The hybridization of rRNA and DNA has furnished the most precise information we have regarding gene multiplicity. The methods of RNA:DNA hybridization described on p. 361 are applicable to the problem, provided precautions are taken to ensure that competition involves use of pure rRNA and that for concentration-dependence studies, extrapolations to the RNA plateau are not made by a subjective evaluation of the asymptotic value. This last error is easily avoided by plotting the reciprocal of hybridization as a function of the reciprocal of RNA concentration. The extrapolation is then to the zero-value of the ordinate axis. The techniques are reviewed by Birnstiel, Chipchase and Spiers (1971).

The number of ribosomal cistrons per genome has been measured for a wide variety of organisms. Only one cistron per genome is present in *Mycoplasma*[12] (Ryan and Morowitz, 1969). The other cases of a single rRNA cistron per genophore[13] relate to rRNA from mitochondria and chloroplasts. These molecules, which are of different nucleotide composition and molecular weight to their cytoplasmic counterparts (pp. 284 and 286) are complementary to the organellar DNA. The hybridization data are reported by Dawid (1970) and Tewari and Wildman (1970) and examples are illustrated in Fig. 12.8.

With the exception of these simple systems, it is generally the case that there are several ribosomal cistrons per genome. In addition to the question of how many cistrons there are, two other questions have been examined: are the two structural genes (for 18 S and 28 S; 16 S and 23 S, etc.) next to one another? and are they bunched in one region of the chromosome or are

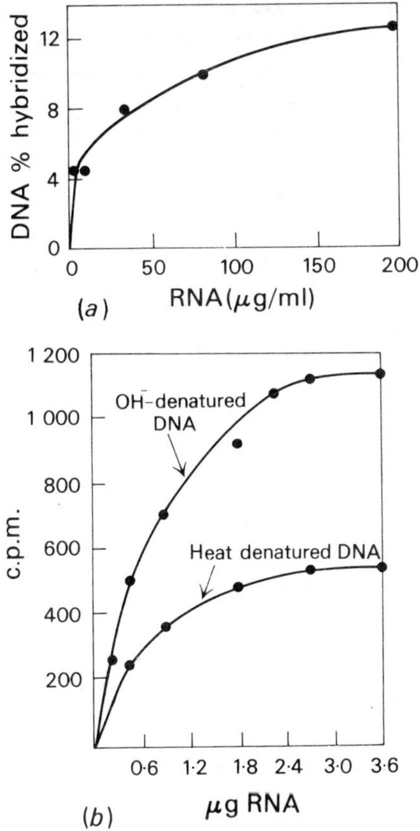

FIG 12.8   Hybridization of organellar DNA with complementary rRNA. (*a*) Hybridization of 21 S + 13 S mitochondrial RNA from *Xenopus* with mitochondrial DNA. Hybridization was in 4 x SSC at 65°C for 12 h with 4 μg DNA on the filter. (*b*) Hybridization of RNA from tobacco chloroplast ribosomes with chloroplast DNA. Hybridization in 2 x SSC at 60° for 18 h with 10 μg DNA on the filter. Data of (*a*) Dawid (1970) and (*b*) Tewari and Wildman (1970).

they scattered? The answer to the first question is that as far as we know these two genes are invariably adjacent.

With eukaryotes, there is transcriptional evidence for it as one of the rRNA precursors contains sequences for both of the molecules (p. 182). With bacteria it is also true; the evidence emerges during the discussion of the other point (whether or not the cistrons are bunched).

In certain animals there is a cytogenetic method of studying rRNA synthesis. It is known that the presence of nucleoli requires a genetic characteristic, the nucleolar organizer (*no*). The evidence that the nucleolus is the site of rRNA synthesis derives from the work of Ritossa and Spiegelman (1965). They showed that by measuring rRNA:DNA hybridization in strains of *Drosophila melanogaster* containing *no* mutations that if the DNA was isolated from a class of mutants called 'bobbed' there were fewer cistrons. The work was extended to *Xenopus* (Birnstiel, Wallace, Sirlin and Fischberg, 1966). This amphibian normally has two nucleoli per nucleus (phenotype 2*n*). Strains with only one (1*n*) are known. These animals are heterozygous with respect to the nucleolar organizer. The genotype is no^+/no^-. The no^- genotype is a classical Mendelian recessive-lethal characteristic. Thus the progeny of a cross (no^+/no^- x no^+/no^-) will consist of 1*n*, 2*n* and 0*n* phenotypes (in the ratio 2:1:1). The 0*n* phenotype is lethal as the animal cannot make any ribosomes. However sufficient ribosomes are 'inherited' in the oöcyte that it can develop as far as the stage of a small tadpole. Birnstiel and others (1966) found that 28 S + 18 S RNA was complementary to between 500 and 800 sites in the $2n$ genome and to no dectable sites in the $0n$ genome. Further it was established that the DNA responsible was the high-GC 'satellite' which is absent in the $0n$ phenotype (p. 151).

This number of cistrons is vastly amplified in the oöcyte which contains 3000 ribosomal cistrons per somatic cell (Evans and Birnstiel, 1968).

Such amplification is not restricted to amphibians. As an example from insect biochemistry, the ovaries of *Acheta domesticus*[14] contain a large 'DNA body' in the nucleus of the oöcytes. This body is larger in oöcytes in interphase before meiosis than in the resting (pachytene-diplotene) phase and is completely absent in the testes of males. The DNA body consists of heterochromatin (p. 279). Lima-de-Faria, Birnstiel and Jaworska (1969) compared the profiles of DNA produced by analytical CsCl centrifugation of the DNA isolated from these various tissues. A significant zone of DNA with a buoyant density corresponding to that expected for ribosomal cistrons (56% GC in this organism) was identified in the oöcyte material. The hybridization of DNA from this region of the chromosome showed that DNA which would hybridize with rRNA was in fact found in this region (Fig. 12.9).

The remaining question is whether the cistrons are scattered in the chromosome or are grouped. The evidence from the animals which have a ribosomal DNA satellite indicates some sort of specific grouping. A general method for studying the grouping of the cistrons which do not naturally form such a satellite, is to band DNA in CsCl and to establish (by hybridization of individual fractions with rRNA) the density of the fractions that contain the cistrons. The experiment is performed using DNA variously fragmented by shearing (p. 363).[15] If the DNA is of very high molecular weight, the rRNA may form a hybrid with a peak corresponding to the

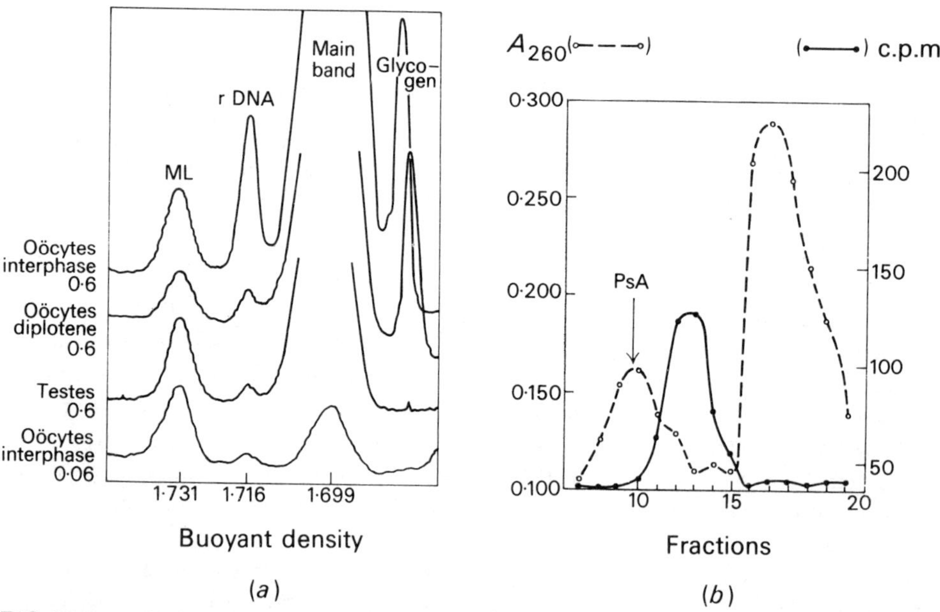

FIG 12.9   DNA from tissues of *A. domesticus*. On the left: traces obtained by scanning analytical CsCl profiles of total DNA. The numbers in brackets refer to the total A_{260} units in the gradient. On the right: Preparative CsCl gradient of oöcyte interphase DNA; the fractions were collected, the absorbance measured and challenged with [32P]rRNA. ML (*Micrococcus lysodeikticus*) and PsA (*Pseudomonas aeruginosa*) are DNA standards employed for calibrating the gradients (p. 267). Data of Lima-de-Faria, Birnstiel and Jaworska (1969).

absorbance maximum, implying that the ribosomal cistrons are in the middle of a fragment consisting of mainly 'typical' DNA. When the fragments are about the size of the cistrons, of course, the hybridization zone will be coincident with that of 'ribosomal DNA'. If there are blocks of cistrons, the shift to this density will occur with much larger fragments. The results of such an analysis for *Proteus mirabilis* are summarized in Fig. 12.10 (Purdom, Bishop and Birnstiel, 1970).

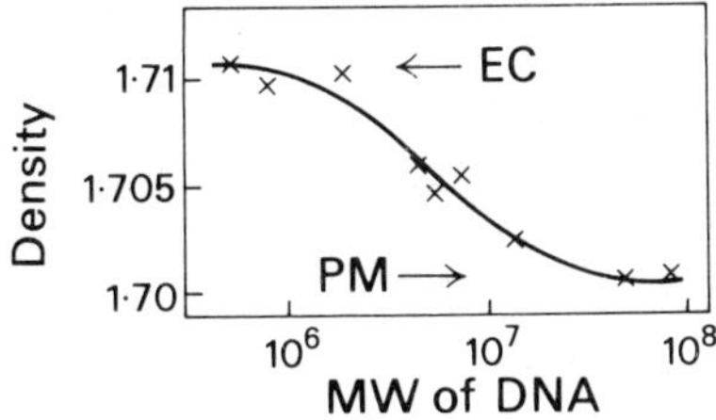

FIG 12.10   Banding position of DNA complementary to *P. mirabilis* rRNA as a function of DNA MW. The arrows indicatd the densities of (PM) *P. mirabilis* DNA and (EC) *E. coli* DNA (which is essentially of the same nucleotide composition as *Proteus* rRNA). Data of Purdom, Bishop and Birnstiel (1970).

From an analysis of results such as these, Birnstiel, Chipchase and Speirs (1971) have compared the ribosomal cistrons of *X. laevis* with those of *P. mirabilis*. Schematically, they are shown in Fig. 12.11. The figures for the amphibian should be compared with the data for eukaryotic rRNA precursors (Fig. 5.4, p. 182). Note that the cistrons are significantly more scattered in the bacterium but that there is no reason (from studies of 16 S and 23 S rRNA separately hybridized to fragments of DNA) to believe that the 16 S and 23 S cistrons are more separated in the bacterium than are the 18 S and 28 S cistrons in the animal even though the transcription of the bacterial cistrons does not yield a giant precursor containing both rRNA sequences (p. 184).

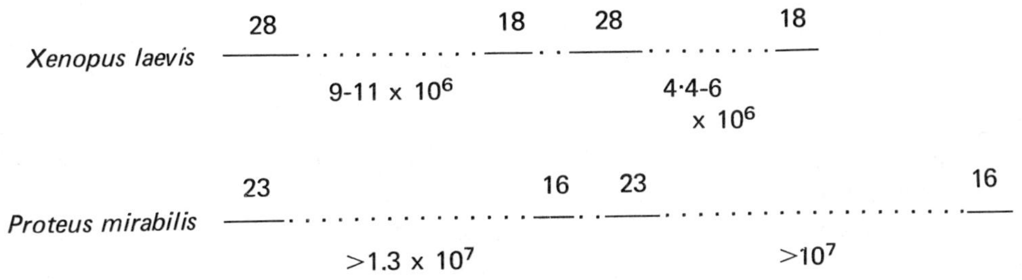

28   18   28   18

Xenopus laevis

9-11 × 10^6 4·4-6 × 10^6

23   16 23   16

Proteus mirabilis

>1.3 × 10^7 >10^7

FIG 12.11   Comparison of the intercistronic gaps between rRNA cistrons in a bacterium and an animal (from Birnstiel, Chipchase and Speirs, 1971).

Transfer RNA   For reasons discussed by Kennell (1971), accurate quantitative studies on tRNA:DNA hybridization present special difficulties. It is possible however (using competition experiments) to unambiguously identify tRNA complementarity with DNA.

Evidence that some chloroplast tRNA is complementary to chloroplast DNA is discussed by Tewari and Wildman (1971). Hybridization was also used by von Heyden and Zachau (1971) to establish that tRNA occurs in thymus chromatin. This result is important as it is certain that this tRNA has been confused in the past with certain 'low-molecular-weight chromatin RNA' molecules to which a variety of hypotheticl functions have been ascribed. It is not my purpose to stir up trouble by making an exact comparison with the earlier work; suffice it to say that von Heyden and Zachau's paper should be essential reading for anyone embarking upon a study of 'chromatin RNA'.

Isolation of tRNA and rRNA cistrons   The principle behind the use of RNA:DNA hybridization for the isolation of cistrons is that the DNA is sheared (p. 363) and hybridized with the RNA under conditions (choice of temperature and mt—p. 357) in which there is only of the order of 10% DNA renaturation. The mixture is then separated into s ingle-strand and double-strand nucleic acids (typically with hydroxyapatite) and the double-strand fraction is repeatedly re-cycled through the denaturation and renaturation until there is no significant double-strand DNA left. Using this method several partial purifications of genes have been achieved; Brenner, Fournier and Doctor (1970) have obtained the *E. coli* tRNA cistrons and used them to demonstrate considerable homology between *E. coli* tRNA (unfractionated) and that of several other Enterobacteria. As an example of the method applied to rRNA cistrons, Udvardi and Venetianer (1971) have succeeded in partially purifying the rRNA cistrons of *Salmonella typhimurium*.

Characterization of RNA synthesized *in vitro*

The *lac* operon again   In a different context, I have referred to evidence for the existence of the CAP factor in *lac* transcription (p. 226). The same experiment can be performed using as the assay system not β-galactosidase but rather hybridization of the newly formed RNA with *lac* DNA in a transducing phage (p. 367). The system is a much cleaner one, as in the former case, ribosomes and *E. coli* supernatant are required to effect the coupled translation of the mRNA. Using the hybridization assay system, de Crombrugghe *et al.* (1971) have shown that the only factors required for the controlled transcription of *lac* are *lac* DNA, its repressor and inducer, RNA polymerase (including σ and its four NTP substrates), cyclic AMP and CAP. This paper describes the most complete dissection of the transcriptional system that is possible.

The transcription of rRNA   The successful synthesis *in vitro* of rRNA has been a surprisingly delayed experiment. The answer came when Travers, Kamen and Cashal (1970) examined the capacity of fractions of a post-ribosomal supernatant of *E. coli* to stimulate the overall incorporation of NTP into RNA by RNA polymerase and σ. They were able to demonstrate, by competition-hybridization studies, that their product included rRNA. Their deduction that the synthesis of rRNA requires a specific factor (called Ψ_r) was possibly invalid.

Transcription of chromatin   This is an example of a process that requires great caution in interpretation. The question that is being asked is, 'If chromatin from different tissues is transcribed *in vitro*, are tissue-specific RNA molecules produced as a consequence of different proteins in the chromatin?' The method of performing the assay is to hybridize RNA transcribed on (say) liver chromatin as template and see whether there is RNA whose hybridization to DNA is competed for by liver RNA but not by RNA from (for example) kidney. The difficulty arises from the fact that the labelled RNA and the competing RNA consist of a mixture of an unknown number of different molecules and these components cover an unknown (but probably extremely wide) concentration range. With the proviso that errors in interpretation are possible (p. 376), there is nevertheless evidence that specificity of transcription does occur. Smith, Church and McCarthy (1968) used chromatin from three mouse tissues (liver, kidney and brain) as template for *E. coli* RNA polymerase. The maximum competition obtained with homologous RNA resulted in hybridization being reduced to about 10%. Less competition was obtained with heterologous RNA. As an example, hybridization of the RNA transcribed on the liver chromatin was reduced (maximally) to 30% by kidney RNA, to 45% with brain RNA and to 90% with *E. coli* RNA (control).

. . . and DNA   The process catalysed by the viral RNA-dependent DNA polymerases (p. 224) results in the synthesis of DNA complementary to an RNA template. The methods used for the identification of the product are described by Spiegelman *et al.* (1970).

Transcription of phage

The identification of classes of RNA synthesized during phage development is ideally suited to hybridization studies. The number of species of RNA is limited and phage DNA is easily purified and employed as selective template. Moreover, in many of the coliphage, mutants with genetically defined deletions are available for comparative studies. There is no room to discuss every type of phage in this way and I have just chosen examples of three very different phage.

Phage λ   As a result of a combined attack using genetic and hybridization studies, the transcription of phage λ is understood in enormous detail. The topic is the subject of a monumental review by Szybalski *et al.* (1970). Here I have limited my discussion to a brief reference to Szybalski's work. The review contains references to other workers in the field. Another very authoritative review of the physiology of λ infection and lysogeny is by Signer (1968).

In the discussion of Szybalski *et al.* (1970), a unit of transcription, the 'scripton' is defined. A scripton is a gene cluster under the grand control of a single autonomous promoter. Within the scripton there may be independent control systems but the expression of any genes in the scripton, requires (in addition to the fulfilment of any local requirements for transcription) initiation of this promoter. The operon (p. 19) is thus a special case of the simplest possible scripton in which the 'autonomous' promoter (or promoter/operater in the case of an operon in which these functions are not discrete) in the only control point for the scripton. A more complex scripton differs from an operon in that the transcription products consist of several mRNA molecules. Having said this, an outline genetic map of λ is shown in Fig. 12.12.

Notice that, not only are both DNA strands employed as transcribing strand in different parts of the genome, but that there is extensive overlap between L2 and R1/R2 (which cover the same ground for much of their length). We do not know how widespread this arrangement will turn out to be but it has interesting consequences in the mutual restriction of RNA (and hence protein) sequences in the gene products from the two overlapping scriptons. One apparent consequence of the arrangement is that it would lead toevolutionary conservation in such a region as it is unlikely that an 'allowed' mutation in one product will be allowed in the other. When the phage is integrated (as prophage) the only functioning scripton is L1.

As example of the sort of evidence that has been used to construct this genetic-and-transcriptional map is shown in Fig. 12.13. The data illustrate the principal technique, in which the RNA synthesized as different times during the early phases of lysogeny is challenged with DNA from a variety of phage which differ from the input phage in established regions.

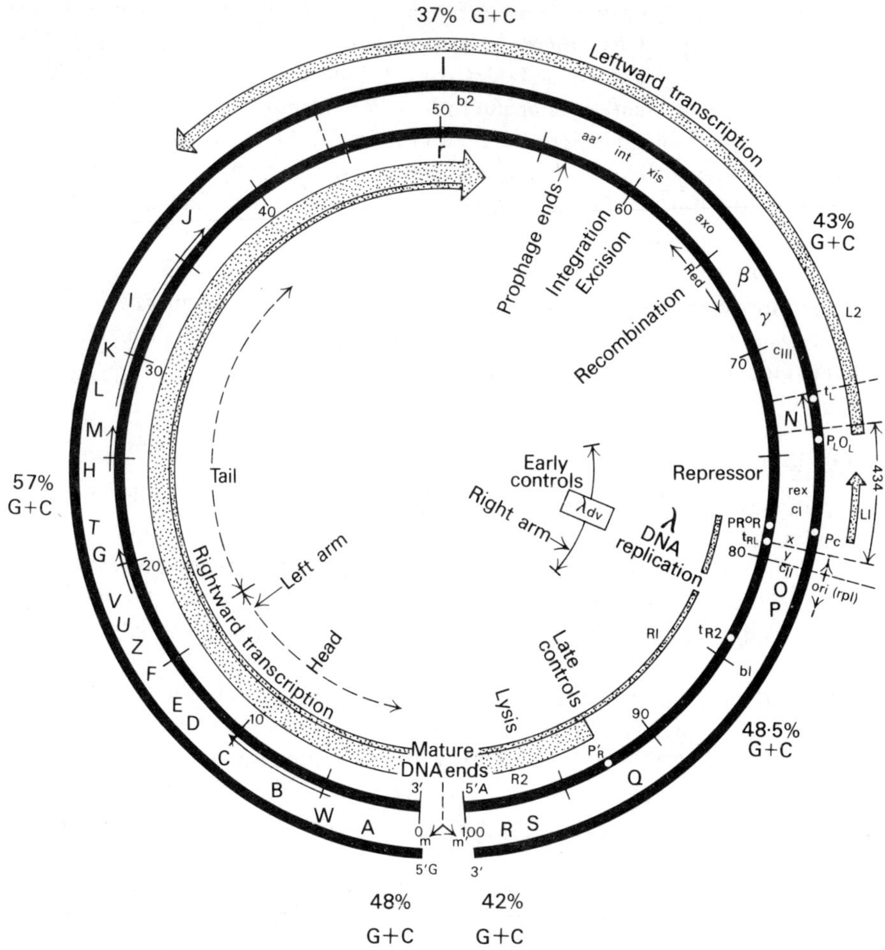

FIG 12.12   Genetic map of λ (after Szybalski *et al.*, (1970). The 'sticky-endedness' (p. 280) occurs between A and R. Complementarity between this genome and *E. coli* is in the egion aa'. The corresponding site (λ_att) in *E. coli* is denoted bb' and lies between bacterial markers *blu* and *bio*. Thus if you wish to envisage the above chromosome incorporated into the *E. coli* chromosome, the outline is as follows:

. . . . *blu* ba' . . . . N . . . . R . . A . . . . J . . b2 . . . ab' *bio* . . . . .

The four scriptons that have been identified so far (there could be another) are shown (L1, L2, R1 and R2) against their transcribing strands.

Phage T4   A virulent phage whose genetics are understood in great detail is T4. The timetable of gene expression in this phage have been examined by Jayaraman and Goldberg (1970). Their methodology is very ingenious and is applicable to any system in which a suitable transformation system is available.

The principle is illustratd with reference to a specific problem, the kinetics of the synthesis of rIIB mRNA. The r phenotype is rapid lysis; a convenient assay is available and the r-region has been mapped in great detail. *E. coli* was infected with T4rIIB and RNA was extracted at given times after infection. The RNA was hybridized with DNA (from the same phage) and was hybridized with single-strand (separated strands are used) DNA and the hybrids were digested with a single-strand specific nuclease.[16] The hybridized DNA segments are protected. These fragments were used to effect a transformation of a T4 strain defective in rIIB (T4Dr$_{73}$).[17] By scoring for transformants in a suitable indicator strain (*E. coli* K12) and total p.f.u. on a non-selective strain (*E. coli* B) it was possible to deduce which preparations contain protected rIIB DNA and hence which preparations of RNA contain rIIB mRNA. In this case it was found that the peak occurs 10 min after infection. Using suitable selective strains eight T4 genes were ordered in this way.

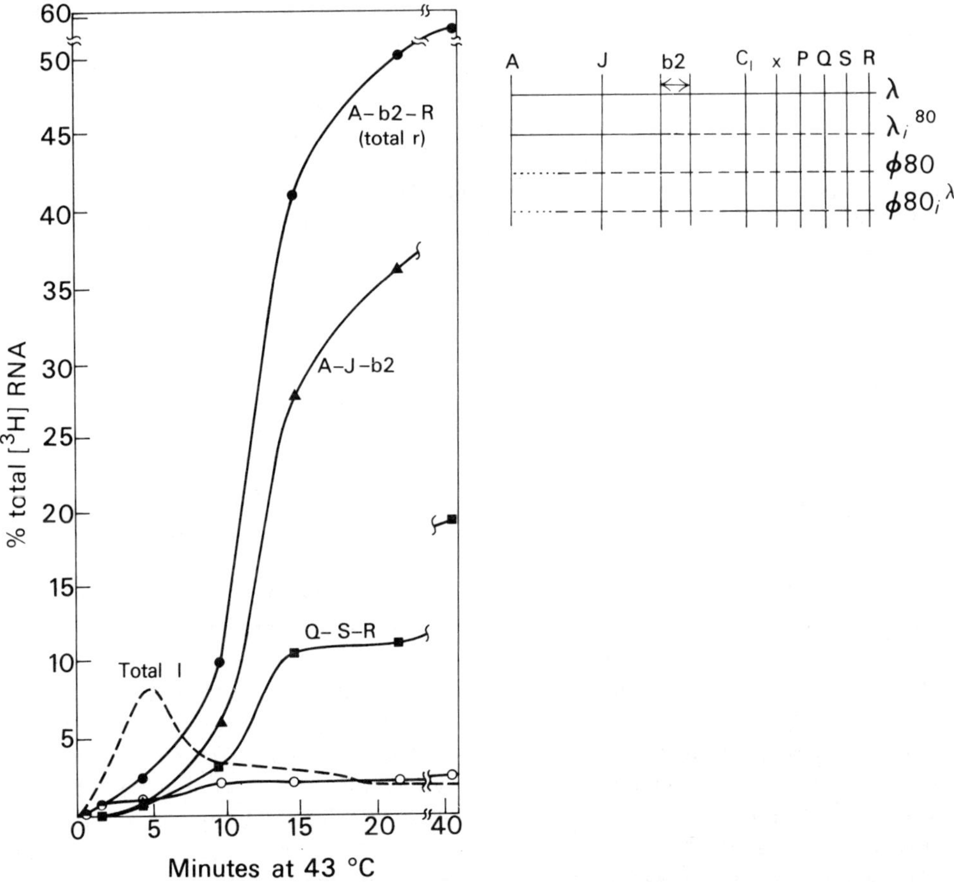

FIG 12.13   Kinetics of rightwards transcription of phage λ. [³H]RNA was isolated from cells infected with wild-type phage for the times indicated. The rather high temperature (43°C) was employed to induce lysogeny (p. 30). DNA was isolated from various strains of λ (and φ80) in which there were only certain specific regions of DNA showed homology with the wild-type DNA. The DNA was denatured and the strands were separated (p. 153). Thus 'total l' indicates the extent of hybridization with l strand from wild-type DNA; 'total r' is the hybridization with r strand from wild-type DNA. A-J-b2 is the extent of hybridization with r-strand from λ*i*⁸⁰ in which only the stretch indicated is homologous with the wild-type; similarly r-strand from φ80 and φ80*i*^λ was used. From the maps of these phage (inset) it can be seen that values for hybridization with λ minus those for λ*i*⁸⁰ give the x-P-Q values and λ minus λ*i*⁸⁰ minus φ80*i*^λ plus φ80 give the values for Q-S-R. In the inset the chromosomes are shown in their linear form with each duplex shown as a single line. Regions drawn in the same form (continuous lines, dots and dashes) are homologous. Data of Szybalski *et al.* (1970).

Phage SP01   SP01 is a phage for *Bacillus subtilis*. It is of particular interest in having one of the largest genomes of any phage studied so far (2×10^5 base pairs, equivalent to between 200 and 400 proteins). It has been extensively studied by E. P. Geiduschek and his colleagues. The timetable for the transcription of six classes of phage-directed RNA (from early to late) has been characterized and defective mutants have been isolated which require mutants of *B. subtilis* for successful infection. Grau, Ohlsson-Wilhelm and Geiduschek (1970) have explored the nature of phage–host cell relationship responsible for the control of transcription. Cell-free extracts (containing RNA polymerase activity) have been employed to synthesize RNA *in vitro*. This newly synthesized RNA has been assigned to classes in the developmental timetable of the phage competition–hybridization experiments. For example it is found that enzyme preparations from infected and uninfected cells transcribe (*in vitro*) exogenous DNA (T4, salmon sperm) with comparable efficiency. There is however considerable specificity in the transcription of SP01 DNA. Host-cell polymerase will not transcribe one of the classes of genes (m_1l–'medium (1) late') but that the enzyme from infected cells primarily transcribes this region *in vitro*. The data reported in this paper form the basis of what should prove to be an extremely valuable study of the nature of the modification to host-cell polymerase by this phage. It is already clear that the mechanism is qualitatively different to the early events during infection of *E. coli* with a phage such as T4 (p. 374).

Cellular RNA

How can RNA:DNA hybridization be used to evaluate the proportion of cellular RNA that is 'stable' (e.g. tRNA and rRNA) as opposed to 'unstable' (mRNA perhaps ?)? Can hybridization be used to identify regions of DNA in complex organisms that are being transcribed *in vivo*? These are probably the most difficult applications of hybridization that are commonly used. Methods which are most readily accessible to semi-quantitative analysis are exemplified below.

Total bacterial RNA   Kennell (1968) devised a method of determining what proportion of a bacterial chromosome was being transcribed *in vivo*. It involves thinking of the DNA as a surface with a lot of sites, some of which are complementary to the RNA in solution and others are not. Clearly there is an equilibrium between the two states of RNA (bound and unbound). The question that is being asked really is, 'How do we know when the RNA concentration is high enough to saturate all the available sites ?'. Kennell devised a procedure for titrating these RNA sites. He applied the Langmuir equation for the equilibrium for absorption to a surface. In the present context, if D is the DNA concentration, R the RNA concentration, S the concentration of bound RNA and a the fraction of DNA competent to receive the RNA (i.e. what was transcribed at the time the RNA was labelled), then the equilibrium constant for the reaction (K) is defined as:

$$K = \frac{(R - S)(aD - S)}{S}$$

rearrangement gives:

$$\frac{D}{S} = \frac{K}{a(R - S)} + \frac{1}{a}$$

Thus if D/S is plotted as a function of $1/(R - S)$, the intercept (on the D/S axis) gives us $1/a$. These results yielded the conclusion that 10% of *E. coli* DNA was complementary to RNA synthesized during the growth of this organism in a defined medium. Assuming that there is no (or little) overlap between operons in opposite transcribing strands (see however p. 373) this means that 20% of the gene sites are transcribed. From an estimate of the proteins required for growth, this represents something over 2000-fold excess of transcription. The significance of this apparent overtranscription (which has a mammalian counterpart, p. 378) remains uncertain.

To establish the proportion of RNA that is unstable, the usual method (Kennell, 1971; Pigott and Midgley, 1968) is to perform a similar series of experiments (i.e. to vary R/D) and to extrapolate hybridization to infinite RNA concentration. If the results of RNA labelled in different ways are compared it is usually possible to work out the proportion of unstable RNA in the preparation. One method is illustrated diagrammatically in Fig. 12.14.

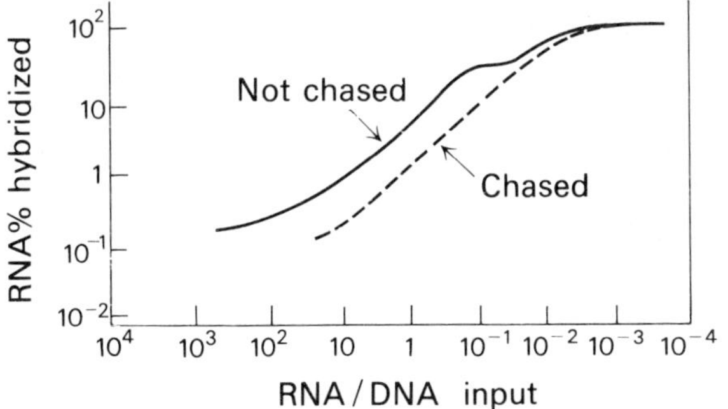

FIG 12.14   Method of distinguishing between stable and unstable RNA components in total cellular RNA. RNA is labelled uniformly (for many generations) with an isotope (say ^{32}P). The hybridization is measured against a variety of R/D inputs and plotted as percentages of the asymptopic maximum against R/D (logarithmic scale). The same experiment is performed with RNA labelled and then 'chased' (with excess inorganic phosphate in the case above) and a similar curve is drawn. Obviously both eventually will reach 100%. The reason the continuously labelled RNA shows greater hybridization at lower R/D inputs is that there is less competition for the same sites among the more dilute mRNA species. From the difference between these two curves it is possible to construct a curve for unstable RNA components in the mixture (after Kennell, 1971).

Mammalian transcription *in vivo*   A method for estimating the extent of transcription in a complex genome has been devised by Geldermann, Rake and Britten (1971). This procedure is an extension of the 'Cot' method of p. 364. Their paper describes several technical procedures which are undoubtedly of wide application. To take one example. DNA from L-cells[18] was sheared, denatured and annealed at 60°C to the extent that mt (p. 357) became 1500 nucleotide-moles s l^{-1}. The preparation was passed through a column of hydroxapatite at 60°C (in 0·12 M phosphate buffer); 54% was unbound, i.e. non-repetitive. This DNA was concentrated by binding to cellulose nitrate and was eluted with water at 90°C. This DNA was used for RNA:DNA hybridization (0·15 ml of a solution of 10 mg/ml in 6 x SSC containing RNA, 60°C for 16 h). The hybrids were identified by three independent methods. In one case the solution was treated with ribonuclease and then fractionated on Sephadex G-100; alternatively it was passed through cellulose-nitrate filters (after ribonuclease). In the third procedure hydroxyapatite was used to bind the hybrids. In general this is a difficult procedure as RNA binds to hydroxyapatite anyway. Conditions were employed in which this background binding was minimized and additionally [14C] DNA was used to identify the hybrids.

The conclusion from these data was that 8% of the repetitive DNA isolated in this experiment was transcribed. When corrected for the recovery of DNA from the hydroxyapatite column, this figure implies that 5·6% of the genome is covered by mRNA synthesized in these cells. The hybrids were found to have a T_m of 77°C in a buffer in which the native L-cell DNA had a T_m of 83°C. This figure implies that the hybrids were essentially perfect as an RNA:DNA duplex has typically a T_m about 5 deg lower than the DNA:DNA analogue (Kennell, 1971). This figure of 5·6% of the genome is again vastly in excess of the transcription anticipated from the protein synthesis required by the cells.[19]

Thus it seems to be generally the case that hybridization studies imply that far too much of the genome of cells are transcribed. It is not possible to argue one's way out by postulating extensive gene duplication. It is known that the duplication in *E. coli* is far less than that implied by the data of Kennell (p. 377) and in the case of the L-cells (above) the DNA had been specifically fractionated to ensure that only non-repetitive DNA was employed in the assay. There is thus a big problem. Is there some process which leads to apparently true, nuclease resistant and correctly melting hybrids which involves some type of nucleic acid secondary to tertiary structure which is confusing hybridization data of this type? Alternatively is there some feature of the kinetics and equilibria of these processes which has been missed in theoretical discussion (such as those on pp. 355–360)? Until these questions have been resolved, hybridization must be regarded as an excellent complement to other biochemical or genetic studies but (so far as total RNA is concerned) it must be treated with considerable caution.

Hybridization as a histochemical method   Potentially, hybridization is an extremely precise method of localizing genes under the electron microscope. I have made it clear that ultrastructural studies are outside the scope of this book. However the prospect of being able to identify genes under the electron microscope is clearly of great importance. A technique which involves the use of [³H] RNA hybridized to ultrathin sections (fixed in glutaraldehyde) of tissue has been described by Jacob, Todd, Birstiel and Bird (1971). They used *Xenopus* 28 S rRNA and were able to confirm the localization of the ribosomal genes in the nucleolar cap. This paper contains several references to earlier work (from Birnstiel's laboratory and elsewhere) to the application of hybridization to cytological studies.

Notes to chapter 12

[1] Note that hybridization (or renaturation of any long duplex) is much more complex than duplex-formation during DNA-replication. The new DNA helix grows in one direction only. Hybridization presumably starts at one or more points in the middle of the complementary molecules.

[2] This value was calculated by comparing the densities of λ- and λ_{dlac}- DNA.

[3] These figures can be calculated by the 'titration' of RNA sites—see Kennell (1971) and also p. 376.

[4] This is the reverse of the hybridization reaction. It is not the same as the dissociation process for the reason explained in note 1 (above).

[5] Millipore type HA filters are suitable.

[6] It is not possible to assume that the absorption of DNA is quantitative and the DNA must be assayed (after the discs have been washed and dried). Undoubtedly the simplest method is to use DNA of accurately known specific activity labelled with a different isotope to the RNA. The DNA label can then be assayed in each disc at the same time as the RNA. Otherwise DNA can be removed from the filter with deoxyribonuclease and assayed by the Burton method (p. 173).

[7] P22 is a *Salmonella* phage.

[8] Note there are two genetic consequences of this fact for a phage such as T4. One is that although the chromosome is linear, the linkage map is circular. The other is that as (we shall see in a moment) the molecular weight of the DNA is invariant, extensive deletions result in gene duplication. For example, if the stretch between (d) and (f) was missing, the molecule (neglecting the two strands) would be a b c g h a b c g h and its permutations.

[9] The reason I follow Kennell (1971) in eschewing the other nomenclature is that $(mt)_{\frac{1}{2}}$ has become, not $(C_0 t)_{\frac{1}{2}}$ which is acceptable (if a little cumbersome), but $Cot_{\frac{1}{2}}$ which is deplorable. Cot is a baby's bed; otherwise (as mathematical symbol) it means $C \times o \times t$.

[10] Establishing where the operon has arrived (and which way round it is) is not too difficult. The transposition mutants are of necessity Hfr strains and by interrupted mating with well marked F⁻ recipients

it is possible to map them fairly simply. The problem of selecting transposition mutants in which the operon is adjacent to a phage attachment site is less straightforward (see references cited in text).

11   The purpose of these deletions was to ensure that no adjacent genes (or parts of such genes) were present in the final preparation of the operon.

12   Primitive small bacteria, formerly 'pleuropneumonia-like organisms'.

13   Word used to describe a genetic system such as the nuclear chromosomes (nuclear genome) or (in this case) the DNA of mitochondria or chloroplasts.

14   The house cricket.

15   This is one of the few experiments in molecular biology which cannot be done with *E. coli*, as it requires an organism for which the nucleotide compositions of rRNA and the bulk of the DNA differ significantly.

16   In this case a shark-liver single-strand-specific endodeoxyribonuclease was used. Reference to its isolation and purification is given in the paper.

17   A method of effecting T4 transformation is discussed on p. 193. In this case, spheroplasts of *Klebsiella aerogenes* (referred to by Jayaraman and Goldberg under its earlier generic name of *Aerobacter*) were infected with the defective T4 in the presence of the shark-liver digest. It was found that separation of the DNA:mRNA hybrids from the fragments was unnecessary.

18   L-cells are tissue-culture cells derived from mice.

19   The use of very low RNA/DNA input ratios as an aid to simplifying hybridization studies on heterogeneous RNA has been considered in detail by Melli, Whitfield, Rae, Richardson and Bishop (1971). Their methods should prove of particular value in the analysis of eukary otic nuclear RNA.

Addendum

Tumour viral RNA   The search for viral genomes in tumour cells constitutes one of the most exciting areas of current cancer research. Spiegelman's use of hybridization to test for the presence of DNA complementary to the RNA of tumour viruses is referred to on p. 373. More recently, R. Axel, J. Schlom and S. Spiegelman (*Proc. Nat. Acad. Sci. U.S.* 1972, **69**, 535) have employed hybridization methods to establish that viral RNA is present in polysomes from a mouse mammary carcinoma.

13   Size and shape

It is difficult to separate a discussion of molecular weight from some understanding of the overall shape of the molecules. I have omitted any detailed description of the secondary and tertiary structure from this discussion, and by 'shape' I just refer to the general geometry of the molecule. We can consider a duplex structure as being rod-shaped and a denatured one as behaving as a sphere. (The sphere is the shape obtained in the average spatial distribution of atoms or other steric parameters of a random coil.) A molecule containing limited duplex structure is somewhere between these extremes. If there is considerable rotation about the bonds between the helical segments, the molecule will consist of semi-flexible structure with rigid (helical) sections within it, rather like a hinged ruler. If on the other hand the helical segments are packed together in a tertiary structure, the molecule will behave as a sphere or other 'hard' solid.

Hydrodynamic properties

The ultracentrifuge

The analytical centrifuge—machine or monster?   The analytical ultracentrifuge is a device for observing the optical properties of a solution during its exposure to high gravitational fields. It can thus be used to determine the rate of sedimentation of boundaries during sedimentation-velocity centrifugation, or (with more difficulty) the sedimentation of zones during rate-zonal centrifugation (see Fig. 4.2, p. 132). Equally it is possible to determine the distribution of molecules at equilibrium, following either sedimentation-equilibrium or buoyant-density centrifugation. As the design features unique to analytical (as opposed to preparative) ultracentrifuges largely derive the optical requirements, it is worth briefly commenting on the optical methods that are involved. For a detailed introduction to the design (and theory) of analytical centrifuges, see Bowen (1970).

The sample is contained in an optical cell which is shaped as a cylinder mounted in a hole drilled in an analytical rotor. The volume occupied by the solution is sector-shaped. This is to avoid the 'edge diffusion' that occurs in parallel-sided centrifuge tubes (see Fig. 4.10, p. 143). A collimated light source shines through the sample and through reference holes either drilled in the rotor or present in the 'counterpoise' which balances the cell on the opposite side of the rotor. The reference holes are at defined distances from the rotor centre so that the magnification factor(s) in the optical system and the photographic or other detection system can be calculated. The general arrangement of the cell is shown in Fig. 13.1.

There are three methods of measuring the optical properties of the solution in a centrifuge.

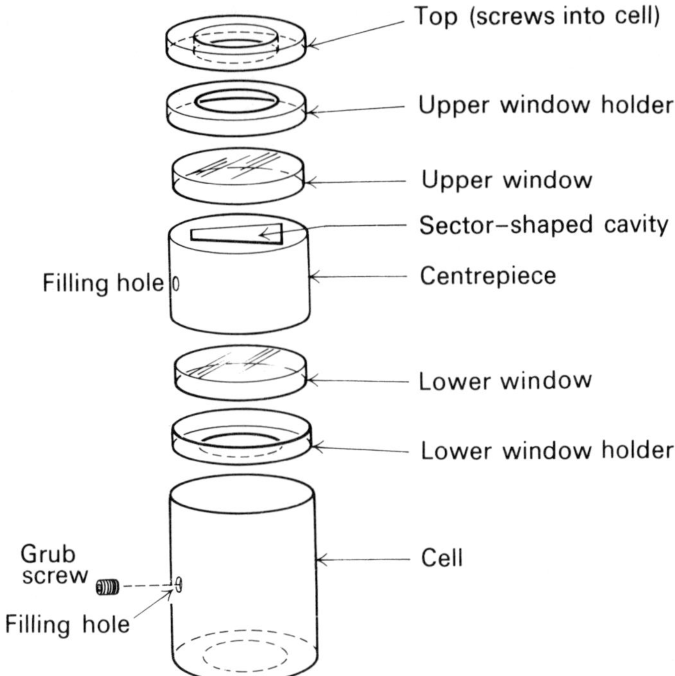

FIG 13.1   General arrangement of a simple sector-shaped cell for an analytical centrifuge rotor.

The simplest, in principle, is the UV-absorbance system. In this case a beam of UV light passes through the sample and the difference in absorption along the sample is recorded. The record can either be made photographically or by scanning the emergent light with photomultipliers. The difficulty with this system derives from quantitating the absorbance with amounts of solute. To make these correlations reliably it is essential to use a centrifuge fitted with a monochromatic light source and a double-beam UV-system. The second technique, which is the easiest to operate, but the least easy to understand in detail is the schlieren system.[1] Schlieren patterns are produced by differences in refractive index. Schlieren optical systems contain many components (see Bowen, 1970, for a very lucid explanation of the principles) but the effect itself is very familiar. If two liquids of different refractive index (like sugar solution and water) are mixed, the streaks can be seen at the interface between regions or zones of different concentration. The third system consists of creating interference fringes between light that has passed through the sample solution and a reference sample containing only solvent. The two samples are usually side-by-side in a double-sector cell.

Again see Bowen (1970) or some other suitable text for a detailed description of these methods. Let us just assume that photographs have been taken with each of these systems of the three types of concentration distribution obtained in the centrifuge and see what they reveal (Fig. 13.2). The last two methods (schlieren and interference) produce a picture with lines on them suitable for direct measurement. UV-pictures do not produce such a picture, but rather a little rectangle with varying degrees of blackening in it. Therefore in Fig. 13.2, the UV-picture is presented as a trace obtained by scanning such a picture with a densitometer. Anyone with the luxury of a photomultiplier unit on their centrifuge can obtain such traces as a direct read-out from the machine.

An analytical centrifuge is, to the outsider, a fairly fearful piece of equipment covered with

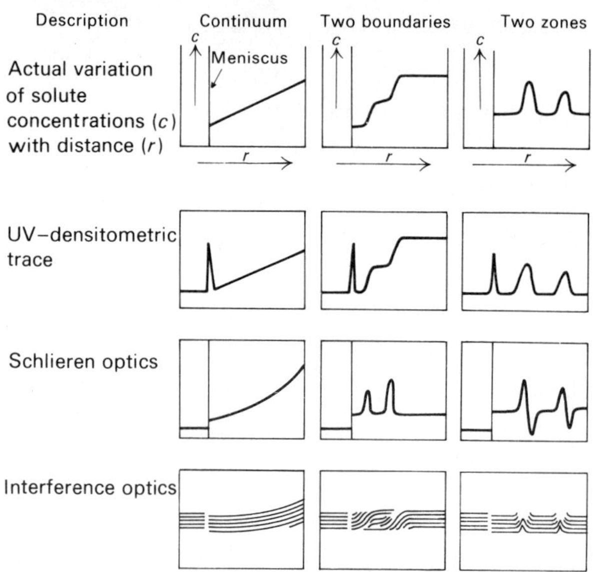

FIG 13.2   General appearance of pictures taken in an analytical centrifuge. The gravitational field is from left to right throughout.

knobs and dials either standing aloof, unloved and unused in a special room, or (hopefully) quietly droning away, with all its inspection panels stripped off and with a keen centrifuger dashing in and out of a dark room with photographic films and plates, dripping developer and fixative all over the floor. The most familiar machine is the Beckman model E. In it, the rotor is suspended by a piano wire in a chamber maintained at a high vacuum with a dibutyl-phthallate diffusion pump backed off by a rotary diffusion pump. The chamber is cooled by a refrigerator and the temperature is maintained by a thermocouple which controls the current in a heating coil in the bottom of the chamber. The motor and drive are above the chamber and the optical bench extends along the top of the instrument and at its end is the camera (or photomultipliers). The controls on the instrument (in addition to normal ultracentrifuge controls) are the calibration for the RTIC (temperature control) system, a manual arrangement for acceleration and an automatic device for operating the camera at pre-determined time intervals. In modern versions, the speed is maintained by an electronic mechanism; a mechanical stopping and starting mechanism was employed in the older versions, in which it is essential to ensure that the drive voltage is so close to that required to keep the centrifuge at the pre-set speed that the amount of switching (acceleration, followed by coasting back to the 'cut-in speed') is kept to a minimum.

Recently, designs for analytical rotors driven from below have been perfected, so that stable, bottom-drive systems can be made. An instrument employing such a drive and incorporating many ingenious design features is the MSE analytical Ultracentrifuge. The other consequence of this technical development has been the development by centrifuge manufacturers of the 'hybrid centrifuge'. This is an instrument with the full range of preparative rotors and as easy to use as an ordinary preparative ultracentrifuge, but with an analytical facility, consisting of an analytical rotor and an optical system and camera that can be mounted above the centrifuge bowl. For velocity-sedimentation studies, this arrangement is apparently as accurate as the 'full' analytical machine.

Starting from scratch, the analytical centrifuge is an expensive method of measuring the molecular weight of nucleic acids. However, if there is one available to you, it is well worth spending a few days learning how to use it, for it is a quick method for analysing nucleic-acid (or nucleoprotein) mixtures and it remains the instrument whose output is required ultimately to calibrate most other semi-empirical estimations of hydrodynamic properties of macromolecules.

Sedimentation-velocity centrifugation   This technique is the direct method of estimating sedimentation coefficients. The rate at which a particle sediments through a gravitational field in a centrifuge is given by the expression:

$$\frac{dr}{dt} = s\omega^2 r$$

where t is time, r is radius and ω is the angular velocity (see p. 131). The sedimentation coefficient s has the dimension of time. The unit is the Svedberg (S) and $1\,S = 10^{-13}$ s. The method of calculating the value of s for a macromolecular solute is extremely simple. The equation above is merely re-written as follows:

$$s = \frac{dx/dt}{r\omega^2} = \frac{d(\ln r)}{\omega^2\,dt}$$

Thus if a boundary (maximum of schlieren peak or the point of inflexion of a UV-trace—see Fig. 13.2, p. 382) is measured at various times and r (peak to rotor centre) is calculated, all that is necessary is to plot $\ln r$ against time (s). The slope is then $\omega^2 s$ from which s (units are s) can be calculated. The intercept on the $\ln r$ axis can in principle be used as an additional point on the graph. A boundary must necessarily start at the meniscus, so the distance from meniscus to rotor centre may be employed as a value for r at time zero. What is 'time zero'? The centrifuge does not start instantaneously, it is progressively accelerated; during the acceleration, the sedimentation commences. A rule that is sufficient for most purposes is to count time zero, not as the time at which the rotor reaches full speed, but rather the time at which it reaches $\frac{2}{3}$ speed.

The value of the sedimentation coefficient depends on the identity of the sedimenting macromolecule, its concentration, the temperature and the nature of the solvent. In order to obtain values characteristic of the macromolecule, all the other variables are standardized to obtain $s^0_{20,w}$. This is the sedimentation coefficient at zero concentration (superscript 0), at 20°C (subscript 20) and in water as solvent (subscript w).

The corrections for temperature and solvent are made using the relationship:

$$s_{20,w} = s \times \frac{\eta}{\eta_{20,w}} \times \frac{(1 - \bar{v}\rho)_{20,w}}{(1 - \bar{v}\rho)}$$

where η is viscosity and $(1 - \bar{v}\rho)$ is a function (pronounced 'one minus vee-bar rho') that crops up frequently in centrifugation theory; ρ is the density of the solution and $\bar{v}$ is the partial specific volume of the macromolecule. Provided the conditions do not differ too widely from the standard conditions (i.e. the solvent is dilute salt solution and the temperature is within 10 deg of 20°C), the corrections for $(1 - \bar{v}\rho)$ can be neglected and the simplified relationship:

$$s_{20,w} = s \times \eta/\eta_{20,w}$$

where the viscosities of water at the experimental temperature and 20°C are employed. For more accurate work, in which it is difficult to estimate the values of $\bar{v}$, the alternative expression:

$$s_{20,w} = s \times \frac{\eta}{\eta_{20,w}} \times \frac{\rho_p - \rho_{20,w}}{\rho_p - \rho}$$

may be employed: in this relationship, ρ_p is the buoyant density of the particle or macromolecule (see p. 389).

The relationship between s^0 and s is as follows:

$$s = s^0 [1 - f(c)]$$

where $f(c)$ is a function of concentration. It is frequently assumed that $f(c)$ is a simple linear proportionality and one sees the relationship written down as:

$$s = s^0 (1 - \text{const.} \times c)$$

This simplification is only justified in cases in which concentration ranges are narrow and the effective magnitude of 'constant' is small. With nucleic acids (in contradistinction to most proteins) this last assumption is not justified. The dependence of s on concentration is extremely marked. Moreover, if the solution is so strong (of the order of 1 mg/ml) that schlieren optics can be used, the former assumption is clearly invalid as well. There are two

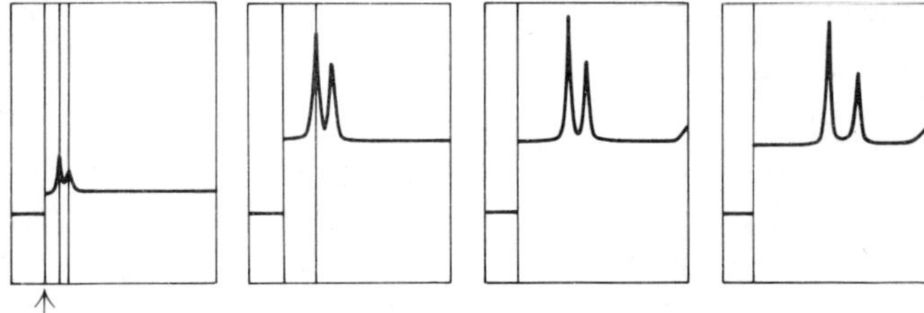

Meniscus

FIG 13.3   Schlieren boundaries (traced from an enlargement of a photographic plate by the author). The sample was in a cell of the An-D rotor of the Beckman model E ultracentrifuge. The vertical lines are caused by the fact that very steep boundaries result in the light being deflected right out of the optical system. The variations in base line are due to the fact that as the centrifugation progressed, one of the optical components of the system (phase-plate angle) was adjusted during the run. The sample consisted of *E. coli* rRNA dissolved in 0·1 M acetate (pH 8); the rotor (Beckman An-D in the model E centrifuge) was run at 59 000 rpm at 20°C.

consequences: one is that to measure sedimentation coefficients of nucleic acids, it is essential to employ a UV optical system and to carefully extrapolate to zero concentration using as many readings at different concentration as possible. Secondly, if rate sedimentation centrifugation using the schlieren system is used to check up on the integrity of RNA preparations,[2] the peaks, which correspond to the boundaries in the cell, are not the gaussian peaks of text books. Rather they have the curious spiky appearance shown in Fig. 13.3. The explanation is that the boundaries are 'self sharpening'. A molecule at the tail of a boundary is in much less concentrated solution of RNA than is one in the solution below the boundary. It therefore has a higher sedimentation coefficient and moves faster, until it catches up with the boundary. The net result is that the boundary becomes extremely sharp.

Measurement of the sedimentation coefficient of DNA by zonal centrifugation   I have pointed out on p. 148 that the zonal centrifugation of DNA is limited by the anomalous hydrodynamic properties of DNA at high concentration. Zonal centrifugation of DNA is only reproducible and meaningful if the DNA is at zero concentration. This means using radioactive DNA. In practice sucrose-gradient analysis of DNA preparations and analytical centrifugation complement one another rather well, as the sucrose gradients are useful for DNA preparations (such as certain viral DNA preparations) that are only obtainable readily in radioactive amounts. A good empirical procedure for general use is that of Burgi and Hershey (1963). A linear gradient of 5–20% (w/v) sucrose in 0·2 M Na$^+$ ions pH 7·0 is established in a 5 ml centrifuge tube (volume of gradient is 4·8 ml). DNA (less than 0·2 μg in 0·2 ml buffer) is layered over the gradient and centrifugation is for such a time that the DNA zone appears about half way down the gradient. If d is the distance (cm) from the mid-point of the zone to the meniscus, the sedimentation coefficient is given by the relationship:

$$s^0_{20,w} = \frac{6·45 \times 10^{10} d}{\omega^2 \times t}$$

where ω is the 'angular velocity' in rpm and t is time in hours. For more accurate work, an internal standard of DNA (from a phage) of known sedimentation coefficient and molecular weight is included in the same gradient. If the subscript 'a' refers to the standard and 'x' to the unknown, then the relationships are:

$$\frac{d_x}{d_a} = \frac{s_x}{s_a} = \left(\frac{MW_x}{MW_a}\right)^{0·35}$$

The reason for the '0·35-th power' will appear shortly.

Measurement of the sedimentation coefficient of RNA from sucrose gradients   The literature contains several references to novel RNA species identified by sucrose-gradient centrifugation, with estimated 'S-values' allocated to them. The procedure is a perfectly legitimate one provided no great claims are made as to the significance of the values that are recorded. The dependence of sedimentation coefficient on distance travelled down the centrifuge tube is, in general, non-linear. With most of the gradients that are employed for RNA fractionation, the plot of d (defined as in the preceding section) against sedimentation coefficient is convex; that is to say the molecules of high sedimentation coefficient are more bunched up at the bottom of the tube than are the ones nearer the meniscus (of low sedimentation coefficient). The reason is that there are two factors contributing to the position to which a zone will sediment in a given time. One is the gravitational force on the molecules (which increases linearly down the tube, p. 383); the other is the viscous drag on the molecule which also increases down the tube as the sucrose becomes more concentrated. Typically the latter effect (which slows the molecules up) overtakes the increasing effect of gravity. For accurate estimates of S-value (and also incidentally for optimizing the resolving power of the gradient), a gradient should be designed in which the two effects are absolutely balanced. In such a gradient an RNA molecule (or other particle) will sediment at a completely uniform speed. Gradients of this type have been termed by Noll 'isokinetic gradients'. Methods of calculating the shape of isokinetic gradients for particles (including RNA) of known density are described by Noll (1967). The relationships which are presented in this paper are worked out in such a way that the isokinetic gradients can be made up directly using the device for creating exponential gradients illustrated in Fig. 4.3(*b*) (p. 134) and employing the parameters defined in the caption to that figure. The advantage of employing such a gradient for the estimation of sedimentation coefficient is, of course, that the sedimentation coefficient is a linear function of d and consequently the S-value of an unknown RNA can be read off with in principle only one reference point, in practice with two (say 16 S and 23 S rRNA).

The linearity of the above method (or the empirical S/d relationship deduced for a non-isokinetic gradient) depend upon the assumption that the RNA molecules employed for calibration are of identical density and consequently of the same (or similar) nucleotide composition. For much work, the RNA species (e.g. different rRNA types and their precursors) are of sufficiently similar nucleotide composition for the approximation to be valid. However, great care is needed in interpreting results obtained with RNA of extremely anomalous nucleotide composition (such as mitochondrial rRNA—see p. 286).

Sedimentation coefficient and molecular weight   The equations that relate sedimentation coefficient to molecular weight are of two types. In one treatment, familiar to anyone who has measured the molecular weight of a protein by the sedimentation-velocity method, the parameter that is required in addition to s is the diffusion coefficient D. The diffusion coefficients of high-molecular-weight nucleic acids are so small that they cannot be measured with sufficient accuracy by any techniques that are generally available. Therefore a different type of calculation (originally designed in a slightly different form for proteins too) is employed. In this case, the other parameter in the equation is 'intrinsic viscosity' $[\eta]$. Methods of measuring this function are discussed on p. 391. However in the meantime, it is worth carrying the argument on a little further as the intrinsic viscosity is not actually needed for some quite reliable approximate solutions to the equations.

The equation that relates sedimentation coefficient of nucleic acids to molecular weight is due to Mandelkern and Flory (1952). The relationship is as follows:

$$MW = \left\{ \frac{s^0_{20,w}\,[\eta]^{1/3}\eta_0 N}{\beta(1-\overline{v}\rho)} \right\}^{3/2}$$

where the units of $s^0_{20,w}$ are seconds (not Svedbergs—see p. 383); η_0 is the viscosity of water at 20°C and ρ is the density of water at 20°C; N is Avogadro's number (the number of molecules is a mole). Thus η_0, N and ρ are all universal constants:

$\eta_0 = 0.01005$ poise

$\rho\ = 0.998$ g/ml

$N = 6.02 \times 10^{23}$

The values of $s^0_{20,w}$, $[\eta]$, $\overline{v}$ and β are constant for a particular nucleic acid: $[\eta]$ is the intrinsic viscosity (see p. 391 for an explanation of the units); $\overline{v}$ is the partial specific volume and β is a conformation-dependent term.[3]

From here we must consider DNA and RNA separately.

Sedimentation of linear DNA.   Here I am only concerned with linear DNA. The extension of sedimentation studies to circular and supercoiled DNA is discussed on p. 426. The values of β and $\overline{v}$ are remarkably constant for all DNA molecules (single- and double-stranded); the 'best' value for $\overline{v}$ is 0.55 ml/g (Hearst, 1962); that for β is 2.4×10^6 (Eigner and Doty, 1965). If these values, together with those for η_0, N and the factor 10^{-13} (see p. 391; to enable $s^0_{20,w}$ to be expressed in S-units) are put into the Mandelkern–Flory equation, the final form is:

$$MW = 13\ 200\ (s^0_{20,w}\,[\eta]^{1/3})^{3/2}$$

In practice it can be made simpler than this. It has been established empirically that it is not necessary to measure $[\eta]$ provided standard conditions are employed. The sedimentation is observed at 22°C, pH 7.0 in 0.2 M Na^+ ions. From these measurements $s^0_{20,w}$ is determined. Then the following equation applies for double-stranded DNA for the molecular weight range $0.2 \times 10^6 - 100 \times 10^6$ (Eigner and Doty, 1965; Crothers and Zimm, 1965).

$$0.445 \log (MW) = 1.819 + \log (s^0_{20,w} - 2.7)$$

For single-stranded DNA of molecular weight up to 4×10^6, Eigner and Doty (1965) derived the relationship:

$$s^0_{20,w} = 0.056\ (MW)^{0.36}$$

For high-molecular-weight single-strand DNA ($2-70 \times 10^6$), Studier (1965) derived the following:

$$s^0_{20,w} = 0.053\ (MW)^{0.40}$$

A more recent analysis of available data (Freifelder, 1970) suggests that a more accurate relationship is the following:

$$s^0_{20,w} = 2{\cdot}8 + 0{\cdot}00834\,(MW)^{0{\cdot}479}$$

For phage DNA studies, this equation is the best available to date.

If polydisperse DNA is used for such measurements, the sedimentation coefficient that is computed is at first sight the weight average $(\overline{s^0_{20\,w}})_W$. Likewise the intrinsic viscosity values should be $(\overline{[\eta]})_W$. In practice we cannot assume that these values are the same as the values for a single species of molecular weight $(\overline{MW})_W$, as s and $[\eta]$ are not linear functions of MW. For many purposes it is legitimate to neglect this complication. To take the question further it is necessary to analyse the distribution of apparent molecular weights by calculating the distribution of sedimentation coefficients for various points chosen along a densitometric trace of the boundary in the ultracentrifuge. The arithmetic involved is fairly formidable but a simplified procedure is very clearly described by Schumaker and Schachman (1957). From such a distribution it is equally possible to calculate the number-average sedimentation coefficient, $(\overline{s^0_{20,w}})_N$. To establish whether the 'weight-average' values of s and $[\eta]$ are appropriate for calculating $(\overline{MW})_W$ (and/or estimating the error implicit in an assumption that they are), Hayes and Zimm (1970), have described the following procedure. Equally spaced points are taken along the concentration axis of a plot of $s^0_{20,w}$ versus concentration. The values are weighted so that $(\overline{s_{20,w}})_W$ and $(\overline{[\eta]})_W$ could be computed. A comparison of the predicted and experimental dependences of $s^0_{20,w}$ upon $[\eta]$ can then be made.

Sedimentation of RNA   The Mandelkern–Flory equation (p. 386) is appropriate for the determination of the molecular weight of RNA. It will be recalled from the previous discussion of DNA, that the first job in approximating the equation is to select appropriate values for β and $\bar{v}$ (p. 386). With RNA, the literature contains conflicting estimates of 'good' values for these parameters. The problem is discussed by Boedtker (1968a). She points out that although the relationship is sensitive to changes in the estimates of these individual parameters, the consistency in RNA molecular-weight estimations in the literature is due to the fact that the various authors have fixed on different values of $\bar{v}$ and have estimated β accordingly. If the data are put together, the 'best' approximation for the molecular weight of RNA is obtained from the equation:

$$MW = 14\,300\,\{s^0_{20,w}\,[\eta]^{\,1/3}\}^{3/2}$$

where the units are the same as those for the corresponding DNA equation (p. 386).

Unfortunately, we cannot approximate this equation any further. There is no universal relationship between S-value and molecular weight for RNA. The factors which determine the variability are unknown. Thus various 'families' of RNA are recognized (Boedtker, 1968a) in which $s^0_{20,w}$ is a linear function of molecular weight. However there is no means of allocating an uncharacterized RNA to one of these families on the basis of nucleotide composition, estimated structure, function (viral, ribosomal, etc.) or anything else that has been recognized so far.

Certainly far too much emphasis has been placed on 'S-values' in the past. They are only of value if one is sure one is comparing like with like. Sedimentation coefficients are very valuable for detecting qualitatively aggregation in RNA. An example of this application is the work of Martin *et al.* (1971) on oligo A : oligo U complexes (see p. 343) in which it was essential to establish that true duplices (not triplices) were being studied. The nature of the oligomers was described on p. 343. Consider the oligomers in which a string of A-residues is followed by a string of U-residues. For simplicity, let us represent the A-stretch as - - - - - - - and the U-stretch as . . . . The hairpin conformation is then:

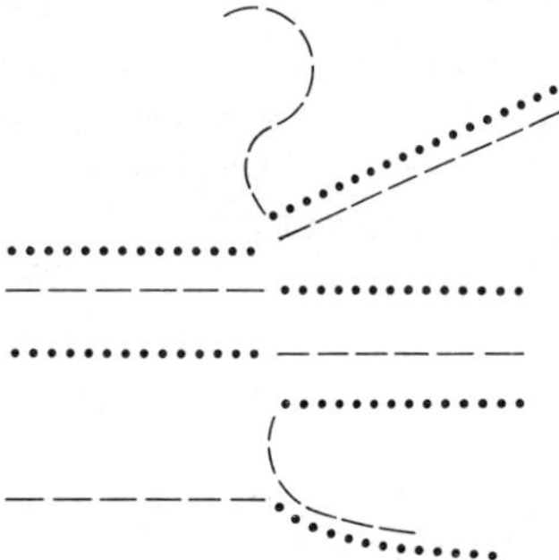

A linear duplex dimer is:

If triplices are formed, large aggregates will form such as:

The formation of such complexes if favoured at low temperatures if the oligomers contained more U's than A's. Thus for the substance (using the above 'convention') - - - . . . ., the uncorrected sedimentation coefficient (s_{20}^0) was 1 at 36°C but 3·1 (in the same solvent) at 6°C.

The problems of correlating S with MW can be overcome by using random-coil RNA. Standard conditions for achieving this denaturation with formaldehyde (see p. 65) have been described by Boedtker (1968b). The method consists of adding $\frac{1}{5}$th volume of 0·45 M Na_2HPO_4, 0·05 M NaH_2PO_4 to 37% aqueous formaldehyde and adding 1 volume of this to 9 volumes of RNA solution (about 0·1 mg/ml). This mixture is heated to 63°C for 15 min and cooled to 0°C. The sedimentation coefficient is then measured. The sedimentation coefficients are less than those for the unreacted molecules; for example '23 S RNA' gives $s_{20,w}$ = 13·9 in this system. Using these 'formaldehyde S-values', the relationship is as follows:

$$0.40 \log (MW) = -\log 0.05 + \log s_{20,w}$$

Alternatively, the zonal techniques using organic solvents (p. 147) may be employed for estimating (rather approximately) the molecular weight.

Sedimentation-equilibrium ultracentrifugation   Enough has been written above to convince even the newcomer to ultracentrifugation, that sedimentation-velocity untracentrifugation suffers from the disadvantage that in order to calculate a molecular parameter such as molecular weight from the S-value, other physicochemical parameters have to be measured or guessed. The disadvantages of this procedure can be overcome by applying one of the methods of sedimentation-equilibrium ultracentrifugation. The techniques have been widely applied to proteins. For a survey and guide to the calculations involved, see Bowen (1971). Qualitatively, the methods can be summarized by saying that if the solution of macromolecules is put in a centrifugal field insufficiently strong to sediment them, a gradient of concentration will be formed (like a CsCl gradient) at equilibrium. The attractive feature of the method is that in the theoretical treatment of the results, the effects of sedimentation and diffusion cancel out. Thus by an analysis of the shape of the gradient that is produced, it is possible to calculate the molecular weight of the protein. There are two variants to the procedure. In the 'high-speed' or 'meniscus-depletion' method, the centrifugation speed is such that the equilibrium distribution results in zero concentration at the top of the column of liquid in the cell. In the 'low-speed' method, this is not the case. The calculations are simpler in the former case. The technical drawback to the equilibrium centrifugation of very large molecules (with very low diffusion coefficients) is that extremely long times are required to reach equilibrium. The problem can be partly overcome by employing very short column lengths in the ultracentrifuge cell but at some expense as accurate measurements are made more difficult. Thus the technique falls short of the requirements for nucleic acid research. It is an excellent method of measuring the molecular weight of tRNA (Lindahl, Henley and Fresco, 1965). However, we now have precise knowledge of the molecular weight of tRNA from sequencing studies. It is possible to measure the molecular weight of large molecules (e.g. *E. coli* 23 S RNA, Stanley and Bock, 1965) by the same method employing very short (1 mm) columns of solution but the method has not found general application.

Buoyant density centrifugation Several references to the use of caesium-salt centrifugation as a preparative technique have appeared elsewhere (pp. 149, 186 and 261). It remains to discuss the analytical centrifugation of caesium gradients.

The resolving power of the system is so great that with macromolecules, it is possible to do with the analytical ultracentrifuge an experiment which requires a mass spectrometer in the chemistry of low-molecular-weight compounds, namely the separation of molecules that differ only in the identity of an isotope. In the classical experiment of Meselson and Stahl (1958) the separation of $[^{14}N]$ DNA and $[^{15}N]$ DNA was exploited to demonstrate the semi-conservative replication of DNA in *E. coli* (p. 17). The experiment consisted of growing the bacteria in medium containing only $[^{15}N]$ nitrogen compounds. The bacteria were transferred to $[^{14}N]$ medium and after successive periods of time, DNA was extracted and banded in a CsCl gradient. The gradient resolved three components: DNA containing ^{15}N in both strands ('heavy'), DNA composed of one strand of $[^{15}N]$ DNA ('intermediate') and DNA containing $[^{14}N]$ DNA in both strands ('light'). The distribution of DNA between these three zones at various fractions or multiples of a mean generation time was consistent with the semi-conservative mode of replication. For example at time zero (before the transfer to $[^{14}N]$ medium), only heavy DNA was found. After one generation, only intermediate DNA was found, after two generations, intermediate and light in equal amounts and so on.

As a method of measuring nucleotide composition indirectly, the technique has been described on p. 267. The simplest method of making the measurements is with an analytical centrifuge with UV optics.

Extensive reviews of caesium-salt centrifugation have been written by Vinograd and Hearst (1962) and Szybalski (1968). The former is an extremely useful summary of the theoretical aspects of the method; the latter is especially useful as a summary of the use of caesium salts other than CsCl (notably Cs_2SO_4) which have certain advantages over the choride.

The zone of DNA in a CsCl (or other salt gradient) forms a gaussian distribution. The nature of this distribution can be employed to determine certain parameters of the DNA molecules. Qualitatively, the zone is broader if the diffusion coefficient is higher (i.e. if the molecular weight is lower).

The method is technically and theoretically one of the most esoteric procedures for measuring the molecular weight of macromolecules. The calculation requires knowledge of the precise gradient of CsCl concentration across the zone as well as the variance of the Gaussian distribution. Methods of determining the former quantity are described by Ifft, Voet and Vinograd (1961) and the approaches to the calculation are described by Vinograd and Hearst

(1962). The method is limited by the fact that any zone-broadening due to microheterogeneity in the DNA renders the calculation totally invalid. In other words it is really only useful for phage DNA. An empirical procedure of Thomas and Pinkerton (1963) was devised to simplify the arithmetic. Unfortunately, their procedure is theoretically unsound and can lead to misleading results. The whole question has been thoroughly re-examined by Schmid and Hearst (1969). The procedure they describe is probably too complicated for routine use but it does yield what are probably the best molecular weight data for some phage DNA molecules, which thus form excellent standards for calibrating semi-empirical methods such as correlations between molecular weight and sedimentation coefficient (although Schmid and Hearst themselves deplore such procedures).

The methods of Schmid and Hearst are based on a recognition of the thermodynamic non-ideality of DNA solutions that cause a concentration dependence of band width. Their procedure requires perfectly aligned double-beam UV optics with a photoelectric scanner. Apparent molecular weight is measured at a variety of concentrations employing the procedures described in the review of Vinograd and Hearst (1962). These apparent values are plotted versus average absorbance in the total band $\langle A \rangle$ (this last value is computed from each trace by numerical integration) and the values are extrapolated to zero $\langle A \rangle$. The results that are obtained (converting from CsDNA to NaDNA) are for T7 $(23 \cdot 1 \pm 0 \cdot 4) \times 10^6$, for T5 $(64 \pm 6) \times 10^6$ and for T4 $(105 \pm 6) \times 10^6$.

The review by Vinograd and Hearst (1962) includes reference to two different applications of CsCl gradients to studies on DNA which are of great interest (in both cases the data come from Vinograd's own laboratory). One is the determination of DNA hydration. The method is based on the comparison of the apparent density of DNA in various salt gradients and extrapolation of the data to a water activity of 1·000. The net solvation of CsDNA was estimated to be 30% by weight. The other technique consists of measuring buoyant density as a function of pH. Denatured DNA has a higher density than native DNA (see for example p. 154). Thus if the buoyant density of DNA is measured at a number of pH-values, it is possible to obtain a titration curve with a point of inflexion at the pH at which native duplex is in equilibrium with single-stranded DNA (around 11). The interesting point is that single-strand viral DNA (from φX 174) produces a transition of the sort characteristic of (for example) *E. coli* or T4 DNA. The conclusion is that there is considerable base-paired secondary structure in this DNA.

Viscometry

The viscosity of a fluid is determined by measuring the frictional resistance experienced by a layer of liquid as it moves past an adjacent layer. There are two viscometric phenomena which can be quantitated for measuring viscosity. Either one can measure the rate at which the liquid flows through a narrow tube (this is the basis of the capillary viscometers) or alternatively one can measure the torque required to maintain the rotation of a cylinder at constant speed if the cylinder is inside a concentric cylinder with the space between them filled with the liquid (this is the basis of the 'Couette viscometer'). The most familiar type of viscometer is of the former type. In practice, it is usual to employ one of the various models of the 'Ubbelohde capillary viscometer' which is designed in such a way that the liquid which has passed through the capillary (under gravity) spills into a chamber so that there is no correction required for the rising meniscus on the 'dead' side of the apparatus. A design suitable for RNA and DNA of low molecular weight (up to 2×10^6) and also a very low-shear Ubbelohde viscometer (designed by J. and J. J. Hermans) suitable for high-molecular-weight DNA and a Couette viscometer (also suitable for high-molecular-weight DNA designed by B. H. Zimm and D. M. Crowthers) are described by Eigner (1968). The same article describes the experimental procedures employed in viscometry. I have avoided describing the methods here as they are very specialized and are not suitable for routine work in a biochemistry laboratory. One of the great drawbacks of viscometry is that large amounts of material are required. The technique is thus one for specialists to fully characterize a few select molecules which can be used a reference substances by the rest of us. However, it is worth being conversant with the underlying principles of viscometry, if only to understand what 'intrinsic viscosity', which is one of the fundamental parameters of any macromolecule, really means.

Consider a layer of liquid flowing past another layer. Its free flow is opposed by the frictional force (symbol F). This force per unit area (F/A) is a function of the relative translational velocity between the layers (dx/dt) and the distance between them (r). In general terms this relationship may be written:

$$F/A = f \left\{ \frac{(dx/dt)}{r} \right\}$$

For an ideal or 'Newtonian' liquid, the function f is a simple linear relationship so that the above equation becomes:

$$F/A = \eta \times \frac{(dx/dt)}{r}$$

The constant η is called the coefficient of viscosity; its units are dynes, s, cm^{-2} or poises. If this equation is applied to a capillary viscometer, the coefficient of viscosity is determined by Poiseuille's equation:

$$\eta = \frac{hg\rho \, R^4 \pi t}{8lv}$$

where t is the time taken for volume v to pass through a capillary of length l, radius R under a hydrostatic pressure of $hg\rho$ (h is the height of the liquid, ρ its density and g the acceleration due to gravity). The device is thus suitable for measuring coefficients of viscosity in Newtonian liquids. In non-Newtonian liquids, such as DNA solutions, the calculations require a knowledge of F/A ('the shear stress') and $(dx/dt)/r$ ('the shear gradient'). The calculations with a capillary viscometer require a mathematical treatment of the problems that arise from the fact that the shear gradient varies with R. From this point of view, the Couette viscometer is theoretically easier to use, as the shear gradient is constant throughout the sample and can be directly calculated from the speed at which the inner cylinder is rotating and the geometry of the instrument. The aim of viscometric measurements with non-Newtonian liquids is to extrapolate the readings to zero shear stress and zero shear gradient. The other point is that the value we want is to be characteristic of one particular component (the nucleic acid) of the viscous solution. Viscometry differs fundamentally from other hydrodynamic measurements in that the solvent (say water) is itself viscous and the measurements are on the effect of altering a pre-existing parameter.

The viscosity of nucleic acid solution is measured relative to the viscosity of the solvent lacking nucleic acid thus:

$$\frac{\text{Viscosity of solution}}{\text{Viscosity of solvent}} = \text{Relative viscosity } (\eta')$$

The 'reduced viscosity' (η'') is defined as:

$$\eta'' = (\eta' - 1)/c$$

where c is concentration. The 'intrinsic viscosity' is the extrapolated value of reduced viscosity at zero shear stress, zero shear gradient and zero concentration thus:

$$[\eta] = \lim_{c \to 0} \quad \lim \, (\eta'')$$

$$= \lim_{c \to 0} \quad \lim \left(\frac{\eta' - 1}{c}\right)$$

$$= \lim_{c \to 0} \quad \lim \left(\frac{\ln \eta'}{c}\right)$$

In these equations, the second limit refers to shear stress and shear gradient reduced to zero. They have been separated from the concentration limit to take advantage of the mathematical theorem that the limit to take advantage of the mathematical theorem that the limit, as x approaches zero, of $(y - 1)/x$ is the same as the limit of $\ln y/x$.

The values of intrinsic viscosity can be put into the Mandelkern–Flory equation (together with sedimentation coefficients) to work out the molecular weights. We have seen (p. 387) that this is essential if the equation is to be employed for the determination of the molecular weight of RNA. For DNA, we noted that the molecular weight could be reliably computed for an empirical equation in which sedimentation coefficient is the only independent variable (p. 386). It is also possible to write such an equation in terms of intrinsic viscosity (Crothers and Zimm, 1965):

$$0.665 \log (MW) = 2.863 + \log ([\eta] + 5)$$

A recent application of the sedimentation-viscosity method to RNA, is the work of Cox and others (1970), referred to elsewhere (p. 392) on the double-strand RNA from Penicillium phage. It must be rare these days for newly discovered RNA to be characterized so thoroughly. The molecular weights of the two (p. 396) molecules were found to be 1.2×10^6 and 0.14×10^6. It will be interesting to learn whether RNA (also double-stranded) isolated from the mycophage PS 1 (host fungus *Penicillium stoloniferum*) by Frank, Ellis and Kleinschmidt (1971) and that of the *Aspergillus foetidus* phage of Banks *et al.* (1970) have similar properties. With Cox's data there are excellent reference molecules for comparative or semi-empirical molecular-weight determinations on double-stranded RNA.

Conformation of nucleic acids and hydrodynamic properties

Whatever the shape of a macromolecule, in solution it is tumbling around so that the average volume that it occupies during finite time is a sphere. The radius of this sphere is an important molecular parameter, the 'radius of gyration' R_G. We shall see in the next section (p. 398) that light scattering forms the direct method of measuring R_G. However, if we accept that R_G can be determined, we are left with two questions. What is the actual shape of the molecule that is gyrating and to what extent is this shape constant? Simplified models are required to analyse the data satisfactorily. First we shall consider only three possible models of the shape of a nucleic acid: random coil, real sphere and rod. A random coil conformation is one in which there are no intramolecular associations of any sort so that the shape of the molecule (irrespective of its gyration) is constantly changing: the time-average shape of such a conformation is spherical. A real sphere is the spherical conformation assumed by a molecule such as rRNA. The 'flexibility' of such a molecule is then envisaged as the limited conformational variance (if any) which arises as a result of limited, permitted transitions within the conformation. The question of flexibility in this case can be reduced to the question 'is the radius of gyration the radius of the real sphere?'. With duplices the problem is a somewhat different one. The radius of gyration for a stiff rod is readily calculated, but if the rod bends, the radius will be smaller. There are two separate models for the flexing of such rods. One envisages there is continuum of flexible distortions of the rod: this is known as the 'wormlike coil model'. The other imagines that the chain consists of a number of stiff elements linked by joints. The positions of the links are randomized and the statistical analysis of such a model is analogous the 'random walk problem'. We shall refer to such a model as the 'gaussian chain'. We shall see shortly that the gaussian chain and the worklike coil are in fact reconcilable with one another.

RNA   There is every reason to believe that high-molecular-weight rRNA does adopt an overall spherical conformation (see p. 127). The question that remains is whether or not the molecule is a real sphere or random coil. It is difficult to come down hard in favour of the real sphere. Boedtker (1960) examined the molecular weight dependence of R_G in fragmented TMV RNA and deduced that the data fitted the random coil. On the other hand, RNA is considerably hypochromic (p. 340), is extensively base-paired and presumably has some rigid structure; the optical melting of RNA is associated with a dramatic reduction in sedimentation coefficient in sedimentation and increase in intrinsic viscosity (Cox and Littauer, 1959). There is a small decrease in the sedimentation coefficient of *E. coli* rRNA which accompanies a change to more dilute salt. These conditions do not result in any detectable hyperchromic effect and so there is no melting of duplices.

It thus seems that in solution of medium ionic strength, the real sphere is fairly tightly held together but that in low salt there is a disordering of 'random' regions of the molecules so that the duplices are free to move relative to one another (Möller and Boedtker, 1962).

DNA   There is an excellent study of flexibility in DNA by Gray and Hearst (1968). They show, in the introduction to their paper, that the wormlike coil model can be regarded as a special case of the gaussian chain as follows. The simple version of the gaussian chain states that the mean square end-to-end distance ($\langle r^2 \rangle$) is related to the number of consecutive units of the chain, n, and the length of each, l by the relationship:

$$\langle r^2 \rangle = nl^2$$

This formula assumes that every possible conformation of the chain is permitted. Clearly this is not the case for a real chain. If the chain is to have some limiting angle (θ) between the segments, the equation becomes:

$$r^2 = nl^2(1 + \cos \theta)/(1 - \cos \theta)$$

In these formulations, it is sometimes the practice to replace nl^2 by $L \times l$ where $L(= nl)$ is the total contour length. In the argument of Gray and Hearst (1968), the wormlike coil is derived from the above by saying that it is the limiting case of the restricted gaussian chain in which l and θ approach zero in such a manner that $[l \times (1 + \cos \theta)/(1 - \cos \theta)]$ remains constant.

The conformational rigidity of a wormlike coil is determined by the length of the projection of the end-to-end distance on to an axis tangential to one end of the coil. This parameter is known as the persistence length. Gray and Hearst calculated the persistence length (or more precisely, twice the persistence length which is known as 'the Kuhn statistical length') as a function of temperature and recognized that if the sedimentation coefficient of a worklike coil were measured as a function of length and temperature (as independent variables) it should be possible to calculate the thermodynamic parameters of flexing. The 'wormlike coils' they studied were four phage DNA molecules of differing molecular weight. The sedimentation coefficients were measured between 5 and 49°C (they even made one series of measurements at

$60^\circ C$). From these results, the entropy change of flexing and the heat content change of flexing were evaluated. The actual magnitude of these values require a detailed understanding of the vectors used to define the flexing process. The interesting qualitative point is that although the process is endothermic (ΔH positive), ΔS is a small, negative value. The implication is that the process is radically different to denaturation, and that DNA (if it is a wormlike coil) readily assumes flexed conformations. Gray and Hearst discuss various models for the process and come down in favour of an idea proposed previously by Printz and von Hipple (1965) that randomized local unstacking of segments of DNA result in the continuous appearance and disappearance of 'flexing positions' in the molecule. The flexing process is of more than academic interest as DNA is normally extensively folded *in vivo* (in the highly condensed conformation that must exist in chromatin). Moreover, we know that in the case of DNA phage, what must originally be a 'soluble' molecule in the host cell has to be folded up in order to get into the head of the newly assembled phage (p. 448).

Hays and Zimm (1970) have extended this treatment to examining the effect of single-strand breaks ('nicks') upon the flexibility of DNA. Polydisperse calf-thymus DNA was used and the breaks were introduced with pancreatic deoxyribonuclease (p. 210). The extent of hydrolysis was assayed by measuring the number-average molecular weights (p. 387) in neutral and alkaline solution. If M_7 is the value of $(\overline{MW})_N$ at pH 7, M_{12} the value at pH 12 and M_R is the mean-residue molecular weight, then the frequency of single-strand breaks, f, is given by:

$$f = \frac{(M_7/2M_{12}) - 1}{M_7/2M_R}$$

The persistence length was found to be the same for both 'nicked' and 'unnicked' samples. This conclusion was based on a comparison of values of $s^0_{20,w}$ and $[\eta]$ for control and nicked samples of differing $(\overline{MW})_N$.[4] An analysis of the results implied that the view that a single-strand discontinuity results in the production of a 'universal joint' in the molecule is untenable. In other words, the stacking interaction overcomes any weakness introduced by the break.

An effect of the nicks was found when the DNA was heated. A decrease in $[\eta]$ was obtained when the nicked preparation was heated: some of this decrease was reversible and presumably reflects the partial denaturation of a 'nicked region'. Most of this decrease was irreversible however and is supposed to be due to the complete removal of a segment (between two breaks, relatively close to one another, in the same strand).

Gel electrophoresis

General methods   The techniques for gel electrophoresis in polyacrylamide described on p. 155 are suitable for molecular-weight estimation. The calculations are based on the assumption that if the mobility is treated as the independent variable, the molecular weight is related to it by a simple exponential relationship, so that a plot of log (MW) against mobility should be a straight line. The methods based on this assumption are the simplest method of estimating the molecular weight of RNA. The experiment must always be performed with internal markers as the parameters such as potential difference and the properties of the gel are extremely difficult to standardize.

In the method of Peacock and Dingman (1968) composite gels of polyacrylamide plus agarose (see p. 160) are employed. Their work is worth studying as they demonstrated the linear relationship between migration distance and time (which is essential for these techniques to be valid) and have also accumulated useful information regarding the relationship between gel-concentration and the electrophoretic mobility of molecules of different molecular weights. In their procedure for calculating molecular weight, the measurements were all made at a standard temperature (6°C) and the mobilities are expressed as $cm^2 V^{-1} s^{-1}$.

A more empirical treatment is that of Loening (1968a). His procedure is the more convenient for most purposes. The gels are 2·4% polyacrylamide run in a buffer containing detergent at room temperature (described on p. 157). Provided the molecular weights of the marker RNA are not too dissimilar from the experimental sample (in other words rRNA should be run on a gel in the presence of a different rRNA as marker and so on) there is a reproducible linear relationship between long (MW) and distance of migration.

Cytoplasmic ribosomal RNA   Loening (1968) estimated the molecular weights of the two large rRNA moleculaes from a variety of sources. In all prokaryotes studied the molecular weight of the smaller molecule (16 S) was $0·56 \times 10^6$ with the exception of *Rhodopseudomonas* (see p. 139) in which it was $0·59 \times 10^6$. Somewhat more variation was found in the larger (23 S) molecule; in most bacteria and blue-green algae it was $1·07 \times 10^6$ but in the two Actinomycetes (see note 2, p. 103) studied it was larger ($1·11$ and $1·13 \times 10^6$). Between eukaryotic organisms (with the exception of protozoa—see below) there is little variation in the size of the smaller (18 S) molecule; it is around $0·7 \times 10^6$. The biggest variation is apparently in fung; the two extremes of molecular weight for this species are shown by the fungi *Aspergillus* ($0·73 \times 10^6$) and *Betrytis (0·68 \times 10^6$). Of animals *Drosophila* is anomalous in having a value of $0·73 \times 10^6$ (most are $0·70 \times 10^6$). It is possible that this result is due to the fact that the anomalous nucleotide composition of this RNA (see p. 285) renders the calibration invalid. The larger rRNA molecule in eukaryotes ('25 S' or '28 S') show considerable variation in molecular weight. Among plants and fungi, the molecular weight is fairly consistently around $1·30 \times 10^6$. The lower limit in this group is $1·28 \times 10^6$ and the upper $1·30 \times 10^6$ with the fern *Dryopteris* ($1·34 \times 10^6$) an apparent exception. Animals show great variation and apparently the '28 S' rRNA has got larger during their recent evolution. *Arbacia* (sea urchins) and *Drosophila* have values of $1·40 \times 10^6$ (although the *Drosophila* figure could be unreliable—see above); among the vertebrates that were studied the figures were, for *Xenopus* (amphibian) $1·51 \times 10^6$,[5] for chicken $1·58 \times 10^6$, for rabbit and mouse $1·72 \times 10^6$ and for man (HeLa) and rat $1·75 \times 10^6$. Protozoa appear to form a special group of their own. The data for the larger molecule are difficult to interpret as in the examples studied the RNA appeared to be especially unstable, but in an amoeba (*Acanthamoeba castellanii*) the value appears to be $1·53 \times 10^6$, which puts it in the middle of the animal range. However in both the amoeba and in *Euglena* (a photosynthetic protozoan) the smaller ('18 S') molecules' molecular weights ($0·89 \times 10^6$ and $0·85 \times 10^6$) are higher than those found in the comparable molecule from any other source.

Ribosomal RNA precursors   The molecular weights of rRNA-precursors have been most reliably estimated by gel electrophoresis (p. 183).

The recent data have been brought together by Grierson, Rogers, Sartirana and Loening (1970) who have measured the molecular weights of the precursors and their various intermediate products in a variety of eukaryotes. Their conclusions are summarized in Fig. 13.4.

Organellar rRNA   The molecular weight of mitochondrial rRNA remains something of a mystery. The problem has been extensively studied and well reviewed by Dawid (1970). He estimated the molecular weight of the two species from *Xenopus* mitochondria both by

	Larger rRNA	Segment I	Segment II	Smaller rRNA	Segment III
Birds and mammals	1·6–1·7	0·3–0·5	0·3	0·7	0·8–1·0
Other eukaryotes	1·3–1·5	0·1–0·15	0·3	0·7	0·1–0·45

FIG 13.4   Evolution of the rRNA precursor. All the sections are identified by their molecular weight $\times 10^{-6}$. It is not known which end of the system is $5'$, nor where the segment III occurs relative to the rest. Segment I is a region that has increased in size during the evolution of warm-blooded animals. Segment II has been conserved; III is variable and no generalizations are apparent.

measuring the sedimentation coefficients on sucrose gradients (see p. 385) and by gel electrophoresis. In both cases the standard was *E. coli* rRNA. The data from the sucrose gradients suggested that the sedimentation coefficients are 21 S and 13 S. They are therefore more slowly sedimenting than any other rRNA species that have been studied. Both centrifugation and electrophoresis imply that the 13 S molecule is of MW $0·35–0·4 \times 10^6$. From the sedimentation coefficient, the 21 S molecule should be $0·7–0·8 \times 10^6$. However the gels implied a significantly higher value ($0·95 \times 10^6$). It is likely that in this case, the extremely anomalous nucleotide composition of mitochondrial rRNA (see p. 286) results in the empirical assumptions of the calibrations being invalid.

Chloroplast rRNA presents anomalies of a different type. The results of Ingle, Possingham, Wells, Leaver and Loening (1970) and Leaver and Ingle (1971) can be qualitatively summarized as follows. The rRNA from the chloroplasts of many higher plants consists of a smaller component and a larger one (as with other rRNA preparations). The larger component ('23 S') is degraded *in vivo*. This degradation accompanies the ageing of the plant tissue. The degradation products can be characterized if the RNA is extracted by the procedure of p. 121 and the gels are run in the 'standard' electrophoresis system of p. 156. On the other hand, if the RNA is extracted in the presence of magnesium ions and electrophoresed in the presence of these ions, the degradation is masked. The degradation apparently involves the very specific cleavage of a particular region of the 23 S molecule.

Examples from their data are presented in Fig. 13.5.

The significance of these results is still somewhat obscure. It is clear that in the 'old ribosomes' of these (and many other plant) chloroplasts, the larger molecule ($1·06 \times 10^6$) is specifically broken to two fragments ($0·7 \times 10^6$ and $0·4 \times 10^6$). Is this breakdown essential for the control of protein synthesis? We do not know as yet. An even more puzzling feature (at least to a chemist) is that the damage can be hidden specifically by magnesium-binding. If it was (say) a break in a single loop of RNA structure, one would expect high concentrations of monovalent concentrations (and low temperatures) would be effective in maintaining the overall secondary structure of the molecule. The implication is that either this RNA contains some features of secondary structure (maintained by Mg^{2+} ions) not previously described, or alternatively the damage is in some region other than a loop and that the fragments are artefactually aggregated in the presence of Mg^{2+}. Studies on the optical properties of this RNA should prove extremely rewarding.

Segmented viral genomes   We have already discussed Cox's work on the *Penicillium* phage which apparently contains two molecules of double-strand RNA. The best established example of a segmented viral genome is that of reovirus. Recent work is described (and earlier results are reviewed) in the paper by Millward and Nonoyama (1970). Fractionation of the RNA by gel

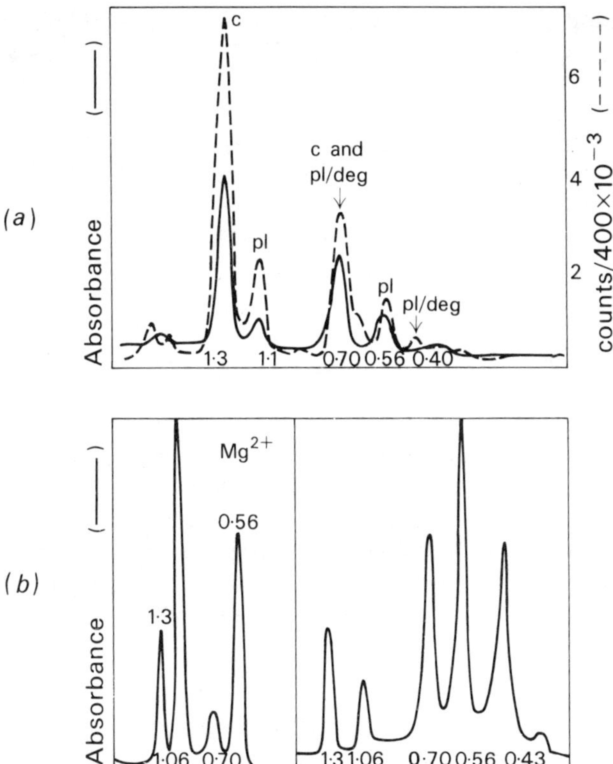

FIG 13.5   Chloroplast rRNA. (*a*) Total RNA from radish cotyledons. The cotyledons were excised from three-day seedlings and incubated for 6 h in [^{32}P] phosphate. The radioactivity (broken line) represents the newly synthesized RNA: c, cytoplasmic rRNA positions, pl, chloroplast rRNA positions and pl/deg degradation products from the 23 S Chloroplast RNA.

(*b*) Similar experiment to the above except that unlabelled tissue was employed and the chloroplasts were purified (so the level of cytoplasmic rRNA is much smaller). In this case the RNA extraction was performed in the presence of Mg^{2+} ions. In one case (Mg^{2+}) the electrophoresis was run in the presence of a magnesium salt. Note that the molecular weight of one of the degradation products of chlorplast 23 S RNA is the same as that of the smaller cytoplasmic rRNA molecule. Data taken from Ingle *et al.* (1970). Numbers refer to molecular weights (x 10^{-6}).

electrophoresis yields ten molecules of molecular weights between 0.72×10^6 and 2.72×10^6. The viral genome is similar to DNA in its behaviour, that is to say it is transcribed by an enzyme present in the virion and mRNA molecules complementary to each of the ten segments are produced (p. 255). Studies on the effect of exonucleases (p. 208) on the RNA suggest that although there is an unpaired section at the end of each segment, these unpaired regions are all purine-rich and not complementary so that the genome is apparently not held together by 'sticky ends' (p. 280).

It is probable that several findings of this type will come to light as more viruses are studied. Influenza virus (strain X-31) is an example of the presence of several RNA species of RNA in a single-stranded RNA virus. In this case the RNA molecules were identified by sucrose-density-gradient centrifugation (p. 285) as well as by electrophoresis (Skehl, 1971). The viral genome consists of six dissimilar molecules of molecular weights ranging from 0.34×10^6 to 0.98×10^6. In this particular case there is excellent circumstantial evidence that any one virion contains one of each of the six molecules. This provides a total influenza genome size of 3.9×10^6. This figure is in excess of values obtained by the 'particle count method' (p. 398) as determined in earlier work and serves as a warning to the reliability that can be placed on such indirect procedures.

Messenger RNA   The haemoglobin mRNA (p. 197) is probably the only cellular messenger RNA which is known in a pure form at the moment. The molecular weight of this 9 S RNA has been estimated by electrophoresis to be of molecular weight 2×10^5 (Gaskill and Kabat, 1971). This figure implied that there are considerable stretches of 'non-informational' sequence.

Direct methods of measuring molecular weight

Light scattering

Light scattering is a phenomenon normally associated with particles (the Tyndall effect) although it is observable with macromolecules. The experiment consists of shining light through a solution of the sample and measuring the light scattered at various angles to the incoming light. Technical details and precautions that must be taken are described by Boedtker (1968a). Measurements are made at a variety of angles and a variety of concentrations. In this section I only present sufficient information for the details of any paper involving light scattering to be understood. The technique is one of the absolute calibration methods used for establishing the size of a few carefully selected reference molecules. Its main limitation is the requirement for several mg of a sample.

The light scattering equation   The following equation is applicable for the scattering of light by particles (or molecules) with dimensions comparable to that of the incident (visible) light. In this equation, observations are made at angle θ to the incident beam, R_θ is the ratio of scattered-light intensity to incident-light intensity, $\overline{MW}_w$ is the weight and c is concentration (conventionally in g/ml). K, B and α are constants that will be discussed shortly. The equation is:

$$\frac{Kc}{R_\theta} = \frac{1}{\overline{MW}_w} + 2Bc + \frac{\sin^2(\theta/2)}{\overline{MW}_w}$$

This equation does make certain assumptions about the factorization of the 'particle scattering factor' but is sufficient for the present discussion.

Treatment of results   The constant K is determined solely by the geometry of the cell and the wavelength of the incident light. It can thus be calculated for any set of experiments. The other two are thermodynamic parameters of the solute. B is actually the second virial coefficient for the solution and α is a function of the radius of gyration.

The usual procedure for analysing light-scattering data is to plot them on a graph in which Kc/R_θ is the y-axis and the x-axis is $[\sin^2(\theta/2) + kc]$. In this plot, k (the 'spreading constant') is arbitrarily chosen to spread out the x-axis nicely. The value of k for most nucleic acid plots is of the order of a few thousand. At first sight this (known as a 'Zimm plot') is a weird method of presenting results as the experimental values form a non-rectilinear grid on the graph. In practice the Zimm plot is extremely convenient as $\overline{MW}_w$ and R_G may be read off directly. Figure 13.6 is a hypothetical Zimm plot for a solution of RNA or DNA. The intersection of the two extrapolated lines (Fig. 13.6) gives the value of $1/(\overline{MW})_{Mw}$. The radius of gyration is measured from the slope of the $c = 0$ line as:

$$R_G = \frac{\sqrt{3}\,\lambda\,(Kc/R_0)^{1/2}}{4\,(\text{slope})^{1/2}}$$

where λ is the wavelength in solution.

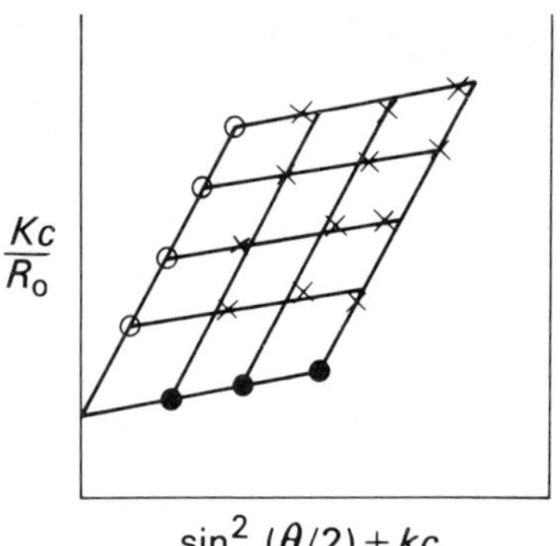

FIG 13.6   Hypothetical Zimm plot for a nucleic acid solution. The crosses are experimental points. In this case three concentrations have each been measured at five angles. The circles are the extrapolated values. Thus ●————————● is the plot of the function when $c = 0$ and ○————————○ is the plot for $\theta = 0$.

The R_G obtained in this way is a number-average value. Therefore great care is required in comparing MW and R_G unless the system is strictly monodisperse (e.g. viral RNA). The value of R_G for a random coil or sphere of MW 2×10^6 is of the order of 350 Å. However the value is affected by flexibility, although it may give the wrong impression of flexibility (p. 393).

Direct measurement

The DNA molecule may be indirectly visualized by autoradiography (p. 178) or by electron microscopy (p. 188). Similarly RNA may be visualized in the electron microscope. These lengths may be converted to molecular-weight values. Schmid and Hearst (1969) are unimpressed with the method and point out several problems associated with it, in particular one has to make assumptions regarding the secondary structure of the nucleic acid. This could be overcome by calibrating the method with nucleic acids of known molecular weight. According to Schmid and Hearst, the groups who have been responsible for these determinations and correlations have not on the whole included the reference sample in the same preparation as the test sample so that variation in drying-out between different slide or grid preparations is not aken into account. The effect of ionic strength (in the solvent from which the grids are prepared) on the contour lengths has been studied by Inman (1967).

Despite the difficulties encountered in interpreting the data, several estimates of length and molecular weight appear in the literature and represent the only knowledge we have of some of the sizes of nucleic acids. The essential point about such data is that they are valueless unless the authors make it clear how many measurements have been made and that they quote the standard error of the mean. A stringent statistical test for the accuracy of histograms employed for chain-length estimates of nucleic acids from electron microscope pictures is discussed by Inman (1965).

According to Highton and Beer (1963), for single polynucleotide chains (denatured DNA or RNA) the molecular weight is given by the formula:

$$\text{MW} = \frac{\text{length (Å)}}{5 \cdot 2} \times \text{mean residue MW}$$

For a discussion of the relationships for duplex viral DNA, see Schmid and Hearst (1969).

The value for the autoradiographs of *E. coli* DNA obtained by Cairns (see p. 178) was 1100

μ. For large DNA molecules, the value of 1.97×10^6 daltons/μ seems to be reasonably accurate. Thus the molecular weight of the *E. coli* chromosome is about 2.2×10^9.

Indirect and chemical methods

This section is admittedly something of a hotchpotch of procedures, some of which are not entirely satisfactory. All do share one great advantage over the preceding methods; no assumptions are made about the conformation of the molecules. Before referring to these procedures it is necessary (for completeness) to cross-refer to the 'Cot method' (p. 364) which is an indirect method of measuring the molecular weight of DNA.

End-group methods

On the whole end-group methods are accurate for small nucleic acids and oligomers but rather inaccurate for large nucleic acids. They thus fill the gap extremely well as the hydrodynamic properties of the small molecules are difficult to correlate accurately with molecular weight. There are two methods of performing the experiment; as both of them involve no new principles they can be dismissed very briefly.

The terminal phosphate may be removed with alkaline phosphatase; the ratio of inorganic/organic phosphate in the products is a measure of chain length. The simplest procedure is either to use uniformly labelled [^{32}P] nucleic acid and to follow the radioactivity that appears in inorganic phosphate, or to remove the terminal phosphate with the enzyme and to rephosphorylate the 5′-terminus with polynucleotide kinase (p. 228) using [^{32}P] ATP of known specific activity. There is a check on the reliability of this method as the product can be digested with alkaline phosphatase again and the quantitative release of ^{32}P can be examined.

The alternative (for RNA or RNA fragments) is to use the glycol splitting reaction (p. 287). The main drawback is that there is no method of checking that the oxidation (and the formation of derivatives) is quantitative. Another method only suitable for small RNA fragments, is to digest the oligomer with alkali. The ratio of nucleotide/nucleoside in the product is the chain length (see p. 289).

A possible end-group method for DNA is described on p. 281.

Chemical analysis

Methods for the assay of DNA are described on p. 173. Alternatively the DNA in viruses or bacteria can be determined from the radioactivity of DNA-phosphate in cells (or viruses) that have been labelled with [^{32}P] phosphate *in vivo*.

If the DNA content of a viral preparation is known, the total weight of the viral genome may be determined either by counting the viruses as plaque forming units or by measuring the total protein in the virus (assuming that nucleic acid and protein are the only components of the virion. The former method is suitable only if the plaque-forming efficiency is 100% but it has been reliably employed in the past for estimating the genome size of phage.

With bacteria, the same general method applies. Some assumption about the number of chromosomes per cell has to be made. This is usually done by staining the cells for nuclear bodies and assuming that there is one replicating chromosome per 'nucleus'. This is in all probability invalid as several replication forks can exist in the same chromosome. The number of bacteria can be counted directly in a Petroff-Hauser chamber and no assumptions of clone-forming efficiency need to be made.

Suicide and stars

Suicide rates   This method is only applicable to phage. The principle is very simple. A uniformly labelled radioactive viral chromosome is inactivated if a radioactive decay occurs in it. The method then consists of just keeping a radioactive preparation and assaying the phage titre at a series of times. The suicide rate is given by the relationship:

$$-\frac{\mathrm{d}S}{\mathrm{d}t} = \alpha\, N^*vs$$

where S is the surviving virus titre at time t, α is a constant, N^* is the number of radioactive atoms and v the fractional rate of decay. This equation is integrated and re-written to eliminate αN^* and hence becomes an exponential function of $(-vt)$. The empirical values of constants that are required if the isotope is ^{32}P (Hershey, Kamen, Kennedy and Gest, 1951) or ^{3}H (Caro, 1965) are available in the literature. The second reference describes an accurate value for the molecular weight of T4 obtained in this way.

Star gazing   The principle of this method is that the DNA (typically phage DNA) is labelled with ^{32}P at the time of its synthesis. The preparation is embedded in a photographic emulsion and after a given time the film is developed and examined under the light microscope. What is seen is a set of tracks (few hundred microns long) each produced by the β-particle emitted from a ^{32}P-atom at the time of its decay. As the ^{32}P-atoms are located in DNA molecules, the pattern that is seen is a set of 'stars' of these tracks. The centre of each star corresponds to the presence of a DNA molecule. The molecular weight is given by the relationship:

$$MW = \frac{N^*m}{3\cdot4 \times 10^{-6}\, A_0}$$

where N^* has the meaning defined above, m is the mean-residue molecular weight, A_0 is the specific activity of phosphorus in the growth medium in Ci/g and $3\cdot4 \times 10^{-6}$ is a proportionality constant whose magnitude is derived from the definition of the curie and the value of v (see above). The attractive feature of this procedure is that it does not matter whether the 'preparation' is isolated DNA or a virus particle and hence it may be used to show that the molecular weight of isolated DNA is the same as that in the virus (Rubinstein, Thomas and Hershey, 1961). There are problems however; if there are too many rays in the star, the true number of disintegrations may be in excess of that scored (on account of the superimposition of two or more tracks). If the number is too small, the statistical analysis of the results becomes more difficult. Kahn (1964) has made a critical appraisal of the method.

Notes to chapter 13

[1]   I do not write 'schlieren' with a capital S because, although it is a German noun, it is not the name of a physical chemist, but simply means 'streaks'.

[2]   If you have an analytical centrifuge in your department, this is by far the quickest way of checking that an RNA preparation is undegraded. Allow 30 min to get the cell ready, the rotor installed and the centrifuge evacuated and so on. Then run it at top speed (60 000 rpm). Boundaries for a mixture of (for example) 16 S and 23 S RNA are seen after 15 min or so. Allow 10 min to develop and fix the plates (assuming you need a photographic record).

[3]   Strictly β is the ratio of two different conformation-dependent terms, which arise in the theoretical expressions for s and $[\eta]$; these two expressions are combined in the Mandelkern–Flory equation.

[4]   These preparations were obtained by sonicating the original DNA.

[5]   This figure relates to tadpole RNA. For some reason not so far explained, the adult tissues contain a 28 S component of higher molecular weight ($1\cdot54 \times 10^6$).

14 Interaction between nucleic acids and small molecules

The nucleic acids contain a variety of polar groups and systems of delocalized electrons; consequently, in addition to the extensive chemistry of DNA and RNA (chapter 8), a variety of intermolecular interactions are possible. In this chapter, I have selected three classes of low-molecular-weight compounds which form stable complexes with nucleic acids. Each of these classes is important as the interactions are of biological importance and additionally form useful probes into the structure of nucleic acids.

Metal ions

Considerable interest attaches to the study of the interactions between metal ions and nucleotides and polynucleotides. For one thing many enzymes are dependent upon the presence of a polyvalent metal ion for their activity. Thus every enzyme for which ATP or any other nucleoside triphosphate is a cofactor or substrate have an absolute requirement for a divalent metal ion (typically Mg^{2+}). The binding of metals to such nucleotides is outside the scope of this book. However the interaction of metal ions with polynucleotides is vital to the metabolism of nucleic acids and their interactions with protein. Additionally, the interaction of DNA with certain transition metal ions forms the basis of a method of fractionating DNA on the basis of its nucleotide composition.

Some general considerations of the low-molecular-weight complexes

The ligand properties of metal ions I refer the reader to textbooks of general or inorganic chemistry for an account of ligand field theory (the generalizations that predict the number of coordination positions around metal cations and the geometry of these positions). The fact that a metal has six octahedral positions of this type (the commonest arrangement), does not mean, of course, that these positions are all filled with the ligand in which we are interested; some of the positions may be occupied with low-molecular-weight anions or water molecules. However, the general layout of the coordination complexes of the cations important in studies on nucleic acids are as follows.

Mg^{2+} has six octahedral positions, Mn^{2+} the same but the octahedron is somewhat distorted. The same situation arises in some Zn^{2+} complexes; others have four tetrahedral positions. The coordination chemistry of cobalt is complicated by valency considerations: Co^{2+} forms either four (tetrahedral) or six (octahedral) coordination positions; Co^{3+} forms six (octahedral); moreover octahedrally coordinated Co^{2+} is more readily oxidized than is the free

Co^{2+} ion. Ni^{2+} has four planar coordination positions. Fe^{2+} and Fe^{3+} form six octahedral ligand positions. The stereochemistry of coordinated Cu^{2+} is surprisingly complex: essentially it has four planar coordination positions with the possibility (at least in the solid state) of forming two weak ligand associations perpendicular to the plant ('completing' an octahedron). Ag^+ and Hg^{2+} both have two (linear) ligand positions in most 'complexes' (just 'compounds' in the latter case). However Ag^+ can also form tetrahedral coordination complexes and Hg^{2+} forms some complexes in which the metal lies at the centre of a distorted octahedron.

Complexes between metal ions and bases or nucleosides   It is uncertain what significance, if any, the complexes between metal ions and bases or nucleosides may have for the complex formation with the nucleic acids. Several data on the stability of these complexes are compiled by Weser (1968). In summary, purines form more stable complexes than pyrimidines and bases

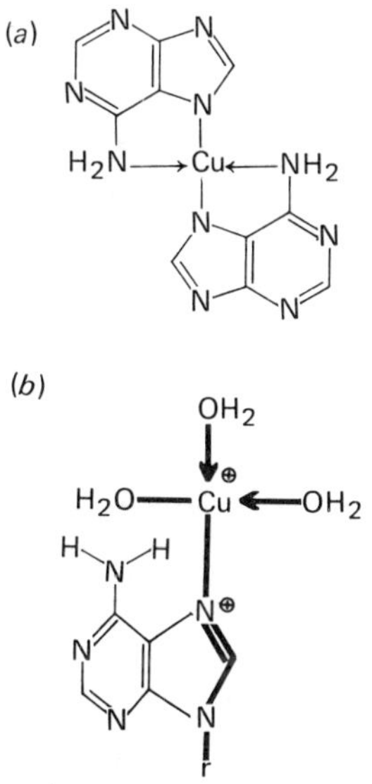

FIG 14.1   Proposed structures of the planar complexes between Cu^{2+} and adenine (*a*—after Harkins and Freiser, 1958) and adenosine (*b*—after Schneider, Brintzinger and Erlenmeyer, 1964).

stronger ones than nucleosides. Indeed no pyrimidine nucleoside complexes are formed with any metal except Hg^{2+}. The reason for the greater stability of the base complexes has been ascribed to the availability of the 'glycosidic' nitrogen for chelation in the bases. The point is illustrated by a comparison of two of the stronger complexes, viz. those between Cu^{2+} and Ade and Ado (Fig. 14.1).

The situation with nucleotides is rather different. The effort has largely been put into understanding the biologically important interactions between metal ions ATP. The difficulty attaching to the interpretation of the metal-ion binding by nucleic acids in terms of these (and other) data is that the freedom of conformational states in the nucleotides is so great that equilibria can exist in which various parts of the nucleotide are ligand donors. The present formulations are derived from the work of Brintzinger (1963), who recognized the equilibria were between three classes of complex; those in which the heterocyclic system formed the donor ('metal-ring complexes'), those in which the phosphates did so ('metal-chain complexes') and those in which both types of group participate ('intermediate').

Complexes with nucleic acids

For the purposes of a brief review such as the present, it is convenient to group the metals into groups. Finally, the possible biological significance of some of the findings is reviewed.

Mercury and silver   These two metals form linear complexes with ligands (p. 403). Both bind strongly to DNA and have a marked preference for single-strand DNA. Moreover they have considerable selectivity in terms of primary structure (Ag for high-GC DNA and Hg for high-AT DNA). As the complexes have markedly different properties to those of uncomplexed DNA and the metal ions can be readily removed by any suitable anion (such as EDTA, CN^- or Cl^-)[1] it is clear that these metals can be used for the fractionation of DNA. In practice the techniques have been relatively little exploited. It is probable that the whole question of applying these salts to DNA fractionation would be well worth thorough re-examination.

References to the spectrophotometric and other assays for the bindings are to be found in Weser (1968). A model for the Hg^{2+}-complex involving only T-residues is due to Katz (1963). The binding of Ag^+ ions by GC-pairs has been postulated by Jensen and Davidson (1966). They point out that if the ion is trapped between a base pair, less distortion of the Watson–Crick geometry is involved in the case of the insertion in GC-pair (see below) than is required for the analogous insertion between an AT-pair.

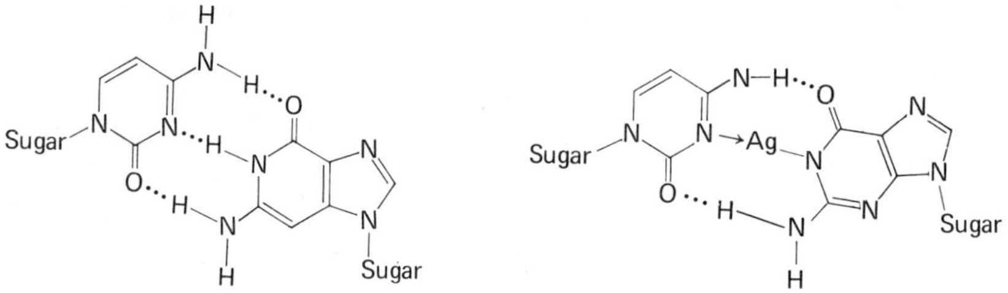

FIG 14.2   Watson–Crick GC-pair compared with the model for a G-Ag-C pair proposed by Jensen and Davidson (1966).

The separation of DNA fractions on the basis of their metal binding capacities is performed by buoyant density centrifugation (p. 149) using Cs_2SO_4.[2] The most familiar use of the technique is in the preparation of crab poly d(AT) (p. 152; the following method is from Davidson et al., 1965). Crab DNA is dialysed against 5 mM sodium borate pH 9·2; $HgCl_2$ (0·1 mole/mole DNA-P) and Cs_2SO_4 to a final density of 1·48 are added and the solutions are centrifuged to equilibrium at 20°C. The poly d(AT)/Hg^{2+} forms a heavier zone than the bulk DNA.

Copper and zinc   Both these metals bind to DNA but with very different effects; this is hardly surprising as the geometry of the complexes is very different (p. 402). Zinc has a stabilizing effect on the duplex. The melting temperature is increased in the presence of zinc ions (1 mole Zn^{2+}/mole DNA-P). More remarkably the melting is reversible. Thus DNA preparations that do not renature faithfully normally (p. 363) anneal apparently perfectly in the presence of the zinc, as though they had been cross-linked (p. 240; Shin and Eichorn, 1968).

There is an extensive (and somewhat conflicting) literature on the interaction between DNA and Cu^{2+} ions. Fortunately, the uncertainties have been cleared up by an excellent definitive paper by Zimmer and Venner (1970). They examined the interaction between cupric salts and DNA of different nucleotide composition, and discovered the effects (a strange sort of hyperchromicity and additionally reduced melting temperature) were greater, the higher the GC-content. The contention that it is interaction with GC-pairs that is responsible for the effects was confirmed by the observation that the replacement of G-residues by me^{7}G reduced the effects.

Magnesium and manganese   There is no reason to suppose that the interaction of Mg^{2+} ions with nucleic acids involves any nucleotide-ligands other than phosphate. The role of magnesium in stabilizing nucleic acid and nucleoprotein complexes thus seems to be one of bridging phosphates or (presumably bridging a phosphate and a carboxylic acid side chain in a protein. It is equally true of Mn^{2+} that there is no real evidence that complexes other than of the same ('metal-chain') type are formed (Weser, 1968).

Biological implications   There are two possible biological implications of the interactions of heavy metals and nucleic acids. These metals are trace requirements for the growth of cells; conversely they are toxic when present in high concentrations. Are the requirements and/or the toxicities due to interactions with nucleic acids? The questions are extremely hard to answer as few satisfactory experimental systems have been devised. It is the view of Wacker (1962) that a crucial consequence of zinc deficiency is the inhibition of DNA synthesis and transcription. No clear explanation for this supposed requirement for zinc in DNA-dependent processes is available.

A more productive approach is likely to be the study of the effect of inhibitory amounts of these ions on nucleic acid metabolism. D. G. Wild's laboratory is engaged on studies of heavy-metal salts on RNA metabolism in micro-organisms. A survey of several salts (Blundell and Wild, 1969) revealed that $CoCl_2$ is a selective inhibitor of protein synthesis. During treatment of *E. coli* with the salt (up to 0·3 mM) protein synthesis is stopped but RNA synthesis continues with the result that 'incomplete ribosomes' are assembled. The extension of this type of experiment to ribosome assembly *in vitro* (p. 461) should prove of great value. In the meantime there is every reason for a thorough study of the interaction between Co^{2+} and RNA to be carefully characterized.

Polyamines

The polyamines are a widely distributed class of natural products which interact strongly with nucleic acids. They undoubtedly play a role, not yet fully understood, in maintaining the integrity of ribosomes structure *in vivo* (p. 125) and are possibly involved in the control of transcription. There follows a brief survey of the polyamines and their interaction with nucleic acids. For a fuller review see Stevens (1970).

Description of the polyamines

Occurrence and nomenclature   The major, unsubstituted polyamines that occur in nature are the following. The shorthand formulation is not a standard convention but makes the structural relationship of the substances immediately apparent.

1,3-Diaminopropane	$H_2N(CH_2)_3NH_2$	n c$_3$ n
Putrescine	$H_2N(CH_2)_4NH_2$	n c$_4$ n
Cadaverine	$H_2N(CH_2)_5NH_2$	n c$_5$ n

bis(3-Aminopropyl)amine	$H_2N(CH_2)_3NH(CH_2)_3NH_2$	n c₃ n c₃ n
Spermidine		n c₃ n c₄ n
Spermine		n c₃ n c₄ n c₃ n

On the whole, spermine is only found in eukaryotic cells. In most animal tissues, the major polyamines are spermidine and spermine; plants contain these two and putrescine as well. Most bacteria contain putrescine and spermidine as their major polyamines and on the whole, less polyamine is found in Gram positive bacteria than in Gram negative species.

Chemical properties   The polyamines are either liquids or low-melting-point solids with mildly unpleasant smells. They are strongly basic; the pK-values for the primary amino groups are around 10–11; those for the secondary amine groups are around 9–10. They form crystalline salts with acids. The phosphates are of very low solubility.[3]

Fractionation of the polyamines is relatively simple. They are volatile, stable and differ from one another in ionic properties. They may thus be separated by gas chromatography (Smith, 1971), ion-exchange chromatography (Dubin and Rosenthal, 1960) or paper chromatography and paper electrophoresis (Herbst, Keister and Weaver, 1958). The only precautions that are necessary in handling them derive from the fact that (in common with many aliphatic amines) the free bases slowly react with CO_2 in air to form the substituted ammonium carbamates. Small quantities of the free bases should therefore be handled under nitrogen.

Interaction between polyamines and nucleic acids

DNA   Polyamines stabilize the DNA duplex in that they increase the melting temperature. Surveys of the effectiveness of different polyamines have been made (Tabor, 1962; Stevens, 1967). Spermine is the most effective (concentrations as low as 2×10^{-6} M spermine will cause an increase in T_m of as much as 5°). The stoicheiometry of the binding of spermine to DNA under saturation concentrations has been found to correspond to one amino group per 5 DNA-phosphate residues (Hirschman, Leng and Felsenfield, 1967). Details of the nature of the association remain to be clarified. No gross conformational changes occur in solution, as the ORD of the DNA is unchanged in spermine solution (Maestre and Tinoco, 1967). It seems unlikely therefore that models in which the polyamine skeleton cross-links the helix are tenable.

A different type of association is obtainable out of solution. DNA may be precipitated from solution with very high concentrations of spermine. The complex that is precipitated contains much higher ratios of spermine to DNA (about one amino group per DNA-phosphate residue). X-ray fibre diagrams of these complexes have been interpreted in terms of two categories of spermine molecule; some that do span the strands of the helix and others that span the minor groove (Suwalsky, Traub, Shmueli and Subirana, 1969). It is tentatively proposed that in solution all the spermine/DNA interactions are of the latter type.

Polynucleotides and RNA   Spermidine and spermine have a striking effect upon the optical properties of poly U. The addition of one of these polyamines to the nucleotide results in a pronounced hypochromic effect which can be 'melted out' co-operatively. From an analysis of these data and the production in the poly U solution of a large Cotton effect (p. 89), Rogers, Ulbricht and Szer (1967) proposed that the effect of the polyamines was to produce a duplex conformation of the poly U. This conclusion might explain earlier results in which it was established that spermine protects rRNA and viral RNA against thermal denaturation and also ribonuclease (see p. 122 and also references cited by Stevens, 1970). Certainly rRNA has more spermine-binding sites than does DNA (Stevens, 1969).

Effects of polyamines on the biochemistry of nucleic acids

DNA polymerase and RNA polymerase   The effects of polyamines on the reaction catalysed (*in vitro*) by RNA polymerase have been extensively studied. A paper which describes many of these effects (using the *E. coli* polymerase) and contains extensive references to earlier work is by Abraham (1968). In summary, the addition of either spermidine or spermine to the reaction in which double-stranded DNA is employed as template is stimulatory (polyamine concentrations up to about 1 mM—higher concentrations are inhibitory). The effect is not found if single-stranded DNA is employed as template. There are several possible explanations for the stimulation but it is most probably due an interaction of the polyamine with the RNA chain. Thus polyamines will reduce the inhibitory effect of added exogenous RNA on the reaction; they will allow the reaction to proceed for a longer time than occurs in the control system and, if they are added to a system which has become exhausted (reached the plateau of synthesis), they will cause a new surge of synthesis. Moreover it has been shown by Gumport and Weiss (1969) that with the Micrococcal polymerase the reaction is more asymmetric; that is to say the enzyme then behaves more as it does *in vivo* in choosing faithfully the transcribing strand.

The effect on DNA polymerase is different. At very low concentrations spermine is marginally stimulatory but above about 10^{-4} M it is markedly inhibitory (O'Brien, Olenick and Halm, 1966). Their data may be of pharmacological interest as they point to parallels with the effects of antimalarial drugs on the same reaction.

It is most reasonable that the former of these effects (the one involving RNA polymerase) is of significance *in vivo*. There is an extensive literature (reviewed by Stevens, 1970) on the intracellular levels of polyamines and the activities of enzymes required for their biosynthesis and the possible correlations of these levels with the phase of growth of the cells. In *E. coli* the ratio of polyamine-N:RNA-P is approximately constant (about 0·22). The direction of the control appears to be that it is the polyamines that operate a positive control on RNA synthesis, for if the amount of polyamine in the medium is increased, the synthesis of RNA (but not of DNA nor protein) is increased (Raina, Jansen and Cohen, 1967). However it is not necessarily the case that the effect is a direct one. Another possibility is that the control is due to the effect of the polyamines on the availability of amino acids for protein synthesis (Stevens, 1970). The review also refers to analogous work on mammalian cells in culture and to the indirect evidence that the synthesis of polyamines which accompanies liver regeneration may be associated with the RNA synthesis in the tissue.

Ribosomes and protein synthesis   In view of the fact that polyamines bind to RNA (p. 406) it is hardly surprising that they have a considerable effect on ribosomes. They promote the association of subunits and (in high concentration) they cause aggregation of ribosomes. The effects on protein synthesis differ considerably from system to system. However in certain instances they produce mis-coding. There is probably some confusion in the literature owing to the fact that the effects of different polyamines (notably putrescine and spermidine) may be antagonistic in this reaction. The review of Stevens (1970) contains a discussion of this point, together with many references to the studies on cell-free systems for protein synthesis.

Aryl-substituted triamines

These substances are of great importance in structural studies on both nucleic acids and nucleoprotein (p. 455) as the spectral properties of the aromatic chromophore are sensitive to the environment of the chromophore. When they are bound to nucleic acids, the conformation of the molecule can be deduced from the shifts in the spectral (and optical) properties of the chromophore. They are thus examples of 'reporter molecules'.[4]

The examples quoted in this section are all taken from the paper by E. J. Gabbay (1970). He synthesized a wide range of reporters which differed both in chain length and in the nature of the chromophore. The structures of these referred to here are shown in Fig. 14.3.

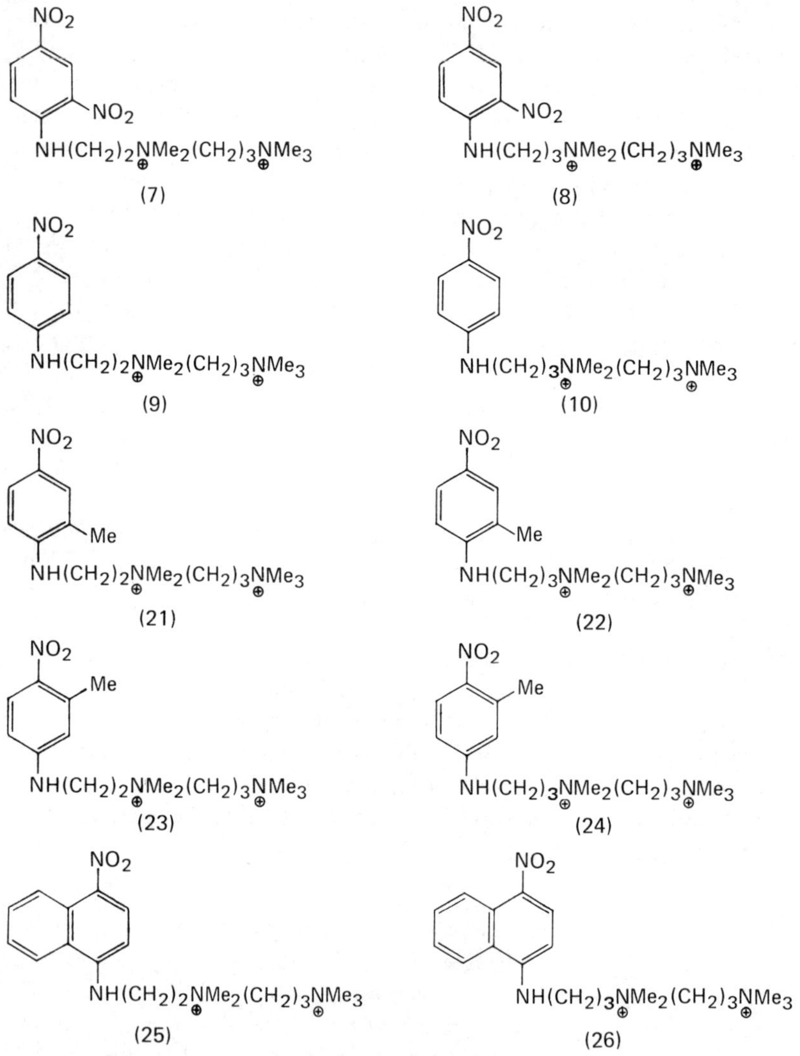

$NH(CH_2)_2\overset{\oplus}{N}Me_2(CH_2)_3\overset{\oplus}{N}Me_3$

(7)

$NH(CH_2)_3\overset{\oplus}{N}Me_2(CH_2)_3\overset{\oplus}{N}Me_3$

(8)

$NH(CH_2)_2\overset{\oplus}{N}Me_2(CH_2)_3\overset{\oplus}{N}Me_3$

(9)

$NH(CH_2)_3\overset{\oplus}{N}Me_2(CH_2)_3\overset{\oplus}{N}Me_3$

(10)

$NH(CH_2)_2\overset{\oplus}{N}Me_2(CH_2)_3\overset{\oplus}{N}Me_3$

(21)

$NH(CH_2)_3\overset{\oplus}{N}Me_2(CH_2)_3\overset{\oplus}{N}Me_3$

(22)

$NH(CH_2)_2\overset{\oplus}{N}Me_2(CH_2)_3\overset{\oplus}{N}Me_3$

(23)

$NH(CH_2)_3\overset{\oplus}{N}Me_2(CH_2)_3\overset{\oplus}{N}Me_3$

(24)

$NH(CH_2)_2\overset{\oplus}{N}Me_2(CH_2)_3\overset{\oplus}{N}Me_3$

(25)

$NH(CH_2)_3\overset{\oplus}{N}Me_2(CH_2)_3\overset{\oplus}{N}Me_3$

(26)

FIG 14.3   Structures of various nucleic acid reporters. See text for discussion. The anions (not shown) are Br⁻ in every case. The numbers of the molecules are employed for reference in the text. They are the same as those used by Gabbay (1970) in order to facilitate cross-reference to this very important paper for those who wish to read further into the subject.

Using the convention of p. 405 (slightly modified) to simplify these formulae, it can be seen that the molecules with odd numbers are all of the form Ar n c_2 ṅ c_3 ṅ (where Ar is an aryl group and ṅ is a quarternary amonium centre, the remaining valencies being occupied by methyl groups). The molecules with even numbers are all of the type Ar n c_3 ṅ c_3 ṅ.

The rigorous argument relating to the interaction of these substances with nucleic acids are discussed in Gabbay's paper. In general terms, the chain lies down the outside of the helix and the chromophore interacts with the bases.

UV spectra of bound reporters   All the reporters in Fig. 14.3 have a strong UV absorption spectra. For those in which the chromophore is a nitrobenzene group (7–10; 21–24), the maximum is at around 430 nm with a molar absorptivity of around 15 000. When bound to nucleic acid, there is a marked red shift and a concomitant hypochromic effect. The red shift can be analysed in terms of charge-neutralization; thus the aliphatic chain of the reporter lies along phosphate residues. The hypochromic effect is due to a parallel orientation of the transition moments of the reporter chromophore with the bases in the nucleic acid. In other words the chromophore is in some way stacked parallel to the bases. This effect is greater with DNA than RNA. The geometry of the association between the chromophore and the base could be of one of two types. Either the chromophore could be sandwiched between bases ('intercalated'—see below) or it could form an edge-to-edge association with a base. Gabbay showed that the latter alternative is the correct one. It is sterically impossible to build a model in which the chromophore is intercalated and the aliphatic chain is aligned along the sugar-phosphate backbone. That the aliphatic chain is so aligned is established partly by an analysis of the UV data and also by the fact that diamines (such as cadaverine and putrescine) compete with reporter for DNA binding sites.

CD of bound reporters   None of the reporters is an optically active compound. However, when a chromophore is associated with an optically active matrix, a Cotton effect (and associated dichroic maximum—see p. 90) is induced. The sign and magnitude of this CD maximum was measured for several cases. The simplest reporters are (9) and (10). The dichroic maxima are positive for either DNA or RNA;[5] if the compound with the shorter chain length (9) is compared with the longer (10) it is found that, for DNA, the molar ellipsisity is much larger in the latter case than the former. No such dramatic effect is seen with the RNA. These data confirm that the sugar-phosphate backbone is involved in the binding reaction. In DNA there is only one sort of (duplex) conformation; in the RNA there were various types of binding site available.

When the disubstituted reporters (7) and (8) were used, the results were different. The CD peak for RNA was positive but that for DNA was negative. This has been analysed by Gabbay

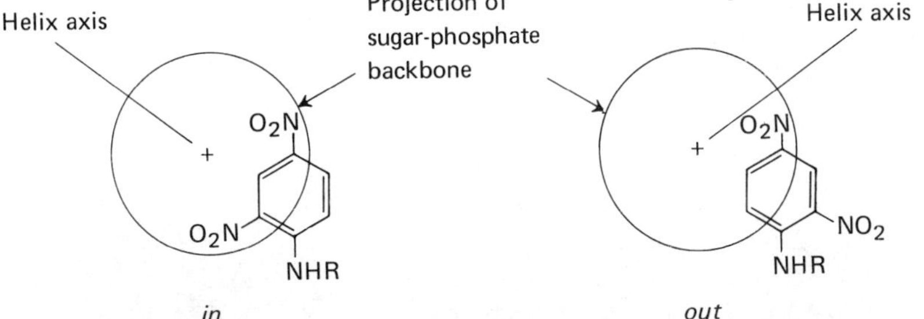

FIG 14.4   Definitions of the *in* and *out* conformations of the reporter/nucleic acid complexes illustrated with reference to dinitro reporters (7) and (8). After Gabbay (1970).

in terms of the two possible conformations of the chromophore in the helix, referred to as '*in*' and '*out*'. The conformations are defined in Fig. 14.4.

A systematic study of the effects of chain-length in the reporters (7) and (8) and (21)–(26) was undertaken. The data were analysed in terms of the steric restrictions imposed by the extra methyl groups (in T-residues) in DNA and the 2'-OH groups in RNA. In model-building, strictly parallel alignments of transition moments were regarded as axiomatic. The data were entirely consistent with the following generalizations. The chromophore/base interaction is a genuine case of an edge-to-edge complex; the chain lies in the small groove of the duplex; the RNA/reporter complex is invariably in the *out* conformation. With DNA, the *in* conformation is allowed in certain cases (including (7) and (8)).

The importance of this work is that there is now a set of completely defined conformational probes for examining details of nucleic-acid conformation. An application of their use is given on p. 405.

Intercalation

Intercalation is the name given to the sandwiching of a flat molecule in between the stacked bases in a nucleic acid. The process is accompanied by local unwinding of the helix, to make the distance between adjacent base pairs large enough for the intercalated species to be slipped in. The phenomenon is important for the light it throws on the biological properties of an important class of dyes. It is also important in that it can be used to probe details of the tertiary structure of DNA (see p. 427) and as the basis of a fractionation procedure for DNA of differing tertiary structure (p. 424). The discussion is subdivided according to the nature of the intercalating species.

Substituted acridines and phenanthridines

Intercalation was first proposed (by L. S. Lerman) as a model for the binding of acridine dyes to DNA and the bulk of the theoretical and experimental work relates to this group. The structure of the compounds most commonly used are shown in Fig. 14.5.

Experimental evidence for intercalation The binding of acridine dyes to DNA may be followed spectrophotometrically (see below). The results show that with duplex DNA or homopolymers such as poly A : poly U or poly I : poly C the binding isotherm is biphasic. That

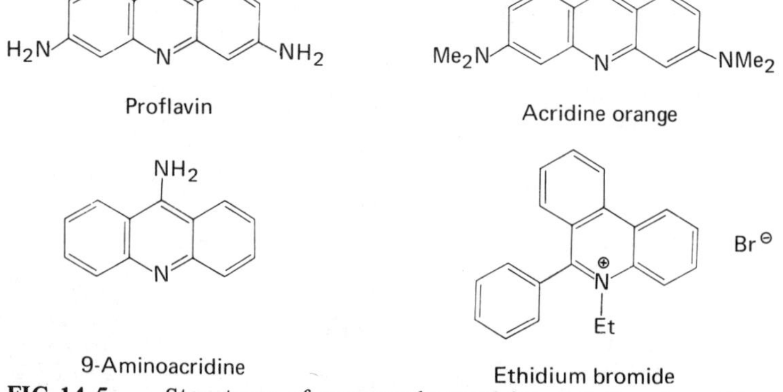

FIG 14.5 Structures of commonly used intercalating dyes. The first three are substituted acridines; ethidium is a substituted phenanthridine.

is to say that if the ratio of moles of bound dye per mole DNA-P (symbol r) plotted against moles of unbound dye, the curve plateaus at r = approximately 0·25 (at unbound concentrations of the order of 10^{-5} M). At higher concentrations of dye the value increases to r = 1·0 (or even above). Such data are summarized in the article by Jordan (1968). Similar data can be obtained with ethidium bromide. The first phase of binding is referred to as the formation of complex I; the second phase (high r-values) as the formation of complex II. The following evidence supports the view that complex I is intercalated.

Bound acridine chromophores (in complex I) exhibit a marked red shift. Bradley and Wolf (1959) pointed out that this shift would be characteristic of a conformation in which the chromophore was parallel to a different π-electron system. The stacking of acridine chromophores with one another should produce a blue shift, to be expected if the dye stacks on the surface of the polyanion. Indeed, when the dye is in excess (complex II) such a blue shift is observed. Fluorescence spectra of bound and unbound proflavin were compared by Weill and Calvin (1965) and were found to be consistent with the intercalated model. The chemical properties of acridines in complex I are in agreement with the model (Lerman, 1964) as are sedimentation and electron microscopic data to be referred to in the next chapter (p. 427).

Additional evidence that the bases are genuinely separated in the intercalated states comes from the observation that proflavin, bound to DNA, inhibits the UV-induced production of pyrimidine dimers (p. 245; Setlow and Carrier, 1967).

Theoretical aspects   The theories have to account for the spectral and other data on the bound dyes. Before considering the probable geometry of the intercalated structure, it is worth considering whether there are other models that ought to be considered. Bradley and Lifson (1968) have employed a statistical mechanical procedure, the sequence-generating function (SGF) to test the plausibility of various models of acridine binding. According to them there are three plausible structures: at low dye-concentrations complex I (intercalated dyes); at high dye-concentrations complex II (stacked dyes on the outside of the helix); at very low concentrations of dye a third type (? complex 0) in which the dye molecules lie along one of the grooves of the helix. Experimental support for the existence of 'complex 0' is said to come from details of the Cotton effect induced (see p. 409) in the bound dye and from the fact that r never reaches its theoretical maximum of 0·5 in complex II.

Jordan (1968) has analysed the free energy of interaction of intercalated dyes by a method analogous to that described for base-pair interactions on p. 77. Thus this free energy is considered as being composed of six terms:

$$F_{\text{total}} = F_{\rho\rho} + F_{\rho\mu} + F_{\rho\alpha} + F_{\mu\mu} + F_{\mu\alpha} + F_L$$

In this equation (compare with p. 77), ρ is charge, μ dipole and α polarizability. F_L is the contribution from the London dispersion forces.

Gilbert and Claverie (1968) have considered the changes in free energy brought about by the intercalation process. There are several types of intercalation site in DNA (those illustrated in a different context in Table 2.8, p. 81). For a site of category j, the average change in free energy/site ($\Delta\bar{F}_j$) is given by the expression;

$$\Delta\bar{F}_j = \Delta F_j^{\text{dis}} + \Delta F_j^{\text{int}} + \Delta F_j^{\text{oth}}$$

In this expression, the first term is the free energy of distortion of the helix accompanying the intercalation process, ΔF_j^{int} is the interaction term (comparable with F_{total} above) and ΔF_j^{oth} represents all the other terms.[6] ΔF_j^{oth} is composed of terms to account for the restriction of thermal movement of the dye once it has been bound, the interaction between dye and water molecules and the fact that the 'hole' in the solvent created for the dye is filled with water molecules when the dye is removed from free solution and inserted in the helix. Similarly the shape of the 'DNA hole' is altered as the geometry of the helix changes to accommodate the intercalated base.[7]

Of the two versions, Gilbert and Claverie's paper should be read as a more rigorous and definitive treatment. Jordan comes to qualitatively the same conclusions and his nomenclature is employed in the following description of the intercalated state of aminoacridines.

All models for intercalation require a separation of base pairs to 6·7 Å. There is sufficient slack in the sugar-phosphate backbone of the helix for several orientations of the base-pairs with respect to one another to be possible. The parameter for these models are defined in Fig. 14.6.

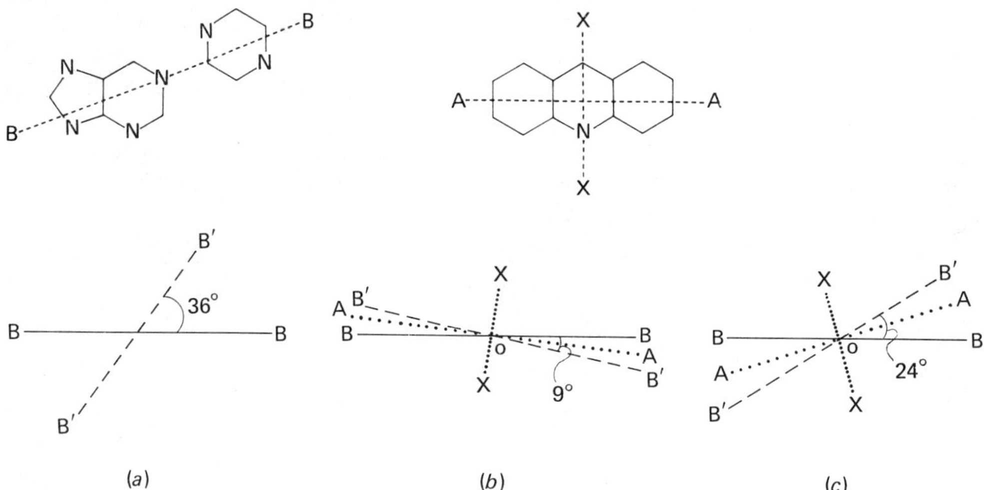

(a)   (b)   (c)

FIG 14.6   Models for the intercalation of an acridine dye. At the top, the axes for a base pair and an acridine nucleus are defined. XX (the sense of the dipole vector) is perpendicular to AA. Below are shown the orientations of adjacent base pairs in B-DNA. B' - - - - - - - B' is below B————————B; the acridine axes are shown thus . . . . Orientations are shown in (a) DNA, (b) complex Ia (Lerman, 1961) and (c) complex Ib (Fuller and Waring, 1964). Point o (at the intersection of the axes) is referred to in the text.

The theoretical analysis of these models consists of calculating the free energy of interaction of the complexes in cases in which the base-pair sequence differs, and in which the angle BOX (Fig. 14.6) is regarded as a dependent variable. In this way it is possible to predict the relative stabilities for the different model complexes in DNA molecules of differing sequence. The experimental and theoretical orders of stability fit if model Ia applies to the binding of acridines. Thus it is generally assumed that this is the configuration of intercalated acridines. Note that the model implies that the helix has unwound through 45° (9 + 36). It is probable that Ib is the configuration for ethidium complexes (Fuller and Waring, 1964).

A spanner in the works   All the models for intercalation imply that a gap is created between base pairs by unwinding the helix. On the basis of model building, Crawford and Waring (1967) calculated that a 12° unwinding of the helix is required to create a gap the right size to slide in an ethidium. It has been recently pointed out by Paoletti and le Pecq (1971) that the appropriate gap can equally well be generated by winding the helix through 12°. Ones first reaction to this may be disbelief, but remember that there is considerable 'slack' in the sugar-phosphate backbone. If you were to take a model of a duplex segment and try and screw it down, movement would be possible. As the backbones tighten up, there is a conformational change in the centre of the helix which results in base pairs coming apart. Photographs which illustrate the point are contained in their paper. What is more, they present experimental evidence which is in support of the winding (as opposed to unwinding) model. The methods they employ rely on the fluorescence properties of the bound ethidium.[8] The fluorescence emission of ethidium is vastly increased by its association with DNA in complex I. This increase is due to 'energy transfer' between the singlet states of the ethidium and the base pairs. Moreover, the orientation of the ethidium (at right angles to the helix) results in the bound chromophore exhibiting 'fluorescence polarization'. This is the phenomenon exhibited by oriented fluorescent chromophores when the excitation light is polarized. If the chromophore is firmly oriented, the emitted light is also polarized.[9] The analysis of the results is complicated by the fact that the relative positions of the ethidium chromophores are randomly distributed along the length of the DNA duplex. Paoletti and le Pecq used a computer simulation of the predicted emissions for different intercalation modes for their conclusion that their winding model is correct. They claim additional support for their model by a fresh analysis of X-ray diffraction data on the ethidium-DNA complex previously reported by Neville and Davies (1969).

The importance of this finding (if the deductions are valid) is that it has profound consequences for one of the most important methods available for the analysis of the tertiary structure of DNA (p. 427).

Biological consequences of intercalation Acridines are routinely employed in experimental bacteriology for two purposes. One is to induce 'frameshift' mutations. The other is to release from bacteria episomes (p. 271) of one sort or another. This latter process is referred to as 'plasmid curing'.

Frameshift mutations are of enormous value in microbial genetics. They are mutations that arise from the insertion of deletion of one base pair in the DNA. The result is that mRNA contains one too many or one too few nucleotides. The message from the frameshift onwards is totally wrong. An example of a frameshift is discussed on p. 318. The precise molecular mechanism whereby frameshifting comes about is still not completely certain. A survey of different models (including his own) is made by Lerman (1964). A detailed study of the effect of 9-aminoacridine on DNA synthesis is phage T4 has been undertaken by Lerman and Altman (1968).

Plasmid curing is an even more puzzling phenomenon in many ways. The data of Lerman and Altman (above) might have some bearing on it as they demonstrared that the phage synthesizes abnormally short lengths of DNA when the dye is bound. An interesting facet of plasmid curing is that specific co-segregation of genes occurs during the process (demonstrated for the F' gal CI 857 episome in *E. coli* by Hohn and Korn, 1969).

Antibiotics

Actinomycin D Actinomycin D is a potent inhibitor of transcription in all cells.[10] It is secreted by *Streptomyces chrysomallus*. It has been established for some time that the inhibition of RNA synthesis is due to the binding of the drug to DNA.[11]

Actinomycin D consists of a chromophore of a substituted phenoxazine with two (identical) cyclo-peptide side chains. The structure is shown in Fig. 14.7.

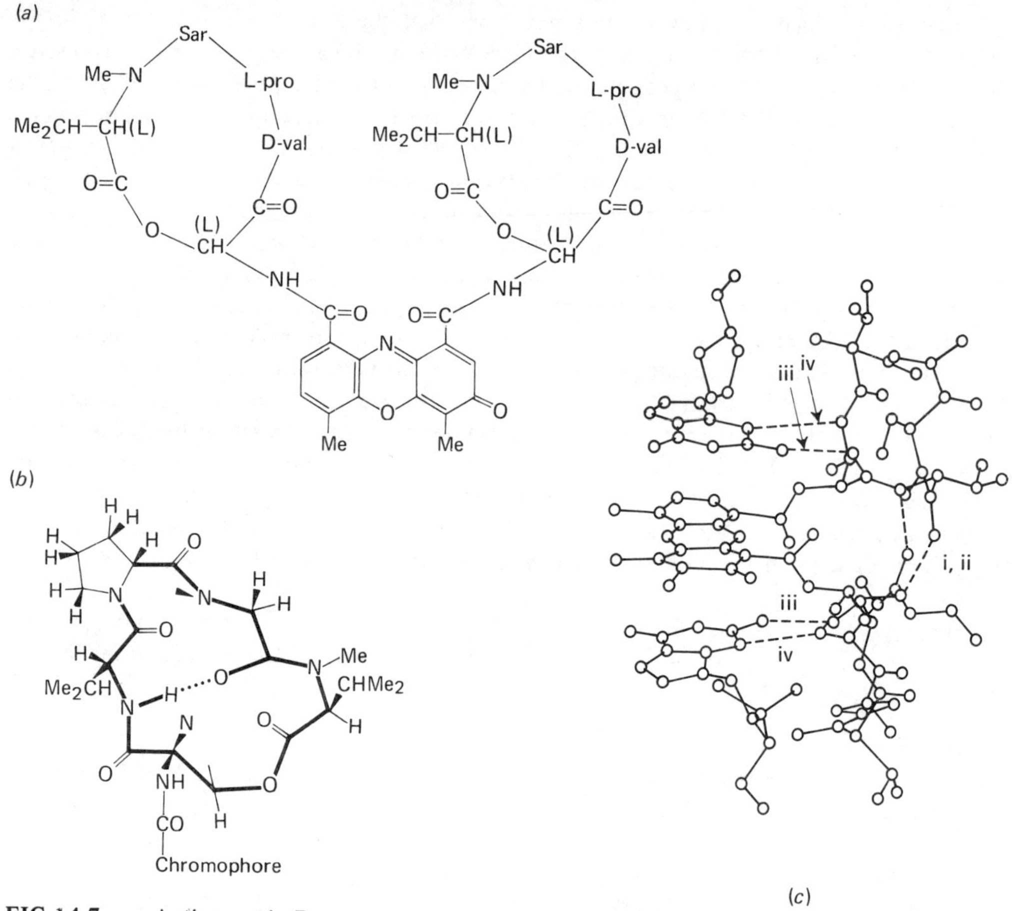

(a)

(b)

Chromophore

(c)

FIG 14.7   Actinomycin D.

(a) Structure of actinomycin D. Two of the amino-acid residues of the peptide have been drawn out to illustrate the bonding more clearly. The N-C direction of these chains is anti-clockwise. The bottom amino acid residue is in fact L-thr, its OH group having formed a lactone with the carboxyl end of the amino acid to its left, which is N-methyl valine.

(b) Conformation of a peptide side-chain in actinomycin (from Lackner, 1970). This structure contains elements of the β polypeptide conformation.

(c) Model of the intercalation of actinomycin D between two dGuo-residues (from Sobell *et al.*, 1971). This is almost a side view of the complex. Note the following features of this model. The dGuo conformations are *anti*. There are three strong H-bonds. Two (i and ii) link the C—O of one valine with the N—H of the other (and vice versa). The other links the C—O of Thr with the —NH$_2$ of Gua. A weak H-bond (iv) links the N—H of Thr to the N-3 of Gua.

The conformation of the drug has two-fold symmetry. The configuration of the peptide chain is shown in Fig. 14.7(b). The binding of actinomycin to DNA occurs in G-rich regions. The maximum binding occurs with poly dG : poly dC with one molecule bound per 12 base pairs (Wells and Larson, 1970).

From X-ray diffraction studies on the complex formed between deoxyguanosine and actinomycin, Sobell, Jain, Sakore and Nordman (1971) deduced that this particular complex has the structure shown in Fig. 14.7(c). Using this stereochemistry as axiomatic, they constructed a model for the actinomycin/DNA complex. This model is consistent with all the data that have been accumulated regarding this complex. In essence, the phenoxazine is intercalated between the guanines in one chain. The two peptide chains are accommodated in the small groove of the helix. The fit is extremely good and precisely covers the region (six bases in each direction away from the phenoxazine) predicted from the experimental data of Wells and Larson (above).

Further evidence for the intercalation of the actinomycin chromophore will be mentioned on p. 428.

Other antibiotics   Actinomycin D is by no means unique in being the only drug with a flat chromophore which inhibits DNA-dependent RNA synthesis. There is considerable evidence that the members of two groups of drugs do intercalate. These groups are the anthracyclines and the tetrahydroanthracene antibiotics. The general structural features of the chromophores are shown in Fig. 14.8. The side chains of all these compounds are sugars.

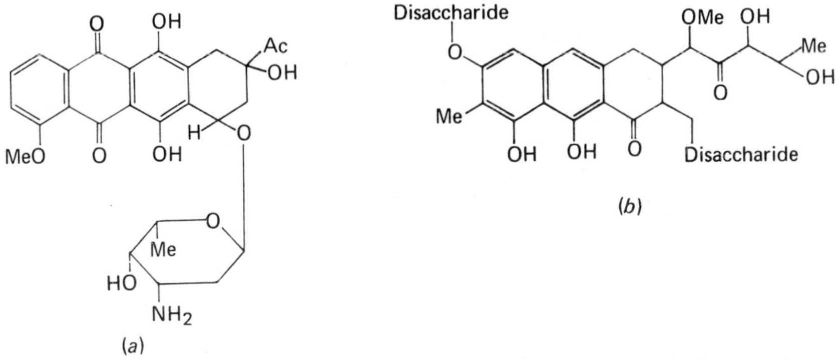

FIG 14.8   (*a*) The structure of daunomycin (an anthracycline). The chromophores in the other anthracyclines (cinerubin and nogalamycin) are similar. (*b*) The structures of chromomycin A3 and mithramycin (tetrahydroanthracene antibiotics).

Studies on these antibiotics have been made by W. and H. Kersten. Much useful information is summarized in their review (Kersten and Kersten, 1968).

The anthracyclines appear to be 'classical' intercalating substances. When these drugs complex with DNA, the sedimentation behaviour (p. 428), viscosity and melting temperature of the DNA all change in a manner characteristic of DNA containing type I intercalated dyes. Kersten and Kersten suggest that the geometry of these complexes may differ from that of the acridine complexes as the anthracyclines are not mutagenic. One chemical difference between them and the acridine dyes is immediately obvious; these drugs are not ionic; the consequences of this are not yet clear.

The nature of the association between DNA and the tetrahydroanthracenes is probably a different character. The binding can be assayed spectrophotometrically (compare p. 411). It appears that the effect on the spectrum of the dyes is absolutely dependent upon the presence of Mg^{2+} ions. However, even in the absence of Mg^{2+} there is an effect upon the T_m of DNA. Thus it appears that there are at least two modes of binding; one dependent on Mg^{2+} and the other not.

Aromatic hydrocarbons   The possibility that the polycyclic hydrocarbons can intercalate into the DNA structure has been considered for some time. There is really no direct experimental approach to the problem. These substances are so insoluble in water, that the maximal 'binding' to DNA is too small for any physico-chemical measurements (in which normal DNA is compared with DNA containing bound hydrocarbon) to be meaningful. There have been two approaches to the problem. One has been to calculate the theoretical stability of such structures on the basis of change-transfer complexes;[12] the other has been to examine the efficiency with which purines and pyrimidines will solubilize the hydrocarbons in water. The principle behind this latter experiment is that if the substances are solubilized it is probably because base stacks will intercalate the hydrocarbon molecule. Neither approach has proved too useful. The solubilization experiments have been made to see whether it is possible to correlate the effectiveness of the hydrocarbons in this assay with their carcinogenicity. Such correlations are not apparent. If the nucleic acids are the targets for the biological effects of these substances, it is more likely that the arylation reactions of their derivatives (see p. 248) are the crucial processes. For interesting discussions, see Bergmann (1968) and Caillet and Pullman (1968).

Notes to chapter 14

[1] All these reagents are suitable for the removal of mercury salts; silver cyanide and silver chloride are both extremely insoluble and hence for the removal of silver from a nucleic acid preparation, EDTA is clearly more suitable.

[2] CsCl is obviously unsuitable as Cl^- complexes the metal.

[3] It is essential that phosphate buffers are not employed in experiments involving polyamines.

[4] A reporter molecule is one which is bound to a macromolecule (or more complex structure such as a membrane) to 'report' on the configuration of the macromolecule in the vicinity of its (the reporter's) binding site. The nature of the report depends on the reporter system that is used. In the cases to be discussed, absorption spectra and CD are used. For other reporters, employed in determinations of protein conformation, luminescence spectra of the reporters are employed.

[5] The DNA employed in this work was native thymus or salmon sperm DNA. The RNA was a commercial preparation of unfractionated RNA, which must have contained hairpins and single-strand regions. These studies would be of great interest if they were extended to cover one of the double-stranded viral RNA molecules.

[6] I am sorry about this—I cannot think of a better superfix; it looks better in the original French.

[7] The technical phrase for these terms is 'cavitation energy'.

[8] An earlier study (Le Pecq and Paoletti, 1967) of the fluorescence of bound ethidium points to a use for the greatly magnified emission (following intercalation) for the analysis of DNA in crude tissue extracts, etc.

[9] This technique will be familiar to anyone acquainted with the determination of protein conformation. It may be used to study the conformational stability of fluorescent groups (such as tryptophan residues) as the extent of depolarization is a measure of the half-life of the conformational state(s) in the molecule.

[10] Provided it can enter the cell. Many Gram negative bacteria are not inhibited by the drug as it cannot enter the cell (note 12 p. 265). However it will inhibit RNA synthesis *in vitro* in every system that has been examined so far.

[11] This in contrast to the more recently discovered inhibitors of transcription (such as the drugs of the rifamycin family) which bind to the polymerase. Thus there is a genetic characteristic of rifamycin resistance, due to an altered RNA polymerase. There is no such thing as actinomycin resistance (see however footnote 10, above).

[12] Charge-transfer complexes: for non-chemists (this account will make any chemist writhe), imagine two molecules A and B both rich in π-electrons (if you were doing this particular sum, you might make A benzpyrene and B guanine). The molecules of A are an equilibrium mixture of those in their ground state (the vast majority) and those in the first excited state (p. 82). Likewise, the molecules B. The interaction between A and B can be considered as consisting of four terms; the dispersion forces between the two ground states, the interaction between 'ground A' and 'excited B', vice versa (excited A and ground B) and the two excited ones. If any of the last three terms are significant, the result is referred to as charge transfer, because a consequence of the calculation is that an excited electron from (say) A may spend some of its time within the province of B and the effect will be a transfer of negative charge from A to B.

15 Tertiary structure

What is tertiary structure? In protein structure, secondary structure refers to the existence of regions of H-bonded conformation (typically α helix and the various forms of β pleated sheet), tertiary structure to the further folding of this structure (and maintained by cystine bridges and 'hydrophobic' interactions), and quaternary structure to the association of more than one polypeptide chain to form a multi-chain complex. The divisions are somewhat arbitrary but are of some value in discussing the generalities of protein conformation. With nucleic acids, we know rather little about the conformations other than the nature of the H-bonded (duplex) and stacked conformations. I have arbitrarily grouped details of conformation other than these features into the category 'tertiary structure'.

Transfer RNA

Nothing is known of the tertiary structure of species of RNA other than tRNA. All we can do with larger RNA molecules is to estimate the degree of secondary structure (duplex content) and attempt to guess which regions are associated with which (see p. 329 and the following). With tRNA there is every reason to suppose that there is a common overall shape. With an understanding of the tertiary structure it should be possible to rationalize the two functions of tRNA (p. 15) and the astonishing evolutionary conservatism of various aspects of its function (see for example, p. 32) in terms of a common shape. The features of the sequence and secondary structure which are common to all tRNA molecules examined to date have already been summarized (Fig. 10.8, p. 305).

Physico-chemical properties common to all tRNA molecules

Most of the examples in this section relate to experiments in which either mixed tRNA preparations are used or alternatively which have been performed on several different species. Some of the methods are only applicable to certain tRNA species but can reasonably be regarded as bearing on other types of tRNA.

tRNA is not a real sphere nor a random coil The simplest demonstration of the overall non-spherical shape of tRNA comes from estimates of the radius of gyration. We have seen that R_G is measured from light-scattering data (p. 398). For this particular purpose, the most reliable data come from low-angle X-ray scattering. The technique is formally analogous to light scattering. The instrumentation is different, of course, and more accurate measurements are possible, especially for small molecules such as tRNA. The procedure consists of plotting I/I_0 as a function of the logarithm of the scattering angle: the shape of the plot is then fitted to the curves predicted for various conformations of the molecules. Excellent studies on yeast tRNA^{Phe} by Pilz, Kratky, Cramer, von der Haar and Schlimme (1970) have established that the curve fits that predicted for a distorted 'cigar' with a fatter cross-section at one end than the other. The length of the model is 92 Å and the diameter 22 Å (thin end) and 36 Å (thick end). The mean radius of gyration is 24 Å; on heating the solution, the radius increases and the curve fits that characteristic of a random coil. Similar studies on the low-angle X-ray scattering of different tRNA molecules are in broad agreement with those of Pilz *et al.* (above). The data have been reviewed (Cramer, 1971).

Another direct proof of the non-spherical geometry of tRNA comes from the observation of Tsetkov *et al.* (1965) that, unlike rRNA, tRNA demonstrates flow birefringence, that is to say, in a stream, the molecules can be oriented in a preferred direction (like logs of wood in a river). This property is also found in DNA.

Certain tRNA molecules contain the highly fluorescent Y-base in their sequence in the anti-codon loop. Were it possible to introduce another equally strong fluorescent group (or 'probe') in the structure, it would be possible to estimate the distance between the two fluorescent groups by the alteration in the spectrum of one by the presence of the other (fluorescent energy transfer). So far, success has been limited to the introduction of the probe (an acridine derivative) at the 3′-end of the molecule (after formation of the dialdehyde (p. 287)). The effects to be expected for any energy-transfer are quenching of the Y-base fluorescence and enhancement of the acridine fluorescence. The results (Beardsley and Cantor, 1970) were that a very slight effect (implying a distance from Y to 3′-terminal A of more than 40 Å) was observed. It is to be hoped that probes can be placed in the molecule closer to Y so that more accurate estimates of the geometry can be made in future.

Certain 'single-strand regions' are highly ordered   Evidence for the ordered structure of the anti-codon loop comes from a fluorometric study of the codon–anticodon interaction. This experiment is discussed in a different context on p. 439.

The other region that is suitable for spectrophotometric study is the 3′-terminal . . . C-C-A (p. 13). Aminoacylation of the A-residue does not affect the ORD or CD spectra of the molecule (Adler and Fasman, 1970). The amino acid can therefore be used as a probe for the tRNA conformation. Hoffman, Schofield and Rich (1969) employed a 'spin label' to probe the aminoacyl configuration. The technique involves the attachment of a free radical to the amino acid present in the charged tRNA. The splitting of the ESR signal (p. 99) is used to probe the conformation of the part of the molecule to which the radical is attached. The method was to attach the nitroxide derivative shown below to val-tRNA^{Val}. Although there was some reaction between the reagent, employed for forming this derivative, and other parts of the tRNA molecule, this secondary binding was small enough to be neglected in an analysis of the spectra.

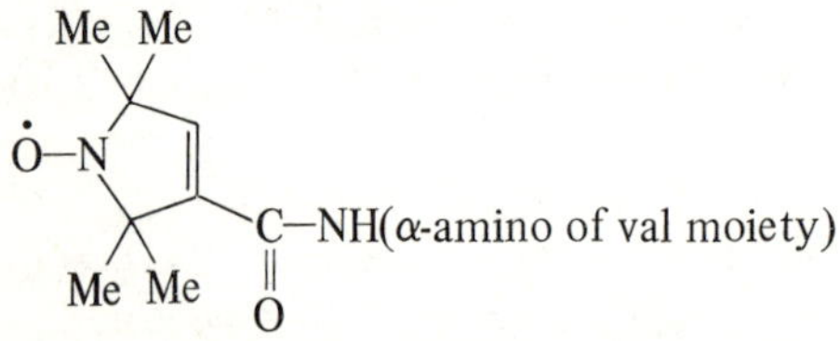

$$\text{Me} \quad \text{Me}$$
$$\dot{O}-N$$
$$\text{C}-\text{NH}(\alpha\text{-amino of val moiety})$$
$$\text{Me} \quad \text{Me} \quad \overset{\|}{O}$$

The splitting of the label was measured as a function of temperature. A transition corresponding approximately to the hyperchromic melting temperature was found. However the transition was much sharper and implied that conformational degrees of freedom could be assumed by the CCA-end while the bulk of the molecule retained its native structure.

All tRNA molecules are the same shape   It is generally assumed that this statement is an axiom for tRNA-model building. It is difficult to envisage a ribosome working unless many of the molecular parameters of the different tRNA species are in fact the same. For example, it is presumably the case that the distance from the $3'$(aminoacyl-CCA) end to the anti-codon is constant. Several pieces of evidence are commonly adduced to prove this point; for example, the sedimentation coefficient of unfractionated tRNA has been measured as a function of temperature (Henley, Lindahl and Fresco, 1966). The result is that a profile roughly 'superimposable' with the melting profile is obtained. It is difficult to estimate the extent to which this result is to be expected from the common clover-leaf secondary structures. A more convincing (and remarkable) result is that unfractionated tRNA can be crystallized (see p. 422; Blake, Fresco and Langridge, 1970).

Chemical and enzymic evidence for tertiary structure   The specificity of nucleases for double-stranded structure can be used to establish the extent of H-bonded secondary structure in RNA and, in the case of a molecule of known sequence, may be used to deduce the details of this structure (p. 315). It is equally possible to employ the same techniques to obtain evidence for H-bonded tertiary structure. Chemical reactions have been employed to probe tertiary structure in two distinct ways. On the one hand, chemical reactivity of nucleosides is dependent upon their conformations. By examining the reactivity of residues in a known sequence, it is possible to explore the possibility that residues, apparently unpaired in the clover-leaf, are in fact H-bonded. The alternative is to attempt to cross-link the molecule selectively; if such an experiment is successful, it suggests that the two residues are close together in the native conformation. In this section, I have chosen illustrative examples of these three techniques.

Limited nucleolysis: the $5'$-terminal phosphate of tRNA is not removed by alkaline phosphatase at normal temperatures, but only when the RNA is heated to a temperature close to T_m (Stern, Zutra and Littauer, 1969). It thus appears to be in some way buried in the structure. When attempts are made to erode tRNA with venom phosphodiesterase (p. 207), the reaction is held up after the removal of the terminal A- and C-residues (Zubay, 1968); thus the next residue (C) is tightly bonded in some way (even though it is never H-bonded in a clover-leaf structure (see Fig. 10.8, p. 305). Attempts to correlate the action of endonucleases with possible tertiary structures have not provided any unambiguous assignments of tertiary conformation so far.

Chemical modification: Several reagents have been employed for such studies. Here I only choose the reagents which have provided fairly clear-cut results. The data are summarized (Fig. 15.1) for yeast tRNA[Phe] for which the most extensive experiments have been performed. The picture, which looks so simple in the figure, in fact represents the accumulated results of extremely time consuming and painstaking experiments, as the RNA must be hydrolysed by an endonuclease and fragments containing the modified bases must then be sequenced to establish the nature of the reactive sites. The reagents that have been employed are (i) perphthallic acid which forms *N*-oxides of A- (and Y-) residues (p. 63; Cramer, Erdmann, von der Haar and Schlimme, 1969), (ii) reduction with NaB^3H_4 (p. 69; Igo-Kemenes and Zachau, 1969) and (iii) kethoxal which attacks G-residues (p. 65; Litt, 1969). In the figure, the groups of bases in the loops of the clover-leaf and not attacked by these reagents (and presumably present in a tertiary H-bonded conformation) are indicated.

Cross-linking: Two cross-linking reactions of tRNA have been exploited. One is the observation that *E. coli* tRNA[Tyr] reacts with I_2 to yield sUrd-disulphide bridges (p. 242). Unfortunately, this reaction is of little interest as the two residues in question are adjacent (in the base of the stem of the 'dihydroU loop'—see Fig. 10.9, p. 308). A more interesting

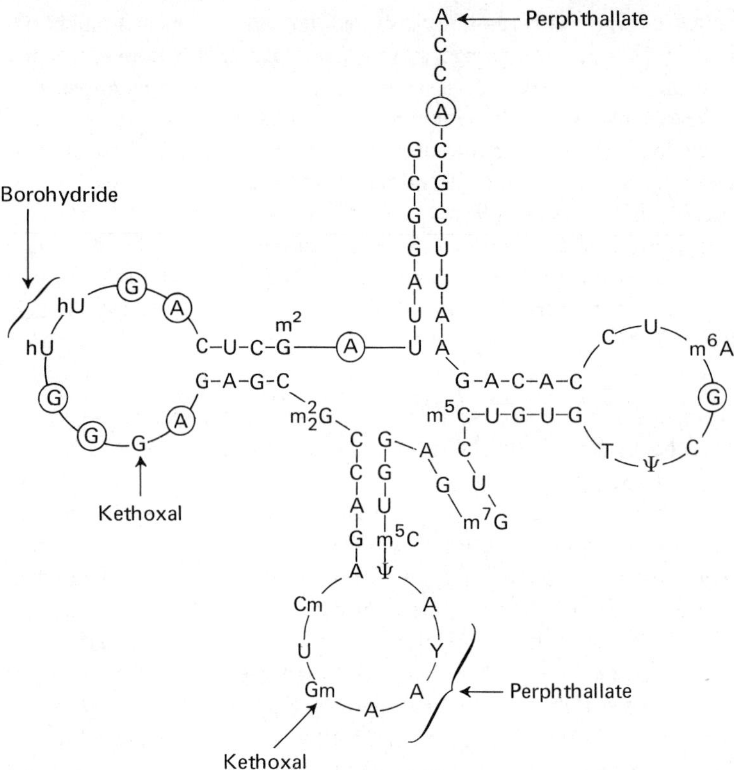

FIG 15.1   Sequence of yeast tRNA^{Phe} (see also Fig. 10.9f, p. 307) with the positions attacked by three base-specific reagents indicated. References are in the text. This picture is adapted from Cramer (1971). Residues with circles round them are ones which (from the standpoint of the clover-leaf) are 'unexpectedly' unreactive and have to be accommodated in the tertiary structure.

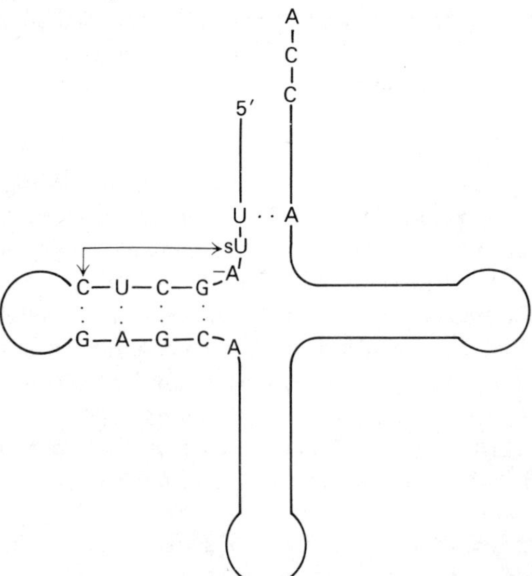

FIG 15.2   Partial sequence of *E. coli* tRNA^{Val} showing the positions of the two pyrimidines between which a photocatalysed reaction occurs (from Yaniv, Favre and Barrell, 1969). The complete sequence of this molecule is given in Fig. 10.9a, p. 306.

reaction is the photochemical reaction between an sU-residue and a C-residue in *E. coli* tRNA^{Val} (Yaniv, Favre and Barrell, 1969). The positions of these two residues in the clover-leaf are shown in Fig. 15.2. Thus it is presumed that these residues are either close together or that a simple conformational change is possible that brings them into proximity.

How many tertiary structures are there?   We have already pointed out there is no apparent change in conformation upon the charging of tRNA with amino acids. At least one model for protein syntheses (see p. 440) requires conformational changes in the anti-codon loop during translation. The fluorescence of the Y-base may be used for the probing of conformational changes in this region. Unfortunately, it has not been possible so far to examine conformational changes during translation using this probe. On the other hand, it has been possible to examine the conformation of a synthetic complex between tRNA^{Phe} and 'mRNA fragments' using this probe (p. 439). Moreover, it is known that divalent cations affect the quenching of this fluorescence and consequently, it can be assumed that they effect a conformational change in this region. Two papers describe work on the fluorescence of this base in tRNA^{Phe} in different buffers. Robinson and Zimmerman (1971) demonstrated that Mg^{2+} and (to a lesser extent) polyamines (notably spermidine—p. 406) enhance the fluorescence of this base. Beardsley, Cantor and Tao (1971) have interpreted these differences in terms that the Y-base is more shielded from water molecules in the 'Mg^{2+}-conformation' than in the 'normal' state. Robinson and Zimmerman extended their studies to the effect of temperature on the enhanced fluorescence in the Mg^{2+}-RNA. The result was that there was a loss of the enhanced fluorescence at temperatures below the absorbancy melting temperature. The conclusion from these studies is that there are two states for the conformation of the anti-codon loop. In the presence of Mg^{2+}, a state in which Y is shielded from solvent is stabilized; the denaturation of this Mg^{2+}-stabilized state is distinct from (and more facile than) the overall denaturation of the duplex strands of the molecule.

Additional evidence for the existence of two conformations for tRNA comes from the work of H. G. Zachau on polynucleotide kinase (p. 228). Hänggi, Streeck, Voigt and Zachau (1970) studied the kinetics of the reaction of the phosphorylation of 5′-dephosphorylated tRNA^{Phe} with the enzyme. The reaction rate was found to be biphasic; after a rapid reaction 0·2 mole phosphate incorporated per mole of RNA was achieved. Following this a very slow increase (essentially a plateau—see Fig. 15.3(*a*) was observed. These data imply that there are two forms of the rRNA; one (about 20% in the equilibrium mixture) has a 5′-end accessible to this enzyme and the other (80%) has not. The interconversion of these forms is apparently very slow. This observation was extended in a very imaginative way to show that the two major halves of the molecule assemble the native equilibrium mixture of tertiary structures on mixing. The molecule tRNA$^{Phe}_{yeast}$ contains the anomalous nucleoside 'Y' in the anti-codon loop. This nucleoside is acid-labile and it is possible to cleave the molecule (Wintermeyer *et al.*, 1969) at this point to yield two 'half-molecules', each with a 5′-phosphate end. The 5′-phosphate of the '5′-half is, of course, the 5′-pG ... present in the intact molecule. It was shown that in a mixture of the dephosphorylated half molecules could be phosphorylated in the assay and showed similar biphasic kinetics but with the plateau at a much higher value (about 1·1 moles phosphate/mole RNA). Analysis of the data established that this was due to about 0·8–0·9 moles P/mole of 5′-terminus of 3′-half and 0·2–0·3 moles P/mole of 5′-terminus of 5′-half. They then examined the effect of adding increasing amounts of one phosphorylated half to the other dephosphorylated half on the re-phosphorylation reaction (Fig. 15.3(*b*)). The results showed that addition of the 3′-half inhibited the phosphorylation reaction progressively until the ratio of 5′-half/3′-half was 1:1, when the 'native' value of about 0·2 moles P/mole RNA was achieved.

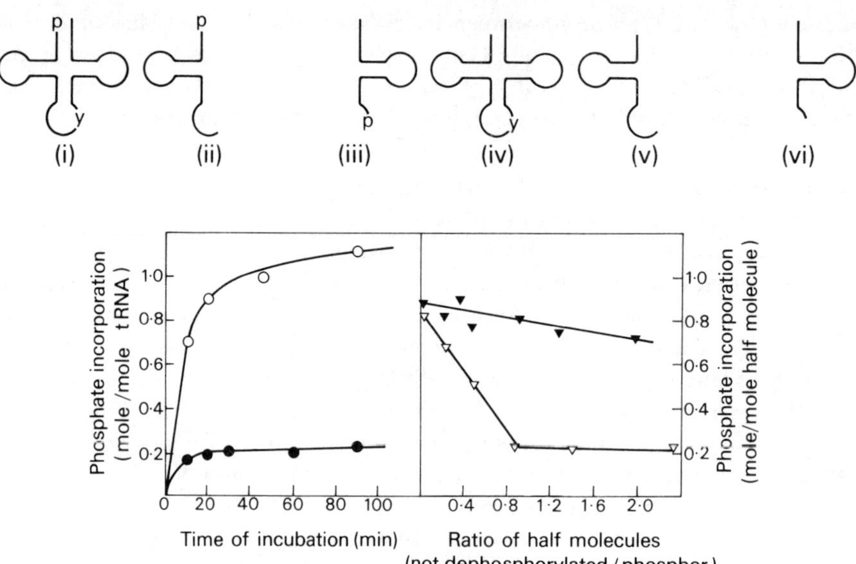

FIG 15.3   The re-phosphorylation of dephosphorylated $tRNA_{yeast}^{Phe}$ and its halves. At the top of the figure: simplified structures of (i) the intact RNA (see Fig. 0.00, p. 000 for the complete sequence, (ii) and (iii) the 5′- and 3′-halves and (iv, v and vi) the dephosphorylated tRNA, 5′-half and 3′-half respectively.

(*a*) Kinetics of re-phosphorylation reaction. Dephosphorylated tRNA (iv above, ●———●) and dephosphorylated halves (mixture of iv and v, ○———○) were incubated with polynucleotide kinase (3 units/nmole RNA) and [α-³²P] ATP of known specific activity.

(*b*) Re-phosphorylation of mixtures of half molecules. A standard incubation was employed: 2·2 units of enzyme/nmole of half molecule for 30 min. The mixture consisted of either dephosphorylated 3′-half and increasing amounts of 5′-half (vi and ii, ▼———▼) or dephosphorylated 5′-half and increasing amounts of 3′-half (v and iii, ▽———▽). Data of Hänggi *et al.* (1971).

These results tell us a great deal about tRNA. There are two (or more) tertiary structures in equilibrium with the '5′-protected conformation' in excess and with very slow equilibration between the forms. Admixture of the two halves of the molecule results in the formation of a stoicheiometric complex with the two native conformations in the same equilibrium mixture. The 5′-end of the 3′-half is not present in this complex in a conformation in which it is protected from the polynucleotide kinase reaction. Further studies by Zachau and his colleagues on these half molecules has thrown a great deal of light of a different aspect of tRNA structure, namely the nature of the recognition reaction of aminoacyl tRNA synthetase (see p. 453).

X-ray diffraction studies   Several tRNA species (and unfractionated tRNA—see p. 419) have been crystallized. The various techniques of crystallization and the variety of crystal types that have been obtained are summarized by Cramer (1971). So far, insufficient reflections have been analysed for a complete synthesis of the structure of tRNA to be possible. However, we can say with reasonable confidence (Young *et al.*, 1969) that there appears to be an extended helical stretch through the molecule about 80–85 Å long (compare with the low-angle X-ray scattering estimate, p. 418). Thus such X-ray data as are available support the view that the tertiary structure of tRNA is a cigar-shaped molecule about 80 Å or so long. The aim of model-building is to fold the clover-leaf up in some way so that base-stacking and additional H-bonds stabilize a rigid molecule. The folding and H-bonding possibilities are restricted by two requirements: they must be reconcilable with the chemical data that are summarized in the preceding section and they must provide a molecule with much the same overall shape and definitely the same length for all the tRNA sequences that are established to date.

Models for the tertiary structure of tRNA

Many models for tRNA have been constructed. There are two good reviews of these models together with an analysis of their fit with experimental data (Levitt, 1969; Cramer, 1971). Both authors have their own proposed models. Cramer's article is probably the better review as he discusses other peoples' models very objectively. The models can be grouped into three general categories, based on the position of the TΨC and dihydroU arms. In the first category (for which there is one example—'the Cramer model' (Cramer *et al.*, 1969; Cramer, 1971)—neither of these is stacked up to contribute to the overall length of the helix. Rather the CCA stem through to the anti-cocodon are stacked up to form a continuous duplex with the other arms looped back in the direction of the CCA terminus. In the second category the TΨC stem is stacked in such a way that it contributes to the overall length. Thus the CCA-anticodon distance is longer than in the Cramer model by the length (or a portion of the length) of the TΨC stem. There are three versions of this: the 'Connors model' (Connors, Labanouskas and Beeman, 1969), the 'Fuller model' (Fuller, Arnott and Creek, 1969) and the 'Levitt model' (Levitt, 1969). In the final category (Ninio, Favre and Yaniv, 1969), the dihydroU loop contributes to the length of the structure and the TΨC loop is apart from the structure (though in a very different position to that in the Cramer model). In summary, the Cramer model does not account for the photoreaction between sU and C in tRNA^{Val} (p. 421) unless one accepts that the reaction occurs as a result of rare conformational states of the RNA. It is possible that the CCA-anticodon distance is too short in this model. Otherwise the fit with chemical and enzymic properties of tRNA is excellent. The Ninio-Favre-Yaniv model was really designed to account for the above reaction and does so better than any other version. However there are serious problems in that many other reactivities of tRNA are unaccounted for. Of the 'TΨC stacked' models, I confess to be insufficiently expert to compare them critically.

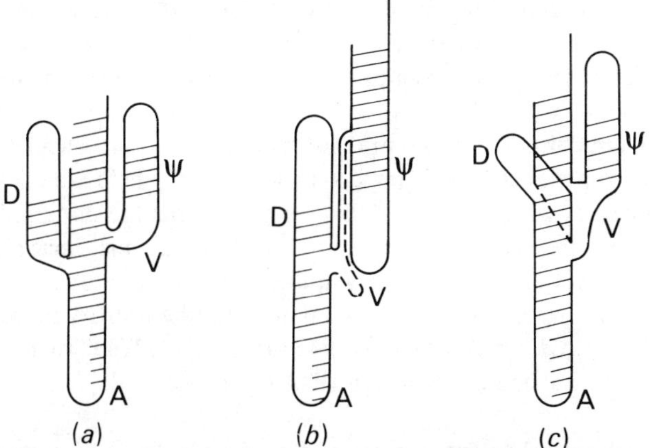

FIG 15.4   Schematic drawings of the layout of the arms of three models for tRNA (after Cramer, 1971—see text for other references). The vertical dotted lines represent the axis of a continuous duplex stack. A—anti-codon loop; V—variable arm; Ψ—TΨC loop; D-dihydroU loop. (a) Cramer model; (b) Levitt model; (c) Ninio-Favre-Yaniv model.

Certainly none explains the chemical reactivities of tRNA as well as the Cramer model but they probably fit the biophysical data rather better. Examples of these models are drawn in a highly schematic form in Fig. 15.4. For good three-dimensional representations of these models, refer to the original literature.

Small continuous DNA molecules

We have encountered several types of small 'circular' DNA already in this book; examples are the intermediates of sticky-ended phage (p. 280), plasmids of various kinds (p. 27) and RF'S for single-strand DNA phage. Additionally, organellar DNA (from mitochondria and chloroplasts) and many animal viral DNAs are of this class. In this section, I discuss the methods of establishing the tertiary structure of these molecules. I have not attempted a catalogue of the types described so far; it would be a pointless exercise as the number is growing all the time. It is more important to question which methods provide satisfactory procedures for their isolation and characterization.

General and qualitative considerations

Open circles and superhelices   Consider a model of a completely unstrained helix. If the two double ends are brought together, the structure (allowing for a certain wormlike flexing–p. 393) should be quite stable. In other words the molecule will look like a bracelet made from two strands (of copper or something of the sort) entwined. Now imagine that before the joint is made, the helix is distorted either by winding or unwinding. Thus it is constrained in a deformed conformation, the ends are joined and then it is left alone. The primary (Watson–Crick) helix can adopt its stable conformation by rotation. The consequence of this rotation will be that a secondary (or super-) helix will be imposed on the structure. The result will be a tangled coil (as a bracelet–no good). Now imagine cutting one of the strands in one place (with an endonuclease so to speak). The constraint can now be relieved by rotation about the backbone bonds of the continuous strand. The result is that the superhelix will unwind and a 'Watson–Crick bracelet' will be produced (but with a break in one of its strands). Small continuous DNA duplices are apparently universally in the superhelical conformation. The bracelet structure is known as an open circle and the transition from superhelix to open circle is effected, as implied above, by an endonuclease. An example of two molecules related to one another in this way or φX RF I and RF II (see p. 154).

Now let us imagine intercalating ethidium (p. 410) into the superhelix. What happens depends on (1) which way round the superhelix is with respect to the primary helix and (2) whether the ethidium intercalation involves winding or unwinding. These points are discussed below. For the moment, just accept that the effect must be profound. For example, imagine that the superhelix arises as a result of relief from overwinding in the 'initial' bracelet (above). Now just for the sake of argument, let us suppose that the ethidium unwinds the primary helix. The result will be that with intercalated ethidium there will be less supercoiling. We have now got far enough to understand how superhelical and open circular DNA can be isolated.

Isolation   The isolation of DNA from mitochondria, viruses and so on requires no special techniques (p. 123) and the peculiar properties of the superhelices do not need to be employed. However, it is sometimes required to isolate small closed-circle DNA molecules which are minor components of a preparation. The classical example is the isolation of plasmids from bacteria.

Sometimes the plasmids differ in density from the chromosomal DNA (p. 150) and can be separated by ordinary buoyant density centrifugation. In other instances, the densities of plasmid and chromosomal DNA may be the same. If the DNA is isolated carefully, the small plasmid molecules will remain unbroken; the chromosomes will be broken up a bit (p. 105). So in solution there should be a mixture of superhelices (plasmids) and linear DNA molecules (chromosome fragments). Ethidium bromide is now added and intercalates the DNA. The superhelices will change shape, the linear molecules will just change in length a little bit. Consequently the densities will change differently (typically the plasmids will be less effected and hence will be heavier than the linear molecules—see below). Thus the plasmids and chromosome fragments may be fractionated by CsCl centrifugation in the presence of ethidium bromide. With care it is possible to separate really big plasmids by this technique. As an example of a technical tour-de-force, Goebel and Helinski (1968) isolated a giant episome F ColV ColB *trp cys* from *Proteus mirabilis*[1] in this fashion. Their recipe gives a good idea of the amounts of ethidium used in the procedure. DNA (400 μg) in very dilute buffer (3 ml) and ethidium bromide (1·5 ml of a solution 1 mg/ml) were mixed. To the mixture were added 10·5 ml buffer and the appropriate amount of CsCl (15 g in this case) and the mixture was centrifuged to equilibrium (p. 149).

This technique will not work for open circles; their density changes (when intercalated) in the same fashion as that of linear DNA. An ingenious method of isolating open circles is due to Fuke and Thomas (1970). They employed it for isolating the open circles of λ-DNA obtained by incubating linear, sticky-ended (p. 424) phage DNA in the absence of ligase (see Fig. 5.9, p. 280). The DNA was set in agar gel; the gel was broken into little pieces and eluted with buffer. The linear pieces were eluted. When the gel was melted, the closed circular DNA was obtained considerably enriched. How did it work? The author's idea (for which there is no direct physico-chemical evidence) is that some linear fragments become anchored to the gel at more than one point and circles become latched in like the ring in a bull's nose.

Separation of fragments from open circles   The procedures for this experiment have been discussed on p. 154. They can now be seen in a different context. The procedure is exemplified in an interesting assay for *E. coli* endonuclease I (p. 209) by Goebel and Helinski (1970). They found that the enzyme forms a stable complex with tRNA and that this complex has high specificity for single-strand breaks in duplex DNA. The substrate was the supercoiled φX RF II. Following the single-strand break ('nicking') the molecule opens out (see pp. 154 and 424) to an open circle (RF I). The RF II product was centrifuged through an alkaline sucrose-gradient (p. 188) and the two classes of molecules (intact single-strand circles and linear molecules) were isolated as two fractions. The two fractions were analysed by alkaline CsCl centrifugation; in each case two peaks were obtained (i.e. from the single-strand circles, + circles and − circles; from the linear molecules, linear + and linear − strands). It was thus shown that the enzyme (as its tRNA-complex) lacks strand-specificity.

Quantitative and structural aspects

Molecular weights   The equations relating the molecular weight of DNA to sedimentation coefficient (pp. 386 and 387) apply to linear DNA. For continuous DNA molecules, different equations are required. The methods employed, consist of correlating (on a semi-empirical basis) the molecular weights, deduced either by sedimentation studies on the linear forms of these molecules or from electron micrographs, with corrected sedimentation coefficients.

Several relationships are in the literature. Recently, Böttger, Bierwolf, Wunderlich and Graffi (1971) published data on a superhelical DNA (that from the oncogenic Papova virus). During their work, they thoroughly re-examined the validities of the earlier assumptions and derived what are 'the best' equations for general use. Their paper should be carefully read by anyone intending to work on superhelical viral DNA as they show that many literature values are over-estimated. For superhelical DNA (form I), their equation is:

$$s^0_{20,w} = +5 \cdot 16 + 0 \cdot 00439 \, (MW)^{0 \cdot 553}$$

and, for open-circle (or in their terminology, the relaxed ring, form II):

$$s^0_{20,w} = 2 \cdot 50 + 0 \cdot 0219 \, (MW)^{0 \cdot 435}$$

Definitions of topological terms   The conventions used in this book are those of Bauer and Vinograd (1968) and Vinograd, Lebowitz and Watson (1971). Three topological parameters are defined. The duplex winding number β is the number of duplex turns of the molecule in the primary helix (for B-DNA—p. 327, $\beta = N/10$, where N is the total number of base pairs). The superhelix winding number τ is the number of turns made by the duplex round the axis of the superhelix. The topographical winding number α is the number of complete revolutions made by one of the duplex strands around the duplex axis when the superhelix is twisted so that the axis of the superhelix is forced to lie flat (in one plane). The relationship between these terms (Glaubiger and Hearst, 1967) is as follows:

$$\tau = \alpha - \beta$$

The superhelix density (σ) is defined as the number of superhelical turns per 10 base pairs; clearly $\sigma = 10\tau/N$.

Superhelices can arise in more than one way. The alternatives are most easily seen with respect to an extremely simple model (Fig. 15.5).

In Fig. 15.5(c and d), topologically equivalent states have been arrived at (so τ has the same sign and magnitude in each case). If you ignore the axes of the superhelices (which are at right angles to one another) the sense of the supercoils is opposite in the two cases. The sign-convention for τ is universally used, so it can be said that τ-values are *negative* for

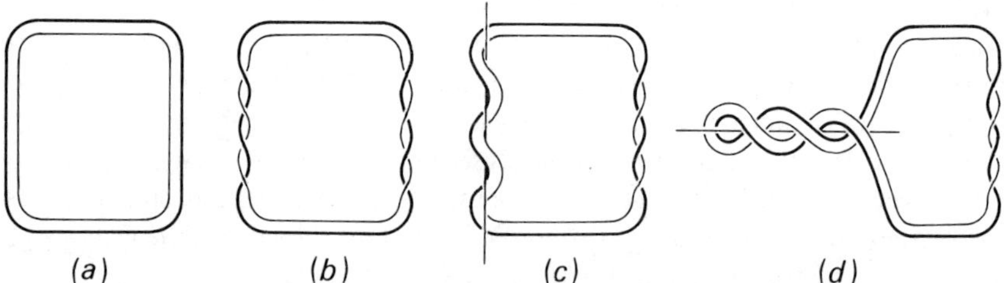

(a)   (b)   (c)   (d)

FIG 15.5   Simplified model to illustrate superhelix types.

(a) A hypothetical molecule of two non-entwined strands (drawn thick and thin for convenience in later identification). This is not a Watson–Crick duplex; the strands are just lying side-by-side.

(b) Duplices have been introduced into the molecule in (a) by winding three turns of right- and left-handed duplex. The helices cancel one another so that $\beta = \tau = 0$.

(c) and (d) Three right-hand and three left-hand duplex turns have been introduced (as in b) but in this case in such a way that the two helical sections do not cancel one another. In (c) a 'toroidal superhelix' has been created. In (d) they do not rotate around each other but intertwine. In both cases, $\alpha = 0$, $\beta = 3$, $\tau = -3$. See text for discussion. Diagram adapted from Vinograd, Lebowitz and Watson (1968).

left-handed toroidal superhelices (Fig. 15.5(*c*)) and for *right-handed interwound* superhelices. It is generally assumed that superhelical DNA is interwound.

Determination of superhelix density–'direct methods'   By direct methods, I mean anything that does not rely upon the use of the intercalating reaction of ethidium bromide (or other dyes). Of course, the most direct way of looking at the superhelical structure is looking at is (under the electron microscope). Recent pictures of this form are to be found in the paper by Böttger *et al.* (1971) mentioned on p. 426. Another interesting example is the attractive visualization of mitochondrial DNA by van Bruggen *et al.* (1968). This is a good method of measuring the length of the DNA ($5 \cdot 2 \pm 0 \cdot 4$ μ for chick-liver mitochondria for this reference). Twisted circles can be seen and cross-overs counted. It is not a particularly suitable technique for determining the sign of τ for (as Paoletti and le Pecq, 1971, point out) the decision about which duplex is on top of which is a difficult one. The appearance of the cross-over is a function, not only of the truth about the superhelix but also of the shadowing technique employed.

Another approach is that of Vinograd, Lebowitz and Watson (1968). They studied the sedimentation coefficients of (1) superhelical DNA (from the oncogenic Polyoma virus), (2) nicked open-circle DNA from Polyoma and (3) DNA as in (1) but at increasingly high pH. A helix-coil transition was obtained such that the superhelical DNA was converted to form with the same sedimentation properties as the nicked form when $3 \cdot 2\%$ of the bases had been titrated with alkali. The process was accompanied by a reduction (slight but significant) in the buoyant density. They calculated from their data that $\tau = -15 \pm 1$, and σ $-0 \cdot 32 \pm 0 \cdot 002$ (this corresponds to one turn per 300 base pairs).

There is possibly a flaw in this calculation. It assumes that the native (state I) superhelix has proper base-pairing. This assumption may be wrong. Dean and Lebowitz (1971) have employed the reaction between DNA and formaldehyde to explore the secondary structure of superhelical φX RF II (p. 154). The reagent is specific for non-paired bases (see pp. 388 and 438). The extent of reaction of the following was compared: single-strand φX DNA (reacted), open-circle nicked φX RF II (no reaction—it's pure Watson–Crick) and RF I. In this case there was limited reaction (although less than with the single-strand DNA). Thus there are apparently some unpaired regions in the superhelix.

Intercalation and the superhelix   Let us assume that we actually know what's going on; i.e. that intercalation requires a $12°$ unwinding of the helix (p. 412). Additionally, let us consider a negative superhelix. What is the effect of intercalation? The answer is straightforward: the superhelix unwinds. It is difficult to see this without models but the position is shown diagrammatically in Fig. 15.6.

The effects described in Fig. 15.6 can be monitored in three ways. The buoyant densities of superhelical and open circle DNA can be plotted as a function of ethidium concentration. Using such data, Bauer and Vinograd (1968, 1970, 1970b) have derived equations to relate the binding parameters and dye concentration. The other experiments consist of measuring either sedimentation coefficient or viscosity of DNA as a function of dye-concentration. The effect on sedimentation coefficient will be that as the superhelix opens the S-value will fall until a minimum (corresponding to the $\tau = 0$ state—see Fig. 15.6) is reached. Then (as τ reaches increasing positive values) the S-value will start rising again. With intrinsic viscosity, the effect will be converse of this (i.e. a maximum intrinsic viscosity will be characteristic of $\tau = 0$). In both cases it is necessary to use an unconstrained (open-circle or straight-chain) molecule as control as the binding of dye to DNA affects the hydrodynamic properties slightly in any case.

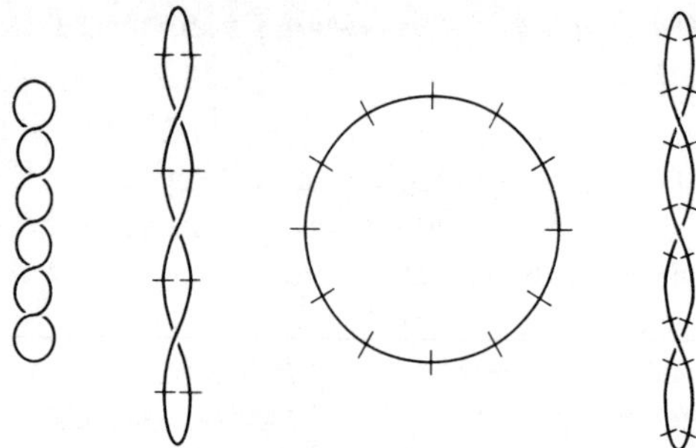

FIG 15.6   Hypothetical situation that arises as more ethidium chromophores (−) are bound to a negative superhelix, if the intercalation involves unwinding. Progressively, the helix is unwound, achieves an open circular form and then starts winding up again in the opposite sense (positive superhelix).

The sedimentation coefficient method has been employed by Crawford and Waring (1967) to survey several closed DNA molecules, all of which appeared to have negative τ-values (using ethidium as intercalating species). More recently, Waring (1970) has employed the intercalation of φX RF (τ apparently -13) to survey a number of dyes to establish whether they intercalate. His data demonstrated unwinding (and hence intercalation) with ethidium, acridines, actinomycin D (see p. 413), the anthracyclins (p. 415) and other drugs (chloroquin and propidium) but not with the tetrahydroanthracenes (p. 415) nor various other controls including spermine and streptomycin. According to Révet, Schmir and Vinograd (1971) the viscometric data are easier to analyse and will presumably be employed in future publications from Vinograd's laboratory.

Obviously, if intercalation involves a winding of the helix (p. 412) these conclusions are invalid. If the effects of intercalation on hydrodynamic properties are due to the basic models produced by Vinograd (and I think it is inconceivable that they are not), the implication of the winding model is that all these τ-values are positive. For the moment the problem must remain unresolved. Clearly many data must be re-analysed and further more definitive experiments planned. The CD properties of superhelices during 'titration' with ethidium would seem to be a possible experimental approach to the problem.

Proposed intermediates of nucleic acid metabolism

It is impossible to envisage DNA replication, transcription or (probably) translation without predicting structures different to conventional Watson–Crick or denatured configurations. Many ingenious models have been developed to account for these processes. There is no room for me to discuss them all or even to do proper justice to the more important ones. Rather I have merely selected those models which are accessible to experimental testing or which might become testable in the foreseeable future.

Models for DNA replication and recombination

The central problem of models for DNA synthesis is to overcome the problem imposed by the fact that DNA polymerase is a undirectional enzyme which extends the growing chain in only one direction (from 3′-end—see pp. 17 and 222). Any model for this process can make use of DNA polymerase, ligase (p. 230) and deoxyribonucleases. The theory should also account for several facts which are summarized below (not all these facts are relevant to all DNA systems).

Data relating to DNA synthesis   Of points that have been referred to already in this book, the following are (or could be) relevant to DNA synthesis: the existence of Okazaki fragments (p. 185), the semi-conservative mode of chromosome replication (p. 289), the existence of a replication fork working its way round the bacterial chromosome, the strange permuted characteristics of phage DNA (p. 362). A few additional experimental points may be mentioned. Okazaki, Okazaki, Sakube, Sugimoto and Sugino (1968) have shown that if replicating T4 DNA is lysed carefully (following very rapid pulse-labelling), 'Okazaki fragments' may be isolated from the growing region of the chromosome without denaturation of total

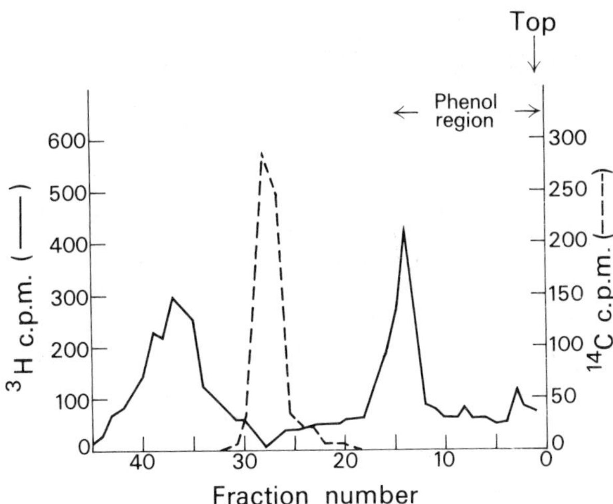

FIG 15.7   Slow-sedimenting, newly synthesized HeLa DNA. The cells were labelled with [^{3}H] dThd for 5 min, lysed as described on p. 146 and centrifuged through a gradient of the type shown in Fig. 4.13 (p. 147). Precautions were taken to avoid shearing or aggregation. Thus the cells were lysed in the bottom of the centrifuge tube and the gradient was put in underneath them (displacing the lysate) to avoid pipetting the lysate and the concentration was very low (4 x 10^4 cells altogether); [^{14}C]T4 DNA was included as a molecular-weight reference. The gradient was centrifuged for 15 h at 18°C and 17 000 rpm in the MSE Superspeed 50 ultracentrifuge. The gradient was sampled from the top (on the right). The slowly sedimenting zone (peak in fraction 15) was only found in 'pulsed' preparations. With longer labelling times, the size of this zone (relative to the very-high-molecular weight HeLa DNA around fraction 37) is diminished.

Unpublished data generously provided by Dr. J. R. B. Hastings.

DNA. This implies that there are short single-strand regions of DNA in the replication fork. Similar data have been obtained with bacterial DNA (Oishi, 1968). The important feature of these experiments is that the lysis should be achieved as gently as possible. The 'phenol gradient' of Hastings (p. 146) is particularly suited to this type of experiment. An application of this technique to the demonstration of the existence of a slow-sedimenting, rapidly labelled species in animal DNA is illustrated in Fig. 15.7.

During replication of phage DNA, there are intermediates much larger than the phage genome (see p. 188 for evidence in φX 174). The same findings apply to many double-stranded DNA phage (see for example the work of Botstein (1968) on P22^2 and Stone (1970) for an extensive bibliography). These structures are referred to as 'concatenates' and have sedimentation coefficients about twice those of the corresponding DNA isolated from phage particles (implying that the concatenates are many times as large as the phage DNA, p. 386).

Models for the general mode of DNA synthesis The models are briefly described in the following paragraphs. There are essentially three types of model. The first type seeks to explain the existence of phage concatenates; the second is designed to overcome the problems raised by the single polarity of the polymerase (p. 222); the third type is designed to overcome the especial problems raised by eukaryotic DNA replication.

First of all: phage concatenates. The rolling circle has been discussed already (p. 187) in the context of φX 174. I choose this example for detailed discussion as it is the system for which the most direct experimental evidence is available. The version of the model that applies to a duplex chromosome is shown in Fig. 15.8.

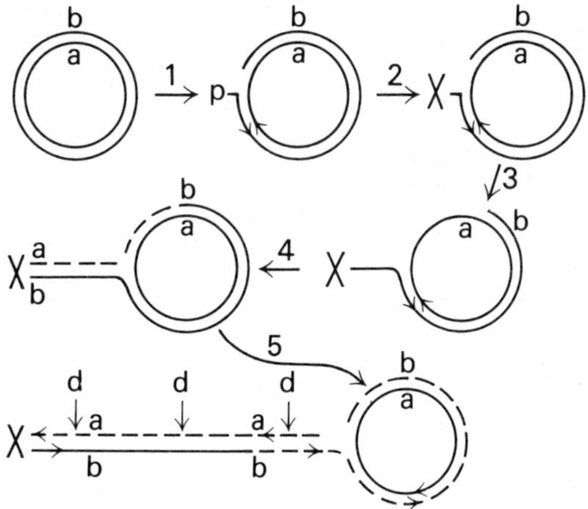

FIG 15.8 The rolling circle as applied to a continuous duplex DNA. The strands are labelled (a) and (b) for convenience. Arrows (>) indicate the 5$'$ to 3$'$ direction. The stages are: (1) introduction of a single-strand break ('nick'); (2) anchorage of the free 5$'$-end—see text for a discussion of 'X'; (3) the circle starts to roll; (4) DNA synthesis starts—newly synthesized DNA - - - - - - - ; (5) after several more rolls—discontinuities in (a)—shown d—are going to be linked up with ligase. After Gilbert and Dressler (1968).

The supposition is that the '5$'$-anchor' X in the figure is an attachment site on the cell membrane. What is the evidence for this model? It is a simple model that explains the presence of Okazaki fragments and the production of concatenates in the same system. However, the only direct (electron microscopic) evidence for the rolling circle has been obtained with the φX RF (see frontispiece *b*). So far there is no direct proof of the existence of this intermediate from the double-strand phage. Ideas regarding the further processing of the concatenates are discussed on p. 187.

The *E. coli* chromosome is not replicated by the rolling circle method. Very exact methods of studying chromosome replication in bacteria require methods of holding up the process and synchronizing replication (e.g. by thymine starvation of a Thy$^-$ organism). Re-initiation starts at a specific chromosomal locus, 'the origin' (Pritchard and Lark, 1964). The kinetics of DNA synthesis in rapidly growing *E. coli* have been analysed by Helmstetter and Cooper (1968). Their work provides an analysis of the situation that arises when several replication forks are present in the same molecule and are inconsistent with the rolling circle. Recent data of Fritsch and Worcel (1971) have been obtained by creating a 'Meselson-and-Stahl density shift' (p. 289) in rapidly growing cells, in which multi-fork chromosomes must be present. The results demonstrate that there is a symmetric transfer of label which is incompatible with multiple rolling circles and imply symmetric multi-fork replication (see Fig. 15.9).

Eukaryotic chromosomes: Several theories have been advances for an explanation of some of

FIG 15.9   Possible consequences of density transfer in multi-fork chromosomes. For simplicity, the chromosome is regarded as a linear structure with the origin at the top. 'Heavy DNA' ———————— is present in the bacteria at time 0. The symmetric labelling (above) was confirmed by measurement of the densities of DNA following transfer to light medium (light DNA - - - - - - -). Rolling circles would predict the asymmetric labelling shown below. The situation is shown after density-transfer and growth for $\frac{1}{2}$ and for 1 generation. From Fritsch and Worcel (1971).

the puzzling facts regarding eukaryotic-chromosome replication. One of these (the conservation of genotype among multiple copies of the same gene) is discussed in the context of recombination below. Another problem is that there is some evidence (outside the scope of this book) that chromosome replication can result in a conservation of the heterozygous state. To account for this there are models in which the two strands of the parent duplex are assigned the roles of master-strand and passive-strand. In one model (Jehle, 1965) the 5'-to-3' strand is read as master by DNA polymerase to produce a single daughter strand. This daughter is then the template for the synthesis of the other 'daughter'. The scheme is summarized in Fig. 15.10.

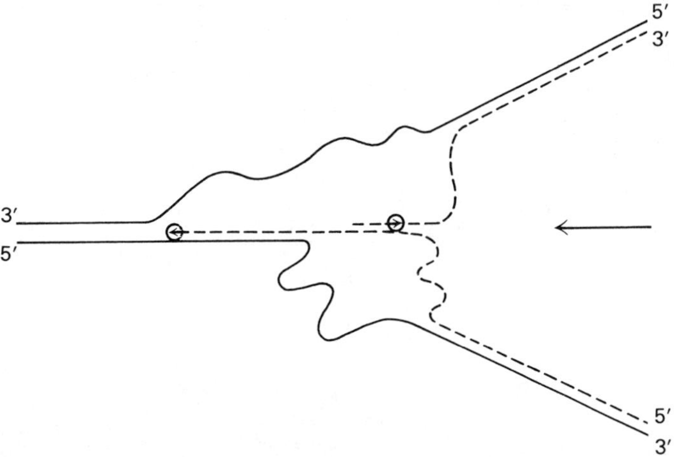

FIG 15.10   Tertiary structure at the replication fork in the 'master-and-passive-strand' model of Jehle (1965). The strands are labelled thus: Original (parent) duplex ————————, daughters - - - - - - -. The big arrow shows the replication direction and the little circles with arrows are DNA polymerase extending the chains in those directions.

Note that in this model, the master/first daughter relationship is the same as that between the transcribing strand and nascent RNA during transcription. The model apparently fulfils one requirement of a scientific theory—it should be experimentally verifiable, as it implies that rather extensive regions of single-strand DNA occur in the region of the replication fork. It is certain that in replication of phage (and probably all prokaryotic replications) this arrangement cannot apply as heterozygotes segregate after only one round of replication (Russo, Stahl and Stahl, 1970; see however p. 436).

Knife-and-fork and broken-parent models   These two models represent more intimate pictures of the replication fork in the growing chromosome and are designed to overcome the 'priming problem' that arises in any version of the replication of DNA by a Kornberg-type enzyme (p. 428).

The knife-and-fork model is the easier to understand. It is illustrated in Fig. 15.11. It is fully described by Guild (1968). Once the 'knife' (nuclease) has cut the fork, synthesis may continue and some steric factor is assumed to determine the distance the polymerase moves up before it turns round and starts coming down. There is no direct physico-chemical evidence for the unique triangular tertiary structure implied as an intermediate in this model. Indirect evidence is discussed by Becker and Hurwitz (1971).

If synthesis of the type required to account for Okazaki fragments occurs by a discontinuous mechanism other than the knife-and-fork model, it is necessary to postulate single-strand DNA

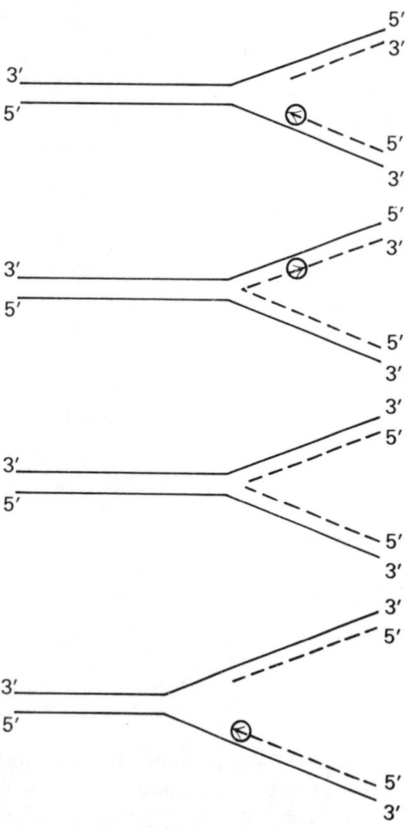

FIG 15.11   The knife-and-fork model. The conventions are those employed in Figs. 15.8 and 15.9. The sequence of events should be self-explanatory.

in the region of the replication fork. In other words, the fork has a single-strand point of attachment to the prong in which discontinuous synthesis is occurring. Inman and Schnös (1971) have made electron micrographs of replicating λ-DNA[3] and have discovered that a very high proportion of the replication forks do contain single-strand DNA. These single-strand stretches are of the order of a few hundred nucleotides long. A few pictures contained apparently 'pure duplex forks'. However Inman and Schnös regard their data as supporting a model for the λ replication fork containing single-strand connexions.

The 'broken-parent' (my name) is due to Haskell and Davern (1969). It is summarized in Fig. 15.12. Here again the model should be directly testable as one of its consequences is that in one region (ahead of the replication fork) the nascent (daughter) is covalently linked to parent. Tentative evidence for this model is discussed by Becker and Hurwitz (1971). The attractive feature of this model is that it simplified the problem of unwinding of the parental strands.

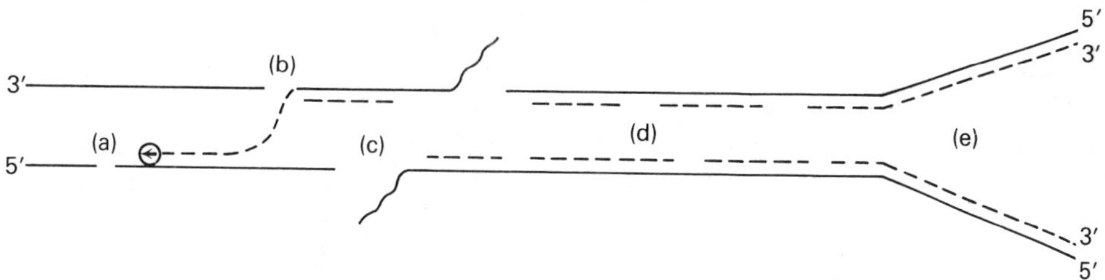

FIG 15.12   The broken-parent model of Haskell and Davern (1971). Again the parents are represented by continuous lines and the daughter by broken lines. Overall replication is from right to left. Successive stages are represented by (*a*) to (*e*). (*a*) A single-strand break in a parent; (*b*) The free 3'-end of such a break is a primer for synthesis; (*c*) such a product is detached from its parent; (*d*) further extension of such fragments. By the time we get to (*e*), breaks in the parents and discontinuities in the daughter have been sealed with ligase.

I have omitted all reference to models which account for the events involved in the initiation and completion of the replication of a continuous chromosome as at present it is difficult to see how the techniques covered by this book may be used to test them. The review by Becker and Hurwitz (1971) contains a summary of the various ideas.

Recombinational events   The models described here refer to the processes that occur when two DNA duplices with regions of homology line up and a cross-over occurs. There is very little structural evidence for the models and I summarize them in order (hopefully) to stimulate the reader into thinking of experiments that might confirm the presence of intermediates predicted by them. For an informative and extremely entertaining review of the subject, see Davern (1971).

First let us briefly recapitulate the biological events that are used for the study of recombination. There is the insertion of a small circular DNA molecule into a larger one (such as the establishment of lysogeny by phage—p. 29). In this case the genetic systems involved is λ-specific; there are several, genetically unrelated systems: *Int*, which specifies insertion and excision; *Red*, a complex of systems, one of whose gene-products is λ-exonuclease (p. 211) which determines vegetative λ recombination, β which is related to *Red* system in some obscure way and *Ter* which is involved in the production of sticky-endedness (p. 280). The recombination between large genomes has been extensively studied in T4 (Moshig, 1970); two genes (*x* and 47) control both recombination and repair. In *E. coli*, the genetic loci for recombination are called *rec*. *Rec⁻* mutants are both deficient at recombination (thus mating between an Hfr and an F⁻ *rec*⁻ strain is abortive) and are UV-sensitive. There are several classifications of *rec*-characteristics. One is meaningful in molecular terms; the mutants are regarded as 'cautious' (*rec* B, *rec* C), in which case there is less-than-the-normal amount of DNA-breakdown following UV-irradiation, or 'reckless' (*rec* A), in which case there is more than the usual amount of breakdown. The cautious mutants apparently lack nucleases although the precise ones that are missing are unidentified. The reckless (*rec* A⁻) mutants belong to a quite separate class. It is now clear that the limited repair that is possible in *pol* A⁻ mutants (p. 259) is effected by gene-product(s) from the *rec* A system; moreover *Pol* A⁻ rec A⁻ is a completely lethal genotype. It thus looks as though *rec* A and *Pol* A will to some extent take over one anothers work in the repair process but that *rec* A is uniquely required for recombination. Moreover one or other of these genes is required for normal vegetative growth; whether this is because some repair is always required or whether there is some part of normal chromosome replication that requires these activities is not clear (Monk, Peacey and Gross, 1971; Gross, Grunstein and Witkin, 1971).

Recombination in eukaryotic organisms is the essential feature of meiosis. Despite the incredible complexity of the system, models are accessible to genetic testing. By a careful analysis of the segregants occurring following an interallelic cross, it is possible to probe the nature of the intermediate stage. As an example of the methodology see the work of Kitane and Olive (1970) using the spore-colour markers in the fungus *Sordaria*. The other approach is to try and correlate the cytological effects of meiosis (synapsis and chiasma-formation) with the models. The procedure raises more questions that it answers as there is no indication of the timing of the molecular events relative to the appearance of the phenomena; Pritchard (1955) has made a critical appraisal of the problems.

Recognition   Prior to recombination, the chromosomes must line up so that identical DNA duplices are in perfect alignment. Probably most people imagine that the proteins in chromatin play an important role in this recognition. This is certainly possible and there are no data to use to test any hypothesis regarding this process. However, it has been pointed out to me by Dr. W. Fuller (personal communication) that it is at least possible that two duplices might associate through specific hydrogen bonding between bases (Fig. 15.13). The base-pairing schemes shown in this figure have actually be found in the crystal structures of mixtures of derivatives of the

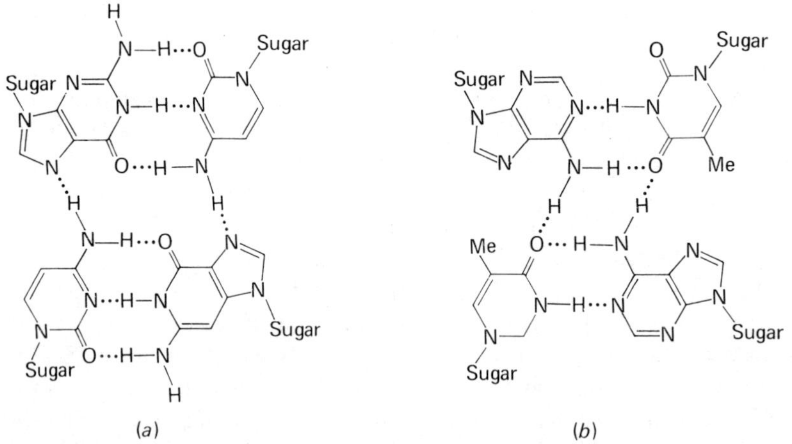

FIG 15.13   Structures of proposed quartets (*a* G:C/G:C and *b* A:T/A:T) to account for duplex/duplex recognition prior to recombination. Based on figures kindly supplied by Dr. W. Fuller. See text for discussion.

bases (Fig. 2.29(*b*) and (*c*), p. 102). The attractive feature of these particular quartets is that the parameters defining equivalent positions in the four sugars (say C-1') are the same in the G:C/G:C quartet as in A:T/A:T.

Recombinational models   Four possible models for recombination are summarized in Fig. 15.14. In all cases, a single cross-over is illustrated. Such a cross occurs in meiosis and in the insertion of a temperate viral (or plasmid) DNA molecule into a chromosome. A genetic cross of the type that occurs during bacterial or meiotic recombination requires two such events.

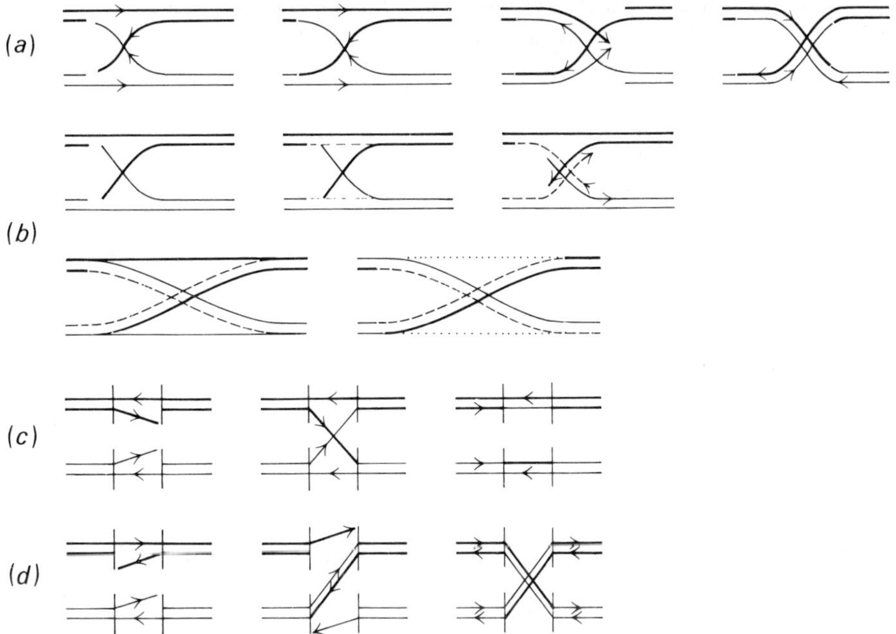

FIG 15.14   Models of cross-over in DNA. Little arrows represent 5'-to 3' directions; the two original duplices are drawn in heavy and light lines to distinguish them. Discontinuous lines (- - - - - - -) represent DNA that is newly synthesized during recombination; dotted lines (. . . . . . .) represent DNA that is eroded during recombination. See text for discussion. The models are due to (*a*) Holliday (1964), (*b*) Whitehouse and Hastings (1965) and (*c* and *d* Kitane and Olive (1969).

The model of Holliday (1964; Fig. 15.14(*a*)) is based on the idea that following the production of single-strand breaks, a hybrid DNA is produced and subject to 'mis-match repair'. Only nucleases and ligases are required in this model. The idea of Whitehouse and Hastings (1965) is similar except the repair involves nucleolysis and repair synthesis. The model is a more attractive one, because it resembles other types of repair process (p. 259) in that DNA synthesis is required. The idea that repair synthesis is a function of the *Rec* A system is discussed on p. 434. Moreover the other *Rec* systems (B and C) seem to be involved in DNA degradation (p. 434).

The two models of Kitane and Olive (1969—Fig. 15.14(*c*) and (*d*) are specifically designed to account for recombination in eukaryotic cells. The idea is that there exist fixed-pair regions (? cistrons) linked by metabolically labile positions at which single-strand breaks can occur. The two alternative versions account for those occasions on which there is asymmetry between the frequencies of parental and recombinational events for outside markers and those for which there is symmetry. Thus model (*c*) would produce no recombination outside the cross whereas (*d*) would. It is proposed that in fungi, both types of cross-over are possible.

Cycloid chromomeres   Whitehouse (1969) and Holliday (1969) have proposed a novel type of cross-over during eukaryotic chromosome duplication. The model accounts for three separate facts: one is the existence of extensive tandem genes (e.g. for rRNA—see p. 370) which are apparently not evolutionarily divergent; another is that many closely related animals (especially amphibians) contain differing amounts of DNA/cell (range over a factor of as much as ten) thus implying that tandem genes or multiple gene copies are common (see also p. 365); finally there are the special properties associated with 'lampbrush chromosomes' (see Whitehouse, 1969, for an account of these).

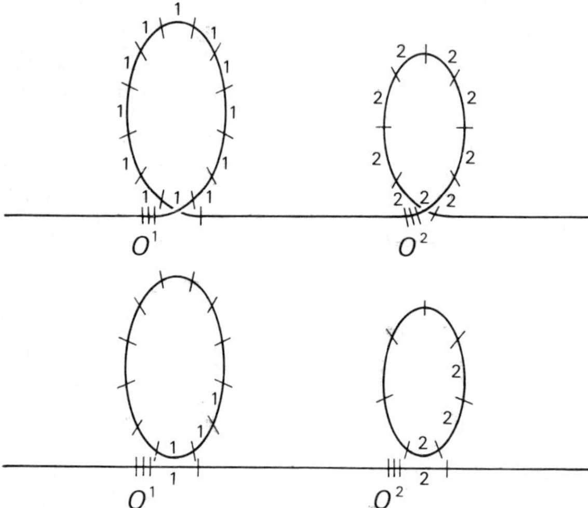

FIG 15.15   A simplified version of Whitehouse's cycloic model of the chromosome (see text for discussion). Lines in this picture are DNA duplices. In the upper picture, two master genes (1 and 2) and their slave copies are arranged in a cycloid conformation. The loci o^1 and o^2 are hypothetical 'recombination-operator genes'. In the lower picture all the slaves have been excised as loop-chromomeres. This is the state presumed to exist prior to DNA replication, when a new cycloid chromosome will be synthesized with masters as templates for the synthesis of new slave copies.

The suggestion is that the chromosomes contain two classes of genes 'masters and slaves'. A simplified version of the model is shown in Fig. 15.15. The process involved in this figure (the excision of the slaves as cyclic chromomeres) does not involve any mechanism different to those discussed above. The tertiary structures implied by the copying of the masters to form new strings of slave concatenates during the next round of replication must be ectremely complex although Whitehouse has discussed possible models. The slave-loop chromomeres are (in principle at least) entities with a characteristic structure which should be identifiable in a DNA preparation.

Transcriptional intermediates

Two models   There are two possibilities for transcription. Either the DNA must unwind locally for a Watson–Crick structure to be possible between the nascent RNA chain (and the incoming pppN molecules prior to condensation) and the transcribing strand or alternatively the DNA is not denatured and some kind of local triplex is used for the recognition reaction. Recent versions of the two alternatives appear conveniently as consecutive articles in *Nature* (Florentiev and Ivanov, 1970; Riley 1970). Both are extremely ingenious and take thermodynamic and structural parameters into consideration.

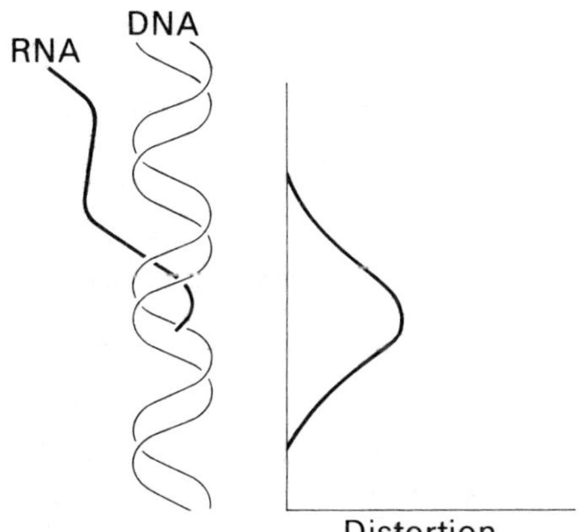

FIG 15.16   The Riley model for transcription (see text and Riley, 1970).

(*a*) Left: An A:T base pair (Watson–Crick), and a picture of the orientation following the necessary distortion (about 40° rotation of one base with respect to the other) and the formation of a triplex with an A-residue (above) in the wide groove. The right-hand DNA base (Thy) is in the transcribing strand.

Right: An analogous case illustrating the incorporation of a pyrimidine (G-residue) by a G:C pair. Equivalent pictures can be drawn for the insertion of a T-residue and a G-residue.

(*b*) Distortion of the DNA duplex to form a 'transcriptional triplex'. The graph illustrates the extent of angular distortion of the duplex in the region of the complex.

The Florentiev–Ivanov model assumes the following sequence of processes in the region of the RNA polymerase. The DNA locally undergoes the following changes B-form to A-form to unwound (local conformation) in which the non-transcribing strand is stabilized by stacking. The incoming pppN is bound (requires energy) and the pyrophosphate is released (p. 12) with a subsequent re-winding of the DNA and 'release' of the nascent RNA from the complex. The suggestion is that these processes are concerted; in other words the exergonic pyrophosphate-release accompanying the nucleotide-bound formation of residue *n* in the RNA chain is coupled to (or 'drives') the unfavourable process of binding the triphosphate for residue ($n + 1$) together with the conformational alignment of this residue for condensation.

The 'Riley model' assumes that the incoming pppN lies in the wide groove of a slightly distorted DNA duplex. The three-base complexes and tertiary structure involved in this model are shown in Fig. 15.16.

Evidence for transcriptional intermediates   Evidence for the second (Riley) model should be found by the isolation of the RNA:DNA:DNA triplex structure. There is no direct evidence for the presence of this material in a well-defined medium containing RNA polymerase *in vitro*. There is considerable evidence for the existence of RNA/DNA hybrids *in vivo*. For a description of the isolation of such a complex from *E. coli* (and reference to earlier papers) see Tongur, Wladytchenskey and Kotchina (1968). A more recent, and better defined, experiment is by Avadhani, Schuit, Róe and Buetow (1970). RNA was isolated from the photosynthetic flagellate, *Euglena gracilis*[4] by a method which used Macaloid (p. 122) and tri-*iso*propyl-naphthalene sulphonate (p. 110). However the product was essentially RNA, insoluble in 1 M NaCl. The preparation was found to contain some DNA and protein. The DNA/RNA/protein material could be precipitated by addition of pre-cooled ($-20°$C) *iso*propanol (p. 107). Analysis by polyacrylamide-gel electrophoresis (p. 155) showed that the DNA, RNA and protein existed as a complex with an estimated sedimentation coefficient of 13 S. The DNA was detected in this complex by the diphenylamine reaction (p. 173). The complex was found to be stable to ribonuclease; the effect of deoxyribonuclease was to produce an apparent hyperchromicity.[5] Treatment of this complex with urea or heat resolved the RNA and DNA components (on the gels). The simplest explanation of these results is that the 13 S complex contains duplex DNA and H-bonded RNA together with some protein. The nature of the protein component was not studied. A possible explanation of these data is that a rather large triplex region (Fig. 15.16, p. 437) exists in the transcriptional complex of *E. gracilis* and this complex was isolated. Alternatively, a special tertiary structure may be formed between DNA and RNA which is in some way stored for subsequent metabolism.

On the other hand there is direct evidence for the local unwinding of DNA during transcription *in vitro*. Kosaganov *et al.* (1971) employed formaldehyde (p. 427) as a probe for locally unwound regions of DNA. The principle of the method they employed is that formaldehyde, in relatively high concentration, reacts with denatured regions of DNA and then works its way along the duplex, unwinding and reacting as it goes.

It is possible with this method to detect discontinuities in a duplex separated by as much as several thousand nucleotide pairs. The procedure has to be carefully calibrated with sheared DNA of known weight-average molecular weight in order to calculate the unwinding parameters. The extent of discontinuities in native T2 DNA was found to be unchanged when RNA polymerase is bound to the enzyme. However if the triphosphates are added, so that transcription may commence, the extent of local unwinding (as reflected in the increased rate of formaldehyde-induced unwinding) was markedly increased. Of course it is possible that the helix distortions implied by Riley's model might create conformations accessible to the formaldehyde reaction, but at first sight the data seem to favour the Florentiev–Ivanov model.

Translational intermediates

The codon–anti-codon complex   Fuller and Hodgson (1967) presented a model of the anti-codon loop of rTNA which accounted for codon-binding in a simple and attractive form.

The model is shown in Fig. 15.17; essentially, the A-structure of the stem of this loop is envisaged as extended beyond the base-paired region. The two pyrimidines which universally occur on the 5′-side of the loop (Fig. 10.8, p. 305) are hooked back out of the way and the anti-codon bases and their 3′-neighbours are stacked in a continuation of the helix. The attractive feature of this model is that the codon of mRNA can fit into place in this helix. Moreover the 'wobble position' is (of the three interactions) in the most flexible conformation.

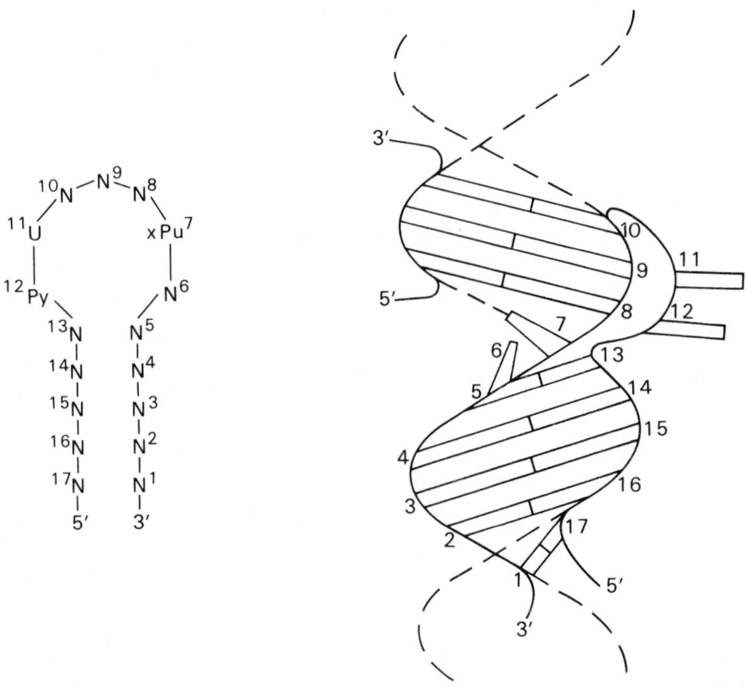

FIG 15.17   Model of the codon–anti-codon interaction according to Fuller and Hodgson (1967). On the left, the generalized structure of the anti-codon region of all tRNA molecules is redrawn from Fig. 10.8, p. 305. The numbers are drawn to help to identify the residues in the model. On the right is the codon–anti-codon complex. The continuation of the A-helix is shown thus — — — —. The sugar-phosphate chains are drawn as continuous lines and the bases thus ▭. The model is discussed in the text.

In this particular form the model has been tested and found wanting. Although it is not possible to study the conformational changes occurring during protein synthesis directly, it has been possible to study the effect of binding oligomers to tRNA. The ideal tRNA to use is tRNA^{Phe} which contains its own fluorescent probe (Y) in the anti-codon region (see p. 302). By studying the blue shift in the fluorescence (p. 418), Eisinger, Feuer and Yamane (1971) were able to estimate the binding constants for various codon–anti-codon interactions. From such measurements at different temperatures, it was possible to calculate ΔG, ΔH and ΔS for such reactions. Measurements were made with U-U-U, U-U-U-U, U-U-U-U-U, poly U and U-U-C. The results were as follows: U-U-C binds more strongly than U-U-U (binding constant six-fold higher). This is perhaps not too surprising as the wobble base in the anti-codon is Gua (attached to methyl ribose—Fig. 10.9f, p. 307). The binding constants for the remaining substances were all about the same. The sequence in the anti-codon loop of this molecule is as follows (the asterisk for reference later, the line covers the anti-codon) . . . Å-Y-A-A-G̲m̲-̲U̲-̲m̲C . . . . This implies that Å is not free for base-pairing whereas the model in Fig. 15.17 implies that it is. The thermodynamic parameters themselves are of great interest. Rounding the figures off a bit, the values are, for ΔH, -11, for $T\Delta S$, -8, for ΔG, -3. These data correspond to the complex formed between two triplets, one in a locked conformation (the anti-codon is tRNA) and the other to the free trinucleotide conformation. It is known that if both components are free trinucleotides, no complex is formed. Thus ΔG is apparently positive; assuming that ΔH is of the same order of magnitude, the assumption is that in the case of two free triplets, no interaction occurs as $T\Delta S$ swamps the contribution of ΔH.[6] One may assume that on the ribosome, the mRNA is in a locked conformation so that the rRNA-anti-codon interaction is much stronger. How much? If the ribosome works perfectly, we should be able to equate the entropy change (and hence $T\Delta S$) with zero. This puts $\Delta G = \Delta H$. From this value ($\Delta G = -11$ kcal/mole) an affinity constant of the order of 10^{12} mole^{-1} is implied; this is sufficient to account for the fidelity of protein synthesis.

The conformation of the mRNA A complex model for protein synthesis has been proposed by Woese (1970). In essence, the model allows for protein synthesis to follow from allosteric changes in the two tRNA molecules (aminoacyl tRNA and polypeptidyl tRNA) and does not require the hypothesis that there are two dissimilar sites (A and P) on the 50 S subunit. From the point of view of the present discussion, the valuable point is that it involves the constraint of the mRNA in a system similar to that of Fuller and Hodgson (above) but which requires a specific messenger conformation. Thus the implication from the work of Eisinger (above) that the conformation of the tRNA is not that in which the anti-codon duplex is extended (in the way described by Fuller and Hodgson) could be irrelevant, if it were confirmed that the specific 'translational conformations' of mRNA (attached to a ribosome) are required to lock the anti-codon into this form. Beyond this it is not necessary to go; it is impossible at present to imagine ways in which Woese's tertiary structure might be explored experimentally.

Notes to chapter 15

[1]   This is a complicated system genetically. The episome which carries an F'-factor, genes for two colicins (p. 31) and two amino-acid markers, is derived from an *E. coli* donor and had been introduced into *Proteus* (a *trp⁻ cys⁻* strain to make scoring the transfer simpler).

[2]   A *Salmonella* phage.

[3]   This is not a rolling circle; DNA has two replicating forks moving in opposite directions. Thus the RF has the following form: ⊂▭▭▭⊃ .

[4]   Most of these experiments were performed on *E. gracilis* cells lacking chloroplasts or by employing a wild-type strain in the presence of streptomycin that inhibits chloroplast metabolism.

[5]   This phenomenon was measured indirectly. The ratio of absorbance at 260 nm to amounts of diphenylamine-positive material in the 13-S zone on the gels was compared for sampled untreated and treated with deoxyribonuclease.

[6]   ... because from the second law of thermodynamics, $\Delta G = \Delta H - T\Delta S$; and for any feasible process ΔG must be negative.

16 Interaction between nucleic acids and proteins

This book is largely a source of recipes, background material and references and I imagine that the average reader will be dipping into sections here and there in a search for particular topics. However I may have a few faithful readers who are reading through it to see the whole subject in the perspective which I have drawn. To such a reader: we are approaching the end of our journey. We have ended up at the point at which the biologist would start. The essential functions of nucleic acids depend on their unique interactions with protein. This is also the point at which a historical account of the subject would start as nucleic acids were first recognized as a component of Miescher's 'nuclein' (nucleoprotein). However, to a chemist, the problems of the interaction of one class of polyelectrolyte (nucleic acids) with a different class (proteins) are the most intractable of all. Ideally we should divide such a discussion into two parts. The first would answer the question, what are the structures of nucleoprotein particles (chromosomes, viruses, ribosomes)? The second would explain the mechanisms of those dynamic processes in which nucleic acids react with proteins specifically (conformational changes in ribosomes during protein synthesis, the aminoacyl-tRNA-synthetase reaction, the reaction between repressors, polymerases and so on, and DNA). Our knowledge of nucleoprotein at the moment is so fragmentary that such a classification would be over ambitious. However the subject is expanding at a fast rate and one can look forward to the possibility of such a sub-division in the future.

At present it is only possible to survey the methods available for studying nucleoprotein. Basically one can do three types of experiment. The overall structure of nucleoprotein can be examined by structural, chemical and spectroscopic means; the structure can be selectively broken down to try and deduce the specific associations in the complex; finally the nucleoprotein can be built up from its components. There are many variants of this last procedure that are possible. Modified components may be used to enquire which structural features are essential for the final quarternary structure; the kinetics of the association may be examined and the thermodynamic parameters of the association can be determined.

This chapter is divided into sections dealing with viruses, cellular deoxyribonucleoprotein, aminoacyl tRNA synthetases and ribosomes.

I have not devoted any significant space to the determination of the molecular weights of the nucleoprotein particles. Essentially the methods are those hydrodynamic procedures suitable for very large proteins. If the meniscus-depletion equilibrium method (p. 389, Yphantis, 1964) is used, it is necessary to know the value of $(1 - \bar{v}\rho)$ (see p. 383). The value of $\bar{v}$ can be measured by very accurate pycnometry. If the molecular weight is to be determined from the sedimentation coefficient, either $[\eta]$ or D must be determined (p. 386). Probably the method employing D is the more accurate for the present purpose. An extremely accurate method of measuring D for large nucleoproteins (T-phage) using optical-mixing spectroscopy[1] has been described by Dubin, Benedek, Bancroft and Freifelder (1970). A comparison of the molecular weights of several T-phage estimated by the equilibrium method (Bancroft and Freifelder, 1970) and the sedimentation/diffusion method (Dubin *et al.*, 1970) are in excellent agreement.[2]

Viruses

Clearly it is impossible to attempt even a superficial survey of virus structure here. I have chosen particular examples of viruses to illustrate methods of examining the structural interactions between their macromolecules. I have restricted the examples to ones where the virion is a pure nucleoprotein (i.e. have not included any 'envelope viruses' that contain fragments of host-cell membrane—p. 32).

There are essentially two groups of viruses, 'simple' and 'complex'. A simple virus is one which has very few different types of protein molecule and exhibits geometric symmetry; a complex virus is one which contains many different protein components and has an asymmetric complex morphology. It was pointed out by Crick and Watson (1956) that viruses with very small genomes must be simple as there is not enough genetic information for the many different classes of proteins required for complex morphology. They predicted (quite correctly) that the simple viruses should contain a nucleic-acid core with the protein subunits arranged around it in either helical or cubic[3] symmetric array. The confirmation of these predictions and the detailed knowledge of the overall structure of viruses comes from electron microscopy.

Simple viruses

The assembly of simple viruses is a fascinating subject as a wide variety of experimental techniques for studying the process which can be related to models in solid geometry. These processes are the simplest of all examples of morphogenesis and are the only morphogenetic assemblages which can be studied *in vitro* at the moment.

Icosahedral viruses   The icosahedron is a regular solid composed of twenty faces (each face an equilateral triangle). It possesses 5:3:2 axial symmetry. This fact and other details of icosahedral structures are most simply observed with a solid model. One method of folding paper to make an icosahedron (which is instructive in a separate connexion—see p. 447) and an explanation of the symmetry are shown in Fig. 16.1.

Crick and Watson (1956) pointed out that it was unlikely that the protein molecules themselves that were packed together in the icosahedron. It is the case (Caspar and Klug, 1962) that for close packing in an icosahedron, the packed units (capsomeres) should be either pentagons or hexagons. Thus the assembly of the virus consists of the assembly of the coat protein into capsomeres which then close-pack into the faces of an icosahedron. The geometric arrays that make this assembly possible are reviewed by Almeida and Ham (1965).

Two classes of icosahedral RNA viruses have been used extensively for studies on assembly *in vitro*, the RNA coliphage (such as R17 and Qβ) and the small plant viruses (such as cowpea chlorotic mosaic virus, CCMV and bromegrass mosaic virus, BMV). The work is reviewed in two excellent concurrent articles (Hohn and Hohn, 1970; Bancroft, 1970). Different experimental systems are involved in the two cases. A phage such as R17 is infective in Hfr strains of *E. coli*;

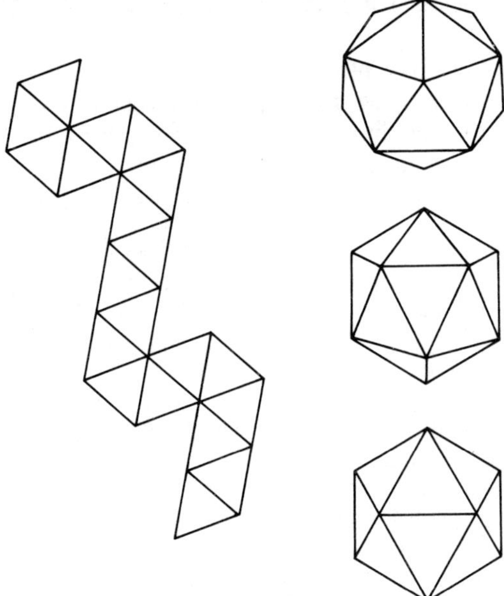

FIG 16.1   Icosahedra. On the left, a strip of paper which can be folded into an icosahedron. On the right, views of an icosahedron down its three axes of symmetry. The axis runs through the centre of the picture in each case. The upper one is the five-fold axis, as if this picture is rotated through 360° the identical view will be obtained five times. The middle is the three-fold axis and the bottom one the two-fold axis. There are four axes of the middle type in the icosahedron (defining the 'cubic symmetry'—see p. 466).

its RNA is not infective (although it can be assayed in spheroplasts—p. 193). Small plant viruses and their RNA are both infective. It is possible that only damaged plant cells are infected by either virus or RNA. However in this case it is not at all possible to study the unique infectivity of the reconstituted particles.

Under favourable circumstances, the phage proteins aggregate to produce particles of the same approximate sedimentation and electrophoretic properties as the complete phage. It thus seems that specific protein:protein interactions are important (see Hohn and Hohn, 1970, for references). In the presence of A-protein (p. 313), coat protein and RNA form RNP particles

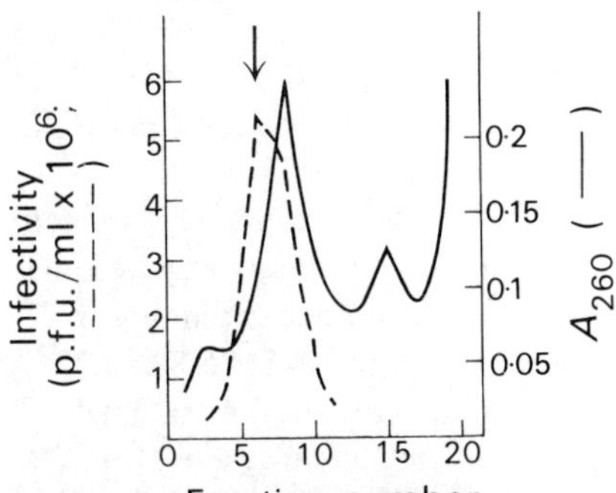

FIG 16.2   Reconstitution of phage R17 *in vitro*. Sucrose-gradient profile (sedimentation from right to left) of the results of an incubation of RNA, coat protein and A-protein. The arrow shows the position of viable R17 in a separate gradient. Data of Roberts and Steitz (1967).

which are predominantly lighter than R17 but which contain a shoulder (in sucrose gradients) of infective particles (see Fig. 16.2).

The apparent low efficiency of this process (and the production of the inactive 70 S particles) is not surprising as R17 assembly is an extremely inefficient process *in vivo*. By using non-optimal buffer composition it is possible to produce uninfective small complexes containing only a few subunits per RNA chain (there are 180 coat-protein subunits and one A-protein per virus particle: the nature of the clustering of the subunits into capsomeres and

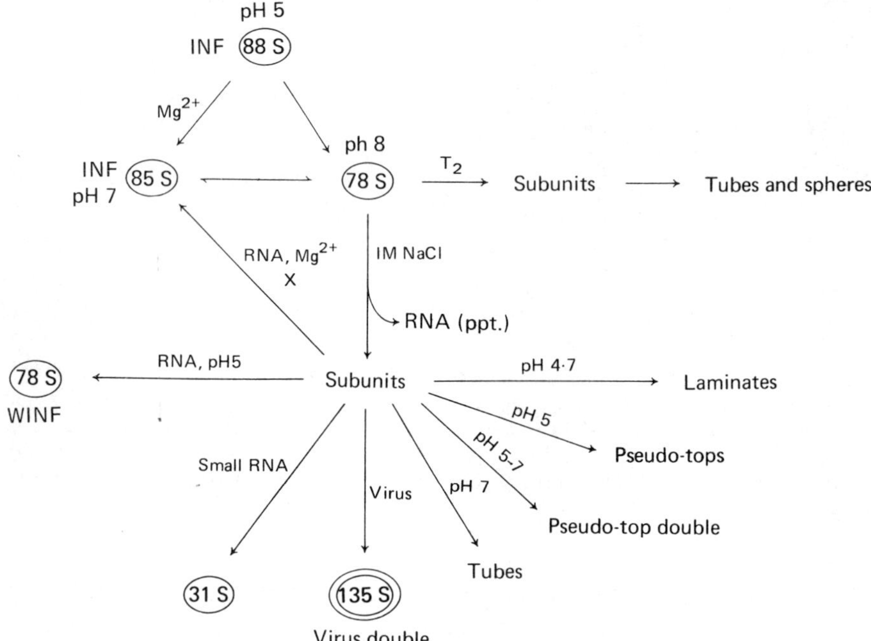

FIG 16.3   Map of the assembly and dissociation reactions of CCMV (after Bancroft, 1970). The molarities are strengths of buffer used in the experiment: Mg^{2+} ions present in some incubations are typically 1 mM. See text for discussion and Fig. 16.4 for an example of a reconstitution experiment such as process X above. INF indicates an infective preparation; WINF indicates weakly infective.

the location of the A-protein are not precisely certain). It is probable that some sort of linear array is formed. Certainly protein monomers aggregate to form long strands under certain conditions. An electron micrograph of these filamentous aggregates appears in Hohn and Hohn (1970).

The dissociation and association properties of CCMV have been extensively studied. Infectivity of reconstituted particles is assayed by demonstrating that the infectivity becomes resistant to snake venom phosphodiesterase. Some of the reactions are summarized in Fig. 16.3.

Several points emerge from these studies. The three forms of the virus (88 S, 85 S and 78 S) reflect differences in the protein:protein interactions in the coat. Bancroft has rationalized these changes in terms of interactions between —CO$_2$H side chains (in glu- and asp-residues). It is proposed that at low pH (88 S form) H-bonding exists between two neutral residues thus:[4]

At higher pH (78 S form) these residues are charged and repel one another:

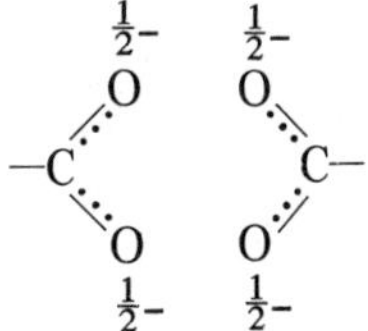

In the presence of Mg^{2+} (85 S form), this structure is stabilized by chelation:

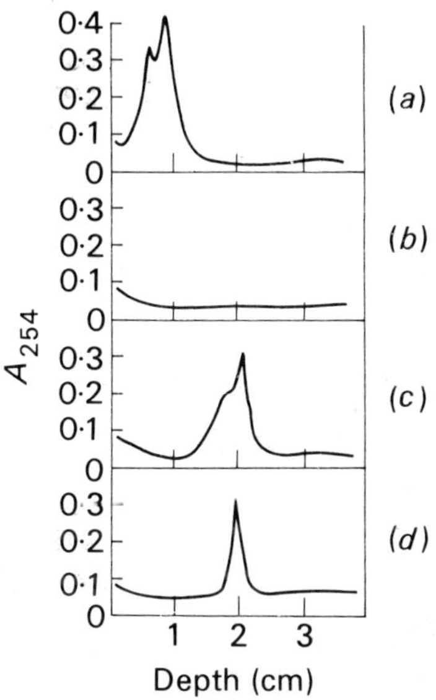

In the reassembly reaction it is clear that Mg^{2+}-chelation is essential for the faithful locating of these 'carboxylic dimers'. An example of such a reconstitution is shown in Fig. 16.4.

FIG 16.4   The reconstitution of CCMV *in vitro*. Sucrose-gradient analysis of (*a*) RNA, (*b*) protein, (*c*) reconstituted particles and (*d*) 85 S virus. In (*a*) there is evidence (in my view) of aggregation probably due to the high NaCl used for isolating the RNA (see p. 139). The material in (*c*) was obtained by mixing RNA and protein in 0·5 NaCl and dialysing (4°C) against 10 mM KCl, 10 mM tris HCl pH 7·4, 5 mM $MgCl_2$. Data of Bancroft (1970).

In the absence of RNA (or in the presence of oligonucleotides) the virus proteins self-associate to form a variety of structures identified by electron microscopy (Fig. 16.3, p. 444) depending on the pH and ionic conditions of the medium. The 'pseudo tops' are particles which are probably cup-shaped with positive charges on the inside. They thus may reflect a structure in which the subunits are packed in a conformation similar to half a virus capsid.

The evidence from two quite different types of virus both suggest that the RNA is required to assemble the virion in its correct icosahedral morphology, but that there exists considerable specificity among the protein subunits for the formation of quarternary structure. There are two alternatives. One is that the RNA forms a regular tertiary structure first which is the core of the virion and the protein then stacks round it. The second is that the two types of macromolecule interact to form assemblage intermediates prior to the folding of a nucleoprotein complex. All the evidence supports the latter view. The ability of CCMV to form

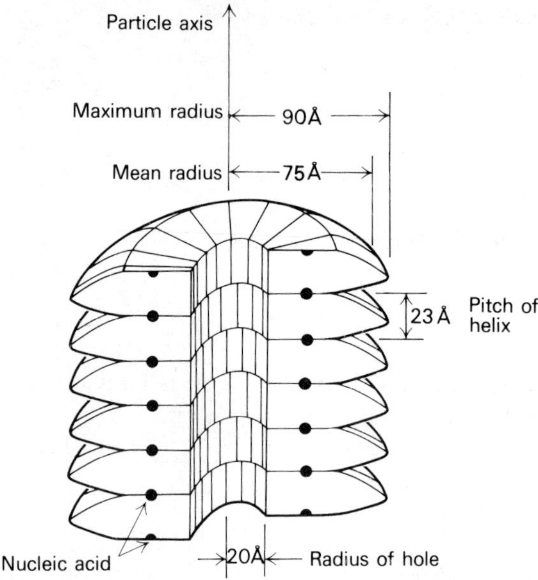

Fig 16.5   Dimensions of TMV (from Franklin, Klug and Holmes, 1957).

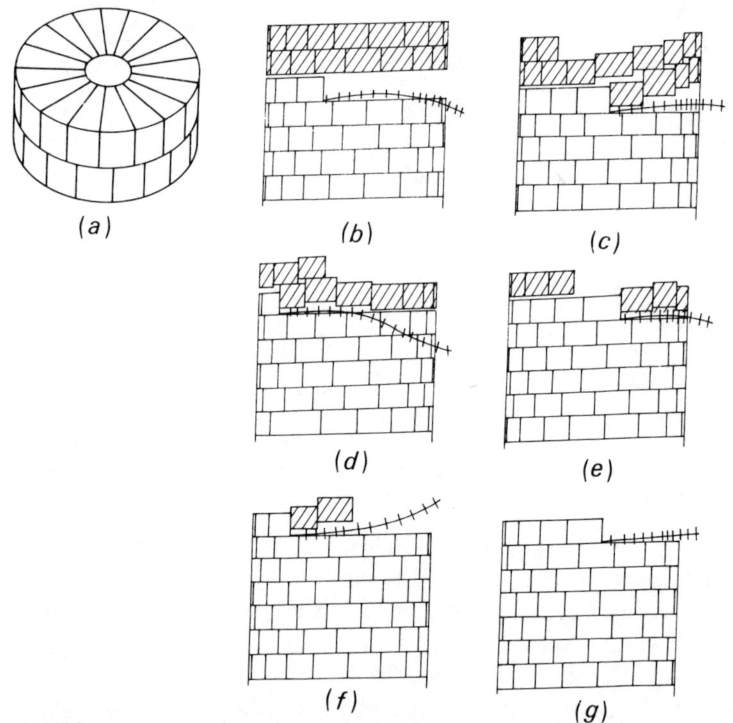

FIG 16.6   Intermediates in TMV assembly (from Durham *et al.*, 1971).
  (*a*) Schematic drawing of a 20 S disc
  (*b*) Such a disc lying on top of an incomplete helix (+++++ = RNA)
  (*c*) to (*f*) Successive dislocations resulting in the rearrangements of protein in the disc to extend the helix
  (*g*) An incomplete helix ready for a new disc to approach (as in *b*).
In (*b*) to (*f*) the shaded subunits are those not present in the helical quaternary structure.

a diesterase-sensitive structure (78 S) as a result of reactions, which must surely be protein-conformational changes, argues against the RNA being buried in the centre of the structure. The existence of RNP intermediates in R17 assembly (p. 443) likewise supports such a view. Hohn and Hohn (1970) present a model for the assembly of R17 and similar phage in which the subunits bind to the growing RNA chain (during synthesis) and that the linear RNP structure then collects more subunits and folds into an icosahedron. At least a paper strip may be folded into an icosahedron (Fig. 16.1, p. 442) so why not an RNP strip? The role of the A-protein may be to form some kind of nucleus for the process.

Helical viruses One helical virus (tobacco mosaic virus, TMV) has been studied far more extensively than any other and in this section I just report some recent work on the self-assembly of this structure which is probably of wide significance in helix-assembly in general. The virus particle contains one RNA molecule (MW 2×10^6 27 S) and protein subunits (A protein, MW 17 420). The overall picture of a section of the intact virion is illustrated in Fig. 16.5. The overall length of the virion is 3 000 Å.

Until recently, it was generally assumed that the helix was created by the continuous winding of the linear RNP particle. Recent work by A. Klug and his colleagues (Durham, Fincham, Butler and Klug, 1971) has shown the situation to be very much more complex than this. They have carefully characterized (with the electron microscope and the ultracentrifuge) the various states of aggregation of TMW protein. The natural state of aggregation at neutral pH and ionic strength of 0·1–0·2 (or at higher ionic strength and higher pH) is a 20 S disc. The structure of such a disc and the probable way in which they might act as intermediates, in the extension of the helix in the presence of TMV RNA, are shown in Fig. 16.6. The essential feature of this process is that protein forms a quaternary structure on its own with the correct diameter and these discs then form a helix which can be regarded as a dislocated stack.

Complex viruses

Complex helical viruses Many animal RNA viruses consist of an inner core of helical ribonucleoprotein with a sheath of surrounding protein (of a different type). Examples are vesicular stomatitis virus[5] (VSV) and rabies virus.

VSV is a bullet-shaped particle. The central core consists of a nucleoprotein helix and it is surrounded by layers of protein sheath. There are only four different species of viral protein (Kang and Prévec, 1969). Of these one is present in much smaller amounts than the other three. Nakai and Howatson (1968) have proposed an attractive model for the formation of the virion. It envisages a chain of RNA with the protein subunits arranged along it and wound into a broad hollow helix with the nose of the bullet created by turns of diminishing diameter. It is unfortunate that the assembly of such a particle cannot be studied *in vitro*; it would be extremely interesting to examine how such an open helical structure can originate.

According to Nakai and Howatson (1968) rabies virus has a similar internal structure but is surrounded by two protein envelopes whose contours closely follow those of the internal helical region.

T-even phage The T-even phage have a complex morphology that is idealized in Fig. 16.7.

These T-phage cannot be re-assembled from their components *in vitro*. The problem of their morphogenesis might appear to be insurmountable. In fact there are two experimental approaches to it. Both involve the assemblages of aberrant structures. These structures can be

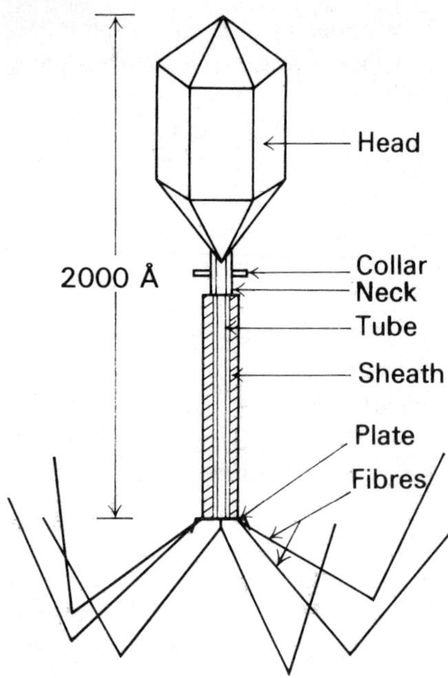

FIG 16.7   Structure of a T-even phage (after Cummings, Couse and Forrest, 1970). It is not absolutely certain that the head has the shape (short hexagonal tube capped with hexagonal pyramids) shown here.

induced in the phage by infecting in the presence of one of several drugs or chemicals (such as dimethylsylphoxide). Alternatively many amber mutants of T4D (p. 302) are available in genes which determine morphogenesis. See the review by Cummings, Couse and Forrest (1970) for references to these two techniques. The amber mutants are maintained in amber-suppressor hosts; when used to infect non-permissive (su^-) hosts the erroneously constructed structures may be seen in the hosts by electron microscopy of thin sections of the bacteria. Structures of this type that are found include long tubular sheaths of head protein called 'polyheads'. The packing of capsomeres in the polyheads can be visualized more clearly than it can in the bacteriophage themselves. From the observations that have been made, several models of the packing of these capsomeres have been made. The significant point is that no specific nucleic acid–protein interaction is important in these structures. The question of direct interest is 'how does the DNA fold up to get into the phage head'? It seems that there is sufficient flexibility in native DNA for the process to occur without single-strand discontinuities of any kind (p. 394).

Aminoacyl-tRNA synthetases

The reactions catalysed by these enzymes are described on p. 14. The second part of the reaction (reaction of aminoacyl adenylate with tRNA) is the crucial reaction of protein synthesis, in the sense that it is the one stage of the whole process in which a specific amino-acid nucleotide recognition process occurs. In all the other processes (nucleic-acid synthesis and peptide-bond formation) the specific recognition is achieved by a nucleic acid–nucleic acid interaction. Thus there is great interest attaching the problem of which part of the tRNA sequence is recognized by the synthetase in its selection of the tRNA appropriate to its aminoacyl adenylate. For example, is one of the sequences common to all tRNA molecules (Fig. 10.8, p. 305) responsible for a binding reaction common to all synthetase/

tRNA interactions? Which of the variable regions in the sequence is responsible for the specific binding of the correct tRNA species to the synthetase?[6] What role (if any) does the anti-codon play in these processes? We do not really know the answers to any of these questions in any detail. However there is considerable fragmentary evidence and the purpose of this section is to illustrate the main ways in which the problem can be approached.

Isoacceptors and heterologous synthesis

'Isoacceptors' are tRNA molecules of different sequence found in the same cell and specific for the same amino acid. A comparison of the sequences of isoacceptors should make it possible to rule out certain regions of tRNA as candidates for R-site.[6] A rather different type of information has been obtained by comparing an 'unnatural isoacceptor system' in which the effectiveness of synthetases from one organism in charging (i) its own tRNA molecules and (ii) tRNA molecules from a different species are compared. As it turns out, neither of these approaches has provided satisfactory answers to our questions but experiments of type (ii) have provided interesting data in their own right.

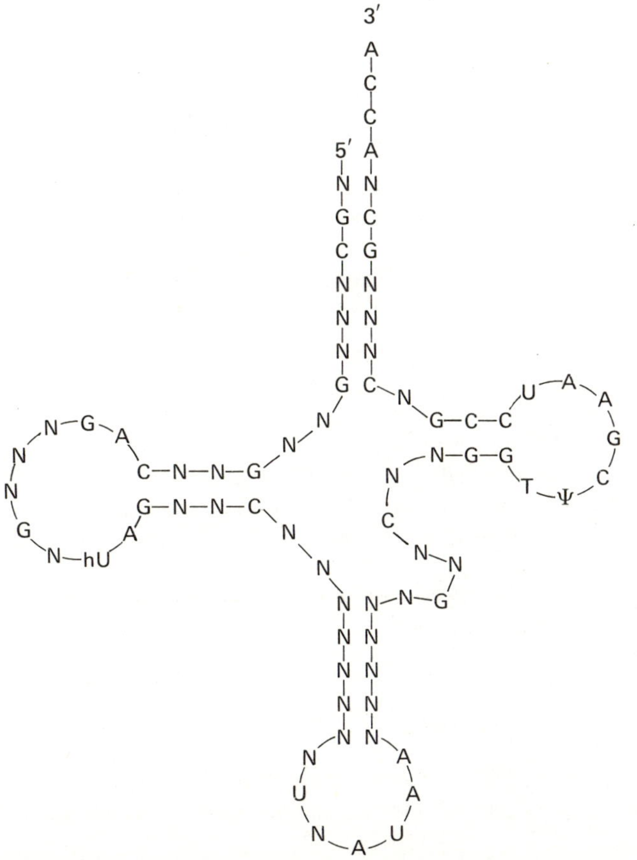

FIG 16.8   Comparison of the sequences of tRNA$_F^{Met}$ and tRNA$_M^{Met}$ from *E. coli*. The residues N differ in the two molecules and presumably are not involved in the R-site (see text for discussion). This type of presentation is due to Chambers (1971). The complete sequences of these two molecules are shown in Fig. 10.9, p. 308.

Sequences of isoacceptor tRNA molecules Rather few isoacceptor sequences are known and some of them are extremely unhelpful. See, for example, the two sequences of $tRNA_{yeast}^{Ser}$ (Fig. 10.9d, p. 307) which differ in sufficiently few places for no sensible guess to be possible. A more instructive case are the $tRNA_M^{Met}$ and $tRNA_F^{Met}$ of *E. coli* (Fig. 10.9, p. 308). Both of these molecules are charged by the same synthetase (p. 24) and the kinetics are similar; they differ considerably in sequence. The composite sequence is illustrated in Fig. 16.8.

Certain sequences in the figure can be discounted immediately as candidates for the R-site: ...C-C-A is common to all tRNA sequences and ...A-C-C-A is common to most of them; T-Ψ-C is common to them all and the whole stretch ...G-G-T-Ψ-C-G-A-A-U-C-C... is found in one other *E. coli* tRNA ($tRNA^{Tyr}$). Similarly ...hU-A-G... is found in $tRNA^{Phe}$. In any case there are strong reasons for thinking that the R-site is not present in the loops (see following paragraphs). Chambers (1971) thus concludes that the R-site lies in the acceptor stem. There is an amusing consequence of this conclusion, which will have occurred to the perceptive reader. We shall just leave it for the moment and return to the point on p. 454.

Heterologous charging reactions There are several possible consequences of heterologous charging processes. The charging reaction may just not work at all; it may work but with less efficiency than the homologous; it may work with greater efficiency; finally it may result in mis-matching. There is a comprehensive survey of these processes by Jacobson (1971). As he points out, many of the data on the 'efficiency' of correct heterologous charging are not really suitable for comparison, as it has been the common practice to compare the extents of charging under standard conditions. What is really required is a comparison of the kinetics of the reaction in the various systems. In general, the remarkable feature of the process is the large number of correct (not mis-matched) heterologous reactions that are possible, even between species widely separated in evolution (like *E. coli* and man) again substantiating the conservation of features of tRNA sequence (see p. 417). The heterologous charging of $tRNA^{Ser}$ in the rat/yeast system presents a particular problem for theories of the R-site and is discussed further on p. 454.

The phenomenon of mis-matching was first demonstrated by Barnett and Jacobson (1964) who found that homogeneous phenylalanyl synthetase from *Neurospora crassa*[7] would charge two different heterologous *E. coli* molecules to produce Phe-$tRNA^{Val}$ and Phe-$tRNA^{Ala}$. Obviously this enzyme recognizes the R-site in its own $tRNA^{Phe}$ but cannot be recognizing two different R-sites in the two *E. coli* molecules (although it is possible that one of them has the same R-site as *N. crassa* $tRNA^{Phe}$, the other one must be different). Thus mis-matching must sometimes occur when there is no specific enzyme/R recognition at all. Perhaps the non-specific binding site (p. 14) is identical in all three molecules.

Reference to other mis-matching systems are to be found in the reviews of Jacobson (1971a) and Chambers (1971).

Acceptor properties of chemically modified tRNA

One type of chemical modification is hydrolysis. The acceptor activity of tRNA fragments is discussed in the next section. Here we are concerned with the modifications in which there is no alteration in chain-length.

Some nucleosides not involved in the R-site Several selective reagents have been employed which have no effect on the charging reaction. Some of the results and a key to the reactions employed are summarized in Fig. 16.9.

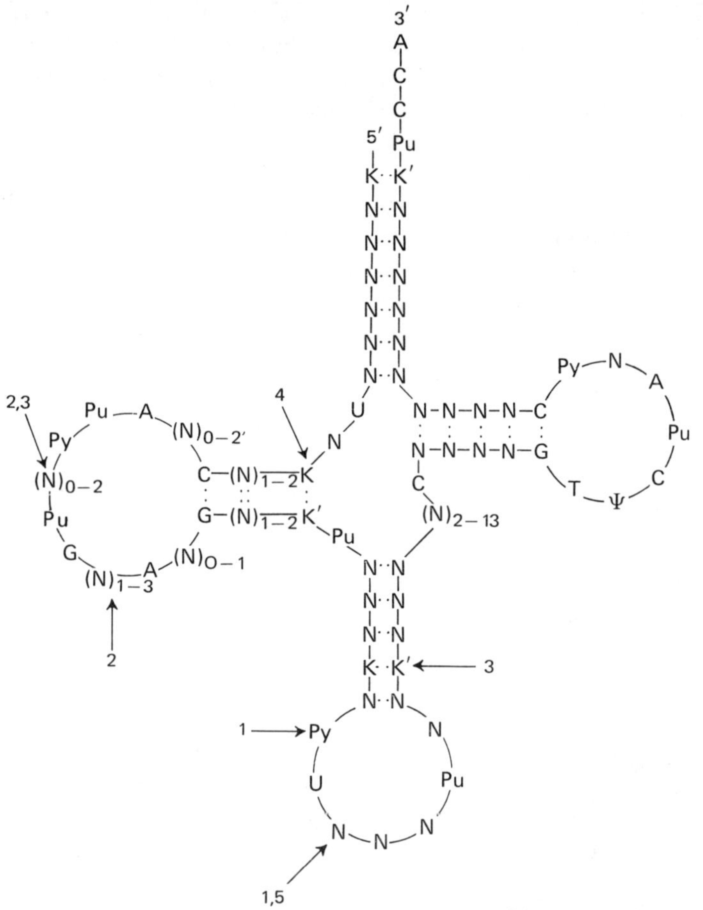

FIG 16.9   Diagram showing positions in tRNA molecules which can be modified without affecting the charging reaction.

Key: (1) Reaction between acrylonitrile and I and Ψ in the 'labile' $\text{tRNA}^{Ala}_{yeast}$ (see p. 452; Yoshida, Kaziro and Ukita, 1968)

(2) Reaction between $NaBH_4$ and hU in $\text{tRNA}^{Ser}_{yeast}$ (Molinaro, Sheiner, Neelon and Cantoni, 1968)

(3) Reaction between $NaBH_4$ and hU and m^5C in $\text{tRNA}^{Phe}_{yeast}$ (Igo-Kemenes and Zachau, 1969)

(4) De-thiation of sU in tRNA^{Tyr}_{coli} (Walker and RajBhandary, 1970)

(5) Reaction between kethoxal (p. 65) and Gm in $\text{tRNA}^{Phe}_{yeast}$ (Litt, 1969).

Obviously the scoring of all these results in the same figure implies that the R-site of all tRNA species is the same. There is no absolute reason why this should be the case but the common clover-leaf secondary structure of all tRNA sequences leads one to expect some kind of universal pattern. However, at this stage the generalizations implied by this figure should be regarded as tentative.

UV-induced changes in tRNA   Extensive work has been done on the photo-inactivation of tRNA and attempts have been made to correlate the uv-inactivation with modification of pyrimidines (p. 63) in the sequence. R. W. Chambers and his colleagues have made the major contribution to this subject and many data are reviewed in Chambers (1971). The techniques may be exemplified by two sets of experiments.

Schulman and Chambers (1968) examined the photo-inactivation of the 'labile' yeast $tRNA^{Ala}$ ($tRNA^{Ala}_{lab}$).[8] The kinetics of the process were found to differ in Na^+ and Mg^{2+} salts. The latter reaction was found to show no isotope effect in D_2O (i.e. kinetics were the same in D_2O and H_2O) from which it was deduced that the inactivating event was not a photohydration (p. 65). It was thus either a pyrimidine-dimerization (p. 243) or a photolytic fragmentation of a Ψ-residue.[9] They developed a method of fractionating the inactivated from the normal tRNA and identified the oligomers by T_1-digestion of these fragments. Similar experiments have been performed (Kućan, Kućan, Gartland and Chambers, 1970; reported in more detail by Chambers, 1971) and have identified UV-modified nucleosides in irradiated and not inactivated $tRNA^{Tyr}_{yeast}$. These results are summarized in Fig. 16.10.

As Chambers (1971) points out, these data can be regarded as pointers to the identity of the R-site. It is not valid to assume that the inactivation region is necessarily the same as the R-site.

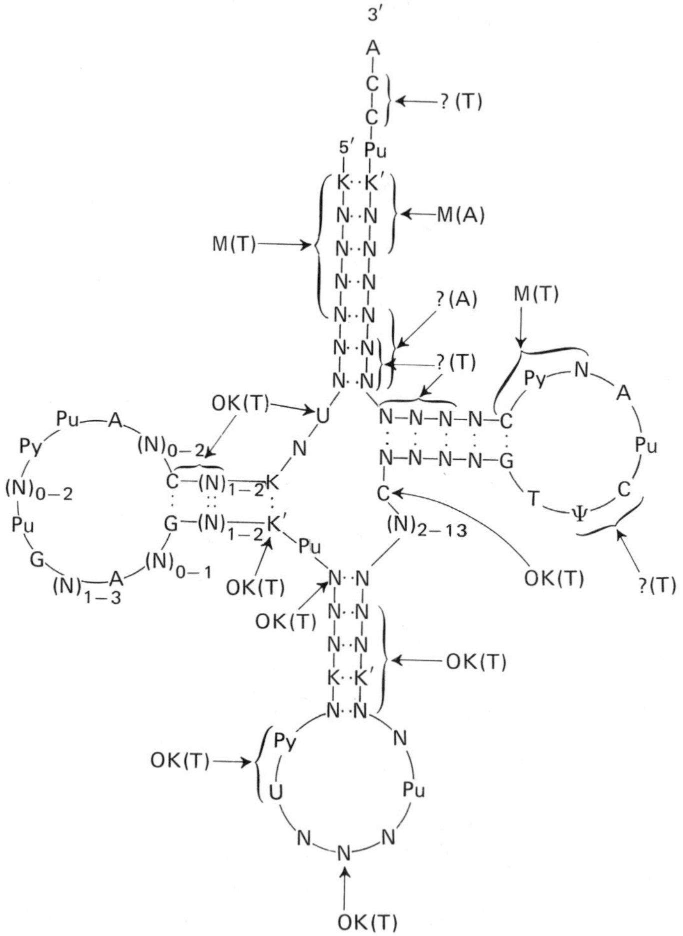

FIG 16.10   Photoinactivation of two tRNA species from yeast. The generalized picture of tRNA is that shown in Fig. 10.8, p. 305 and Fig. 16.9, p. 451.
Key: M(A) Major inactivation target in $tRNA^{Ala}_{lab}$
  ?(A) Possible inactivation target in $tRNA^{Ala}_{lab}$
  OK(T) Regions of $tRNA^{Tyr}$ that can be damaged by UV with no lack of activity
  M(T) Major inactivation targets in $tRNA^{Tyr}$
  ?(T) Possible inactivation targets in $tRNA^{Tyr}$.
Data from various papers of Chambers—see text for references and discussion.

It is possible that a distortion in the secondary or tertiary structure consequent upon a reaction at a site other than R might inhibit binding of tRNA to the synthetase.

Charging of tRNA fragments

Several pieces of work have been reported on the charging of fragments of tRNA. There is not room to refer to every experiment of this type. The review of Chambers (1971) contains complete references up to mid-1970. I have just chosen two cases which exemplify the conclusions that have been reached.

Yeast tRNA^{Phe}   Yeast tRNA^{Phe} can be easily split into two approximately equal 'halves' by excision of the Y-nucleoside (p. 421). Thiebe and Zachau (1971) have surveyed the effects of removing the Y-nucleoside from the tRNA on its optical properties and charging reactions. CD and other spectral studies imply a gross change in tertiary structure in tRNA^{Phe} when the Y-nucleoside is excised (compare the data on p. 421). The product of this reaction (two half-molecules, Fig. 15.3, p. 422) can be charged with the yeast synthetase but require higher Mg^{2+}-concentrations than did the intact molecule. However unlike the intact molecule, it could not be charged with the heterologous *E. coli* synthetase. These data imply that the overall conformation of the molecule is significant in determining the effectiveness of the R-site.

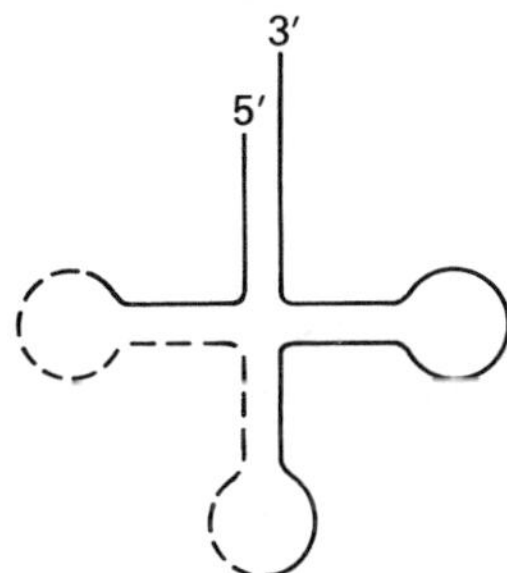

FIG 16.11   A reconstituted part of tRNA$_2^{Ala}$ which can be charged with Phe (by the homologous synthetase). The outline of the complete structure is shown thus - - - - - - - . The continuous line represents the reconstituted fragment employed in the assay. Data from Imura, Weiss and Chambers (1969) and Chambers (1971).

Imura, Weiss and Chambers (1969) have been able to isolate large fragments from yeast tRNA$_2^{Ala}$ by partial digestion with ribonuclease (p. 300). Although the complete sequence of this molecule has not been determined, there are sufficient data available for the position of the fragments to be unambiguously determined. None of the fragments had charging activity on its own. However various combinations of these fragments produced reconstituted structure with limited but significant charging capacity. The smallest of these reconstituted structures that could be charged was the one containing the 5'- and 3'-termini of the sequence (see Fig. 16.11).

Is the R-site in the acceptor stem?

Well, probably—this is the message of Chambers (1971). The work on photo-inactivation (p. 452), the data shown in Fig. 16.10, the work on the tRNA$_2^{Ala}$ (above) all point in this direction. The data of Zachau (p. 451) do not contradict this view but emphasize the possible role of tertiary structure in the effectiveness of the interaction. If this is the case, can we guess which region of the stem is involved; is there a 'second code' of sequences in this region that could account for the fact? To code for the 20 different amino acids, the sequences must contain three (or more) nucleotides. As I hinted on p. 450, in the case of the $tRNA_{coli}^{Met}$ isoacceptors, there is no triplet sequence common to the two molecules. Perhaps the nucleotides in both sides of the stem are responsible for the specificity of the R-site. Of course, it is not necessary for this particular code to be universal. Some amino acids may use a sequence and others a pattern of nucleotides in both strands. The explanation for non-compatibility of heterologous systems could be that these are particular amino acids whose coding mechanisms differ in the two species.

In fact there is circumstantial evidence that a triplet code may operate for certain yeast interactions. Chambers (1971) has pointed out that the yeast tRNA molecules for six amino acids (ala, ser, val, ile, phe, tyr) have a stem sequence as follows:

... N-N-N-N-N-N-W-5′

... N-N-N-N-Q-R-X-Pu-C-C-A 3′

In this sequence, W and X are each G or C (such that W:X is a G:C or C:G pair) and the sequence QRX is uniquely different for each tRNA. Thus it is tentatively proposed that in this family of molecules such a code is operating.

If this is the case, serine poses a bit of problem, although there is a neat way out of it. Rat liver $tRNA^{Ser}$ and yeast $tRNA^{Ser}$ are charged by the liver enzyme with identical rates (both K_m and V_{max} are the same). This implies that the R-site is the same in the two molecules. The sequences in the region shown above are apparently quite different:

... U-C-A-A-C-G-G-5′   ... G-C-U-G-A-U-G-5′

... A-G-U-U-G-U-C-G-C-C-A3′   ... C-G-A-C-U-A-C-G-C-C-A 3′

    Yeast   Rat

In the light of the above 'code' the 'codon' in $tRNA_{yeast}^{Ser}$ should be GUC; the corresponding sequence in $tRNA_{rat}^{Ser}$ is UAC. There is a GUC sequence in the rat molecule. It is in the upper strand (remember to read from right to left) starting with the fourth nucleotide residue. Moreover, Chambers (1971) has shown with model building that there is an axis of symmetry perpendicular to the page and dividing the strands immediately to the right of this fourth nucleotide. As a consequence it is claimed that the synthetase would see the 'codon' in the same relation to the CCA terminus in both molecules.

So much for aminoacyl-tRNA synthetase. We can look forward, over the next few years to an accumulation of more evidence regarding the sequences of tRNA molecules which should enable us to pin down the R-site with more confidence. Knowledge of primary, secondary, tertiary and quaternary structures of one of the enzymes might make it possible to determine what peptide configurations are involved and the role of the binding reaction and the CCA sequence in these reactions.

Deoxyribonucleoprotein

DNA : protein interactions determine the specificities of both DNA synthesis and transcription. The polymerase reactions are discussed elsewhere in the book. Here it is appropriate to discuss the binding of proteins to DNA which control these processes. The subject is a vast one and I do not pretend to be making anything like a systematic survey of the experiments that have been performed. For example I have omitted all reference to the dissociation and reassociation of eukaryotic nucleohistone and the transcription of nucleohistone *in vitro*; these experiments have yielded a lot of results but unambiguous interpretation is difficult, and I believe that a new breakthrough is required in the quantitation of the kinetics of these processes before a realistic assessment of their significance for nucleoprotein structure can be made. Rather I have chosen examples which illustrate experimental methods which, I believe, are at present available for the further study of DNA-protein interactions in any new system that the reader might wish to examine.

Chromatin

Chromatin (p. 116) consists of DNA, basic proteins (histones) and acidic proteins. In addition to the processes which I have omitted (see above) there are three methods of studying chromatin structure. The overall geometry of the chromatin complex may be examined; the conformation of the DNA may be studied; and the conformation of the proteins may be studied. I have chosen (I hope) experiments, each of which does give an unambiguous description of an aspect of chromatin structure.

Overall geometry   Calf-thymus DNP (p. 116) or nucleohistone can be drawn into a fibrous forms and an X-ray diffraction pattern can be obtained. These patterns have been studied by Luzatti and Nicolaieff (1963) and by Pardon, Wilkins and Richards (1967). The conclusions from these data are that the reflections from the DNA duplex are poorly aligned (implying the DNA-helix-axis does not run along the axis of the DNP fibres). The reflections that are obtained include one corresponding to a spacing of 110 Å.

Davies (1967) studied electron micrographs of a similar preparation (this time from chick erythrocytes—see p. 182) and concluded that the DNP consisted of hollow tubes with an outside diameter of 150 Å. Although other interpretations may be possible (as pointed out by Pardon, Wilkins and Richards, 1967) the simplest explanation of the data is that the conformation of the DNP is that of a giant helix with a pitch of 120 Å and a diameter of between 100 Å (X-ray data) and 120 Å (EM data[10]).

Thus DNP is probably a coiled coil; the DNA duplex is wrapped up in protein; the resulting rope is coiled in a giant single helix with the geometry described above.

Circular dichroism of chromatin   Matsuyama, Tagashira and Nagata (1971) have analysed the CD spectrum of chromatin. Their most important finding is that as the pH is increased there is a CD transition at that pH (10·5) at which the nucleoprotein is disaggregated. The CD above pH 10·5 (but below 12·5 when the DNA unfolds) is that characteristic of B-DNA. They suggest that in native chromatin, the DNA is in the C-form (p. 328).

Interaction between chromatin-DNA and a reporter   Gabbay (p. 408) has surveyed the binding of aryl polyamines (used as 'reporters') to DNA. The consequences of the binding are that there is a hypochromic effect on the aryl chromophore and induced dichroic effect in this chromophore. His conclusion was that the aliphatic chain of the reporter lies in the small groove of the duplex.

Simpson (1970) studied the binding of one particular reporter (no. 7 in Fig. 14.3, p. 408) to calf-thymus chromatin. He made three types of measurement: the induced CD maximum; the hypochromism; and (using the latter parameter) the reporter–chromatin binding parameters.

The dichroic maximum (negative in this case—p. 409) was identical with that obtained in the DNA–reporter complex. The association constant for reporter and DNA was $7·5 \times 10^{-4}$; that for the reporter and chromatin (in an identical buffer) was of the same order of magnitude

(3.75×10^{-4}). Thus both the nature and (essentially) the extend of reporter-binding was the same in both cases.

This is certainly the most important information we have about chromatin structure as it clearly points to a conclusion: the minor groove of the DNA duplex in chromatin is empty and its phosphate residues are available for polymine-binding. Therefore the proteins are presumably in the major groove. These data do not preclude the possibility of some limited filling or shielding of the minor groove in particular regions but if one could see the chromatin complex, there would be long stretches of empty minor groove.

Interactions between histones and DNA   Histones consist of a mixture of a fairly limited number of different proteins (about ten in all); the classes of histones are remarkably constant in different eukaryotic organisms. The nomenclature of the various histone fractions is odd at first sight and reflects the historical fractionation and sub-fractionation of the classes. Three histones have been sequenced: F2A1 from two different sources (DeLange, Fambrough, Smith and Bonner, 1969; Ogawa *et al.*, 1969), F2B (Iwai, Ishikawa and Hayashi, 1970) and F1 (Bustin, Rall, Stellwagen and Cole, 1969). The sequences show a curious type of polarity. They all contain a large excess of basic residues (lys and arg). In the F2 histones, these basic residues are concentrated in the N-terminal half of the molecule; in F1 they are concentrated in the C-terminal half. The distribution of the other classes of amino-acid residues is extremely asymmetric too. The basic regions contain residues (such as pro or gly) that are regarded as de-stabilizing the α-helical conformation of proteins. In contrast, the 'neutral' regions contain those residues (such as ala) regarded as helix-stabilizers together with high proportions of aromatic and hydroxy-acids. The proposal was therefore made (see above references) that the basic halves of the molecules (which will have little if any tendency to form their own secondary structure) are involved in binding to the DNA whereas the non-basic halves (with native protein secondary structure) are involved in whatever processes you think histones are specific for.

These predictions have been experimentally tested by Boubliak, Bradbury, Crane-Robinson and Rattle (1971) by studying the PMR signals (p. 95) due to the amino-acid residues in histone in their free and DNA-bound states (in D_2O solution). Any detailed discussion of the results requires a consideration of the PMR spectra of proteins and would be out of place in this book. Suffice it to say that characteristic signals for such groups as the CH_3-protons in ala, the CH_3-protons in val, leu and ilv (all superimposed) and two classes of Ch_2-protons in lys can be assigned in these spectra and the collapse of these signals can be used as a measure of interaction between these residues in secondary, tertiary and quaternary structures. The conclusion they reached was that (as predicted) only the basic (N-terminal) half of F2B is bound to DNA but that for F1 (in which the C-terminal half is basic) the whole molecule is involved in interaction with the DNA.

Repressors

Two highly purified repressors (p. 19) have been prepared; these are the phage repressors for the temperate coliphage λ (p. 373) and the similar phage 434 and the *lac* repressor. The kinetics of repressor-binding can be followed by making use of the fact that the DNA–repressor complex (like other nucleoproteins—see p. 213) is absorbed to cellulose nitrate. Thus radioactive DNA is incubated with a known titre of repressor and the mixture is 'filtered' through cellulose nitrate and the bound radioactivity (in the filter) is counted. The conclusions from two such sets of experiments are summarized below.

λ and 434 phage repressors   Chadwick, Pirrotta, Steinberg, Hopkins and Ptashne (1970) have examined the kinetics of the binding of these repressors to phage DNA. Their paper contains references to the isolation of repressor and a description of the isolation of double-lysogen strains of *E. coli* W3102 containing the λ and 434 prophages which can be mutagenized to yield strains that overproduce one or other repressor.

They observed a sigmoidal binding isotherm for the repressor/DNA reaction. The kinetics can be interpreted in a variety of ways but they favour a model in which there is an equilibrium in solution between repressor and repressor-dimer. If this interpretation is correct, the conclusion from their data is that the dimer is the form that binds to the operator (p. 19). They also investigated the effect of RNA polymerase on the reaction and concluded that RNA polymerase (attached to the *p*-gene—p. 19) inhibited binding of the repressor to *o*. They do not report any direct evidence for the converse (and physiologically more interesting process), i.e. the inhibition of binding of polymerase to *p* by the presence of repressor on *o*.

The *lac* repressor   Extensive studies on the *lac* repressor have been made by A. D. Riggs, Suzanne Bourgeois and their colleagues. Earlier work (including the isolation of the repressor

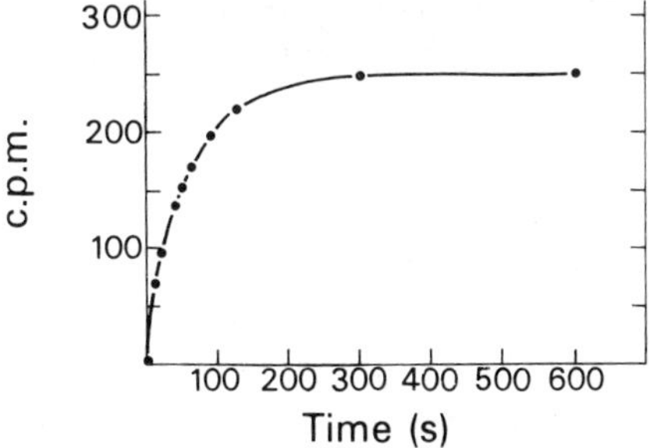

FIG 16.12   Binding of *lac* repressor to DNA from φ80*dlac* (see p. 198). [^{32}P]DNA (effective operator concentration of 2·4 x 10^{-12} M) and repressor (2·4 x 10^{-12} M) were incubated in 10 mM magnesium acetate, 10 mM KCl, 0·1 mM dithiothreitol, 5% dimethyl-sulphoxide, 50 μg/ml albumin, 10 mM tris HCl pH 7·4, for the times shown. The reaction was 'stopped' by adding excess unlabelled DNA (to remove the unbound repressor from solution). The mixture was filtered (cellulose nitrate) and the membrane-bound radioactivity was assayed. A background due to the residual binding of free DNA has been subtracted. Data of Riggs, Bourgeois and Cohn (1970).

and details of their assay method) is referred to in their most interesting publication, Riggs, Bourgeois and Cohn (1970). From the data in this paper, we know more about the details of the *lac* repressor DNA interaction than we do about the phage systems (above).

The dissociation of the complex follows first-order kinetics; its half life is shorter the higher the ionic strength (compare p. 461). The association reaction follows second-order kinetics. Under their 'standard conditions' (see Fig. 16.12) and at 24°C, the association constant is 7 x 10^9 M^{-1} s^{-1}, the equilibrium is 1 x 10^{-13} M^{-1}; ΔG is −18 kcal/mole, ΔH is +8·5 kcal/mole and ΔS + 90 cal/mole/deg. Thus it is the large positive entropy change that 'drives' the reaction.

From the facts that the affinity of non-specific DNA for repressor is very weak (dissociation constant is more than 10-fold greater than that for the repressor–operator complex) and sonicated φ80 *dlac* DNA competes for repressor as well as undegraded DNA, it is deduced that this very fast association reaction involves initial specific binding of repressor to operator and that the DNA does not bind randomly to the DNA and then 'walk along the helix' until it reaches its *o*-gene.

From the extremely rapid reaction and the fact that the repressor requires the *o*-gene DNA to be double-stranded, Riggs *et al.* deduce that in all probability the repressor binds to native Watson–Crick DNA, i.e. the DNA is not denatured in the repressor–operator complex. It would be extremely interesting if it could be confirmed (p. 456) that the repressor lies in the wide groove of the duplex. It would then imply a 'reading' of a sequence of base-edges by a protein molecule and perhaps a novel 'code'.

Ribosomes

As with the other sections of this chapter, there is not space to discuss this subject comprehensively. I have chosen merely to exemplify the techniques employed to determine features of ribosome structure with exclusive reference to bacterial ribosomes.

Ribosome conformation

The purpose of studying ribosome conformation is primarily to determine the tertiary structure of the ribosome. In fact we are a long way from being able to do this with certainty at the moment. However there is another use of these techniques which is to monitor changes in ribosome conformation. Although it is still premature to relate these conformational changes to the changes which occur in the ribosome during translation, it may be possible in time to check the feasibility of such models of translation as do exist (such as the ingenious 'lock and key model' which is fully described by Spirin, 1970).

Optical properties   Thermal denaturation or the effect of removal of cations on ribosome structure can be examined by following hyperchromic effects on the UV spectrum. A more satisfactory probe for these conformational changes is circular dichroism (Adler, Fasman and Tal, (1970).[11]

From previous work, it was known that complete removal of metal ions (excess EDTA followed by extensive dialysis against tris buffer) results in a change in ribosome conformation known as 'unfolding' and characterized by a reduction in sedimentation coefficient. Similar results are obtained if ribosomes are heated and cooled. Adler, Fasman and Tal (1970) employed CD measurements to determine which components (RNA or protein) are involved in the conformational changes involved in these processes. The CD of *E. coli* ribosomes consists of a small negative band at 297 nm and a large positive one at 264 nm; both of these are characteristic of RNA—see p. 341; additionally there is a negative shoulder at 220–225 nm due to the α-helix in protein and large bands (negative at 208 nm and positive at 190 nm) due to both RNA and protein contributions and not employed in their analysis. In both unfolded and heated ribosomes, the 'α-helical shoulder' is diminished, the '297-RNA band' disappears and the '264-RNA band' is red-shifted. The loss of α-helix is evidence for protein-denaturation; both the effects on RNA bands are characteristic of RNA-denaturation (p. 341). Divalent cations were added back to these structures and the 'reconstituted' ribosomes were found to have RNA-CD spectra intermediate between the unfolded (or heated) and native conformation; the α-helical shoulder had even further diminished, indicating that although some restoration of secondary structure in the RNA had occurred, there was no such restoration of protein conformation. Using the criterion of restoration of the RNA-CD spectrum, it was found that of the divalent cations tested Mg^{2+} was the most effective in the case of the unfolded ribosomes but that it was notably more successful than (for example Ni^{2+}) in the case of the heated ribosomes. Thus it appears that there is a difference in the nature of the disruption of native structure in these two processes.

None of these effects could be mimicked in free rRNA which renatures faithfully under these conditions. The RNA-CD spectrum of ribosomes resembles that of free rRNA (first noted by Busch and Scheraga, 1967) so it looks as though rRNA is present in ribosomes in a secondary and tertiary configuration very similar to that in the isolated molecule but that in the ribosome, it is so intermeshed with protein that following denaturation, it can never reassemble its specific H-bonding faithfully (see also p. 458).

Hydrogen exchange   Exchange of hydrogen ions (or tritium ions) between the ribosome and its aqueous environment may be used as a monitor of the extent to which the ribosome is 'open'. The technique has been exploited in a most interesting way to suggest a mechanism by

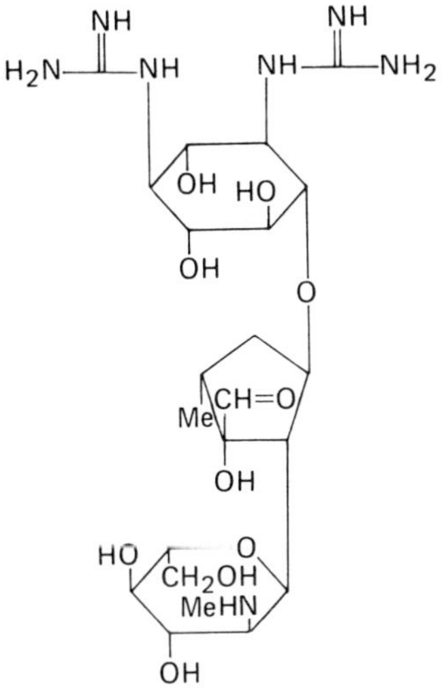

FIG 16.13   Structure of streptomycin.

which streptomycin might work. Streptomycin is an aminoglycoside antibiotic (structure in Fig. 16.13) which is known to bind to one of the proteins of the 30 S subunit (p. 465). The effect of streptomycin is to result (*in vitro* at least) in the misreading of the genetic code. Streptomycin-resistant mutants of *E. coli* can be isolated. It is found that in these organisms the streptomycin-protein of the 30 S subunit is different (different electrophoretic and chromatographic properties).

Sherman and Simpson (1969) employed two strains of *E. coli*, MRE 600 (see p. 314) and an isogenic streptomycin-resistant strain obtained from it by mutagenesis. Ribosomes were equilibrated in ^3H_2O and the unbound ^3H_2O was removed by Sephadex and the ribosomes were allowed to equilibrate with buffer in a rapid-dialysis apparatus (see p. 345). The experiment was performed on ribosomes from streptomycin-sensitive and streptomycin-resistant cells in the presence of varying concentrations of streptomycin. The results with the ribosomes from streptomycin-resistant cells showed no effects whatever and acted as a perfect control. The ribosomes from strain MRE 600 showed a striking effect. With increasing amounts of streptomycin, the H-exchange increased (the ribosome was 'opening') reached a maximum and then dropped to (eventually) a level below that of control (the ribosome was 'closing up'). These ribosomes (in the presence of different amounts of streptomycin) were tested for the capacity to catalyse the poly U-directed synthesis of poly Phe. The incorporation of leu into

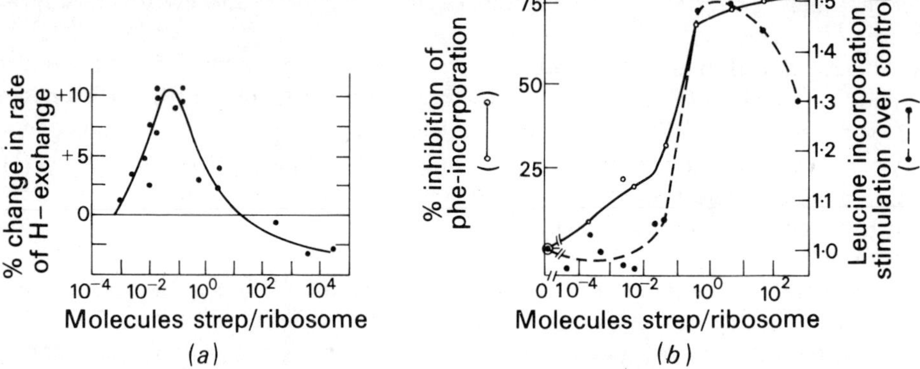

FIG 16.14   Effect of streptomycin on (a) H-exchange in 30 S subunits and (b) poly U-directed protein synthesis by ribosomes from a streptomycin-sensitive bacterium (*E. coli* MRE 600). The data plotted in (b) were obtained from cell-free systems (see p. 194) containing [³H]phe and [¹⁴C]leu. Data from Sherman and Simpson (1969).

the polypeptide product was employed as a measure of mis-coding. It was found that the 'opening' of the ribosome could be correlated with a limited inhibition of protein synthesis and that the 'closing up' could be correlated with considerable inhibition of protein synthesis and with extensive mis-reading of the code (leu-incorporation). The results are summarized in Fig. 16.14.

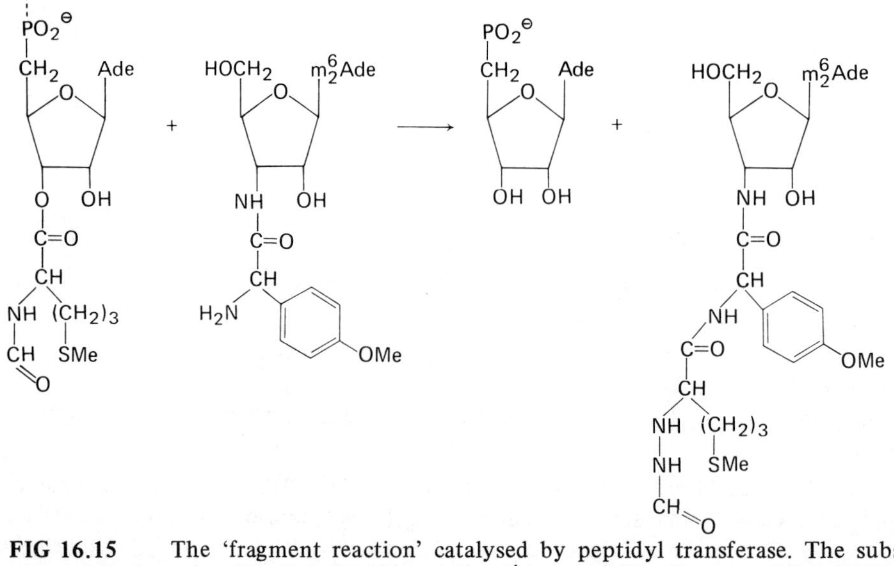

FIG 16.15   The 'fragment reaction' catalysed by peptidyl transferase. The substrates are puromycin (see also Fig. 2.4, p. 42) and the 3′-terminal T_1-oligomer of fMet-tRNA$_F^{Met}$. The product of the reaction are the free oligomer (with 3′-OH) and f-Met-puromycin. The usual way to assay this reaction is to employ a fragment isolated from tRNA charged with [¹⁴C]met. As fMet-puromycin is soluble in organic solvents (such as ethyl acetate) the transfer of radioactivity to a solvent-soluble compound can be easily measured. The conditions were determined by Monro and Marcker (1967), who showed that, providing alcohol is present in the reaction mixture, the reaction will occur on 50 S subunits (from *E. coli*) in the absence of 30 S subunits, or mRNA. It is equally possible to employ undegraded fMet-tRNA$_F^{Met}$ as substrate (Monro, Cerná and Marcker, 1968). This latter procedure was adopted in the experiments referred to in the text. The same reaction can be observed in a more complete system containing 30 and 50 S subunits, A-U-G (as 'messenger') and initiation factors F1 and F2 (p. 25). In this case, alcohol is not required in the incubation mixture.

It would be extremely useful to employ the CD methods of Adler and her colleagues (previous section) to enquire whether these opening and closing reactions could be correlated with changes in the RNA spectrum. For example, does the RNA skeleton expand (like a rib cage) when the whole structure inflates?

Peptidyl transferase   In this paragraph I am straying from my declared aim of restricting the discussion to experiments that are relevant to the study of the RNA/protein interaction. However we have said something in the previous paragraph about probing conformational changes in the 30 S subunit and it is appropriate to say something about such changes in the 50 S subunit. The characteristic biological activity of the 50 S subunit is the peptidyl transferase activity. This is the enzyme responsible for the peptide-bond forming reaction (p. 15). It can be assayed using a system described in Fig. 16.15. This reaction is variously referred to as the 'fragment reaction' or the 'alcohol reaction'.

Employing the various versions of this reaction (see figure) Miskin, Zamir and Elson (1970) demonstrated that the removal of monovalent cations from the medium inhibits the peptidyl transferase reaction. The activity can be slowly restored by adding such cations back. The most effective cation is NH_4^+; Na^+ and Li^+ are totally ineffective although the remaining alkali metals have some effectiveness. That the reactivation process (and consequently the preceding inactivation) are due to an overall change in ribosome conformation, and not merely to a limited change in the enzyme structure, is established by the considerable activation energy (40–50 kcal/mole at 30°C).

Again, one looks forward to a correlation of these changes with optical parameters.

Disruption and reconstitution of ribosomes

I have made no attempt to put this subject in any sort of historical perspective. Reference to the extensive earlier literature will be found in the references that are cited below. The whole subject of ribosome structure was revolutionized by the remarkable work of M. Nomura and his colleagues, who demonstrated that ribosomes can be faithfully reconstituted from their constituent parts (RNA and proteins). Therefore I start with Nomura's work, and explain subsequently how degradative and chemical studies confirm his conclusions and may in future complement reconstitution experiments.

Ribosomal components   Ribosomal subunits are required in a pure state for these studies as large-scale preparations are frequently required; this is a field in which the zonal ultracentrifuge rotor (p. 146) has exceptionally proved its worth. One method of removing proteins from ribosomes is to treat them with phenol (p. 123); however for the purpose of characterizing proteins this is not a particularly convenient procedure. It is more useful to use techniques in which the proteins are maintained in an aqueous phase. Nucleoprotein is disrupted by strong salt (p. 123). Strong salt alone is insufficient to remove all the proteins from ribosomes; rather it results in the release of some proteins and leaves others attached as a ribonucleoprotein 'core'. Strong salt is effective in the presence of urea. Thus 2 M LiCl in 4 M urea removes all the proteins from *E. coli* ribosomes and precipitates the rRNA.[12] Alternative methods are to employ strong acetic acid or detergents (see p. 123).

There are many different ribosomal proteins that fall into two classes, unitary and transitory (p. 22). For an authoritative review of the proteins, see Wittmann (1969). The proteins may be fractionated by ion-exchange chromatography or gel-electrophoresis. As ribosomal protein consists of a mixture of largely acidic and basic proteins, there is a great tendency for them to aggregate. The procedures are therefore normally conducted in the presence of high concentrations of urea to eliminate aggregation. Alternatively (p. 160), electrophoresis may be operated in the presence of a detergent. For a review of methods of fractionating ribosomal proteins see Bloemendahl and Vennegoor (1969). For the routine analysis of mixtures of these proteins, the most versatile system is undoubtedly the electrophoretic method of Leboy, Cox and Flaks (1965).

Methods of fractionating ribosomal RNA are discussed on pp. 137 and 158.

Reconstitution of the 30 S subunit of *E. coli* The conditions for the assembly of these subunits from RNA and protein are described in detail by Traub and Nomura (1969). Their procedure involves the use of certain Mg^{2+} buffers with the following compositions (their nomenclature);

Buffer I: 20 mM $MgCl_2$, 30 mM NH_4Cl, 6 mM β-mercaptoethanol, 10 mM tris HCl pH 7·8 (at 4°C)

Buffer II: as buffer I but $MgCl_2$ 0·3 mM

Buffer III: 5 mM potassium phosphate buffer pH 7·8 (at 4°C), 2 mM $MgCl_2$, 1 M KCl, 6 mM β-mercaptoethanol

Buffer IV: as buffer III but KCl 0·25 M.

Ribosomes were dissociated by suspending the subunits in buffer I and adding an equal volume of 4 M LiCl, 8 M urea. After 24 h at 0°C, the precipitate (RNA) was removed and the supernatant (protein) was dialysed against buffer III. The reconstitution experiments used either RNA isolated as above or phenol-extracted RNA (p. 123); both types of RNA preparation were equally effective.

The reconstitution experiment itself is extremely simple. The RNA (2 mg) and 1·2 protein-equivalents are mixed in 10 ml of a solution with the final composition of a mixture of IV and III (20:1 by volume). The proteins are added (in III) to the RNA at 0°C. The mixture is then incubated at 40°C for 20 min, cooled to 0°C and dialysed against III. Alternatively, the dialysis may be omitted and the reconstituted RNP is harvested by centrifugation and resuspended in II; $MgCl_2$ is then added to a final concentration of 10 mM. The reconstituted particles were found to be indistinguishable from native subunits by the criteria of sedimentation coefficient, electrophoretic 'spectrum' of proteins and their activity (in the presence of 50 S subunits and necessary co-factors) to catalyse the poly U-dependent synthesis of poly Phe.

Traub and Nomura (1969) also demonstrated that the same conditions would reconstitute unfolded ribosomes (see p. 458—this refers to EDTA-treated particles).

From studies on the time-course of the reconstitution experiments at different temperatures, it was possible to deduce the thermodynamic activation parameters of the process. The kinetics were essentially the same for the reconstitution from RNA and protein or from unfolded particles. The activation energy is around 38 kcal, $\Delta H^{\ddagger}$ 37 kcal, $\Delta S^{\ddagger}$ 50 cal/deg/mole and $\Delta G^{\ddagger}$ 22 kcal/mole.[13] Thus the reaction appears to have a rate-limiting step identical in both cases.

These experiments have been extended (Mizushima and Nomura, 1970) to examine the binding of fractionated ribosomal proteins to RNA. They were thus able to construct an 'assembly map' for the subunit and locate the different proteins in the ribosome on the basis of their interactions with RNA in the presence or absence of other proteins. The results are summarized in Fig. 16.16.

Reconstitution with anomalous RNA Sypherd (1971) has attempted the reconstitution of the *E. coli* 30 S subunit (employing Nomura's conditions, p. 462) but with either denatured 16 S

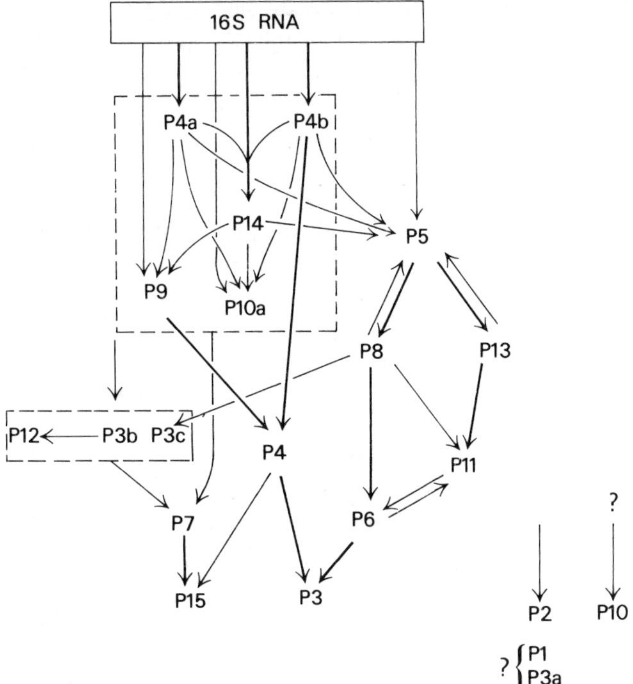

FIG 16.16   Assembly map of the *E. coli* 30 S ribosomal subunit. The nomenclature for the proteins is that employed by Nomura (see reference here and in text). P10 is the streptomycin-binding protein (see text).

Explanation of the figure. An arrow pointing from one protein to another indicates that it helps the binding of the latter protein to the RNP complex. Arrows from 16 S RNA indicate the assembly processes. Thick arrows are the principal interactions, weak ones are shown by thin arrows. Thus the thick arrow from 16 S RNA to P4a means that P4a binds strongly to RNA in the absence of other proteins; likewise P5 binds weakly in the absence of other protein. The arrows converging on P5 show that its binding is weakly helped by P4a, P4b, P14, P8 and P13. The thick arrow from P5 to PP13 shows that P13 will only bind in the presence of P5 irrespective of the presence or absence of other proteins. The arrow from the large box (- - - - - - -) to P7 shows that P7-binding requires some of the proteins in the box (the precise details are not known). P2 binds after the assembly of the whole and is a 'peripheral' protein like P15. The positions of P10 and P3a are uncertain because of difficulties in obtaining homogeneous preparations of these proteins. It seems that they are both peripheral, although it is possible that P3a is not bound in the assembly reaction. P1 is apparently not bound and may be a non-ribosomal contaminant. From Mizushima and Nomura (1970).

RNA or 17 S precursor RNA. The denatured RNA was obtained by heating and very rapidly cooling the RNA. It was established that, under the conditions of the incubation at 40°C, this RNA was not renatured. However it formed 30 S subunits in the presence of 30 S proteins. These subunits were correctly assembled according to the criteria of p. 462. Precursor (17 S) RNA was equally effective in assembling 30 S ribosomal subunits but was not hydrolysed during the assembly. The significance of this latter result is not clear; the former implies that ribosomal protein 'helps' rRNA to assume its native conformation.

The 50 S subunit   The assembly of the 50 S subunit is more complex than that of the 30 S particle. There are two RNA molecules (23 S and 5 S) and more proteins (p. 22). Furthermore there is evidence that this particle is not self-assembling *in vivo* but requires the presence of 30

S subunits. Before considering the method by which this assembly can be achieved *in vitro*, let us briefly consider the evidence for the statement regarding the assembly *in vivo*.

Nashimoto and Nomura (1970) surveyed 'ribosome assembly mutants' of *E. coli*. At first sight this is an impossible phenotype to score directly. However, as the assembly process (at least for 30 S particles) has a high activation energy, it was assumed that some cold-sensitive mutants would demonstrate poor assembly at the non-permissive (low) temperature. This was found to be the case; moreover a survey of streptomycin-resistant strains demonstrated that they included a considerable number that were also cold-sensitive. Thus protein P10 (Fig. 16.16, p. 463) controls to some extent the fidelity of the assembly reaction. Cells of the cold-sensitive strains, were transferred to $20°C$ (non-permissive) and labelled with $[^{14}C]$Ura. Ribosomes were extracted and analysed by sucrose-gradient centrifugation. The wrongly assembled subunits were identified by slow-moving zones (lighter than the corresponding 'proper' subunits). In no case was a mutant found which was sensitive to 50 S assembly and not 30 S assembly. Moreover genetic analysis of a cold-sensitive, spectinomycin-resistant[14] mutant of an Hfr strain of *E. coli* demonstrated that a single mutation had effected all phenotypic changes (resistance to the antibiotic, cold-sensitivity and inability at the low temperature to correctly assemble both 30 S and 50 S subunits). Thus it seems that 50 S assembly is under the control of 30 S assembly.

Nomura concluded that the difficulty of 50 S assembly was essentially that the activation energy was so high that an allosteric role for the 30 S particle was required for 50 S assembly. However he guessed that this activation energy could probably be overcome in the absence of the 30 S particles if the temperature was sufficiently high. The difficulty with *E. coli* is that the incubation mixture cannot be made much hotter than $40°C$ or the proteins are irreversibly denatured. Nomura and Erdmann (1970) therefore switched to a bacterium which grows at $65°C$ (*Bacillus stearothermophilus*), in which evolutionary pressures have ensured that all the macromolecules are heat-stable. The experiment works well at $60°C$; using the buffers described on p. 462 and 30 min incubation times there is faithful assembly of 50 S subunits, active in the poly-U assay (p. 459). The system differs from the 30 S assembly in that phenol-extracted RNA is only weakly active. The LiCl–urea precipitate is required. (5 S RNA was found to be essential for the process; about 70% of the 5 S RNA is obtained, together with the proteins, in the LiCl–urea supernatant.) This 5 S RNA could be replaced with purified 5 S RNA from either *B. stearothermophilus* or *E. coli*.

Chemical and degradative studies on ribosomes   Several degradative procedures have been recommended in the past as probes for the architecture of ribosomes. In future, those methods that have been checked against Nomura's assembly map (p. 463), should be exploited. Two such procedures have been evaluated in details.

Chang and Flaks (1970) digested *E. coli* 30 S subunits with trypsin. Trypsin (between $0·2$ and $4·0$ g/ml) was added to a suspension of subunits ($5·5$ A_{260}-units in $0·05$ ml) in 10 mM $MgCl_2$, 50 mM KCl, 50 mM tris HCl pH $7·8$. Incubation was for 15 min at $30°C$ and was stopped with trypsin inhibitor.[15] Proteins were isolated by the LiCl–urea method and analysed by gel electrophoresis. The relative sensitivity of the proteins was assessed. The data are summarized in Table 16.1.

Craven and Gupta (1970) employed a specific reagent for cys-residues, iodoacetic acid. In this case $[^{14}C]$iodoacetate was incubated with the subunits and the proteins were again isolated and fractionated. They also employed an NH_2-specific reagent, 2-methoxy-5-nitro-topone and performed trypsin digests. Their conditions for this last experiment were different to those of Chang and Flaks (above) in that the trypsin was attached to a cellulose matrix.

Comparison of reactivities was not possible. Craven and Gupta's data correlate well with both Chang and Flaks and with the map of Mizushima and Nomura (Fig. 16.16). They unfortunately employ a different nomenclature for the proteins. In my summary of these results (Table 16.1) I have translated all the data to the Nomura terminology.

The table demonstrates that, on the whole, there is a good correlation between assembly and reactivity. Thus the early proteins of Nomura are apparently 'deep' in the ribosome and not sensitive to the reagents. The data seeem to enable us to locate P10 (streptomycin protein) somewhere deep in the structure. The experiments of Chang and Flaks throw an interesting light on the role of P10. They examined the capacity for the partly digested subunits to bind radioactive dihydrostreptomycin. It was found that removal of the most sensitive proteins to trypsin (P7, P3 and P15) and partial removal of P5 and P13 had no effect on streptomycin binding. However the partial removal of P8 and P11 was associated with a reduction in the binding of the antibiotic and total removal of these proteins eliminated binding completely.

TABLE 16.1   Corelation of reactivities of *E. coli* 30 S proteins with their assembly positions.

Protein[a]	Assembly order[a]	Reaction with iodoacetate[b]	Reaction with MNT[c]	Trypsin[d]	Trypsin[e]
P4a	E	—	—	—	—
P4b	E	—	—	—	—
P14	E	N	—	—	—
P9	E	N–[f]	—[f]	—[f]	—
P10a	E	—	—	—	—
P4	I	N	—	—	—
P5	I	N	—	—	4
P13	I	—	—	—	4
P8	I	N	—	—	2
P12	I			—	—
P3b	I	X[g]	X[g]	X[g]	?
P11	I	—		—	2
P6	I	N	—		?
P7	L			?	6
P3	L	N			5
P15	L				6
P2	L(?)	—	—		?
P1	?				3
P3a	?	N	—		1
P10	?	—	—	—	—

Key: a. Numbers and assembly position of proteins according to Mizushima and Nomura (see Fig. 16.15): early (E), intermediate (I) and late (L)

  b. From Craven and Gupta (1970): – no reaction, + reaction, N no reaction because of no cys-residue(s)

  c. 2-methoxy-5-nitrotopone (same ref. as b)

  d. Same ref. as b

  e. From Chang and Flaks (1970): – no reaction; 6 more reactive than 5, which is more reactive than 4, etc.

  f. This is a mixture of two proteins according to Craven and Gupta

  g. Craven and Gupta do not report any data.

Under all these conditions, P10 is untouched. Thus for P10 to bind streptomycin, a particular conformation dependent upon the presence of P8 and P11 is essential.

Ribosomes can be degraded not only by trypsin as above but also by ribonuclease. This procedure has been employed to isolate specific fragments of rRNA (p. 321). For the dissection of the tertiary structure of the ribosome, a more valuable experiment would be to obtain from such a limited digestion, not specific RNA fragments but rather specific RNP fragments. Preliminary results of such an experiment have been reported by Brimacombe, Morgan, Oakley and Cox (1971). Their method consists of digesting 30 S subunits with ribonuclease T_1, and fractionating the RNP fragments by gel electrophoresis. So far, they have purified one such fragment and have identified the proteins attached to it. This paper is technically interesting in that it reports a simple method of analysing ribosomal protein in a buffer containing the detergent *N*-lauryl sarcosine. Clearly the extension of this work should give us an alternative view of the organization of the ribosome, based, not on the layers of protein in the whole structure, but the association of particular proteins with specific regions of the RNA skeleton.

Notes to chapter 16

[1] This technique relies on the fact that if highly monochromatic (laser) light is used for light scattering (p. 398) the scattered light shows a spectrum of emission. This spectrum is determined by an equation in which D is an independent variable.

[2] If methods can be made to work for these particles, they will work for anything. These phage have a complex morphology (p. 448) and T4 with a molecular weight of 190×10^6 has a sedimentation coefficient of 890!

[3] Cubic symmetry: This does not refer to the type of symmetry exhibited by a cube. A regular solid demonstrates cubic symmetry if it has a set of four three-fold axes of symmetry arranged in space as are the

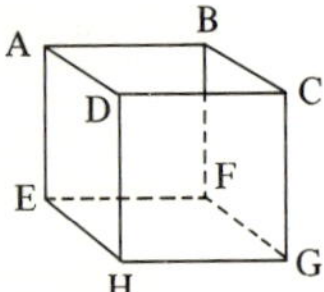

lines joining AG, BH, CE and DF in the cube illustrated here. The three regular solids with cubic symmetry are the tetrahedron, the octahedron and the icosahedron.

[4] This is the type of H-bonding that exists in fatty acids and results in (for example) acetic acid having the colligative properties characteristic of the molecular formula $(CH_3CO_2H)_2$.

[5] Vesicular stomatitis is an epidemic disease among cattle and horses in central America. VSV is the best studied of a large group of viral pathogens in both animals and plants. Some are undoubtedly of potential medical importance as a monkey pathogen of this group (Marburg Virus) has been accidentally transmitted to laboratory workers (using monkeys as experimental animals) with fatal consequences.

[6] The region of tRNA responsible for this recognition process is called the 'R-site'. Thus the question may be re-worded, 'Where is the R-site?'.

[7] *N. crassa* contains three Phe-tRNA synthetase isoenzymes. Only one of them (Sync) effects this mis-reading and was purified to homogeneity for the experiments referred to here.

[8] This is the $tRNA^{ala}$ of Fig. 1.17, p. 24.

[9] There is a complex photolysis of Ψ-residues in which the sugar-ring is cleaved and the chain is broken. Otherwise, any reaction not showing a solvent effect (dimerization) must be located in a region in two (or more) Py-residues are in sequence.

[10] The X-ray data produce an estimate of the diameter of the superhelix measured between points on the axial centre of the threads. Pardon, Wilkins and Richards (1967) assumed that the threads were about 30 Å thick and thus reduced Davies' value from 150 to 120 Å.

[11] The data on UV-hyperchromic effects and also the hydrodynamic properties of 'denatured' ribosomes (which are not discussed here) are referred to in this paper.

[12] This should not be regarded as a universal recipe. H. A. Foster, working in the author's laboratory on the ribosomal proteins from *Myxococcus xanthus*, finds that the precipitate obtained in this case is merely an RNP core and stronger reagents are required for quantitative release of the protein.

13   The rate-constant for a unimolecular process (k) is related to activation energy (E) by the relationship: ln k = $-E/RT$ + const. Hence by measuring k at a variety of temperatures, E can be calculated. An alternative factorization is $k = (K_B T/h) \times \exp(\Delta G^{\ddagger}/RT)$, where h is Planck's constant and K_B Boltzmann's constant. From E, we can calculate $\Delta H^{\ddagger}$ (as E is an internal-energy change and hence, $\Delta H^{\ddagger} = E - RT$) for this system and from $\Delta H^{\ddagger}$ and $\Delta G^{\ddagger}$ we can calculate $\Delta S^{\ddagger}$ ($\Delta G^{\ddagger} = \Delta H^{\ddagger} - T\Delta S^{\ddagger}$).

14   Spectinomycin is an aminoglycoside antibiotic, which like streptomycin, binds to the 30 S subunit of ribosomes.

15   Trypsin inhibitor is a protein obtained from soy beans.

Addenda

Degradation of ribosomes with ribonuclease   The technique of using particles derived from degraded ribosomes for characterizing the internal organization of ribosomal subunits pioneered by Cox (p. 466) has already been extended by P. Schendel, P. Maeba and G. R. Craven (*Proc. Nat. Acad. Sci. U.S.* 1972, **69**, 544).

Viroids   A major recent discovery in virus research is that potato spindle tuber virus contains no protein but only RNA of very low molecular weight (T. O. Diener, *Virology*, 1971, **45**, 411; **46**, 498). Two infective molecules have been identified of molecular weights 2.5×10^4 and 1.1×10^5. The molecules were fractionated by gel electrophoresis, identified by their infectivity, and their molecular weights assigned by the method of p. 394. Obviously such a small genome can only carry enough 'information' to trigger off its own replication by host-cell gene products. In a recent paper, Diener (*Nature New Biology*, 1972, **235**, 218) proposes that such molecules (viroids) may be implicated in the aetiology of certain puzzling animal diseases such as Scrapie (in sheep) and Kuru (in humans).

Discontinuity of rRNA   In addition to the examples from chloroplast RNA (p. 396), it is now clear that there is a 'hidden break' in the large rRNA of *Acanthamoeba castellanii* (U. E. Loening, personal communication; A. R. Stevens and E. F. Pachler, *J. molec. Biol.* 1972, **66**, 225).

Abbreviations and symbols

Symbols that only occur in specific contexts (and are defined in those contexts) are not listed here.

a	arabino
A	adenosine (or deoxyadenosine) in a sequence
A	absorbance; $[A]$ ORD amplitude
ac	acetyl
Ade	adenine
Ado	adenosine
ADP	adenosine diphosphate
ala	alanine
AMP	adenosine monophosphate
arg	arginine
asp	aspartic acid
asn	asparagine
ATP	adenosine triphosphate
Bu	butyl
Bz	C_6H_5CO-
C	cytidine (or deoxycytidine) in a sequence
c	velocity of light
cm	carboxymethyl
CMCT	a water soluble carbodiimide (see p. 63)
cys	cysteine
Cyd	cytidine
Cyt	cytosine
CMP, CDP, CTP	(see AMP, etc.)
d	deoxyribo or (in structures) deoxyribose moiety
DCC	dicyclohexyl carbodiimide
DNA	deoxyribonucleic acid
dnb	dinitro benzene
E	energy
E_D, E_μ, E_M, E_L	interaction energies (p. 77)
EDTA	ethylene diamine tetracetic acid
Et	ethyl
f	function (mathematical)

G   guanosine (or deoxyguanosine) in a sequence
glu   glutamic acid
gln   glutamine
gly   glycine
Gua   guanine
Guo   guanosine
GMP, GDP, GTP (see AMP, etc.)
h   'hydrogen' (hU = dihydrouridine residue)
h Planck's constant
his   histidine
Hyp   hypoxanthine
I intensity
I   inosine in a sequence
Ino   inosine
ile   isoleucine
J coupling constant (NMR)
k rate constant
K equilibrium constant
leu   leucine
lys   lysine
m   methyl
Me   methyl
met   methionine
mam   methylaminomethyl
ms   methyl thio
mRNA   messenger RNA
MW   molecular weight
N   any (or an unknown) nucleoside in a sequence
n refractive index
NEM N-ethylmaleimide
nm   nanometres (= 'millimicrons')
NMG N-methyl-N-nitroso-N'-nitro-guanidine
p   -$\log_{10}$
p momentum
phe   phenylalanine
PCMB p-chloromercuribenzoic acid
pro   proline
Pu   any (or an unknown) purine nucleoside in a sequence
Py   any (or an unknown) pyrimidine nucleoside in a sequence
r   ribo or (in structures) ribose moiety
rp   ribose phosphate (in structures)
RNA   ribonucleic acid
rRNA   ribosomal RNA
s   thio
SDS   sodium dodecyl sulphate (= 'sodium lauryl sulphate')
ser   serine
T   thymidine residue in a sequence
T temperature

T_M melting temperature
Thy   thymine
Thd   thymine riboside ('thymidine' = dThd)
TMP   thymine riboside monophosphate ('thymidine monophosphate' = dTMP)
TDP and TTP (see TMP; the DNA precursor is dTTP)
thr   threonine
tris $(HOCH_2)_3CNH_2$
trp   tryptophan
tyr   tyrosine
val   valine
X   xantosine residue (in sequence and poly X, etc.)
Xao   xanthosine
Xan   xanthine
Y   a minor nucleoside (see p. 39)
Z   carbobenzoxy

Greek

α optical rotation
α, β subunits of RNA polymerase
δ small increment
Δ large increment
ϵ absorptivity ('extinction coefficient')
μ micron
λ wavelength
λ a transducing phage of *E. coli*
Ψ pseudouridine in a sequence
Ψrd   pseudouridine
ψ wave function
ν frequency *or* wave number

References

Abelson, J. and Thomas, C. A., jun. (1966) *J. molec. Biol.* **18**, 262.

Abraham, K. A. (1968) *Europ. J. Biochem.* **5**, 143.

Adams, J. M., Jeppeson, P. G. N., Sanger, F. and Barrell, B. G. (1969) *Nature*, **223**, 1009.

Adler, A. J. and Fasman, G. D. (1970) *Biochim biophys. Acta*, **204**, 183.

Adler, A. J., Fasman, G. D. and Tal, M. (1970) *Biochim. biophys. Acta*, **213**, 424.

Agarwal, K. L., Büchi, H., Caruthers, M. H., Gupta, N., Khorana, H. G., Kleppe, K., Kumar, A., Ohtsuka, E., RajBhandary, U. L., van de Sande, J. H., Sgaramella, V., Weber, H. and Yameda, T. (1970) *Nature*, **227**, 27.

Albertsson, P.-A. (1962) *Arch. Biochem. Biophys. Supplement 1*, 264.

Allfrey, V., Stern, H., Mirsky, A. E. and Saetren, H. (1951) *J. gen. Physiol.* **35**, 529.

Almeida, J. D. and Ham, A. W. (1965) *Tumour Research*, **6**, 1.

Altener, Č. and Maďarič, A. (1966) *Coll. Czech. chem. Comm.* **31**, 4009.

Altman, S. and Smith, J. D. (1971) *Nature New Biology*, **233**, 35.

Amaldi, F. and Attardi, G. (1968) *J. molec. Biol.* **33**, 737.

Anderson, N. G. (1966) *National Cancer Institute Monograph 21*, 9.

Anderson, N. G., Barringer, H. P., Bakelay, E. F., Nunley, C. E., Barthus, M. J., Fisher, W. D. and Rankin, C. T., jun. (1966) *National Cancer Institute Monograph 21*, 137.

Apgar, J., Holley, R. W. and Merrill, S. H. (1961) *Biochim. biophys. Acta*, **53**, 220.

Applequist, J. and Damle, V. (1965) *J. Amer. Chem. Soc.* **87**, 1450.

Applequist, J. and Damle, V. (1966) *J. Amer. Chem. Soc.* **88**, 3895.

Aposhian, H. V., Friedman, M., Nishihara, M., Heimer, E. P. and Nussbaum, A. L. (1970) *J. molec. Biol.* **49**, 367.

Aposhian, H. V. and Kornberg, A. (1962) *J. biol. Chem.* **237**, 519.

Arber, W. (1968) in *Molecular Biology of Viruses* (eds. Crawford, L. V. and Stoker, M. G. P.), London: Cambridge University Press, p. 295.

Ariana, T., Uchida, T. and Egami, F. (1968) *Biochem. J.* **106**, 601 (two consecutive papers).

Armstrong, D. J., Evans, P. K., Burrows, W. J., Skoog, F., Petit, J.-F., Dahl, J. L., Steward, T., Strominger, J., Leonard, N. J., Hecht, S. M. and Occolowitz, J. (1970) *J. biol. Chem.* **245**, 2922.

Arnott, S. (1970) *Prog. Biophys. molec. Biol.* **21**, 267.

Arnott, S., Dover, S. D. and Wonacott, A. J. (1969) *Acta cryst.* **B21**, 2192.

Arnott, S., Fuller, W., Hodgson, A. and Prutton, I. (1968) *Nature*, **220**, 561.

Attardi, G., Parnas, H., Hwang, M.-I. H. and Attardi, B. (1966) *J. molec. Biol.* **20**, 145.

August, J. T., Ortiz, P. J. and Hurwitz, J. (1962) *J. biol. Chem.* **237**, 3786.

Avadhani, N. G., Schuit, K. E., Roe, D. H. and Buetow, D. E. (1970) *Physiol. Chem. Phys.* **2**, 34.

Avery, O. T., Macleod, C. M. and McCarty, M. (1944) *J. exptl. Med.* **79**, 137.

Aviv, H., Boine, I. and Leder, P. (1971) *Proc. Nat. Acad. Sci. U.S.* **68**, 2303.

Ayad, S. R. and Blamire, J. R. (1968) *Biochem. biophys. Res. Comm.* **30**, 207.

Ayad, S. R. and Blamire, J. R. (1969) *J. Chromatog.* **42**, 248.

Ayad, S. R. and Blamire, S. R. (1970) *J. Chromatog.* **48**, 456.

Baczynskyj, L., Biemann, K. and Hall, R. H. (1968) *Science*, **159**, 1481.

Balassa, R. (1960) *Nature*, **188**, 246.

Ball, C. R. and Roberts, J. J. (1970) *Chem. Biol. Interacts.* **2**, 321.

Baldwin, R. L. (1968) in *Molecular Associations in Biology* (ed. Pullman, B.), New York and London: Academic Press, p. 145.
Bancroft, F. C. and Freifelder, D. (1970) *J. molec. Biol.* **54**, 537.
Bancroft, J. B. (1970) *Adv. Virus Res.* **16**, 99.
Banks, G. T., Buck, K. W., Chain, E. B., Darbyshire, J. E., Himmelweit, F. and Ratti, G. (1970) *Nature*, **227**, 505.
Barnett, W. E. and Jacobson, K. B. (1964) *Proc. Nat. Acad. Sci. U.S.* **51**, 642.
Baron, E. and Brown, D. M. (1955) *J. chem Soc.*, 6263.
Barringer, H. P. (1966) *National Cancer Institute Monograph 21*, 77.
Barringer, H. P., Anderson, N. G., Nunley, C. E., Ziehlke, K. T. and Dritt, W. S. (1966) *National Cancer Institute Monograph 21*, 165.
Bauer, W. and Vinograd, J. (1968) *J. molec. Biol.* **33**, 141.
Bauer, W. and Vinograd, J. (1970a) *J. molec. Biol.* **47**, 419.
Bauer, W. and Vinograd, J. (1970b) *J. molec. Biol.* **54**, 281.
Bautz, E. K. F. and Bautz, F. A. (1970) *Nature*, **226**, 1219.
Bayev, A. A., Venkstern, V. T., Mirsabekov, A. D., Krutlina, A. I., Li, T. and Axelrod, V. D. (1967) *Molekularna Biologiya*, **1**, 757 (in Russian).
Beardsley, K. and Cantor, C. R. (1970) *Proc. Nat. Acad. Sci. U.S.* **65**, 39.
Beardsley, K., Cantor, C. R. and Tao, T. (1970) *Biochemistry*, **9**, 3524.
Becker, A. and Hurwitz, J. (1971) *Prog. Nuc. Acid. Res. mol. Biol.* **11**, 423.
Becker, E. F., Zimmerman, B. K. and Geiduschek, E. P. (1964) *J. molec. Biol.* **8**, 377.
Beckwith, J. R., Singer, E. R. and Epstein, W. (1966) *Cold Spring Harb. Symp. quant. Biol.* **31**, 391.
Beers, R. (1957) *Biochem. J.* **66**, 686.
Belozersky, A. N. and Spirin, A. S. (1960) in *The Nucleic Acids vol. 3* (eds. Chargaff, E. and Davidson, J. N.), New York and London: Academic Press, p. 147.
Bergmann, E. D. (1968) in *Molecular Associations in Biology* (ed. Pullman, B.), New York and London: Academic Press, p. 207.
Berman, A. S. (1966) *National Cancer Institute Monograph 21*, 41.
Bernardi, G. and Cordonnier, C. (1965) *J. molec. Biol.* **11**, 141.
Berthod, H., Giessner-Prettre, C. and Pullman, A. (1967), *Internat. J. Quantum Chem.* **1**, 123.
Bial, M. (1903) *Deut. med. Wochschr.* **29**, 253, 477. (two papers; in German).
Billeter, M. A., Dahlberg, J. E., Goodman, H. M., Hindley, J. and Weissman, C. (1969) *Nature*, **224**, 1083.
Birnie, G. D. and Harvey, D. R. (1968) *Analyt. Biochem.* **22**, 171.
Birnstiel, M. L. (1967) in *Cell Differentiation* (a Ciba Symposium), London: Churchill, p. 178.
Birnstiel, M. L., Chipchase, M. and Speirs, J. (1971) *Prog. Nuc. Acid Res. molec. Biol.* **11**, 351.
Birnstiel, M. L., Speirs, J., Purdom, I., Jones, K. and Loening, U. E. (1968) *Nature*, **219**, 454.
Birnstiel, M. L., Wallace, H., Sirlin, J. L. and Fischberg, M. (1966) *National Cancer Institute Monograph 23*, 431.
Blackburn, G. M., Brown, M. J. and Harris, M. R. (1967) *J. chem. Soc. C*, 2438.
Blackburn, G. M. and Davies, R. J. H. (1965) *Chem. Comm.* **11**, 215.
Blackburn, G. M. and Davies, R. J. H. (1966) *Biochem. biophys. Res. Comm.* **22**, 704.
Blake, R. D. and Fresco, J. R. (1966) *J. molec. Biol.* **19**, 145.
Blake, R. D., Fresco, J. R. and Langridge, R. (1970) *Nature*, **225**, 35.
Blake, R. D., Massoulié, J. and Fresco, J. R. (1967) *J. molec. biol.* **30**, 291.
Bloemendahl, H. and Vennegoor, C. (1969) in *Techniques in Protein Synthesis* (eds. Campbell, P. N. and Sargent, J. R.), London and New York: Academic Press, p. 181.
Blundell, M. R. and Wild, D. G. (1969) *Biochem. J.* **115**, 207.
Bock, R. M. and Cherayil, J. D. (1967) in *Methods in Enzymology*, vol **12A** (eds. Grossman, L. and Moldave, K.), New York and London: Academic Press, p. 638.
Boedtker, H. (1960) *J. molec. Biol.* **2**, 171.
Boedtker, H. (1968a) in *Methods in Enzymology*, vol. **12B** (eds. Grossman, L. and Moldave, K.), New York and London: Academic Press, p. 429.
Boedtker, H. (1968b) *J. molec. Biol.* **35**, 61.
Boling, M. E. and Setlow, J. K. (1966) *Biochim. biophys. Acta*, **123**, 26.
Börrensen, H. C. (1963) *Acta Chem. Scand.* **17**, 921, 2359 (two papers).
Botstein, D. (1968) *J. molec. Biol.* **34**, 621.
Böttger, M., Bierwolf, D., Wunderlich, V. and Graffi, A. (1971) *Biochim. biophys. Acta*, **232**, 21.

Boubliak, M., Bradbury, E. M., Crane-Robinson, C. C. and Rattle, H. W. E. (1971) *Nature New Biology,* **229**, 149.
Bowen, T. J. (1970) *An Introduction to Ultracentrifugation* (with additional material by Rowe, A. J.), London: Wiley Interscience.
Boyce, R. P. and Howard-Flanders, P. (1964) *Proc. Nat. Acad. Sci. U.S.* **51**, 293.
Boyland, E. (1964) *Brit. Med. Bull.* **20**, 1211.
Bradley, D. F. and Lifson, S. (1968) in *Molecular Associations in Biology* (ed. Pullman, B.) New York and London: Academic Press, p. 261.
Bradley, D. F. and Wolf, M. K. (1959) *Proc. Nat. Acad. Sci. U.S.* **45**, 944.
Brahms, J. (1963) *J. Amer. chem. Soc.* **85**, 3298.
Brahms, J., Maurizot, J. C. and Michelson, A. M. (1967) *J. molec. Biol.* **25**, 481.
Brakke, M. K. (1964) *Arch. Biochem. Biophys.* **107**, 388.
Brenner, D. J., Fournier, M. J. and Doctor, B. P. (1970) *Nature,* **227**, 448.
Brent, T. P., Butler, J. A. V. and Crathorn, A. R. (1966) *Nature,* **210**, 393.
Brimacombe, R., Morgan, J., Oakley, D. G. and Cox, R. A. (1971) *Nature New Biology,* **231**, 209.
Brink, J. J. and Schein, A. H. (1963) *J. med. Chem.* **6**, 563.
Brintzinger, H. (1963) *Biochem. biophys. Acta,* **77**, 343.
Britten, R. J. and Kohne, D. E. (1968) *Science,* **161**, 529.
Brookes, P. (1968) *Biochim. biophys. Acta,* **157**, 646.
Brookes, P. and Heidelberger, C. (1969) *Cancer Res.* **29**, 157.
Brookes, P. and Lawley, P. D. (1960) *J. chem. Soc.* 3923.
Brookes, P. and Lawley, P. D. (1961) *Biochem. J.* 496.
Brookes, P. and Lawley, P. D. (1964) *Nature,* **202**, 781.
Brown, D. M. and Phillips, J. H. (1965) *J. molec. Biol.* **11**, 663.
Brown, G. B. (1968) *Prog. Nuc. Acid Res. molec. Biol.* **8**, 209.
Brown, G. B. and Weliky, V. S. (1953) *J. biol. Chem.* **204**, 1019.
Brownlee, G. G. (1971) *Nature New Biology,* **229**, 147.
Brownlee, G. G. and Cartwright, E. (1971) *Nature New Biology,* **232**, 50.
Brownlee, G. G. and Sanger, F. (1969) *Europ. J. Biochem.* **11**, 395.
Brownlee, G. G., Sanger, F. and Barrell, B. G. (1968) *J. molec. Biol.* **34**, 379.
Burgess, R. R. (1969) *J. biol. Chem.* **244**, 6160, 6180 (two papers).
Burgess, R. R., Travers, A. A., Dunn, J. J. and Bautz, E. K. F. (1969) *Nature,* **221**, 43.
Burgi, E. and Hershey, A. D. (1963) *Biophys. J.* **3**, 309.
Burness, A. T. H. (1970) *J. gen. Virol.* **6**, 373.
Burness, A. T. H. and Clothier, F. W. (1970) *J. gen. Virol.* **6**, 381.
Burr, H. and Lingrel, J. B. (1971) *Nature New Biology,* **233**, 41.
Burton, K. (1956) *Biochem. J.* **62**, 315.
Burton, K. (1967) *Biochem. J.* **104**, 686.
Busch, C. A. and Scheraga, H. A. (1967) Biochemistry, **6**, 3036.
Busch, H. (1967) in *Methods in Enzymology,* vol. **12A** (eds. Grossman, L. and Moldave, K.) New York and London: Academic Press, p. 426.
Busch, H. and Desjardins, R. (1965) *Exptl. Cell Res.* **40**, 353.
Bustin, M., Rall, S. C., Stellwagen, R. H. and Cole, R. D. (1969) *Science,* **163**, 391.

Caillet, J. and Pullman, B. (1968) in *Molecular Associations in Biology* (ed. Pullman, B.), New York and London: Academic Press, p. 217.
Cairns, J. (1963) *J. molec. Biol.* **6**, 208.
Caldwell, I. C. (1969) *J. Chromatog.* **44**, 331.
Campbell, P. N. and Sargent, J. R. (1967, 1969), editors of *Techniques in Protein Synthesis,* London and New York: Academic Press (two volumes).
Cantor, C. R. (1967) *Nature,* **216**, 513.
Cantor, C. R., Jaskunas, S. R. and Tinoco, I., jun. (1966) *J. molec. Biol.* **20**, 39.
Carbon, J. and David, H. (1968) *Biochemistry,* **7**, 3851.
Caro, L. G. (1965) *Virology,* **25**, 226.
Cashell, M. (1970) *Cold Spring. Harb. Symp. quant. Biol.* **35**, 407.
Caspar, D. L. D. and Klug, A. (1962) *Cold Spring Harb. Symp. quant. Biol.* **27**, 1.

Cassini, G., Burgess, R. R., Goodman, H. M. and Gold, L. (1971) *Nature New Biology,* **230,** 197.

Cavalieri, L. F., Tinber, J. F. and Bendich, A. (1949) *J. Amer. chem. Soc.* **71,** 533.

Černý, R., Černá, E. and Spencer, J. H. (1969) *J. molec. Biol.* **46,** 145.

Cerutti, P. and Miller, N. (1967) *J. molec. Biol.* **26,** 55.

Chadwick, P., Pirrotta, V., Steinberg, R., Hopkins, N. and Ptashne, M. (1970) *Cold Spring Harb. Symp. quant. Biol.* **35,** 283.

Chamberlin, E. M., Baldwin, R. L. and Berg, P. (1963) *J. molec. Biol.* **7,** 334.

Chamberlin, E. M., Zambito, A. J., Sohar, P. and Jenkins, S. (1963) *Biochemical Preps.* **10,** 143.

Chamberlin, E. M. and Berg, P. (1962) *Proc. Nat. Acad. Sci. U.S.* **48,** 81.

Chamberlin, M., McGrath, J. and Waskell, L. (1970) *Nature,* **228,** 1160.

Chambers, R. W. (1971) *Prog. Nuc. Acid Res. molec. Biol.* **11,** 489.

Chan, S. I., Schweitzer, M. P., Ts'o, P. O. P. and Helmkamp, G. K. (1964) *J. Amer. chem. Soc.* **86,** 4182.

Chang, F. N. and Flaks, J. G. (1970) *Proc. Nat. Acad. Sci. U.S.* **67,** 1321.

Chappell, J. B. and Hansford, R. G. (1969) in *Subcellular Components* (eds. Birnie, G. D. and Fox, S. M.), London: Butterworths, p. 43.

Chargaff, E. (1968) *Prog. Nuc. Acid Res. molec. Biol.* **8,** 297.

Chargaff, E., Buchowicz, J., Türler, H. and Shapiro, H. S. (1964) *Nature,* **206,** 145.

Chargaff, E., Rüst, E., Temperli, P., Morisawa, S. and Danon, A. (1963) *Biochim. Biophys. Acta,* **76,** 149.

Chauveau, J., Moulé, Y. and Rouillet, C. (1956) *Exptl. Cell Res.* **11,** 317.

Cherayil, J. D., Hamperl, A. and Bock, R. M. (1968) in *Methods in Enzymology,* vol. **12B** (eds. Grossman, L. and Moldave, K.), New York and London: Academic Press, p. 166.

Chládek, S., Pulkrábek, P., Sonnenbilcher, J., Žemlička, J. and Rychlík, I (1970) *Coll. Czec. chem. Comm.* **35,** 2296.

Clark, L. B. and Tinoco, I., jun. (1965) *J. Amer. chem. Soc.* **87,** 11.

Clark, V. M., Todd, A. R. and Zussman, J. (1951) *J. chem. Soc.* 2952.

Click, R. E. and Hackett, D. P. (1966) *Biochim. biophys. Acta,* **129,** 74.

Cohn, W. E. (1949) *Science,* **109,** 377.

Cohn, W. E. (1950) *J. Amer. chem. Soc.* **72,** 1471.

Cohn, W. E. (1955) in *The Nucleic Acids,* vol. **1** (eds. Chargaff, E. and Davidson, J. N.), New York and London: Academic Press, p. 211.

Cohn, W. E. (1960) *J. biol. Chem.* **235,** 1488.

Cohn, W. E. and Uziel, M. (1967) reported by Cohn in *Chromatography* (2nd edition) by Heftmann, E.; New York: Reinhold, p. 627.

Cohn, W. E. and Volkin, E. (1951) *Nature,* **167** 483.

Colby, C., jun. (1971) *Prog. Nuc. Acid Res. molec. Biol.* **11,** 1.

Cole, R. S. (1970) *Biochim. biophys. Acta,* **217,** 30.

Connors, P. G., Labanouskas, M. and Beeman, W. W. (1969) *Science,* **166,** 1528.

Cook, J. S. and McGrath, J. R. (1970) *Proc. Nat. Acad. Sci. U.S.* **58,** 1359.

Cory, S., Spahr, P. F. and Adams, J. M. (1970) *Cold Spring Harb. Symp. quant. Biol.* **35,** 1.

Cory, S., Marcker, K. A., Dube, S. K. and Clark, B. F. C. (1968) *Nature,* **218,** 232.

Coudray, Y., Quetier, F. and Guille, E. (1970 *Biochim. biophys. Acta,* **217,** 259.

Coukell, M. B. and Yanofsky, C. (1970) *Nature,* **228,** 633.

Cox, R. A. (1966) *Biochemical Preps.* **11,** 104.

Cox, R. A. (1970) *Biochem. J.* **120,** 539.

Cox, R. A. and Kanagalingham, K. (1967) *Biochem. J.* **103,** 431.

Cox, R. A., Kanagalingham, K. and Sutherland, E. S. (1970) *Biochem. J.* **120,** 549.

Cox, R. A. and Littauer, U. Z. (1959) *Nature,* **184,** 818.

Craddock, V. M. and Magee, P. N. (1965) *Biochim. biophys. Acta,* **95,** 671.

Cramer, F. (1966) *Angew. Chem.* **78,** 186 (in German).

Cramer, F. (1971) *Prog. Nuc. Acid Res. molec. Biol.* **11,** 391.

Cramer, F., Erdmann, V. A., von der Haar, F. and Schlimme, E. (1969) *J. Cell Physiol.* **74,** 163.

Cramer, F., Helbig, R., Hettler, H., Scheit, K.-H. and Seliger, H. (1966) *Angew. Chem. Internat. Edition,* **5,** 601.

Cramer, F., Rittesdorf, W. and Böhm, W. (1962) *Liebigs Ann. Chem.* **654,** 180 (in German).

Cramer, F., Wittmann, R., Daneck, K. and Weimann, G. (1963) *Angew. Chem.* **75,** 92 (in German).

Crampton, C. F., Frankel, F. R., Benson, A. M. and Wade, A. (1960) *Analyt. Biochem.* **1**, 249.

Crampton, C. F. and Rodeheaver, J. L. (1967) reported in (by Cohn, W. E.) *Chromatography* (2nd edition) by Heftmann, E.; New York: Reinhold, p. 636.

Crathorn, A. R. and Roberts, J. J. (1966) *Nature,* **211**, 150.

Craven, G. R. and Gupta, V. (1970) *Proc. Nat. Acad. Sci. U.S.* **67**, 1329.

Crawford, L. V. and Waring, M. J. (1967) *J. molec. Biol.* **25**, 23.

Crick, F. H. C. and Watson, J. D. (1956) *Nature,* **177**, 473.

Crosbie, G. W., Smellie, R. M. S. and Davidson, J. N. (1953) *Biochem. J.* **54**, 287.

Crothers, D. M. and Zimm, B. H. (1965) *J. molec. Biol.* **12**, 525.

Cummings, D. J., Couse, N. L. and Forrest, G. L. (1970) *Adv. Virus Res.* **16**, *1.*

Dahlberg, A. E., Dingman, C. W. and Peacock, A. C. (1969) *J. molec. Biol.* **41**, 139.

Dahlberg, J. E. (1968) *Nature,* **220**, 548.

Daneholt, D., Edström, J.-E., Egyházi, E., Lambert, B. and Ringborg, U. (1970) *Cold Spring Harb. Symp. quant. Biol.* **35**, 513.

Daniel, V. and Littauer, U. Z. (1963) *J. biol. Chem.* **238**, 2102.

Darby, G., Dumas, L. B. and Sinsheimer, R. L. (1970) *J. molec. Biol.* **52**, 227.

Davern, C. I. (1966) *Proc. Nat. Acad. Sci. U.S.* **55**, 792.

Davern, C. I. (1971) *Prog. Nuc. Acid Res. molec. Biol.* **11**, 229.

Davidson, N., Widholm, J., Naudi, O. S., Jensen, R., Olivera, B. M. and Wong, J. C. (1965) *Proc. Nat. Acad. Sci. U.S.* **53**, 111.

Davies, H. G. (1967) *Nature,* **214**, 208.

Davies, J. W. (1969) *Biochim. biophys. Acta,* **174**, 689.

Davis, R. C. and Tinoco, I., jun. (1968) *Biopolymers,* **6**, 223.

Dawid, I. B. (1970)   *Symp. Soc. Exp. Biol.* **24**, 227.

Deboer, G., Klinghoffer, O. and Johns, H. E. (1970) *Biochim. biophys. Acta,* **213**, 253.

de Crombrugghe, B., Chen, B., Anderson, W., Nissley, P., Gottesman, M., Pastan, I. and Perlman, R. (1971) *Nature,* **231**, 139.

Dekker, A., Michelson, A. M. and Todd, A. R. (1953) *J. chem. Soc.* **947**.

de Lucia, P. and Cairns, J. (1969) *Nature,* **224**, 1164.

DeLange, J. R., Fambrough, D. M., Smith, E. L. and Bonner, J. (1969) *J. biol. Chem.* **244**, 319.

Dertinger, H. and Nicolau, C. (1970) *Biochim. biophys. Acta,* **199**, 316.

de Wachter, R., Vandenbergh, A., Merregaert, J., Contreras, R. and Fiers, W. (1971) *Proc. Nat. Acad. Sci. U.S.* **68**, 585.

de Wachter, R. and Fiers, W. (1971) in *Methods in Enzymology,* vol. **21** (eds. Grossman, L. and Moldave, K.), New York and London: Academic Press, p. 167.

Diecker, H. R. G., Holmann, H. R. and Kunkel, H. G. (1959) *J. exptl. Med.* **109**, 97.

Dingman, C. W. and Peacock, A. C. (1968) *Biochemistry,* **7**, 659.

Dipple, A., Lawley, P. D. and Brookes, P. (1968) *Europ. J. Cancer,* **4**, 493.

Dische, Z. (1930) *Mikrochemie* **8**, 4 (in German).

Dische, Z. and Schwartz, K. (1937) *Mikrochim. Acta,* **2**, 13 (in German).

Doctor, B. P., Loebel, J. E., Sodd, M. A. and Winter, D. E. (1970) *Science,* **163**, 693.

Doi, R. H. and Igarashi, R. H. (1964) *J. Bact.* **87**, 323.

Dole, V. P. and Cootzias, G. C. (1951) *Science,* **113**, 552.

Donahue, J. (1969) *Science,* **165**, 1091.

Donahue, J. and Trueblood, K. M. (1960) *J. molec. Biol.* **2**, 363.

Doskočil, J. and Šormová, Z. (1965) *Coll. Czech. chem. Comm.* **30**, 479, 490 (two papers).

Doty, P., Boedtker, H., Fresco, J. R., Haselkorn, R. and Litt, M. (1959) *Proc. Nat. Acad. Sci. U.S.* **45**, 482.

Dounce, A. L. (1955) in *The nucleic Acids,* vol. 2, (eds. Chargaff, E. and Davidson, J. N.), New York and London: Academic Press, p. 93.

Drake, J. W. (1966) *Proc. Nat. Acad. Sci. U.S.* **55**, 738.

Dressler, D. (1970) *Proc. Nat. Acad. Sci. U.S.* **67**, 1934.

Dressler, D. and Wolfson, J. (1970) *Proc. Nat. Acad. Sci. U.S.* **67**, 456.

Dube, S. K., Marcker, K. A., Clark, B. F. C. and Cory, S. (1968) *Nature,* **220**, 1039.

Dubin, S. B., Benedek, G. B., Bancroft, F. C. and Freifelder, D. (1970) *J. molec. Biol.* **54**, 547.

Dubin, D. T. and Günlap, A. (1968) *Biochim. biophys. Acta,* **134**, 146.

Dubin, D. T. and Rosenthal, S. M. (1960) *J. biol. Chem.* **235**, 776.
DuBoy, B. and Weissmann, S. M. (1971) *J. biol. Chem.* **246**, 747.
Dudock, B. J., Katz, G., Taylor, E. A. and Holley, R. W. (1969) *Proc. Nat. Acad. Sci. U.S.* **62**, 941.
Dunn, J. J., Bautz, F. A. and Bautz, E. K. F. (1971) *Nature New Biology,* **230**, 94.
Durham, A. C. H., Finch, J. T., Butler, P. J. G. and Klug, A. (1971) *Nature New Biology,* **229**, 37 (three consecutive papers by A. C. H. D., J. T. F. and A. K.; A. C. H. D. and A. K.; and P. J. G. B. and A. K.).
Duschinsky, R., Plevin, E. and Heidelberger, C. (1957) *J. Amer. Chem. Soc.* **79**, 4559.

Eckstein, H. and Grindl, F. (1970) *Europ. J. Biochem.* **13**, 558.
Eckstein, H., Paduch, V. and Hilz, H. (1967) *Europ. J. Biochem.* **3**, 224.
Edmonds, M. and Abrams, R. (1962) *J. biol. Chem.* **237**, 2636.
Eigner, J. (1968) in *Methods in Enzymology,* vol. **12B** (eds. Grossman, L. and Moldave, K.), New York and London: Academic Press, p. 386.
Eigner, J. and Doty, P. (1965) *J. molec. Biol.* **12**, 549.
Einsenstadt, J. and Brawerman, G. (1964) *J. molec. Biol.* **10**, 392.
Einsenstadt, J. and Brawerman, G. (1967) in *Methods in Enzymology,* vol. **12A** (eds. Grossman, L. and Moldave, K.), New York and London: Academic Press, p. 424.
Eisinger, J. and Shulman, R. G. (1963) *Proc. Nat. Acad. Sci. U.S.* **58**, 439.
Eisinger, J., Feure, B. and Yamane, Y. (1971) *Nature New Biology,* **231**, 126.
Emerson, T. R., Swan, R. J. and Ulbricht, T. L. V. (1966) *Biochem. biophys. Res. Comm.* **22**, 505.
Emmerson, P. T. and Howard-Flanders, P. (1965) *Biochem. biophys. Res. Comm.* **18**, 24.
Englander, S. W. (1968) in *Methods in Enzymology,* vol. **12B** (eds. Grossman, L. and Moldave, K.), New York and London: Academic Press, p. 379.
Eoyang, L. and August, J. T. (1968) in *Methods in Enzymology,* vol. **12B** (eds. Grossman, L. and Moldave, K.), New York and London: Academic Press, p. 530.
Evans, D. and Birnsteil, M. L. (1968) *Biochim. biophys. Acta,* **166**, 274.
Eyring, E. J. and Opengard, J. (1967) *Biochemistry,* **6**, 2500.

Fauman, M., Rabinowitz, M. and Getz, G. M. (1969) *Biochim. biophys. Acta,* **182**, 355.
Fedorcsák, I., Natarajan, A. T. and Ehrenberg, L. (1969) *Europ. J. Biochem.* **10**, 450.
Falaschi, A. and Kornberg, A. (1966) *J. biol. Chem.* **241**, 1478.
Fellner, P. and Ebel, J. P. (1970) *Nature,* **225**, 1131.
Fellner, P., Ehresmann, C. and Ebel, J. P. (1970) *Cold Spring Harb. Symp. quant. Biol.* **35**, 29.
Fellner, P. and Sanger, F. (1968) *Nature,* **219**, 236.
Felsenfield, G. (1968) in *Methods in Enzymology,* vol. **12B** (eds. Grossman, L. and Moldave, K.), New York and London: Academic Press, p. 247.
Felsenfield, G. and Hirschmann, S. Z. (1965) *J. molec. Biol.* **13**, 407.
Fiske, C. H. and Subbarow, Y. (1925) *J. biol. Chem.* **66**, 375.
Fixman, M. (1963) *J. molec. Biol.* **6**, 36.
Flamm, W. G., Birnstiel, M. L. and Walker, P. M. B. (1969) in *Subcellular Components* (eds. Birnie, G. D. and Fox. S. M.), London: Butterworths, p. 125.
Flamm, W. G., Bord, H. E. and Burr, H. E. (1966) *Biochim. biophys. Acta,* **129**, 310.
Flessel, C. P., Ralph, P. and Rich, A. (1967) *Science,* **158**, 658.
Florentiev, V. L. and Ivanov, V. I. (1970) *Nature,* **228**, 519.
Forget, B. G. and Weissmann, S. M. (1968) *J. biol. Chem.* **243**, 3709.
Forget, B. G. and Weissmann, S. M. (1969) *J. biol. Chem.* **244**, 3148.
Fox, C. F., Robinson, W. S., Haselkorn, R. and Weiss, S. B. (1964) *J. biol. Chem.* **239**, 186.
Fox, J. J. and Wempen, I. (1965) *Tetrahedron Letters,* 643.
Fox, J. J., Wempen, I., Hampton, A. and Doerr, I. C. (1958) *J. Amer. chem. Soc.* **80**, 1669.
Fraenkel-Conrat, H., Singer, B. and Tsugita, A. (1961) *Virology,* **14**, 54.
Frank, R. M. van, Ellis, L. F. and Kleinschmidt, W. J. (1971) *J. gen. Virol.* **12**, 33.

Franklin, R. E., Klug, A. and Holmes, K. C. (1957) in *The Nature of Viruses* (a Ciba Symposium), London: Churchill, p. 39.
Franklin, R. M. (1966) *Proc. Nat. Acad. Sci. U.S.* **55**, 1504.
Franze de Fernandez, M. T., Eoyang, L. and August, J. T. (1968) *Nature,* **219**, 588.
Freifelder, D. (1966a) *Virology,* **28**, 742.
Freifelder, D. (1966b) *Proc. Nat. Acad. Sci. U.S.* **54**, 128.
Freifelder, D. (1970) *J. molec. Biol.* **54**, 567.
Fremlin, J. H. (1964) *Applications of Nuclear Physics,* London: English Universities Press.
Fried, A. H. and Rappaport, H. P. (1970) *Biochim. biophys. Acta,* **204**, 91.
Fritsch, A. and Worcel, A. (1971) *J. molec. Biol.* **59**, 207.
Fry, M. and Artman, M. (1969) *Biochem. J.* **115**, 287.
Fuke, M. and Thomas, C. A., jun. (1970) *J. molec. Biol.* **52**, 395.
Fuller, W., Arnott, S. and Creek, J. (1969) *Biochem. J.* **114**, 26P.
Fuller, W. and Hodgson, A. (1967) *Nature,* **215**, 817.
Fuller, W. and Waring, M. J. (1964) *Ber. Bunsenges Physik. Chem.* **68**, 805.

Gabbay, E. J. (1970) *J. Amer. chem. Soc.* **91**, 5136.
Galibert, F., Larsen, C. J., Lelong, J. C. and Boison, M. (1966) *Nature,* **207**, 1039.
Gaskill, P. and Kabat, D. (1971) *Proc. Nat. Acad. Sci. U.S.* **68**, 72.
Geiduschek, E. P. (1961) *Proc. Nat. Acad. Sci. U.S.* **47**, 950.
Gelderman, A. H., Rake, A. V. and Britten, R. J. (1971) *Proc. Nat. Acad. Sci. U.S.* **68**, 172.
Gennis, R. B. and Cantor, C. R. (1970) *Biochemistry,* **9**, 4714.
Georgiev, G. P. (1967) *Prog. Nuc. Acid. Res. molec. Biol.* **9**, 259.
Gesteland, R. F., Salser, W. and Bolle, A. (1970) *Proc. Nat. Acad. Sci. U.S.* **58**, 2036.
Ghosh, H. P., Söll, D. and Khorana, H. G. (1967) *J. molec. Biol.* **25**, 275.
Gilbert, M. and Claverie, P. (1968) in *Molecular Associations in Biology* (ed. Pullman, B.), New York and London: Academic Press, p. 245 (in French).
Gilbert, W. and Dressler, D. (1968) *Cold Spring Harb. Symp. quant. Biol.* **33**, 473.
Gillam, I., Millwall, S., Blew, D., von Tigerstrom, M., Wimmer, E. and Tener, G. M. (1967) *Biochemistry,* **6**, 3043.
Gillespie, D. and Spiegelman, S. (1965) *J. molec. Biol.* **12**, 829.
Glaubiger, D. and Hearst, J. E. (1967) *Biopolymers,* **8**, 691.
Glitz, D. G., Bradley, A. and Fraenkel-Conrat, H. (1968) *Biochim. biophys. Acta,* **161**, 1.
Goebel, W. and Helinski, D. R. (1968) *Proc. Nat. Acad. Sci. U.S.* **61**, 1406.
Goebel, W. and Helinski, D. R. (1970) *Biochemistry,* **9**, 4793.
Goodman, H. M., Abelson, J. N., Landy, A., Zadrazil, S. and Smith, J. D. (1970) *Europ. J. Biochem.* **13**, 461.
Gorbach, G. and Henke, J. (1968) *J. Chromatog.* **37**, 225 (in German).
Gordon, A. H. (1969) *Electrophoresis of Proteins in Polyacrylamide and Starch Gels* (in the series *Laboratory Techniques in Biochemistry and Molecular Biology*, eds. Work, T. S. and Work, E.), Amsterdam and London: North Holland.
Gottesman, M. E., Canellakis, Z. N. and Canellakis, E. S. (1962) *Biochim. biophys. Acta,* **61**, 34.
Goulian, M., Kornberg, A. and Sinsheimer, R. L. (1967) *Proc. Nat. Acad. Sci. U.S.* **58**, 2321.
Goulian, M., Lucas, Z. J. and Kornberg, A. (1968) *J. biol. Chem.* **243**, 627.
Grady, L. J. and Pollard, E. C. (1967) *Biochim. biophys. Acta,* **145**, 837.
Grau, O., Ohlsson-Wilhelm, B. M. and Geiduschek, E. P. (1970) *Cold Spring Harb. Symp. quant. Biol.* **35**, 221.
Gray, H. B., jun. and Hearst, J. E. (1968) *J. molec. Biol.* **35**, 111.
Green, M. and Cohn, W. E. (1957) *J. biol. Chem.* **228**, 601.
Greenberg, J. (1967) *Genetics,* **55**, 193.
Greenman, D. C., Huang, R.-C. C., Smith, M. and Farrow, L. M. (1969) *Analyt. Biochem.* **31**, 348.
Gressel, J. and Wolowelsky, J. (1968) *Analyt. Biochem.* **22**, 352.
Grierson, D., Rogers, M. E., Sartirana, M. L. and Loening, U. E. (1970) *Cold Spring Harb. Symp. quant. Biol.* **35**, 589.
Griffin, B. E., Jarman, M. and Reese, C. B. (1968) *Tetrahedron,* **24**, 639.
Griffith, J., Huberman, J. A. and Kornberg, A. (1971) *J. molec. Biol.* **55**, 209.

Griswold, B. C., Humoller, F. L. and McIntyre, A. R. (1957) *Analyt. Chem.* **23**,192.

Gross, J. D., Grunstein, J. and Witkin, E. M. (1971) *J. molec. Biol.* **58**, 631.

Grummt, F. and Bielka, H. (1968) *Biochim. biophys. Acta,* **161**, 253.

Grunberg-Manago, M., Ortiz, P. and Ochoa, S. (1956) *Biochim. biophys. Acta,* **20**, 269.

Guild, W. R. (1968) *Cold Spring Harb. Symp. quant. Biol.* **33**, 142.

Gumport, R. F. and Weiss, S. B. (1969) *Biochemistry,* **8**, 3618.

Gupta, N. K., Ohtsuka, E., Sgaramella, V., Büchi, H., Kumar, A., Wever, H. and Khorana, H. G. (1968) *Proc. Nat. Acad. Sci. U.S.* **60**, 1338.

Gupta, N. K., Ohtsuka, E., Weber, H., Chang, S. H. and Khorana, H. G. (1968) *Proc. Nat. Acad. Sci. U.S.* **60**, 285.

Gurdon, J. B., Lane, C. D., Woodland, H. R. and Marbaix, G. (1971) *Nature,* **233**, 177.

Guthrie, G. D. and Sinsheimer, R. L. (1960) *J. molec. Biol.* **2**, 297.

Hadjivassiliou, A. and Brawerman, G. (1966) *J. molec. Biol.* **20**, 1.

Haines, J. A., Reese, C. B. and Todd, Lord (1964) *J. chem. Soc.* 1406.

Hall, R. H. (1967) in *Methods of Enzymology,* vol. **12A** (eds. Grossman, L. and Moldave, K.), New York and London: Academic Press, p. 305.

Hall, J. B. and Sinsheimer, R. L. (1963) *J. molec. Biol.* **6**, 115.

Hamlin, R. M., Lord, R. C. and Rich, A. (1965) *Science,* **148**, 1734.

Handbook of Chemistry and Physics (1964 edition) International Rubber Co.

Hänggi, U. J., Streeck, R. E., Voigt, H. P. and Zachau, H. G. (1970) *Biochim. biophys. Acta,* **217**, 278.

Hansbury, E., Kerr, V. N., Mitchell, W. E., Ratliff, R. L., Smith, D. A., Williams, D. J. and Hayes, F. N. (1970) *Analyt. Biochem.* **21**, 322.

Harada, F., Gross, H. J., Kimura, F., Chang, S. H., Nishimura, S. and RajBhandary, W. L. (1968) *Biochem. biophys. Res. Comm.* **33**, 299.

Harkins, T. R. and Freiser, H. (1958) *J. Amer. chem. Soc.* **80**, 1132.

Haruna, I. and Spiegelman, S. (1965) *Proc. Nat. Acad. Sci. U.S.* **54**, 579; *Science,* **150**, 884 (two papers refer to similar material).

Haschemeyer, A. E. V. and Sobell, H. M. (1965) *Acta Cryst.* **18**, 525.

Haselkorn, R. and Fox, C. F. (1965) *J. molec. Biol.* **13**, 780.

Hashizume, T. and Sasaki, Y. (1966) *Analyt. Biochem.* **15**, 199.

Haskell, E. H. and Davern, C. I. (1969) *Proc. Nat. Acad. Sci. U.S.* **64**, 1065.

Hastings, J. R. B. and Kirby, K. S. (1966) *Biochem. J.* **100**, 532.

Hastings, J. R. B., Parish, J. H., Kirby, K. S. and Cook, E. A. (1968) *Biochim. biophys. Acta,* **155**, 603.

Hastings, J. R. B., Parish, J. H., Kirby, K. S. and Klucis, E. S. (1965) *Nature,* **208**, 665.

Hayatsu, H. and Khorana, H. G. (1967) *J. Amer. chem. Soc.* **89**, 3880.

Hayes, D. H., Michelson, A. M. and Todd, A. R. (1955) *J. chem. Soc.* 808.

Hayes, F., Hayes, D., Fellner, P. and Ehresmann, C. (1971) *Nature New Biology,* **232**, 52.

Hays, J. B. and Zimm, B. H. (1970) *J. molec. Biol.* **48**, 297.

Hearst, J. E. (1962) *J. molec. Biol.* **4**, 415.

Hecht, C., Zamecnik, P. C., Stephenson, M. L. and Scott, J. F. (1958) *J. biol. Chem.* **233**, 954.

Hecht, N. B. and Woese, C. R. (1968) *J. Bact.* **95**, 986.

Hecht, S. M. (1970) *Biochim. biophys. Acta,* **213**, 269.

Helmstetter, C. E. and Cooper, S. (1968) *J. molec. Biol.* **31**, 507

Henley, D. D., Lindahl, T. and Fresco, J. R. (1966) *Proc. Nat. Acad. Sci. U.S.* **55**, 191.

Heppel, L. A., Whitfield, P. R. and Markham, R. (1955) *Biochem. J.* **60**, 8.

Herak, J. N. and Gordy, W. (1965) *Proc. Nat. Acad. Sci. U.S.* **54**, 1287.

Herbst, E. J., Keister, D. L. and Weaver, R. H. (1958) *Arch. Biochem. Biophys.* **75**, 178.

Herschberger, C., Mickel, S. and Rownd, R. (1971) *J. Bact.* **106**, 238.

Hershey, A. D., Kamen, M. D., Kennedy, J. W. and Gest, H. (1951) *J. gen. Physiol.* **34**, 305.

Herzog, A., Ghysen, A. and Bollen, A. (1971) *FEBS Letters,* **15**, 291.

Heywood, S. M. (1970) *Proc. Nat. Acad. Sci. U.S.* **67**, 1782.

Higashi, K., Shankar Narayan, K., Adams, H. R. and Busch, H. (1966) *Cancer Res.* **26**, 582.

Highton, P. J. and Beer, M. (1963) *J. molec. Biol.* **7**, 70.

Hill, R. (1958) *Biochim. biophys. Acta,* **30**, 636.

Hindley, J. and Staples, D. H. (1969) *Nature,* **224**, 964.

Hirsch, C. A. (1966) *Biochim. biophys. Acta,* **123,** 246.

Hirsch, O. (1971) *J. molec. Biol.* **58,** 439.

Hirschmann, S. Z. and Felsenfield, G. (1966) *J. molec. Biol.* **16,** 347.

Hirschmann, S. Z., Leng, M. and Felsenfield, G. (1967) *Biopolymers,* **5,** 227.

Ho, N. W. Y., Uchida, T., Egami, F. and Gilham, D. T. (1969) *Cold Spring Harb. Symp. quant. Biol.* **34,** 647.

Hoagland, M. B., Stephenson, M. L., Scott, J. F., Hecht, L. and Zamecnik, P. C. (1958) *J. biol. Chem.* **231,** 241.

Hoard, D. E. (1960) *Biochim. biophys. Acta,* **40,** 62.

Hoffman, N. E. (1970) *Analyt. Biochem.* **33,** 209.

Hoffman, B. M., Schofield, B. and Rich, A. (1969) *Proc. Nat. Acad. Sci. U.S.* **62,** 1195.

Hofschneider, P. H. and Delius, H. (1968) in *Methods in Enzymology,* vol. **12B** (eds. Grossman, L. E. and Moldave, K.), New York and London: Academic Press, p. 880.

Hohn, B. and Korn, D. (1969) *J. molec. Biol.* **45,** 385.

Hohn, T. and Hohn, B. (1970) *Adv. Virus Res.* **16,** 43.

Holcomb, D. N. and Tinoco, I., jun. (1965) *Biopolymers,* **3,** 121.

Holley, R. W. (1963) *Biochem. biophys. Res. Comm.* **10,** 186.

Holley, R. W. (1968) *Prog. Nuc. Acid. Res. molec. Biol.* **8,** 37.

Holley, R. W., Madison, J. T. and Zamir, A. (1964) *Biochem. biophys. Res. Comm.* **17,** 389.

Holliday, R. (1964) *Genet. Res.* **5,** 282.

Holliday, R. (1969) *Symp. Soc. Gen. Microbiol.* **20,** 359.

Holy, A. and Scheit, K.-H. (1967) *Biochim. biophys. Acta,* **138,** 230.

Hoogsteen, K. (1959) *Acta Cryst.* **12,** 822.

Hoogsteen, K. (1963) *Acta Cryst.* **16,** 907.

Hotchkiss, R. D. and Gabor, M. (1970) *Ann. Rev. Gen.* **4,** 193.

Hotz, G. and Reuschel, H. (1967) *Molec. gen. Genetics,* **99,** 5.

Howard-Flanders, P. (1968) *Ann. Rev. Biochem.* **38,** 175.

Howard-Flanders, P. and Boyce, R. P. (1966) *Radiation Res. Supplement 6,* 156.

Howard-Flanders, P., Simpson, E. and Theriot, L. (1966) *Proc. Nat. Acad. Sci. U.S.* **53,** 1119.

Huang, R. C. C. and Bonner, J. (1962) *Proc. Nat. Acad. Sci. U.S.* **48,** 1216.

Huang, R. C. C. and Bonner, J. (1965) *Proc. Nat. Acad. Sci. U.S.* **54,** 960.

Huberman, J. H. and Attardi, G. (1966) *J. Cell Biol.* **31,** 95.

Hudson, C. S. (1909) *J. Amer. chem. Soc.* **31,** 66.

Huh, T. Y. and Helleiner, C. W. (1967) *Analyt. Biochem.* **19,** 150.

Hünig, S., Müller, H. R. and Thier, W. (1965) *Angew. Chem. International Edition,* **4,** 271.

Hurwitz, J., Gold, M. and Anders, M. (1964a) *J. biol. Chem.* **239,** 3462, 3474 (two papers); (1964b) *ibid.* 3858.

Ifft, J. B., Voet, D. H. and Vinograd, J. (1961) *J. phys. Chem.* **65,** 1138.

Igo-Kemenes, T. and Zachau, H. G. (1969) *Europ. J. Biochem.* **10,** 549.

Imura, N., Weiss, G. B. and Chambers, R. W. (1969) *Nature,* **222,** 1147.

Ingle, J., Possingham, J. V., Wells, R., Leaver, C. J. and Loening, U. E. (1970) *Symp. Soc. Exptl. Biol.* **24,** 303.

Inman, R. B. (1965) *J. molec. Biol.* **13,** 947.

Inman, R. B. (1967) *J. molec. Biol.* **25,** 209.

Inman, R. B. and Baldwin R. L. (1962) *J. molec. Biol.* **5,** 185.

Inman, R. B. and Schnös, M. (1971) *J. molec. Biol.* **56,** 319.

Inoue, Y., Agoyagi, S. and Nakanishi, K. (1967) *J. Amer. chem. Soc.,* **89,** 5701.

Isaacs, A. and Lindenmann, J. (1957) *Proc. Roy. Soc.* **B147,** 258.

IUPAC-IUB (Commission on Biochemical Nomenclature) (1970) *Europ. J. Biochem.* **15,** 203.

Iwai, K., Ishikawa, K. and Hayashi, H. (1970) *Nature,* **226,** 1056.

Iwanami, Y. and Brown, G. M. (1968) *Arch. Biochem. Biophys.* **126,** 8.

Iyer, V. N. and Szybalski, W. (1963) *Proc. Nat. Acad. Sci. U.S.* **50,** 355.

Jacob, K., Todd, K., Birnstiel, M. L. and Bird, A. (1971) *Biochim. biophys. Acta,* **228,** 761.

Jacob, T. M. and Khorana, H. G. (1964) *J. Amer. chem. Soc.* **86,** 1630.

Jacob, T. M., Narang, S. A. and Khorana, H. G. (1967) *J. Amer. chem. Soc.* **89**, 2177.
Jacobson, K. B. (1971a) *Prog. Nuc. Acid Res. molec. Biol.* **11**, 461.
Jacobson, K. B. (1971b) *Nature New Biology,* **231**, 17.
Jacobson, M., O'Brien, J. F. and Hedgcoth, C. (1968) *Analyt. Biochem.* **25**, 363.
Jaenicke, L. and Volbrechtshausen, I. (1952) *Naturwissenschaften,* **39**, 86.
Jardetzky, C. D. (1961) *J. Amer. chem. Soc.* **83**, 2919.
Jardetzky, C. D. and Jardetzky, O. (1960) *J. Amer. chem. Soc.* **82**, 222.
Jayaraman, R. and Goldberg, E. B. (1970) *Cold Spring Harb. Symp quant. Biol.* **35**, 197.
Jehle, H. (1965) *Proc. Nat. Acad. Sci. U.S.* **53**, 1451.
Jensen, R. H. and Davidson, N. (1966) *Biopolymers,* **4**, 17.
Jeppesen, P. G. N., Steitz, J. A., Gesteland, R. F. and Spahr, P. F. (1970) *Nature,* **226**, 230.
Jeppesen, P. G. N., Nichols, J. L., Sanger, F. and Barrell, B. G. (1970) *Cold. Spring Harb. Symp. quant. Biol.* **35**, 13.
John, M., Skrabei, H. and Dellweg, H. (1969) *FEBS Letters,* **5**, 185.
Johnson, F. B., Setterfield, G. and Stern, H. (1959) *J. biochem. biophys. Cytol.* **6**, 53.
Jones, A. S., Letham, D. and Stacey, M. (1956) *J. chem. Soc.* **2573 (three consecutive papers).**
Jones, J. W. and Robins, R. K. (1963) *J. Amer. chem. Soc.* **85**, 193.
Jones, D. S., Nishimura, S. and Khorana, H. G. (1966) *J. molec. Biol.* **16**, 454.
Jordan, D. O. (1968) in *Molecular Associations in Biology* (ed. Pullman, B.), New York and London: Academic Press, p. 221.
Jorgensen, S. E. and Koerner, J. F. (1966) *J. biol. Chem.* **241**, 3090.
Josse, J., Kaiser, A. D. and Kornberg, A. (1964) *J. biol. Chem.* **236**, 251.

Kahn, P. L. (1964) *J. molec. Biol.* **8**, 392.
Kaiser, A. D. and Hodgness, D. S. (1960) *J. molec. Biol.* **2**, 392.
Kallen, R. G., Simon, M. and Marmur, J. (1962) *J. molec. Biol.* **5**, 248.
Kallenbach, N. R. (1968) *J. molec. Biol.* **37**, 445.
Kamen, R. (1970) *Nature,* **228**, 527.
Kang, C. Y. and Prévec, L. (1969) *J. virol.* **3**, 404.
Kaplan, H. S. (1966) *Proc. Nat. Acad. Sci. U.S.* **55**, 1442.
Kaplan, H. S., Earle, J. D. and Howsden, F. L. (1964) *J. cell. Comp. Physiol.* **64**, Supplement 1, 69; also Kaplan and Howsden, *Proc. Nat. Acad. Sci. U.S.* **51**, 181.
Kaplan, H. S. and Zavarine, R. (1962) *Biochem. biophys. Res. Comm.* **8**, 432.
Kapuler, A. M. and Michelson, A. M. (1971) *Biochim. biophys. Acta,* **232**, 136.
Karplus, M. (1959) *J. chem. Phys.* **30** 11.
Kates, J. (1970) *Cold Spring Harb. Symp. quant. Biol.* **35**, 743.
Kates, J. and Beeson, J. (1970) *J. molec. Biol.* **30**, 19.
Kato, K., Goncalves, J. M., Houts, G. E. and Bollum, F. J. (1967) *J. biol. Chem.* **242**, 2780.
Katz, S. (1963) *Biochim. biophys. Acta,* **68**, 240.
Katz, L. and Penman, S. (1966) *J. molec. Biol.* **15**, 220.
Kennell, D. (1968) *J. molec. Biol.* **34**, 85.
Kennell, D. E. (1971) *Prog. Nuc. Acid Res. molec. Biol.* **11**, 259.
Kennell, D. and Kotoulas, A. (1968) *J. molec. Biol.* **34**, 71.
Kenner, G. W. (1957) in *The Chemistry and Biology of Purines* (eds. Wolstenholme, G. E. W. and O'Connor, C. M.; a Ciba Symposium), Boston, Mass.: Little, Brown & Co., p. 312.
Kersten, W. and Kersten, H. (1968) in *Molecular Associations in Biology* (ed. Pullman, B), New York and London: Academic Press, p. 289.
Keyl, H.-G. (1964) *Chromosoma,* **17**, 139.
Khorana, H. G., Turner, A. F. and Vizsolyi, J. P. (1961) *J. Amer. chem. Soc.* **83**, 686.
Kidson, C. S. and Kirby, K. S. (1963) *Biochim. biophys. Acta,* **76**, 674.
Kimball, R. F. (1966) *Radiation Res. Supplement* **6**, 51.
Kirby, K. S. (1956) *Biochem. J.* **64**, 405.
Kirby, K. S. (1957) *Biochem. J.* **66**, 495.
Kirby, K. S. (1957) *Biochim. biophys. Acta,* **41**, 338.
Kirby, K. S. (1962) *Biochim. biophys. Acta,* **61**, 506.
Kirby, K. S. (1965) *Biochem. J.* **96**, 266.
Kirby, K. S. and Cook, E. A. (1967) *Biochem. J.* **104**, 254.

Kirby, K. S. (1968) in *Methods in Enzymology*, vol. **12B** (eds. Grossman, L. and Moldave, K.), New York and London: Academic Press, p. 93.

Kirby, K. S., Fox-Carter, E. and Guest, M. (1967) *Biochem. J.* **104**, 749.

Kirby, K. S., Hastings, J. R. B. and Parish, J. H. (1969) in *Subcellular Components* (eds. Birnie, G. D. and Fox, S. M.), London; Butterworths, p. 117.

Kit, S. (1960) *Arch. Biochem. Biophys.* **87**, 318.

Kitane, Y. and Olive, C. S. (1969) *Genetics,* **62**, 23.

Klee, C. B. and Singer, M. F. (1968) *Biochem. biophys. Res. Comm.* **29**, 356.

Kleinschmidt, A. K. and Zahn, R. K. (1959) *Z. Naturforschg.* **14b**, 770.

Klemperer, H. G. (1963) *Biochim. biophys. Acta,* **72**, 416.

Kloos, W. E. (1969) *J. gen. Microbiol.* **59**, 247.

Knippers, R. (1970) *Nature,* **228**, 1051.

Knippers, R., Komano, T. and Sinsheimer, R. L. (1968) *Proc. Nat. Acad. Sci. U.S.* **59**, 577.

Knippers, R. and Strätling, W. (1970) *Nature,* **226**, 713.

Knippers, R., Whalley, J. and Sinsheimer, R. L. (1969) *Proc. Nat. Acad. Sci. U.S.* **64**, 275.

Knorre, D. and Shamovsky, G. (1967) *Biochim. biophys. Acta,* **142**, 555.

Kondo, M., Gallerani, R. and Weissman, C. (1970) *Nature,* **228**, 525.

Kornberg, A. (1969) *Science,* **163**. 1410.

Kornberg, A., Bertsch, L. L., Jackson, J. F. and Khorana, H. G. (1964) *Proc. Nat. Acad. Sci. U.S.* **51**, 315.

Kornberg, A., Zimmerman, J. B., Kornberg, S. R. and Josse, J. (1959) *Proc. Nat. Acad. Sci. U.S.* **45**, 772.

Kosaganov, Yu. N., Zaruduaja, M. I., Lazurkin, Yu. S., Frank-Kameneskii, M. D., Beabealashvilli, R. Sh. and Savochkina, L. B. (1971) *Nature New Biology,* **231**, 212.

Kotchekov, N. K. and Budowsky, E. I. (1969) *Prog. Nuc. Acid. Res. molec. Biol.* **9**, 403.

Kriek, E. and Emmelot, P. (1964) *Biochim. biophys. Acta,* **91**, 59.

Kriek, E., Miller, J. A., Johl, U. and Miller, E. C. (1967) *Biochemistry,* **6**, 177.

Kućan, I., Kućan, Ź., Gartland, W. J. and Chambers, R. W. (1970) *Feder. Proc.* **29**, 467.

Kull, F. J. and Soodak, M. (1968) *Analyt. Biochem.* **32**, 10.

Küntzel, H. and Noll, H. (1967) *Nature,* **215**, 1340.

Kuo, T.-T., Huang, T.-c. and Teng, M.-H. (1968) *J. molec. Biol.* **34**, 375.

Kwiatkowski, J. F. (1968) *Theoret. Chim. Acta,* **11**, 167.

Lackner, H. (1970) *Tetrahedron Letters,* 3189.

Lal, B. M. and Burdon, R. H. (1967) *Nature,* **213**, 1134.

Lamb, R. D. and Dukes, G. R. (1964) *Analyt. Biochem.* **7**, 152.

Lamola, A. A., Guéron, M., Yamane, T., Eisinger, J. and Shulman, R. G. (1966) *Proc. Nat. Acad. Sci. U.S.* **55**, 1015.

Langridge, R. and Rich, A. (1963) *Nature,* **198**, 725.

Larsen, C. J., Lebowitz, P., Weissmann, S. M. and DuBoy, B. (1970) *Cold Spring Harb. Symp. quant. Biol.* **35**, 35.

Lawford, G. R., Langford, P. and Schachter, H. (1966) *J. biol. Chem.* **241**, 1835.

Lawley, P. D. (1961) *J. chem. Soc.* **1011**.

Lawley, P. D. (1967) *J. molec. Biol.* **24**, 75.

Lawley, P. D. and Brookes, P. (1963) *Biochem. J.* **89**, 127.

Lawley, P. D. and Brookes, P. (1964) *Biochem. J.* **92**, 19C.

Lawley, P. D. and Brookes, P. (1965) *Nature,* **206**, 480.

Lawley, P. D. and Brookes, P. (1967) *J. molec. Biol.* **25**, 143.

Lawley, P. D. and Brookes, P. (1968) *Biochem. J.* **109**, 433.

Lawley, P. D. and Orr, J. J. (1970) *Chem. Biol. Interacts.* **2**, 147.

Lawley, P. D. and Thatcher, C. J. (1970) *Biochem. J.* **116**, 693.

Leaver, C. J. and Ingle, J. (1971) *Biochem. J.* **123**, 235.

Leboy, P. S., Cox, E. C. and Flaks, J. G. (1965) *Proc. Nat. Acad. Sci. U.S.* **52**, 1367.

Lee, K. Y., Lijinsky, W. and Magee, P. N. (1964) *J. Nat. Cancer Inst.* **32**, 65.

Lee, Hung, S. and Cavalieri, L. F. (1963) *Proc. Nat. Acad. Sci. U.S.* **50**, 1116.

Leffler, A. T. I., Creskoff, E., Lubowsky, S. W., MacFarlane, V. and Mora, P. T. (1970) *J. molec. Biol.* **48**, 455.

Lehman, I. R. and Nussbaum, A. L. (1964) *J. Biol. Chem.* **239**, 233.

Lehman, I. R., Roussos, G. G. and Platt, E. A. (1962) *J. biol. Chem.* **237**, 829.

Leonard, N. L., McDonald, J. J. and Reichmann, M. E. (1970) *Proc. Nat. Acad. Sci. U.S.* **67**, 93.

Le Pecq, J.-B. and Paoletti, J. (1967) *J. molec. Biol.* **27**, 87.

Leppla, S. H., Bjoraker, B. and Bock, R. M. (1968) in *Methods in Enzymology,* vol. **12B** (eds. Grossman, L. and Moldave, K), London and New York: Academic Press, p. 236.

Lerman, L. S. (1961) *J. molec. Biol.* **3**, 18.

Lerman, L. S. (1964) *J. cell. comp. Physiol.* **64**, Supplement 1, 1.

Lerman, L. S. and Altman, S. (1968) in *Molecular Associations in Biology* (ed. Pullman, B.), New York and London: Academic Press, p. 271.

Levene, P. A. and Bass, L. W. (1931) *The Nucleic Acids,* New York: Chemical Catalog Co.

Levene, P. A. and Jacobs, W. A. (1910) *Ber.* **43** 3150.

Levin, Ö. (1962) in *Methods in Enzymology,* vol. **5**, (eds. Colowick, S. P. and Kaplan, H. S.), New York and London: Academic Press, p. 27.

Levitt, M. (1969) *Nature,* **224**, 759.

Lima-de-Faria, A., Birnstiel, M. and Joworska, H. (1969) *Genetics Supplement 61*, 1.

Lin, H. J. and Chargaff, E. (1964) *Biochim. biophys. Acta,* **91**, 691.

Lindahl, T., Henley, D. D. and Fresco, J. R. (1965) *J. Amer. chem. Soc.* **87**, 4961.

Ling, V. (1971) *FEBS Letters,* **19**, 50.

Lingens, F. and Schneider-Bernlöhr, H. (1965) *Liebigs Ann. Chem.* **686**, 134.

Linn, S. and Lehman, I. R. (1965) *J. biol Chem.* **240**, 1287. (two consecutive papers).

Lipkin, D., Howard, F. B., Nowotny, D. and Sano, M. (1963) *J. biol. Chem.* **238**, PC2249.

Lipsett, M. N. (1965) *J. biol. Chem.* **240**, 3975.

Lipsett, M. N. and Doctor, B. P. (1967) *J. biol. Chem.* **242**, 4067 (two consecutive papers, 1st one by Lipsett alone).

Lisý, V., Eckstein, F. and Škoda, J. (1968) *Coll. Czech. chem. Comm.* **33**, 2734.

Litt, M. (1969) *Biochemistry,* **8**, 3249.

Littauer, U. Z. and Kornberg, A. (1957) *J. biol. Chem.* **226**, 1077.

Little, J. W. (1967) in *Methods in Enzymology,* vol. **12A** (eds. Grossman, L. and Moldave, K.), New York and London: Academic Press, p. 263.

Littlefield, J. W. and Dunn, D. B. (1958) *Biochem. J.* **70**, 642.

Lodemann, E. and Wacker, A. (1967) *Z. Naturforschg.* **22b**, 42.

Lodish, H. and Robertson, H. D. (1969) *Cold Spring Harb. Symp. quant. Biol.* **34**, 655.

Loening, U. E. (1968a) *J. molec. Biol.* **38**, 355.

Loening, U. E. (1968b) in *Chromatographic and Electrophoretic Techniques,* vol. 2,(*Electrophoresis*; 2nd edition, ed. Smith, I.), London: Heinemann, p. 437.

Lohrmann, R., Söll, D., Hayatsu, H., Ohtsuka, E. and Khorana, H. G. (1966) *J. Amer. chem. Soc.* **88**, 819.

Longuet-Higgins, H. C. and Zimm, B. H. (1960) *J. molec. Biol.* **2**, 1.

Longworth, J. W., Rahn, R. O. and Shulman, R. G. (1966) *J. chem. Phys.* **45**, 2930.

Lord, R. C. and Thomas, C. A., jun. (1967) *Biochim. biophys. Acta,* **142**, 1.

Loring, H. S., Fairley, J. L., Bortner, H. W. and Seagram, H. L. (1951) *J. biol. Chem.* **197**, 809.

Loring, H. S., Luthry, N. G., Bortner, H. W. and Levy, L. W. (1950) *J. Amer. chem. Soc.,* **72** 2811.

Losick, R. Shorenstein, R. G. and Sonnenshein, A. L. (1970) *Nature,* **227**, 910.

Lovett, J. S. and Leaver, C. J. (1969) *Biochem. biophys. Acta,* **195**, 319.

Loveless, A. (1969) *Nature,* **223**, 206.

Löwdin, P. O. (1965) *Adv. quantum Chem.* **2**, 213.

Lowry, C. V. and Dhalberg, J. E. (1971) *Nature New Biology,* **232**, 54.

Lowry, O. H., Rosebrough, N. J., Farr, N. L. and Randall, R. J. (1951) *J. biol. Chem.* **193**, 265.

Lozeron, H. A. and Szybalski, W. (1967) *J. molec. Biol.* **30**, 277.

Lubos, B. and Wilczok, T. (1970) *Biochim. biophys. Acta,* **224**, 1.

Ludlam, D. B. (1965) *Biochim. biophys. Acta,* **95**, 674.

Luzzati, V. and Nicolaieff, A. (1963) *J. molec. Biol.* **7**, 142.

McCarthy, J. R., jun., Robins, M. J., Townsend, L. B. and Robins, R. K. (1966) *J. Amer. chem. Soc.* **88**, 1549.

McConnell, J. F., Sharma, B. D. and Marsh, R. E. (1964) *Nature,* **203,** 399.
McCoy, T. A. and Carter, E. A. (1968) *J. Chromatog.* **37,** 458.
MacGee, J. (1966) *Analyt. Biochem.* **14,** 305.
McGrath, R. A. and Williams, R. W. (1966) *Nature,* **212,** 534.
MacHattie, L. A., Ritchie, D. A., Thomas, C. A., jun. and Richardson, C. C. (1967) *J. molec. Biol.* **23,** 355.
Macintyre, W. M. (1964) *Biophys. J.* **4,** 495.
Madison, J. T., Everett, G. A. and Kung, H. 1966) *Science,* **153,** 531.
Madison, J. T. and Holley, R. W. (1965) *Biochem. biophys. Res. Comm.* **18,** 153.
Madison, J. T., Holley, R. W., Poucher, J. S. and Connett, P. H. (1967) *Biochim. biophys. Acta,* **145,** 825.
Maestre, M. F. and Tinoco, I., jun. (1967) *J. molec. Biol.* **23,** 323.
Magee, P. N. and Barnes, J. M. (1956) *Brit. J. Cancer,* **10,** 114.
Malbon, R. M. and Parish, J. H. (1971) *Biochim. biophys. Acta,* **246,** 542.
Malt, R. A. (1966) *Biochim. biophys. Acta,* **120,** 461.
Mandel, L. R. and Borek, E. (1963) *Biochemistry,* **2,** 555.
Mandelkern, L. and Flory, P. J. (1952) *J. chem. Phys.* **20,** 212.
Mandell, J. D. and Hershey, A. D. (1960) *Analyt. Biochem.* **1,** 66.
Mangiarotti, G. and Schlessinger, D. (1966) *J. molec. Biol.* **20,** 123.
Mans, R. J. and Novelli, G. D. (1961) *Analyt. Biochem.* **94,** 48.
Marcot-Queiroz, J. and Monier, R. (1965) *J. molec. Biol.* **14,** 490.
Marcus, L., Bretthauer, R. K., Bock, R. M. and Halvorsen, H. O. (1963) *Proc. Nat. Acad. Sci. U.S.* **50,** 782.
Markham, R. and Smith, J. D. (1952) *Biochem. J.* **52,** 552.
Markham, R., Matthews, R. E. F. and Smith, J. D. (1954) *Nature,* **173,** 537.
Marmur, J. (1961) *J. molec. Biol.* **3,** 208.
Marmur, J. (1963) in *Methods in Enzymology,* vol. **6** (eds. Colowick, S. P. and Kaplan, H. S.), New York and London: Academic Press, p. 729.
Marmur, J. and Doty, P. (1962) *J. molec. Biol.* **5,** 109.
Marshak, A. and Vogel, H. J. (1951) *J. biol. Chem.* **189,** 597.
Martelo, O. J., Woo, S. L. C., Reimann, E. M. and Davie, E. W. (1970) *Biochemistry,* **9,** 4807.
Martin, F. H., Uhlenbeck, O. C. and Doty, P. (1971) *J. molec. Biol.* **57,** 201.
Marvin, D. A., Spencer, M., Wilkins, M. H. F. and Hamilton, L. D. (1961) *J. molec. Biol.* **3,** 547.
Mason, S. F. (1954) *J. chem. Soc.* **2071.**
Massoulié, J., Michelson, A. M. and Pochon, F. (1966) *Biochim. biophys. Acta,* **114,** 14.
Matsuyama, A., Tagashira, Y. and Nagata, C. (1971) *Biochim. biophys. Acta,* **240,** 184.
Matthews, F. S. and Rich, A. (1964) *J. molec. Biol.* **8,** 89.
Matthaei, J. H. and Nirenberg, M. W. (1961) *Proc. Nat. Acad. Sci. U.S.* **47,** 1588.
Mattoccia, E. and Comings, D. E. (1971) *Nature New Biology,* **229,** 175.
Maurshige, K. and Bonner, J. (1966) *J. molec. Biol.* **15,** 160.
Maxwell, I. H. (1969) *J. gen. Microbiol.* **58,** p. v.
Melli, M., Whitfield, C., Rao, K. V., Richardson, M. and Bishop, J. O. (1971) *Nature New Biology,* **231,** 8.
Merrifield, R. B. (1965) *Science,* **150,** 178.
Merrill, C. R., Geier, M. R. and Petricciani, J. C. (1971) *Nature,* **233,** 398.
Meselson, M. and Stahl, F. (1958) *Proc. Nat. Acad. Sci. U.S.* **43,** 581.
Michelson, A. M. (1959). *J. chem. Soc.* **3665.**
Michelson, A. M., Dondon, J. and Grunberg-Manago, M. (1962) *Biochim. biophys. Acta,* **55,** 592.
Michelson, A. M., Massoulié, J. and Guschlbauer, W. (1967) *Prog. Nuc. Acid Res. molec. Biol.* **6,** 83.
Michelson, A. M., Szabo, L. and Todd, A. R. (1956) *J. chem. Soc.* 546.
Michelson, A. M., Ulbricht, T. L. V., Emerson, T. R. and Swan, R. J. (1966) *Nature,* **209,** 878.
Miles, H. T. (1956) *Biochim. biophys. Acta,* **22,** 247.
Millar, D. B. and Mackenzie, M. (1970) *Biochim. biophys. Acta,* **204,** 82.
Miller, E. C., Miller, J. A. and Hartman, H. A. (1961) *Cancer Res.* **21,** 815.
Miller, O. L., jun. and Beatty, B. R. (1969) *Science,* **164,** 985.

Miller, O. L., jun., Beatty, B. R., Hamkalo, B. A. and Thomas, C. A., jun. (1970) *Cold Spring Harb. Symp. quant. Biol.* **35**, 505.
Miller, R. A. and Kirkpatrick, J. W. (1969) *Analyt. Biochem.* **27**, 306.
Millward, S. and Nonoyama, M. (1970) *Cold Spring Harb. Symp. quant. Biol.* **35**, 773.
Milman, G., Chamberlin, M. and Langridge, R. (1967) *Proc. Nat. Acad. Sci. U.S.* **57**, 1804.
Miskin, R., Zamir, A. and Elson, D. (1970) *J. molec. Biol.* **54**, 355.
Mitsui, Y., Langridge, R., Grant, R. C., Kodama, M., Wells, R. D. and Cantor, C. R. (1970) *Nature,* **228**, 1166.
Mizushima, S. and Nomura, M. (1970) *Nature,* **226**, 1214.
Mizutani, S., Temin, H. M., Fodana, M. and Wells, R. T. (1971) *Nature New Biology,* **230**, 232.
Moffatt, J. G. (1967) in *Methods in Enzymology,* vol. **12A** (eds. Grossman, L. and Moldave, K.), New York and London: Academic Press, p. 182.
Moffatt, J. G. and Khorana, H. G. (1961) *J. Amer. chem. Soc.* **83**, 649.
Moldave, K. and Grossman, L. (1971) editors of *Methods in Enzymology,* vol. **20**, New York and London: Academic Press.
Molinaro, M., Sheiner, L. B., Neelon, S. A. and Cantoni, G. L. (1968) *J. biol. Chem.* **243**, 1277.
Möller, W. and Boedtker, H. (1962) in *Acides Ribonucléiques et Polyphosphates, Structure, Synthèse, et Fonction,* Paris: CNRS, p. 99.
Monk, M., Peacey, M. and Gross, J. D. (1971) *J. molec. Biol.* **58**, 623.
Monro, R. E., Cerná, J. and Marcker, K. A. (1968) *Proc. Nat. Acad. Sci. U.S.* **61**, 1042.
Monro, R. E. and Marcker, K. A. (1967) *J. molec. Biol.* **25**, 347.
Montgomery, J. A. and Hewson, K. (1960) *J. Amer. chem. Soc.* **82**, 43.
Morrison, J. F. (1968) *Analyt. Biochem.* **24**, 106.
Moses, R. E. and Singer, M. F. (1970) *J. biol. Chem.* **245**, 2414.
Moshig, G. (1970) *Adv. Genetics,* **15**, 1.
Muench, K. H. and Berg, P. (1966) *Biochemistry,* **5**, 970.
Mushynski, W. and Spencer, J. H. (1970) *J. molec. Biol.* **52**, 91.

Nakai, T. and Howatson, A. F. (1968) *Virology,* **35**, 268.
Nakanishi, K., Furutachi, N., Funamiza, M., Grunberg, D. and Weinstein, I. B. (1970) *J. Amer. chem. Soc.* **92**, 7617.
Nash, H. A. and Bradley, D. F. (1966) *J. chem. Phys.* **45**, 1380.
Nashimoto, H. and Nomura, M. (1970) *Proc. Nat. Acad. Sci. U.S.* **67**, 1440.
Nathans, D. (1965) *J. molec. Biol.* **13**, 521.
Némethy, G. and Scheraga, H. A. (1962) *J. phys. Chem.* **66**, 1773.
Neville, D. M. and Davies, D. R. (1966) *J. molec. Biol.* **17**, 57.
Ninio, J., Favre, A. and Yaniv, M. (1969) *Nature,* **223**, 1333.
Nirenberg, M., Caskey, T., Marshall, R., Brimacombe, R., Kellog, D., Doctor, B., Hatfield, D., Levin, J., Rottman, F., Pestka, S., Wilcox, M. and Anderson, F. (1966) *Cold Spring Harb. Symp. quant. Biol.* **31**, 11.
Noll, H. (1967) *Nature,* **215**, 360.
Noll, H. (1969) *Analyt, Biochem.* **27**, 130.
Noll, H. and Stutz, E. (1968) in *Methods in Enzymology,* vol. **12B** (eds. Grossman, L. and Moldave, K.), New York and London: Academic Press, p. 140.
Nomura, M. and Erdmann, V. A. (1970) *Nature,* **228**, 744.

Oberg, B. (1970) *Biochim. Biophys. Acta,* **204**, 430.
O'Brien, E. J. (1967) *Acta Cryst.* **23**, 92.
O'Brien, E. J. and MacEwan, A. (1970) *J. molec. Biol.* **48**, 243.
O'Brien, R. L., Olenick, J. G. and Halm, F. E. (1966) *Proc. Nat. Acad. Sci. U.S.* **55**, 1511.
Ochoa, M., jun. and Weinstein, I. B. (1964). *J. biol. Chem.* **239**, 3834.
Odinstova, M. S., Golubeva, E. U. and Sissiakin, N. M. (1964) *Nature,* **204**, 1090.
Ogawa, H., Shimada, K. and Tomizawa, J. (1968) *Molec. gen. Genet.* **101**, 227.
Ogawa, Y., Quagliarotti, G., Jordan, J., Taylor, C. W., Starbuck, W. C. and Busch, H. (1969) *J. biol. Chem.* **224**, 4387.

Ohtsuka, E., Moon, M. W. and Khorana, H. G. (1965) *J. Amer. chem. Soc.* **87**, 2956.

Oishi, M. (1968) *Proc. Nat. Acad. Sci. U.S.* **60**, 329, 691.

Okazaki, R., Okazaki, T., Sakube, K., Sugimoto, K. and Sugino, A. (1968) *Proc. Nat. Acad. Sci. U.S.* **59**, 598.

Olivera, B. M. and Lehman, I. R. (1967) *Proc. Nat. Acad. Sci. U.S.* **57**, 1426, 1700.

Ono, J., Wilson, R. G. and Grossman, L. (1965) *J. molec. Biol.* **11**, 600.

Oppenheim, J., Scheinbuchs, J., Biava, C. and Marcus, L. (1968) *Biochim. biophys. Acta,* **161**, 386.

Otaka, E., Mitsui, H. and Osawa, S. (1962) *Proc. Nat. Acad. Sci. U.S.* **48**, 425.

Ott, D. G., Keir, V. N., Hansbury, E. and Hayes, F. N. (1967) *Analyt. Biochem.* **21** 469.

Palmiter, R. D., Christensen, A. K. and Schimke, R. T. (1970) *J. biol. Chem.* **245**, 833.

Paoletti, J. and le Pecq, J.-B. (1971) *J. molec. Biol.* **59**, 43.

Pardon, J. F., Wilkins, M. H. F. and Richards, B. M. (1967) *Nature,* **215**, 508.

Parish, J. H. (1968) *Biochim. biophys. Acta,* **169**, 14.

Parish, J. H. (1969) *Biochim. biophys. Acta,* **182**, 454.

Parish, J. H. and Hastings, J. R. B. (1966) *Biochim. biophys. Acta,* **123**, 202.

Parish, J. H., Khairul Bashar, S. A. M., Brown, N. L. and Brown, M. (1971) *Biochem. J.* **125**, 643.

Parish, J. H. and Kirby, K. S. (1966) *Biochim. biophys. Acta,* **129**, 554.

Parish, J. H. and Kirby, K. S. (1967) *Biochim. biophys. Acta,* **142**, 273.

Parish, J. H. and Whiting, M. C. (1964) *J. chem. Soc.* 4713.

Payne, P. I. and Loening, U. E. (1970) *Biochim. biophys. Acta,* **224**, 128.

Peacock, A. C. and Dingman, W. C. (1968) *Biochemistry,* **7**, 668.

Pearson, R. L. and Kelmers, A. D. (1966) *J. biol. Chem.* **241**, 767.

Pearson, R. L., Weiss, J. F. and Kelmers, A. D. (1971) *Biochim. biophys. Acta,* **228**, 770.

Pelling, C. (1970) *Cold Spring Harb. Symp. quant. Biol.* **35**, 251.

Pene, J. J., Knight, E. and Darnell, S. E. (1968) *J. molec. Biol.* **33**, 609.

Penman, S., Scherrer, K., Becker, Y. and Darnell, J. E. (1963) *Proc. Nat. Acad. Sci. U.S.* **49**, 654.

Penman, S., Vesco, C., Weinberg, R. and Zylker, E. (1969) *Cold Spring Harb. Symp. quant. Biol.* **34**, 535.

Peterson, G. G. and Burton, K. (1964) *Biochem. J.* **92**, 666.

Pettijohn, D. E. and Hanawalt, P. C. (1963) *Biochim. biophys. Acta,* **72**, 127.

Pettijohn, D. E. and Hanawalt, P. C. (1964) *J. molec. Biol.* **9**, 395.

Pfitzner, K. E. and Moffatt, J. G. (1964) *Biochem. biophys. Res. Comm.* **17**, 146.

Phillips, J. H. and Brown, D. M. (1967) *Prog. Nuc. Acid. Res. molec. Biol.* **7**, 349.

Phillips, J. H., Brown, D. M., Adman, R. and Grossman, L. (1965) *J. molec. Biol.* **12**, 816.

Phillips, J. H., Brown, D. M. and Grossman, L. (1966) *J. molec. Biol.* **21**, 405.

Piggott, G. H. and Midgley, J. E. M. (1968) *Biochem. J.* **110**, 251.

Pilz, I., Kratky, O., Cramer, F., von der Haar, F. and Schlimme, E. (1970) *Europ. J. Biochem.* **15**, 401.

Pinck, L., Hirth, L. and Bernardi, G. (1968) *Biochem. biophys. Res. Comm.* **38**, 481.

Pinck, M., Yot, P., Chapeville, F. and Duranton, H. M. (1970) *Nature,* **226**, 254.

Plescia, O. J. (1968) in *Methods in Enzymology,* vol. 12B (eds. Grossman, L. and Moldave, K.), New York and London: Academic Press, p. 893.

Pochon, F., Michelson, A. M., Grunberg-Manago, M., Cohn, W. E. and Dondon, L. (1964) *Biochim. biophys. Acta,* **80**, 141.

Pochon, F., Pascal, Y., Pitha, P. and Michelson, A. M. (1970) *Biochim. Biophys. Acta,* **213**, 273.

Pochon, F. and Michelson, A. M. (1965) *Proc. Nat. Acad. Sci. U.S.* **53**, 1425.

Pochon, F. and Michelson, A. M. (1967) *Biochim. biophys. Acta,* **145**, 321.

Pochon, F. and Michelson, A. M. (1969) *Biochim. biophys. Acta,* **182**, 17.

Poland, D. and Scheraga, H. A. (1970) *Theory of Helix-Coil Transitions in Biopolymers,* New York and London: Academic Press, p. 188.

Pollard, C. J. (1964) *Biochem. biophys. Res. Comm.* **17**, 171.

Preiss, J., Dieckmann, M. and Berg, P. (1961) *J. biol. Chem.* **236**, 1748.

Prestayko, A. W., Tonato, M. and Busch, H. (1970) *J. molec. Biol.* **47**, 505.
Printz, M. P. and von Hipple, P. H. (1965) *Proc. Nat. Acad. Sci. U.S.* **53**, 363.
Pritchard, R. H. (1955) *Heredity,* **9**, 343.
Pritchard, R. H. and Lark, K. G. (1964) *J. molec. Biol.* **9**, 288.
Pruden, B., Snipes, W. and Gordy, W. (1965) *Proc. Nat. Acad. Sci. U.S.* **53**, 917.
Pullman, B., Claverie, P. and Caillet, J. (1966) *Proc. Nat. Acad. Sci. U.S.* **55**, 904.
Pullman, B. and Pullman, A. (1968) *Adv. quant. Chem.* **4**, 267.
Pullman, B. and Pullman, A. (1969) *Prog. Nuc. Acid. Res. molec. Biol.* **9**, 327.
Purdom, I., Bishop, J. O. and Birnstiel, M. L. (1970) *Nature,* **227**, 239.

Radding, C. M. and Shreffer, D. C. (1966) *J. molec. Biol.* **18**, 255.
Raina, A., Jansen, M. and Cohen, S. S. (1967) *J. Bact.* **96**, 1684.
RajBhandary, U. L., Chang, S. H., Stuart, A., Faulkner, R. D., Hoskinson, R. M. and Khorana, H. G. (1967) *Proc. Nat. Acad. Sci. U.S.* **57**, 751.
RajBhandary, U., Young, R. and Khorana, H. G. (1964) *J. biol. Chem.* **239**, 3875.
Ralph, R. K. and Bellamy, A. R. (1964) *Biochim. biophys. Acta,* **87**, 9.
Ralph, R. K., Connors, W. J., Schaller, H. and Khorana, H. G. (1963) *J. Amer. chem. Soc.,* **85**, 1983.
Ralph, R. K. and Khorana, H. G. (1961) *J. Amer. chem. Soc.* **83**, 2926.
Randerath, K. and Randerath, E. (1967) in *Methods in Enzymology,* vol. 12A (eds. Grossman, L. and Moldave, K.), New York and London: Academic Press, p. 323.
Randolph, M. L. and Setlow J. K. (1971) *J. Bact.* **106**, 221.
Rau, J. and Lingens, F. (1967) *Naturwissenschaften,* **54**, 517.
Razzell, W. E. and Khorana, H. G. (1961) *J. biol. Chem.* **236**, 1144.
Rechler, M. M. and Martin, R. G. (1970) *Nature,* **226**, 908.
Reddi, K. K. (1959) *Biochim. biophys. Acta,* **36**, 132.
Reese, E. T. and Maguire, A. H. (1968) *J. Bact.* **96**, 1696.
Reeves, W. J., jun., Seid, A. S. and Greenberg, D. M. (1969) *Analyt. Biochem.* **30**, 474.
Regan, J. D. and Cook, J. S. (1967) *Proc. Nat. Acad. Sci. U.S.* **58**, 2274.
Reiter, H. and Strauss, B. (1965) *J. molec. Biol.* **14**, 179.
Rekosh, D., Lodish, H. F. and Baltimore, D. (1969) *Cold Spring Harb. Symp. quant. Biol.* **34**, 747.
Révet, B. M. J., Schmir, M. and Vinograd, J. (1971) *Nature New Biology,* **229**, 10.
Rho, J. H. and Chipchase, M. I. (1962) *J. Cell Biol.* **14**, 183.
Rice, J. M. (1964) *J. Amer. chem. Soc.* **86**, 1444.
Rich, A. (1958) *Biochim. biophys. Acta,* **29**, 502.
Rich, A., Davies, D. R., Crick, F. H. C. and Watson, J. D. (1961) *J. molec. Biol.* **3**, 71.
Richardson, C. C., Lehman, I. R. and Kornberg, A. (1964) *J. biol. Chem.* **239**, 251.
Richardson, C. C., Schildkraut, C. L., Aposhian, H. V. and Kornberg, A. (1964) *J. biol. Chem.* **239**, 222.
Richardson, C. C. (1965) *Proc. Nat. Acad. Sci. U.S.* **54**, 158.
Rifkin, M. R., Wood, D. D. and Luck, D. J. (1967) *Proc. Nat. Acad. Sci. U.S.* **58**, 1025.
Riggs, A. D., Bourgeois, S. and Cohn, M. (1970) *J. molec. Biol.* **53**, 401.
Riley, P. A. (1970) *Nature,* **228**, 522.
Ritchie, D. A., Thomas, C. A., jun., MacHattie, C. A. and Wensink, P. C. (1967) *J. molec. Biol.* **23**, 365.
Ritossa, F. M. and Spiegelman, S. (1965) *Proc. Nat. Acad. Sci. U.S.* **53**, 737.
Riva, S., Polsinelli, M. and Falaschi, A. (1968) *J. molec. Biol.* **35**, 347.
Ro, T. S. and Busch, H. (1964) *Cancer Res.* **24**, 1630.
Roberts, J. J., Brent, T. P. and Crathorn, A. R. (1968) in *Interaction of Drugs and Subcellular Components in Animal Cells* (ed. Campbell, P. N.), London: Churchill, p. 5.
Roberts, J. J., Pascoe, J. M., Plant, J. E., Sturrock, J. E. and Crathorn, A. R. (1971) *Chem. Biol. Interacts,* **3**, 29.
Roberts, J. J., Pascoe, J. M., Smith, B. A. and Crathorn, A. R. (1971) *Chem. Biol. Interacts.* **3**, 49.
Roberts, J. J. and Warwick, G. P. (1966a) *Internat. J. Cancer,* **1**, 107; (1966b) *ibid.* p. 179.
Roberts, J. W. and Steitz, J. A. (1967) *Proc. Nat. Acad. Sci. U.S.* **58**, 1416.
Robins, M. J., McCarthy, J. R., jun. and Robins, R. K. (1966) *Biochemistry,* **5**, 224.
Robinson, B. and Zimmerman, T. P. (1971) *J. biol. Chem.* **240**, 110.

Rodighiero, G., Musajo, L., Dall'acqua, F., Marciani, S., Caporale, G. and Ciavatta, I. (1970) *Biochim. biophys. Acta,* **217**, 40.

Roeder, R. G. and Rutter, W. J. (1969) *Nature,* **224**, 235.

Rogers, G. T., Ulbricht, T. L. V. and Szer, W. (1961)   *Biochem. biophys. Res. Comm.* **27**, 372.

Rohrer, D. and Sundaralingham, M. (1968) *Chem. Comm.* 746.

Rossett, T., Smith, J. G., jun., Matsuo, I., Bailey, P. I., Smith, D. B. and Surakiat, S. (1970) *J. Chromatog.* **49**, 308.

Roth, J. S. (1956) *Biochim. biophys. Acta,* **21**, 34.

Rubinstein, I., Thomas, C. A., jun. and Hershey, A. D. (1961) *Proc. Nat. Acad. Sci.* **43**, 581.

Rudner, R., Karkas, J. D. and Chargaff, E. (1968) *Proc. Nat. Acad. Sci. U.S.* **60**, 630 (two consecutive papers).

Rudner, R., Karkas, J. D. and Chargaff, E. (1969) *Proc. Nat. Acad. Sci. U.S.* **63**, 152.

Rudner, R., Shapiro, H. S. and Chargaff, E. (1966) *Biochim. biophys. Acta,* **129**, 85.

Rupprecht, A. (1970) *Biochim. biophys. Acta,* **199**, 277.

Russell, G. J., Follett, E. A. C., Subak-Sharpe, J. H. and Harrison, B. D. (1971) *J. gen. Virol.* **11**, 129.

Russo, V. E. A., Stahl, M. M. and Stahl, F. W. (1970) *Proc. Nat. Acad. Sci. U.S.* **65**, 363.

Rutenberg, A. M., Persky, L. and Friedman, O. M. (1952) *Cancer,* **5**, 354.

Ryan, J. L. and Morowitz, H. J. (1969) *Proc. Nat. Acad. Sci. U.S.* **63**, 1282.

Sadowski, P. D. and Hurwitz, J. (1969) *J. biol. Chem.* **244**, 6182 (two consecutive papers).

Sakakibara, Y., Tanooka, H. and Terano, H. (1970) *Biochim. biophys. Acta,* **199**, 549.

Salas, M., Smith, M. A., Stanley, W. M., jun., Wahba, A. J. and Ochoa, S. (1965) *J. biol. Chem.* **54**, 1167.

Salser, W., Gesteland, R. G. and Ricard, B. (1969) *Cold Spring Harb. Symp. quant. Biol.* **34**, 771.

Samejima, T., Hashizume, H., Imahori, K., Fujii, I. and Miura, K. (1968) *J. molec. Biol.* **34**, 39.

Samejima, T. and Yang, J. T. (1965) *J. biol. Chem.* **240**, 2094.

Sanchez, R. A., Ferris, J. R. and Orgel, L. E. (1967) *J. molec. Biol.* **30**, 223.

Sanger, F., and Brownlee, G. G. (1967) in *Methods in Enzymology,* vol. **12A** (eds. Grossman, L. and Moldave, K.), New York and London: Academic Press, p. 361.

Sanger, F., Brownlee, G. G. and Barrell, B. G. (1965) *J. molec. Biol.* **13**, 375.

Sarker, P. K. and Yang, J. T. (1963) *Biochem. biophys. Res. Comm.* **20**, 346.

Schachman, H. K., Adler, J., Radding, C. M., Lehman, I. R. and Kornberg, A. (1960) *J. biol. Chem.* **235**, 3242.

Scheid, B., Srininivasan, P. R. and Borek, E. (1968) *Biochemistry,* **7**, 280.

Scheit, K.-H. and Cramer, F. (1964) *Tetrahedron Letters,* 2764.

Scheit, K.-H. and Gaertner, E. (1969) *Biochim. biophys. Acta,* **182**, 1 (two consecutive papers).

Scherrer, K., and Marcaud, L. (1968) *J. Cell Physiol.* **72**, 181.

Scherrer, K., Marcaud, L., Zajdela, F., London, I. and Gros, F. (1966) *Proc. Nat. Acad. Sci. U.S.* **56**, 1571.

Schildkraut, C. L., Marmur, J. and Doty, P. (1962) *J. molec. Biol.* **4**, 430.

Schmid, C. W. and Hearst, J. E. (1969) *J. molec. Biol.* **44**, 143.

Schneider, P. W., Brintzinger, H. and Erlenmeyer, H. (1968) *Helv. chim. Acta,* **47**, 992 (in German).

Schulman, L. H. and Chambers, R. W. (1968) *Proc. Nat. Acad. Sci. U.S.* **61**, 308.

Schumaker, V. N. and Schachman, H. K. (1957) *Biochim. biophys. Acta,* **23**, 628.

Schuster, H. and Wilhelm, R. C. (1963) *Biochim. biophys. Acta,* **68**, 554.

Schweitzer, M. P., Chan, S. I., Helmkamp. G. K. and Ts'o, P. O. P. (1964) *J. Amer. chem. Soc.* **86**, 696.

Scolnick, E. M., Aaronson, S. A. and Todaro, G. J. (1970) *Proc. Nat. Acad. Sci. U.S.* **67**, 660.

Scohnick, E. M., Aaronson, S. A., Todaro, G. J. and Parks, W. P. (1971) *Nature,* **229**, 316.

Sedat, J., Lyon, A. and Sinsheimer, R. L. (1969) *J. molec. Biol.* **44**, 418.

Seeman, N. C., Sussman, J. L., Berman, H. M. and Kim, S. H. (1971) *Nature New Biology,* **233**, 90.

Sela, M., Ungar-Waron, H. and Shechter, Y. (1964) *Proc. Nat. Acad. Sci. U.S.* **52**, 285.

Seno, T., Kobayishi, M. and Nishimura, S. (1968) *Biochim. biophys. Acta,* **169**, 80.

Setlow, R. B. (1966) *Science,* **153**, 379.

Setlow, R. B. (1968) *Prog. Nuc. Acid. Res. molec. Biol.* **8**, 257.

Setlow, J. K. and Bollum, F. (1968) *Biochim. biophys. Acta,* **157**, 233.

Setlow, R. B. and Carrier, W. L. (1964) *Proc. Nat. Acad. Sci. U.S.* **51**, 226.

Setlow, R. B. and Carrier, W. L. (1966) *J. molec. Biol.* **17**, 237.

Setlow, R. B. and Carrier, W. L. (1967) *Nature,* **213**, 906.

Setlow, R. B., Carrier, W. L. and Setlow, J. K. (1969) *Biophys. J.* **9**, A-57.

Setlow, R. B., Carrier, W. L. and Bollum, F. J. (1965) *Proc. Nat. Acad. Sci. U.S.* **53**, 1111.

Setlow, R. B. and Setlow, J. K. (1962) *Proc. Nat. Acad. Sci. U.S.* **48**, 1250.

Setlow, R. B. and Setlow, J. K. (1965) *Photochem. Photobiol.* **4**, 939.

Shankar Narayan, K. and Birnstiel, M. L. (1969) *Biochim. biophys. Acta,* **190**, 470.

Shapiro, J., MacHattie, L., Eron, L., Ihler, G., Ippen, K. and Beckwith, J. (1969) *Nature,* **224**, 768.

Shapiro, H. S. and Chargaff, E. (1957) *Biochim. biophys. Acta,* **26**, 206.

Shapiro, H. S. and Chargaff, E. (1963) *Biochim. biophys. Acta,* **68**, 9.

Shapiro, R., Cohen, B. I. and Clagett, D. C. (1970) *J. biol. Chem.* **245**, 2633.

Shapiro, R. (1964) *J. Amer. Chem. Soc.* **86**, 2948.

Shapiro, R. and Hachmann, J. (1966) *Biochemistry,* **5**, 2799.

Shaw, E. (1950) *J. biol. Chem.* **185**, 439.

Shaw, E. (1958) *J. Amer. chem. Soc.* **80**, 3899.

Shen, T. Y., Lewis, H. M. and Ruyle, W. V. (1965) *J. org. Chem.* **30**, 835.

Sherman, M. I. and Simpson, M. V. (1969) *Proc. Nat. Acad. Sci. U.S.* **64**, 1388.

Shields, H. and Gordy, W. (1959) *Proc. Nat. Acad. Sci. U.S.* **45**, 269.

Shimanouchi, T., Tsuboi, M. and Kyogoku, Y. (1964) *Adv. chem. Phys.* **7**, 435.

Shin, Y. A. and Eichorn, G. C. (1968) *Biochemistry,* **7** 1026.

Short, E. C., jun. and Koerner, J. F. (1969) *J. biol. Chem.* **244**, 1487.

Signer, E. R. (1968) *Ann. Rev. Microbiol.* **22**, 451.

Simpson, R. T. (1970) *Biochemistry,* **9**, 4814.

Simundza, G., Sakore, T. D. and Sobell, H. M. (1970) *J. molec. Biol.* **48**, 263.

Sinanoğlu, O. (1968) in *Molecular Associations in Biology* (ed. Pullman, B.), New York and London: Academic Press, p. 427.

Singer, H. and Fraenkel-Conrat, H. (1969) *Prog. Nuc. Acid Res. molec. Biol.* **9**, 1.

Skehl, J. J. (1971) *J. gen. Virol.* **11**, 103.

Skinner, D. M. (1967) *Proc. Nat. Acad. Sci. U.S.* **58**, 103.

Slater, D. W. and Spiegelman, S. (1966) *Biophys. J.* **6**, 385.

Smith, D. W., Schaller, H. E. and Bonhoeffer, F. J. (1970) *Nature,* **226**, 711.

Smith, I. editor of *Chromatographic and Electrophoretic Techniques*; vol. 1 *Chromatography* (3rd edition, 1969); vol. 2 *Electrophoresis* (2nd edition, 1968), London: Heinemann.

Smith, J. D. (1967) in *Methods in Enzymology,* vol. **12A** (eds. Grossman, L. and Moldave, K.), New York and London: Acacemic Press, p. 350.

Smith, J. D. and Markham, R. (1950) *Biochem. J.* **46**, 509.

Smith, K. D., Church, R. B. and McCarthy, B. J. (1968) *Biochemistry,* **8**, 4271.

Smith, M., Moffatt, J. G. and Khorana, H. G. (1958) *J. Amer. chem. Soc.* **80**, 6204.

Smith, T. A. (1971) *Annals N.Y. Acad. Sci.* **171**, 988.

Smrt, J. and Šorm, F. (1964) *Coll. Czech. chem. Comm.* **29**, 2971.

Sobell, H. M., Jain, S. C., Sakore, T. D. and Nordman, C. E. (1971) *Nature New Biology,* **231**, 200.

Sobell, H. M., Tomita, K. and Rich, A. (1963) *Proc. Nat. Acad. Sci. U.S.* **49**, 885.

Sogin, M., Pace, B., Pace, N. R. and Woese, C. R. (1971) *Nature New Biology,* **232**, 48.

Söll, D. and Khorana, H. G. (1965) *J. Amer. chem. Soc.* **87**, 360.

Solymosy, F., Fedorcsák, I., Gulyás, A., Farkas, G. L. and Ehrenberg, L. (1968) *Europ. J. Biochem.* **5**, 520.

Somes, C. E., Cole, A. and Hsu, T. C. (1963)   *Expt. Cell Res. Supplement 9* 200.

Southern, E. M. (1970) *Nature,* **227**, 794.

Southern, E. M. and Mitchell, A. R. (1971) *Biochem. J.* **123**, 613.

Spadari, S., Sgaramella, V., Mazza, G. and Falaschi, A. (1971) *Europ. J. Biochem.* **19**, 294.

Spencer, J. H. and Chargaff, E. (1963) *Biochim. biophys. Acta,* **68**, 9.

Spiegelman, S. (1961) *Cold Spring Harb. Symp. quant. Biol.* **26**, 75.

Spiegelman, S., Burney, A., Das, M. R., Keydar, J., Schlom, J., Travnicek, M. and Watson, K. (1970) *Nature,* **227**, 563.

Spirin, A. S. (1969) *Cold Spring Harb. Symp. quant. Biol.* **35**, 197.

Spizizen, J., Reilly, B. E. and Evans, A. H. (1966) *Ann. Rev. Microbiol.* **20**, 371.

Spragg, S. P. and Rankin, C. T., jun. (1967) *Biochim. biophys. Acta,* **141**, 164.

Staehelin, M., Rogg, H., Baguley, B. C., Ginsberg, T. and Wehrli, W. (1968) *Nature,* **219**, 1363.

Stahl, J., Lawford, G. R., Williams, B. and Campbell, P. N. (1968) *Biochem. J.* **109**, 155.

Stanier, R. Y., Douderoff, M. and Adelberg, E. A. (1968)   *General Microbiology* (2nd edition), London: Macmillan, p. 169.

Stanley, W. M., jun. and Bock, R. M. (1965) *Biochem. J.* **4**, 1302.

Steele, W. J. and Busch, H. (1963) *Cancer Res.* **23**, 1153.

Steiner, R. F. (1960) *J. biol. Chem.* **235**, 2946.

Steiner, R. F. and Beers, R. F. (1968) *Polynucleotides: Natural and Synthetic Nucleic Acids,* Amsterdam: Elsevier.

Steinhart, W. L. and Herriott, R. M. (1968) *J. Bact.* **96**, 1718.

Steinschneider, A. and Fraenkel-Conrat, H. (1966) *Biochemistry,* **5**, 2729.

Steitz, J. A. (1969) *Nature,* **224**, 957.

Stern, R., Zutra, L. E. and Littauer, U. Z. (1969) *Biochemistry,* **8**, 313.

Steuart, C. D., Anand, S. K. and Bessman, M. J. (1968) *J. biol. Chem.* **243**, 5319.

Stevens, L. (1967) *Biochem. J.* **103**, 811.

Stevens, L. (1969) *Biochem. J.* **113**, 117.

Stevens, L. (1970) *Biol. Rev.* **45**, 1.

Stevens, M. A., Magrath, D. I., Smith, H. W. and Brown, G. B. (1958) *J. Amer. chem. Soc,* **80**, 958.

Stevens, M. A., Smith, H. W. and Brown, G. B. (1958) *J. Amer. chem. Soc.* **81**, 1734.

Stewart, B., Roberts, R. J. and Strominger, J. L. (1971) *Nature,* **230**, 36.

Stone, A. B. (1970) *Prog. Biophys. molec. Biol.* **20**, 133.

Stone, A. B. and Burton, K. (1962) *Biochem. J.* **85**, 600.

Stuart, A. and Khorana, H. G. (1964) *J. biol. Chem.* **239**, 3885.

Studier, F. W. (1965)   *J. molec. Biol.* **11**, 373.

Subak-Sharpe, H., Bürk, R. R., Crawford, L. V., Morrison, J. M., Hay, J. and Keir, H. M. (1966) *Cold Spring Harb. Symp. quant. Biol.* **31**, 737.

Sueoka, N. (1961) *J. molec. Biol.* **3**, 31.

Summers, W. C. and Siegel, R. B. (1969) *Nature,* **223**, 1111.

Sundaralingam, M. and Jensen, L. H. (1965) *J. molec. Biol.* **13**, 930.

Sutherland, B. M., Carrier, W. L. and Setlow, R. B. (1967) *Science,* **158**, 1699.

Suwalsky, M., Traub, W., Shmueli, U. and Subirana, J. A. (1969) *J. molec. Biol.* **42**, 363.

Suzuki, J. and Haselkorn, R. (1968) *J. molec. Biol.* **36**, 47.

Suzuki, K., Moriguchi, E. and Hori, Z. (1966) *Nature,* **212**, 1265.

Swartz, M. N., Trautner, T. A. and Kornberg, A. (1962) *J. biol. Chem.* **237**, 1961.

Sweetman, L. and Nyhan, W. L. (1968) *J. Chromatog.* **32**, 662.

Swenson, P. A. and Setlow, R. B. (1963) *Photochem. Photobiol.* **2**, 419.

Sypherd, P. S. (1971) *J. molec. Biol.* **56**, 311.

Székely, M. and Sanger, F. (1969) *J. molec. Biol.* **43**, 607.

Szulmajster, J., Arnaud, M. and Yang, F. E. (1969) *J. gen. Microbiol.* **57**, 1.

Szybalski, W. (1967) *Radiation Research Supplement* **7**, 147.

Szybalski, W. (1968) in *Methods in Enzymology,* vol. **12B** (eds. Grossman, L. and Moldave, K.), New York and London: Academic Press, p. 330.

Szybalski, W., Bøvre, K., Fiandt, M., Hayes, S., Hradecna, Z., Kumar, S., Lozeron, H. A., Nijkamp, H. J. J. and Stevens, W. F. (1970) *Cold Spring Harb. Symp. quant. Biol.* **35**, 341.

Szybalski, W. and Iyer, V. N. (1964) *Fed. Proc.* **23**, 946.

Szybalski, W., Kubinski, H., Hradecna, Z. and Summers, W. C. (1971) in *Methods in Enzymology,* vol. **21** (eds. Grossman, L. and Moldave, K.), New York and London: Academic Press, p. 383.

Szybalski, W. and Menningman, H.-D. (1962) *Analyt. Biochem.* **3**, 267.

Szybalski, W. and Opera-Kubinska, Z. (1965) in *Cellular Radiation Biology,* Baltimore: Williams and Wilkins, p. 223.

Szylagyi, J. F. (1968) *Biochem. J.* **109**, 191.

Tabor, H. (1962) *Biochemistry,* **1**, 496.
Takahashi, I. and Marmur, J. (1963) *Nature,* **197**, 794.
Takeisha, K. and Ukita, G. J. (1966) *Biochim. biophys. Acta,* **123**, 445.
Takemura, S., Mizutani, T. and Miyazaki, M. (1968) *J. Biochem. (Japan)* **63**, 277.
Temin, H. M. and Mizutani, S. (1970) *Nature,* **226**, 1211.
Temperli, A., Türler, H., Rüst, P., Danon, A. and Chargaff, E. (1964) *Biochem. biophys. Acta,* **91**, 462.
Tener, G. M., Khorana, H. G., Markham, R. and Pol, E. H. (1958) *J. Amer. chem. Soc.* **80**, 6223.
Terry, C. E. and Setlow, J. K. (1967) *Photochem. Photobiol.* **3**, 799.
Tewari, K. K. and Wildman, S. G. (1970) *Symp. Soc. Exptl. Biol.* **24**, 147.
Thach, R. E. and Doty, P. (1965) *Science,* **148**, 632.
Thang, M. N., Graffe, M. and Grunberg-Manago, M. (1965) *Biochim. biophys. Acta,* **108**, 125.
Thiebe, R. and Zachau, H. G. (1970). *Biochim. biophys. Acta,* **217**, 294.
Thiebe, R., Zachau, H. G., Baczynskyj, L., Biemann, K. and Sonnenbilcher, J. (1971) *Biochim. biophys. Acta,* **240**, 163.
Thomas, C. A., jun. (1970) *Biochim. biophys. Acta,* **213**, 417.
Thomas, C. A., jun. and Pinkerton, T. C. (1962) *J. molec. Biol.* **5**, 356.
Tinoco, I., jun., Uhlenbeck, O. C. and Levine, M. D. (1971) *Nature,* **230**, 362.
Toji, L. and Cohen, S. S. (1970) *J. Bact.* **103**, 323.
Tomlinson, R. V. and Tener, G. M. (1962) *J. Amer. chem. Soc.* **84**, 2644.
Tomlinson, R. V. and Tener, G. M. (1963) *Biochemistry,* **2**, 697.
Tongur, V. S., Wladytchenskya, N. S. and Kotchina, W. M. (1968) *J. molec. Biol.* **33**, 451.
Traub, P. and Nomura, M. (1969) *J. molec. Biol.* **40**, 391.
Travers, A. A. (1969) *Nature,* **223**, 1107.
Travers, A., Kamen, R. and Cashel, M. (1970) *Cold Spring Harb. Symp. quant. Biol.* **35**, 415.
Travers, A. A., Kamen, R. I. and Schleif, R. F. (1971) *Nature,* **228**, 748.
Tsetkov, V. N., Kisselev, L. L., Lyurbina, S. Y., Fralova, L. Y., Klenin, S. I., Skaska, V. S. and Nikitin, N. A. (1965) *Biokhimiya,* **30**, 302 (in Russian).
Ts'o, P. O. P. and Lu, P. (1964) *Proc. Nat. Acad. Sci. U.S.* **51**, 272.
Tsuboi, M., Kyogoku, Y. and Shimanouchi, T. (1962) *Biochim. biophys. Acta,* **55**, 1.
Turba, F., Pelzer, H. and Schuster, H. (1954) *Hoppe-Seylers Zeit. physiol. Chem.* **296**, 7. (in German).
Turner, J. C. (1967) *Sample Preparation for Liquid Scintillation Counting,* Amersham (Bucks., U.K.): The Radiochemical Centre.

Uchida, T. and Egami, F. (1967) in *Methods in Enzymology,* vol. **12A** (eds. Grossman, L. and Moldave, K.), New York and London: Academic Press, p. 228.
Udenfriend, S. and Zaltman, P. (1962) *Analyt. Biochem.* **3**, 49.
Udvardy, A. and Venetianer, P. (1971) *Europ. J. Biochem.* **20**, 513.
Uhlenbeck, O. C., Martin, F. H. and Doty, P. (1971) *J. molec. Biol.* **57**, 217.
Ulbricht, T. L. V., Emerson, T. R. and Swan, R. J. (1965) *Biochem. biophys. Res. Comm.* **19**, 643.
Ulbricht, T. L. V., Emerson, T. R. and Swan, R. J. (1966) *Tetrahedron Letters,* **14**, 1561.

van Bruggen, E. F. J., Rimmer, C. M., Borst, P., Ruttenberg, C. J. C. M., Kroon, A. M. and Schuurmans Stekhoven, F. M. A. H. (1968) *Biochim. biophys. Acta,* **161**, 402.
van de Putte, P., Zwenk, H. and Rorsch, A. (1966) *Mutation Res.* **3**, 381.
van Dijk-Salkinoja, M. S., Stoof, T. J. and Planta, J. (1970) *Europ. J. Biochem.* **12**, 474.
van Holde, K. E., Brahms, J. and Michelson, A. M. (1965) *J. molec. Biol.* **12**, 726.
van Holde, K. E. and Rossetti, G. P. (1967) *Biochemistry,* **6**, 272.
van Ravenswaay Claasen, J. C., van Leeuwen, A. B. J., Duijts, G. A. H. and Bosch, L. (1967) *J. molec. Biol.* **23**, 535.
Vanyushin, B. F., Belozersky, A. N., Kokurina, N. A. and Kadirova, D. X. (1968) *Nature,* **218**, 1066.
Vanyushin, B. F., Tkacheva, S. G. and Belozersky, A. N. (1969) *Nature,* **225**, 948.
Vanyushin, B. F., Buryanov, Ya. I. and Belozersky, A. N. (1971) *Nature New Biology,* **230**, 25.
Varghese, A. J. (1970) *Biochemistry,* **9**, 4781.

Veldhuisen, G. and Goldberg, E. B. (1968) in *Methods in Enzymology,* vol. **12B** (eds. Grossman, L. and Moldave, K.), New York and London: Academic Press, p. 858.

Verley, W. G., Dewandre, A., Moutschen-Damen, J. and Moutschen-Damen, M. (1962) *J. molec. Biol.* **11**, 216.

Verwoerd, D. W., Zillig, W. and Kohlhage, H. (1963) *Hoppe Seylers Zeit. physiol. Chem.* **332**, 184.

Vinograd, J. and Hearst, J. E. (1962) *Fortschr. Chem. org. Naturstoffe,* **20**, 372.

Vinograd, J., Lebowitz, J. and Watson, R. (1968) *J. molec. Biol.* **33**, 173.

Vischer, E. and Chargaff, E. (1948) *J. biol. Chem.* **176**, 715.

Vizsolyi, J. P. and Tener, G. M. (1962) *Chem. and Industry,* 263.

Voet, D. and Rich, A. (1970) *Prog. Nuc. Acid Res. molec. Biol.* **10**, 183.

Vold, B. S. (1970) *J. Bact.* **102**, 711.

von Heyden, H. W. and Zachau, *Biochim. biophys. Acta,* **232**, 651.

Wacker, A., Mennigmann, H. D. and Szybalski, W. (1962) *Nature,* **191**, 635.

Wacker, W. E. C. (1962) *Biochemistry,* **1**, 859.

Walker, J. R. (1969) *J. Bact.* **99**, 713.

Walker, R. T. and RajBhandary, U. L. (1970) *Biochem. biophys. Res. Comm.* **38**, 907.

Walter, T. J. and Mans, R. J. (1970) *Biochim. biophys. Acta,* **217**, 72.

Walter, G., Seifert, W. and Zillig, W. (1968) *Biochem. biophys. Res. Comm.* **30**, 240.

Walton, E., Holley, R. W., Boxer, G. E. and Nutt, R. F. (1966) *J. org. Chem.* **31**, 164.

Wang, S. Y. (1957) *Nature,* **180**, 91.

Waring, M. J. (1968) *Ann. Rep. chem. Soc.* **65B**, 551.

Waring, M. J. (1970) *J. molec. Biol.* **54**, 247.

Warner, J., Rich, A. and Hall, C. E. (1962) *Science,* **138**, 1399.

Watson, J. D. (1970) *The Molecular Biology of the Gene* (2nd edition), New York: Benjamin.

Watson, J. D. and Crick, F. H. C. (1953) *Nature,* **171**, 737.

Watson, D. G., Sweet, R. M. and Marsh, R. (1965) *Acta Cryst.* **19**, 573.

Watts, R. L. and Mathias, A. P. (1967) *Biochim. biophys. Acta,* **145**, 828.

Weber, K. (1967) *Biochemistry,* **6**, 3144.

Webster, R. E., Engelhardt, D. L., Zinder, N. D. and Konigsberg, W. (1967) *J. molec. Biol.* **29**, 27.

Weill, G. and Calvin, M. (1963) *Biopolymers,* **1**, 401.

Weinberg, R. A., Loening, U. E., Willems, M. and Penman, S. (1967) *Proc. Nat. Acad. Sci. U.S.* **58**, 1088.

Weinstein, I. B. (1963) *Cold Spring Harb. Symp. quant. Biol.* **28**, 579.

Weinstein, I. B. and Schechter, A. N. (1962) *Proc. Nat. Acad. Sci. U.S.* **48**, 1686.

Weiss, J. J. (1964) *Prog. Nuc. Acid Res. molec. Biol.* **3**, 103.

Weiss, J. F. and Kelmers, A. D. (1967) *Biochemistry,* **6**, 2507.

Wells, R. D. and Larsen, J. E. (1970) *J. molec. Biol.* **49**, 319.

Wempen, I. and Fox, J. J. (1967) in *Methods in Enzymology,* vol. **12A** (eds. Grossman, L. and Moldave, K.), New York and London: Academic Press, p. 59.

Wempen, I., Ueda, T. and Fox, J. J. (1963) *Biochem. Preps,* **10**, 98.

Werner, R. (1971) *Nature New Biology,* **223**, 99.

Weser, U. (1968) in *Structure and Bonding,* vol. **5**, Berlin: Springer Verlag, p. 41.

Westmoreland, B. C., Szybalski, W. and Ris, H. (1969) *Science,* **163**, 1343.

Wettstein, F. O., Staehelin, T. and Noll, H. (1963) *Nature,* **197**, 430.

Wheeler, H. L. and Merriam, H. F. (1903) *Amer. chem. J.* **29**, 478.

Whitehouse, H. L. K. (1969) in *Towards an Understanding of the Mechanisms of Heredity* (2nd edition) London: Arnold.

Whitehouse, H. L. K. and Hastings, P. J. (1965) *Genet. Res.* **6**, 27.

Wierzchowski, K. L., Litonska, E. and Shugar, D. (1965) *J. Amer. chem. Soc.* **87**, 4621.

Wilkins, M. H. F., Arnott, S., Marvin, D. A. and Hamilton, L. D. (1969) *Nature,* **167**, 1693; and consecutively, Crick, F. H. C. (ibid. p. 1694), Arnott, S. (ibid. p. 1694) and Donahue, J. (ibid. p. 1700).

Williams, F. R. and Grunberg-Manago, M. (1964) *Biochim. biophys. Acta,* **89**, 66.

Wilson, G. A. and Bott, K. K. (1968) *J. Bact.* **95**, 1439.

Wilson, R. G. and Caicuts, M. J. (1965) *J. biol. Chem.* **241**. 1725.

Wintermeyer, W., Thiebe, R., Zachau, H. G., Riesner, D., Römer, R. and Maas, G. (1969) *FEBS Letters,* **5**, 23.
Wittmann, H. G. (1969) *Symp. Soc. gen. Microbiol.* **20**, 55.
Woese, C. R. (1970) *Nature,* **226**, 817.
Wood, D. D. and Luck, D. J. L. (1969) *J. molec. Biol.* **41**, 211.
Wu, R. and Taylor, E. (1971) *J. molec. Biol.* **57**, 491.
Wulff, D. L. (1962) *Biophys. J.* **3**, 355.
Wyatt, G. R. (1955) in *The Nucleic Acids,* **vol.** 1 (eds. Chargaff, E. and Davidson, J. N.), New York and London: Academic Press, p. 243.

Yang, J. T. and Samejima, T. (1969) *Prog. Nuc. Acid Res. molec. Biol.* **9**, 223.
Yang, J. T., Samejima, T. and Sakar, P. K. (1966) *Biopolymers,* **4**, 623.
Yaniv, M., Favre, A. and Barrell, B. G. (1969) *Nature,* **223**, 1331.
Yarus, M. and Sinsheimer, R. L. (1967) *Biophys. J.* **7**, 267.
Yoneda, M. and Bollum, F. J. (1965) *J. biol. Chem.* **240**, 3385.
Yoshida, M., Kaziro, Y. and Ukita, T. (1968) *Biochim. biophys. Acta,* **166**, 646.
Young, J. D., Bock, R. M., Nishimura, S., Ishikura, H., Yamada, Y., RajBhandary, U. L., Labanouskas, M. and Conners, P. G. (1969) *Science,* **166**, 1527.
Young, E. T. (1970) *J. molec. Biol.* **51**, 591.
Young, F. E. and Jackson, A. P. (1966) *Biochem. biophys. Res. Comm.* **23**, 490.
Yourno, J. (1970) *J. molec. Biol.* **40**, 437.
Yphantis, D. A. (1964) *Biochemistry,* **3**, 297.
Yung, N. C., Burchenal, J. M., Fecher, R., Duschinsky, R. and Fox, J. J. (1961) *J. Amer. chem. Soc.* **83**, 4060.

Zachau, H. G. (1965) *Hoppe Seylers Zeit. physiol. Chem.* **342**, 98.
Zachau, H. G., Acs, G. and Lipmann, F. (1958) *Proc. Nat. Acad. Sci. U.S.* **44**, 885.
Zachau, H. G., Dütting, D. and Feldman, H. (1966) *Hoppe Seylers Zeit. physiol. Chem.* **347**, 212.
Zahn, R. Z. (1970) *FEBS Letters,* **6**, 141.
Zamecnik, P. C. and Keller, E. B. (1954) *J. biol. Chem.* **209**, 337.
Zamir, A., Holley, R. W. and Marquisee, M. (1964) *J. biol. Chem.* **240**, 1267.
Zillig, W., Zechel, K., Rabussay, D., Schachner, M., Sethli, V. S., Palm, P., Heil, A. and Seifert, W. (1970) *Cold Spring Harb. Symp. quant. Biol.* **35**, 47.
Zimmer, Ch. and Venner, H. (1970) *Europ. J. Biochem.* **15**, 40.
Zimmermann, F., Kröger, H., Hager, U. and Keck, K. (1964) *Biochim. biophys. Acta,* **87**, 160.
Zubay, G. (1968) in *Methods in Enzymology,* vol. **12B** (eds. Grossman, L. and Moldave, K.), New York and London: Academic Press, p. 227.
Zubay, G. and Chambers, D. A. (1969) *Cold Spring Harb. Symp. quant. Biol.* **34**, 753.
Zubay, G., Schwartz, D. and Beckwith, J. (1970) *Cold Spring Harb. Symp. quant. Biol.* **35**, 433.

Author Index

Numbers in parenthesis refer to pages on which the author's work is referred to, although his name does not appear.

Subject Index